Lecture Notes in Mathematics

Edited by A. Dold and B. Eckmann

964

Ordinary and Partial Differential Equations

Proceedings of the Seventh Conference
Held at Dundee, Scotland, March 29 – April 2, 1982

Edited by W. N. Everitt and B. D. Sleeman

Springer-Verlag
Berlin Heidelberg New York 1982

Editors

W. N. Everitt
Department of Mathematics, University of Birmingham
P.O. Box 363, Birmingham B15 2TT, England

B. D. Sleeman
Department of Mathematics, University of Dundee
Dundee DD1 4HN, Scotland

AMS Subject Classifications (1980): 34-06, 35-06

ISBN 3-540-11968-X Springer-Verlag Berlin Heidelberg New York
ISBN 0-387-11968-X Springer-Verlag New York Heidelberg Berlin

Printed in Germany

Printing and binding: Beltz Offsetdruck, Hemsbach/Bergstr.
2146/3140-543210

D E D I C A T I O N

These Proceedings are dedicated to

The University of Dundee

on the occasion of the centenary celebrations

1882 - 1982

PREFACE

These Proceedings form a record of the lectures delivered at the seventh Conference on Ordinary and Partial Differential Equations which was held at the University of Dundee, Scotland during the week Monday 29 March to Friday 2 April 1982.

The Conference formed part of the centenary celebrations for the University of Dundee 1882-1982. The University of Würzburg was officially represented at the Conference, for this purpose, by Professor Dr H W Knobloch.

The Conference was attended by 120 mathematicians from more than 25 countries.

The Conference was organised by a Committee with membership: E R Dawson, W N Everitt, B D Sleeman.

Following again the tradition set by the earlier Dundee Conferences the Organising Committee named as Honorary Presidents of the 1982 Conference

Professor Jean Mawhin (Belgium)

Professor Tung Chin-Chu (P R China).

The Committee expresses its gratitude to all mathematicians who took part in the work of the Conference.

The Committee thanks : the University of Dundee for generously supporting the Conference; the Warden, Domestic Bursar and Staff of West Park Hall for all their help in providing accommodation for the participants; the Department of Mathematical Sciences for continuing support for the Conference; the Bursar of Residences, the Accommodation Officer, the Information Officer and the Finance Officer of the University.

The Committee offers special thanks to the Vice-Principal of the University, Professor P D Griffiths, for representing the University at the opening ceremony, and at the University reception held for all participants.

The Committee records with gratitude financial support, in the form of travel and subsistence funds, from

The Royal Society of London
and
The European Research Office of the United States Army.

The Professors Everitt and Sleeman record special thanks to their colleague Commander E R Dawson R N whose services, from retirement, were generously offered in helping to organise the Conference, and who carried the main responsibility for the accommodation of all participants in West Park Hall. Commander Dawson also edited a number of the manuscripts submitted for publication in these Proceedings.

Likewise special thanks are due to all the Secretaries in the Department of Mathematical Sciences of the University; in particular to Mrs Caroline Peters for her invaluable contribution.

W N Everitt B D Sleeman

CONTENTS

Lectures given at the Conference which are not represented by contributions to these Proceedings.

(Results may be presented elsewhere)

A U Afuwape:

On Lyapunov Razumikhin approach to certain third-order equations with delay

A H Azzam:

Boundary value problems for elliptic and parabolic equations in domain with corners

I Bihari:

Distribution of the zeros of Bôcher's pairs with respect to second order homogeneous differential equations

T R Blows:

Limit cycles of polynomial differential equations

L Brüll:

A new abstract existence theory for non-linear Schrödinger and wave equations

L Collatz:

Inclusion theorems for singular and free boundary value problems

P A Deift:

Some recent results in ordinary differential equations and approximation techniques

J Dönig:

Positive eigensolutions and the lower spectrum of Schrödinger operators

W D Evans:

On the distribution of eigenvalues of Schrödinger operators

W N Everitt:

Two examples of the Hardy-Littlewood type of integral inequalities

R E Kleinman:

Recent developments in modified Green's functions

M K Kwong:

Second order linear and nonlinear oscillation results

A G Ramm:

Basisness property and asymptotics of spectrums of some nonselfadjoint differential and pseudo-differential operators

B P Rynne:

Bloch waves and multiparameter spectral theory

B D Sleeman:

An abstract multiparameter spectral theory

R A Smith:

Poincaré's index theorem and Bendixson's negative criterion for certa: differential equations of higher dimension

R Villella-Bressan:

Functional equations of delay type in L^1-spaces

W L Walter:

Generalized Volterra prey-predator systems

Address list of authors and speakers

A U Afuwape:	Department of Mathematics, University of Ife, ILE-IFE, Nigeria
J M Amillo Gil:	Universidad Politechnica De Madrid, MADRID, Spain
F V Atkinson:	Department of Mathematics, University of Toronto, TORONTO, Ontario, Canada
A H Azzam:	Department of Mathematics, King Saud University, RIYADH, Saudi Arabia
P W Bates:	Department of Mathematics, Heriot-Watt University, Riccarton, Currie, EDINBURGH, EH14 4AS, Scotland
H Behncke:	Fachbereich Mathematik, University of Osnabrüch, 45 OSNABRÜCH, West Germany
V Benci:	Istituto de Matematica Applicata, Via Re David 200, 70125 BARI, Italy
I Bihari:	Mathematical Institute, BUDAPEST, Realtanoda-U 13/15, H-1053, Hungary
T R Blows:	Pure Mathematics Department, University College Wales, ABERYSTWYTH, Dyfed, SY23 3BZ, Wales
R C Brown:	Department of Mathematics, University of Tennessee, KNOXVILLE, Tennessee, U.S.A.
P J Browne:	Department of Mathematics & Statistics, University of Calgary, CALGARY, Alberta, T2N 1N4, Canada
L Brüll:	Mathematisches Institut der Universität zu Köln, 5000 KÖLN 41, Weyertal 86, West Germany
J A Burns:	Department of Mathematics, Virginia Tech University, BLACKSBURG, VA 24061, U.S.A.
P A Clarkson:	The Queen's College, OXFORD, OX1 4AW, England
E M Cliff:	Department of Mathematics, Virginia Tech University, BLACKSBURG, VA 24061, U.S.A.
L Collatz:	Eulenkrugstrasse 84, 2000 HAMBURG 67, Germany
D L Colton:	Department of Mathematical Sciences, University of Delaware, NEWARK, Delaware 191711, U.S.A.
P A Deift:	Courant Institute of Mathematical Sciences, 251 Mercer Street, NEW YORK, NY 10012, U.S.A.

J Donig: Technische Hochschule Darmstadt, Fachbereich Mathematik, Schlossgartenstrasse 7, D6100 DARMSTADT, West Germany

N X Dũng: Texas Tech University, Department of Mathematics, LUBBOCK, Texas 79409, U.S.A.

M S P Eastham: Department of Mathematics, Chelsea College, Manresa Road, LONDON, SW3 6LX, England

Á Elbert: Mathematical Institute, BUDAPEST, Realtanoda-U 13/15, H-1053, Hungary

D E Edmunds: School of Mathematical & Physical Sciences, University of Sussex, FALMER, Brighton, BN1 9QH, England

W D Evans: Department of Pure Mathematics, University College, P O Box 78, CARDIFF, CF1 1XL, Wales

W N Everitt: Department of Mathematics, University of Birmingham, P.O. Box 363, Birmingham B15 2TT, England

P C Fife: Department of Mathematics, University of Arizona, TUCSON, Arizona 85721, U.S.A.

J Fleckinger: Université Paul Sabatier, UER MIG, 118 Route de Narbonne, 31062 TOULOUSE, Cedex, France

C T Fulton: Florida Institute of Technology, Department of Mathematical Sciences, MELBOURNE, Florida 32901, U.S.A.

I M Gali: Mathematics Department, Qatar University, P O Box 2713, DOHA, Qatar, Arabian Gulf

R C Grimmer: Institut für Mathematik, Elisabethstrasse 11, A-8010 GRAZ, Austria

D B Hinton: Mathematics Department, University of Tennessee, KNOXVILLE, Tenn. 37916, U.S.A.

F A Howes: Department of Mathematics, University of California, DAVIS, CA 95616, U.S.A.

H G Kaper: Applied Mathematics Division, Argonne National Laboratory, ARGONNE, Ill 60439, U.S.A.

R M Kauffman: Department of Mathematics, Western Washington University, BELLINGHAM, WA 98225, U.S.A.

R E Kleinman: Department of Mathematics, University of Strathclyde, Livingstone Tower, 26 Richmond Street, GLASGOW, G1 1XH, Scotland

H W Knobloch: Mathematisches Institut, Am Hubland, D-8700 WÜRZBURG, Germany

I W Knowles: Department of Mathematics, University of Alabama in Birmingham, BIRMINGHAM, Alabama 35294, U.S.A.

M A Kon: Department of Mathematics, Boston University, BOSTON, MA 02215, U.S.A.

A M Krall: McAllister Building, Pennsylvania State University, UNIVERSITY PARK, PA 16802, U.S.A.

K Kreith: Mathematics Department, University of California, DAVIS, CA 95616, U.S.A.

M K Kwong: Department of Mathematics, Northern Illinois University, DEKALB, Illinois, U.S.A.

R T Lewis: Department of Mathematics, University of Alabama in Birmingham, BIRMINGHAM, Alabama 35294, U.S.A.

L L Littlejohn: Department of Mathematics, Computer Science & Systems Design, The University of Texas at San Antonio, SAN ANTONIO, Texas 78285, U.S.A.

S-O Londén: Institute of Mathematics, Helsinki University of Technology, ESPOO 15, Finland

J Mawhin: Institut Mathematique, Université de Louvain, Chemin du Cyclotron 2, B-1348 LOUVAIN LA NEUVE, Belgium

A C McBride: Department of Mathematics, University of Strathclyde, Livingstone Tower, 26 Richmond Street, GLASGOW G1 1XH, Scotland

P A McCoy: Department of Mathematics, U S Naval Academy, ANNAPOLIS, Maryland 21402, U.S.A.

J R McLaughlin: Department of Mathematical Sciences, Rensselaer Polytechnic Institute, TROY, New York 12181, U.S.A.

J B McLeod: Wadham College, Oxford University, 24-29 St Giles, OXFORD, OX1 3LB, England

A B Mingarelli: Department of Mathematics, University of Ottawa, OTTAWA, Ontario, K1N 9B4, Canada

S E A Mohammed: Department of Pure Mathematics, University of Khartoum, Khartoum, the Sudan

P A Morales: Departément de Mathématiques et d'Informatique, Université de Sherbrooke, SHERBROOKE, Québec, J1K 2R1, Canada

M N M Nakao: Department of Mathematics, Heriot-Watt University, Riccarton, Currie, EDINBURGH, EH14 4AS, Scotland

F Neuman: Mathematical Institute, Czechoslovak Academy of Sciences, Janáckovo Nám 2a, 66295 BRNO, Czechoslovakia

D Pascali: Fachbereich Mathematik, Technische Hochschule Darmstadt, Schlossgartenstrasse 7, D-6100 DARMSTADT, West Germany

D Race: Mathematics Department, Witwatersrand University, Jan Smuts Avenue, JOHANNESBURG, South Africa

A G Ramm: Department of Mathematics, Kansas State University, MANHATTAN, Kansas 66506, U.S.A.

L A Raphael: Howard University, Department of Mathematics, WASHINGTON, DC 20059, U.S.A.

K A R Rautmann: Department of Mathematics, The University, Warburger Strasse 100, D-4790 PADERBORN, West Germany

T T Read: Department of Mathematics, Western Washington University, BELLINGHAM, Washington 98225, U.S.A.

H Röh: Department of Mathematics, Heriot-Watt University, Riccarton, Currie, EDINBURGH, EH14 4AS, Scotland

B P Rynne: Department of Mathematical Sciences, The University, DUNDEE, DD1 4HN, Scotland

D A Sánchez: Department of Mathematics, University of New Mexico, ALBUQUERQUE, NM 87131, U.S.A.

R A Saxton: Department of Mathematics, Brunel University, UXBRIDGE, Middlesex, UB8 3PH, England

P W Schaefer: Department of Mathematics, University of Tennessee, KNOXVILLE, Tennessee, U.S.A.

K Seitz: Technical University of Budapest, 1111 Muegyetem, RAKPART 9, H. EP. V EM. 5, Hungary

J K Shaw: Department of Mathematics, Virginia Tech University, BLACKSBURG, VA 24061, U.S.A.

B D Sleeman: Department of Mathematical Sciences, The University, DUNDEE, DD1 4HN, Scotland

R A Smith: Department of Mathematics, University of Durham, South Road, DURHAM, DH1 3LE, England

K Soni: Department of Mathematics, University of Tennessee, KNOXVILLE, Tennessee, U.S.A.

E Stephan: Fachbereich Mathematik, Technische Hochschule, Schlossgartenstrasse 7, D-6100 DARMSTADT, West Germany

R L Sternberg: Office of Naval Research, Bld 114 Section D, 666 Summer Street, BOSTON, Mass 02210, U.S.A.

A Z-A M Tazali: The University of Mosul, College of Science, Department of Mathematics, MOSUL, Iraq

Tung Chin-Chu: Graduate School, Chinese Academy of Sciences, P O Box 3908, BEIJING, P R China

L Turyn: Department of Mathematics & Statistics, University of Calgary, CALGARY, Alberta, T2N 1N4, Canada

A Vanderbauwhede: Instituut voor Theoretische Mechanica, Rijksuniversiteit Gent, Krijgslaan 281-S9, B-9000 GENT, Belgium

R Villella-Bressan: Istituto de Analisi E Meccanica, Università di Padova, Via Belzoni 7, 35100 PADOVA, Italy

J Walter: Institut für Mathematik, Templergraben 55, D-5100 AACHEN, Germany

W L Walter: Mathematisches Institut I, Universität Karlsrühe, Kaiserstrasse 12, D-7500 KARLSRÜHE, Germany

A D Wood: School of Mathematical Sciences, N.I.H.E., Ballymun Road, DUBLIN 9, Ireland

S D Wray: Department of Mathematics, Royal Roads Military College, F.M.O., VICTORIA, British Columbia, VO5 1BO, Canada

E M E Zayed: Department of Mathematics, Zagazig University, Faculty of Science, ZAGAZIG, Egypt

A Zettl: Department of Mathematics, Northern Illinois University, DEKALB, Illinois, U.S.A.

On the asymptotic behaviour of the Titchmarsh-Weyl m-coefficient and the spectral function for scalar second-order differential expressions.

F. V. Atkinson

Summary: It is shown that a modified version of the Stieltjes inversion formula for Nevanlinna-type functions, together with improved estimates of the Titchmarsh-Weyl function $m(\lambda)$, constitute an effective approach to the problem of finding approximations to spectral functions for the case of $-y'' + q(x)y = \lambda y, \quad 0 < x < b \leq \infty$.

CHAPTER I. Preliminary discussion and definitions.

1.1. Introduction.

We discuss here the asymptotics of two functions, both basic to the theory of boundary-value problems for

$$-y'' + q(x)y = \lambda y, \quad 0 < x < b \leq \infty ; \tag{1.1.1}$$

here $q(x)$ is to be real-valued, and should satisfy certain integral conditions, specified later in §3.2. The fundamental roles of these functions are illustrated by the variety of ways in which they may be introduced. Partipular ways, presented in outline only, might be the following. We denote by θ, ϕ solutions of (1.1.1) determined by

$$\theta(0, \lambda) = 0, \ \theta'(0, \lambda) = -1, \ \phi(0, \lambda) = 1, \ \phi'(0, \lambda) = 0 . \tag{1.1.2}$$

For the Titchmarsh-Weyl m-coefficient, we claim that there is at least one function $m(\lambda)$, defined and holomorphic in the open upper half-plane, with the Nevanlinna property

$$\operatorname{Im} m(\lambda) > 0 \quad \text{if} \quad \operatorname{Im} \lambda > 0, \tag{1.1.3}$$

and such that

$$\int_0^b |\theta(x, \lambda) + m(\lambda)\phi(x, \lambda)|^2 \, dx \leq \operatorname{Im} m(\lambda)/\operatorname{Im} \lambda , \tag{1.1.4}$$

and in particular such that $\theta + m\phi \in L^2(0, b)$. For the spectral function, we claim that there is at least one non-decreasing function $\tau(t)$ from the real line to itself, such that if

$$f \in L^2(0, b), \quad g(\lambda) = \int_0^b f(x)\phi(x, \lambda)\, dx, \tag{1.1.5}$$

then

$$\int_0^b |f|^2 \, dx = \int_{-\infty}^{\infty} |g|^2 \, d\tau \quad . \tag{1.1.6}$$

The asymptotics of these functions, of $m(\lambda)$ for large complex λ, and of $\tau(t)$ for large positive or negative t, have tended to be developed separately, at widely different times. In the case of $\tau(t)$, the basic results were obtained in the early 50's by Marčenko ([17], [18]), and developed by him and by Levitan ([15],[16]). So far as leading terms are concerned, it is known that

$$\tau(-\infty) > -\infty, \quad \tau(t) \sim 2\pi^{-1} t^{\frac{1}{2}} \quad \text{as } t \to +\infty. \tag{1.1.7-8}$$

In the case of $m(\lambda)$, the leading term was obtained much later by Everitt, in [6], who showed in particular that

$$m(\lambda) \sim i \lambda^{-\frac{1}{2}}, \tag{1.1.9}$$

as $\lambda \to \infty$ in any sector of the form

$$0 < \epsilon \leq \arg \lambda \leq \pi - \epsilon . \tag{1.1.10}$$

Earlier order-results of this nature were found by Hille (see [3], [10])

Results of the forms (1.1.7-8) and (1.1.9-10) are, however, not independent, being linked by formulae such as (see e.g. [20])

$$m(\lambda) = \int_{-\infty}^{\infty} (t - \lambda)^{-1} d\tau(t), \quad \operatorname{Im} \lambda > 0; \tag{1.1.11}$$

in cases of non-uniqueness, we must here have for each m the appropriate choice of τ. It is clear that information such as (1.1.7-8) can be fed into the right of (1.1.11), so as to obtained results such as (1.1.9-10).

It is however our purpose here, following a suggestion of Everitt, to explore the opposite path. We show that (1.1.7-8) and certain refinements can be obtained from developments of (1.1.9-10). As is well-known, the Stieltjes inversion formula (see e.g. [5], [13])

$$\tau(t_2) - \tau(t_1) = \pi^{-1} \lim_{\delta \to 0+} \int_{t_1}^{t_2} \operatorname{Im} m(t + i\delta) \, dt , \tag{1.1.12}$$

valid except possibly at points of discontinuity of $\tau(t)$, shows that τ is essentially determined by m. However this formula cannot be applied for asymptotic purposes in the present setting; the limiting process called for in (1.1.12) is specifically excluded by (1.1.10).

We attack this problem on two fronts. Firstly, we prove a more quantitative version of (1.1.12), not involving any limiting process. This is the subject of Chapter II, and utilises the Nevanlinna property (1.1.3), without specific reference to differential equations. Secondly, in Chapter III, we extend the range of formulae such as (1.1.9) beyond (1.1.10). Applications to the proof of results such as (1.1.7-8), and refinements thereof, will be the subject of Chapter IV.

1.2. Constructive definitions of m and τ.

The characterization (1.1.4) of a class of functions such as $m(\lambda)$ is not the only way of introducing them. We rely in the sequel on the following construction. For any $X \in (0, b)$ and any ζ we define

$$m(\lambda, \zeta, X) = - \{\theta(X, \lambda) - \zeta\theta'(X, \lambda)\}/\{\phi(X, \lambda) - \zeta\phi'(X, \lambda)\}. \qquad (1.2.1)$$

For fixed X, ζ, this will be a meromorphic function of λ, whose poles will be the eigenvalues of the problem $\phi(X, \lambda) - \zeta\phi'(X, \lambda) = 0$. In particular, if $\zeta = \tan\gamma$, where γ is real, and $\lambda_n = \lambda_n(\gamma, X)$ are the roots of $\phi(X, \lambda)\cos\gamma - \phi'(X, \lambda)\sin\gamma = 0$, then the λ_n will be the poles of $m(\lambda, \tan\gamma, X)$; a simple calculation shows that the residue at λ_n is

$$\left\{\int_0^X \phi^2(x, \lambda_n)\, dx\right\}^{-1}. \qquad (1.2.2)$$

One is then led to a representation such as (1.1.11), namely

$$m(\lambda, \tan\gamma, X) = \int_{-\infty}^{\infty} (t - \lambda)^{-1}\, d\tau(t, \gamma, X), \qquad (1.2.3)$$

where $\tau(t, \gamma, X)$ is a step-function with jumps (1.2.2) at the λ_n.

Actually, we will make only incidental use of (1.2.2), but will rely heavily on (1.2.1), together with the nesting-circle property. The logical order adopted here will be that for any $X \in (0, b)$, and fixed λ with $\text{Im}\,\lambda > 0$, (1.2.1) defines a map of the real ζ-axis to a circle located in the upper half-plane, to be denoted by $C(X, k)$ where $k = \lambda^{\frac{1}{2}}$; for any $X \in (0, b)$, we can thus define a family $\mathfrak{M}(X)$ of functions $m(\lambda)$, holomorphic in the open upper half-plane, and such that $m(\lambda) \in D(X, k)$ for $\text{Im}\,\lambda > 0$, where $D(X, k)$ is the closed disc bounded by $C(X, k)$. The family $\mathcal{M}(X)$ will include the left of (1.2.3) as special cases, and also many others, such as convex linear combinations of (1.2.3) for various γ. More importantly, $\mathcal{M}(X)$ will include $\mathcal{M}(X')$ for any $X' \in (X, b)$, and so also the common intersection of all such $\mathcal{M}(X')$, in particular the unique $m(\lambda)$ associated with the limit-point case. Again for any $X \in (0, b)$, the elements of $\mathcal{M}(X)$ will be functions of Nevanlinna type and will admit a representation (1.1.11); with the formulation based on (1.1.2) and the hypotheses of §3.2, we can use the form (1.1.11) rather than the more general form discussed in Chapter II (see [3], p. 91). Thus, from the present point of view, as in [3], the various spectral functions are seen as derived objects, defined via the various m-coefficients, rather than by means of (1.1.5-6) or (1.2.2). Even though the latter can be used to derive asymptotic information about spectral functions, as for example in [4], we shall here derive such information exclusively from similar information about m-coefficients.

1.3. Variation in initial dqta.

The above formulae are of course heavily dependent on the choice (1.1.2) of initial data. However a variation in this choice simply occasions a linear-fractional transformation in m (see [3], p. 66); the consequent change in τ is less easy to visualise. Here we note only the case when, as in [15] - [18], the data for ϕ are replaced by

$$\phi(0, \lambda) = 1, \quad \phi'(0, \lambda) = h, \tag{1.3.1}$$

with real finite h. The effect on (1.2.1) is then to replace m by $m/(1 + hm)$. Assuming (1.1.9) to hold with error $o(\lambda^{-1})$, which will in fact be the case, we will then have that

$$m = i\lambda^{-\frac{1}{2}} + h\lambda^{-1} + o(\lambda^{-1}), \tag{1.3.2}$$

subject to (1.1.10), and possibly elsewhere. As will be noted in §4.6, this leads to an additive term $-h$ in certain approximations to the spectral function.

1.4. Further remarks.

An extensive account of the m-coefficient, together with a full survey of the literature, is to be found in the recent monograph [3] of Bennewitz and Everitt. In the case of the vibrating string a full theory, dealing also with spectral functions in detail, is given by Kac and Krein in [14], where the letter "Ω" is used in place of "m".

Asymptotics of the spectral function for generalized second-order operators have been developed by Kac [12]; these imply, by way of relations such as (1.1.11), asymptotic results for m subject to (1.1.10), though only so far as leading terms are concerned.

Matrix extensions of m-coefficients are of current interest, important contributions to the basic theory being due to Hinton and Shaw [11]; see also their recent paper [8] with Everitt.

To Everitt is due a remarkable connection between the behaviour of the m-coefficient and the constants appearing in certain integral inequalities which extend classical results of Hardy, Littlewood and others; we cite in this connection [3], [9], and the recent article [5] for references.

1.5. Acknowledgements.

It is a pleasure to acknowledge many stimulating discussions on these matters with Professor W. N. Everitt. Appreciation is also expressed for the support of the Science and Engineering Research Council of the U.K., for the hospitality of the Department of Mathematics of the University, Dundee, and for the continuing support of the National Science and Engineering Research Council of Canada, under Grant A-3979.

CHAPTER II. A modified Stieltjes inversion formula.

2.1. Nevanlinna functions and their representations.

In this chapter we are concerned with functions holomorphic in the open upper half-plane, satisfying (1.1.3), but not necessarily arising from constructions such as (1.2.1). We have the general representation (see e.g. [3], [5] and [13] for discussion and references)

$$m(\lambda) = A + B\lambda + \int_{-\infty}^{\infty}\{(t-\lambda)^{-1} - t(1+t^2)^{-1}\}d\tau(t), \tag{2.1.1}$$

where A , B are real and $B \geqslant 0$, and $\tau(t)$ is real-valued and non-decreasing, and such that

$$\int_{-\infty}^{\infty}(1+t^2)^{-1}\, d\tau(t) < \infty . \tag{2.1.2}$$

The function $\tau(t)$ may be rendered unique by fixing its value at some point, and by fixing its value at points of discontinuity in some way, for example by right-continuity. In any case we have (1.1.12), except possibly at points of discontinuity. If $\tau(t)$ satisfies the stronger condition

$$\int_{-\infty}^{\infty}(1+|t|)^{-1}\, d\tau(t) < \infty , \tag{2.1.3}$$

the term $t(1+t^2)^{-1}$ in (2.1.1) can be dropped, its contribution being absorbed into the constant A ; if additionally $m(i\nu) \to 0$ as $\nu \to \infty$, we can set $A = B = 0$, thus arriving at the form (1.1.11). The replacement of (2.1.1) by (1.1.11) when (1.1.9) holds is discussed in ([3], p. 91).

Our purpose in this Chapter is to replace (1.1.12) by a result with specific bounds instead of limits.

2.2. A quantitative inversion formula.

Theorem 2.2.1. Let $-\infty < \Lambda_1 < \Lambda_2 < \infty$, $\delta > 0$. Then

$$\left| \operatorname{Im}\int_{\Lambda_1+i\delta}^{\Lambda_2+i\delta} m(\lambda)\,d\lambda - \pi\Big[\tau(\mu)\Big]_{\Lambda_1}^{\Lambda_2} - \delta\Big[\operatorname{Re} m(\mu+i\delta)\Big]_{\Lambda_1}^{\Lambda_2}\right| \leqslant$$

$$\leqslant \tfrac{1}{2}\pi\delta\left\{\operatorname{Im} m(\Lambda_1+i\delta) + \operatorname{Im} m(\Lambda_2+i\delta)\right\} . \tag{2.2.1}$$

Proof: On the left of (2.2.1), the integral may be taken along any path joining the two limits in the upper half-plane. For the purposes of the proof we suppose it taken along a straight line, so that "$d\lambda$" will be real, and the "Im" may be taken under the integral sign, as in (1.1.12).

We prove the result first for the special form (1.1.11), with the additional restriction that $\tau(t)$ should be constant outside some finite interval. Integrals occurring in the following manipulations will then be absolutely convergent. We have

$$\int_{\Lambda_1+i\delta}^{\Lambda_2+i\delta} \operatorname{Im} m(\lambda)\,d\lambda = \int_{-\infty}^{\infty} d\tau(\mu) \int_{\Lambda_1}^{\Lambda_2} f(\lambda,\mu)\,d\lambda,$$

where

$$f(\lambda,\mu) = \delta\left\{(\mu-\lambda)^2 + \delta^2\right\}^{-1} .$$

We note that

$$\int_{-\infty}^{\infty} f(\lambda,\mu)\,d\lambda = \pi ,$$

and introduce the function

$$F(\Lambda) = \int_{\Lambda}^{\infty} d\tau(\mu) \int_{-\infty}^{\Lambda} f(\lambda,\mu)\,d\lambda - \int_{-\infty}^{\Lambda} d\tau(\mu) \int_{\Lambda}^{\infty} f(\lambda,\mu)\,d\lambda .$$

Then, as is easily verified,

$$\int_{\Lambda_1+i\delta}^{\Lambda_2+i\delta} \operatorname{Im} m(\lambda)\,d\lambda = \Big[\pi\tau(\Lambda) + F(\Lambda)\Big]_{\Lambda_1}^{\Lambda_2} .$$

Thus, in order to complete the proof of (2.2.1) for this special case, it will be sufficient to prove that

$$\left|F(\Lambda) - \delta \operatorname{Re} m(\Lambda + i\delta)\right| \leqslant \tfrac{1}{2}\pi\,\delta \operatorname{Im} m(\Lambda + i\delta). \qquad (2.2.2)$$

Now

$$\delta \operatorname{Re} m(\Lambda + i\delta) = \int_{-\infty}^{\infty} (\mu - \Lambda) f(\Lambda,\mu)\,d\tau(\mu),$$

and so the left of (2.2.2) equals

$$\Bigg|\int_{\Lambda}^{\infty} d\tau(\mu)\left\{\int_{-\infty}^{\Lambda} f(\lambda,\mu)\,d\lambda - (\mu-\Lambda) f(\Lambda,\mu)\right\} -$$

$$- \int_{-\infty}^{\Lambda} d\tau(\mu)\left\{\int_{\Lambda}^{\infty} f(\lambda,\mu)\,d\lambda - (\Lambda-\mu) f(\Lambda,\mu)\right\}\Bigg| . \qquad (2.2.3)$$

We write, for $\sigma \geqslant 0$,

$$h(\sigma) = \int_{\sigma}^{\infty} (1+t^2)^{-1}\,dt - \sigma(1+\sigma^{-2})^{-1} \qquad (2.2.4)$$

$$= \tfrac{1}{2}\pi - \tan^{-1}\sigma - \sigma(1+\sigma^{-2})^{-1} , \qquad (2.2.5)$$

and can then replace (2.2.3) by

$$\left| \int_{\Lambda}^{\infty} d\tau(\mu)\, h\{\delta^{-1}(\mu - \Lambda)\} - \int_{-\infty}^{\Lambda} d\tau(\mu)\, h\{\delta^{-1}(\Lambda - \mu)\} \right| . \quad (2.2.6)$$

We claim next that $0 \leqslant h(\sigma) \leqslant \frac{1}{2}\pi(1 + \sigma^2)^{-1}$. The first fact follows from (2.2.4). For the second inequality we introduce the function $j(\sigma) = h(\sigma)(1 + \sigma^2)$, and observe that $j(0) = \frac{1}{2}\pi$, and that $j'(\sigma) < 0$ if $\sigma > 0$. Thus (2.2.6) is bounded by

$$\tfrac{1}{2}\pi \int_{-\infty}^{\infty} d\tau(\mu) \left\{1 + \delta^{-2}(\mu - \Lambda)^2\right\}^{-1} ,$$

and this, in view of (1.1.11), completes the proof of (2.2.2), and so of Theorem 2.2.1 for this special case.

It is a simple matter to extend the result to the general case. We remark first that (2.2.1) is unaffected if to $m(\lambda)$ we add any real constant A, and so we can apply the result to the form (2.1.1) if $B = 0$ and $\tau(t)$ is still required to be constant outside some finite interval.

In the next stage we remove the latter restriction. We assume now just that $\tau(t)$ is non-decreasing and satisfies (2.1.2). For any T_1, T_2 with $-\infty < T_1 < T_2 < \infty$, we define $\tau(t;\, T_1,\, T_2)$ as equal to $\tau(t)$ in $[T_1,\, T_2]$ and constant in $(-\infty,\, T_1]$ and in $[T_2,\, \infty)$. We then define an associated function $m(\lambda;\, T_1,\, T_2)$ by taking $A = B = 0$ in (2.1.1) and replacing $\tau(t)$ by $\tau(t;\, T_1,\, T_2)$, and to this function apply Theorem 2.2.1. Making $T_1 \to -\infty$, $T_2 \to +\infty$ then yields the required result for the general case of (2.1.1) with $A = B = 0$, or for that matter for any real A.

Finally, we must allow for a positive B in (2.1.1). In fact, adding such a term $B\lambda$ to the previous case leaves the left of (2.2.1) unaffected, on account of cancellation between the first and third terms, while the right is increased. This completes the proof of Theorem 2.2.1.

2.3. A weakened form of the inversion formula.

Although the full form of Theorem 2.2.1 was needed for its proof, our application will treat the term in $\operatorname{Re} m(\mu + i\delta)$ as an error term. If we combine this with the right of (2.2.1) we get a simpler result, namely

Theorem 2.3.1. Under the conditions of Theorem 2.2.1, we have

$$\left| \pi^{-1} \operatorname{Im} \int_{\Lambda_1 + i\delta}^{\Lambda_2 + i\delta} m(\lambda)\, d\lambda - \left[\tau(\mu)\right]_{\Lambda_1}^{\Lambda_2} \right| \leqslant$$

$$\leqslant (\pi^{-2} + \tfrac{1}{4})^{\frac{1}{2}} \delta \left\{ |m(\Lambda_1 + i\delta)| + |m(\Lambda_2 + i\delta)| \right\}. \quad (2.3.1)$$

A convenient bound for the numerical factor on the right will be 0.6.

CHAPTER III. Asymptotics of $m(\lambda)$.

3.1. Preliminary remarks.

We now deal with the other main component of the investigation, that of improving (1.1.9-10); since (1.1.7-8) imply somewhat more than (1.1.9-10), it is clear that the latter cannot serve to prove sharpened forms of (1.1.7-8). The reasoning of this chapter is very close to that of [2], modified for the present purpose. We write, without loss of generality,

$$\lambda = k^2 \ , \quad k = \alpha + i\beta \ , \quad \text{where } \alpha \geqslant 0 \ , \ \beta > 0, \quad (3.1.1)$$

and, with $C(X, k)$, $D(X, k)$ as in § 1.2, obtain a bound for elements of $D(X, k)$ by combining a bound for some element of $D(X, k)$ with a bound for its diameter. Bounds for $D(X, k)$ will then automatically be bounds also for

$$D(b, k) = \bigcap D(X,k) \text{ over } X \in (0, b), \quad (3.1.2)$$

provided that $\alpha > 0$.

3.2. Hypotheses on q .

In addition to the reality of q , we assume:

(i) $q \in L(b_1, b_2)$ for all b_1, b_2 with $0 < b_1 < b_2 < b$,

(ii) the limit

$$Q(x) := \lim_{t \to 0} \int_t^x q(s)\,ds \quad (3.2.1)$$

exists and is finite,

(iii) the function

$$G(X, \rho) = \sup \left| \int_0^x e^{2ki(x-t)} q(t)dt \right|, \tag{3.2.2}$$

with "sup" over the domain

$$0 \leq x \leq X, \quad \operatorname{Im} k \geq 0, \quad |k| \geq \rho, \tag{3.2.3}$$

satisfies, for each fixed $X \in (0, b)$,

$$G(X, \rho) \to 0 \text{ as } \rho \to \infty. \tag{3.2.4}$$

The above hypotheses will certainly hold under the usual condition

$$q \in L(0, b') \text{ for all } b' \in (0, b). \tag{3.2.5}$$

However we do not need to assume this; as an example in which (i) - (iii) hold, but not (3.2.5) we cite the case $q(x) = x^{-1} \sin (x^{-1})$.

We verify briefly that (i), (ii) are sufficient to ensure the existence of θ, ϕ satisfying the initial data (1.1.2). If ϕ does exist, and we define $\psi = \phi'/\phi$, we shall have $\psi' = q - \lambda - \psi^2$, $\psi(0) = 0$, and so $\psi(x) = Q(x) - \lambda x - \int_0^x \psi^2(t)dt$. We can reverse this argument and claim that this integral equation has a solution $\psi(x)$ in a neighbourhood of $x = 0$, and then define $\phi(x) = \exp \int_0^x \psi(t)dt$, $\theta(x) = - \phi(x) \int_0^x dt/\phi^2(t)$. These solutions can be extended to the right by means of standard theorems.

3.3. An auxiliary function.

This will provide an element of $D(X, k)$ on which the approximation will be based. We assume $X \in (0, b)$, and that λ, k are as in (3.1.1), with however $\alpha > 0$. We denote by $z(x) = z(x, X)$ the solution of (1.1.1) such that

$$z(X) = 1, \quad z'(X) = ik. \tag{3.3.1}$$

We note that $(z \bar{z}' - z' \bar{z})' = 2i \alpha \beta z \bar{z}$, so that $\operatorname{Im} (z \bar{z}' - z' \bar{z})$ is non decreasing. Since this function is negative at $x = X$, by (3.3.1), we have that z, z' do not vanish in $[0, X]$. We can therefore define

$$n(\lambda, X) = - z(0)/z'(0). \tag{3.3.2}$$

Combining these observations with remarks from [2], p. 347, we have

Lemma 3.3.1. For fixed $X \in (0, b)$, the function $n(\lambda, X)$ is regular and of Nevanlinna type in $\operatorname{Im}\lambda > 0$. Any $m \in D(X, k)$ satisfies

$$|m - n(\lambda, X)| \leq \left\{2\alpha\beta\int_0^X |\phi|^2 dx\right\}^{-1} . \qquad (3.3.3)$$

3.4. Approximation to $n(\lambda, X)$.

Again as in [2], we have recourse to a Riccati equation technique.

Lemma 3.4.1. Let there be a number η such that, with k as in (3.1.1),

$$\left|\int_{x_1}^{x_2} \exp\{2ki(x_2 - t)\}\, q(t)dt\right| \leq \eta , \quad 0 \leq x_1 \leq x_2 \leq X , \qquad (3.4.1)$$

and also

$$\eta < \tfrac{1}{8}\beta . \qquad (3.4.2)$$

Then $n(\lambda, X)$ is defined, and

$$\left|n(\lambda, X) - ik^{-1} - k^{-2}\int_0^X e^{2kit}\, q(t)dt\right| \leq 8\eta^2|k|^{-2}\beta^{-1} . \qquad (3.4.3)$$

Here the finiteness of $n(\lambda, X)$ is noted only since we now allow $\alpha \geqslant 0$, instead of $\alpha > 0$ as in § 3.3. As in [2], we write $v = ik - z'/z$ and have the integral equation

$$v(x) = -\int_0^x e^{2ki(x-t)}\{v^2(t) - q(t)\}dt ,$$

from which, together with (3.3.1), we have

$$|v(x)| \leqslant 2\eta < \tfrac{1}{4}\beta , \quad 0 \leq x \leq X , \qquad (3.4.4)$$

and so

$$\left|v(0) - \int_0^X e^{2kit} q(t)dt\right| \leq 2\eta^2\beta^{-1} . \qquad (3.4.5)$$

Arguing as in [2], p. 350, we have

$$n(\lambda, X) = \{v(0) - ik\}^{-1} = ik^{-1} + v(0)k^{-2} + R ,$$

say, where

$$|R| \leqslant |v^2(0)|\,|k|^{-2}\{|k| - |v(0)|\}^{-1} \leqslant (16/3)\eta^2\,|k|^{-3} .$$

We then obtain (3.4.3) on combining this with (3.4.5-6).

We use (iii) of § 3.2 to bring about (3.4.2). We can take

$$\eta = 2G(X, |k|), \qquad (3.4.6)$$

and then (3.4.2) will hold if

$$G(X, |k|) < \beta/16. \qquad (3.4.7)$$

In particular, we may derive

Lemma 3.4.2. For any $X \in (0, b)$ there is a real $\lambda^*(X) \in (-\infty, 0)$ such that for real $\lambda < \lambda^*(X)$ the function $n(\lambda, X)$ is analytic and real-valued, and so can be continued into the lower half-plane.

Considered as functions of λ, for fixed X, $z(0, X)$ and $z'(0, X)$ can be continued analytically from the upper half-plane across the negative real axis, and so this is true also for $n(\lambda, X)$ so long as it is finite. This is ensured by Lemma 3.4.1 if, as in (3.4.7), $\lambda = -\beta^2$ and $G(X, \beta) < \beta/16$. That $n(\lambda, X)$ is real-valued, if finite, when λ is real and negative follows from the reality in that case of (1.1.1) and the data (3.3.1).

3.5. Estimation of ϕ.

To deal with the right of (3.3.3) we need a lower bound for ϕ.

Lemma 3.5.1. Let the conditions of Lemma 3.4.1 hold. Then

$$|\phi(x)| \geq \tfrac{1}{4} e^{\beta' x} - (11/8)e^{-\beta' x}, \quad \beta' = \beta - 2\eta, \quad 0 \leq x \leq X. \quad (3.5.1)$$

It will be sufficient to prove this when $x = X$; the definition of ϕ is independent of X, and the hypotheses (3.4.1-2), if true for some X, remain true when X is replaced by any lesser positive number. We can express ϕ in terms of z by

$$\phi(x) = z(x)/z(0) - z(x)z'(0)\int_0^x dt/z^2(t),$$

and so have

$$|\phi(X)| \geq \left| z'(0)\int_0^X dt/z^2(t) \right| - |1/z(0)|. \quad (3.5.2)$$

From (3.4.4) we have that $\mathrm{Re}\,(z'/z) \leq -\beta'$, and so

$$|z(x)| \geq \exp\{\beta'(X - x)\}, \quad (3.5.3)$$

which deals with the last term in (3.5.2). For the first term on the right of (3.5.2) we set

$$\int_0^X dt/z^2(t) = I_1 + I_2,$$

where

$$I_1 = (2ki)^{-1}\int_0^X 2z'(t)\, dt/z^3(t) = (2ki)^{-1}\{z^{-2}(0) - 1\},$$

and

$$I_2 = (ik)^{-1} \int_0^X v(t)\, dt/z^2(t) \ .$$

In I_1 we use (3.5.3), which yields

$$|I_1| \geqslant \tfrac{1}{2}|k|^{-1}(1 - \exp(-2\beta' X)). \qquad (3.5.4)$$

In I_2 we use (3.4.4) and (3.5.3), to get

$$|I_2| \leqslant \tfrac{1}{4}\, \beta\, |k|^{-1} \int_0^X \exp(-2\beta'(X-x))\, dx < |6k|^{-1} \ ,$$

since $\beta' \geqslant \frac{3}{4}\beta$. Combining this with (3.5.4) we have

$$\left| \int_0^X dt/z^2(t) \right| \geqslant |3k|^{-1} - |2k|^{-1} \exp(-2\beta' X) \ . \qquad (3.5.5)$$

We note also that

$$|z'(0)| = |z(0)|\, |z'(0)/z(0)| \geqslant |z(0)|(|k| - |v(0)|) \geqslant \tfrac{3}{4}\, |k| \exp(\beta' X),$$

and so

$$\left| z'(0) \int_0^X dt/z^2(t) \right| \geqslant \tfrac{1}{4} \exp(\beta' X) - \tfrac{3}{8} \exp(-\beta' X) \ .$$

Using this in (3.5.2) along with (3.5.3) we get (3.5.1) with $x = X$, as required.

We now use this on the right of (3.3.3), to get

Lemma 3.5.2. Under the conditions of Lemma 3.4.1, we have, if $\beta X \geqslant 4$,

$$\beta \int_0^X |\phi|^2 dx \geqslant 76^{-1} \exp(2\beta' X) \ . \qquad (3.5.6)$$

We use the remark that if a_1, a_2, a_3 are non-negative, and $a_1 \geqslant a_2 - a_3$, then $a_1^2 \geqslant \frac{1}{2}a_2^2 - a_3^2$. Thus (3.5.1) yields

$$\int_0^X |\phi|^2 dx \geqslant (1/32) \int_0^X \exp(2\beta' x)\, dx - (121/64) \int_0^X \exp(-2\beta' x)\, dx \ .$$

Hence, noting that $\beta' \leqslant \beta$, $\beta' X \geqslant 3$, we have

$$\beta \int_0^X |\phi|^2\, dx \geqslant (1/64)(\exp(2\beta' X) - 1) - 121/128,$$

$$\geqslant (1/64) \exp(2\beta' X)\{1 - (123/2)e^{-6}\},$$

which establishes (3.5.6).

3.6. The main result.

In what follows we will assume that $X \in (0, b)$ and λ as in (3.1.1) satisfy

$$\beta X \geq 4 , \quad XG(X,|k|) \leq \tfrac{1}{8} . \qquad (3.6.1\text{-}2)$$

For fixed X, we can arrange (3.6.2) by taking $|k|$ large enough, by (iii) of §3.2; in the case $b = \infty$, an alternative method will be to make $X \to \infty$ and also $|k| \to \infty$, subject to (3.6.2). Always with the assumptions of §3.2, we have

Theorem 3.6.1. Let (3.6.1-2) hold, and also $\alpha > 0$. Then, for any $m \in D(X, k)$

$$\left| m - ik^{-1} - k^{-2} \int_0^X e^{2kit} q(t)dt \right| \leq 32G^2(X,|k|)|k|^{-2}\beta^{-1} + 104\alpha^{-1}e^{-2\beta X} . \qquad (3.6.3)$$

In particular,

$$|m - ik^{-1}| \leq \tfrac{1}{4}X^{-1}|k|^{-2} + 108\alpha^{-1}e^{-2\beta X} . \qquad (3.6.4)$$

We apply Lemma 3.4.1. It follows from (3.6.1-2) that

$$\eta \leq 2G(X, |k|) \leq \tfrac{1}{4}X^{-1} \leq \beta/16 , \qquad (3.6.5\text{-}7)$$

so that the requirement (3.4.2) is satisfied. We then combine (3.3.3) and (3.4.3) to get a result of the form (3.6.2), with the right replaced by

$$8\eta^2 |k|^{-2}\beta^{-1} + 38\alpha^{-1}\exp(-2\beta X), \qquad (3.6.8)$$

where the last term comes from (3.5.6). We now have to replace these two terms by those on the right of (3.6.3). For the first term in (3.6.8) we use (3.6.5). In the case of the last term we note that

$$-2\beta' X = -2\beta X + 4\eta X \leq -2\beta X + 1 ,$$

by (3.6.5-6), and that $38e < 104$. This proves (3.6.3).

For the simpler but weaker form (3.6.4) we note that

$$\left| k^{-2} \int_0^X e^{2kit} q(t)dt \right| \leq |k|^{-2} G(X,|k|) \leq \tfrac{1}{8}|k|^{-2} X^{-1} ,$$

by (3.6.6), and that the first term on the right of (3.6.3) satisfies the same bound, again by (3.6.6). This completes the proof of the Theorem.

For λ near to the negative real axis we use the simpler

Theorem 3.6.2. Under the hypotheses of Theorem 3.6.1,

$$|m - n(\lambda, X)| < 104\alpha^{-1} e^{-2\beta X} . \qquad (3.6.9)$$

This follows from (3.3.3) by the same arguments.

3.7. Extension of some results of Hille and Everitt.

In this section we apply the results of §3.6 with fixed $X \in (0, b)$, to obtain bounds which will apply to any $m \in D(X', b)$, $X \leq X' < b$, and also to points in the limit-disc or limit-point $D(b, k)$ (3.1.2). We take first the Hille order result corresponding to (1.1.9), which will serve for the right of (2.3.1). It is convenient to separate the cases of the first and second quadrants, more precisely those of

$$\text{Re}\,\lambda \geqslant 0, \quad \text{Im}\,\lambda > 0 , \qquad (3.7.1)$$

$$\text{Re}\,\lambda \leqslant 0, \quad \text{Im}\,\lambda \geqslant 0. \qquad (3.7.2)$$

All the following statements refer to $m \in D(X, k)$ with fixed $X, \lambda \to \infty$.

Theorem 3.7.1. We have, in (3.7.1),

$$m = O(|k|^{-1}) , \qquad (3.7.3)$$

provided that

$$\text{Im}\,\lambda^{\frac{1}{2}} \geqslant 4X^{-1} . \qquad (3.7.4)$$

This requires, of course, that λ should lie above a certain parabola with focus at the origin and vertex on the negative real axis; it yields an essentially larger region than the sectorial requirement (1.1.10).

Theorem 3.7.2. In (3.7.2) we have that (3.7.3) holds, provided that

$$\liminf \quad (\text{Im}\,\lambda)|\lambda|^{-1} \exp(2X|\lambda|^{\frac{1}{2}}) > 0. \qquad (3.7.5)$$

Thus in the second quadrant λ may come, in a sense, exponentially close to the negative real axis; this is linked to exponentially small behaviour of $\tau(t) - \tau(-\infty)$. For the proof we suppose first that $\text{Im}\,\lambda \geqslant 1$, so that $\alpha^{-1} = O(|\lambda|^{\frac{1}{2}})$, $\beta \geqslant |\tfrac{1}{2}\lambda|^{\frac{1}{2}}$, and the last term in (3.6.4) is neglgible. If again $0 < \text{Im}\,\lambda \leq 1$, we have

$$\alpha^{-1} = O(|\lambda|^{\frac{1}{2}} |\text{Im}\,\lambda|^{-1}), \qquad \beta = |\lambda|^{\frac{1}{2}} + o(1) ,$$

and so we are led to the requirement (3.7.5).

We consider next the problem of extending the range (1.1.10) of the Everitt asymptotic formula (1.1.9). It is only necessary to strengthen (3.7.4-5). Still with fixed X we have

Theorem 3.7.3. We have

$$m \sim ik^{-1} \tag{3.7.6}$$

as $\lambda \to \infty$ in the first quadrant if also

$$\operatorname{Im} \lambda^{\frac{1}{2}} \to \infty , \tag{3.7.7}$$

and in the second quadrant if

$$(\operatorname{Im} \lambda)|\lambda|^{-1} \exp (2X|\lambda|^{\frac{1}{2}}) \to \infty . \tag{3.7.8}$$

In the case of $n(\lambda, X)$ it is not necessary to stay away from the negative real axis. We have easily from (3.4.3)

Theorem 3.7.4. As $\lambda \to \infty$ in the region (3.7.2), we have

$$n(\lambda, X) = ik^{-1} + o(|k|^{-2}). \tag{3.7.9}$$

3.8. The case $b = \infty$.

As indicated in § 3.6, it seems advantageous here to make X and $|k|$ tend to ∞ together. We determine $X(\rho)$ subject to $X(\rho) \to \infty$ and

$$X(\rho)G(X(\rho), \rho) \leqslant \tfrac{1}{8} . \tag{3.8.1}$$

For example, we may determine $X(\rho)$ for large ρ by equality in (3.8.1). The resulting bounds will then apply to $D(\infty, k)$, the limit-point or limit-disc. In the inequalities of § 3.6 we take $X = X(|k|)$, so that (3.6.2) holds, and in view of (3.6.1) must have

$$(\operatorname{Im} k)X(|k|) \geqslant 4. \tag{3.8.2}$$

For example, let $q(x)$ be monotone, with $q(0) = 0$, so that

$$G(X, \rho) \leqslant |q(X)/\rho| . \tag{3.8.3}$$

We then may require that

$$X(\rho) \to \infty, \quad |X(\rho)q(X(\rho))| \leqslant \tfrac{1}{8}\rho . \tag{3.8.4-5}$$

More specially, let us take

$$q(x) = x^N , \quad N > 0. \tag{3.8.6}$$

Choosing equality in (3.8.5) then yields

$$X(\rho) = (\tfrac{1}{8}\rho)^{1/(N+1)} . \qquad (3.8.7)$$

Subject to (3.7.4), these choices for $X(|k|)$ can be inserted on the right in the estimates of Theorems 3.6.1-2, 3.7.1-2, with the result applying to any $m \in D(\infty, k)$. We can also use this choice of $X(|k|)$ on the left in Theorem 3.7.4. We can also modify Theorem 3.7.3 in this sense. For example, in the special case (3.8.6), we would have (3.7.6) as $\lambda \to \infty$ in the first quadrant, provided that

$$|\lambda|^{1/(2N+2)} \operatorname{Im} \lambda^{\frac{1}{2}} \to \infty .$$

CHAPTER IV. Estimates for the spectral function.

4.1. Preliminaries.

We now put together the results of the last two chapters in order to arrive at approximations to $\tau(t)$, the non-decreasing function on the real line appearing in (1.1.11), (1.2.3) and elsewhere. The bounds of the last chapter were for arbitrary points $m \in D(X, k)$, or in $D(b, k)$ or $D(\infty, k)$ as the case may be. We will now confine attention to the situation that $m = m(\lambda)$ is an analytic function, from the open upper half-plane to itself, satisfying such inclusion properties, with $k = \lambda^{\frac{1}{2}}$ as in (3.1.1). As a Nevanlinna function obeying (1.1.9-10), or even only the corresponding Hille-type order bound, $m(\lambda)$ will (see (3), p. 91) admit a representation (1.1.11), essentially determining the associated $\tau(t)$. We use (2.3.1) and estimates for $m(\lambda)$ to estimate $\tau(t)$.

As in Chapter III, there are two procedures. We may use bounds for $m(\lambda)$ satisfying $m(\lambda) \in D(X, k)$ with fixed $X \in (0, b)$; the results will then apply to any $\tau(t)$ associated with an $m(\lambda)$ satisfying

$$m(\lambda) \in D(X', k), \text{ for some } X' \in [X, b), \qquad (4.1.1)$$

and also to any $\tau(t)$ associated with an $m(\lambda)$ satisfying

$$m(\lambda) \in D(\infty, k), \qquad (4.1.2)$$

if of course $b = \infty$. Again, if $b = \infty$, we can use the procedure indicated in § 3.8, in which we make $X \to \infty$, so as possibly to obtain better results for the case (4.1.2).

We assume throughout the hypotheses (i) - (iii) of §3.2 concerning $q(x)$. Our results are consistent with the standard ones for the case that q satisfies the more special requirement (3.2.5).

4.2. Estimation of $\tau(t) - \tau(-\infty)$ as $t \to -\infty$.

It is a question of showing that this is exponentially small as $t \to -\infty$, and in particular that $\tau(-\infty)$ is finite (and could be taken to be zero). Since much larger error-terms, such as $O(1)$ or $o(1)$, occur in the estimation of $\tau(t)$ as $t \to +\infty$, it will be sufficient to work with a fixed $X \in (0, b)$.

Theorem 4.2.1. Under the assumptions of §3.2, for the spectral function $\tau(t)$ associated with any Nevanlinna function satisfying (4.1.1), we have that $\tau(-\infty)$ is finite and that, as $t \to -\infty$,

$$\tau(t) - \tau(-\infty) = O\left\{ |t| \exp(-2X|t|^{\frac{1}{2}})\right\}. \tag{4.2.1}$$

We use Theorem 2.3.1 with $\Lambda_2 = t < -1$, $\Lambda_1 \in (-\infty, t)$, and

$$\delta = |t| \exp(-2X|t|^{\frac{1}{2}}). \tag{4.2.2}$$

We write

$$K_1(\Lambda, t) = \operatorname{Im} \int_{\Lambda+i\delta}^{t+i\delta} m(\lambda)\, d\lambda, \tag{4.2.3}$$

and then have from (2.3.1) that

$$\tau(t) - \tau(\Lambda) = \pi^{-1} K_1(\Lambda, t) + O(\delta |t|^{-\frac{1}{2}}), \tag{4.2.4}$$

by Theorem 3.7.2, or directly from (3.6.4). We need to show that $K_1(\Lambda, t)$ is bounded as $\Lambda \to -\infty$, and that it admits a bound of the form appearing in (4.2.1).

As an approximation to K_1 we use

$$K_2(\Lambda, t) = \operatorname{Im} \int_{\Lambda+i\delta}^{t+i\delta} n(\lambda, X)\, d\lambda. \tag{4.2.6}$$

Clearly

$$|K_1 - K_2| \leqslant \left| \int_{\Lambda+i\delta}^{t+i\delta} \{m(\lambda) - n(\lambda, X)\} \, d\lambda \right|.$$

We estimate this by deforming the contour so as to pass through the points $\Lambda + i$, $t + i$, taking it that t is so large that $\delta < 1$. On the segment $(t + i, \; t + i\delta)$ we write $\lambda = t + i\nu$, and have

$$m(\lambda) - n(\lambda, X) = O\left\{ \nu^{-1}|t|^{\frac{1}{2}} \exp(-2X|t|^{\frac{1}{2}}) \right\}, \tag{4.2.6}$$

by (3.6.9). Integrating over $\delta \leqslant \nu \leqslant 1$, we get an amount of order $O\left\{ |\ln\delta| \, |t|^{\frac{1}{2}} \exp(-2X|t|^{\frac{1}{2}}) \right\}$; this order-bound will also apply to the contribution of the segment $(\Lambda + i\delta, \; \Lambda + i)$. On the segment $(\Lambda + i, \; t + i)$ we use (4.2.6) with $\nu = 1$ and with t replaced by s, where $\Lambda \leqslant s \leqslant t$, to get an amount of order

$$O\left\{ \int_{\Lambda}^{t} |s|^{\frac{1}{2}} \exp(-2X|s|^{\frac{1}{2}}) \, ds \right\} = O\left\{ (|t| X^{-1} + X^{-3}) \exp(-2X|t|^{\frac{1}{2}}) \right\}. \tag{4.2.7}$$

Hence, uniformly in Λ,

$$K_1 - K_2 = O\left\{ |t| \exp(-2X|t|^{\frac{1}{2}}) \right\}. \tag{4.2.8}$$

Turning to K_2 , we deform the contour in (4.2.5) so as to pass along the real axis from Λ to t ; this is permissible in view of Lemma 3.4.2, if t is suitably large and negative. Since $n(\lambda, X)$ is then real-valued along (Λ, t), this segment of the contour contributes zero to (4.2.5). Using Theorem 3.7.4, or (3.4.3), we see that the contributions of the vertical segments of the resulting contour to (4.2.5) are of orders $O(\delta \, |\Lambda|^{-\frac{1}{2}})$, $O(\delta \, |t|^{-\frac{1}{2}})$. Hence, uniformly in Λ ,

$$K_2(\Lambda, t) = O(|t|^{\frac{1}{2}} \exp(-2X|t|^{\frac{1}{2}})) . \tag{4.2.9}$$

It follows that $K_2(\Lambda, t)$, and so also $K_1(\Lambda, t)$, are bounded as $\Lambda \to -\infty$. Since $\operatorname{Im} m(\lambda) > 0$, we have that the integral in (4.2.3) must tend to a finite limit as $\Lambda \to -\infty$, and so, by

(4.2.4), the non-decreasing function $\tau(\Lambda)$ must also tend to a finite limit as $\Lambda \to -\infty$. We then derive the required result (4.2.1) from (4.2.4) and (4.2.8-9).

If we apply this result to the case $m(\lambda) \in D(b, k)$, we deduce that

$$\int_{-\infty}^{1} \exp(\sigma |t|^{\frac{1}{2}})\, d\tau(t) < \infty \ , \text{ if } \sigma < 2b \ . \qquad (4.2.10)$$

For the case that q satisfies (3.2.5), this is a well-known result of Marčenko (see, e.g., the discussion in (15) (English translation), p. 193).

The results apply uniformly to any $m(\lambda)$ satisfying the conditions of the Theorem. For an individual function $m(\lambda) \in C(X, k)$, of the form given by (1.2.2-3), the result (4.2.1) would be trivial since the set of points of increase of $\tau(t)$ would be bounded below. However the statement is non-vacuous when applied to the collection of such $m(\lambda)$. The case $q \equiv 0$ will serve here to show that the left of (4.2.1) can be of order at least $|t|^{\frac{1}{2}} \exp(-2X|t|^{\frac{1}{2}})$.

4.3. Approximation to $\tau(t)$ as $t \to +\infty$.

It is here that we meet a recognisable asymptotic behaviour, as given by (1.1.8). We first separate off the tail of the distribution by writing, for large $t > 0$,

$$\tau(t) - \tau(-\infty) = \{\tau(-t) - \tau(-\infty)\} + \{\tau(t) - \tau(-t)\} . \qquad (4.3.1)$$

We choose some fixed $X_0 \in (0, b)$, and use (4.2.1) to estimate the first term on the right, and thus have

$$\tau(-t) - \tau(-\infty) = O\{t \exp(-2X_0 t^{\frac{1}{2}})\} , \qquad (4.3.2)$$

provided that this $\tau(t)$ arises from an $m(\lambda)$ satisfying

$$m(\lambda) \in D(X, k), \quad X_0 \leq X < b \ . \qquad (4.3.3)$$

For such X, we then proceed to find an approximation to the last term in (4.3.1), though with a much large error-term than that in (4.3.2).

We impose the restrictions

$$Xt^{\frac{1}{2}} > 10 \ , \quad XG(X, t^{\frac{1}{2}}) < \tfrac{1}{8} \ , \qquad (4.3.4\text{-}5)$$

which will, for given $X \in [X_0, b)$, be satisfied for large t, by the assumptions of § 3.2. We have then

Theorem 4.3.1. Under the above conditions

$$|\tau(t) - \tau(-t) - 2\pi^{-1} t^{\frac{1}{2}}| \leq 20X^{-1} . \qquad (4.3.6)$$

We use Theorem 2.3.1 with

$$\delta = 10X^{-1} t^{\frac{1}{2}}, \quad \Lambda_1 = -t, \quad \Lambda_2 = t . \qquad (4.3.7)$$

We note that $\delta < t$, and take the contour in the integral on the left of (2.3.1) through the points

$$-t + i\delta, \quad -t + it, \quad t + it, \quad t + i\delta . \qquad (4.3.8)$$

We need to bound a number of error-terms by numbers $J_1, \ldots, J_7$; these will in turn be bounded by various multiples of X^{-1} .

The first two of these error-terms arise from the right of (2.3.1). Noting that the numerical factor there is less than 0.6, we set

$$J_r = (0.6)\, \delta\, |m(\Lambda_r + i\delta)|, \quad r = 1, 2. \qquad (4.3.9)$$

We write $(-t + i\delta)^{\frac{1}{2}} = \alpha + i\beta$ and have, if $\beta X \geq 4$, by (3.6.3),

$$J_1 \leq (0.6)\, \delta \{t^{-\frac{1}{2}} + (4Xt)^{-1} + 108\alpha^{-1} e^{-2\beta X}\}. \qquad (4.3.10)$$

Now

$$\alpha^2 = \tfrac{1}{2}\{(t^2 + \delta^2)^{\frac{1}{2}} - t\} \geq \tfrac{1}{2}\delta^2 t^{-1}(2^{\frac{1}{2}} - 1), \qquad (4.3.11)$$

since $\delta > t$, and so

$$\alpha \geq t^{-\frac{1}{2}}(0.455). \qquad (4.3.12)$$

Similarly,

$$\beta^2 = \tfrac{1}{2}\{\sqrt{(t^2 + \delta^2)} + t\}, \qquad (4.3.13)$$

and so

$$\beta > t^{\frac{1}{2}} ; \qquad (4.3.14)$$

in particular, (4.3.4) ensures that $\beta X \geq 4$, as required for (3.6.3). Proceeding to bound the individual terms on the right of (4.3.10), we have that $\delta t^{-\frac{1}{2}} = 10X^{-1}$ by (4.3.7), that $\delta X^{-1} t^{-1} = 10X^{-2} t^{-\frac{1}{2}} \leq X^{-1}$ by (4.3.4), and that

$$\delta \alpha^{-1} e^{-2\beta X} \leq (0.455)^{-1} t^{\frac{1}{2}} e^{-2Xt^{\frac{1}{2}}} \leq (0.455)^{-1} 10e^{-20} X^{-1} ,$$

by (4.3.4) and (4.3.12). Combining these estimates we get that

$$J_1 < 7X^{-1} . \qquad (4.3.15)$$

In the estimation of J_2 we set $\alpha + i\beta = (t + i\delta)^{\frac{1}{2}}$, and have to interchange the bounds (4.3.12), (4.3.14), so that

$$\alpha > t^{\frac{1}{2}} , \quad \beta > \delta t^{-\frac{1}{2}}(0.455) = (4.55)X^{-1} , \qquad (4.3.16)$$

and again $\beta X > 4$. For the last term in (3.6.3-4) we thus have $\alpha^{-1} e^{-2\beta X} < t^{-\frac{1}{2}} e^{-9}$. Hence, from a bound of the form (4.3.10),

$$J_2 < 7X^{-1} . \qquad (4.3.17)$$

We now consider the contribution of the main term ik^{-1} in (3.6.3-4) to the integral in (2.3.1); we show that this approximates to the main term in (4.3.6). We write

$$J_3 = 2\pi^{-1} \left| t^{\frac{1}{2}} - \mathrm{Re}\,(t + i\delta)^{\frac{1}{2}} + \mathrm{Re}\,(-t + i\delta)^{\frac{1}{2}} \right| . \qquad (4.3.18)$$

Using (4.3.11) we have

$$\left| \mathrm{Re}\,(-t + i\delta)^{\frac{1}{2}} \right| < \tfrac{1}{2}\delta t^{-\frac{1}{2}} = 5X^{-1} ,$$

by (4.3.7) and, in a similar way,

$$0 < \mathrm{Re}\,(t + i\delta)^{\frac{1}{2}} - t^{\frac{1}{2}} < \tfrac{1}{8}\delta^2 t^{-3/2} < 5/(4X) ,$$

by (4.3.4), (4.3.7). Hence

$$J_3 < 4X^{-1} . \qquad (4.3.19)$$

Next we denote by J_4 a bound for the error arising from integrating the first term on the right of (3.6.4) around the contour (4.3.8). We can take

$$J_4 = (4t)(4Xt)^{-1} \pi^{-1} < X^{-1} ; \qquad (4.3.20)$$

here $4t$ is a bound for the length of the contour, and the factor π^{-1} comes from the left of (2.3.1).

In the last stage of the proof of Theorem 4.3.1 we write J_5, J_6 and

J_7 for the contributions, in modulus, of the last term in (3.6.3-4), when integrated along the three segments of the contour (4.3.8). For J_5 we write $\lambda = -t + is$, $\delta \leq s \leq t$, and have, using (4.3.12) with δ replaced by s, and (4.3.14), that

$$J_5 \leqslant 108(0.455\pi)^{-1} \int_\delta^t t^{\frac{1}{2}} s^{-1} \exp(-2Xt^{\frac{1}{2}})\, ds$$

$$< 76t^{\frac{1}{2}} \exp(-2X t^{\frac{1}{2}}) \ln (Xt^{\frac{1}{2}}/10) .$$

Since $Xt^{\frac{1}{2}} \geqslant 10$, $\ln v < v$ if $v > 0$, we find that

$$J_5 < \left\{(7.6)X^2 t \exp(-2Xt^{\frac{1}{2}})\right\} X^{-1} .$$

Since $u^2 e^{-2u}$ is decreasing when $u > 1$, and $Xt^{\frac{1}{2}} \geqslant 10$, we have that

$$J_5 \leqslant 7600\, e^{-20} X^{-1} < 10^{-4} X^{-1} . \tag{4.3.21}$$

For J_6, we set $\lambda = s + it$, $-t \leqslant s \leqslant t$, $\alpha + i\beta = (s + it)^{\frac{1}{2}}$, so that $\alpha^2 \geqslant \frac{1}{2}\left\{(s^2 + t^2)^{\frac{1}{2}} + s\right\} \geqslant \frac{1}{2}t(2^{\frac{1}{2}} - 1)$, and hence $\alpha > t^{\frac{1}{2}}(0.455)$, and likewise for β. Hence

$$J_6 \leqslant (2t)(108\pi^{-1})(0.455)^{-1} t^{-\frac{1}{2}} \exp\left\{-(0.9)Xt^{\frac{1}{2}}\right\}$$

$$< 152\, t^{\frac{1}{2}} \exp\left\{-(0.9)Xt^{\frac{1}{2}}\right\}.$$

Since $Xt^{\frac{1}{2}} \geqslant 10$, and since $u \exp\left\{-(0.9)u\right\}$ is decreasing for $u > 10/9$, we deduce that

$$J_6 \leqslant 1520\, e^{-9} X^{-1} < 5^{-1} X^{-1} . \tag{4.3.22}$$

Finally, for J_7 we put $\lambda = t + is$, $\delta \leqslant s \leqslant t$, and have (4.3.17) with s in place of δ. This gives

$$J_7 < 108\pi^{-1} t^{-\frac{1}{2}} \int_\delta^t \exp\left\{-(0.9)sXt^{-\frac{1}{2}}\right\} ds$$

$$< 108\pi^{-1}(0.9)^{-1} X^{-1} e^{-9} < 10^{-2} X^{-1} . \tag{4.3.23}$$

Our estimates for $J_1, \ldots, J_4$ above total $19X^{-1}$, and so with the smaller bounds (4.3.21-23) for J_5, J_6 and J_7 we have proved (4.3.6).

4.4. Asymptotic formulae for the case of fixed X.

We summarize the results of §§4.2-3 using the first of the two procedures indicated in §4.1, with $b \leqslant \infty$. Always with the hypotheses of §3.2, we have

Theorem 4.4.1. Let $X_0 \in (0, b)$ be fixed and, for some $X \in [X_0, b)$ let $m(\lambda)$ be a Nevanlinna function satisfying $m(\lambda) \in D(X, k)$ for all λ with $\mathrm{Im}\,\lambda > 0$. Let $\tau(t)$ be an associated spectral function, satisfying (1.1.11). Let $T = T(X)$ satisfy

$$XT^{\frac{1}{2}} > 10, \quad XG(X, T^{\frac{1}{2}}) \leqslant \tfrac{1}{8}. \tag{4.4.1-2}$$

Then, for $t \geqslant T(X)$,

$$\tau(t) - \tau(-\infty) = 2\pi^{-1}t^{-\frac{1}{2}} + E_1(t) + E_2(t), \tag{4.4.3}$$

where

$$E_1(t) < 20X^{-1}, \tag{4.4.4}$$

and

$$E_2(t) = O(t \exp(-2X_0 t^{\frac{1}{2}})). \tag{4.4.5}$$

Here (4.4.5) holds uniformly in X, for $X \in [X_0, b)$. In particular, one has

$$\tau(t) = 2\pi^{-1}t^{\frac{1}{2}} + O(1), \tag{4.4.6}$$

as $t \to \infty$.

Looking more closely at (4.4.3), one sees an $O(1)$ term (4.4.4), and an exponentially small term (4.4.5). In the first of these, the numerical factor "20" is of course too large; however this factor cannot be reduced below "2". We see this by considering the case $q \equiv 0$ with boundary condition $\phi'(X, \lambda) = 0$, when the jumps (1.2.2) in the spectral function are all, except for the first, equal to $2X^{-1}$.

For similar results, with an $O(1)$ or $o(1)$ error-term, obtained by entirely different methods, we refer to (15) - (18).

4.5. Asymptotic formulae with an $o(1)$ error-term.

We now concentrate on the case $b = \infty$, though without any hypothesis as to whether the limit-point or limit-circle case holds; the result will apply to an $m(\lambda)$ satisfying (4.1.2), whether this is unique or not.

Theorem 4.5.1. Let the assumptions of § 3.2 hold, with $b = \infty$. Let $X = X(t)$ be determined so that

$$X(t) \to \infty \quad \text{as} \quad t \to \infty , \quad X(t)G(X(t), t^{\frac{1}{2}}) \leq \tfrac{1}{8}, \qquad (4.5.1\text{-}2)$$

for large $t > 0$. Then for any $\tau(t)$ associated with an $m(\lambda)$ satisfying (4.1.2) we have

$$\tau(t) - \tau(-\infty) = 2\pi^{-1}t^{\frac{1}{2}} + O(X^{-1}) + O(\exp(-\sigma t^{\frac{1}{2}})), \qquad (4.5.3)$$

as $t \to \infty$, for any $\sigma > 0$. In particular,

$$\tau(t) - \tau(-\infty) = 2\pi^{-1}t^{\frac{1}{2}} + o(1) \quad \text{as} \quad t \to +\infty . \qquad (4.5.4)$$

The last error-term in (4.5.3) will almost certainly be negligible compared to the term $O(X^{-1})$. The latter can be estimated in particular cases, as in § 3.8, however with a variation in notation from (3.8.4-5). If $q(x)$ is monotone, with $q(0) = 0$, we need that

$$X(t) \to \infty , \quad \limsup \; t^{\frac{1}{2}}X(t)q(X(t)) < \tfrac{1}{8} .$$

In the case $q(x) = x^N$, $N > 0$, this gives

$$\tau(t) - \tau(-\infty) = 2\pi^{-1}t^{\frac{1}{2}} + O(t^{-1/(2N+2)}).$$

4.6. The case $\phi'(0, \lambda) = h$.

The same arguments can be applied if the last of (1.1.2) is modified in this way. As indicated in § 1.3, one gets a correction term $h\lambda^{-1}$ in the approximations to $m(\lambda)$. When this is fed into the integral in (2.3.1) there results an additional term h in the approximation to $\tau(t)$, so that, if $b = \infty$, and (4.1.2) holds,

$$\tau(t) - \tau(-\infty) = 2\pi^{-1}t^{\frac{1}{2}} - h + o(1).$$

More substantial changes take place if the roles of θ, ϕ in (1.1.1-2) are interchanged, so that, roughly speaking, $m(\lambda)$ behaves as $\lambda^{\frac{1}{2}}$ and $\tau(t)$ as $t^{3/2}$. It seems likely that the present arguments can be used for this case, but the details will be very different.

REFERENCES

1. F. V. Atkinson, "Discrete and Continuous Boundary Problems" Academic Press, New York, 1964.

2. F. V. Atkinson, On the location of the Weyl circles, Proc. Roy. Soc. Edin. A 88, 345-356 (1981).

3. C. Bennewitz and W. N. Everitt, Some remarks on the Titchmarsh-Weyl m-coefficient, in "Tribute to Åke Pleijel", Department of Mathematics, University of Uppsala, Uppsala, 1980.

4. E. A. Coddington and N. Levinson, "Theory of Ordinary Differential Equations", McGraw-Hill, New York, 1955.

5. W. D. Evans and W. N. Everitt, A return to the Hardy-Littlewood inequality, Proc. Roy. Soc. London, A 380, 447-486 (1982).

6. W. N. Everitt, On a property of the m-coefficient of a second-order linear differential equations, Jour. London Math. Soc. (2), 443-457 (1972).

7. W. N. Everitt and S. G. Halvorsen, On the asymptotic form of the Titchmarsh-Weyl m-coefficient, Applicable Analysis, 8, 153-169, (1978).

8. W. N. Everitt, D. B. Hinton and J. K. Shaw, The asymptotic form of the Titchmarsh-Weyl coefficient for systems, (preprint).

9. W. N. Everitt and A. Zettl, On a class of integral inequalities, Jour. London Math. Soc. (2), 17, 291-303, (1978).

10. E. Hille, "Lectures on Ordinary Differential Equations", Addison-Wesley, London, 1969.

11. D. B. Hinton and J. K. Shaw, On Titchmarsh-Weyl $M(\lambda)$-functions for linear Hamiltonian systems, Jour. of Differential Equations, 40, 316-342 (1981).

12. I. S. Kac, A generalization of the asymptotic formula of V. A. Marčenko for the spectral function of a second-order boundary-value problem, Izv. Akad. Nauk SSSR Ser. Mat. 37, 422-436 (1973), also Mathematics Izvestija USSR 7, 424-436 (1973).

13. I. S. Kac, R-functions - analytic functions mapping the upper half-plane into itself, Amer. Math. Soc. Transl. (2), 103, 1-18, (1974), (Supplement I to reference no. 1 above, "Mir", Moscow, 1968)

14. I. S. Kac and M. G. Krein, On the spectral functions of the string, Amer. Math. Soc. Transl. (2),103, 19-102 (1974), (Supplement II to reference no. 1 above, "Mir", Moscow, 1968).

15. B. M. Levitan, On the asymptotic behaviour of the spectral function of a selfadjoint second-order differential equation, Izv. Akad. Nauk SSSR Ser. Mat. 16, 325-352 (1952), also Amer. Math. Soc. Transl.(2) 101, 192-221 (1973).

16. B. M. Levitan and I. S. Sargsyan,"Introduction to Spectral Theory for Selfadjoint Ordinary Differential Equations", "Nauka", Moscow, 1970, also in "Translations of Mathematical Monographs", Vol. 39, American Mathematical Society, Providence, R. I., 1975.

17. V. A. Marčenko, Some questions in the theory of one-dimensional second-order linear differential operators. I, Trudy Moskov. Mat. Obšč. 1, 327-340 (1952), also Amer. Math. Soc. Transl. (2) 101, 1-104 (1973).

18. V. A. Marčenko, "Sturm-Liouville Operators and their Applications", Naukova Dumka, Kiev, 1977.

19. E. C. Titchmarsh, "Eigenfunction Expansions", Part I, 2nd. edn., Oxford University Press, 1962.

20. S. D. Wray, On Weyl's function m(λ), Proc. Roy. Soc. Edin. A 74, 41-48, 1974/75.

F V Atkinson
Department of Mathematics
University of Toronto
Toronto
Ontario
Canada

Some Limit Circle Eigenvalue Problems and

Asymptotic Formulae for Eigenvalues†

by

F V Atkinson and C T Fulton

§1 Introduction

Asymptotic properties of eigenvalues of the second order equation

$$-y'' + qy = \lambda y$$

on the half line $[0,\infty)$ with a regular boundary condition at $x = 0$ have been obtained by a number of authors under various conditions on $q(x)$ in the case when $q(x) \to +\infty$ as $x \to \infty$. We mention, in particular, the work of Titchmarsh [24,26, 27], McLeod and Titchmarsh [21], McLeod [19,20], Giertz [9], K Jörgens [17], Levitan and Sargsjan [18, Chap. 12] and Hartman [11,12,13,14,15]. The work on this case goes back as far as 1929 when W E Milne [22] obtained an asymptotic estimate for the number of eigenvalues in the interval $[0,\lambda]$ (the so-called Milne formula). The case when $q(x)$ increases to ∞ belongs to the limit point case at ∞, has a discrete spectrum, and was perhaps the simplest situation to investigate since no boundary condition at ∞ was required.

In contrast to the amount of effort which has been devoted to the asymptotics of eigenvalues in the limit point case, considerably less has been done in the case when $q(x) \to -\infty$ faster than $-x^2$ and belongs to the limit circle case at ∞. In 1954, P Heywood [16], was the first to obtain a limit circle analogue of the Milne formula, obtaining density formulae for both the positive and negative eigenvalues. Heywood's results were rediscovered much later by Belograd and Kostyuchenko [6] in 1973, who were unaware of his work. Subsequently, A G Alenitsyn [1] in 1976 improved on Heywood's density formulae, obtaining 'quantum conditions' for both the positive and negative eigenvalues in which the parameter indexing the choice of boundary condition at ∞ made an appearance. Alenitsyn's quantum conditions were sharp enough to permit reversion for the eigenvalues, but his error bounds for them were only good enough to restrict the large positive and large negative eigenvalues to lie in disjoint intervals.

Recently, the present authors have studied the asymptotics of eigenvalues for problems with limit circle endpoints on both finite and infinite intervals, which cover the following cases:

(i) $[0,\infty)$, $q(x) \to -\infty$ faster than $-x^2$,

(ii) $(0,b]$, $q(x) = C/x^\alpha$, $C \neq 0$, $1 \leq \alpha < 2$,

(iii) $(0,b]$, $q(x) = -1/x^\alpha$, $\alpha > 2$.

The full results of these investigations will be published separately in a sequence of three papers [3,4,5]. For problem (i) an iterative procedure of Atkinson [2] enables lower order terms in the quantum condition for the positive eigenvalues to be produced algorithmically, so that higher order terms, as many as desired, can be produced in the asymptotic expansion of the nth positive eigenvalue as $n \to \infty$. For the negative eigenvalues, a piecewise turning point analysis is employed which improves on the strength of Alenitsyn's error bounds, but which still falls short of locating the eigenvalues with error term tending to zero. Here, there is still room for improvements, and possibly a different method of approach will prove more successful.

In the present paper we restrict attention to the Heywood density formulae, and show that his results are already sharp enough to draw conclusions about the exponent of convergence of the positive and negative eigenvalues. We also show that the Heywood formula for the positive eigenvalues can be obtained very easily by following the approach the authors have given in [3]. Finally, we also make use of the results on the exponent of convergence of the eigenvalues to deduce results on the order of the entire $w(\lambda)$-function whose zeros are the eigenvalues of a given limit-circle eigenvalue problem. The $w(\lambda)$-function was introduced by Fulton [8, p.56, Equa. (2.14)(i)] as a replacement for the Weyl-Titchmarsh $m(\lambda)$-function (of which it is simply the denominator), and is known to have order ≤ 1. A conjecture of Professor W N Everitt to the effect that $w(\lambda)$ can have order larger than 1/2, in contrast to regular Sturm-Liouville problems, is answered in the affirmative, and, in fact, it is shown by examples that all possible values of the order, 1/2 up to 1, occur.

One might be inclined to ask why the asymptotics of limit circle eigenvalues was not investigated much earlier than 1982. While there are several intricate problems which have to be overcome, one matter seems to stand out: Previous authors make their analysis rest on the formulation of limit circle boundary conditions given by Titchmarsh [26] in his classic book on eigenfunction expansions. Heywood, for example, was lead to use Titchmarsh's solutions $\{\phi_\lambda,\theta_\lambda\}$ of the basic equation defined at $x = 0$, analyze its distribution of zeros over $[0,b]$, and then pass $b \to \infty$ in order to apply a limit circle boundary condition of the form

$$\lim_{x\to\infty} W_x\big(y,\, \theta_\lambda + m(\lambda)\phi_\lambda\big) = 0 ,$$

where $m(\lambda)$ is the limit of Titchmarsh's $\ell_b^\beta(\lambda)$-function as $b \to \infty$. Since this limit is not uniquely defined in the limit circle case, it is well known that one must let $b \to \infty$ through an appropriate sequence of values of b, or make β to vary with b in such a way that the limit will exist as $b \to \infty$ continuously. Heywood chose the latter method, linking his analysis to the 1950 paper of Sears and Titchmarsh [23], the only existing piece of literature at that time where the limit circle $m(\lambda)$ functions had been parametrized in the form

$$m(\lambda) = -\,\frac{a(\lambda)\cot K + b(\lambda)}{c(\lambda)\cot K + d(\lambda)}\,, \qquad K \in [0,\pi) ,$$

with K being the index on the choice of boundary condition at ∞ and $a(\lambda)$, $b(\lambda)$, $c(\lambda)$, $d(\lambda)$ being entire functions of λ. Similarly, A G Alenitsyn used a Titchmarsh-type boundary condition at ∞, using a parametrization of the limit circle $m(\lambda)$-functions equivalent to that given by Fulton [8, p.52, Equa. (1.9)]. Alenitsyn's reliance on the $m(\lambda)$-function to formulate the boundary condition at ∞ seems to have given rise to the less than optimal '$O(1)$'-error bound, cf. [1; p.300, Equa. (1.9)].

The work of the present authors, in contrast, dispenses with the use of $m(\lambda)$-functions to formulate limit circle boundary conditions, and employs instead the formulation used by Fulton [8, p.57, Equa. (2.18)], the limit circle boundary condition at ∞ being parametrized in the form

$$\lim_{x\to\infty} W_x(f, v \cos\gamma + v \sin\gamma) = 0 , \quad \gamma \in [0,\pi) ,$$

where $\{u,v\}$ is a fundamental system of the basic equation for $\lambda = 0$. Relying on the same Prüfer equation as Heywood, the reader will note (Theorem 1 below) that this formulation of boundary conditions at ∞ affords a considerably simpler proof of the Heywood formula for the positive eigenvalues, than that given by Heywood. This seems to suggest that results on asymptotics of limit circle eigenvalues would have been obtained much earlier if Titchmarsh had formulated the limit circle boundary conditions in the above form at the outset. The above mentioned work of the present authors seems therefore to fill a long neglected gap in the Weyl-Titchmarsh theory for second-order ordinary differential equations.

§2 Heywood Density Formulae

Putting $q(x) = -F(x)$ we consider the singular eigenvalue problem on the interval $[0,\infty)$,

$$(*) \quad \begin{cases} -y'' - Fy = \lambda y & (1) \\ y(0)\cos\alpha + y'(0)\sin\alpha = 0 , \quad \alpha \in [0,\pi) & (2) \\ [\lim_{x\to\infty} W_x(y,v)]\cos\gamma + [\lim_{x\to\infty} W_x(y,u)]\sin\gamma = 0, \quad \gamma \in [0,\pi) , & (3) \end{cases}$$

where $\{u,v\}$ is a fundamental system of the equation for $\lambda = 0$, $y'' + Fy = 0$, satisfying

$$W_x(u,v) = uv' - u'v = 1 . \qquad (4)$$

This formulation the limit circle boundary condition at ∞ was the choice used by Fulton in [8], where the equivalence to Titchmarsh's form of boundary conditions is established. The only essential difference is that the above form of boundary conditions makes use of limits of Wronskian combinations of y with solutions of the basic equation for a real value of λ, while Titchmarsh's form of boundary conditions makes use of limits of Wronskian combinations of y with solutions of the basic equation for nonreal values of λ. In applications it is usually easier to apply the above form of boundary conditions because the solutions $\{u,v\}$ can often be found explicitly for many of the equations of mathematical physics. As we shall see in Theorem 1 below the above form also has some theoretical advantages.

For the case of the positive spectrum we make the following assumptions on F:

$$F(x) \geq 0 \quad \text{for} \quad x \in [0,\infty), \text{ and } \quad F'(x) > 0 \quad \text{for} \quad x > 0 \,, \tag{5)(i}$$

$$\lim_{x\to\infty} F(x) = \infty \tag{ii}$$

$$F(x) \in C^2[0,\infty) \tag{iii}$$

$$\int^{\infty} \frac{1}{\sqrt{F}}\, dx < \infty \tag{iv}$$

$$\lim_{x\to\infty} \frac{F'(x)}{F(x)^{3/2}} = 0 \tag{v}$$

$$\frac{F''}{F^{3/2}}, \frac{(F')^2}{F^{5/2}} \in L_1(x_0,\infty) \,, \quad 0 < x_0 < \infty \tag{vi}$$

$$F'' \,, \quad (F')^2 \in L_1(0,x_0) \,, \quad 0 < x_0 < \infty \,. \tag{vii}$$

Under these assumptions we prove:

Theorem 1 (Heywood/Belogrud-Kostyuchenko)

Let $E(\lambda)$ denote the number of nonnegative eigenvalues of (*) in $[0,\lambda]$. Then

$$E(\lambda) = \frac{1}{\pi} \int_0^{\infty} \frac{\lambda}{\sqrt{\lambda+F} + \sqrt{F}}\, dx + O(1) \,, \quad \text{as} \quad \lambda \to +\infty \,. \tag{6}$$

Under these same assumptions we can also prove the following theorems:

Theorem 2 (Alenitsyn, $\alpha = 0$)

Let the positive eigenvalues of (*) be ordered by $0 \leq \lambda_1 < \lambda_2 < \ldots . < \lambda_n < \ldots$.

Then the nth positive eigenvalue satisfies the equation

$$n\pi = \int_0^{\infty} \frac{\lambda_n}{\sqrt{\lambda_n+F} + \sqrt{F}}\, dx - \gamma + O(1) \,, \quad \text{as} \quad \lambda_n \to \infty \,. \tag{7}$$

Theorem 3 $\alpha \in (0,\pi)$, $\alpha \neq 0$, and $\gamma \in [0,\pi)$.

The nth positive eigenvalue of (*) is characterized implicitly as the unique root of the equation

$$(n+\tfrac{1}{2})\pi = \int_0^\infty \frac{1}{\sqrt{\lambda+F} + \sqrt{F}}\,dx - \gamma + \frac{\cot\alpha}{\left(\lambda+F(0)\right)^{\frac{1}{2}}}$$

$$+ O\left[\int_0^\infty \frac{|F''|}{(\lambda+F)^{3/2}} + \frac{|F'|^2}{(\lambda+F)^{5/2}} + \frac{1}{\lambda^{3/2}}\right], \text{ as } \lambda \to \infty . \qquad (8)$$

In the case of $\alpha = 0$, this holds with $(n+\frac{1}{2})\pi$ replaced by $n\pi$ and the term involving $\cot\alpha$ deleted.

Theorem 1 is contained in Theorem 2 and Theorem 2 is contained in Theorem 3. Indeed, putting $\lambda = \lambda_n$ in (5) gives (6), and (7) just provides a sharper error bound to replace Alenitsyn's O(1)-error term in (6). The Heywood formula (5) is a density formula which estimates only the number of eigenvalues in $[0,\lambda]$ irrespective of the choice of boundary condition parameters α and γ. The Alenitsyn formula (6), on the other hand, incorporates the dependence on γ, but the dependence of the eigenvalues on the boundary condition at $x = 0$ is still contained in the O(1)-error term. This 'quantum condition' can be shown to restrict the eigenvalues to lie in disjoint intervals for the case of the example $F(x) = x^{2\varepsilon}$, $\varepsilon > 1$, but is not sharp enough to locate the eigenvalues with error term tending to zero. The quantum condition (8) obtained by the present authors incorporates the dependence of the nth eigenvalue, $\lambda_n = \lambda_n^{\alpha,\gamma}$ on both α and γ and the dependence on γ is seen to be more dominant than the dependence on α. In examples, a reversion for the asymptotics of λ_n as $n \to \infty$ gives an asymptotic formula for λ_n with error term tending to zero, thus "locating" the large eigenvalues.

Under the additional assumptions

$F''(x)$ is ultimately of one sign (9)(i)

$F'(x)/\left(F(x)\right)^d$ is monotone decreasing for x sufficiently large and for some $d < 3/2$ (ii)

$F''(x) = O\left(\left(F'(x)\right)^\gamma\right)$ where $1 < \gamma < 4/3$. (iii)

P Heywood obtained the following result for the density of the negative eigenvalues:

<u>Theorem 4</u> Let $\mu = -\lambda$, and let $G(\mu)$ be the number of negative eigenvalues of (*) contained in $[-\mu,0]$. Let $x_0 = x_0(\mu)$ be the turning point of $F(x)$ defined

by $F(x_0(\mu)) = \mu$ for all $\mu > 0$. Then

$$G(\mu) = \frac{1}{\pi}\left[\int_0^{X_0(\mu)} \sqrt{F}\,dx + \int_{X_0(\mu)}^{\infty} \frac{\mu}{\sqrt{F}+\sqrt{F-\mu}}\,dx\right] + O(1)\ , \text{ as } \mu \to \infty\ . \tag{10}$$

Proof of Theorems 1-3:

For $\lambda > 0$ we make in (1) the modified Prüfer transformation,

$$\begin{aligned} y &= \frac{r \sin\theta}{(\gamma+F)^{1/4}} \\ y' &= (\lambda+F)^{1/4} r \cos\theta\ , \end{aligned} \tag{11}$$

which gives rise to the first order system for r and θ,

$$\begin{aligned} \theta' &= \sqrt{\lambda+F} + \frac{F'}{4(\lambda+F)} \sin 2\theta \\ \frac{r'}{r} &= -\frac{1}{4}\frac{F'}{(\lambda+F)} \cos 2\theta\ . \end{aligned} \tag{12}$$

Writing (12) in the form

$$1 = \frac{\theta'}{\sqrt{\lambda+F}} - \left(\frac{F'}{4(\lambda+F)^{3/2}}\right) \sin 2\theta$$

we multiply the second term in (12) by the right hand side of the above and perform an integration by parts on the term $\frac{F'}{4(\lambda+F)} (\sin 2\theta\cdot)\left(\frac{\theta'}{\sqrt{\lambda+F}}\right)$, to obtain the equation

$$\begin{aligned} \theta' = \sqrt{\lambda+F} &- \frac{1}{8}\left(\frac{F'}{(\lambda+F)^{3/2}} \cos 2\theta\right)' - \frac{1}{32}\left(\frac{(F')^2}{(\lambda+F)^{5/2}}\right) \\ &+ \frac{\cos 2\theta}{8}\left(\frac{F'}{(\lambda+F)^{3/2}}\right)' + \frac{\cos 4\theta}{32}\left(\frac{(F')^2}{(\lambda+F)^{5/2}}\right)\ . \end{aligned} \tag{13}$$

A similar manipulation on the right hand side of the r-equation brings the r-equation into the form

$$r'/r = \frac{1}{8}\left(\sin 2\theta \frac{F'}{(\lambda+F)^{3/2}}\right)' - \frac{\sin 2\theta}{8}\left(\frac{F'}{(\lambda+F)^{3/2}}\right)' + \frac{(F')^2}{32(\lambda+F)^{5/2}} \sin 4\theta\ . \tag{14}$$

Equation (13) is the same equation used by Hewood to prove theorem 1, cf. [16, p.459, Equa. (2.3)]. Heywood integrates this equation over $[0,b]$ to obtain an estimate on the number of zeros of the solution satisfying the boundary condition at $x = 0$ ([16, p.459, Equa. (2.8)]), and then makes use of this result to obtain a Milne-type formula for the number of eigenvalues in $[0,\lambda]$ for a regular Sturm-Liouville problem on $[0,b]$, e.g. [16; p.464, Equa. (5.2)],

$$E_b(\lambda) = \frac{1}{\pi} \int_0^b \left(\sqrt{\lambda+F(x)} - \sqrt{F(x)}\right)dx + O(1) . \tag{15}$$

The rest of his analysis is concerned with passing $b \to \infty$, so as to obtain equation (6) for the density of positive eigenvalues on the infinite interval. In order to fix a choice of boundary condition at ∞ he relies on results of Sears-Titchmarsh [23], who formulated sufficient conditions for the occurrence of the limit circle case at ∞, and provided a formula for $\beta = \beta(b,K)$ such that the limit circle $m(\lambda)$-functions were obtainable in the form

$$m(\lambda,K) = \lim_{b\to\infty} - \frac{\theta(b,\lambda) \cos \beta(b,K) + \theta'(b,\lambda) \sin \beta(b,K)}{\phi(b,\lambda) \cos \beta(b,K) + \phi'(b,\lambda) \sin \beta(b,K)}$$

$$= - \frac{a(\lambda) \cos K + b(\lambda) \sin K}{c(\lambda) \cos K + d(\lambda) \sin K}, \text{ with } b \to \infty \text{ continuously} .$$

The basic idea was to show that for large b the number of poles of $\ell_b^{\beta(b,K)}(\lambda)$ for fixed K in $[0,\Lambda]$ did not differ by more than two from the number of poles of $m(\lambda,K)$ in $[0,\Lambda]$. Once established, this permits passage to the limit $b \to \infty$ in (15) to give a proof of Theorem 1. The choice of K fixes the boundary condition at ∞, but since the dependence of $E_b(\lambda)$ on the boundary conditions at 0 and b is already contained in the $O(1)$ error term in (15), Heywood's approach rules out any hope of obtaining the dependence of $E(\lambda)$ on K. Also, his assumptions,

$$F''(x) \text{ ultimately of one sign ,} \tag{16)(i}$$

and

$$F''(x) = O\left(\left(F'(x)\right)^{\gamma}\right), \text{ for some } 1 < \gamma < 4/3 , \tag{ii}$$

seem to have been imposed primarily to guarantee that the Sears-Titchmarsh criteria for occurrence of the limit circle case were satisfied, cf. [16, p.457]. Theorem 1 actually holds under the weaker assumptions listed in (5) above.

To prove Theorem 1 more efficiently, and obtain Theorems 2 and 3 at the same time, we first apply the boundary condition (3) at ∞, relying on the parameter γ instead of the Sears-Titchmarsh parameter K, to index the boundary conditions at ∞. Indeed the main advantage of (13) over (12) is that (13) enables us to replace the boundary condition (3) by an equivalent growth condition on $\theta(x,\lambda)$ as $x \to \infty$, while (12) does not. To this end we prove:

Claim 1 Let $\gamma \in [0,\pi)$. Under the assumptions in (5), there exists, for each $\lambda \in (0,\infty)$, a unique solution of (13) satisfying

$$\lim_{x\to\infty} [\theta(x,\lambda) - \int_0^x \sqrt{F(x)}\ ] = \gamma . \tag{17}$$

Proof: Subtracting $\sqrt{F}$ from both sides of (13) and integrating over $[x_0,x]$ we have

$$\theta(x,\lambda) - \int_0^x \sqrt{F}\,dx = \theta(x_0,\lambda) - \int_0^{x_0} \sqrt{F}\,dt + \int_{x_0}^x \frac{\lambda}{\sqrt{\lambda+F} + \sqrt{F}}\,dt$$

$$- \frac{1}{8}\frac{F'}{(\lambda+F)^{3/2}} \cos 2\theta \Big|_{x_0}^{x} - \frac{1}{32}\int_{x_0}^x \frac{(F')^2}{(\lambda+F)^{5/2}}\,dt \tag{18}$$

$$+ \frac{1}{8}\int_{x_0}^x \cos 2\theta \left(\frac{F'}{(\lambda+F)^{3/2}}\right)'dt + \frac{1}{32}\int_{x_0}^x \cos 4\theta \left(\frac{(F')^2}{(\lambda+F)^{5/2}}\right)dt .$$

But under the assumptions (5)(iv), (v), (vi) the limit as $x \to \infty$ of the right hand side exists. The fact that (17) defines a unique solution of (13) then follows from a standard result on asymptotic integration (cf. Hartman [10; p.273, Theorem 1.1]).

Claim 2 For each $\lambda \in (0,\infty)$ and each solution (r,θ) of (13), (14), $\lim_{x\to\infty} r(x,\lambda)$ exists.

Proof: An integration of (14) over $[x,\infty)$ obtains the result because of (5)(v), (vi). q.e.d.

Claim 3 Let $\theta^\gamma(\cdot,\lambda)$ denote the solution of (13) which is defined by (17), and let $r^\gamma(x,\lambda)$ be any corresponding solution of (14). Then if $y^\gamma(\cdot,\lambda)$ denotes the corresponding solution (11) of the basic equation (1) we have for $\gamma_1 = \gamma_2$ and $\lambda,\lambda' \in [0,\infty)$,

$$\lim_{x\to\infty} W_x\left(y^{\gamma_1}(\cdot,\lambda), y^{\gamma_2}(\cdot,\lambda')\right) = r^{\gamma_1}(\infty,\lambda)\cdot r^{\gamma_2}(\infty,\lambda')\sin(\gamma_1-\gamma_2) \ . \tag{19}$$

Proof: The result follows by a computation of the Wronskian using (11) and a passage to the limit which takes advantage of (17). q.e.d.

Definition 1 We now fix the normalization of $y^\gamma(\cdot,\lambda)$ by fixing the initial condition defining $r^\gamma(\cdot,\lambda)$ in (14). For convenience we fix $r(0,\lambda)$ so as to have

$$r^\gamma(\infty,\lambda) \equiv 1 \quad \text{for all} \quad \gamma \in [0,\pi) \quad \text{and} \quad \lambda \in [0,\infty) \ . \tag{20}$$

Claim 4 Letting the solutions $\{u,v\}$ in the boundary condition at ∞ be chosen by

$$u(x) := \frac{r^{\pi/2}(x,0)\sin\theta^{\pi/2}(x,0)}{F(x)^{1/4}} \ , \qquad v(x) := \frac{r^0(x,0)\sin\theta^0(x,0)}{F(x)^{1/4}} \ , \tag{21}$$

we have

$$\lim_{x\to\infty} \begin{pmatrix} W_x\left(y^\gamma(\cdot,\lambda),v\right) \\ \\ -W_x\left(y^\gamma(\cdot,\lambda),u\right) \end{pmatrix} = \begin{pmatrix} \sin\gamma \\ \\ \cos\gamma \end{pmatrix}, \text{ for all } \lambda \in (0,\infty) \ . \tag{22}$$

Proof: Use equation (19) to compute the limits.

It follows from (22) that the solution $y^\gamma(\cdot,\lambda)$ obtained by fixing the initial conditions on θ and r at ∞ by (17) and (20) satisfies the boundary condition (3) at $x = \infty$ (under the choice of $\{u,v\}$ made in (21)).

Having obtained a solution satisfying the boundary condition at ∞, we can proceed to obtain a 'quantum condition' for the positive eigenvalues by applying the boundary condition at $x = 0$. In terms of $\theta^\gamma(x,\lambda)$ the boundary condition at $x = 0$ may be expressed as

$$\tan\theta(0,\lambda) = \begin{cases} -\sqrt{\lambda+F(0)}\ \tan\alpha \ , & \alpha \in [0,\pi), \quad \alpha \neq \frac{\pi}{2} \\ \infty \ , & \alpha = \pi/2 \ , \end{cases}$$

or

$$\theta(0,\lambda_n) = \begin{cases} -n\pi + \operatorname{Arctan}\left(-\sqrt{\lambda+F(0)}\ \tan\alpha\right) \ , & \alpha \neq \frac{\pi}{2} \\ -n\pi - \frac{\pi}{2} \ , & \alpha = \frac{\pi}{2} \ . \end{cases} \tag{23}$$

Claim 5 The nth positive eigenvalue of (*) is characterized as the unique root of the quantum condition (8). This proves Theorem 3, under the assumptions (5), and therefore also the weaker versions stated as Theorems 1 and 2.

Proof: Put $x_0 = 0$ in (18), let $x \to \infty$, use (17) and (23), and replace the Arctan using

$$\text{Arctan}\left(-\sqrt{\lambda+F(0)}\ \tan\alpha\right) = -\frac{\pi}{2} + \frac{1}{\sqrt{\lambda+F(0)}\ \tan\alpha} + O\left(\frac{1}{\lambda^{3/2}}\right) .$$

q.e.d.

Example $F(x) = x^{2\varepsilon}$, $1 < \varepsilon < \infty$, $\alpha = 0$.

In this case the quantum condition (8) becomes

$$\gamma + n\pi = C_1\lambda^{\varepsilon+1/2\varepsilon} + O\left(\frac{1}{\lambda^{\varepsilon+1/2\varepsilon}}\right) , \quad \lambda \to \infty ,$$

where

$$C_1 = \int_0^\infty \frac{1}{\sqrt{1+x^{2\varepsilon}} + x^\varepsilon}\, dx .$$

A reversion for the eigenvalues gives

$$\begin{aligned}\lambda_n &= \left(\frac{\gamma + n\pi}{C_1} + O\left(\frac{1}{n}\right)\right)^{2\varepsilon/\varepsilon+1} \\ &= \left(\frac{\gamma + n\pi}{C_1}\right) + O\left(\frac{1}{n^{2/\varepsilon+1}}\right) ,\end{aligned} \tag{25}$$

which shows that Theorem 3 locates the eigenvalues with error term tending to zero. In contrast, the result obtained by Alenitsyn in Theorem 2 yields

$$\lambda_n = \left(\frac{\gamma + n\pi}{C_1} + O(1)\right)^{2\varepsilon/\varepsilon+1} = \left(\frac{\gamma + n\pi}{C_1}\right)^{2\varepsilon/\varepsilon+1} \left(1 + O\left(\frac{1}{n}\right)\right) , \tag{26}$$

which is not sharp enough to locate the eigenvalues with error term tending to zero.

An iterative procedure for obtaining successively more terms of lower order in λ in the quantum condition (8), and thereby higher order terms in asymptotic expansions like (25) has been given by the authors in [3], as well as improvements and refinements of Theorem 4 for the negative spectrum.

§3 The $w(\lambda)$-Function

According to Fulton's formulation of the basic expansion theory for limit circle problems in [8], the eigenvalues of the problem (*) are characterized as the zeros of an entire function $w(\lambda)$, which is defined as the Wronskian of two solutions of the basic equation, one which is defined by initial conditions at $x = 0$, so as to satisfy the boundary condition there, and one which is defined by initial conditions, or 'end' conditions at ∞, so as to satisfy the boundary condition at ∞. The solution, $X_\lambda = X_\lambda^\gamma(x)$, which is entire in λ and satisfies the boundary condition at ∞, is uniquely defined by the requirements

$$\begin{aligned} &\lim_{x\to\infty} W_x(X_\lambda,v) = \sin\gamma \\ &\lim_{x\to\infty} W_x(X_\lambda,u) = -\cos\gamma\ , \text{ for all } \lambda \in \mathbb{C}\ . \end{aligned} \tag{27}$$

This follows from Fulton [8, Theorem 1 and Equa. (2.5)].

Similarly, the solution $\phi_\lambda = \phi_\lambda^\alpha(x)$, which is entire in λ and satisfies the boundary condition at $x = 0$, may be uniquely defined by the initial conditions

$$\begin{aligned} &\phi_\lambda(0) = \sin\alpha \\ &\phi_\lambda'(0) = -\cos\alpha\ , \text{ for all } \lambda \in \mathbb{C}\ . \end{aligned} \tag{28}$$

As in [8] the transformation

$$Y = \begin{pmatrix} u & v \\ u' & v' \end{pmatrix} \begin{pmatrix} y(x,\lambda) \\ y'(x,\lambda) \end{pmatrix} = \begin{pmatrix} W_x(y,v) \\ -W_x(y,u) \end{pmatrix} \tag{29}$$

carries solutions $y(x,\lambda)$ of the basic equation into solutions $Y = Y(x,\lambda)$ of the modified Sturm-Liouville equation,

$$\frac{dY}{dX} = \lambda \begin{pmatrix} uv & v^2 \\ -u^2 & -uv \end{pmatrix} Y\ . \tag{30}$$

Since $u,v \in L^2(0,\infty)$ by virtue of the limit circle case occuring at ∞, the coefficient matrix is in $L^1(0,\infty)$, and a Caratheodory initial value problem may be posed at ∞. Indeed this justifies the definition of solutions of the basic equation by means of the "transformed" initial conditions (27) for X_λ. By means of the determinantal identity ([8; Equa. 2.9])

$$W_x(\phi_\lambda,X_\lambda) = \begin{vmatrix} W_x(\phi_\lambda,v) & W_x(X_\lambda,v) \\ -W_x(\phi_\lambda,u) & -W_x(X_\lambda,u) \end{vmatrix} ,$$

and passage to the limit $x \to \infty$, the entire function $w(\lambda)$, whose zeros determine the eigenvalues of (*) may be written in the various forms,

$$\begin{aligned} w(\lambda) = w^{\alpha,\gamma}(\lambda) &= W_x(\phi_\lambda,X_\lambda), \quad x \in (0,\infty) \\ &= W_0(\phi_\lambda,X_\lambda) = (\sin\alpha)X_\lambda'(0) + (\cos\alpha)X_\lambda(0) \\ &= W_\infty(\phi_\lambda,X_\lambda) = \cos\gamma\, W_\infty(\phi_\lambda,v) + \sin\gamma\, W_\infty(\phi_\lambda,u) . \end{aligned} \tag{31}$$

We have the following theorems.

Theorem 5 The order of $w(\lambda)$ is less than or equal to 1.

Proof: Let

$$Y^\phi = Y^\phi(x, \ ) = \begin{pmatrix} W_x(\phi_\lambda,v) \\ -W_x(\phi_\lambda,u) \end{pmatrix} \tag{32}$$

be the solution of (30) corresponding to the solution ϕ_λ of the basic equation. For fixed $b < \infty$,

$$w_b(\lambda) = \cos\gamma\, Y_1^\phi(b,\lambda) - \sin\gamma\, Y_2^\phi(b,\lambda)$$

is an entire function of order 1/2, since $w_b(\lambda)$ is the entire function whose zeros are the eigenvalues of a regular Sturm-Liouville problem on $[0,b]$. By Theorem 1(i) of [8] we also have the estimate for all $x,b \in (0,\infty)$:

$$|Y^\phi_{(X,\lambda)}| \le |Y^\phi_{(b,\lambda)}| \exp\{|\lambda| \cdot \int_b^X (u^2+v^2)ds\} . \tag{33}$$

Letting $X \to \infty$ we therefore obtain

$$|Y^\phi(\infty,\lambda)| \le |Y^\phi(b,\lambda)| e^{|\lambda| \int_b^\infty (u^2+v^2)ds} . \tag{34}$$

It suffices to show that for every $\epsilon > 0$, the right hand side of the above inequality is $\le e^{\varepsilon|\lambda|}$ for $|\lambda|$ sufficiently large. To this end we make use of the fact that $|Y^\phi(b,\lambda)|$ has order 1/2 for fixed b. Hence, there exist constants K and $c = c_b$ such that

$$|Y^\phi(b,\lambda)| \le Ke^{c_b|\lambda|^{1/2}}$$

for $|\lambda|$ sufficiently large. Given $\varepsilon > 0$, we first pick b so large that

$$\int_b^\infty (u^2+v^2)ds < \varepsilon/2 .$$

Then for $|\lambda|$ so large that $c_b < \frac{\varepsilon}{2} |\lambda|^{1/2}$, we have

$$|Y^{\phi}(\infty,\lambda)| \le Ke^{\frac{\varepsilon}{2}|\lambda|} \cdot e^{\frac{\varepsilon}{2}|\lambda|} = Ke^{\varepsilon|\lambda|} \quad .$$

This proves the theorem since $w(\lambda)$ is representable in terms of the components of $Y^{\phi}(\infty,\lambda)$ by (31). q.e.d.

Let the zeros of $w(\lambda)$ be ordered according to their absolute value,

$$0 \le |\lambda_1| \le |\lambda_2| \le |\lambda_3| \le \dots \quad . \tag{35}$$

Suppose that α and γ are such that $|\lambda_1| > 0$, that is, $\lambda = 0$ is not an eigenvalue of (*). Then $w(\lambda)$ has a Hadamard factorization of the form

$$w(\lambda) = e^{Q(\lambda)}\left[\prod_{n=1}^{\infty} \left(1 - \frac{\lambda}{\lambda_n}\right) e^{\lambda/\lambda_n + \frac{1}{2}(\lambda/\lambda_n)^2 + \dots + \frac{1}{p}(\lambda/\lambda_n)^p}\right] \tag{36}$$

where $Q(\lambda)$ is a polynomial of degree less than or equal to ρ, and p is the order of $w(\lambda)$. The order ρ and the exponent of convergence ρ_1 of the zeros of $w(\lambda)$ are defined by

$$\rho = \inf\{\beta |\ w(\lambda) = O(e^{|\lambda|^{\beta}}) \quad \text{as} \quad |\lambda| \to \infty\} \tag{37}$$

$$\rho_1 = \inf\{\alpha |\ \sum_{n=1}^{\infty} \left|\frac{1}{\lambda_n}\right|^{\alpha} < \infty\} \ . \tag{38}$$

The following lemmas relating ρ, ρ_1 and p (which is independent of n in (36)) are well known, cf. Titchmarsh [25, pp.248-254]:

<u>Lemma 1</u> $\rho_1 \le \rho$.

<u>Lemma 2</u> If the degree q of $Q(\lambda)$ is zero, then $\rho_1 = \rho$, and p is the greatest integer $\le \rho_1$. If ρ_1 is an integer, then

$$p = \begin{cases} \rho_1 & \text{if} \quad \sum_{n=1}^{\infty} |\lambda_n|^{-\rho_1} = \infty \\ \rho_1 - 1 & \text{if} \quad \sum_{n=1}^{\infty} |\lambda_n|^{-\rho_1} < \infty \end{cases} \quad . \tag{39}$$

<u>Lemma 3</u> If ρ is not an integer, then $\rho_1 = \rho$.

Since $\rho \le 1$ by Theorem 5, it follows from the above that there are three different possibilities for the Hadamard factorization of $w(\lambda)$:

Since $\rho \leq 1$ by Theorem 5, it follows from the above that there are three different possibilities for the Hadamard factorization of $w(\lambda)$:

Case 1 If $0 < \rho < 1$, then $\rho_1 = \rho < 1$, $q = 0$ and $p = 0$. In this case

$$w(\lambda) = w(0) \prod_{n=1}^{\infty} (1 - \frac{\lambda}{\lambda_n}) \quad . \tag{40}$$

Case 2 If $\rho = 1$ and $\rho_1 = 1$, then q may be 0 or 1 and p may be 0 or 1. In this case

$$w(\lambda) = e^{q_0 + q_1\lambda} [\prod_{n=1}^{\infty} (1 - \frac{\lambda}{\lambda_n}) e^{0 + p(\lambda/\lambda_n)}] \tag{41}$$

where q_1 and p may be 0 or 1. The determination of p is governed by

$$p = \begin{cases} 1 & \text{if} \quad \sum_{n=1}^{\infty} |\lambda_n|^{-1} = \infty \\ 0 & \text{if} \quad \sum_{n=1}^{\infty} |\lambda_n|^{-1} < \infty \end{cases} .$$

Case 3 If $\rho = 1$ and $0 < \rho_1 < 1$, then $p = 0$ and $q = 1$. In this case

$$w(\lambda) = e^{q_0 + q_1\lambda} [\prod_{n=1}^{\infty} (1 - \frac{\lambda}{\lambda_n})] , \tag{42}$$

where $q_1 \neq 0$.

In all cases the basic relation $p = \max(q, \rho_1)$ holds. In the next section we use Heywood's density formulae to find ρ_1 in several special cases, and show that case 1 may occur with any $\rho \in [1/2, 1)$ and that Case 2 may occur with $p = 1$. Examples to show that Case 3, or Case 2 with $p = 0$, may occur, are not known.

§4 Exponent of Convergence of Eigenvalues

The exponent of convergence ρ_1 of the zeros of $w(\lambda)$ can be determined by computing the exponent of convergence of the positive and negative eigenvalues separately. For if

$$\rho_1^+ = \inf\{\alpha \mid \sum_{n=1}^{\infty} \left| \frac{1}{\lambda_n} \right|^{\alpha} < \infty \} \tag{43}$$

where λ_n are the positive eigenvalues ordered by $0 < \lambda_1 < \lambda_2 < \ldots$, and

$$\rho_1^- = \inf\{\alpha \mid \sum_{n=1}^{\infty} \left| \frac{1}{\mu_n} \right|^{\alpha} < \infty \} \tag{44}$$

where $\mu_n = -\lambda_n$ are the negative eigenvalues ordered by $0 < \mu_1 < \mu_2 < \ldots$, then

$$\rho_1 = \max\{\rho_1^+, \rho_1^-\} \ . \tag{45}$$

To determine ρ_1^+ and ρ_1^- we put

$$E_1(\lambda,F) = \frac{1}{\pi}\int_0^{\infty} \frac{\lambda}{\sqrt{\lambda+F} + \sqrt{F}}\, dx \ , \tag{46}$$

and

$$G_1(\mu,F) = \frac{1}{\pi}\Big[\int_0^{x_0(\mu)} \sqrt{F}\, dx + \int_{x_0(\mu)}^{\infty} \frac{\mu}{\sqrt{F} + \sqrt{F-\mu}}\, dx \Big] \ , \tag{47}$$

where $x_0 = x_0(\mu)$ and $x_1 = x_1(\mu)$ are defined by the conditions

$$F\big(x_0(\mu)\big) = \mu \quad \text{for all} \quad \mu > 0 \tag{48}$$

$$F\big(x_0(\mu) + x_1(\mu)\big) = 2\mu \quad \text{for all} \quad \mu > 0 \ . \tag{49}$$

We then have the following theorems:

Theorem 6 Under the assumptions of Theorem 1 we have for all $\lambda > 0$,

$$\frac{1}{2^{3/2}\pi} A(\lambda) \le E_1(\lambda,F) \le \frac{1}{\pi} A(\lambda) \tag{50}$$

where

$$A(\lambda) = \Big[\int_0^{x_0(\lambda)} \sqrt{\lambda}\, dx + \int_{x_0(\lambda)}^{\infty} \frac{\lambda}{\sqrt{F}}\, dx \Big] \ . \tag{51}$$

Theorem 7 Let $\mu = -\lambda$. Then under the assumptions of Theorem 4 we have for all $\mu > 0$,

$$\frac{1}{2^{3/2}\pi} B_1(\mu) \le G_1(\mu,F) \le \frac{1}{\pi} B_2(\mu) \tag{52}$$

where

$$B_1(\mu) = \Big[\int_0^{x_1(\mu)} \sqrt{\mu}\, dx + \int_{x_2(\mu)}^{\infty} \frac{\mu}{\sqrt{F(x)}}\, dx \Big] \ , \tag{53}$$

$$B_2(\mu) = [\int_0^{x_2(\mu)} \sqrt{\mu}\, dx + \int_{x_2(\mu)}^{\infty} \frac{\mu}{\sqrt{F(x)}}\, dx] , \tag{54}$$

and

$$x_2(\mu) = x_0(\mu) + x_1(\mu) . \tag{55}$$

<u>Proof of Theorem 6:</u> Since $F(x)$ is strictly increasing and positive in $[0,\infty)$ we have

$$0 \le F(x) \le \lambda \quad \text{for} \quad x \in [0,x_0(\lambda)]$$

and

$$\lambda \le F(x) < \infty \quad \text{for} \quad x \in [x_0(\lambda),\infty] .$$

Hence

$$E_1(\lambda,F) \le \frac{1}{\pi} [\int_0^{x_0(\lambda)} \frac{\lambda}{\sqrt{\lambda+F}}\, dx + \int_{x_0(\lambda)}^{\infty} \frac{\lambda}{\sqrt{\lambda+F}}\, dx] \le \frac{1}{\pi} A(\lambda) ,$$

and

$$E_1(\lambda,F) \ge \frac{1}{2\pi} [\int_0^{x_0(\lambda)} \frac{\lambda}{\sqrt{\lambda+F}}\, dx + \int_{x_0(\lambda)}^{\infty} \frac{\lambda}{\sqrt{\lambda+F}}\, dx]$$

$$\ge \frac{1}{2\pi} [\int_0^{x_0(\lambda)} \frac{\lambda}{\sqrt{2\lambda}}\, dx + \int_{x_0(\lambda)}^{\infty} \frac{\lambda}{\sqrt{2F}}\, dx] = \frac{1}{\pi 2^{3/2}} A(\lambda) . \qquad \text{q.e.d.}$$

<u>Proof of Theorem 7:</u> Letting $x = x_0(\mu) + t$ and putting

$$F_1(t) = F\big(x_0(\mu)+t\big) - F\big(x_0(\mu)\big) = F\big(x_0(\mu)+t\big) - \mu ,$$

we have

$$0 \le F_1(t) \le \mu \quad \text{for} \quad t \in [0,x_1(\mu)] ,$$

and

$$\mu \le F_1(t) \quad \text{for} \quad t \in [x_1(\mu),\infty) .$$

Estimating as before we have

$$G_1(\mu,F) \le \frac{1}{\pi} [\int_0^{x_0(\mu)} \sqrt{\mu}\, dx + \int_{t=0}^{x_1(\mu)} \frac{\mu}{\sqrt{\mu + F_1(t)}}\, dt + \int_{x_1(\mu)}^{\infty} \frac{\mu}{\sqrt{\mu + F_1(t)}}\, dt$$

$$\le \frac{1}{\pi} B_2(\mu) ,$$

and

$$G_1(\mu,F) \ge \frac{1}{2\pi}\Big[\int_{t=0}^{x_1(\mu)} \frac{\mu}{\sqrt{\mu+F_1(t)}}\,dt + \int_{x_1(\mu)}^{\infty} \frac{\mu}{\sqrt{\mu+F_1(t)}}\,dt\Big]$$

$$\ge \frac{1}{2\pi}\Big[\int_{t=0}^{x_1(\mu)} \frac{\mu}{\sqrt{2\mu}}\,dt + \int_{x_1(\mu)}^{\infty} \frac{\mu}{\sqrt{\mu+F_1(t)}}\,dt\Big]$$

$$\ge \frac{1}{2^{3/2}\pi} B_1(\mu)\ . \qquad \text{q.e.d.}$$

Corollary Under the assumptions of Theorem 1, the exponent of convergence of the zeros of $w(\lambda)$ is bounded below by 1/2. Hence, by lemma 1, the order of $w(\lambda)$ is also bounded below by 1/2.

Proof: Since $x_0(\lambda) \to \infty$ as $\lambda \to \infty$ we have

$$A(\lambda) \ge \lambda^{1/2}\, x_0(\lambda) \ge \lambda^{1/2}$$

whenever $x_0(\lambda) \ge 1$. By Theorem 1 we have $E(\lambda) - E_1(\lambda,F) = O(1)$, so applying Theorem 6 it follows that there exists a positive constant C such that

$$C\lambda^{1/2} \le E(\lambda)$$

for λ sufficiently large. Letting λ_n be the nth positive eigenvalue, we therefore have

$$C\lambda_n^{1/2} \le n\ ,$$

or

$$\frac{C}{n} \le \frac{1}{\lambda_n^{1/2}}\ .$$

Hence the series $\sum_{n=1}^{\infty} \left|\frac{1}{\lambda_n}\right|^{1/2}$ is always divergent. We may therefore conclude that $\rho_1 \ge \rho_1^+ \ge 1/2$. q.e.d.

We now apply Theorems 6, 7 and Theorems 1, 4 to calculate ρ_1^+ and ρ_1^- for 3 examples:

Example 1 $F(x) = X^{2\varepsilon}$, $1 < \varepsilon < \infty$: $\rho_1^+ = \rho_1^- = \rho_1 = \frac{\varepsilon+1}{2\varepsilon}$, which assumes all values in (1/2, 1) except for the endpoints. In this case we have $X_0(\lambda) = \lambda^{1/2\varepsilon}$, and a calculation gives

$$A(\lambda) = \Big(1 + \frac{1}{\varepsilon-1}\Big)\lambda^{(\varepsilon+1)/2\varepsilon}\ ,$$

$$B_1(\mu) = [(2^{1/2\varepsilon} - 1) + \frac{1}{\varepsilon - 1}]\mu^{\varepsilon+1/2\varepsilon} \quad ,$$

and

$$B_2(\mu) = [2^{1/2\varepsilon} + \frac{1}{\varepsilon - 1}]\mu^{\varepsilon+1/2\varepsilon} \quad .$$

Since $E(\lambda) - E_1(\lambda,F) = O(1)$ by Theorem 1 it follows from Theorem 6 and the above formula for $A(\lambda)$ that there exist positive constants $C_1 < C_2$ such that for sufficiently large λ

$$C_1\lambda^{\varepsilon+1/2\varepsilon} \leq E(\lambda) \leq C_2\cdot\lambda^{\varepsilon+1/2\varepsilon} \quad . \tag{56}$$

For $\lambda = \lambda_n$, the nth positive eigenvalue, this gives

$$C_2\lambda_n^{\varepsilon+1/2\varepsilon} \leq n \leq C_1\cdot\lambda_n^{\varepsilon+1/2\varepsilon} \quad ,$$

or

$$\left(\frac{C_2}{n}\right)^{2\varepsilon/\varepsilon+1} \leq \left|\frac{1}{\lambda_n}\right| \leq \left(\frac{C_1}{n}\right)^{2\varepsilon/\varepsilon+1} \quad , \tag{57}$$

for sufficiently large n.

It follows that $\sum_{n=1}^{\infty} \left|\frac{1}{\lambda_n}\right|^{\alpha} < \infty$ if and only if $\alpha > (\varepsilon+1)/2\varepsilon$. A similar argument using Theorem 4 and Theorem 7, together with the above formulae for $B_1(\mu)$ and $B_2(\mu)$ shows that $\rho_1^- = (\varepsilon+1)/2\varepsilon$.

<u>Example 2</u> $F(x) = e^{2x}$: $\rho_1^+ = \rho_1^- = 1/2$.

In this case we have $x_0(\lambda) = 1/2 \ln \lambda$, and a calculation gives

$$A(\lambda) = \lambda^{1/2} \ln \lambda^{1/2}(1 + \frac{2}{\ln \lambda}) \quad ,$$

$$B_1(\mu) = \left(\frac{\ln 2}{2} + \frac{1}{\sqrt{2}}\right) \mu^{1/2} \quad ,$$

and

$$B_2(\mu) = \mu^{1/2}(\ln \mu^{1/2} + \ln \sqrt{2} + \frac{1}{\sqrt{2}}) \quad .$$

Since $E(\lambda) - E_1(\lambda,F) = O(1)$ by Theorem 1, it follows from Theorem 6 and the above formula for $A(\lambda)$ that there exist positive constants $C_1 < C_2$ such that for sufficiently large λ

$$C_1\lambda^{1/2}\ \ell n(\lambda^{1/2}) \le E(\lambda) \le C_2\lambda^{1/2}\ \ell n(\lambda^{1/2}) . \tag{58}$$

For $\lambda = \lambda_n$, this gives

$$C_1\lambda_n^{1/2}\ \ell n\ \lambda_n^{1/2} \le n \le C_2\lambda_n^{1/2}\ \ell n\ \lambda_n^{1/2} ,$$

or, for some constants $0 < K_1 \le C_1 < C_2 \le K_2$,

$$K_1 \left|\frac{\ell n\ n}{n}\right| \le \frac{1}{\lambda_n^{1/2}} \le K_2\left|\frac{\ell n\ n}{n}\right| , \tag{59}$$

for n sufficiently large. It follows that $\sum_{n=1}^{\infty} \left|\frac{1}{\lambda_n}\right|^{\alpha} < \infty$ if and only if $\alpha > 1/2$ since $\sum_{n=1}^{\infty} \left|\frac{\ell n\ n}{n}\right|^{k}$ is convergent only for $k > 1$. Hence we conclude that $\rho_1^+ = 1/2$. For the negative eigenvalues we find similarly from Theorems 4 and 7 that there exist positive constants $C_1 < C_2$ such that for sufficiently large μ

$$C_1\mu^{1/2} \le G(\mu) \le C_2\mu^{1/2}\ \ell n(\mu^{1/2}) . \tag{60}$$

For $\mu = \mu_m$ ($\mu_m = -\lambda_m$ where λ_m is the mth negative eigenvalue) we have

$$C_1\mu_m^{1/2} \le m \le C_2\mu_m^{1/2}\ \ell n(\mu_m^{1/2}) ,$$

or

$$C_1 \frac{1}{m} \le \frac{1}{\mu_m^{1/2}} \le K_2\left|\frac{\ell n\ m}{m}\right| , \tag{61}$$

for m sufficiently large. It follows that $\sum_{m=1}^{\infty} \left|\frac{1}{\mu_m}\right|^{\alpha}$ is convergent if and only if $\alpha > 1/2$. Hence we conclude that $\rho_1^- = 1/2$.

Example 3 $F(x) = x^2\ \ell n^4 x$: $\rho_1^+ = \rho_1^- = \rho_1 = 1$.

In this case $x_0(\lambda)$ is defined by

$$x_0(\lambda)\ \ell n^2\big(x_0(\lambda)\big) = \sqrt{\lambda} .$$

Squaring and taking the logarithm gives

$$\ell n^2\big(x_0^2(\lambda)\ \ell n^4 x_0(\lambda)\big) = \ell n^2 \lambda .$$

Dividing and applying L'Hospital's rule to get

$$\lim_{x_0 \to \infty} \left| \frac{\ln(x_0^2 \ln^4 x_0)}{\ln x_0} \right|^2 = 4 , \tag{62}$$

we see that

$$x_0(\lambda) = 4 \frac{\sqrt{\lambda}}{\ln^2 \lambda} \left(1 + o(1)\right) \quad \text{as} \quad \lambda \to \infty . \tag{63}$$

A calculation gives

$$A(\lambda) = \sqrt{\lambda}\, x_0(\lambda) + \frac{\lambda}{\ln\left(x_0(\lambda)\right)}$$

$$= \frac{\lambda}{\ln \lambda} \left(1 + O\left(\frac{1}{\ln \lambda} \right)\right) , \quad \text{as} \quad \lambda \to \infty ,$$

$$B_1(\mu) = \sqrt{\mu}\, x_1(\mu) + \frac{\mu}{\ln\left(x_2(\mu)\right)}$$

$$= \sqrt{\mu} \left[4 \frac{\sqrt{2\mu}}{\ln^2 2\mu} \left(1 - \frac{1}{\sqrt{2}} + o(1)\right) \right] + \frac{\mu}{\ln\left(x_2(\mu)\right)}$$

$$= \frac{\mu}{\ln \mu} \left(1 + O\left(\frac{1}{\ln \mu} \right)\right) , \text{ and}$$

$$B_2(\mu) = \sqrt{\mu} \cdot x_2(\mu) + \frac{\mu}{\ln\left(x_2(\mu)\right)}$$

$$= \sqrt{\mu} \left[4 \frac{\sqrt{2\mu}}{\ln^2 2\mu} \left(1 + o(1)\right) \right] + \frac{\mu}{\ln\left(x_2(\mu)\right)}$$

$$= \frac{\mu}{\ln \mu} \left(1 + O\left(\frac{1}{\ln \mu} \right)\right) , \quad \text{as} \quad \mu \to \infty .$$

Since $E(\lambda) - E_1(\lambda,F) = O(1)$ by Theorem 1, it follows from Theorem 6 and the above result for $A(\lambda)$ that there exist positive constants $C_1 < C_2$ such that for sufficiently large λ

$$C_1\left(\frac{\lambda}{\ln \lambda} \right) \le E(\lambda) \le C_2\left(\frac{\lambda}{\ln \lambda} \right) . \tag{64}$$

For $\lambda = \lambda_n$ this gives

$$C_1\left(\frac{\lambda_n}{\ln \lambda_n} \right) \le n \le C_2\left(\frac{\lambda_n}{\ln \lambda_n} \right) ,$$

or

$$\frac{K_1}{n \ln n} \le \frac{1}{\lambda_n} \le \frac{K_2}{n \ln n} , \tag{65}$$

for n sufficiently large, and constants K_1, K_2 with $K_1 \le C_1 < C_2 \le K_2$. It follows that $\sum_{n=1}^{\infty} \left| \frac{1}{\lambda_n} \right|^{\alpha}$ converges if and only if $\alpha > 1$. For $\alpha = 1$ we have

$$\sum_{n=1}^{\infty} \left| \frac{1}{\lambda_n} \right| = \infty . \tag{66}$$

Thus we conclude that $\rho_1^+ = 1$. Similarly, the above estimates for $B_1(\mu)$, $B_2(\mu)$ and application of Theorem 4 gives $\rho_1^- = 1$ and

$$\sum_{n=1}^{\infty} \left| \frac{1}{\mu_n} \right| = \infty , \tag{67}$$

also for the negative eigenvalues.

Since $\rho_1 = \rho_1^- = \rho_2^- = 1$, we conclude from lemma 1 that $\rho = 1$ also. This example therefore belongs to Case 2, and the divergence of the above series determines that $p = 1$ in the Hadamard factorization of $w(\lambda)$. This example, incidentally, was considered by J Weidmann [28, pp.219-220]. Weidmann establishes that for the sum over all eigenvalues,

$$\sum_{\lambda_n \neq 0} \left| \frac{1}{\lambda_n} \right|^{\alpha} = \infty$$

for $\alpha < 1$, and that it is convergent for $\alpha > 1$, cf. [28, p.217, Satz 3.5]. The results (66) and (67) answer his open question, showing that we have divergence of the above sum for $\alpha = 1$. It follows that Weidmann's Satz 3.5(a) is indeed a best possible result. This divergence for $\alpha = 1$ also has another noteworthy consequence: if one considers the Hilbert-Schmidt integral operator T which corresponds to the inverse of the self-adjoint differential operator, then by a lemma of Dunford and Schwartz [7, p.1092, Lemma 6] it follows that

$$\sum_{n=1}^{\infty} \mu_n(T) = \infty ,$$

where $\mu_n(T)$ are the singular values of T. Example 3 therefore represents an oscillatory limit circle problem for which the corresponding Hilbert-Schmidt integral operator does not belong to the trace class. On the other hand, for limit circle problems where the spectrum is bounded below, the corresponding Hilbert-

Schmidt integral operator always belongs to the trace class, cf. Weidmann [28, p.216, Satz 2.6].

It is possible to determine the exponents of convergence for the positive and negative eigenvalues by a more direct appeal to the behaviour of $F(x)$ near ∞. Namely, we have the following theorems:

Theorem 8 (i) Under the assumptions of Theorem 1, and for $1/2 < \alpha < 1$, the series of positive eigenvalues

$$\sum_{n=1}^{\infty} \left| \frac{1}{\lambda_n} \right|^{\alpha} \tag{68}$$

converges and diverges together with

$$\int^{\infty} F^{\frac{1}{2}-\alpha}\, dx \,. \tag{69}$$

(ii) For the case $\alpha = 1$, the above series converges and diverges together with

$$\int^{\infty} \frac{\ell n\, F}{\sqrt{F}}\, dx \,. \tag{70}$$

Proof: Since $E(\lambda)$ is the number of positive eigenvalues in $[0,\lambda]$, it is a step function of λ, which increases by 1 at each eigenvalue. We may therefore write

$$\sum_{1<\lambda_n<\Lambda} \lambda_n^{-\alpha} = \int_1^{\Lambda} \lambda^{-\alpha}\, dE(\lambda) + C_1 + O(\Lambda^{-\alpha})$$

for some constant C_1. The first term can be integrated by parts to give

$$[\lambda^{-\alpha}\, E(\lambda)\,]_1^{\Lambda} + \int_1^{\Lambda} \alpha\, \lambda^{-\alpha-1}\, E(\lambda)\, d\lambda \,.$$

Since the integral defining $E_1(\lambda,F)$ is bounded by $\lambda \int_0^{\infty} \frac{1}{\sqrt{F}}\, dx$, it is uniformly convergent with respect to λ and may be differential under the integral sign to give

$$\frac{d}{d\lambda} \left(\frac{1}{\pi} \int_0^{\infty} (\sqrt{\lambda+F} - \sqrt{F})dx \right) = \frac{1}{2\pi} \int_0^{\infty} \frac{1}{\sqrt{\lambda+F}}\, dx \,.$$

Thus, a similar integration by parts gives

$$\int_1^\Lambda \lambda^{-\alpha} E_1'(\lambda,F)d\lambda$$

$$= \lambda^{-\alpha} E_1(\lambda,F)]_1^\Lambda + \int_1^\Lambda \alpha\, \lambda^{-\alpha-1} E_1(\lambda,F)d\lambda \quad .$$

Putting $E(\lambda) = E_1(\lambda,F) + O(1)$ in the first result and substituting the above for the E_1-part we have

$$\lambda^{-\alpha} E(\lambda)]_1^\Lambda + \int_1^\Lambda \alpha\, \lambda^{-\alpha-1} E(\lambda)d\lambda$$

$$= \int_1^\Lambda \lambda^{-\alpha} E_1'(\lambda,F)d\lambda + O(\Lambda^{-\alpha}) + O(\int_1^\Lambda \alpha\, \lambda^{-\alpha-1} d\lambda) \ .$$

Hence there exists a constant M independent of Λ such that

$$\left| \sum_{1<\lambda_n<\Lambda} \lambda_n^{-\alpha} - \frac{1}{2\pi} \int_1^\Lambda \lambda^{-\alpha} \int_0^\infty \frac{1}{\sqrt{\lambda+F}}\, dx\, d\lambda \right| \le M \ . \tag{71}$$

Interchanging the order of integration we find

$$\int_1^\Lambda \lambda^{-\infty} \int_0^\infty \frac{1}{\sqrt{\lambda+F}}\, dx\, d\lambda$$

$$= \int_0^\infty \int_1^\Lambda \frac{1}{\lambda^\alpha(\lambda+F)^{1/2}}\, d\lambda\, dx$$

$$= \int_0^\infty F(x)^{\frac{1}{2}-\alpha} \left(\int_{1/F(x)}^{\Lambda/F(x)} \frac{1}{\nu^\alpha(1+\nu)^{1/2}}\, d\nu \right) dx \ .$$

Passing to the limit $\Lambda \to \infty$ in the above inequality we thus see that the series $\sum^\infty \left| \frac{1}{\lambda_n} \right|^\alpha$ converges and diverges together with the integral

$$\int_0^\infty F(x)^{\frac{1}{2}-\alpha} \left(\int_{1/F(x)}^\infty \frac{1}{\nu^\alpha(1+\nu)^{1/2}}\, d\nu \right) dx \ . \tag{72}$$

For $\alpha \in (1/2,1)$ we have

$$\int_0^\infty \frac{1}{\nu^\alpha(1+\nu)^{1/2}}\, d\nu < \infty \ ,$$

and the inner integral is therefore bounded above and below for x sufficiently large. This proves (i). For $\alpha = 1$, we may integrate by parts in the inner integral, obtaining

$$\int_{1/F(x)}^{\infty} \frac{1}{\nu(1+\nu)^{1/2}}\, d\nu = \frac{\ln F(x)}{\sqrt{1 + 1/F(x)}} + O(1) \,, \tag{73}$$

as $x \to \infty$. Since the dominant term is $\ln F(x)$ this proves (ii). q.e.d.

Application of Theorem 8 to the above examples yields the same results obtained before. For example 3, in particular, the integral in (70) is found to be divergent for $\alpha = 1$, so we again obtain the result that

$$\sum_{n=1}^{\infty} \left| \frac{1}{\lambda_n} \right| = \infty \,.$$

Theorem 9 (i) Under the assumptions of Theorem 4 and for $1/2 < \alpha < 1$, the series of negative eigenvalues $\mu_m = -\lambda_m$,

$$\sum_{m=1}^{\infty} \left| \frac{1}{\mu_m} \right|^{\alpha} \tag{74}$$

converges and diverges together with

$$\int^{\infty} F^{\frac{1}{2}-\alpha}\, dx \,. \tag{75}$$

(ii) For the case $\alpha = 1$, the above series converges and diverges together with

$$\int^{\infty} \frac{\ln F}{\sqrt{F}}\, dx \,. \tag{76}$$

Proof: We follow the same steps as before using theorem 4 to establish the existence of a constant M independent of Λ such that

$$\left| \sum_{K<\mu_m<\Lambda} \mu_m^{-\alpha} - \frac{1}{2\pi} \int_K^{\Lambda} \frac{1}{\mu^{\alpha}} \left(\int_{x_0(\mu)}^{\infty} \frac{1}{\sqrt{F - \mu}}\, dx \right) d\mu \right| \le M \,. \tag{77}$$

Interchanging the order of integration in the above integral gives

$$\int_K^{\Lambda} \mu^{-\alpha} \left(\int_{x_0(\mu)}^{\infty} \frac{1}{\sqrt{F-\mu}}\, dx \right) d\mu$$

$$= \int_{x_0(\mu)}^{\infty} \left(\int_{K}^{\min\{\Lambda,F(x)\}} \frac{1}{\mu^{\alpha}(F-\mu)^{1/2}} \, d\mu \right) dx$$

$$= \int_{x_0(\mu)}^{\infty} F(x)^{\frac{1}{2}-\alpha} \left\{ \int_{K/F(x)}^{\min(\Lambda/F,1)} \frac{1}{\nu^{\alpha}(1-\nu)^{1/2}} \, d\nu \right\} dx .$$

Passing to the limit $\Lambda \to \infty$ in the above inequality we thus see that the series $\sum^{\infty} \left| \frac{1}{\mu_m} \right|^{\alpha}$ converges and diverges together with the integral

$$\int_{x_0(\mu)}^{\infty} F(x)^{\frac{1}{2}-\alpha} \left\{ \int_{K/F(x)}^{1} \frac{1}{\nu^{\alpha}(1-\nu)^{1/2}} \, d\nu \right\} dx . \tag{78}$$

For $\alpha \in (1/2,1)$ we have

$$\int_0^1 \frac{1}{\nu^{\alpha}(1+\nu)^{1/2}} \, d\nu < \infty ,$$

and the inner integral is therefore bounded above and below for x sufficiently large. This proves (i).

For $\alpha = 1$, an integration by parts in the inner integral gives

$$\int_{K/F(x)}^{1} \frac{1}{\nu(1-\nu)^{1/2}} \, d\nu = \frac{\ell n \, F(x)}{\sqrt{1 - K/F(x)}} + O(1) , \tag{79}$$

as $x \to \infty$. Since the dominant term is $\ell n \, F(x)$ this proves (ii). q.e.d.

References

1. A G Alenitsyn, Asymptotic properties of the spectrum of a Sturm-Liouville operator in the case of a limit circle. Differential Equations 12, No. 2 (1976), 298-305. (Differentsial'nye Uravneniya 12, No. 3 (1976), 428-437.)

2. F V Atkinson, On second-order linear oscillators. Revista Tucuman 8 (1951), 71-87.

3. F V Atkinson and C T Fulton, Asymptotic formulae for eigenvalues of limit circle problems on a half line. Annali di Math. Pur. ed. Appl., to appear.

4. F V Atkinson and C T Fulton, Asymptotics of Sturm-Liouville eigenvalues for problems on a finite interval with one limit circle singularity, I. Proc. Roy. Soc. Edin., Ser.A, to appear.

5. F V Atkinson and C T Fulton, Asymptotics of Sturm-Liouville eigenvalues for problems on a finite interval with one limit circle singularity, II. Submitted.

6. V P Belogrud and A G Kostyuchenko, Usp. Mat. Nauk, 28, No. 2 (170) (1973), 227-228.

7. N Dunford and J T Schwartz, Linear Operators, II. New York: Interscience, 1963.

8. C Fulton, Parametrizations of Titchmarsh's m(λ)-functions in the limit circle case. Trans. A.M.S. 229 (1977), 51-63.

9. M Giertz, On the solutions in $L_2(-\infty,\infty)$ of $y'' + (\lambda - q(x))y = 0$, when q is rapidly increasing. Proc. London Math. Soc. (3) 14 (1964), 53-73.

10. P Hartman, Ordinary Differential Equations. Baltimore: Wiley, 1973 (corrected reprint of original 1964 edition).

11. P Hartman, Differential equations with nonoscillatory eigenfunctions. Duke Math. J. 15 (1948), 697-709.

12. P Hartman, On the eigenvalues of differential equations. Amer. J. Math. 73 (1951), 657-662.

13. P Hartman and C R Putnam, The least cluster point of the spectrum of boundary value problems. Amer. J. Math. 70 (1948), 849-855.

14. P Hartman, On the zeros of solutions of second order linear differential equations. J. London Math. Soc. 27 (1952), 492-496.

15. P Hartman, Some examples in the theory of singular boundary value problems. Amer. J. Math. 74 (1951), 107-126.

16. P Heywood, On the asymptotic distribution of eigenvalues. Proc. London Math. Soc. (3) 4 (1954), 456-470.

17. K Jörgens, Die Asymptotische Verteilung der Eigenwerte singulärer Sturm-Liouville probleme. Annales Acad. Scientiarum Fennicae, Ser. A (1963), 3-23.

18. B M Levitan and I S Sargsjan, Introduction to spectral theory: self-adjoint ordinary differential operators. Providence, R.I., A.M.S. Transl., 1975.

19. J B McLeod, The distribution of the eigenvalues for the hydrogen atom and similar cases. Proc. London Math. Soc. (3) 11 (1961), 139-158.

20. J B McLeod, The determination of phase shift. Quart. J. Math. Oxford (2) 12 (1961), 17-32.

21. J B McLeod and E C Titchmarsh, On the asymptotic distribution of eigenvalues. Quart. J. Math. (Oxford) (2) 10 (1959), 313-320.

22. W E Milne, On the degree of convergence of expansions in an infinite interval. Trans. A.M.S. 31 (1929), 906-913.

23. D B Sears and E C Titchmarsh, Some eigenfunction formulae. Quart. J. Math. Oxford (2) 1 (1950), 165-175.

24. E C Titchmarsh, On the asymptotic distribution of eigenvalues. Quart. J. Math. Oxford (2) 5 (1954), 228-240.

25. E C Titchmarsh, Theory of Functions. (2nd edition) London: Oxford University Press, 1939.

26. E C Titchmarsh, Eigenfunction expansions associated with second-order differential equations, I. (2nd edition) London: Oxford University Press, 1962.

27. E C Titchmarsh, Eigenfunction expansions associated with second-order differential equations, II. London: Oxford University Press, 1958.

28. J Weidmann, Verteilung der Eigenwerte für eine Klasse von Integraloperatoren in $L_2(a,b)$. Journal für Mathematik 276 (1973), 213-220.

F V Atkinson
Department of Mathematics
University of Toronto
Toronto
Ontario
Canada

C T Fulton
Department of Mathematical Sciences
Florida Institute of Technology
Melbourne
Florida 32901
U.S.A.

N.B. This research supported by the National Science Foundation under grant No MCS-7902025.

INTEGRAL INEQUALITIES AND EXPONENTIAL CONVERGENCE OF SOLUTIONS OF DIFFERENTIAL EQUATIONS WITH BOUNDED DELAY

by

F. V. Atkinson

J. R. Haddock

O. J. Staffans

1. Introduction.

In this paper we establish results for integral inequalities that, when applied to certain functional (delay) differential equations, produce information regarding exponential convergence of solutions. Our results are primarily based on an inequality

$$\rho(t) \leq \int_{t-r}^{t} m(t,s)\rho(s)ds \qquad (t \geq t_0), \tag{1.1}$$

where

(i) $r > 0$ and $t_0 \geq 0$;

(ii) ρ is measurable, locally bounded and non-negative on $[t_0 - r, \infty)$; and

(iii) m is non-negative and integrable on $[t_0, T] \times [t - r, t]$ for every $T, t \geq t_0$.

In Section 2 we state the two main theorems of the paper; the proofs are given in Section 3. These theorems combine (1.1) with other integral inequalities to yield exponential convergence properties of $\int_t^\infty \rho(s)ds$ and/or $\rho(t)$. Examples (corollaries) are given in Section 2 to illustrate how the inequalities can be applied to obtain exponential asymptotic constancy of solutions of functional differential equations with bounded delay. As shall be seen, in some cases the convergence of solutions is even faster than exponential.

It should be mentioned that the functional differential equations under consideration often, but not always, possess the property that _each constant function is a solution_. Such equations have provided much of the motivation for the work presented here. Indeed, some of the examples given in Section 2 represent generalizations of recent results of Atkinson and Haddock [2], where the study of (perturbations of) the aforementioned equations is the main theme.

2. Statements of the Theorems and Applications.

We assume throughout the remainder of this paper that inequality (1.1) is always accompanied by conditions (i), (ii) and (iii) of Section 1. We note that a locally integrable function h is said to be in $L^p[t_1,\infty)$, $(t_1 \in R, p > 0)$ if

$$\int_{t_1}^{\infty} |h(t)|^p dt < \infty .$$

THEOREM 2.1. Let (1.1) hold and suppose there exist positive continuous functions ϕ and ψ such that

$$\phi(s) - \int_s^{s+r} \phi(t)m(t,s)dt \geq \psi(s) \qquad (s \geq t_0). \tag{2.1}$$

Then

$$\psi p \in L^1[t_0,\infty). \tag{2.2}$$

A few comments are in order. If ψ in (2.1) is non-decreasing on $[t_0,\infty)$, then it readily follows from (2.2) that $p \in L^1[t_0,\infty)$. On the other hand, if ψ is allowed to be negative, then the conclusion (2.2) can be replaced by

$$\sup_{T \geq t_0} \int_{t_0}^{t} \psi(s)p(s)ds < \infty .$$

Finally it should be pointed out that Theorem 2.1 is often quite simple to apply. It appears that the most fruitful approach is to begin with an increasing function ϕ, impose conditions for which the left-hand side of (2.1) is positive and then define ψ such that (2.1) is satisfied. For example, by setting $\phi(s) = e^{\alpha s}$, $\alpha > 0$, we can deduce:

COROLLARY 1. In addition to the conditions of Theorem 2.1, suppose

$$\limsup_{s \to \infty} \int_s^{s+r} m(t,s)dt = \gamma < 1. \tag{2.3}$$

$$\left.\begin{array}{l}\text{If } \gamma > 0, \text{ let } \alpha \text{ satisfy} \\ 0 < \alpha < -(\ln\gamma)/r.\end{array}\right\} \tag{2.4}$$

$$\left.\begin{array}{l}\text{If } \gamma = 0, \text{ let } \alpha \text{ be any} \\ \text{positive number.}\end{array}\right\} \tag{2.5}$$

Then

$$\int_{t_0}^{\infty} e^{\alpha s} \rho(s)\,ds < \infty . \tag{2.6}$$

PROOF. Let $\phi(s) = e^{\alpha s}$, where α satisfies (2.4) if $\gamma > 0$ and (2.6) if $\gamma = 0$. Then

$$\phi(s) - \int_s^{s+r} \phi(t)m(t,s)\,dt = e^{\alpha s} - \int_s^{s+r} e^{\alpha s} m(t,s)\,dt$$

$$\geq e^{\alpha s}[1 - e^{\alpha r}\gamma] \overset{\text{def}}{=} \psi(s).$$

Since the restrictions on α imply ψ is positive, the conditions of Theorem 2.1 hold. The conclusion $\int_{t_0}^{\infty} e^{\alpha s}\rho(s)\,ds < \infty$ now follows directly from (2.2).

Under the conditions of the above corollary, it is easy to see from (2.6) that

$$\int_t^{\infty} \rho(s)\,ds = O\{e^{-\alpha t}\} \quad \text{as} \quad t \to \infty . \tag{2.7}$$

As shall be seen in the next corollary, this property is useful in determining an exponential rate of convergence of solutions of certain functional differential equations for which each constant is a solution. A prototype of these equations is

$$x'(t) = g(t,x(t)) - g(t,x(t-r)) \quad (r > 0), \tag{2.8}$$

where $g : [0,\infty) \times R^n \to R^n$ is continuous.

COROLLARY 2. Suppose there exists a function p, continuous and non-negative on $[0,\infty)$, such that g in (2.8) satisfies

$$\left.\begin{array}{l} |g(t,u) - g(t,v)| \leq p(t)|u-v| \\ \text{for all } t \geq 0 \text{ and } u,v \in R^n. \end{array}\right\} \tag{2.9}$$

($|\cdot|$) is a norm in R^n) If

$$\limsup_{s\to\infty} \int_s^{s+r} p(t)\,dt = \gamma < 1, \tag{2.10}$$

then each solution x of (2.8) on an interval $[t_1 - r,\infty)$ ($t_1 \geq 0$) tends to a constant as $t \to \infty$. Furthermore, the rate of convergence is exponential. If $\gamma = 0$ (i.e., if

$$\lim_{s\to\infty} \int_s^{s+r} p(t)\,dt = 0), \tag{2.11}$$

then x tends to its limit faster than any exponential.

PROOF. Let x be a solution of (2.8) on $[t_1 - r,\infty)$. It follows from (2.8) that x' is continuous on $[t_0,\infty)$, where $t_0 = t_1 + r$. Thus, for $t \geq t_0$

$$|x'(t)| \leq p(t)|x(t) - x(t-r)| \leq \int_{t-r}^{t} p(t)|x'(s)|\,ds. \tag{2.12}$$

That is, condition (1.1) is satisfied for

$$\rho(t) = |x'(t)| \quad \text{and} \quad m(t,s) = p(t). \tag{2.13}$$

Hence, from (2.13) and the hypothesis of Corollary 2, we see that the conditions of Corollary 1 are satisfied. We conclude from (2.7)

$$\int_t^\infty |x'(s)|ds = O\{e^{-\alpha t}\} \quad \text{as} \quad t \to \infty, \tag{2.14}$$

where α is given by (2.4) or (2.5) depending, respectively, on whether $\gamma > 0$ or $\gamma = 0$. In either case, $|x'| \in L^1[t_0,\infty)$ and it follows that

$$\lim_{t \to \infty} x(t) = x(\infty) \quad \text{exists (and is finite).}$$

We deduce from (2.14) that

$$x(t) - x(\infty) = O\{e^{-\alpha t}\} \quad \text{as} \quad t \to \infty. \tag{2.15}$$

Clearly, if $\lim_{s \to \infty} \int_s^{s+r} \rho(t)dt = 0$, x converges to its limit faster than any exponential since, for this case, (2.15) holds for any $\alpha > 0$. This completes the proof.

It is already known that solutions of (2.8) tend to finite limits under the conditions of Corollary 2 (cf. [2] or, for the linear case only, [1].) Also, each solution is uniformly stable [2, Theorem 4.1]. The above corollary represents a marked extension of the previous results since a rate of convergence is actually determined. That the constant 1 in the condition $\limsup_{s \to \infty} \int_s^{s+r} p(t)dt < 1$ is the best possible can be seen from the equation

$$x'(t) = r^{-1}[x(t) - x(t-r)],$$

where $x(t) = \alpha + \beta t$ is a solution for arbitrary constants α and β.

REMARK. Corollary 2 can be applied directly to the linear system

$$x'(t) = P(t)[x(t) - x(t-r)], \tag{2.16}$$

where P is a continuous $n \times n$ matrix function. Condition (2.10) becomes, for $p(t) = |P(t)|$ ($|\cdot|$ a matrix norm),

$$\limsup_{s \to \infty} \int_s^{s+r} |P(t)|dt = \gamma < 1. \tag{2.17}$$

Thusfar, we have concentrated on providing results regarding convergence to zero of $\int_t^{\infty} \rho(s)ds$; in particular, we have said nothing at all with respect to the behavior of ρ itself at ∞. This is remedied in Theorem 2.2 below, where a "strong" exponential decay property of ρ is given. This result requires only a slight modification of the conditions of Corollary 1, and among its consequences is exponential asymptotic stability of an integral equation.

Theorem 2.2. Let (1.1) hold and suppose

$$\sup_{t \geq t_0} \int_{t-r}^{t} m(t,s)ds < \infty; \tag{2.18}$$

$$\limsup_{t \to \infty} \int_{t-r}^{t} m(t,s)ds = \gamma < 1; \tag{2.19}$$

$$\lim_{h \to 0} \sup_{t \geq t_0} \int_t^{t+h} m(t+h,s)\,ds = 0 . \tag{2.20}$$

Then ρ tends to zero at an exponential rate as $t \to \infty$. More specifically, let (2.4) hold if $\gamma > 0$ and (2.5) if $\gamma = 0$. Then there exists a constant $K(\alpha)$, independent of ρ and $t_1 \geq t_0$, such that

$$\rho(t) \leq K(\alpha) e^{-\alpha(t-t_1)} ||\rho||_{t_1-r}^{t_1} \qquad (t \geq t_1), \tag{2.21}$$

$$||\rho||_u^v \overset{\text{def}}{=} \sup_{u \leq s \leq v} \rho(s)$$

An immediate consequence of Theorem 2.2 is a stability result for the functional integral equation

$$x(t) = \int_{t-r}^{t} G(t,s,x(s))\,ds. \tag{2.22}$$

Here we take x to be continuous and require G to satisfy conditions in order that the integral in (2.21) is well-defined and solutions exist locally (see, e.g. [4, pp. 86-87, 98]).

COROLLARY 3. Suppose

$$\left.\begin{array}{l} |G(t,s,u)| \leq m(t,s)|u| \\ t \geq 0, t-r \leq s \leq t, u \in R^n \end{array}\right\} \tag{2.23}$$

with m satisfying the same conditions as in Theorem 2.2 for $t_0 = 0$. Then the zero solution is exponentially asymptotically stable in the large. That is, there exists a constant $\alpha > 0$ such that:

for every $\delta > 0$ there exists a $M(\delta) > 0$ such that, for each $t_1 \geq 0$ and each initial function ϕ on $[t_1 - r, t_1]$ satisfying $|\phi(s)| \leq \delta$ $(t_1 - r \leq s \leq t_1)$, the solution x of (2.21) satisfies

$$|x(t)| \leq M(\delta) e^{-\alpha(t-t_1)} \sup_{t_1 - r \leq s \leq t_1} |\phi(s)| \quad \text{for all}$$

$t \geq t_1$.

PROOF. Setting $\rho(t) = |x(t)|$, we see that Corollary 3 follows from Theorem 2.2 and the inequality

$$|x(t)| \leq \int_{t-r}^{t} |G(t,s,x(s)|ds \leq \int_{t-r}^{t} m(t,s)|x(s)|ds.$$

We close this section by pointing out that we have chosen in this paper to place the major emphasis on certain integral inequalities. As a result, the applications have not been given in the most general form. This is particularly true since most examples we have considered for functional differential equations have actually been restricted to simple differential-difference equations (2.8), (2.16) and (2.17). By employing the usual notation $x' = f(t,x_t)$ for functional differential equations, one can easily generalize, say Corollary 2, by considering an inequality $|f(t,x_t)| \leq \int_{t-r}^{t} m(t,s)|x'(s)|ds$ (cf. [2]).

3. Proofs of the Theorems.

PROOF OF THEOREM 2.1. This proof is not at all difficult. Multiply both sides of (1.1) by $\phi(t)$ and integrate from t_0 to T $(T > t_0)$ to obtain

$$\int_{t_0}^{T} \phi(t)\rho(t)dt \le \int_{t_0}^{T} \phi(t)\int_{t-r}^{t} m(t,s)\rho(s)dsdt$$
$$\le c + \int_{t_0}^{T} \rho(s)\int_{s}^{s+r} \phi(t)m(t,s)dtds, \tag{3.1}$$

where $c = \int_{t_0-r}^{t_0} \rho(s)\int_{s}^{s+r} \phi(t)m(t,s)dtds$. Changing the variable of integration in the left-hand side of (3.1) to s and applying (2.1), we have

$$\int_{t_0}^{T} \psi(s)\rho(s)ds \le \int_{t_0}^{T} [\phi(s) - \int_{s}^{s+r} \phi(t)m(t,s)dt]\rho(s)ds \le c. \tag{3.2}$$

Since (3.2) holds for arbitrary $T > t_0$, (2.2) follows and the proof is complete.

PROOF OF THEOREM 2.2. Fix $\alpha > 0$ according to (2.4) or (2.5) (whichever is appropriate) and define

$$\varepsilon = (1 - \gamma e^{\alpha r})/(1 + e^{\alpha r}). \tag{3.3}$$

Then $0 < \varepsilon < 1$, and so by (2.20) one can find a positive integer n such that

$$\int_{t}^{t+h} m(t+h,s)ds \le \varepsilon \quad (0 \le h \le r/n; \quad t \ge t_0). \tag{3.4}$$

For the sake of notational convenience, set $b(t) = \int_{t-r}^{t} m(t,s)ds$. From (2.19) and the choice of ε in (3.3), we can find constants $\beta, T \ge 0$ such that

$$\left.\begin{aligned} b(t) &\le (1-\varepsilon)e^{\beta} \quad (t \ge t_0) \text{ and} \\ b(t) &\le (1-\varepsilon)e^{-\alpha r} \quad (t \ge t_0 + T). \end{aligned}\right\} \tag{3.5}$$

It follows from (3.1) that $\gamma+\varepsilon=(1-\varepsilon)e^{-\alpha r}$, which means

$$\left.\begin{aligned} b(t) &\le (1-\varepsilon)e^{\beta} && (t \ge t_0) \\ b(t) &\le (1-\varepsilon)e^{-\alpha r} && (t \ge t_0+T). \end{aligned}\right. \tag{3.5}$$

Let $0 \le h \le r/n$; $t \ge t_1 \ge t_0$ and use (1.1) and (3.4) to obtain

$$\begin{aligned} \rho(t+h) &\le \int_{t+h-r}^{t} m(t+h,s)\rho(s)ds + \int_{t}^{t+h} m(t+h,s)\rho(s)ds \\ &\le b(t+h)\,||\rho||_{t+h-r}^{t} + \varepsilon\,||\rho||_{t}^{t+h} \\ &\le b(t+h)\,||\rho||_{t-r}^{t} + \varepsilon||\rho||_{t}^{t+r/n}, \end{aligned}$$

where $||\rho||_{u}^{v}$ is defined in (2.21). Now, take the supremum over $0 \le h \le r/n$, move the term $\varepsilon||\rho||_{t}^{t+r/n}$ to the other side, divide by $1-\varepsilon$ and employ (3.5) to obtain

$$\left.\begin{aligned} ||\rho||_{t}^{t+r/n} &\le e^{\beta}\,||\rho||_{t-r}^{t} && (t \ge t_1), \\ ||\rho||_{t}^{t+r/n} &\le e^{-\alpha r}\,||\rho||_{t-r}^{t} && (t \ge t_1+T). \end{aligned}\right\} \tag{3.6}$$

First, consider the interval $t_1 \le t \le t_1+T$. For every non-negative integer k, define

$$d_k = ||\rho||_{t_1+kr/n-r}^{t_1+kr/n}. \tag{3.7}$$

Then by (3.6)

$$d_k \le e^{\beta} d_{k-1} \qquad (k \ge 1),$$

which implies

$$d_k \le e^{\beta k} d_0 \qquad (k \ge 1).$$

This inequality combined with (3.7) yields

$$\rho(t) \le K_1 ||\rho||_{t_1-r}^{t_1} \qquad (t_1 \le t \le t_1 + T), \tag{3.8}$$

where $K_1 = e^{\beta(1+nT/r)}$.

Next, consider the interval $t_1 + T \le t < \infty$. By (3.6),

$$||\rho||_{t+r/n-r}^{t+r/n} \le ||\rho||_{t-r}^{t} \qquad (t \ge t_1 + T).$$

Thus, (3.6) with t replaced by $t + r/n$ yields

$$||\rho||_{t+r/n}^{t+2r/n} \le e^{-\alpha r} ||\rho||_{t+r/n-r}^{t+r/n} \le e^{-\alpha r} ||\rho||_{t-r}^{t}.$$

Similarly, for every $k \ge 1$,

$$||\rho||_{t+(k-1)r/n}^{t+kr/n} \le e^{-\alpha r} ||\rho||_{t-r}^{t} \qquad (t \ge t_1 + T).$$

However,

$$||\rho||_{t}^{t+r} = \max_{1 \le k \le n} ||\rho||_{t+(k-1)r/n}^{t+kr/n},$$

and so

$$||\rho||_{t}^{t+r} \le e^{-\alpha r} ||\rho||_{t-r}^{t} \qquad (t \ge t_1 + T) \tag{3.9}$$

In particular, choosing $t = t_1 + T + kr$, $k = 1,2,3,\ldots$

and solving (3.9) recursively leads to

$$||\rho||_{t_1+T-(k-1)r}^{t_1+T+kr} \le e^{-\alpha kr}||\rho||_{t_1+T-r}^{t_1+T}.$$

For any $t \ge t_1 + T$, choose k so that $t_1 + T + (k-1)r \le t \le t_1 + T + kr$ and obtain

$$\rho(t) \le e^{-\alpha kr}||\rho||_{t_1+T-r}^{t_1} \le e^{-\alpha(t-t_1+T)}||\rho||_{t_1+T-r}^{t_1+T}.$$

This combined with (3.8) yields

$$\rho(t) \le K_1 e^{\alpha T} e^{-\alpha(t-t_1)}||\rho||_{t_1-r}^{t_1} \qquad (t \ge t_1).$$

That is, (2.21) holds with $K(\alpha) = K_1 e^{\alpha T}$, $K_1 = e^{\beta(1+nT/r)}$.

The proof of Theorem 2.2 can be shortened by borrowing from the terminology of Volterra equations and, in particular, by applying recent results of [3]. However, we have retained the above proof since it requires only elementary techniques.

REFERENCES

[1] O. Arino and P. Séguier, Publications Mathématiques, Université de Pau; Pau, France, 1980.

[2] F. V. Atkinson and J. R. Haddock, Criteria for asymptotic constancy of solutions of functional differential equations, J. Math. Anal. Appl., to appear.

[3] G. Gripenberg, On the resolvents of nonconvolution Volterra Kernels, Funkcial. Ekvac., 23 (1980), 83-95.

[4] R. K. Miller, "Nonlinear Volterra Integral Equations", W. A. Benjamin, Menlo Park, California, 1971.

BIFURCATION OF PERIODIC TRAVELLING WAVES FOR A REACTION-DIFFUSION SYSTEM

by

D.L. Barrow and P.W. Bates

The second author was supported by an SERC visiting fellowship grant at Heriot-Watt University.

1. INTRODUCTION

In a recent paper we reported on the existence and stability of 2π-periodic steady state and travelling wave solutions of a certain system of reaction-diffusion equations. Here we give conditions which imply the existence of other 2π-periodic travelling wave solutions bifurcating from a circle of degenerate steady state solutions. To show the rich structure of the solution set of the system below, and to set the stage for the new results we first summarize the contents of [1].

The system (1.1) was proposed by J.M. Lasry as a model which may have properties sufficiently similar to the Hodgkin-Huxley equations of nerve conduction that its study will lead to a better understanding of that system.

Consider

$$(1.1)\qquad \begin{cases} u_t = u_{xx} + k(1-\rho)u - \varepsilon\phi(\beta)v \\ v_t = v_{xx} + k(1-\rho)v + \varepsilon\phi(\beta)u \quad , \quad -\infty < x < \infty , \quad t \geqslant 0, \end{cases}$$

where $u = \rho\cos\beta$, $v = \rho\sin\beta$, k and ε are parameters with $k > 0$ and ϕ is a smooth 2π-periodic function which changes sign on $[0,2\pi]$. Thus, the vector field determined by the nonlinearities has a rotationally symmetric radial component directed towards the unit circle and a rotational component which changes direction, vanishing on certain rays from $(u,v) = (0,0)$. There are, therefore, constant solutions of (1.1), viz., $(0,0)$ and points where these rays meet the unit circle.

We seek solutions which are 2π-periodic in x and which travel with time, considering first the case where $\varepsilon = 0$ and the speed of propagation is zero. Then

(1.1) reduces to

$$(1.2)\qquad \begin{cases} u'' + k(1-\rho)u = 0 \\ v'' + k(1-\rho)v = 0 \quad , \quad u,v\ 2\pi\text{-periodic.} \end{cases}$$

The following lemmas are easily established.

Lemma 1.1 All solutions of (1.2) with $k > 0$ satisfy $u^2 + v^2 \leqslant 1$.

Idea of proof Suppose not and use the maximum principle to reach a contradiction.

Lemma 1.2 If $0 < k \leqslant 1$ then (1.2) has only the constant solutions comprising (0,0) and points (u,v) with $u^2 + v^2 = 1$.

Idea of proof Use Lemma 1.1 and a comparison of eigenvalues argument.

Lemma 1.3 The solution set of (1.2) is invariant under the group O(2) acting both in the uv-plane and in the independent variable. Furthermore, any solution of (1.2) can be obtained from a solution (u,v) with u even and v odd by rotating in the uv-plane and translating in x.

Idea of proof The first part follows by inspection. The second part follows from the first part and a uniqueness argument for the corresponding first order system.

There are certain easily found solutions of (1.2) which we call circular and collinear because of their image in the uv-plane. For each nonzero integer j there is a family of circular solutions given by $u(x) = (1 - j^2/k)\cos jx$, $v(x) = (1 - j^2/k)\sin jx$ (and those generated from this pair by the symmetry group action). So we have a branch (surface) of solutions bifurcating from $(u,v) \equiv (0,0)$ at $k = j^2$ and existing for $k > j^2$. In seeking collinear solutions, that is u,v linearly dependent, we can in view of Lemma 1.3 assume that $v \equiv 0$ and u is even. The system (1.2) becomes

$$(1.3)\qquad u'' + k(1 - |u|)u = 0, \quad u\ 2\pi\text{-periodic.}$$

Either by using a bifurcation argument or a shooting method, as was done in [1], one can prove

Theorem 1.4 For each nonzero integer j, problem (1.3) has a continuous family of nontrivial even solutions $\{u_j(\cdot;k) : k > j^2\}$ such that $u_j(\cdot;k)$ has $2j$ zeros in $[0,2\pi]$ and $\lim_{k\to j^2} u_j(\cdot;k) = 0$.

Returning to the circular solutions, if we rewrite (1.2) in polar coordinates and linearize about a point on the jth branch of circular solutions we obtain the operator defined by

$$L(\rho,\beta) = \begin{bmatrix} \rho'' - 2j(k-j^2)\beta'/k - (k-j^2)\rho \\ \beta'' + 2jk\rho'/(k-j^2) \end{bmatrix}.$$

Considered as an operator from $\{\rho,\beta \in C^2(S^1)\colon \rho \text{ even}, \beta \text{ odd}\}$ into $(C(S^1))^2$, L is noninvertible (for $k > j^2$) iff $k = 5j^2 - n^2$, $n = 1,2,\ldots,2|j|-1$. For each such singular value of k, L has a one dimensional kernel and codimensional one range. It is easy to check that the other hypotheses of Theorem 1.7 of Crandall and Rabinowitz [2] are satisfied, thus obtaining secondary bifurcation from the branches of circular solutions at the above values of k.

Now consider (1.1) with $\varepsilon \neq 0$ and look for 2π-periodic solutions depending upon the single variable $z = x - ct$ (waves travelling with speed c), where c is a parameter which may depend upon ε. This gives the system

$$(1.4) \qquad \begin{cases} u'' + k(1-\rho)u + cu' - \varepsilon\phi(\beta)\, v = 0 \\ v'' + k(1-\rho)v + cv' + \varepsilon\phi(\beta)u = 0\,, \quad u,v \quad 2\pi\text{-periodic in } z. \end{cases}$$

Again, changing to polar coordinates and linearizing about the point $(\rho,\beta,c) = (1 - j^2/k, jx, 0)$ on the jth branch of circular solutions yields an operator from $\{\rho,\beta \in C^2(S^1)\colon \beta(0) = 0\} \times \mathbb{R}$ to $(C(S^1))^2$. For $k \neq 5j^2 - n^2$, $n = 1,2,\ldots,2|j|-1$, this operator is an isomorphism, so the implicit function theorem gives the existence of solutions to (1.4) for $|\varepsilon|$ small, near the circular solutions of (1.2). The stability of these travelling waves and steady state soltuions can be determined using linearization, the perturbation theory in Kato [4] and the stability results of Sattinger [5] (or Henry [3]). This has been done in [1].

The question now arises as to the existence of travelling wave solutions of (1.4) with $\varepsilon \neq 0$ in the vicinity of the collinear solutions of (1.2). The next section is devoted to answering this question and these results are not contained in [1].

2. <u>BIFURCATION FROM THE COLLINEAR SOLUTIONS</u>

Recall that for each fixed integer n and $k > n^2$ there are $|n|$ circles of solutions, $(u_i(x;k)\cos\lambda, u_i(x;k)\sin\lambda)$, $\lambda \in [0,2\pi]$, $i = 1,\ldots |n|$. We will consider one of these circles and treat λ as a bifurcation parameter. Let $\bar{u} = u_j(\cdot;k)$ for some j,k and recall that $\bar{u}$ is even and has $2j$ zeros in $[0,2\pi]$.

We would like to use the bifurcation results found in [2] but because of the degeneracy of the collinear solutions, passing through $(u,v) = (0,0)$ with the fact that $\beta = \beta(u,v)$ is not continuous there, we need an abstract theorem with less restrictive hypotheses. The following two theorems are probably known but we have not found them in the literature. The first is an implicit function theorem, the second is a bifurcation theorem proved using the first theorem in much the same way as the Crandall-Rabinowitz result [2].

Let Y,Z be Banach spaces and S a normed linear space and let f be a mapping with domain in $S \times Y$ and range in Z.

<u>Theorem 2.1</u> Suppose $f(s,y) = Ly - N(s,y)$ for $(s,y) \in \mathrm{dom}(f)$ with $N(s,\cdot)$ continuous on Y for all $s \in S$ sufficiently small. Assume

(i) $f(0,0) = 0$

(ii) L is linear with compact inverse $L^{-1}:Z \to Y$.

(iii) $N(0,y) = o(\|y\|)$ as $y \to 0$ and

(iv) for some $R > 0$, $N(s,y) - N(0,y) \to 0$ uniformly for $\|y\| \leq R$ as $s \to 0$.

Then there exists $\delta > 0$ such that for all s with $|s| < \delta$ there exists $y(s) \in Y$ with $\|y(s)\| \leq R$ such that $f(s,y(s)) = 0$. Furthermore, $y(s) \to 0$ as $s \to 0$.

<u>Proof</u> Using (ii) and (iii) we may choose r arbitrarily small in $(0,R]$ such that $\|N(0,y)\| \leq r/2\|L^{-1}\|$ for $\|y\| \leq r$. Solving $f(s,y) = 0$ for $\|y\| \leq r$ is equivalent to finding a fixed point of $Ty \equiv L^{-1}(N(s,y))$ in $\{y:\|y\| \leq r\}$. This map is completely continuous by (ii) and may be written as $L^{-1}(N(s,y) - N(0,y)) + L^{-1}(N(0,y))$. By (iv) there exists $\delta > 0$ such that $\|N(s,y) - N(0,y)\| \leq r/2\|L^{-1}\|$ for $\|y\| \leq r$ if $|s| < \delta$. Hence, for each $s \in S$ with $|s| < \delta$, the completely continuous map T takes $\{y:\|y\| \leq r\}$ into itself and has a fixed point by Schauder's theorem. □

Now, let X and Z be Banach spaces; suppose L_1 is a map from $\mathbb{R}$ into the (possibly unbounded) linear operator from X to Z with $\mathrm{dom}(L_1(\lambda))$ independent of λ. Suppose $M_1:\mathbb{R} \times X \to Z$ is continuous. Define $F(\lambda,x) = L_1(\lambda)x + M_1(\lambda,x)$ for all $\lambda \in \mathbb{R}$ and $x \in \mathrm{dom}\, L_1(\lambda)$.

Theorem 2.2 Suppose that $M_1(\lambda,x) = o(\|x\|)$ as $x \to 0$ uniformly for λ bounded and that $L_1(\lambda)x = L_1(0)x + L_2\lambda x + M_2(\lambda)x$ with L_2 and $M_2(\lambda)$ bounded linear operators from X to Z and $M_2(\lambda)x = o(|\lambda|)$ as $\lambda \to 0$, uniformly for x in bounded sets. Suppose that $M_2(\cdot)x$ is continuous for each fixed x. Suppose $L_1(0)$ has a one dimensional null space spanned by $u_0 \in X$, that $\operatorname{codim} R(L_1(0)) = 1$ and that $L_1(0)^{-1}$ defined from $R(L_1(0))$ into $\overline{X}$ (a topological complement of span$\{u_0\}$) is compact. Finally, suppose that $L_2u_0 \notin R(L_1(0))$. Then there exists $\delta > 0$ such that for all $|s| \leq \delta$ there exists $\lambda(s) \in \mathbb{R}$, $x(s) \in \overline{X}$ such that $\overline{F}(\lambda(s),s(u_0 + x(s))) = 0$ and $\lambda(s) \to 0$, $x(s) \to 0$ as $s \to 0$.

Proof For $s \in \mathbb{R}$ and $y = (\lambda,x) \in \mathbb{R}\times\overline{X} \equiv Y$ define

$$f(s,y) = \begin{cases} L_1(\lambda)(u_0+x) & \text{if } s = 0 \\ \frac{1}{s}\overline{F}(\lambda,s(u_0+x)) & \text{if } s \neq 0. \end{cases}$$

Let $Ly \equiv L_1(0)x + L_2\lambda u_0$ and $N(s,y) \equiv Ly - f(s,y)$

so
$$N(s,y) = \begin{cases} -L_2\lambda x - M_2(\lambda)(u_0+x) & \text{if } s = 0 \\ -L_2\lambda x - M_2(\lambda)(u_0+x) - \frac{1}{s}M_1(\lambda,s(u_0+x)) & \text{if } s \neq 0, \end{cases}$$

and $N(s,\cdot)$ is continuous. Furthermore, $f(0,0) = 0$, $N(0,y) = o(\|y\|)$ as $y \to 0$ and $N(0,y) - N(s,y) = \frac{1}{s}M_1(\lambda,s(u_0+x)) \to 0$ as $s \to 0$, uniformly for (λ,x) bounded. We have $\operatorname{codim} R(L_1(0)) = 1, L_1(0)$ is 1 - 1 on $\overline{X}$ and $L_2u_0 \notin R(L_1(0))$, so L is an isomorphism from Y onto Z.

Finally, L^{-1} is compact by the assumption on $L_1(0)^{-1}$ and Theorem 2.1 gives the conclusion.

Returning to the differential equations, let

$$(2.1) \qquad F(u,v,c,\varepsilon) = \begin{bmatrix} u'' + k(1-\rho)u + cu' - \varepsilon\phi(\beta(u,v))v \\ v'' + k(1-\rho)v + cv' + \varepsilon\phi(\beta(u,v))u \end{bmatrix}$$

where ρ and $\beta = \beta(u,v)$ are given by $u = \rho\cos\beta$, $v = \rho\sin\beta$. We suppose that ϕ is 2π-periodic and differentiable, so that $\phi(\beta(\cdot,\cdot))$ is differentiable on the punctured uv-plane and is uniformly bounded there. Let $\overline{u}$ be the solution of $u'' + k(1-|u|)u = 0$ mentioned at the beginning of this section and let $w(\lambda) = (\overline{u}\cos\lambda, \overline{u}\sin\lambda, 0,0)$, then $F(w(\lambda)) \equiv 0$. We will write $a \perp b$ whenever $\int_{-\pi}^{\pi}ab = 0$. Let $X = \{x = (u,v,c,\varepsilon) \in (H^1(S^1))^2 \times\mathbb{R}^2 : u \perp \overline{u}', v \perp \overline{u}\}$ with norm given by $\|x\|^2 = \|u\|^2_{H^1} + \|v\|^2_{H^1} + c^2 + \varepsilon^2$, then X is a Banach space. Define the (unbounded) operator in $\mathbb{R}\times X$ to $Z \equiv (L^2(S^1))^2$ by

$$\overline{F}(\lambda,x) = F(w(\lambda) + x). \tag{2.2}$$

Let $L_1(\lambda)x = \begin{cases} u'' + k(1 - |\overline{u}|(1+\cos^2\lambda))u - k|\overline{u}|\cos\lambda\sin\lambda v + c\overline{u}'\cos\lambda - \varepsilon\phi(\overline{\beta}(\lambda))\overline{u}\sin\lambda \\ v'' + k(1 - |\overline{u}|(1+\sin^2\lambda))v - k|\overline{u}|\cos\lambda\sin\lambda u + c\overline{u}'\sin\lambda + \varepsilon\phi(\overline{\beta}(\lambda))\overline{u}\cos\lambda \end{cases}$

where $\overline{\beta}(\lambda) = \begin{cases} \lambda & \text{if } \overline{u} > 0 \\ \lambda + \pi & \text{if } \overline{u} < 0. \end{cases}$ Clearly, $\mathrm{dom}(L_1(\lambda))$ is independent of λ.

Let $M_1(\lambda,x) = \overline{F}(\lambda,x) - L_1(\lambda)x$, then since second order terms cancel, it is easy to see that $M_1 : \mathbb{R}\times X \to Z$ is continuous. The first component of $M_1(\lambda,x)$ is

$m_1 \equiv k\cos\lambda[\,|\overline{u}|(u\cos\lambda + v\sin\lambda) + \overline{u}(|\overline{u}| - \sqrt{(\overline{u}\cos\lambda + u)^2 + (\overline{u}\sin\lambda + v)^2})] +$

$+ \varepsilon\overline{u}\sin\lambda[\phi(\overline{\beta}(\lambda)) - \phi(\beta(\overline{u}\cos\lambda + u, \overline{u}\sin\lambda + v))]$ + terms which are $o(\|x\|)$, uniformly in λ, as $x \to 0$. The bracket in the first term can be written (when $\overline{u} \neq 0$) as

$\overline{u}(u\cos\lambda + v\sin\lambda)[1 - 2|\overline{u}|/(|\overline{u}| + \sqrt{(\ \ )^2 + (\ \ )^2})] + \overline{u}(u^2+v^2)/(|\overline{u}| + \sqrt{(\ \ )^2 + (\ \ )^2})$

and both of these terms are $o(\|x\|)$ in L^2 uniformly in λ as $\|x\| \to 0$. The second term of m_1 is seen to be $o(\|x\|)$ in L^2 as $\|x\| \to 0$ by using Lebesgue's Dominated Convergence Theorem (the uniformity in λ is not difficult to prove but we omit the proof to save space).

The second component of $M_1(\lambda,x)$ is similar to m_1.

Lemma 2.3 Let $A, B : H^2(S^1) \to L^2(S^1)$ be given by

$$Au = u'' + k(1 - 2|\overline{u}|)u$$
$$Bv = v'' + k(1 - |\overline{u}|)v\ .$$

Let the eigenvalues of A and B be denoted by $\lambda_0 > \lambda_1 \geqslant \lambda_2 \geqslant \dots$ and $\mu_0 > \mu_1 \geqslant \mu_2 \geqslant \dots$, respectively. Then

(i) $\lambda_{2j-1} = 0$ and $N(A) = \mathrm{span}\{\overline{u}'\}$

(ii) $\mu_{2j} = 0$ and $N(B) = \mathrm{span}\{\overline{u}\}$.

Proof By definition, $\overline{u} \in N(B)$ and since $\overline{u}(\cdot + h)$ is a 2π-periodic solution of $u'' + k(1 - |u|)u = 0$, differentiation shows that $\overline{u}' \in N(A)$. Since $|\overline{u}| > 0$ a.e., standard Sturm-Liouville theory gives (a) $\mu_i > \lambda_i$, $i = 0,1,\dots$. Also, (b) 0 is an eigenvalue for both A and B with the eigenfunctions each having $2j$ zeros in $[-\pi,\pi)$. Again using Sturm theory (c) the eigenfunctions corresponding to

$\lambda_{2j-1}, \mu_{2j-1}, \lambda_{2j}, \mu_{2j}$ are precisely those which have $2j$ zeros in $[-\pi,\pi)$. Statements (a), (b) and (c) imply

$$\lambda_{2j} < \mu_{2j} = 0 = \lambda_{2j-1} < \mu_{2j-1} \text{ and } \lambda_{2j-1} < \lambda_{2j-2} \text{ and } \mu_{2j+1} < \mu_{2j}, \text{ which}$$

proves the lemma.

We now assume

(H1) $\qquad \phi(0) = -\phi(\pi)$.

To find $N(L_1(0))$ set $L_1(0)x = 0$ with $x \in X$. This gives

$$\begin{cases} Au + c\overline{u}' = 0 \\ Bv + \varepsilon\phi(\overline{\beta}(0))\overline{u} = 0. \end{cases}$$

Multiplying the first equation by $\overline{u}'$ and integrating gives $c = 0$. By Lemma 2.3 and the definition of X we have $u = 0$. Since $\phi(\overline{\beta}(0))\overline{u} \perp \overline{u}$ by (H1) and the symmetry of $\overline{u}$, there is, for each ε, a unique solution, v, of the second equation with $v \perp \overline{u}$. Thus, if $\overline{v}$ is that solution corresponding to $\varepsilon = 1$, we see that $N(L_1(0))$ is spanned by $u_0 \equiv (0,\overline{v},0,1)$. One can now check that if $f,g \in L^2(S^1)$ then $(f,g)^T \in R(L_1(0))$ iff $g \perp \overline{u}$, so $\operatorname{codim} R(L_1(0)) = 1$. If $\overline{X}$ is a topological complement of $\operatorname{span}\{u_0\}$ in X one sees that $L_1(0) : \overline{X} \cap \operatorname{dom}(L_1(0)) \to R(L_1(0))$ is invertible. Furthermore, because of the compactness of the embedding of $H^2(S^1)$ in $H^1(S^1)$ the inverse is compact. Define L_2 and $M_2(\lambda) : X \to Z$ by

$$L_2 x = \begin{pmatrix} -k|\overline{u}|v - \varepsilon\phi(\overline{\beta}(0))\overline{u} \\ -k|\overline{u}|u + c\overline{u}' + \varepsilon\phi'(\overline{\beta}(0))\overline{u} \end{pmatrix}, \qquad M_2(\lambda) = L_1(\lambda) - L_1(0) - \lambda L_2.$$

Then, apart from terms which are obviously $o(|\lambda|)$ as $\lambda \to 0$, uniformly for $\|x\|$ bounded, $M_2(\lambda)$ has components $\varepsilon\overline{u}(\phi(\overline{\beta}(0))\lambda - \phi(\overline{\beta}(\lambda))\sin\lambda)$ and $\varepsilon\overline{u}(\phi(\overline{\beta}(\lambda))\cos\lambda - \phi(\overline{\beta}(0)) - \lambda\phi'(\overline{\beta}(0)))$. These too are seen to be $o(|\lambda|)$ as $\lambda \to 0$ uniformly for $|\varepsilon|$ bounded, by Lebesgue's Dominated Convergence Theorem. To be able to apply Theorem 2.2 we need to assume

(H2) $\qquad \phi'(0) \neq -\phi'(\pi)$.

This ensures that $\phi'(\overline{\beta}(0))\overline{u} \not\perp \overline{u}$ and so $L_2 u_0 \notin R(L_1(0))$.

We have proved

Theorem 2.4 Suppose that ϕ is differentiable and 2π-periodic and let θ be any angle such that $\phi(\theta) + \phi(\theta+\pi) = 0$, $\phi'(\theta) + \phi'(\theta+\pi) \neq 0$. Then there exists an interval I containing 0 and mappings $\lambda, c, \varepsilon : I \to \mathbb{R}$, $u, v : I \to C^2(S^1)$ which are continuous at $0 \in I$ and take on the value zero there, such that for each $s \in I$, (1.4) is satisfied with $(u,v,c,\varepsilon) =$

$$(\bar{u}\cos(\theta+\lambda(s)) + s(u(s) - \bar{v}\sin\theta), \bar{u}\sin(\theta+\lambda(s)) + s(v(s)+\bar{v}\cos\theta), sc(s), s+s\varepsilon(s))$$

The only claims in the theorem not discussed above are the regularity, which follows by bootstrapping, and that θ need not be zero which is not really a difficulty since we can rotate in the uv-plane and translate ϕ.

REFERENCES

1. D.L. Barrow and P.W. Bates, Bifurcation and Stability of Periodic Travelling Waves for a Reaction-Diffusion System, to appear, J. Differential Equations.

2. M.G. Crandall and P.H. Rabinowitz, Bifurcation from Simple Eigenvalues, J. Functional Analysis 8 (1971), 321-340.

3. D. Henry, Geometric Theory of Semilinear Parabolic Equations, Lecture Notes in Mathematics vol. 840, Springer-Verlag, New York, 1981.

4. T. Kato, Perturbation Theory for Linear Operators, Springer-Verlag, New York, 1966.

5. D.H. Sattinger, On the Stability of Waves of Nonlinear Parabolic Systems, Advances in Math. 22 (1976), 312-355.

The Dirac Equation with an Anomalous Magnetic Moment II

Horst Behncke

Though it is well known, that the electron possesses an anomalous magnetic moment, this term has not been considered so far in the mathematical investigations of the Dirac equation [3,4,5,6] and it has at most been treated by perturbation methods in the physical literature. Recent investigations of Barut [1] and this author [2] make it desirable to study the interaction due to the anomalous magnetic moment nonpertubatively, because this term is rather singular. This singularity leads to a number of interesting results. Most notable among these is the essential selfadjointness of the Hamiltonian for almost all potentials and the small distance behaviour of the wavefunctions. The essential spectrum of the Dirac Hamiltonian is the main object of this paper. The investigations are carried out mainly for spherically symmetric potentials.

The Dirac Hamiltonian for an electron in an external static electromagnetic potential A and scalar field B is [2] .

$$H = \Sigma \gamma^o \gamma^k (-i\partial_k - e_o A^k - e_1 \Delta A^k) +$$

$$(1) \quad e_o A^o + e_1 \Delta A^o + m_o \gamma^o + \gamma^o B + i f \Sigma \gamma^o \gamma^k \gamma^l (\partial_l A^k) -$$

$$i f \Sigma \gamma^k \partial_k A^o$$

Here e_o represents the charge, e_1 the electrical dipole moment and f the anomalous magnetic moment.

Theoretically one could also include higher moment interaction terms, which arise from radiative corrections in quantum electrodynamics. However, since these terms lead to nonlocal interactions and are of the same order as other effects, we restrict ourselves to (1). In addition we assume that the most singular terms in (1) are spherically symmetric and due to the electric potential A_o. Because ΔA_o corresponds to the charge distribution of the electric field, (1) can be rewritten as

$$(2)\quad H = \alpha \cdot (-i\,\nabla) + V_e + \beta(V_s + m) - i\beta\alpha \cdot \frac{x}{r} V_m + H_1 = H_o + H_1$$

With $V_e = e_o A_o + e_1 \Delta A_o$ and $V_m = fA'_o$ spherically symmetric. H_1 is a symmetric 4 by 4 matrix, whose coefficients are linear combinations of $A_1, \ldots, \partial_3 A_3$. In order for H to be a symmetric operator on $L^2(\mathbb{R}^3) \otimes \mathbb{C}^4$ with domain $\mathcal{D}_H = \mathcal{C}_o^\infty(\mathbb{R}^3_+) \otimes \mathbb{C}^4$, $\mathbb{R}^3_+ = \mathbb{R}^3 \setminus \{0\}$ we require in addition:

(3.1) $V_e, V_s, \ldots, \partial_3 A_3$ are real valued and in $L^2_{loc}(\mathbb{R}^3_+)$

(3.2) $\operatorname{sgn} V_m(r)$ is constant for small r

Here $\mathcal{C}_o^k(\mathbb{R}^3_+)$ is the space of k-times differentiable functions on $\mathbb{R}^3_+$ with compact support. We shall also assume V_m to be more singular than $1/r$, V_e, V_s and H_1 near 0. This is satisfied in all cases of physical interest.

Since H_o is spherically symmetric, it can be separated by using polar coordinates. Thus H_o is unitarily equivalent to a direct sum of operators $K(\varepsilon.l.m)$.

$$H_o \simeq \Sigma^{\oplus} K(\varepsilon,\ell,m) \qquad \ell = 1,2,3,\ldots \quad \varepsilon = \pm 1$$
$$m = -\ell + {}^1/_2, \ldots, \ell - {}^1/_2$$

Each $K(\varepsilon,\ell,m)$ operates on $L^2(0,\infty) \otimes \mathbb{C}^2$ by

$$(4)\quad K(\varepsilon,\ell,m) = \begin{pmatrix} V_e - V_s - m & D + \varepsilon(-1)^\ell \frac{\ell}{r} - V_m \\ -D + \varepsilon(-1)^\ell \frac{\ell}{r} - V_m & V_e + V_s + m \end{pmatrix} \quad ; \; D = \frac{d}{dr}$$

with domain $\mathcal{D} = \mathcal{C}_o^\infty((0,\infty)) \otimes \mathbb{C}^2$.

For simplicity of notation consider now

$$(5)\quad K = \begin{pmatrix} V_e - V_s & D+V_3 \\ -D+V_3 & V_e+V_s \end{pmatrix} \quad \text{with} \quad \mathcal{D}_K = C_0^\infty \ ((0,\infty)) \otimes \mathbb{C}^2$$

Theorem 1:

Let H_0 be as in (2). In addition to (3) assume

i) $\lim_{r\to 0} r\, V_m(r) = \pm \infty$

ii) $V_e = c_e V_m + U_e + W_e$; $V_s = c_s V_m + U_s + W_s$ with $1 > c_e^2 - c_s^2$ and

iii) U_e, U_s are locally integrable near 0 and

$$\lim_{r\to 0} W_e(r)/V_m(r) = \lim_{r\to 0} W_s(r)/V_m(r) = 0$$

Then H_0 is essentially selfadjoint.

Proof:

a) H_0 is essentially selfadjoint, if this is true for all K (ε,ℓ,m). Since K is limit point at infinity [6, theorem 5.1], it is essentially selfadjoint if the equation

$$K \binom{u}{v} = 0$$

has only one nontrivial solution, which is square integrable over $(0,b)$, for some $b > 0$.

The necessary estimates for the solutions of this equation can be found in [1]. Here we sketch another proof. With the aid of the Prüfer transformation [6] $u = \rho\cos\theta$ and $v = \rho\sin\theta$, one obtains

$$(6)\quad \begin{aligned} \theta' &= -\{V_1 \cos^2\theta + V_2 \sin^2\theta + V_3 2 \sin\theta \cos\theta\} \quad \text{and} \\ +(\ln\rho)' &= -{}^1/_2(V_1 - V_2)\sin 2\theta + V_3 \cos 2\theta \end{aligned}$$

Argue now as in [5] and show that (6) has a solution θ_+ with

$$\lim_{r\to 0} - c_s \sin 2\theta_+(r) - \cos 2\theta_+(r) = \sqrt{1 - c_e^2 + c_s^2}$$

Then the corresponding ρ can be estimated by

$$\rho(r)^2 \geqq \rho\,(r_0)^2 \exp 2 \int_r^{r_0} \left(\sqrt{1 - c_e^2 + c_s^2} \right) V_3(t)\, dt$$

Thus there is a solution, which is not square integrable.

Remark:

This result can be extended readily to more general potentials. As in [2] this method also allows to determine the deficiency indices of H_0 in more general situations.

Corollary:

Let H_0 be as above and assume that all matrix coefficients of H_1 are L^∞_{loc} $(\mathbb{R}^3)$. Then H is essentially selfadjoint.

This is shown as in [2] .

In the remainder we shall restrict ourselves to an investigation of the spectrum and essential spectrum of H.

Lemma 1:

Let H be as in (2) with

(7) $A_i(x)$, $V_e(r)$, $V_m(x)$, $V_s(x)$, $H_1(x) \to 0$ as $|x| = r \to \infty$

Then $(-m,m)^c \subset \sigma_{ess}(H)$ for any selfadjoint extension H of H .

Proof:

The free Dirac Hamiltonian H_{00} has the essential spectrum σ_{ess} $(H_{00}) = (-m,m)^c$. For each $\lambda \notin (-m,m)$ there exists therefore infinitely many independent smooth and λ-approximate eigenvectors, whose support is compact and far from the origin. By assumption these functions are also λ-approximate eigenvectors for H and $\tilde{H}$.

Remark:

Let H_o be as in (2) and assume in addition to (3) and (7):

V_e, V_m and V_s are monotonic for large r.

Then it follows from [6, § 8] that there are no eigenvalues of H_o embedded in $(-m,m)^c$. Thus one would expect that the spectrum of H consists of $(-m,m)^c$ and countably many eigenvalues accumulating at $\pm$ m. Under physically reasonable assumptions this is indeed th case.

For any $\phi \in C^1(\mathbb{R}^3)$ with $\|\phi\|$, $\|\partial_i\phi\|_\infty < \infty$ one has $\phi\mathcal{D}_{\bar{H}} \subset \mathcal{D}_{\bar{H}}$ and

$$\bar{H}\phi - \phi\bar{H} = A_\phi = -i\,\alpha\cdot\nabla\phi \qquad \text{with } A_\phi \text{ bounded}$$

Lemma 2:

Let H be as in (2) with

(8) $V_e,\ldots,\partial_3 A_3 \in L^3_{loc}(\mathbb{R}^3_+)$

and let $\phi \in C_o^\infty(\mathbb{R}^3_+)$. Then the operator M_ϕ, multiplication by ϕ, is $\bar{H}$-compact.

Since the proof of this Lemma is rather straight forward we delete it. The conclusions of the Lemma are also true if $L^3_{loc}(\mathbb{R}^3)$ is replaced by suitable Stummel classes.

Corollary:

Let H be as in Lemma 2 and assume $\lim H_1(x) = 0$ for $|x| \to \infty$. Then H_1 is $\bar{H}_o$-compact.

With slightly stronger assumptions we can show a considerable strengthening of Theorem 1.

Lemma 3:

Let H be as in (2) such that the potentials satisfy (7) and (8). Moreover, assume that V_m is absolutely continuous near 0 with

$$\lim_{r\to 0} {}^{V_s(r)}/_{V_m(r)} = \lim_{r\to 0} {}^{V_e(r)}/_{V_m(r)'} = \lim_{r\to 0} {}^{V'_m(r)}/_{V_m^2(r)} = 0$$

and $\lim_{r\to 0} |r\ V_m(r)| = \infty$

Then V_e and V_s are $K(\varepsilon,\ell,m)$-compact and $\sigma_{ess}\ (K(\varepsilon,\ell,m)) = (-m,m)^c$

Then V_e and V_s are $K\ (\varepsilon,1,m)$-compact and

Proof:

Considering the form induced by K on $C^\infty(0,\infty) \otimes \mathbb{C}^2$ one sees

$$K^2 \geqq V_3^2 - |V'_3| - |2V_e\ V_3|$$

This estimate together with Lemma 2 shows the relative compactness of V_e and V_s with respect to K. Hence we may assume $V_e = 0$ and $V_s = 0$ With the aid of the decomposition method the result is shown as in [6].

Remark:

Though Lemma 3 is true for fixed ℓ, one cannot expect it to be valid for H_o, because the estimates are not uniform in ℓ.

Thus we have to impose the following smoothness assumptions near 0.

V_e and V_s are twice differentiable

(10) V_m is differentiable and positive

sgn $V_e + V_s + m$ is constant

$$\frac{V_e' + V_s'}{V_e + V_s + m}(r) = \frac{\alpha(r)}{r} \quad ; \quad \frac{V_e'' + V_s''}{V_e + V_s + m}(r) = \frac{\beta(r)}{r^2} \quad ;$$

(10)

$$\frac{V_m'}{V_m}(r) = \frac{\gamma(r)}{(r)}$$

with α , β and γ continuous and $\alpha(r) \leq 0$ and $\gamma(r) < -1$ near 0. Though, these conditions seem rather restrictive, they hold for most physically interesting potentials, e.g. the Coulomb potential. With these additional assumptions we can now show the main result of this paper.

Theorem 2:

Let H be as in (2) and assume the potentials satisfy (7), (8), (10) and $\lim_{r\to 0} H_1(x) = 0$. Then

$$\sigma_{ess}(H) = (-m,m)^c$$

Proof:

Because of Lemma 2 we may assume $H_1 = 0$. Choose $r_0 > 0$ so small such that the conditions (10) hold for all $r \leq r_0$. Moreover, change the potentials such that $V_e(r) = V_s(r) = V_m(r) = 0$ for $r \geq r_0$. Such a change in the potentials will not affect the essential spectrum of H.

With this one shows

$$\sigma(K(\varepsilon,\ell,m)) = (-m,m)^c \qquad \text{for large } \ell$$

by transforming the eigenvalue equation into a single second order equation for u as in [1]. Since this is rather tedious, this part will be deleted.

In general the computation of the spectrum of $K(\varepsilon,\ell,m)$ will be rather complicated, because V_m is singular at 0. For the neutral particle, however, a complete solution can be given.

Theorem 3:

Assume $V_s \equiv V_e \equiv 0$ and that V_m satisfies (7), (8) and (10). Then $\sigma(K(\varepsilon,\ell,m)) = (-m,m)^c$.

Proof:

Assume $\lambda \in \sigma(K)$ satisfies $\lambda^2 < m^2$. Then the eigenvalue equation fo u becomes

$$(-D^2 + V_m{}^2 - V_m{}' + m^2 - \lambda^2)\, u = 0$$

With $F(r) = \exp(\int_r^{r_o} V_m(t)\, dt)$ this can be rewritten as

$$\{-(F^{-1}D\, F)(FDF^{-1}) + (m^2 - \lambda^2)\} \quad u = 0$$

Since the first summand is positive, this implies $m^2 = \lambda^2$.

If in addition to the above conditions V_m is monotonic for large r, the only eigenvalues of $K(\varepsilon,\ell,m)$ can be $\pm$ m.

References

1. Barut, A.O., Kraus, J.: Solution of the Dirac equation with Coulomb and magnetic moment interaction. J. Mathematical Phys. 17, 506-508 (1976)

2. Behncke, H.: The Dirac equation with an anomalous magnetic moment. Math. Z. 174 (1980) 213-225

3. Jörgens, K.: Perturbations of the Dirac Operator; in: Conference on the Theory of Ordinary and Partial Differential Equations (Dundee 1972), pp. 87-102. Lecture Notes in Mathematics 280. Berlin-Heidelberg-New York: Springer 1972

4. Kalf, H.: A Limit Point Criterion for Separated Dirac Operators and a Little Known Result on Riccati's Equation. Math. Z. 129, 75-82 (1972)

5. Rejto, P.A.: On Reducing Subspaces for One-Electron. Dirac Operators, Israel J. Math. 9, 144-171 (1971)

6. Weidmann, J.: Oszillationsmethoden für Systeme gewöhnlicher Differentialgleichungen. Math. Z. 119 , 349-371 (1971)

PERIODIC SOLUTIONS OF A CLASS OF HAMILTONIAN SYSTEMS

V. BENCI - A. CAPOZZI - D. FORTUNATO

Let $H \in C^1(\mathbb{R}^{2n}, \mathbb{R})$ and consider the Hamiltonian system of 2n ordinary differential equations

$$(1) \qquad \dot{p} = -H_q(p,q), \quad \dot{q} = H_p(p,q),$$

where p and q are n-tuples, $\cdot$ denotes $\frac{d}{dt}$ and $H_q = \mathrm{grad}_q\, H$, $H_p = \mathrm{grad}_p\, H$. This system can be represented more concisely as

$$(2) \qquad -J\dot{z} = H_z(z)$$

where $z = (p,q)$ and $J = \begin{pmatrix} 0 & -I \\ I & 0 \end{pmatrix}$, I being the identity matrix in R^n.

There are many types of questions, both local and global, in the study of periodic solutions of such systems (cf. |10| and its references).

Here we are concerned about the existence of periodic solutions of (2) when the period $T = 2\pi\omega$ is prescribed. Making the change of variable $t \longrightarrow \frac{1}{\omega} t$, (2) becomes

$$(3) \qquad -J\dot{z} = \omega H_z(z)$$

and we seek 2π-periodic solutions of (3), which, of course, correspond to the $2\pi\omega$-periodic solutions of (2). These solutions are the critical points of the functional of the action

$$(4) \qquad f(z) = \int_0^{2\pi} \left(\frac{1}{2}(-J\,\dot{z},z)_{\mathbb{R}^{2n}} - \omega\, H(z)\right) dt.$$

If there exist positive constants k_1, k_2, α such that

$$(5) \qquad |H_z(z)| \leq k_1 + k_2\, |z|^{\alpha}$$

it is easy to see that f is continuously Fréchet-differentiable on the space $W^{1/2}(S^1, \mathbb{R}^{2n})$ of 2n-tuples of 2π-periodic functions, which possess square integrable "derivative of order 1/2".(*)

The spectrum of the linear operator $z \longmapsto -J\,\dot{z}$ (with periodic conditions) consists of infinitely many positive and negative eigenvalues. For this reason the functional (4) is indefinite in a strong sense, i.e.

---

* We set $L^t = L^t(S^1, \mathbb{R}^{2n})$, $t \geq 1$, and for every $s \in \mathbb{R}$ we shall set

$$W^s = \{u \in L^2 \mid \sum_{\substack{j\in \mathbb{Z}\\ k=1,..,2n}} j^{2s}\, |u_{jk}|^2 < +\infty\},$$

where u_{jk} are the Fourier components with respect to the basis $\psi_{jk} = e^{jtJ}\,\phi_k$ ($j \in \mathbb{Z}$; ϕ_k (k=1,...,2n) is the standard basis in $\mathbb{R}^{2n}$).

W^s is an Hilbert space with the inner product $(u|v)_{W^s} = \sum_{j,k} j^{2s}\, u_{jk}\, v_{jk}$.

it is not bounded from above or from below, even modulo weakly continuous perturbations (cf. |2|,|5|).

Using direct methods of the calculus of variations and suitable minimax arguments, Rabinowitz proved that if H(p,q) grows more than quadratically at infinity in both the variables p and q, i.e. there exists $r > o$, $\mu > 2$ s.t.

$$(6) \qquad (H_z(z), z)_{\mathbb{R}^{2n}} \geq \mu H(z) > o \quad \text{for } |z| > r ,$$

then (2) possesses a nonconstant T-periodic solution for each prescribed period T (cf. |9|,|10|).
For "superquadratic" Hamiltonians other results are available (cf. |1|, |2|,|6|,|7|,|8|).

Consider now a mechanical system with constraints which do not depend on time and suppose that it is embedded in a field of forces deriving from a potential $V(q) \in C^1(\mathbb{R}^n,\mathbb{R})$. The Hamiltonian of such a system has the following form

$$(7) \qquad H(p,q) = \sum_{ij} a_{ij}(q)\, p_i p_j + V(q) ,$$

where $\{a_{ij}(q)\}$ is a positive definite matrix for every $q \in \mathbb{R}^n$. If a_{ij} do not depend on q, (1) can be reduced to a second order system of n equations of the form

$$(8) \qquad \ddot{x} = - \frac{\partial U}{\partial x} \qquad U = U(x) \qquad x \in \mathbb{R}^n .$$

Therefore, in this case, the research of the T-periodic solutions is reduced to the study of the critical points of the functional

$$(9) \qquad f(x) = \int_o^{2\pi} \left(\frac{1}{2}\dot{x}^2 - \omega^2 U(x)\right) dt \quad x \in W^1, \ \omega = T/2\pi.$$

The functional (9) is not "strongly" indefinite, because it is bounded from below modulo the weakly continuous perturbation $x \to \omega^2 \int_o^{2\pi} U(x)dt$. In this case it is known that if U grows more than quadratically at infinity (in the sense of (6)), then (8) has a non-constant T-periodic solution for each fixed $T > o$ (cf. |4|,|10|).
However, if a_{ij} depend on q, (1) cannot be reduced to (8). Moreover since (7) is quadratic in p, it does not satisfy the growth condition (6).

Here we are concerned with the Hamiltonian (7), with a_{ij} depending on q.
The following result holds (cf. |3|).

THEOREM.1. Consider the Hamiltonian system (1) with H(p,q) having the form (7). Suppose that $V \in C^1(\mathbb{R}^n,\mathbb{R})$ and $a_{ij} \in C^1(\mathbb{R}^n,\mathbb{R})$ (i,j=1,...,n) satisfy the following assumptions:

i) there are constants $c, r > o$ and $\alpha > 2$ s.t. for $|q| > r$

i_1) $o < \alpha V(q) \leq (V_q, q)_{\mathbb{R}^n}$

i_2) $|V_q(q)| \leq c\,|q|^\alpha$

ii) there exists $\upsilon > o$ s.t.

$$\sum_{ij} a_{ij}(q)\, p_i p_j \geq \upsilon\, |p|^2 \text{ for every } p,q \in \mathbb{R}^n$$

iii) *if* $p, q \in \mathbb{R}^n$

$$\sum_{ij} A_{ij}(q)\, p_i p_j \geq o ,$$

where $A_{ij}(q) = (\text{grad } a_{ij}(q), q)_{\mathbb{R}^n}$

iv) $|\text{grad } a_{ij}(q)|$ is bounded $(i,j = 1,\dots,n)$

Under the above assumptions, (1) *possesses a nonconstant T-periodic solution for each fixed period T.*

In order to prove the Theorem.1 we need the following abstract critical point theorem (cf. |3|) :

THEOREM.2. *Let* X *be a real Hilbert space, on which a unitary representation* T_g *of the group* S^1 *acts. Let* $f \in C^1(X,\mathbb{R})$ *be a functional on* X *satisfying the following assumptions:*

f_1) $f(u) = \frac{1}{2}(Lu\,|u)_X - \psi(u)$, *where* $(\cdot|\cdot)_X$ *is the inner product in* X, L *is a bounded selfadjoint operator and* $\psi \in C^1(X,\mathbb{R})$, $\psi(o) = o$, *is a functional whose Fréchet derivative is compact. We suppose that both* L *and* ψ' *are equivariant with respect to the action of the group* S^1.

f_2) O *does not belong to the essential spectrum of* L.

f_3) *Every sequence* $\{u_n\} \subset X$, *for which* $f(u_n) \to c \in \left]o,+\infty\right[$ *and* $||f'(u_n)||\cdot||u_n|| \to o$, *possesses a bounded subsequence.*

f_4) *There are two closed subspaces* S^1*-invariant* $V, W \subset X$ *and* $R, \delta > o$ *s.t.*

a) W *is L-invariant, i.e.* $LW = W$

b) Fix $(S^1)^*$ $\subset V$ or Fix $(S^1) \subset W$

c) $f(u) < \delta$ for $u \in$ Fix (S^1)

d) f is bounded from above on W

e) $f(u) \geq \delta$ for $u \in V$ s.t. $||u|| = R$

f) codim $(V+W) < +\infty$, dim $(V \cap W) < +\infty$

Under the above assumptions there exists at least

$$\frac{1}{2} (\dim (V \cap W) - \operatorname{codim} (V+W))$$

orbits** of critical points, with critical values greater or equal than δ.

This theorem is a generalization of the Theorem 0.1 in |2|, because only a weak P.S. condition, namely (f_3), is required.

Outline of the proof of the Theorem.1

Obviously the functional (4) satisfies assumption (f_1) with $X = W^{1/2}$, $\psi(z) = \int_0^{2\pi} H(z(t))\,dt$ (we shall take in the sequel $\omega = 1$), and L is the continuous selfadjoint operator in $W^{1/2}$ defined by the bilinear form

(10) $\qquad (Lu\,|v)_X = \langle -J\dot{u}, v \rangle$,

where $u,v \in W^{1/2}$ and $\langle \cdot , \cdot \rangle$ denotes the canonical pairing between X and its dual. The spectrum of L is given by the eigenvalues $\lambda_j = \frac{j}{1+|j|}$, $j \in \mathbb{Z}$. λ_j has multiplicity 2n and Ker L = $\mathbb{R}^{2n}$, so also assumption (f_2) is satisfied.

Now we have to verify that (f_3) is satisfied. Let $\{z_n\} \subset W^{1/2}$ s.t. $f(z_n) \to c \in \left]0,+\infty\right[$ and $||f'(z_n)|| \cdot ||z_n|| \to 0$. The first step,which

---

* $\mathrm{Fix}(S^1) = \{u \in X \mid T_g u = u$ for every $g \in S^1\}$

** If $u \in X$ the "orbit" of u is the set $\{T_g u : g \in S^1\}$

will be not given here, is to prove that, by virtue of (i_1), (ii) and (iii), the sequences

(11) $$||p_n||_{L^2}\ ,\ ||q_n||_{L^\alpha}$$

are bounded. Hence $||z_n||_{L^2 x L^\alpha}$ is bounded.
The second step is to prove that $||z_n||_{W^{1/2}}$ is bounded.
By (i_2), (iv) some computations show that $||H_z(z_n)||_{L^1}$ is bounded. Then, if $\eta > o$, by the Sobolev imbedding theorems we deduce that the sequence

(12) $$||H_z(z_n)||_{W^{-1/2-\eta/2}}$$

is bounded. On the other hand we know that $-J\,\dot{z}_n - H_z(z_n) \to o$ in $W^{-1/2}$, then by (12) we deduce that the sequence

(13) $$-J\,\dot{z}_n$$

is bounded in $W^{-1/2-\eta/2}$. Let us now set X^+ (resp. X^-) the span of the eigenfunctions corresponding to the positive (resp.negative) eigenvalues, and we set X^o = Ker L. Then $z = z^+ + z^- + z^o = \tilde{z} + z^o$ with $z^+ \in X^+$, $z^- \in X^-$, $z^o \in X^o$. It is easy to show that $||\tilde{z}||_{W^{+1/2-\eta/2}} \leq$ const $\cdot$ $||J\,\dot{z}||_{W^{-1/2-\eta/2}}$, then by (13) we deduce that $||\tilde{z}_n||_{W^{+1/2-\eta/2}}$ is bounded. Moreover since Ker L is finite-dimensional, we have also that $||z_n||_{W^{+1/2-\eta/2}}$ is bounded. Hence, as $\eta > o$ is arbitrary, by Sobolev embedding theorems we deduce that for every $t>1$

(14) $$||z_n||_{L^t x L^t}$$

is bounded. Now, as $< f'(z_n), z_n > \to o$, it is easy to see that for $n \in \mathbb{N}$

(15) $$||z_n^+||_{W^{1/2}} \leq \text{const}\ (1+\int_o^{2\pi} (H_z(z_n),\ z_n^+)_{\mathbb{R}^{2n}}\, dt)$$

Moreover by (i_2) and (iv) we deduce that for $n \in \mathbb{N}$

(16) $$\int_o^{2\pi} (H_z(z_n),\ z_n^+)_{\mathbb{R}^{2n}}\, dt \leq \text{const}\ (1+ ||z_n||^t_{L^{2t}} \cdot ||z_n^+||_{L^2}),$$

where t = max {2,α}. By (15), (16) and (14) we conclude that $||z_n^+||_{W^{1/2}}$ is bounded. Analogously we prove that $||z_n^-||_{W^{1/2}}$ is bounded. Finally, because Ker L is finite-dimensional, we have also that $||z_n^o||_{W^{1/2}}$ is bounded.

In order to verify the "geometrical" assumption (f_4) we need more technicality. Here we shall state the steps of this proof only:

I step) Let δ > o be s.t.

(17) $$\delta > 2\pi \cdot \sup\{-H(z) \; | z \in \mathbb{R}^{2n}\}$$

In correspondence to δ we can choose R> o and j ε ℕ sufficiently large in order that f(z) ≥ δ for every $z \in X_k^+$, $||z||_{W^{1/2}} = R$, where X_k^+ is the span of the eigenfunctions corresponding to the eigenvalues λ_j, j> k .

We set $V = X_k^+$.

II step) We set $W = (X_{k+\bar{n}}^+)^{\perp}$, with $\bar{n} \in \mathbb{N}$, $\bar{n} \geq 1$. Using the superquadratic growth of V(q) it can be proved that f is bounded from above on W.

So the assumptions of the Theorem 2 are fullfilled. Now observe that, by the Theorem.2, the critical values, which we find, are greater or equal than δ. So by (17) we have that the corresponding critical points are not constants. Therefore we find $\frac{1}{2}(\dim(V \cap W) - \operatorname{codim}(V+W)) \geq$ $\geq \bar{n}$ non constant critical points.

Observe, moreover, that, by the fact that $\bar{n}$ is arbitrary, we can deduce also that these critical points are infinite. So there exist infinitely many non constant T-periodic solutions of (2).

REFERENCES

|1| A. AMBROSETTI - G. MANCINI, Solutions of minimal period for a class of convex Hamiltonian systems, Preprint.

|2| V. BENCI, On the critical point theory for indefinite functionals in the presence of symmetries, to appear on Trans. Amer.Math.Soc.

|3| V. BENCI - A. CAPOZZI - D. FORTUNATO, Periodic solutions of Hamiltonian systems with a prescribed period, Preprint.

|4| V. BENCI - D. FORTUNATO, Un teorema di molteplicità per un'equazione ellittica non lineare su varietà simmetriche, Proceedings of the Symposium "Metodi asintotici e topologici in problemi diff. non lineari", L'Aquila (1981).

|5| V. BENCI - D. FORTUNATO, Soluzioni periodiche multiple per equazioni differenziali non lineari relative a sistemi conservativi, Proceedings of the Symposium "Metodi asintotici e topologici in problemi diff. non lineari", L'Aquila (1981).

|6| V. BENCI - D. FORTUNATO, The dual method in critical point theory. Multiplicity results for indefinite functionals, to appear on Ann.Mat.Pura e Applicata.

|7| V. BENCI - P.H. RABINOWITZ, Critical point theorems for indefinite functionals, Inv.math., 52, (1979), 336-352.

|8| I. EKELAND, Periodic solutions of Hamiltonian equations and a theorem of P. Rabinowitz, J. Diff.Eq., 34, (1979), 523-534.

|9| P.H. RABINOWITZ, Periodic solutions of Hamiltonian systems,Comm. Pure Appl.Math., 31, (1978), 157-184.

|10| P.H. RABINOWITZ, Periodic solutions of Hamiltonian systems:a survey, Preprint.

MULTIPARAMETER PROBLEMS : THE LAST DECADE

Patrick J. Browne

Dedicated to Richard Guy on the occasion of his 65th birthday.

1. INTRODUCTION

Research in the area known generally as multiparameter spectral theory has been progressing at a steady rate for the last 10 years or so. This lecture will highlight some of the features of that period. I make no claims of this being a complete coverage of the decade; rather I have concentrated on areas with which I am most familiar and where open problems and ideas needing further development can be easily identified.

Why concentrate on the last 10 years? Atkinson's book [2] appeared in 1972 and in many ways set the stage for much of the subsequent development. The book deals with multiparameter eigenvalue problems for matrices, i.e. finite dimensional problems (except for the last chapter where problems with compact operators are introduced). It is tantalizingly entitled "Vol. I".

Two other references which fall outside our 10-year span must also be mentioned. Atkinson's survey lecture [1] is mandatory reading for newcomers to the field. Many of the results of the 10 years in discussion have roots which can be found here. The paper [28] of Faierman is also important (and we shall refer to it later) for it seems to be the first successful attack on the eigenfunction completeness problem for a multiparameter system of ordinary differential equations.

Another useful reference is the 1978 monograph of Sleeman [34] where most of the results of the first half of our decade are collected together and presented in orderly fashion.

2. TYPICAL PROBLEMS

Our aim here is to describe briefly the standard types of multiparameter eigenvalue problems firstly in relation to differential equations and subsequently in terms of an abstract formulation.

Consider the system of ordinary differential equations with unknown functions $y_r(x_r)$, $x_r \in [a_r, b_r]$, $1 \le r \le k$:

$$\frac{d^2 y_r(x_r)}{dx_r^2} + q_r(x_r) y_r(x_r) + \sum_{s=1}^{k} \lambda_s a_{rs}(x_r) y_r(x_r) = 0 , \quad 1 \le r \le k. \tag{1}$$

Here $\lambda_1, \ldots, \lambda_k$ are spectral parameters and the coefficient functions $q_r(x_r)$, $a_{rs}(x_r)$ are real valued and continuous. We subject the equations to separated boundary conditions of the form

$$\begin{aligned} y_r(a_r) \cos \alpha_r + y_r'(a_r) \sin \alpha_r &= 0 \\ y_r(b_r) \cos \beta_r + y_r'(b_r) \sin \beta_r &= 0, \quad 1 \le r \le k. \end{aligned} \tag{2}$$

Thus we have k Sturm-Liouville equations, each with several parameters, linked via these parameters.

We say that a (necessarily real) k-tuple $\underset{\sim}{\lambda} = (\lambda_1, \ldots, \lambda_k)$ is an *eigenvalue* and $y(\underset{\sim}{x}) = y_1(x_1) \ldots y_k(x_k)$, $\underset{\sim}{x} = (x_1, \ldots, x_k)$, a corresponding *eigenfunction* if the functions $y_r(x_r)$ satisfy the boundary conditions and differential equations with the given $\underset{\sim}{\lambda}$.

Multiparameter eigenvalue problems for systems of ordinary differential equations essentially revolve around the investigation of such equations.

An abstract multiparameter spectral problem is formulated as follows. Let $H_1,\dots,H_k$ be Hilbert spaces and in each space consider operators T_r, V_{rs}, $1 \le s \le k$, such that $T_r : D(T_r) \subseteq H_r \to H_r$ is self-adjoint, possibly unbounded, $V_{rs} : H_r \to H_r$ is self-adjoint and continuous.

We work with the operators $W_r(\underset{\sim}{\lambda})$, $\underset{\sim}{\lambda} \in \mathbb{R}^k$, where

$$W_r(\underset{\sim}{\lambda}) = T_r + \sum_{s=1}^{k} \lambda_s V_{rs}$$

and we consider the linked eigenvalue problems

$$W_r(\underset{\sim}{\lambda})\, x_r = 0,\ x_r \ne 0, \quad 1 \le r \le k. \tag{3}$$

It turns out to be appropriate to work in the tensor product space $H = \bigotimes_{r=1}^{k} H_r$ as we shall see later. We now say that $\underset{\sim}{\lambda} = (\lambda_1,\dots,\lambda_k)$ is an *eigenvalue* and $x = x_1 \otimes \dots \otimes x_k$ a corresponding *eigenvector* of the above equations are satisfied.

The connexion between the two problems is easy to see for if in the abstract formulation we take $H_r = L^2([a_r,b_r])$, $(V_{rs} f)(x_r) = a_{rs}(x_r) f(x_r)$ and T_r to be the Sturm-Liouville operator $T_r f = f'' + q_r f$ with domain specified by the above boundary conditions, then we immediately recover the original problem in ordinary differential equations.

3. DEFINITENESS CONDITIONS

For the standard Sturm-Liouville problem with a weight function : viz

$$y'' + qy + \lambda r y = 0$$

a customary assumption is to demand $r(x) > 0$. The corresponding hypothesis for the multiparameter system (1) is

$$A(\underset{\sim}{x}) = \det|a_{rs}(x_r)| > 0,\ \forall\, \underset{\sim}{x} \in [\underset{\sim}{a},\underset{\sim}{b}] = \mathop{\times}_{r=1}^{k} [a_r,b_r].$$

Of course the continuity of the functions a_{rs} yields the existence of $\gamma > 0$ such that

$$A(\underset{\sim}{x}) > \gamma > 0, \quad \forall\, \underset{\sim}{x} \in [\underset{\sim}{a},\underset{\sim}{b}].$$

This condition is called *strong right definiteness.*

Determinants of the type $\det|a_{rs}(x_r)|$ where the rth-row contains functions of the rth variable x_r, are called Stäckel determinants. They have some curious properties. For example, under strong right definiteness, given any k-tuple of sign factors $\varepsilon_r = \pm 1$, $1 \le r \le k$, there exists $\underset{\sim}{\alpha} = (\alpha_1,\ldots,\alpha_k) \in \mathbb{R}^k$ such that

$$\varepsilon_r \sum_{s=1}^{k} \alpha_s\, a_{rs}(x_r) > 0, \quad \forall\, x_r.$$

Further, this sign factor property is sufficient to imply strong right definiteness. This is a result of F.V. Atkinson.

For a 2×2 Stäckel determinant satisfying

$$A(x,y) = \det \begin{vmatrix} a_{11}(x) & a_{12}(x) \\ a_{21}(y) & a_{22}(y) \end{vmatrix} \ge 0,$$

it turns out that the set $\{(x,y) \mid A(x,y) = 0\}$ can be expressed as the disjoint union of 4 Cartesian product sets - i.e. sets of the form $C_1 \times C_2$ where C_1, C_2 are subsets of the domains of x,y respectively. This prevents e.g. the zero set of $A(x,y)$ from being a diagonal of its rectangle of definition. This is an unpublished result of the author. An open question is to determine the analogue of this result for $k \times k$ Stäckel determinants satisfying $A(\underset{\sim}{x}) \ge 0$.

For the system (1) an alternative definiteness condition is to suppose the existence of $\underset{\sim}{\mu} \in \mathbb{R}^k$ so that each of the k determinants formed from $[a_{rs}(x_r)]$ by replacing row r with $\underset{\sim}{\mu}$ is positive for all $\underset{\sim}{x}$; this can be compactly expressed as requiring

$$h_r(\underset{\sim}{x}) = \sum_{s=1}^{k} \mu_s \, a_{rs}^{*}(x) > 0, \quad 1 \le s \le k, \quad \underset{\sim}{x} \in [a,b],$$

where a_{rs}^{*} denotes the co-factor of a_{rs} in $\det|a_{rs}|$. If we further require that $q_r(x_r) > 0$, $x_r \in [a_r,b_r]$, we arrive at the condition known as *strong left definiteness.*

For the abstract problem (3) definiteness conditions can take several forms. For so called *strong right scalar definiteness* we ask

$$SRD_\delta : \det(V_{rs}\, u_r,u_r) \ge \gamma > 0, \ \forall\, u_r \in H_r, \ \|u_r\| = 1.$$

A weaker condition, *right scalar definiteness* is

$$RD_\delta : \det(V_{rs}\, u_r,u_r) > 0, \ \forall\, u_r \in H_r, \ \|u_r\| = 1,$$

and a yet weaker condition, *weak right scalar definiteness,* is

$$WRD_\delta : \det(V_{rs}\, u_r,u_r) \ge 0, \ \forall\, u_r \in H_r, \ \|u_r\| = 1.$$

Strong left scalar definiteness could be formulated as

$$SLD_\delta : (T_r\, u_r,u_r) \ge \tau_r > 0,$$

$$\exists\, \underset{\sim}{\mu} \in \mathbb{R}^k \text{ such that}$$

$$\sum_{s=1}^{k} \mu_s (V_{rs}\, u_r,u_r)^{*} \ge \gamma > 0, \ \forall\, u_r \in H_r, \ \|u_r\| = 1,$$

where $(V_{rs}\, u_r,u_r)^{*}$ denotes the co-factor of $(V_{rs}\, u_r,u_r)$. Conditions LD_δ, WLD_δ take the same form with appropriate relaxations of the inequalities.

We can also formulate operator definiteness conditions as follows. Each V_{rs} induces an operator $V_{rs}^{\dagger}$ in H by means of

$$V_{rs}^{\dagger}(x_1 \otimes \ldots \otimes x_k) = x_1 \otimes \ldots \otimes V_{rs}\, x_r \otimes \ldots \otimes x_k$$

extended to general $x \in H$ by linearity and continuity. We can consider the array of operators $[V_{rs}^{\dagger}]$ in H. Operators from different rows commute and we form the determinantal operator

$$\Delta_0 = \det|V_{rs}^{\dagger}| \; .$$

For $x = \overset{k}{\underset{r=1}{\otimes}} x_r$, $y = \overset{k}{\underset{r=1}{\otimes}} y_r$, decomposable tensors,

$$(\Delta_0 \; x,y) = \det(V_{rs} \; x_r, y_r).$$

Our right operator definiteness conditions run as follows:

$$SRD_{\Delta} \; : \; (\Delta_0 x,x) \geq \gamma > 0, \quad \forall \; x \in H, \quad \|x\| = 1$$

$$RD_{\Delta} \; : \; (\Delta_0 x,x) > 0 \; , \qquad \forall \; x \in H,$$

$$WRD_{\Delta} \; : \; (\Delta_0 x,x) \geq 0 \qquad \forall \; x \in H.$$

An early result of Atkinson showed that in finite dimensions, $SRD_{\delta} = SRD_{\Delta}$ and more recently Binding [4,5] has proved this in infinite dimensions; in fact he has shown that if $(\Delta_0 \; x,x) \geq \gamma\|x\|^2$ for decomposable tensors x, then $(\Delta_0 \; x,x) \geq k^{-1} \; \gamma\|x\|^2$ for all $x \in H$. Binding in [4,5] also shows $SLD_{\delta} = SLD_{\Delta}$ and provides a detailed and thorough discussion of these (and other) definiteness conditions. The implication $RD_{\delta} \Rightarrow RD_{\Delta}$ remains an open question in infinite dimensions.

4. EIGENFUNCTION EXPANSIONS

The following result which incorporates an eigenfunction expansion theorem and a Klein oscillation theorem for the system of differential equations (1) subject to the boundary conditions (2) and strong right definiteness was first established by Faierman [28].

THEOREM 1. *For each k-tuple of non-negative integers* $\underset{\sim}{p} = (p_1,\ldots,p_k)$ *there is an eigenvalue* $\underset{\sim}{\lambda}^{\underset{\sim}{p}}$ *of* (1) *with corresponding eigenfunction* $\underset{\sim}{y}^{\underset{\sim}{p}}(x) = \prod_{r=1}^{k} y_r^{\underset{\sim}{p}}(x_r)$ *such that* $y_r^{\underset{\sim}{p}}$ *has* p_r *interior zeros in* (a_r,b_r).

The eigenvalues λ^p *have no finite point of accumulation in* $\mathbb{R}^k$ *and form a complete orthonormal basis for the weighted Lebesgue space* $L^2([a,b], A(x))$.

Faierman's original approach to this beautiful result was to replace each of the differential equations by difference equations and then use limiting arguments. The problem was also studied by Browne [14,18] from a more abstract point of view.

The left definite problem has been successfully studied by Faierman [29,30,31], Källstrom and Sleeman [32] and Sleeman [34].

The singular problem in which all or some of the variables x_r may range over a half-line has been studied by Browne [15]. Here a spectral function $\rho(t)$, $t \in \mathbb{R}^k$ can be produced. It is positively monotonic on $\mathbb{R}^k$ and the eigenfunction series expansion for the regular case is replaced by an integral with respect to the measure generated by $\rho(t)$ thus providing a direct extension of the well known Titchmarsh-Kodaira theory to $k \times k$ systems.

The doubly singular case when each x_r ranges over $(-\infty,\infty)$ remains unsettled although a spectral theorem should involve a matrix of size $2^k \times 2^k$ of functions of the type of $\rho(t)$ - see [17] for a discussion of this.

Sleeman [35] has provided an introductory study of limit point-limit circle theory, although there is much left to be investigated and understood here.

Abstractly for the problem (2), spectral theory has been studied by Browne [16,18,20,22] in a series of papers. Under the condition of strong right definiteness (the easiest case to present in this lecture), the operator Δ_0 has a bounded inverse $\Delta_0^{-1} : H \to H$. We proceed by investigating the operators $\Gamma_s = \Delta_0^{-1} \Delta_s$ in H where now we use the inner product generated by Δ_0, viz

$$[x,y] = (\Delta_0 x,y), \quad x,y \in H.$$

The operators Δ_s, $0 \le s \le k$, are defined via the equation

$$\sum_{s=0}^{k} \alpha_s \Delta_s = \det \begin{vmatrix} \alpha_0 & \alpha_1 \dots \alpha_k \\ T_1^\dagger & V_{11}^\dagger \dots V_{1k}^\dagger \\ T_k^\dagger & V_{k1}^\dagger \dots V_{kk}^\dagger \end{vmatrix}$$

for arbitrary $\alpha_0,\dots,\alpha_k$. (In the simplest case, $T_1,\dots,T_k$ are bounded.)

The main step is to extend some earlier finite dimensional results of Atkinson to infinite dimensions and to establish

(i) the operators Γ_s are self-adjoint with respect to the $[\cdot,\cdot]$-inner product

(ii) the operators Γ_s are pairwise commutative

(iii) if $\Gamma_s x = \lambda_s x$, $1 \le s \le k$, then x can be taken to be decomposable and as such is an eigenvalue for the problem (2).

Of course when the T_r's are unbounded, commutativity of the Γ_s's must be interpreted as commutativity of their spectral resolutions.

One can then appeal to the functional calculus for several commuting self-adjoint operators and so produce a spectral measure E on the Borel sets of $\mathbb{R}^k$ which is H-projection valued and such that, for example,

$$[x,x] = \int_{\mathbb{R}^k} d[E(\underset{\sim}{\lambda})\, x,x]$$

The papers mentioned above attempt this line of analysis under various weakenings of the definitiness conditions.

Results under left definiteness conditions can be found in Sleeman's monograph [34].

The theory of spectral representation for 1-parameter problems is detailed by Dunford and Schwartz in [27, pp.1205-1222]. Roughly speaking, given a self-adjoint operator $T = \int \lambda\, E(d\lambda)$ on a Hilbert space H and a non-zero vector $x \in H$, one considers the closure of the set of all vectors of the form $k(T)x$ where k is a bounded Borel function. Thus is generated a space $H_x \subset H$ which is isomorphic to the space $L^2(\mu_x)$, $\mu_x(\cdot) = (E(\cdot)x,x)$ by means of the correspondence $k(T)x \leftrightarrow k(t)$. This makes T (at least on H_x) an isomorphic copy of multiplication by the independent variable. T is said to have simple spectrum if $H_x = H$ for some x ; in general, H can always be expressed as a direct sum of such spaces H_x.

To date there is no satisfactory parallel theory for the multiparameter case. It would be desirable to show for example that if each $W_r(\underset{\sim}{\lambda})$ has simple spectrum for each $\underset{\sim}{\lambda}$ then H could be represented as a space of functions on $\mathbb{R}^k$.

5. VARIATIONAL METHODS

Here we consider the abstract problem (3) subject to strong right definiteness and the extra hypothesis that each T_r is bounded below and has compact resolvent. Then each $W_r(\underset{\sim}{\lambda})$ has these properties too and so has a spectrum consisting entirely of eigenvalues

$$\rho_r^0(\underset{\sim}{\lambda}) \le \rho_r^1(\underset{\sim}{\lambda}) \le \quad \ldots$$

each of finite multiplicity and accumulating only at ∞ . The main tool is the minimax principle for eigenvalues for such operators:

$$\rho_r^i(\underset{\sim}{\lambda}) = \text{Max}\{\text{Min}\{(W_r(\underset{\sim}{\lambda})u_r,u_r) \mid \|u_r\| = 1,\ (u_r,y^j) = 0\} \mid$$

$$y^j \in H_r,\quad 1 \le j \le i\} \quad .$$

The first important result to be obtained was an abstract oscillation theorem which states that given $\underset{\sim}{p} = (p_1,\ldots,p_k)$, a k-tuple of non-negative integers, there is an eigenvalue $\underset{\sim}{\lambda}^{\underset{\sim}{p}}$ so that $\rho_r^{p_r}(\underset{\sim}{\lambda}^{\underset{\sim}{p}}) = 0$. All eigenvalues occur in this manner and so they can be systematically numbered according to the various choices of $\underset{\sim}{p}$. These ideas were first discussed by Binding and Browne [9,10].

Given $u_1,\ldots,u_k$, $u_r \in H_r$, $\|u_r\| = 1$, consider the matrix $V(u) = [(V_{rs}\, u_r,u_r)]$, of real numbers and the cone $C \subseteq \mathbb{R}^k$ defined by

$$C = \{\underset{\sim}{\alpha} \in \mathbb{R}^k \mid V(u)\, \underset{\sim}{\alpha} \le \underset{\sim}{0} \quad \text{for some } u_1,\ldots,u_k\} \ .$$

Here for points $\underset{\sim}{\alpha}$, $\underset{\sim}{\beta} \in \mathbb{R}^k$, $\underset{\sim}{\alpha} \le \underset{\sim}{\beta}$ means $\alpha_r \le \beta_r$, $1 \le r \le k$. Descriptions of the geometrical features of the spectrum of the above problem were given by Binding and Browne in [11] using the cone C. Specifically, a key result is

$$\underset{\sim}{i} \ge \underset{\sim}{j} \Rightarrow \underset{\sim}{\lambda}^{\underset{\sim}{i}} \in \underset{\sim}{\lambda}^{\underset{\sim}{j}} + C$$

and in particular, if Σ denotes the totality of all eigenvalues, then

$$\Sigma \subseteq \underset{\sim}{\lambda}^{\underset{\sim}{0}} + C.$$

The cone C was also used by Browne [21] and Browne and Sleeman [24] when discussing a system of ordinary differential equations of the type (1) where the coefficients are periodic and the system is subject to periodic and semi-periodic boundary conditions : 2^k possibilities exist and the interlacing of the 2^k sets of eigenvalues can be described using C.

6. ASYMPTOTICS FOR ONE EQUATION

Important initial work on this topic as it relates to Sturm-Liouville equations with many parameters was done by Turyn [36] and was subsequently put into a more abstract setting by Binding and Browne [12] and Binding, Browne and Turyn [13].

Abstractly, we consider one equation

$$W(\underset{\sim}{\lambda})x = (T + \sum_{s=1}^{k} \lambda_s V_s)x = 0, \quad \underset{\sim}{\lambda} \in \mathbb{R}^k,$$

in a Hilbert space H where T is self-adjoint, bounded below and has compact resolvent and $V_1,\ldots,V_k$ are bounded and self-adjoint. As before $W(\underset{\sim}{\lambda})$ has a sequence of eigenvalues

$$\rho^0(\underset{\sim}{\lambda}) \leq \rho^1(\underset{\sim}{\lambda}) \leq \quad \ldots$$

and we are interested in the nature of the set where $\rho^0(\underset{\sim}{\lambda}) = 0$.

The case $k = 2$ has been given complete treatment in [12,13], and results for $k \geq 3$ are currently under investigation. We shall concentrate on the case $k = 2$. We put

$$Z = \{ \underset{\sim}{\lambda} \mid \rho^0(\underset{\sim}{\lambda}) = 0\}$$

$$P = \{ \underset{\sim}{\lambda} \mid \rho^0(\underset{\sim}{\lambda}) > 0\}$$

$$N = \{ \underset{\sim}{\lambda} \mid \rho^0(\underset{\sim}{\lambda}) < 0\}$$

We have that Z is closed, P and N open; P is convex.
Given $u \in H$, $\|u\| = 1$, we set $\underset{\sim}{v}(u) = ((V_1 u,u), (V_2 u,u)) \in \mathbb{R}^2$, and define, co denoting convex hull,

$$V = \text{co}\{\alpha\, \underset{\sim}{v}(u) \,|\, \alpha > 0, \|u\| = 1\} \subset \mathbb{R}^2 .$$

V is a convex cone in $\mathbb{R}^2$. We also introduce

$$K = \{\underset{\sim}{\lambda} \,|\, \underset{\sim}{v}\cdot\underset{\sim}{\lambda} \geq 0 \;\forall\; \underset{\sim}{u} \in V\} \subset \mathbb{R}^2$$

$$C^- = \{\underset{\sim}{\lambda} \,|\, \underset{\sim}{v}(u)\cdot\underset{\sim}{\lambda} \geq \alpha > 0 \quad \forall \|u\|=1\}$$

$$= \text{int}\, K \text{ if } C^- \neq \emptyset .$$

We introduce two conditions:

$$I : \underset{\sim}{0} \notin V \qquad\qquad I^s : C^- \neq \emptyset .$$

I^s is the stronger of the two and under it $\overline{V}$ is a convex cone contained strictly in a closed half space in $\mathbb{R}^2$. The recession cone of P (or the set of asymptotic directions of P) is

$$0^{+}P = \{\underset{\sim}{\lambda} \mid P + \underset{\sim}{\lambda} \subseteq P\} .$$

Important results are

(1) $I \Rightarrow P \neq \emptyset$

(2) $P \neq \emptyset \Rightarrow 0^{+} = K$

(3) $P \neq \emptyset \Rightarrow Z = \partial P = \partial N$

Thus under I (or I^s) Z is the boundary of an open convex set and asymptotically has directions given by ∂K. Notice that these results are in terms of the operators V_1, V_2 only and hold for any choice of T

These results are useful when we return to a linked 2×2 system for now we know asymptotic directions of the curves Z_1 and Z_2 and can then discuss in geometrical terms conditions under which the curves must intersect or conditions under which the curves may for certain choices of T_1 and T_2, not intersect. This detailed analysis can be found in [13]. Corresponding results for $k \geq 3$ are under investigation.

Space does not permit a discussion of the various results for non-linear multiparameter problems. The interested reader could consult Binding [5,6], Browne [19] or Browne and Sleeman [23,25,26].

BIBLIOGRAPHY

1. F.V. ATKINSON, Multiparameter spectral theory, *Bull. Amer. Math. Soc.*, 74 (1968), 1-27.

2. F.V. ATKINSON, Multiparameter Eigenvalue Problems I : Matrices and Compact Operators, *Academic Press*, New York, 1972.

3. P.A. BINDING, Another positivity result for determinantal operators, *Proc. Roy. Soc., Edinburgh*, 86A (1980), 333-337.

4. P.A. BINDING, Multiparameter definiteness conditions, *Proc. Roy. Soc., Edinburgh*, 89A (1981), 319-332.

5. P.A. BINDING, Multiparameter definiteness conditions II, *Proc. Roy. Soc., Edinburgh*, (to appear).

6. P.A. BINDING, On the use of degree theory for nonlinear multiparameter eigenvalue problems, *J. Math. Anal. Appl.*, 73, (1980), 381-391.

7. P.A. BINDING, Variational methods for one and several parameter non-linear eigenvalue problems, *Can. J. Math.*, 33 (1981), 210-228.

8. P.A. BINDING and P.J. BROWNE, Positivity results for determinantal operators, *Proc. Roy. Soc. Edinburgh*, 81A (1978), 267-271.

9. P.A. BINDING and P.J. BROWNE, A variational approach to multiparameter eigenvalue problems of matrices, *S.I.A.M. J. Math. Anal.*, 8 (1977), 259-273.

Bibliography (Cont'd)

10. P.A. BINDING and P.J. BROWNE, A variational approach to multiparameter eigenvalue problems in Hilbert space, *S.I.A.M. J. Math. Anal.*, **9** (1978), 1054-1067.

11. P.A. BINDING and P.J. BROWNE, Comparison cones for multiparameter eigenvalue problems, *J. Math. Anal. Appl.*, **77** (1980), 132-149.

12. P.A. BINDING and P.J. BROWNE, Spectral properties of two-parameter eigenvalue problems, *Proc. Roy. Soc. Edinburgh*, **89A** (1981), 157-173.

13. P.A. BINDING, P.J. BROWNE and L. TURYN, Existence conditions for two-parameter eigenvalue problems, *Proc. Roy. Soc. Edinburgh*, to appear

14. P.J. BROWNE, A multiparameter eigenvalue problem, *J. Math. Anal. Appl.*, **38** (1972), 553-568.

15. P.J. BROWNE, A singular multiparameter eigenvalue problem in second order ordinary differential equations, *J. Differential Equations*, **12**, (1972), 81-94.

16. P.J. BROWNE, Multiparameter spectral theory, *Indiana Univ. Math. J.*, **24** (1974), 249-257.

17. P.J. BROWNE, Multiparameter problems, *Proceedings 174 Dundee Conference on Ordinary and Partial Differential Equations, Springer-Verlag Lecture Notes in Mathematics*, Vol. 415, 78-84.

18. P.J. BROWNE, Abstract Multiparameter Theory I, *J. Math. Anal. Appl.* **60** (1977), 259-273.

19. P.J. BROWNE, A completeness theorem for a non-linear multiparameter eigenvalue problem, *J. Differential Equations*, **23** (1977), 285-292.

20. P.J. BROWNE, Abstract multiparameter theory II, *J. Math. Anal. Appl.*, **60** (1977), 274-279.

21. P.J. BROWNE, The interlacing of eigenvalues for periodic multiparameter problems, *Proc. Roy. Soc. Edinburgh*, **80A** (1978), 357-362.

22 P.J. BROWNE, Abstract multiparameter theory III, *J. Math. Anal. Appl.*, **73** (1980), 561-567.

Bibliography (Cont'd.)

23. P.J. BROWNE and B.D. SLEEMAN, Non-linear multiparameter Sturm-Liouville problems, *J. Differential Equations*, 34 (1979) 139-146.

24. P.J. BROWNE and B.D. SLEEMAN, Stability regions for periodic multiparameter eigenvalue problems, *Proc. Roy. Soc. Edinburgh*, 84A (1979), 249-257.

25 P.J. BROWNE and B.D. SLEEMAN, Bifurcation from eigenvalues in non-linear multiparameter Sturm-Liouville problems, *Glasgow Math. J.*, 21 (1980), 85-90.

26. P.J. BROWNE and B.D. SLEEMAN, Non-linear multiparameter eigenvalue problems for ordinary differential equations, *J. Math. Anal. Appl.*, 77 (1980), 425-432.

27. N. DUNFORD and J.T. SCHWARTZ, Linear Operators, Part II, *Wiley-Interscience*, New York (1963).

28. M. FAIERMAN, The completeness and expansion theorems associated with the multiparameter eigenvalue problem in ordinary differential equations, *J. Differential Equations*, 5 (1969), 179-213.

29. M. FAIERMAN, Eigenfunction expansions associated with a two-parameter system of differential equations, *Proc. Roy. Soc. Edinburgh*, 81A (1978), 79-93.

30. M. FAIERMAN, An Oscillation theorem for a two-parameter system of differential equations, *Quaestionnes Math.*, 3 (1979), 313-321.

31. M. FAIERMAN, An eigenfunction expansion associated with a two parameter system of differential equations, *Proc. Roy. Soc. Edinburgh* 89A (1981), 143-155.

32. A. KÄLLSTRÖM and B.D. SLEEMAN, A left definite multiparameter eigenvalue problem in ordinary differential equations, *Proc. Roy. Soc. Edinburgh*, 74A (1974/5), 145-155.

33. A. KÄLLSTRÖM and B.D. SLEEMAN, Multiparameter spectral theory, *Ark . für Mat.*, 15 (1977), 93-99.

34. B.D. SLEEMAN, Multiparameter Spectral Theory in Hilbert Space, *Pitman*, London, (1978).

35. B.D. SLEEMAN, Singular linear differential operators with many parameters, *Proc. Roy. Soc. Edinburgh*, 71A (1973), 199-232.

36. L. TURYN, Sturm-Liouville problems with several parameters, *J. Differential Equations*, 38 (1980), 239-259.

AN APPROACH TO THE DIRICHLET INDEX FOR OPERATORS SATISFYING MINIMAL CONDITIONS

Richard C. Brown

§1 Introduction and Notation.

In this note we offer an interpretation of the Dirichlet Index for symmetric differential expressions with non-negative coefficients. Essentially we relate the index to "limit point" properties of a pair of "maximal" and "minimal" vector valued operators $\mathcal{L}$ and $\mathcal{L}_0$, which are introduced in Section 2. By doing this several facts about Dirichlet index become easy to state and prove. Most of our material is familiar (cf. [1], [4], [6], [7]). However, the results are extended to a "minimal condition" setting. Also (Theorem 6) we present a new result on the invariance of the index with respect to perturbations of the associated quadratic form. Other results (cf., Theorem 5) complement recent unpublished work of Hinton and Kauffman [4]. We also hope that our approach may be of some independent interest since it differs from methods previously employed.

The main results are presented in Section 3. Most proofs are at least outlined and should not be difficult for the reader to reconstruct. (An extended version of the paper with full details will appear elsewhere.)

We now fix notation and introduce some preliminary concepts.

In what follows $L^2(a, b)$ signifies the classical Hilbert space generated by the complex square integrable functions on $[a, b)$ where $-\infty < a < b \leq \infty$. $L^1_{loc}(a, b)$ and AC^n are respectively the spaces of complex valued functions which are locally lebesgue integrable or whose n^{th} derivative is absolutely continuous on $[a, b)$ (we write AC^0 as AC).

Let $p_0, p_1, \ldots, p_n$ be real nonnegative functions on $[a, b)$ such that $p_0 > 0$, $p_n \geq \varepsilon > 0$, and $M[f]$ be the quasi-derivative $f^{[2n]}$ as defined in Naimark [9], §15. (Note that if the p_i are sufficiently smooth, then

[1]The author wishes to thank Don B. Hinton for many interesting discussions concerning the subject of this paper, especially relating to Theorem 4.

$M[f] = \sum_{i=0}^{n} (-i)^{(n-i)}(P_i\, f^{(n-i)})^{(n-i)})$. In the usual way we define the maximal and minimal operators $T(M)$, $T_0(M)$ in $L^2(a, b)$. Let $\underline{t}(y)$ denote the quadratic form $\int_a^b \sum_{i=0}^{n} P_i |y^{(n-i)}|^2$ where the integrals are absolutely convergent. We say that $y \in D(T(M))$ is Dirichlet if $\underline{t}(y) < \infty$ and that $T(M)$ is Dirichlet if its domain is Dirichlet. The Dirichlet index (DI) of M is the span of the Dirichlet solutions of $M[f] = 0$. It is known that $DI \geq n$ and a conjecture of Kauffman [6] is that it is exactly n. Although the conjecture is still open, significant partial verifications have been made by Kauffman [6], [7], Robinette [12], and Read [10], [11]. These results imply that the conjecture is "almost always" true. We also note here that if M is limit point then $DI = n$; however Kauffman [7] has found a sixth order non limit point operator of Euler type for which $DI = 3$.

§2 The operators $\mathcal{L}$, $\mathcal{L}_0'$, $\mathcal{L}_0$, $\mathcal{L}^+$, and $\mathcal{L}_0^+$.

Let $H := \overset{n}{\underset{i=1}{X}} L^2(a, b)$. Define $\mathcal{L}: L^2(a, b) \to H$ by

$$\mathcal{L}(y) := \begin{pmatrix} p_0^{1/2}\, y^{(n)} \\ p_1^{1/2}\, y^{(n-1)} \\ \vdots \\ p_n^{1/2}\, y \end{pmatrix}$$

on the domain

$$\mathcal{D} := \{y \in L^2(a, b) \cap AC^{(n-1)} : \mathcal{L}y \in H\} .$$

Further let $\mathcal{L}_0'$ be the restriction of $\mathcal{L}$ to the class $\mathcal{D}_0'$ of functions with compact support in (a, b). Note that $\|\mathcal{L}y\|^2 = \underline{t}(y)$.

Lemma 1. *$\mathcal{L}$ is a normally solvable operator.*

Proof. A demonstration that $\mathcal{L}$ is closed may be found in [2]. It is apparent from the structure of $\mathcal{L}$ that $\mathcal{L}$ has closed range if and only if it is closed.

Remark 1. Lemma 1 is a way of saying that the form $\underline{\underline{t}}$ is closed (cf. Kato [5], ch. VI). Moreover since $\mathcal{L}$ is normally solvable it has a bounded inverse $\mathcal{L}^{-1}$ defined on its range.

Now set $\tilde{z}_j := (z_1, \ldots, z_j) \in \overset{j}{\underset{i=1}{X}} L^2(a, b)$ and define recursively the expressions

$$\ell_1^+(\tilde{z}_1) = p_0^{1/2}\, z \,,$$

$$\ell_2^+(\tilde{z}_2) = -\ell_1^+(z_1)' + p_1^{1/2}\, z_2$$

$$\cdots$$

$$\ell_{n+1}^+(\tilde{z}_j) = -\ell_n^+(z_n)' + p_n^{1/2}\, z_{n+1} \,.$$

We assume that the $\ell_i^+(z_i)$, $1 \leq i \leq n$, are absolutely continuous and define $\mathcal{L}^+ : H \to L^2(a, b)$ by $\mathcal{L}^+(z_{n+1}) = \ell_{n+1}^+(\tilde{z}_{n+1})$. We denote the domain of $\mathcal{L}^+$ by $\mathcal{D}^+$.

Lemma 2 (a Green's formula). For all y in $\mathcal{D}$ and $\tilde{z}_{n+1} \in \mathcal{D}^+$ then

$$[\mathcal{L}(y), \tilde{z}_{n+1}] - [y, \mathcal{L}^+(\tilde{z}_{n+1})] = \{y, \tilde{z}_n\}(b^-) - \{y, \tilde{z}_n\}(a^+) \,.$$

where

$$\{y, \tilde{z}_n\}(s) = \sum_{i=1}^{n} (-1)^{(i-1)}\, y^{(n-i)}(s)\, \ell_j^+(\tilde{z}_j)\,(s) \,.$$

Let $\mathcal{L}_0^+$ be the restriction of $\mathcal{L}^+$ to the domain $\mathcal{D}_0^+$ such that

$$\{y, \tilde{z}_n\}(b^-) = \{y, \tilde{z}_n\}(a^+) = 0 \,,\ \forall\, y \in \mathcal{D}$$

and $\mathcal{L}_0$ be the restriction of $\mathcal{L}$ to the domain $\mathcal{D}_0$ such that

$$\{y, \tilde{z}_n\}(b^-) = \{y, \tilde{z}_n\}(a^+) = 0 \,, \forall\, \tilde{z}_{n+1} \in \mathcal{D}^+ \,.$$

These definitions imply that $y^{(j)}(a) = \ell^+(\tilde{z}_k)(a) = 0,\ 0 \leq j \leq n-1,\ 1 \leq k \leq n$.
The proof of the next result tying all these operators together may be found in [2]

Proposition 1. $\mathcal{L}_0'^* = \mathcal{L}^+$, $\qquad \mathcal{L}^* = \mathcal{L}_0^+$,

$$\mathcal{L}^{+*} = \mathcal{L}_0, \qquad \mathcal{L}_0^{+*} = \mathcal{L} .$$

Thus $\mathcal{L}_0'$, $\mathcal{L}_0$, $\mathcal{L}$, $\mathcal{L}_0^+$, $\mathcal{L}^+$ are densely defined operators. $\mathcal{L}_0$, $\mathcal{L}_0^+$, $\mathcal{L}^+$ are normally solvable and in particular $\mathcal{L}_0^+$, $\mathcal{L}^+$ are onto.

Corollary 1. $\mathcal{L}^*\mathcal{L}$ and $\mathcal{L}_0^*\mathcal{L}_0$ are self-adjoint restrictions of $T(M)$. Their domains are cores respectively of $\mathcal{L}$ and $\mathcal{L}_0$.

Proof. See Kato [5], p. 275. That $\mathcal{L}^*\mathcal{L} \subset T(M)$, etc. follows by a computation: e.g., for $n = 1$

$$\mathcal{L}^*\mathcal{L}(y) = -(p_0^{1/2}(p_0^{1/2}\, y'))' + p_1^{1/2}\, y$$

with boundary conditions $y'(a) = 0$, $(p_0 y'\ f)(b^-) = 0$, $\forall\, f \in \mathcal{D}$.

§3 Dirichlet Index Theory

Lemma 3. The set of Dirichlet functions in $D(T(M))$ is exactly $D(\mathcal{L}^+\mathcal{L})$ and the space of Dirichlet solutions is exactly $R(\mathcal{L}) \cap N(\mathcal{L}^+)$.

Theorem 1. $DI = \dim \mathcal{D}/\mathcal{D}_0$.

Proof. This follows from the easily verified identity

$$(1) \qquad R(\mathcal{L}) = R(\mathcal{L}_0) \oplus (N(\mathcal{L}^+) \cap R(\mathcal{L})) .$$

Corollary 2. $DI = n$ iff $\mathcal{L}$ is a n-dimensional extension of $\mathcal{L}_0$; i.e., $\mathcal{L}$ is "limit point". Equivalently $DI = n$ iff $\{y, \tilde{z}_n\}(b^-) = 0$ $\forall y \in \mathcal{D}$, $\tilde{z}_{n+1} \in \mathcal{D}^+$.

Theorem 2. $DI = \dim D(\mathcal{L}^+\mathcal{L})/D(\mathcal{L}^+\mathcal{L}_0)$.

Proof. Intersect both sides of (1) by $\mathcal{D}^+$.

Remark 2. A calculation shows that if $\tilde{z}_{n+1} \in R(\mathcal{L})$ then

$$\{y, \tilde{z}_n\}(s) = D(f, \overline{g})(s) := \sum_{i=0}^{n-1} y^{[i]} \overline{g}^{[2n-i-1]}$$

where $\mathcal{L}(g) = \tilde{z}_{n+1}$. Consequently Corollary 3 and Theorem 2 show that DI = n iff $D(y, \overline{g})(b^-) = 0$, $\forall y \in \mathcal{D}$ and $y \in D(\mathcal{L}^+\mathcal{L})$.

Theorem 3. Let T_1, T_2 be restrictions respectively of T(M) to the Dirichlet functions such that either (i) $y^{(i)}(a) = 0$, $0 \leq i \leq n - 1$ or (ii) $y^{[i]}(a) = 0$, $n \leq i \leq 2n - 1$ then T_1 and T_2 are self-adjoint iff DI = n ; in which case $D\sqrt{T_1} = \mathcal{D}_0$ and $D\sqrt{T_2} = \mathcal{D}$.

Proof. It is known (cf. Kato [5], p. 334) that $\mathcal{D}_0 = D(\sqrt{\mathcal{L}_0^* \mathcal{L}_0})$ and $\mathcal{D} = D(\sqrt{\mathcal{L}^*\mathcal{L}})$. Further by Corollary 2 if DI = n , $\mathcal{L}_0$ and $\mathcal{L}_0^* = \mathcal{L}_0^+$ are free of boundary conditions at b so $T_1 = \mathcal{L}^+\mathcal{L}_0$ and $T_2 = \mathcal{L}_0^+\mathcal{L}$. On the other hand, $\mathcal{L}^+\mathcal{L}_0 \subset T_1$ and $\mathcal{L}_0^+\mathcal{L} \subset T_2$ so that by Corollary 1 $\mathcal{L}^+\mathcal{L}_0 = T_1$ and $\mathcal{L}_0^+\mathcal{L} = T_2$ if T_1 and T_2 are self-adjoint.

Lemma 4. For j = 0,..., 2n - 1 $\mathcal{D}$ contains a function of compact support in D(T(M)) such that

$$y_j^{(k)}(a) = \begin{cases} 1 & \text{if } k = j \\ 0 & \text{otherwise.} \end{cases}$$

Proof. See Naimark [9], §17.3, Lemma 2.

Theorem 2. DI $\geq$ n .

Proof. Using Lemma 4 we can produce at least an n-dimensional extension of $\mathcal{L}_0$.

Theorem 4. DI = n iff either (i) D_0'(: = functions of compact support on (a, b) in D(T(M)) is a core of the restriction $\mathcal{L}_0$ of $\mathcal{L}$ such that $y^{(i)}(a) = 0$, $0 \leq i \leq n - 1$, or (ii) D_0'' (: = functions y of compact support in [a, b) in D(T(M)) such that $y^{[i]}(a) = 0$, $n \leq i \leq 2n - 1$) is a core of $\mathcal{L}$.

Proof. If DI = n (i) follows from the fact that $\mathcal{L}^+\mathcal{L}_0$ is the unique Friedrichs extension of $T(M)|_{D_0'}$ and Corollary 1. Similar reasoning using the fact that $\mathcal{L}_0^+\mathcal{L}$ has no boundary conditions at b yields (ii). If on the other hand D_0' and D_0' have the stated core property it is easy to show via Lemma 4 that $\mathcal{L}$ is an n-dimensional extension of $\mathcal{L}_0$.

Remark 3. Theorem 4 can be extended to show that subspaces of $D(T(M))$ of compact support in [a, b) satisfying general symmetric boundary conditions at a (see Hinton [3]) are cores of $\mathcal{D}$ iff DI = n . A similar generalization is possible for Theorem 3 (cf. [6]).

The deficiency index of M (DEF) is $1/2 \dim D(T(M))/D(T_0(M))$. (See [11] for discussion of the properties of the deficiency index).

Theorem 5. DI = DEF iff T(M) is Dirichlet.

Proof. $DI = \dim D(\mathcal{L}^+\mathcal{L})/D(\mathcal{L}^+\mathcal{L}_0)$ and $DEF = D(T(M))/D(\mathcal{L}^+\mathcal{L}_0)$ by standard theory. Further $D(\mathcal{L}^+\mathcal{L}) = D(T(M))$ iff T(M) is Dirichlet.

Remark 4. Theorem 4 is a partial generalization (to the case of minimal conditions) of a unpublished result of Hinton and Kauffman [4]. However, in their setting the p_i $1 \le i \le n$ may be negative since $[T_0(M)y, y]$ need only be $\ge \varepsilon > 0$. Also their treatment posits the existence of a closed extension of the closure of $[T_0(M)y, y]|_{C_0^\infty}$.

Suppose $p_0', p_1', \ldots, p_n'$ are real nonnegative measurable functions with $p_i' \in L^1_{loc}(a, b)$, $1 \le i \le n$. Then $(p_0 + p_0')^{-1} \in L^1_{loc}(a, b)$. Let $M[f]$ be $f^{[2n]}$ computed relative to $\{p_i + p_i'\}$, $0 \le i \le n$. Define $\tilde{\mathcal{L}}, \tilde{\mathcal{L}}_0, \tilde{\mathcal{L}}^+, \tilde{\mathcal{L}}_0^+$ relative to $\{p_i + p_i'\}$. Thus for example:

$$\mathcal{L}y = \begin{pmatrix} (p_0 + p_0')^{1/2}\, y^{(n)} \\ \vdots \\ (p_n + p_n')^{1/2}\, y \end{pmatrix} .$$

Further let $\underline{\underline{t}}'(y)$ denote the quadratic form $\sum_{i=0}^{n} \int_a^b p_i' |y^{(n-i)}|^2$. We say that $\underline{\underline{t}}'$ is $\underline{\underline{t}}$ bounded iff $D(\underline{\underline{t}}') \supset D(\underline{\underline{t}})$ and there exist a, b > 0 such that $\underline{\underline{t}}'(y) \le a\|y\|^2 + b\,\underline{\underline{t}}(y)$.

It is natural to investigate the relationship between the Dirichlet indices of M and $\tilde{M}$. (Of course, if the Dirichlet index conjecture is true, the indices are both n). We have the following simple results.

Theorem 6. Suppose $\underline{t}'$ is $\underline{t}$ bounded, then the Dirichlet indices of M and $\tilde{M}$ are the same.

Indication of proof. Since $\underline{t}'$ is positive and $\underline{t}$ bounded one can show that $\tilde{\mathscr{D}} = \mathscr{D}$ and $\tilde{\mathscr{D}}_0 = \mathscr{D}_0$, whence the theorem is a consequence of Theorem 1.

Corollary 3. If $D(\underline{t}') \supset D(\underline{t})$ and $\underline{t}'$ is dominated by $\underline{t}$ (i.e., $\underline{t}'(y) \leq K\,\underline{t}(y))$, $K > 0$ then the indices of M and $\tilde{M}$ are the same.

Corollary 4. If the p_i' are essentially bounded and $p_0 \geq \delta > 0$ then the indices of M and $\tilde{M}$ are the same.

REFERENCES

1. J. Bradley, D.B. Hinton, and R.M. Kauffman. On minimization of singular quadratic functionals. Proc. Roy. Soc. Edinburgh, 87A(1981), 193-208.

2. R.C. Brown. An operator theory approach to inequalities involving Dirichlet functionals, submitted.

3. D.B. Hinton. On the eigenfunction expansions of singular ordinary differential equations. J. Differential Equations, 24(1977), 282-308.

4. D.B. Hinton and R.M. Kauffman. On some properties of the Dirichlet index, in preparation.

5. T. Kato. "Perturbation Theory for Linear Operators", (Springer-Verlag, New York, 1966).

6. R.M. Kauffman. The number of Dirichlet solutions to a class of linear ordinary differential equations. J. Differential Equations, 31(1979), 117-129.

7. R.M. Kauffman. On the limit-n classification of ordinary differential operators with positive coefficients. Proc. London Math. Soc., 35(3) (1977), 495-526.

8. R.M. Kauffman, T.T. Read, and A. Zettl. "The Deficiency Index Problems for Powers of Ordinary Differential Expressions", (Lecture Notes in Mathematics #621, Ed. A. Dold and B. Eckmann, Springer-Verlag, Berlin, Heidelberg, and New York, 1977).

9. M.A. Naimark. "Linear Differential Operators, Part II", (Ungar, New York, 1968).

10. T.T. Read. Dirichlet solutions of fourth order differential operators, "Spectral theory of Differential Operators", Ed. I. Knowles and R. Lewis, (Math. Study Series, North Holland Publishing Co., Amsterdam, 1981).

11. T.T. Read. The number of the Dirichlet solutions of a fourth order differential equation, submitted.

12. J.B. Robinette. On the Dirichlet index of singular differential operators, in preparation.

NONLINEAR NEUTRAL FUNCTIONAL DIFFERENTIAL EQUATIONS IN PRODUCT SPACES

Jose M. Amillo Gil

John A. Burns*
Eugene M. Cliff*

I. INTRODUCTION

In the past few years there has been a rapid development of approximation methods for systems governed by functional differential equations. The literature on this topic is extensive. In this paper we will concentrate on an approximation scheme for systems governed by nonlinear neutral functional differential equations (which include retarded systems as a special case). Moreover, we will confine our discussion to a particular class of approximations that have proven to be especially well suited for the numerical solution of optimal control and parameter identification problems. Since it is impossible in a few pages to give a detailed review of the relevant literature, we list a number of papers in the references ([1-10],[18-22],[24-26,28]) and refer the interested reader to the discussions given there.

The usual Lebesgue space, Sobolev space and space of k continuously differentiable $\mathbb{R}^n$-valued functions defined on $[a,b]$ will be denoted by $L^p(a,b)$, $W^{1,p}(a,b)$ and $C^k(a,b)$, respectively. We shall use Schultz's notation (see [27]) and let $PC^{k,p}(a,b)$ stand for the space of functions x such that x has $k-1$ continuous derivatives, $x^{[k-1]}$ is piecewise continuously differentiable and $x^{[k]} \in L^p(a,b)$. When $a = -r$ and $b = 0$ we simply write L^p, $W^{1,p}$, C^k and $PC^{k,p}$ for the above spaces. The symbol $|\cdot|_X$ denotes the norm in the normed linear space X. If $T > 0$, $r > 0$ and $x:[-r,T] \to \mathbb{R}^n$, then for $0 \le t \le T$ we define the function $x_t : [-r,0] \to \mathbb{R}^n$ by $x_t(s) = x(t+s)$. Note that x_t does not represent a partial derivative.

II. EXAMPLES

It has been known for sometime that certain initial-boundary value problems for hyperbolic partial differential equations could be transformed to equivalent initial value problems for neutral functional differential equations. In [14] Brayton and Miranker observed that the lossless transmission line with nonlinear elements

* This research was supported in part by the National Science Foundation under grant ECS-8109245.

at each end could be realized by an equivalent differential-difference equation. Brayton and Willoughby [15] made use of this equivalence in their study of a numerical scheme for approximating and simulating such circuits. Later, Cooke and Krumme [16] developed a systematic procedure for transforming a general class of hyperbolic problems to neutral equations. More recently, Reid, Russell and Teglas (see [24-26,28]) have shown that neutral equations may be viewed as canonical forms for large classes of linear boundary control systems. This result has a number of interesting ramifications. For example, it implies that many of these systems can be realized as neutral equations that are "minimal" in the sense that the equations contain the fewest possible parameters necessary in order to completely describe the input-output behavior of the system. Clearly, such realizations are ideal for parameter estimation problems. Moreover, as we shall see below there are problems where the neutral equation realization reduces the complexity of the parameter dependence in the model and hence simplifies the theoretical and computational aspects of various identification problems.

The basic ideas can best be illustrated by a few simple examples. Consider first the idealized cable-mechanical system shown in Figure 1. We assume that the cable of length l has lineal density σ, is under tension τ and is connected to a mass-spring-dashpot system at the right end. We take the classical elastic boundary condition at the left end. The purpose of the devices at the ends is to maintain the tension and we assume they provide no impedance to the vertical motion of the cable at the ends.

A simple model for the vertical motion of the cable is given by the wave equation

$$(2.1)\qquad \frac{\partial^2}{\partial t^2} y(s,t) = \alpha^2 \frac{\partial^2}{\partial s^2} y(s,t) \qquad 0 < s < l, \quad t > 0,$$

with initial data

$$y(s,0) = g(s), \qquad \frac{\partial}{\partial t} y(s,0) = h(s), \tag{2.2}$$

where $\alpha = \tau/\sigma$. At the right end we have the system

$$\frac{\partial}{\partial t}[m \frac{\partial}{\partial t}y(1,t)] = -c\frac{\partial}{\partial t}y(1,t) - ky(1,t) - \tau \frac{\partial}{\partial s}y(1,t) + u(t), \tag{2.3}$$

where $u(t)$ is a control function. The classical elastic boundary condition at the left end is written in integral form

$$\frac{\partial}{\partial s} y(0,t) = \gamma[g(0) + \int_0^t \frac{\partial}{\partial t} y(0,\mu)d\mu], \tag{2.4}$$

where we assume $\gamma \neq 0$.

Note that the initial conditions for (2.3) are specified by

$$y(1,0) = g(1), \frac{\partial}{\partial t} y(1,0) = h(1). \tag{2.5}$$

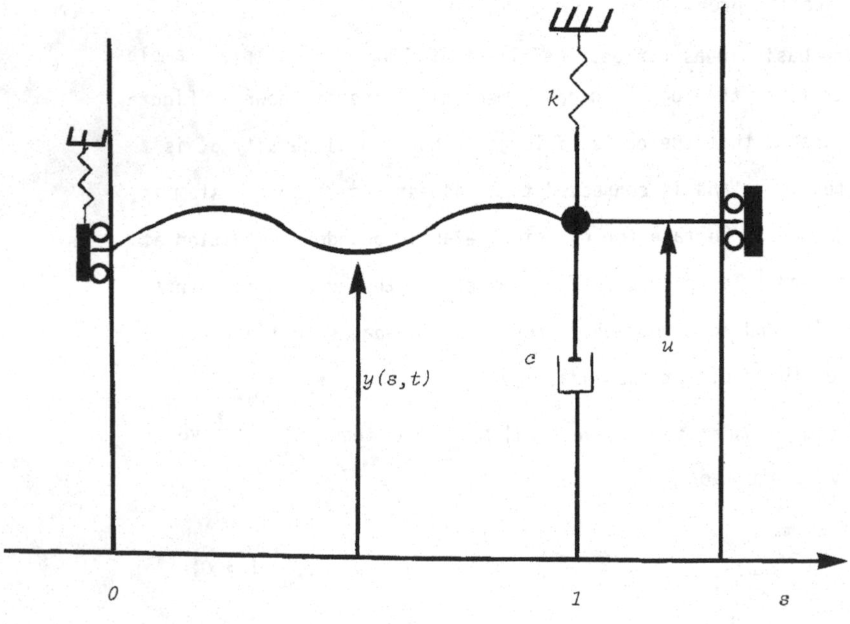

FIGURE 1

In order to construct an equivalent neutral system we follow the general procedure of Cooke and Krumme [16] and first rewrite (2.1)-(2.4) as a system. If $p(s,t)$ and $q(s,t)$ are defined by

$$p = \frac{\partial}{\partial s} y + (1/\alpha) \frac{\partial}{\partial t} y, \quad q = \frac{\partial}{\partial s} y - (1/\alpha)\frac{\partial}{\partial t} y ,$$

then

$$\frac{\partial}{\partial s} y = [p+q]/2, \ \frac{\partial}{\partial t} y = \alpha[p-q]/2 ,$$

and since $\frac{\partial}{\partial s} y(0,t) = \gamma y(0,t)$ it follows that

$$(2.6) \qquad y(s,t) = [p(0,t)+q(0,t)]/2\gamma + \tfrac{1}{2} \int_0^s [p(\mu,t)+q(\mu,t)]d\mu.$$

Problem (2.1)-(2.2) becomes the system

$$(2.7) \qquad \frac{\partial}{\partial t} \begin{bmatrix} p \\ q \end{bmatrix} + \frac{\partial}{\partial s} \begin{bmatrix} -\alpha & 0 \\ 0 & \alpha \end{bmatrix} \begin{bmatrix} p \\ q \end{bmatrix} = 0 \quad 0 < s < 1,\ t > 0,$$

with initial data

$$(2.8) \qquad \begin{aligned} p(s,0) &= g'(s) + (1/\alpha)h(s) \\ q(s,0) &= g'(s) - (1/\alpha)h(s). \end{aligned}$$

The boundary condition (2.3) becomes

$$(2.9) \qquad \begin{aligned} \frac{\partial}{\partial t} \{\tfrac{\alpha}{2}[p(1,t)-q(1,t)]\} = &- \frac{\sigma\alpha}{2m} [p(1,t)-q(1,t)] \\ &- \frac{k}{2m}\{[p(0,t)+q(0,t)]/\gamma + \int_0^1 [p(\mu,t)+q(\mu,t)]d\mu\} \\ &- \frac{\tau}{2m}[p(1,t)+q(1,t)] + \frac{u(t)}{m} \end{aligned}$$

with initial condition

$$\frac{\alpha}{2}\,[p(1,0) - q(1,0)] = h(1). \tag{2.10}$$

Differentiate (2.4) with respect to t to obtain

$$\frac{\partial}{\partial t}\{\tfrac{1}{2}[p(0,t)+q(0,t)]\} = \frac{\gamma\alpha}{2}\,[p(0,t)-q(0,t)]\ , \tag{2.11}$$

with initial data

$$\frac{1}{2}\,[p(0,0) + q(0,0)] = \gamma g(0)\ . \tag{2.12}$$

Observe that the characteristics for (2.7) are straight lines defined by $s = -\alpha[t - \xi] + 1$ and $s = \alpha[t - \xi]$, where ξ is constant. In particular, the functions $t \to p(-\alpha[t - \xi]+ 1,t)$ and $t \to q(\alpha[t - \xi],t)$ are constant. Therefore, it follows that for all $t \geq 0$,

$$\begin{aligned} p(1,t) &= p(0,t + 1/\alpha) \\ q(0,t) &= q(1,t + 1/\alpha). \end{aligned} \tag{2.13}$$

Let $r = (1/\alpha)$ and define $x_i:[-r,+\infty) \to R \quad i = 1,2,$ by

$$x_1(t) = p(0,t+r), \qquad x_2(t) = q(1,t+r), \tag{2.14}$$

respectively. Equations (2.13)-(2.14) imply that if $t \geq 0$, then

$$\begin{aligned} p(1,t) &= x_1(t)\ , \qquad q(0,t) = x_2(t) \\ p(0,t) &= x_1(t-r)\ , \qquad q(1,t) = x_2(t-r)\ . \end{aligned} \tag{2.15}$$

Integrating along characteristics, it follows that for $0 \leq s \leq 1,\ t \geq 0$

(2.16) $$p(s,t) = x_1(t + r(s-1)),\ q(s,t) = x_2(t-rs),$$

and hence

(2.17) $$y(s,t) = [x_1(t-r)+x_2(t)]/2\gamma + \tfrac{1}{2}\int_0^s [x_1(t+r(\mu-1))+x_2(t-r\mu)]d\mu.$$

Therefore, y is completely determined by $x_1(t)$, $x_2(t)$ and their histories $(x_1)_t$, $(x_2)_t$. To determine the governing equations for $x_1(t)$, $x_2(t)$ first note that a simple change of variables yields the equality

(2.18) $$\int_0^1 [p(\mu,t)+q(\mu,t)]d\mu = \frac{1}{r}\int_{-r}^0 [x_1(t+\mu)+x_2(t+\mu)]d\mu\ ,$$

and for $-r \le s \le 0$,

(2.19)
$$\phi_1(s) = g'(\tfrac{s+r}{r}) + r\,h\,(\tfrac{s+r}{r}) = p\,(\tfrac{s+r}{r},\ 0) = x_1(s)$$
$$\phi_2(s) = g'(\tfrac{-s}{r}) - r\,h\,(\tfrac{-r}{s}) = q\,(\tfrac{-s}{r},\ 0) = x_2(s)\ .$$

Substituting equations (2.15)-(2.17) into (2.9) and (2.11) yields the neutral equations

(2.20)
$$\frac{d}{dt}[x_1(t)-x_2(t-r)] = -\frac{c}{m}[x_1(t)-x_2(t-r)]$$
$$-\frac{kr}{m}\{x_1(t-r)+x_2(t) + \frac{\gamma}{r}\int_{-r}^0 [x_1(t+\mu)+x_2(t+\mu)]d\mu$$
$$-\frac{\tau r}{m}[x_1(t) + x_2(t-r)] + \frac{2r}{m}u(t)$$

and

(2.21) $$\frac{d}{dt}[x_1(t-r) + x_2(t)] = \frac{\gamma}{r}[x_1(t-r) - x_2(t)]\ .$$

The appropriate initial data for (2.20)-(2.21) is given by

$$[x_1(0) - x_2(-r)] = 2r\, h(1)$$

(2.22)

$$[x_1(-r) + x_2(0)] = 2\gamma\, g(0)$$

and

(2.23) $$x_1(s) = \phi_1(s)\ , \quad x_2(s) = \phi_2(s) \qquad -r \le s < 0\ .$$

Let $x(t) = col(x_1(t), x_2(t))$, $\eta = col(\eta_1, \eta_2) = col(2rh(1), 2\gamma g(0))$, and $\phi(s) = col(\phi_1(s), \phi_2(s))$. Define the matrices

$$A_0 = \begin{bmatrix} 1 & 0 \\ 0 & 1 \end{bmatrix} \qquad A_1 = \begin{bmatrix} 0 & -1 \\ 1 & 0 \end{bmatrix}$$

$$B_0 = \begin{bmatrix} \frac{-(c+\tau r)}{m} & \frac{-kr}{m} \\ 0 & \frac{-\gamma}{r} \end{bmatrix} \qquad B_1 = \begin{bmatrix} \frac{-kr}{m} & \frac{(c-\tau r)}{m} \\ \frac{\gamma}{r} & 0 \end{bmatrix}$$

$$K = \begin{bmatrix} \frac{-k}{m} & \frac{-k}{m} \\ 0 & 0 \end{bmatrix} \qquad C = \begin{bmatrix} \frac{2r}{m} \\ 0 \end{bmatrix},$$

and let D, L be the linear R^2-valued operators defined on $C(-r,0)$ by

$$D\phi = A_0\phi(0) + A_1\phi(-r)$$

$$L\phi = B_0\phi(0) + B_1\phi(-r) + \int_{-r}^{0} K\phi(\mu)d\mu\ ,$$

respectively. The system (2.20)-(2.23) can now be written in standard form

(2.24) $$\frac{d}{dt}[Dx_t] = Lx_t + Cu(t) \qquad t \ge 0\ ,$$

$$Dx_0(\cdot) = \eta\ , \qquad x_0(s) = \phi(s) \qquad -r \le s < 0\ . \tag{2.25}$$

Equations (2.24)-(2.25) are equivalent to the initial-boundary value problem (2.1)-(2.5) in the sense that given the proper initial data, $y(s,t)$ defined by (2.17) will be a solution (or generalized solution) to (2.1)-(2.5) and conversely. Moreover, as indicated above, the procedure described above does apply to much more general systems, including nonlinear boundary conditions.

As a second example, consider the lossless transmission line shown in Figure 2 with the nonlinear element at the right end. The current through device (in the direction shown) is given by the

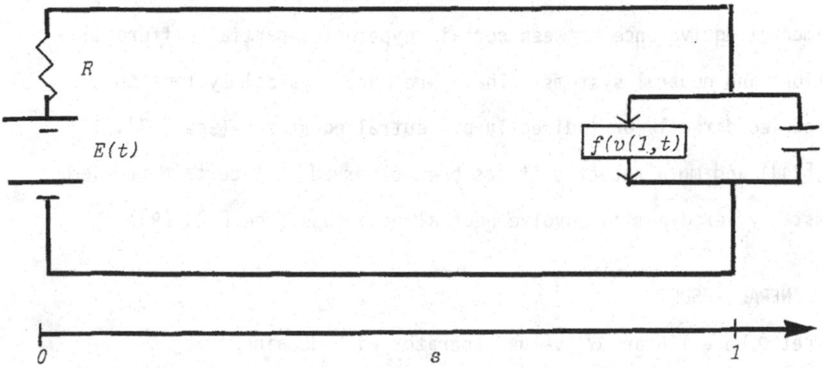

FIGURE 2

nonlinear function $f = f(v(1,t))$ of the voltage at $s = 1$. The equations for the line are

$$\begin{aligned} L_s \frac{\partial}{\partial t} i(s,t) &= -\frac{\partial}{\partial s} v(s,t) \\ C_s \frac{\partial}{\partial t} v(s,t) &= -\frac{\partial}{\partial s} i(s,t) \end{aligned} \tag{2.6}$$

with boundary conditions

$$v(0,t) + Ri(0,t) = E(t)$$

(2.7)

$$C \frac{d}{dt} v(1,t) = i(1,t) - f(v(1,t)) ,$$

where L_s, C_s are the specific inductance and capacitance of the line, respectively.

If one follows the general Cooke-Krumme procedure, then as for the mechanical system above an equivalent nonlinear neutral equation is obtained. For the lossless transmission line, the neutral equation has the form

$$\frac{d}{dt}[Dx_t] = Lx_t + F(t,x(t),x(t-r)) + Cu(t), \tag{2.8}$$

where F is determined by f and $u(t) = \dot{E}(t)$.

These two simple examples are given merely to illustrate the fundamental equivalence between certain hyperbolic partial differential equations and neutral systems. There are many physical systems that are modeled directly or indirectly by neutral equations (see [13],[17], [23],[24]) and more recently it has been observed that certain problems in unsteady aerodynamics involve neutral equations (see [7], [9]).

III. GENERAL RESULTS

Let D be a linear $\mathbb{R}^n$-valued operator with domain $\mathcal{D}(D) \subseteq L^p$ such that $W^{1,p} \subseteq \mathcal{D}(D)$, D restricted to C is continuous and atomic at $s = 0$. In particular, we assume that there is a $n \times n$ matrix valued function μ of bounded variation satisfying $\underset{[-\varepsilon,0]}{Var}(\mu) \to 0$ as $\varepsilon \to 0^+$ and sucl that if $\phi \varepsilon C$, then

$$D\phi = \phi(0) + \int_{-r}^{0} d\mu(s)\phi(s) .$$

We consider the nonlinear system governed by the neutral functional differential equation

$$\frac{d}{dt} Dx_t = f(x(t),x_t) + Cu(t), \quad 0 \leq t \leq T , \tag{3.1}$$

with initial data

(3.2) $$Dx_0 = \eta \ , \qquad x_0 = \phi \ ,$$

where $(\eta,\phi)\varepsilon Z = \mathbb{R}^n \times L^p$, $u\varepsilon L^p(0,T)$ and the functional f has domain $\mathcal{D}(f)$ satisfying $\mathbb{R}^n \times \mathcal{D}(D) \subseteq \mathcal{D}(f) \subseteq Z$.

A *solution* (classical) to the initial value problem (3.1)-(3.2) is a function $x(t) = x(t;\eta,\phi,u)$ such that: i) $x\varepsilon W^{1,p}(-r,T)$; ii) $x_0(s) = \phi(s)$ on $[-r,0]$; iii) the function $y(t) = Dx_t$ is absolutely continuous on $[0,T]$; iv) $y(0) = \eta$ and v) (3.1) holds a.e on $[0,T]$. A *generalized solution* (L^p-solution) to (3.2)-(3.3) is a pair of functions $(y(\cdot),x(\cdot))$ such that: i) $y(t;\eta,\phi,u)$ belongs to $C(0,T)$; ii) $x(t) = x(t;\eta,\phi,u)$ belongs to $L^p(-r,T)$; iii) $x_0(s) = \phi(s)$ a.e on $[-r,0]$ and iv) the pair $(y(\cdot),x(\cdot))$ satisfies the system

(3.3)
$$y(t) = \eta + \int_0^t f(x(s),x_s)ds + \int_0^t Cu(s)ds$$
$$x(t) = y(t) - \int_{-r}^0 d\mu(s)x(t+s) \ ,$$

a.e. on $[0,T]$.

In order to obtain existence, uniqueness, continuous dependence and regularity of solutions, we impose what are now fairly standard (yet clearly not the most general) assumptions on f. For $a > 0$ and $x\varepsilon W^{1,p}(-r,a)$, define $[Q_a(x)](t)$ on $[0,a]$ by

$$[Q_a(x)](t) = f(x(t),x_t) \ .$$

We shall need the following conditions on f:

F1) For each $a\varepsilon[0,T]$, Q_a defines a continuous operator from $W^{1,p}(-r,a)$ into $L^p(0,a)$ and Q_a can be extended to a unique continuous operator $\tilde{Q}_a : L^p(-r,a) \to L^1(0,a)$.

F2) For each bounded set $M \subseteq L^p(-r,a)$, there exists a continuous nondecreasing function α^M such that $|\tilde{Q}_a(x) - \tilde{Q}_a(y)|_{L^1(0,a)} \le \alpha^M(a)|x-y|_{L^p(-r,a)}$ for all $x,y \in M$.

F3) Given $0 < a < r$ and a bounded set $N \subseteq W^{1,p}(-r,a)$, there exists a continuous nondecreasing function α^N, such that $\alpha^N(0) = 0$ and

$$|Q_a(x)-Q_a(y)|_{L^p(0,a)} \le \alpha^N(a)|x-y|_{W^{1,p}(0,a)},$$

for all $x,y \in N$ satisfying $x(s) = y(s)$ for $-r \le s \le 0$.

F4) For each bounded set $K \subseteq W^{1,p}(-r,0)$, there exists a continuous nondecreasing function α^K such that

$$|Q_a(x)-Q_a(y)|_{L^p(0,a)} \le \alpha^K(a)|x-y|_{L^p(-r,a)}$$

for all $x,y \in K$.

The following results may be established using techniques similar to those found in [1] and [2].

THEOREM 3.1. Suppose that conditions F1) - F2) hold. Given $\delta > 0$, there is a $T = T(\delta) > 0$ such that if $(\eta,\phi) \in \mathbb{R}^n \times L^p$ and $|(\eta,\phi)| \le \delta$, then (3.1)-(3.2) has a unique generalized solution on $[-r,T]$.

Let $B_\delta = \{(\eta,\phi,u) \in \mathbb{R}^n \times L^p \times L^p(0,T(\delta)) \,\big|\, |(\eta,\phi)| \le \delta\}$ and define $\Phi : B_\delta \to C(0,T(\delta)) \times L^p(-r,T(\delta))$ by $[\Phi(\eta,\phi,u)](t) = (y(t;\eta,\phi,u), x(t;\eta,\phi,u))$, where (y,x) is the unique generalized solution to (3.1)-(3.2).

THEOREM 3.2. If F1) - F2) hold and $\delta > 0$, then $\Phi : B_\delta \to C(0,T(\delta)) \times L^p(-r,T(\delta))$ is a continuous operator.

THEOREM 3.3. If F1) - F3) hold and $\phi \in W^{1,p}$ satisfies $D\phi = \eta$, then $x(t) = x(t;\eta,\phi,u)$ is the unique (classical) solution to (3.1)-(3.2).

Let $H : W^{1,p} \to \mathbb{R}^n$ be defined by $H(\phi) = f(\phi(0),\phi)$ and define the nonlinear operator $A : \mathcal{D}(A) \subseteq Z \to Z$ by

$$\mathcal{D}(A) = \{(D\phi,\phi) \in \mathbb{R}^n \times L^p \,\big|\, \phi \in W^{1,p}\} \tag{3.4}$$

and

$$A(D\phi,\phi) = (H(\phi),\dot{\phi}) . \tag{3.5}$$

If F1) - F3) hold, $(D\phi,\phi)\varepsilon\mathcal{D}(A)$ and $x(\cdot)$ is the unique solution to (3.1) - (3.2) on $[0,T]$, then it follows that $z(t) = (Dx_t,x_t)$ satisfies the abstract equation

$$\begin{aligned} \frac{d}{dt} z(t) &= Az(t) + (Cu(t),0) \\ z(0) &= z_0 = (D\phi,\phi) . \end{aligned} \tag{3.6}$$

Moreover, the solution of (3.6) satisfies the integral equation

$$z(t) = z_0 + \int_0^t Az(\sigma)d\sigma + \int_0^t (Cu(\sigma),0)d\sigma . \tag{3.7}$$

In order to obtain a representation for generalized solutions, the integral equation must be extended. Define $I:\mathcal{D}(I)\rightarrow C(0,T;Z)$ by $\mathcal{D}(I) = \{z\varepsilon C(0,T;Z) \mid z(t) = (Dx_t,x_t),\ x\varepsilon W^{1,p}(-r,T)\}$ and for $z\varepsilon\mathcal{D}(I)$

$$[I(z)](t) = \int_0^t Az(\sigma)d\sigma .$$

If F4) holds, then one can show that I can be extended to a continuous operator $\bar{I}:\overline{\mathcal{D}(I)}\rightarrow C(0,T;Z)$. Therefore, (3.7) is replaced by the operator equation

$$z(t) = z_0 + [\bar{I}(z)](t) + \int_0^t (Cu(\sigma),0)d\sigma , \tag{3.8}$$

and we have the following result.

<u>THEOREM 3.4</u>. Let $z_0 = (\eta,\phi)\varepsilon Z$ and assume that F1)-F4) hold. If $(y(\cdot),x(\cdot))$ is the unique generalized solution to (3.1)-(3.4), then $z(t)=(y(t),x_t)$ is the unique solution of the abstract equation (3.8).

IV. APPROXIMATING SYSTEMS

We turn now to the development of an approximation scheme for nonlinear neutral equations. The scheme is analogous to the method used by Kappel [19] for linear equations. In order to shorten the presentation, we shall assume a global Lipschitz condition. However, the results are valid if one assumes the appropriate local conditions and properly modifies the statement of the results. We shall concentrate on the nonlinear problem

$$(4.1) \qquad \frac{d}{dt} Dx_t = h(x(t),x(t-r_1),\cdots,x(t-r_v),x_t) + Cu(t)$$

$$(4.2) \qquad Dx_0 = \eta \ , \qquad x_0 = \phi \ ,$$

where $0 = r_0 < r_1 <\cdots< r_v = r$ and

$$D\phi = \phi(0) - \sum_{j=1}^{v} B_j\phi(-r_\delta) - \int_{-r}^{0} B(s)\phi(s)ds \ .$$

Moreover, we assume that h is continuously differentiable, $B(\cdot)$ is an $n \times n$-valued C^1 function and h satisfies the condition:

H1) There is a constant K such that
$|h(x_0,x_1,\cdots,x_v,\phi)-h(y_0,y_1,\cdots,y_v,\psi)| \le K\{\sum_{j=0}^{v}|x_j-y_j|+|\phi-\psi|_{L^p}\}$,
for all $(x_0,x_1,\cdots,x_v,\phi),(y_0,y_1,\cdots,y_v,\psi) \in \mathbb{R}^{n(v+1)}\times L^p$.

Note that (4.1)-(4.2) is a special case of (3.1)-(3.2) where $f(x,\phi) = h(x,\phi(-r_1),\cdots,\phi(-r_v),\phi)$. Consequently, it follows from Theorem 3.4 that $z(t) = (Dx_t,x_t)$ is the unique solution to

$$(4.3) \qquad z(t) = z_0 + \int_0^t Az(\sigma)d\sigma + \int_0^t (Cu(\sigma),0)d\sigma$$

when $z_0 = (D\phi,\phi)\epsilon\mathcal{D}(A)$.

We shall use the Hilbert space $z = Z_g = \mathbb{R}^n \times L^2$ with the (equivalent) weighted norm $|(\eta,\phi)|_g^2 = |\eta|^2 + \int_{-r}^{0}|\phi(s)|^2 g(s)ds$, where g is the piecewise linear function defined on $(-r_j,-\gamma_{j-1})$ by $g(s) = \alpha_j(s+\gamma_{j-1})+v-j+1$ where α_j satisfy $\alpha_j \le -v(v+2)|B_j|/(r_j-r_{j-1})$, $j = 1,2,\cdots,v$. (See [1],[2], [19]). It is easy to show that in this case the nonlinear operator A satisfies

$$\langle Az-Aw,z-w\rangle_g \le \omega|z-w|_g^2$$

for all $z,w \in \mathcal{D}(A)$, where ω is a constant (see [1]).

DEFINITION 4.1. A triple $\{Z^N, P^N, A^N\}$, $N = 1,2,\cdots$, is called an approximation scheme for the Cauchy problem (4.1)-(4.2) if:

i) $\{Z^N\}$ is a sequence of finite dimensional subspaces of Z_q such that $Z^N \subseteq \mathcal{D}(A)$ for all $N = 1,2,\cdots$.

ii) $\{P^N\}$ is a sequence of orthogonal projections from Z_g onto Z^N.

iii) $\{A^N\}$ is the sequence of operators on Z^N defined by $A^N = P^N A P^N$, $N = 1,2,\cdots$.

DEFINITION 4.2. The approximation scheme $\{Z^N, P^N, A^N\}$ satisfies property $P1$ if for each $z \in Z_g$, $P^N z \to z$.

DEFINITION 4.3. The approximation scheme $\{Z^N, P^N, A^N\}$ satisfies property $P2\text{-}k$ if for each $z = (D\phi, \phi)$ with $\phi \in PC^{k,\infty}$, $AP^N z \to Az$.

DEFINITION 4.4. Let $\{Z^N, P^N, A^N\}$ be an approximation scheme that satisfies $P2\text{-}k$. We say that $\{Z^N, P^N, A^N\}$ satisfies $P3$ if for each $x \in PC^{k,\infty}(-r,T)$ there is a function $\Gamma(t)$ and a sequence ε_N such that for $z(t) = (Dx_t, x_t)$ and $0 \le t \le T$,

$$|AP^N z(t) - Az(t)| \le \Gamma(t)\varepsilon_N \; .$$

Note that $\Gamma(t)$ may depend on x.

Given an approximation scheme for (3.1)-(3.2), we consider the approximating system in Z^N given by

$$z^N(t) = P^N z_0 + \int_0^t A^N z^N(\sigma)d\sigma + \int_0^t P^N(Cu(\sigma),0)d\sigma \; , \tag{4.5}$$

which is equivalent to the ordinary differential equation

$$\begin{aligned} \frac{d}{dt} z^N(t) &= A^N z^N(t) + P^N(Cu(t),0) \\ z^N(0) &= P^N z_0 \; . \end{aligned} \tag{4.6}$$

The following result establishes the convergence of the approximating systems. Its proof is quite involved and is essentially contained in [1] and [2].

THEOREM 4.1. Assume that H1) holds, $z_0 = (D\phi,\phi)\varepsilon\mathcal{D}(A)$, $u\varepsilon L^2(0,T)$ and $\{Z^N,P^N,A^N\}$ is an approximating scheme satisfying *P1*, *P2-k* and *P3* If $z^N(t)$ is the solution to (4.5) and $z(t) = (Dx_t,x_t)$ is the solution to (4.3), then $z^N(t)$ converges to $z(t)$ uniformly in t on $[0,T]$ and u in bounded subsets of $L^2(0,T)$.

The general scheme given above can be realized using spaces of spline functions as the approximating spaces. For the case of one discrete delay r, partition $[-r,0]$ by $t_j^N = -jr/N$, $j = 0,1,2,\cdots N$ and define Z_s^N to be the space of all elements $(D\phi,\phi)$ where ϕ is a first order, a cubic or a cubic Hermite spline with knots at t_j^N, respectively. Let P_s^N be the orthogonal projection of $\mathbb{R}^n \times L^2 = Z_g$ onto Z_s^N. It is fairly easy to establish that the resulting approximating scheme satisfies *P1*, *P3* and *P2-k* for some $k \geq 1$ (see [1,2] and [19]).

IV. CONCLUDING REMARKS

Before selecting a particular approximation scheme for any problem, it is important to decide the purpose of the approximation. If one is interested only in simulation, then a high order direct approach might be best. However, if one is interested in numerical methods for identification and control, then rapid convergence may not be as important as stability and robustness of the approximating model. The spline schemes described above work extremely well when applied to identification and control problems.

Converting initial-boundary value problems for PDE's (similar to the examples in Section 2) to initial value problems for neutral equations can in some cases simplify the parameter estimation problem. For example, in the mechanical system if the parameter γ in (2.4) was unknown, then the estimation of γ by a direct approximation of the PDE system is difficult On the other hand, in the corresponding neutral system the parameter γ appears on the "right hand side" of (2.24) and the finite dimensional part of the initial data (2.22). This dependency is easier to handle numerically and theoretically.

REFERENCES

1. J. M. Amillo Gil, J. A. Burns and E. M. Cliff, Approximation of nonlinear neutral functional differential equations on product spaces, preprint.

2. J. Amillo Gil, *Nonlinear Neutral Functional Differential Equations on Product Spaces,* Ph.D. Thesis, Virginia Polytechnic Institute and State University, Blacksburg, Virginia, August 1981.

3. H. T. Banks, Approximation of nonlinear functional differential equation control systems, *J. Optimization Theory Appl., 29* (1979), 383-408.

4. H. T. Banks, Identification of nonlinear delay systems using spline methods, *International Conference on Nonlinear Phenomena in the Mathematical Sciences,* University of Texas, Arlington, Texas, June 16-20, 1980.

5. H. T. Banks and J. A. Burns, Hereditary control problems: Numerical methods based on averaging approximations, *SIAM J. Control and Optimization, 16,* (1978), 169-208.

6. H. T. Banks, J. A. Burns and E. M. Cliff, Spline-based approximation methods for control and identification of hereditary systems, in *Intl. Symp. on Systems Optimization and Analysis,* A. Beusousan and J. L. Lions, Eds., Lecture Notes in Control and Info. Sci. Vol. 14, Springer, Heildeberg, 1979, 314-320.

7. H. T. Banks, J. A. Burns and E. M. Cliff, Parameter estimation and identification for systems with delays, *SIAM J. Control and Optimization, 19* (1981), 791-828.

8. H. T. Banks and F. Kappel, Spline approximations for functional differential equations, *J. Diff. Eqs., 34* (1979), 496-522.

9. J. A. Burns and E. M. Cliff, Hereditary models for airfoils in unsteady aerodynamics, numerical approximations and parameter estimation, *AFWAL Technical Report,* Wright-Patterson AFB, Ohio, 1981, to appear.

10. J. A. Burns and E. M. Cliff, An approximation technique for the control and identification of hybrid systems, *Proc. Third AIAA Symposium on Dynamics and Control of Large Flexible Spacecraft,* L. Meirovitch, Ed., June 1981, 269-284.

11. J. A. Burns, T. L. Herdman and H. W. Stech. Linear functional differential equations as semigroups on product spaces, *SIAM J. Math. Anal.,* in press.

12. J. A. Burns, T. L. Herdman and H. W. Stech, Differential-boundary operators and associated neutral functional differential equations, *Rocky Mountain J. Math.,* in press.

13. J. A. Burns, T. L. Herdman and H. W. Stech, The Cauchy problem for linear functional differential equations, in *Integral and Functional Differential Equations,* T. L. Herdman, S. M. Rankin and H. W. Stech, Eds., Marcel Dekker, 1981, 137-147.

14. R. K. Brayton and W. L. Miranker, A stability theory for nonlinear mixed initial boundary value problems. *Arch. Rational Mech. Anal. 17* (1964), 358-376.

15. R. K. Brayton and R. A. Willoughby, On the numerical integration of difference-differential equations of neutral type, *J. Math. Anal. Appl., 18* (1967), 182-189.

16. K. L. Cooke and D. W. Krumme, Differential-difference equations and nonlinear initial-boundary value problems for linear hyperbolic partial differential equations, *J. Math. Anal. Appl., 24* (1968), 372-387.

17. J. Hale, *Theory of Functional DIfferential Equations*, Springer-Verlag, New York, 1977.

18. F. Kappel, Approximation of neutral functional differential equations in the state space $R^n \times L_2$, Colloquia Mathematica Societatis János Bolyai, 30. Qualitative Theory of Differential Equations, Szeged (Hungary), 1979, 463-506.

19. F. Kappel, An approximation scheme for delay equations, *Proc. Int'l Conference on Nonlinear Phenomena in Mathematical Sciences*, Arlington, Texas, June 1980, in press.

20. F. Kappel and K. Kunisch, Spline approximations for neutral functional differential equations, *SIAM J. Numer. Anal., 18* (1981), 1058-1080.

21. K. Kunisch, Neutral functional differential equations in L^p-spaces and averaging approximations, *J. Nonlinear Anal. TMA 3* (1979), 419-448.

22. K. Kunish, Approximation schemes for nonlinear neutral optimal control systems, *J. Math., Anal. Appl., 82* (1981), 112-143.

23. O. Lopes, Forced occillations in nonlinear neutral differential equations, *SIAM J. Appl. Math., 29* (1975), 196-207.

24. R. M. Reid, Ph.D. Thesis, Department of Mathematics, University of Wisconsin, Madison, Wisconsin, August, 1979.

25. D. L. Russell, "Control canonical structure for a class of distributed parameter systems," Proc. Third IMA Conferençe on Control Theory, Sheffield, September 1980.

26. D. L. Russell, "Closed-loop eigenvalue specification for infinite dimensional systems: augmented and deficient hyperbolic cases," Technical Summary Report #2021, Mathematics Research Center, University of Wisconsin, Madison, August 1979.

27. M. H. Schultz, *Spline analysis*, Prentice Hall, Englewood Cliffs, NJ, 1973.

28. R. G. Teglas, A control canonical form for a class of linear hyperbolic systems, Ph.D. Thesis, Mathematics Department, University of Wisconsin, Madison, Wiscons, June 1981.

"A Connection Formula for the Second Painlevé Transcendent"

by

P.A. Clarkson, The Queen's College, Oxford
and J.B. McLeod, Wadham College, Oxford

Abstract

We consider a particular case of the Second Painlevé Transcendent

$$y'' = xy + 2y^3$$

It is known that if $y(x) \sim kAi(x)$ as $x \to +\infty$, then if $0 < k < 1$,

$$y(x) \sim d|x|^{-\frac{1}{4}}\sin\{\tfrac{2}{3}|x|^{3/2} - \tfrac{3}{4}\, d^2\, \ln\, |x| - c\} \text{ as } x \to -\infty$$

where d(k) and c(k) are the connection formulae for this nonlinear ordinary differential equation.

The lecture shows that

$$d^2(k) = -\pi^{-1}\ln(1 - k^2)$$

which confirms the numerical estimates of Abtowitz and Segar.

I. Introduction

In this lecture we shall consider solutions of

$$\frac{d^2y}{dx^2} = xy + 2y^3 \qquad (1)$$

satisfying

$$y(x) \to 0 \quad \text{as} \quad x \to +\infty . \qquad (2)$$

(1) is a particular case of the Second Painlevé Transcendent. The Painlevé transcendents were first studied by Painlevé and his colleagues around the turn of the century in an investigation into which second order equations have the property that the singularities other than poles of any of the solutions are independent of the particular solution and so dependent only on the equation. In the case of the second transcendent, no solution has any singularity at all except for poles and the point at infinity, (for a survey see Chapter 14 of [4]).

(1) is interesting because it arises from similarity solutions of the Kortweg-de Vries equation

$$u_t + 6uu_x + u_{xxx} = 0 \tag{3}$$

If $f = \frac{dy}{dx} - y^2$, then $u(x,t) = (3t)^{-2/3} f(x/(3t)^{1/3})$.

Hastings and McLeod [3] have proved the following result, in connection with (1) and (2).

The Airy function, $A_i(x)$, is defined to be the solution of

$$A_i'' - xA_i = 0$$

satisfying

$$A_i(x) \sim \pi^{-\frac{1}{2}} |x|^{-\frac{1}{4}} \cos(\tfrac{2}{3}|x|^{3/2} - \tfrac{1}{4}\pi) \quad \text{as} \quad x \to -\infty$$
$$A_i(x) \sim \tfrac{1}{2}\pi^{-\frac{1}{2}} |x|^{-\frac{1}{4}} \exp(-\tfrac{2}{3}x^{3/2}) \quad \text{as} \quad x \to +\infty$$

Theorem A.

Any solution of (1) satisfying (2) is asymptotic to KAi(x), for some k, and conversely, for any k, there is a unique solution of (1) asymptotic to kAi(x). Let this solution be $y_k(x)$, then

If $|k| < 1$,

$$y_k(x) \sim d|x|^{-\frac{1}{4}}\sin\{\tfrac{2}{3}|x|^{3/2} - \tfrac{3}{4}\, d^2 \ln|x| - c\} \text{ as } x \to -\infty \quad (4)$$

where the constants d and c are dependent on k.

If $|k| = 1$,

$$y_k(x) \sim \operatorname{sgn}(k)(-\tfrac{1}{2}\, x)^{\frac{1}{2}} \qquad \text{as } x \to -\infty\,.$$

If $|k| > 1$,

$y_k(x)$ has a pole at a finite x_0, dependent on k, so

$$y_k(x) \sim \operatorname{sgn}(k)(x - x_0)^{-1} \qquad \text{as } x \downarrow x_0\,.$$

Notice that (1) is left unchanged by the transformation $y \to -y$, hence assume without loss of generality that $k > 0$.

It has become a matter of some interest to establish the exact dependence of the constants d, c and x_0 on k. This lecture takes a step in that direction by showing

Theorem 1

$$d^2 = -\pi^{-1}\ln(1 - k^2) \qquad (5)$$

This result was first conjectured, and then verified numerically, by Ablowitz and Segur [1].

The proof of Theorem 1 is dependent upon the fact that, because of the relationship between (1) and (3), the inverse scattering method of solution of (3) leads to an expression for $y_k(x)$ in terms of a linear integral equation.

Ablowitz and Segur [2] show that if the integral equation

$$K(x,y) = k\mathrm{Ai}(\tfrac{x+y}{2}) + \frac{k^2}{4}\int_x^\infty\!\!\int_x^\infty K(x,s)\mathrm{Ai}(\tfrac{s+t}{2})\mathrm{Ai}(\tfrac{t+y}{2})\,ds\,dt \qquad (6)$$

is considered for x sufficiently large (depending on k), say $x > x_1$, then there exists a unique solution for K in $L^2(x_1,\infty)$ and for $x > x_1$

$$y_k(x) = K(x,x) \tag{7}$$

Hastings and McLeod, [3], show further that (6) and (7) hold for all x if $0 < k < 1$.

Ablowitz and Segur [2] also show that K(x,y) satisfies the differential equation

$$\left(\frac{\partial}{\partial x} + \frac{\partial}{\partial y}\right)^2 K(x,y) = \left(\frac{x+y}{2}\right)K(x,y) + 2(K(x,x))^2 K(x,y) \tag{8}$$

for $y \geq x$ and Hastings and McLeod [3] show that (8) holds for all x for $0 < k < 1$.

In (8) let $u = \frac{1}{2}(y + x)$, $v = \frac{1}{2}(y - x)$ and $F(u,v) = K(x,y)$ then (8) becomes

$$\frac{\partial^2 F}{\partial u^2}(u,v) = uF(u,v) + 2(y_k(u - v))^2 F(u,v) \tag{9}$$

which is a linear ordinary differential equation for F(u,v) if v is a fixed positive constant.

To prove Theorem 1, the technique is to investigate the asymptotics of K(x,y) as $x \to -\infty$ with $y - x$ fixed and strictly positive. There are two ways of doing this, either through the integral equation (6), in which case the asymptotic behaviour is expressed in terms of k, or through the asymptotic behaviour of the differential equation (9), which in turn depends upon the asymptotic behaviour as $u \to -\infty$ of $y_k(u - v)$ and this involves d.

A comparison of the two resultant expressions for the asymptotic behaviour of K(x,y) yields Theorem 1.

II The Asymptotics of the linear ordinary differential equation

In the differential equation

$$\frac{d^2y}{du^2}(u) - \{u + 2(y_k(u - v))^2\}y(u) = 0 \qquad (10)$$

make the transformation

$$\eta(u) = q^{\frac{1}{4}}(u)y(u)$$

$$\xi(u) = \int_0^u q^{\frac{1}{2}}(t)dt$$

where $q(u) = -(u + 2y_k^2(u - v))$.

Then (10) becomes

$$\frac{d^2\eta}{d\xi^2} + \eta - \left(\frac{q''}{4q^2} - \frac{5q'^2}{16q^3}\right)\eta = 0 \qquad (11)$$

(where $' \equiv d/du$)

Using (4) one then shows that

$$\frac{d^2\eta}{d\xi^2} + \{1 - \frac{2d^2}{3\xi}\cos(2\xi + 2(\tfrac{3}{2})^{1/3}v\xi^{1/3} + \tfrac{1}{3}d^2\ln|\xi| - c_1) + 0(|\xi|^{-4/3})\}\eta = 0 \qquad (12)$$

where c_1 is a constant and

$$\xi = -\tfrac{2}{3}|u|^{3/2} + \tfrac{1}{2}d^2\ln|u| + c_2 + 0(|u|^{-1})$$

where c_2 is a constant.

Then, since in (12) v is a strictly positive constant, one can show that $\eta(\xi)$ behaves asymptotically like a linear combination of $\cos\xi$ and $\sin\xi$. Hence we obtain

Theorem 2

As $x \to -\infty$ with $y-x$ fixed and strictly positive then

$$|x+y|^{\frac{1}{4}}K(x,y) \sim A(\tfrac{y-x}{2})\cos\{\tfrac{2}{3}|\tfrac{x+y}{2}|^{3/2} - \tfrac{d^2}{2}\ln|\tfrac{x+y}{2}|\}$$
$$+ B(\tfrac{y-x}{2})\sin\{\tfrac{2}{3}|\tfrac{x+y}{2}|^{3/2} - \tfrac{d^2}{2}\ln|\tfrac{x+y}{2}|\} \qquad (13)$$

for some functions A and B.

III <u>The Asymptotics of the integral equation</u>

If

$$K(x,y) = kAi(\frac{x+y}{2}) + \frac{k^2}{4}\int_x^\infty\int_x^\infty K(x,s)Ai(\frac{s+t}{2})Ai(\frac{t+y}{2})dsdt \qquad (6)$$

then K(x,y) can be expressed as a Neumann series

$$K(x,y) = k\sum_{n=0}^{\infty}(\tfrac{1}{2}k)^{2n} I_n(x,y)$$

which is convergent for $0 < k < 1$, where

$$I_0(x,y) = Ai(\tfrac{1}{2}(x+y))$$

and

$$I_n(x,y) = \underbrace{\int_x^\infty \cdots \int_x^\infty}_{2n} Ai(\frac{x+t_1}{2})Ai(\frac{t_1+t_2}{2})\ldots\ldots Ai(\frac{t_{2n}+y}{2})dt_1\ldots dt_{2n}$$

Consider the operator L_x defined by

$$(L_xf)(y) = \frac{1}{2}\int_x^\infty Ai(\frac{y+s}{2})f(s)ds \; .$$

Then as shown by Hastings and McLeod [3], L_x is a compact (indeed Hilbert-Schmidt) self-adjoint operator in $L^2(x,\infty)$ and, at least in a formal sense

$$L^2_{-\infty} = I$$

so

$$f(y) = \frac{1}{4}\int_{-\infty}^\infty Ai(\frac{y+s}{2})\{\int_{-\infty}^\infty Ai(\frac{s+t}{2})f(t)dt\}ds \; .$$

Using this it can be shown that for $y > x$ and $0 < k < 1$ that

$$K(x,y) = \frac{k(2-k^2)}{2(1-k^2)}\sum_{n=0}^{\infty}(\frac{-k^2}{1-k^2})^n A_n(x,y) \qquad (14)$$

where

$$A_0(x,y) = Ai(\tfrac{1}{2}(x+y))$$

and

$$A_n(x,y) = (-1)^n \int_1^\infty \overset{n}{\cdots} \int_1^\infty Ai(x\lambda_n) \prod_{r=1}^{n} \left\{ \frac{Ai(x\lambda_r)Ai'(x\lambda_{r-1}) - Ai'(x\lambda_r)Ai(x\lambda_{r-1})}{\lambda_r - \lambda_{r-1}} \right\} d\lambda_1 d\lambda_2 \ldots d\lambda_n$$

$$\text{with } x\lambda_0 = \tfrac{1}{2}(x+y)$$

These $A_n(x,y)$ are multiple integrals from $-\infty$ to x.

Using the known asymptotic expansions of Ai(z) and Ai'(z), as $z \to -\infty$

$$Ai(z) \sim \pi^{-\frac{1}{2}}|z|^{-\frac{1}{4}}\cos(\tfrac{2}{3}|z|^{3/2} - \tfrac{1}{4}\pi)$$

$$Ai'(z) \sim \pi^{-\frac{1}{2}}|z|^{\frac{1}{4}}\sin(\tfrac{2}{3}|z|^{3/2} - \tfrac{1}{4}\pi)$$

one can show that

$$\int_1^\infty Ai(x\lambda_n)\{Ai(x\lambda_n)Ai'(x\lambda_{n-1}) - Ai(x\lambda_{n-1})Ai'(x\lambda_n)\} \frac{d\lambda_n}{\lambda_n - \lambda_{n-1}}$$

$$= -\frac{1}{2} Ai(x\lambda_{n-1}) + \frac{\sin(\frac{2}{3}|x|^{3/2}\lambda_{n-1}^{3/2} - \frac{1}{4}\pi)}{2\pi|x|^{\frac{1}{4}}\lambda_{n-1}^{\frac{1}{4}}} \ln\left(\frac{\lambda_{n-1}^{\frac{1}{2}}+1}{\lambda_{n-1}^{\frac{1}{2}}-1}\right) + O(|x|^{-7/4})$$

The logarithmic term arises from integration of $\cos^2(x\lambda_n) = \frac{1}{2}(1 + \cos(2x\lambda_n))$.

Then, by induction, this process of multiplying by the "wronskian" and integration produces terms like

$$Re\left[\frac{\exp\{i(\frac{2}{3}|x|^{3/2}\lambda^{3/2} - \frac{1}{4}\pi)\}}{\lambda^{\frac{1}{4}}} \prod_{k=0}^{m-1} \left\{\frac{1}{2\pi i} \ln\left(\frac{\lambda^{\frac{1}{2}}+1}{\lambda^{\frac{1}{2}}-1}\right) + k\right\}\right]$$

Hence (14) becomes, upon doing the summation after obtaining a couple of recurrence relations,

$$K(x,y) = Re\left[\frac{\exp\{i(\frac{2}{3}|x|^{3/2}\lambda_0^{3/2} - \frac{1}{4}\pi)\}}{\pi^{\frac{1}{2}}|x|^{\frac{1}{4}}\lambda^{\frac{1}{4}}} \frac{k}{(1-k^2)^{\frac{1}{2}}} (1-k^2)^{-\frac{1}{2\pi i}\ln\left(\frac{1+\lambda_0^{\frac{1}{2}}}{1-\lambda_0^{\frac{1}{2}}}\right)}\right] + O(|x|^{-7/4})$$

(where $x\lambda_0 = \frac{1}{2}(x+y)$)

$$(1-k^2)^{-\frac{1}{2\pi i}\ln\left(\frac{1+\lambda_0^{\frac{1}{2}}}{1-\lambda_0^{\frac{1}{2}}}\right)} = \left(\frac{1+\lambda_0^{\frac{1}{2}}}{1-\lambda_0^{\frac{1}{2}}}\right)^{-\frac{1}{2\pi i}\ln(1-k^2)}$$

$$\sim \left(\frac{4|x|}{v}\right)^{-\frac{1}{2\pi i}\ln(1-k^2)} \qquad \text{as } x \to -\infty$$

Hence

$$K(x,y) \sim \frac{K}{(x+y)^{\frac{1}{4}}} \cos\{\tfrac{2}{3}|\tfrac{x+y}{2}|^{3/2} + \tfrac{1}{2\pi}\ln(1-k^2)\ln|\tfrac{x+y}{2}| - \beta\} \tag{15}$$

for some constants K and β, dependent on v,

as $x \to -\infty$

and therefore comparison of (13) and (15) yields Theorem 1, that is

$$d^2 = -\pi^{-1}\ln(1-k^2)$$

References

[1] M.J. Ablowitz and H. Segur, "Asymptotic solutions of the Korteweg-de Vries equation", Stud. Appl. Math. 57 pp.13-44 (1977).

[2] M.J. Ablowitz and H. Segur, "Exact solution of a Painlevé Transcendent", Phys. Rev. Lett. 38 pp.1103-1106 (1977).

[3] S.P. Hastings and J.B. McLeod, "A Boundary Value Problem Associated with the Second Painlevé Transcendent and the Korteweg-de Vries equation", Arch. Rat. Mech. Anal. 73 pp.31-51 (1980).

[4] E.L. Ince, "Ordinary Differential Equations", Dover (1944).

The Inverse Scattering Problem for Acoustic Waves*

by

David Colton

All important decisions must be made on the basis of insufficient data - If You Meet the Buddha on the Road, Kill Him! by Sheldon Kopp

I. Introduction.

In this paper we shall survey recent progress and discuss open problems connected with the inverse scattering problem for acoustic waves. However before proceeding it is first necessary to be more precise on what we mean by "the inverse scattering problem" since this phrase has been used to describe a large variety of problems concerned with target identification in acoustic wave propagation. We first make a distinction between "scattering" and "diffraction" and note that the latter is basically a high frequency phenomena whereas the former is more accurately applied to low and intermediate values of the frequency. Hence in this paper we shall not discuss any of the important new results on the "inverse diffraction problem" but instead refer the reader to the recent paper of Brian Sleeman for a survey of these results ([19]). We shall further restrict our attention to the scattering of a plane time harmonic wave by a fixed bounded obstacle situated in a homogeneous medium and in particular to determine information

---

* This research was supported in part by AFOSR Grant 81-0103.

about this obstacle from a knowledge of the asymptotic behavior of the scattered wave. Hence we are excluding such topics as the scattering of waves by moving obstacles, the determination of the speed of sound in a non-homogeneous medium, and the location of equivalent sources. Finally, we shall only be concerned with determining two basic properties of the scattering obstacle, viz. its shape and/or its surface impedance. We note that although the problem of determining the shape of an obstacle from far field measurements has long been recognized as a basic problem in a variety of areas of technology such as radar, sonar, and tomography, the inverse impedance problem has received less attention. Nevertheless the inverse impedance problem is of basic importance in many applications since it gives information on the material composition of the unknown scattering object, e.g. in the case of sonar it can help answer the question of whether or not the scattering obstacle is a whale or a submarine.

The inverse scattering problem, as defined above, is particularly difficult to solve for two reasons: it is (1) nonlinear and (2) improperly posed. Of these two reasons it is perhaps the latter that creates the most difficulty. Indeed we shall see shortly that for a given measured far field pattern in general no solution exists to the inverse scattering problem, and if a solution does exist it does not depend continuously on the measured data. Hence before we can begin to construct a solution to the inverse scattering problem we must answer the question of what we mean by a "solution". At this point it is worthwhile recalling the remark of Lanczos: "A lack of information cannot be remedied by any mathematical trickery".

Hence in order to determine what we mean by a solution it is necessary to introduce "nonstandard" information that reflects the physical situation we are trying to model. Having resolved the question of what is meant by a solution, we then have to actually construct this solution, and this is complicated not only by the fact that the problem is nonlinear, but also the fact that the above mentioned "nonstandard" information has been incorporated into the mathematical model.

We shall now give a mathematical formulation of the inverse scattering problem and outline the specific topics we want to discuss in this paper. We shall formulate our problem in $\mathbb{R}^2$ and state, when appropriate, the necessary modifications that are needed to consider the full three dimensional problem. Let D be a bounded, simply connected domain in $\mathbb{R}^2$ with smooth boundary ∂D and unit outward normal ν. If we let $\lambda = \lambda(\underset{\sim}{x}) \geq 0$ denote the continuous surface impedance of ∂D and $k>0$ the wave number, then the impedance boundary value problem for acoustic waves can be mathematically formulated as the problem of finding a function $u \in C^2(\mathbb{R}^2\setminus\bar{D}) \cap C^1(\mathbb{R}^2\setminus D)$ such that

$$u = u^i + u^s \quad \text{in} \quad \mathbb{R}^2\setminus D \tag{1.1a}$$

$$\Delta_2 u + k^2 u = 0 \quad \text{in} \quad \mathbb{R}^2\setminus\bar{D} \tag{1.1b}$$

$$\frac{\partial u}{\partial \nu} + ik\lambda u = 0 \quad \text{on} \quad \partial D \tag{1.1c}$$

$$\lim_{r\to\infty} r^{1/2} \left(\frac{\partial u^s}{\partial r} - iku^s\right) = 0 \tag{1.1d}$$

where the "incoming wave" u^i is an entire solution of the Helmholtz equation (1.1b) and the "scattered wave" u^s satisfies the radiation condition (1.1d) uniformly with respect to θ where (r,θ) are polar coordinates. We shall also consider the case when λ is infinite, i.e. the boundary condition (1.1c) becomes

$$u = 0 \quad \text{on } \partial D. \tag{1.1c'}$$

In this case we shall refer to the scattering obstacle as being "soft". The existence and uniqueness of a solution to (1.1a)-(1.1d) and (1.1a), (1.1b), (1.1c'), (1.1d) is well known (c.f. [10]). We shall see shortly that if u is a solution of (1.1a)-(1.1d) (or (1.1a), (1.1b), (1.1c'), (1.1d) then u^s has the asymptotic behavior

$$u^s(r,\theta) = e^{ikr}\, r^{-1/2} F(\theta;k) + 0(r^{-3/2}) \tag{1.2}$$

where F is known as the <u>far field pattern</u> corresponding to the given incoming wave u^i. The inverse scattering problems we shall discuss in this paper are (1) given a knowledge of u^i, F and D, find λ, or (2) given a knowledge of u^i, F and the boundary condition (1.1c'), find D. We shall be more precise on what we mean by a "knowledge" of F shortly, but as mentioned above we in general only know F from measurements which are by definition inexact and this fact makes both of the above inverse scattering problems improperly posed. The basic fact we begin with is that the existence of a unique solution to the direct scattering problem (1.1a)-(1.1d) or (1.1a), (1.1b), (1.1c'), (1.1d) by the method of integral equations (c.f. [10]) defines a (nonlinear) mapping from D or λ to F. Hence from an abstract point of view we can :ate our tasks as follows:

.) Determine the range of $\underset{\sim}{T}$ (denoted by $R(\underset{\sim}{T})$) as a subset of $L^2[0,2\pi]$.

!) Establish the existence of $\underset{\sim}{T}^{-1}$ on $R(\underset{\sim}{T})$, i.e. show the uniqueness of the solution to the inverse scattering problem.

!) Determine a subset $X \subset R(\underset{\sim}{T})$ and an operator $\hat{\underset{\sim}{T}}^{-1}$ defined on $L^2[0,2\pi]$ such that $\hat{\underset{\sim}{T}}^{-1} = \underset{\sim}{T}^{-1}$ on X and $\hat{\underset{\sim}{T}}^{-1}$ is continuous on $L^2[0,2\pi]$, i.e. stabilize the inverse scattering problem (In order to do this it is necessary to assume a priori information about D or λ).

:) Give a constructive method for determining $\hat{\underset{\sim}{T}}^{-1}x, x \in X$.

We shall examine what is known about the above four ·oblems in the following sections and in addition give directions .at should be taken if further progress is to be achieved.

. <u>The Mapping $\underset{\sim}{T}$ and its Range.</u>

As mentioned in the Introduction we can obtain the solution ' (1.1a)-(1.1d) (or the corresponding problem with Dirichlet ·undary condition (1.1c$'$)) by the method of integral equations. . particular we can represent the solution u^s of (1.1a)-(1.1d) . the form of a modified single layer potential

$$u^s(\underset{\sim}{x}) = \frac{1}{\pi} \int_{\partial D} \phi(\underset{\sim}{y}) G(\underset{\sim}{x},\underset{\sim}{y}) ds(\underset{\sim}{y}) \quad ; \quad \underset{\sim}{x} \in \mathbb{R}^2 \backslash \bar{D} \tag{2.1}$$

where G is an appropriately chosen fundamental solution ([22]) and ϕ is the unique solution of an integral equation of the form

$$\phi + \underset{\sim}{K}_1\phi + \lambda\underset{\sim}{K}_2\phi = \frac{\partial u^i}{\partial\nu} + \lambda u^i \tag{2.2}$$

where $\underset{\sim}{K}_1$ and $\underset{\sim}{K}_2$ are compact integral operators that are independent of λ and defined on $C(\partial D)$. The solution of the corresponding Dirichlet problem can be represented in the form of a double layer potential ([22])

$$u^s(\underset{\sim}{x}) = \frac{1}{\pi}\int_{\partial D} \psi(\underset{\sim}{y}) \frac{\partial}{\partial\nu(\underset{\sim}{y})} G(\underset{\sim}{x},\underset{\sim}{y})\,ds(\underset{\sim}{y}) \quad ; \ \underset{\sim}{x}\in\mathbb{R}^2\backslash\bar{D} \tag{2.3}$$

where ψ is the unique soluton of an integral equation of the form

$$\psi + \underset{\sim}{K}_3\psi = u^i \tag{2.4}$$

where $\underset{\sim}{K}_3$ is a compact integral operator defined on $C(\partial D)$. If in (2.1), (2.2) or (2.3), (2.4) we now let $r=|\underset{\sim}{x}|$ tend to infinity and use the asymptotic behavior of $G(\underset{\sim}{x},\underset{\sim}{y})$ we obtain the relationship (1.2) and the mapping $\underset{\sim}{T}:\lambda\to F$ or $\underset{\sim}{T}:\partial D\to F$. In particular such a calculation establishes the validity of the following theorem:

Theorem 1: The far field pattern is an entire function of θ and a continuous function of k for $k>0$.

The above analysis shows both the nonlinear nature of the operator $\underset{\sim}{T}$ as well as the fact that $R(\underset{\sim}{T})\neq L^2[0,2\pi]$. Furthermore since it is not possible to determine the analyticity of a function from inexact measurements, Theorem 1 implies that in general for a given measured far field pattern no solution exists to the inverse scattering problem unless further assumptions are made.

From (2.2) or (2.4) it is clear that the operator $\underset{\sim}{T}$ depends on u^i, i.e. $\underset{\sim}{T}=\underset{\sim}{T}(u^i)$. We can therefore pose the following question: If u^i ranges over all entire solutions of the Helmholtz equations where λ or D is kept fixed, is the corresponding set of far field patterns dense in $L^2[0,2\pi]$, or more concisely, does $\overline{R(\underset{\sim}{T})}=L^2[0,2\pi]$. The following example shows that this is not true in general.

Example: Consider problem (1.1a), (1.1b), (1.1c'), (1.1d) when D is the unit disk. Then since u^i is an entire solution of the Helmholtz equation we can expand u^i in the form

$$u^i(r,\theta) = \sum_{n=0}^{\infty} J_n(kr)[a_n\cos n\theta + b_n \sin n\theta] \tag{2.5}$$

where J_n denotes Bessel's function and the series (2.5) is uniformly convergent on any compact subset of $\mathbb{R}^2$. Then for $r\geq 1$ we can expand u^s in the uniformly convergent series

$$u^s(r,\theta) = - \sum_{n=0}^{\infty} H_n^{(1)}(kr) \frac{J_n(k)}{H_n^{(1)}(k)}[a_n \cos n\theta + b_n \sin n\theta] \tag{2.6}$$

where $H_n^{(1)}$ denotes Hankel's function of the first kind and from the asymptotic behavior of Hankel's function we have that the far field pattern for u^s is given by

$$F(\theta;k)= e^{i\pi/4} \sqrt{\frac{2}{\pi k}} \sum_{n=0}^{\infty} \frac{(-i)^n J_n(k)}{H_n^{(1)}(k)}[a_n \cos n\theta + b_n \sin n\theta] \tag{2.7}$$

If k_o^2 is an eigenvalue of the interior Dirichlet problem then $J_n(k_o)=0$ for some integer $n=n_o$ and hence in this case $F(\theta;k_o)$ is orthogonal to $\cos n_o\theta$ and $\sin n_o\theta$ for all incident fields u^i. Hence the class of far field patterns for such values of k is not dense in $L^2[0,2\pi]$.

It is an open question to determine if similar examples are valid for arbitrary domains D. If such examples exist this would establish an interesting relationship between the far field patterns of exterior boundary value problems for the Helmholtz equation and the (interior) eigenvalue problem for Laplace's equation.

The validity of the above example is based on the fact that the set

$$\begin{matrix} J_n(kr)\cos n\theta \\ J_n(kr)\sin n\theta \end{matrix} ;\ n = 0,1,2,\ldots \qquad (2.8$$

is incomplete in $L^2(\partial D)$ if k^2 is an eigenvalue of the interior Dirichlet problem. However it can be shown ([5]; a simpler proof can be based on the ideas of [13]) that the set

$$\begin{matrix} (\frac{\partial}{\partial\nu} + ik\lambda)\ J_n(kr)\cos n\theta \\ (\frac{\partial}{\partial\nu} + ik\lambda)\ J_n(kr)\sin n\theta \end{matrix} ;\ n=0,1,2,\ldots \qquad (2.9)$$

is complete in $L^2(\partial D)$ for arbitrary bounded domains provided $0<\lambda<\infty$ where λ is a constant. This fact implies the following theorem ([5]):

Theorem 2: Let λ be a constant such that $0<\lambda<\infty$. Then $\overline{R(\underset{\sim}{T})} = L^2[0,2\pi]$.

A wealth of questions concerning $R(\underset{\sim}{T})$ remain unanswered and this provides a major mathematical challenge of basic importance to the inverse scattering problem. We have already mentioned one of these questions in connection with the Dirichlet problem. Another is the following: Determine the compact set of far field patterns in $L^2[0,2\pi]$ that corresponds to a fixed incident field and wave number, but with (constant) surface impedance lying in a given compact subset of the positive real axis. As opposed to the problem answered by Theorem 2, this problem is complicated by the fact that it is nonlinear. A similar problem can be posed for λ fixed but ∂D lying in a compact set of closed surfaces.

III. The Existence of $\underset{\sim}{T}^{-1}$.

As we have already mentioned in the Introduction, the existence of $\underset{\sim}{T}^{-1}$ is equivalent to establishing the uniqueness of solutions to the inverse scattering problem. We first consider the inverse scattering problem of determining D from a knowledge of u^i and F where we assume that the boundary data is given by (1.1c') ([17]).

<u>Theorem 3</u>: D is uniquely determined by a knowledge of the far field pattern F for all angles θ and k on any interval of the positive real axis.

<u>Proof</u>: Suppose there existed two obstacles D_1 and D_2 having the same far field pattern F. Consider first the case when $\bar{D}_1$ and $\bar{D}_2$ are disjoint. Then from the analyticity of solutions to the Helmholtz equation, F uniquely determines u^s outside a disk containing D_1 and D_2 in its interior and we can conclude by analytic continuation that u^s is an entire solution of the Helmholtz equation satisfying the radiation condition. But this implies that $u^s \equiv 0$ ([15]), i.e. from (1.1c') we have $u^i=0$ on ∂D_1. Hence u^i is an eigenfunction of the Laplacian for an interval of k values and this is a contradiction since the set of eigenvalues for the Laplacian is discrete. A similar argument holds when

$D=\bar{D}_1\cap\bar{D}_2$ is nonempty, in particular by analytic continuation one can show that the Laplacian defined in $D_1\backslash\bar{D}$ or $D_2\backslash\bar{D}$ has a continuous spectrum, which is a contradiction.

We now consider the inverse scattering problem of determining λ from a knowledge of u^i, F and D, i.e. the boundary value problem is given by (1.1a)-(1.1d). The following theorem is due to Colton and Kirsch ([8]).

Theorem 4: λ is uniquely determined by a knowledge of the far field pattern F for all angles θ and any fixed positive value of the wave number k.

Proof: Let λ_1 and λ_2 be two solutions of the inverse scattering problem corresponding to the same far field pattern F. Then by analytic continuation we can conclude that $u_1=u_2$ and $\partial u_1/\partial\nu = \partial u_2/\partial\nu$ on ∂D where u_1 and u_2 are the solutions of (1.1a)-(1.1d) corresponding to λ_1 and λ_2 respectively. Then from the boundary condition (1.1c) we have

$$(\lambda_1-\lambda_2)u_1 = 0 \quad \text{on } \partial D. \tag{3.1}$$

We now note that u_1 is not identically zero in any neighborhood $S\subset\partial D$ since if this were the case (1.1c) implies that $\partial u_1/\partial\nu\equiv 0$ in S and hence by Holmgren's uniqueness theorem ([3]) $u_1\equiv 0$ in $\mathbb{R}^2\backslash D$. But this is a contradiction since $u_1=u_1^s+u_1^i$ and u_1^s satisfies the radiation condition but u_1^i does not. Therefore if $\underset{\sim}{x}\in\partial D$ there exists a sequence of points $\underset{\sim}{x}_n\to\underset{\sim}{x}$ such that $u_1(\underset{\sim}{x}_n)\neq 0$. From (3.1) we have that $\lambda_1(\underset{\sim}{x}_n)=\lambda_2(\underset{\sim}{x}_n)$ and since λ_1 and λ_2 are assumed to be continuous we have $\lambda_1(\underset{\sim}{x})=\lambda_2(\underset{\sim}{x})$. Since $\underset{\sim}{x}$ is an arbitrary point on ∂D this completes the proof of the theorem.

We note that from Theorem 1 and the identity theorem for analytic functions, it suffices in Theorem 3 and Theorem 4 to assume that F is only known for an interval of θ values contained in $[0,2\pi]$ instead of for all values of θ in $[0,2\pi]$.

An open problem of considerable interest is to determine the validity of the above theorems if instead of the far field pattern we are given the <u>scattering cross section</u> σ defined by

$$\sigma(k;\underset{\sim}{\alpha}) = \int_0^{2\pi} |F(\theta;k,\underset{\sim}{\alpha})|^2 d\theta \tag{3.2}$$

where $F(\theta;k,\underset{\sim}{\alpha})$ is the far field pattern corresponding to the incoming plane wave $u^i = \exp(ik\underset{\sim}{x}\cdot\underset{\sim}{\alpha})$.

IV. The Operator $\hat{\underset{\sim}{T}}^{-1}$.

As pointed out in the Introduction, the inverse scattering problems we are considering in this paper are improperly posed in the sense that in general no solution exists and if a solution does exist it does not depend continuously on the initial data. Hence a major task in a satisfactory treatment of these problems is to restore stability. We shall accomplish this by assuming extra a priori information is available concerning the unknown impedance or scattering obstacle such that it is possible to conclude that the exact far field pattern lies in a compact set $X \subset R(\underset{\sim}{T})$. We shall then show that it is possible to define an operator $\hat{\underset{\sim}{T}}^{-1}$ defined on $L^2[0,2\pi]$ such that $\hat{\underset{\sim}{T}}^{-1} = \underset{\sim}{T}^{-1}$ on X and $\hat{\underset{\sim}{T}}^{-1}$ is continuous on $L^2[0,2\pi]$, thus restoring the continuous dependence of λ or D on the far field pattern. At this point we would like to mention that the term "a priori information" is not meant to imply that such information is impossible to obtain from experimental data, but only that this (in general nonstandard) information is sufficient to stabilize the problem under consideration.

We first consider the problem of determining the impedance λ from a knowledge of the far field pattern F where the incident wave u^i and the scattering obstacle are assumed known ([8]). By using the compactness of the operators $\underset{\sim}{K}_1$ and $\underset{\sim}{K}_2$ in (22) it is possible to show that the mapping $\underset{\sim}{T}:\lambda \to F$ is continuous and in particular if ϕ is the solution of the integral equation (2.2) we can write $F=F(\theta;\phi,\lambda)$ where F is a continuous function of θ,ϕ and λ. Now let $f \in L^2[0,2\pi]$ be the measured far field pattern. Then we can reformulate our inverse scattering problem as an optimization problem, denoted by P_f, as follows: Minimize

$$C_f(\phi,\lambda) = \int_0^{2\pi} |F(\theta;\phi,\lambda)-f(\theta)|^2 d\theta \tag{4.1}$$

subject to ϕ being a solution of (2.2) and $\lambda \in U$ where U is a "control set" to be specified shortly. We first note the following facts:

(1) The set U contains the a priori information we are assuming about λ.

(2) If $\lambda^* \in U$ is a solution of the inverse scattering problem for a given far field pattern f and ϕ^* is the corresponding density, then (ϕ^*,λ^*) is a solution of P_f since $C_f(\phi^*,\lambda^*)=0$.

(3) If (ϕ^*,λ^*) is a solution of P_f and $C_f(\phi^*,\lambda^*)=0$ then λ^* is a solution of the inverse scattering problem. If $C_f(\phi^*,\lambda^*)>0$ then the inverse scattering problem is not solvable for $\lambda^* \in U$, but λ^* is a best approximation in the sense that $||F-f||_{L^2[0,2\pi]}$ is minimal.

We now define our control set U in such a way that P_f has a solution. In particular let

$$U = \{\lambda \in C^+(\partial D): |\lambda(\underset{\sim}{x})| \leq M_1, |\lambda(\underset{\sim}{x}) - \lambda(\underset{\sim}{y})| \leq M_2 |\underset{\sim}{x} - \underset{\sim}{y}|^{\alpha}\}$$

where $C^+(\partial D)$ is the cone in $C(\partial D)$ consisting of all continuous functions λ defined on ∂D such that $\lambda \geq 0$, and M_1, M_2 and $\alpha, 0<\alpha\leq 1$, are fixed constants. The Arzela-Ascoli Theorem implies that U is compact in $C(\partial D)$ and since the functional C_f is continuous we have the following theorem:

<u>Theorem 5</u>: Let $f \in L^2[0,2\pi]$. The P_f has a solution.

In general the solution of P_f is not unique. Let $\Phi^*(f)$ be the set of all solutions (ϕ^*, λ^*) of P_f. Then the compactness of U and the continuity of $\underset{\sim}{T}$ implies the following result on continuous dependence:

<u>Theorem 6</u>: If $f_n \to f$ in $L^2[0,2\pi]$, $(\phi_n^*, \lambda_n^*) \in \Phi^*(f_n)$, then there exists a convergent subsequence of $\{(\phi_n^*, \lambda_n^*)\}$ and every limit point lies in $\Phi^*(f)$.

Note that if P_f is uniquely solvable, i.e. $\Phi^*(f)$ is a single ordered pair (ϕ^*, λ^*), then Theorem 6 simply says that $f \to (\phi^*, \lambda^*)$ is a continuous mapping from $L^2[0,2\pi]$ into $C(\partial D) x U$. In particular the operator $\hat{\underset{\sim}{T}}^{-1}: f \to (\phi^*, \lambda^*)$, $\underset{\sim}{T}U = X$, satisfies the conditions on $\hat{\underset{\sim}{T}}^{-1}$ as set forth in the opening paragraph of this section. If P_f is not uniquely solvable then we must interpret $\hat{\underset{\sim}{T}}^{-1}$ as a set valued mapping and the criteria on $\hat{\underset{\sim}{T}}^{-1}$ as set out in this opening paragraph must be modified in an obvious manner.

For the extension of the above results to the case of the inverse scattering problem for electromagnetic waves see [12].

The inverse scattering problem of determining the shape of a "soft" scattering obstacle from a knowledge of the far field pattern can be handled in a similar manner as the inverse impedance problem provided we can define an appropriate compact family of surfaces U. This problem was discussed in [7] where the set U consisted of those boundaries ∂D contained in a fixed circular annulus such that ∂D has a uniformly bounded Hölder continuous tangent. In the case of the three dimensional inverse scattering problem it was also necessary to assume that D was starlike with respect to the origin ([1]).

In closing we mention that although we have succeeded in using a priori information to stabilize the inverse scattering problem, we have made no statement on the <u>type</u> of continuity that results (i.e. Lipschitz, Hölder, logarithmic, etc.). This is an important problem that remains to be investigated since if the continuity is too weak there is little hope of using the optimization problem in a constructive fashion to actually determine λ or D. For a discussion of this point in the case of <u>linear</u> inverse scattering problems we refer the reader to [2].

V. Constructive Methods for Determining $\hat{\underset{\sim}{T}}^{-1}x, x \in \underset{\sim}{T}U$.

The previous section shows that the inverse scattering problems we are considering in this paper can be reformulated as constrained, nonlinear optimization problems. The numerical solution of these optimization problems is presently being investigated by A. Kirsch ([16]). Related approaches have been previously considered by A. Roger ([18]) and B. Sleeman ([19]). Needless to say there are many open problems connected with such a numerical approach, particularly in connection with the questions of stability and convergence.

The computational problems associated with computing $\hat{T}x^{-1}$, $x \in TU$, are considerably simplified if one has access to low frequency data. We first briefly consider the problem of determining the surface impedance of an obstacle from low frequency far field data ([6], [21]). Let F be the far field pattern arising from the addition of the scattered waves corresponding to two incoming plane waves striking the obstacle from opposite directions. If we expand F in its Fourier series

$$F(\theta;k) = \sum_{n=-\infty}^{\infty} a_n(k)e^{in\theta} \tag{5.1}$$

then the (weighted) low frequency limit of the coefficient $a_n(k)$ determine a sequence of numbers b_n from which we can define the harmonic function

$$h(x) = \sum_{n=-\infty}^{\infty} b_n r^{-|n|} e^{in\theta}. \tag{5.2}$$

It can be shown that the series (5.2) converges for $r \geq a$ where a is the radius of a disk containing the scattering obstacle in its interior. From [6] we now have that the unknown impedance λ is the solution of the integral equation of the first kind

$$h(x) = \int_{\partial D} \lambda(y) N(x,y) ds(y) \quad ; \quad |x| = a \tag{5.3}$$

where N denotes the Neumann function for Laplace's equation in the exterior of D. Since F is assumed to be only approximately known, the computation of the coefficients b_n ($|n| \leq N$ for some positive integer N) must be done through the use of regularization procedures. In particular the left hand side of (5.3) is not known exactly and the problem of solving (5.3) is hence improperly posed. However stabilization can be achieved by assuming a priori that $\lambda \epsilon U$, in agreement with the results of the previous section.

A similar precedure can be used to determine the shape of a soft scattering obstacle from a knowledge of the far field pattern at low values of the frequency ([3], [9]). This method is based on relating the low frequency limit of the Fourier coeffi cients of the far field pattern to the coefficients of the Laurent expansion of the (unique) analytic function f that conformally maps the exterior of the unit disk onto the (unknown) scattering obstacle such that at infinity f has the Laurent expansion

$$f(w) = aw + a_o + \frac{a_1}{w} + \frac{a_2}{w^2} + \cdots \tag{5.4}$$

where a>0 is the mapping radius. In particular if μ_n denotes the (weighted) low frequency limit of the Fourier coefficients then the μ_n are related to the a_n by a relation of the form

$$\mu_n = 2\pi n\, a^{n-1}\, a_{n-1} + \text{lower order coefficients}, \tag{5.5}$$

for n=1,2.,,,. where the mapping radius a can be determined from experimental data. Furthermore, if the coefficients $a_o, a_1, \ldots a_N$ are determined from (5.5) and f_N is defined by

$$f_N(w) = aw + a_o + \frac{a_1}{w} + \dots + \frac{a_N}{w^N} \tag{5.6}$$

then from the Area Theorem in geometric function theory it is possible to deduce the L^2-error estimate

$$\frac{1}{2\pi} \int_o^{2\pi} |f(e^{i\phi}) - f_N(e^{i\phi})|^2 d\phi \le \frac{a^2}{N+1}. \tag{5.7}$$

Note that from (5.5) small values of a can cause large errors in the computation of the coefficients a_n whereas large values of a imply that the error estimate (5.7) is large. Hence in order to achieve stability a must be bounded from above and below. This will be guaranteed if ∂D is known a priori to be contained in a given annulus, in agreement with the results of the previous section. Note however that the error estimate in this case is with respect to the L^2 norm instead of the pointwise estimates of Section IV and hence it is not necessary to assume a priori that ∂D has a uniformly bounded Hölder continuous tangent. This suggests the problem of examining the stability of the inverse scattering problem where continuity is measured with respect to norms different from the maximum norm and determining the appropriate compact sets that are associated with these norms.

The above function-theoretic approach to the inverse scattering problem of determining the shape of the scattering obstacle from a knowledge of the far field pattern was first given by Colton ([4]) and Colton and Kleinman ([9]). Further developments of this method have subsequently been provided by Hariharan ([14]), Sleeman ([20]), and Smith ([21]). A partially successful attempt to extend this approach to the inverse scattering problem in $\mathbb{R}^3$ has been given by Colton and Kress ([11]). The problem in this case is that instead of conformal mapping methods one must rely on the use of level curves in potential theory and such an approach yields much weaker results.

The discussion provided in this section obviously only touches the surface of the computational problems inherent in constructing approximate solutions to the inverse scattering problem and it is hoped that the future will yield new developments and insights in this direction. Indeed the effective numerical solution of the inverse scattering problem is the basic open problem in the field and is intimately linked to all of the problems we have previously mentioned. Our hope is that this paper has indicated recent progress and possible new directions to a sufficient extent to encourage others to enter this important and fascinating area of research.

References

1. T. S. Angell, D. Colton, and A. Kirsch, The three dimensional inverse scattering problem for acoustic waves, J. Diff. Eqns., to appear.

2. M. Bertero, C. de Mol, and G. A. Viano, The stability of inverse problems, in Inverse Scattering Problems in Optics, H. P. Baltes, editor, Springer-Verlag, Berlin, 1980, 161-214.

3. D. Colton, Analytic Theory of Partial Differential Equations, Pitman Publishing, London, 1980.

4. D. Colton, The inverse scattering problem for a cylinder, Proc. Royal Soc. Edinburgh 84A (1979), 135-143

5. D. Colton, Runge's theorem and far field patterns for the impedance boundary value problem in acoustic wave propagation, SIAM J. Math. Anal., to appear.

6. D. Colton, Stable methods for determining the surface impedance of an obstacle from low frequency far field data, submitted for publication.

7. D. Colton and A. Kirsch, Stable methods for solving the inverse scattering problem for a cylinder, Proc. Royal Soc. Edinburgh 89A (1981), 181-188

8. D. Colton and A. Kirsch, The determination of the surface impedance of an obstacle from measurements of the far field pattern, SIAM J. Applied Math. 41 (1981), 8-15.

9. D. Colton and R. Kleinman, The direct and inverse scattering problems for an arbitrary cylinder: Dirichlet boundary conditions, Proc. Royal Soc. Edinburgh 86A (1980), 29-42

10. D. Colton and R. Kress, Integral Equation Methods in Scattering Theory, John Wiley, New York, to appear.

11. D. Colton and R. Kress, Iterative methods for solving the exterior Dirichlet problem for the Helmholtz equation with applications to the inverse scattering problem for low frequency acoustic waves, J. Math. Anal. Appl. 77 (1980), 60-72.

12. D. Colton and R. Kress, The impedance boundary value problem for the time harmonic Maxwell equations, Math. Methods in Applied Sciences 3 (1981), 475-487.

13. D. Colton and R. Kress, The unique solvability of the null field equations of acoustics, Quart. J. Mech. Applied Math., to appear.

14. S. I. Hariharan, Inverse scattering for an exterior Dirichlet problem, Quart. Applied Math., to appear.

15. G. Hellwig, Partial Differential Equations, Blaisdell, New York, 1964.

16. A. Kirsch, Habilitation Dissertation, University of Göttingen, in preparation.

17. P. D. Lax and R. S. Phillips, Scattering Theory, Academic Press, New York, 1967.

18. A. Roger, Newton-Kantorovitch algorithm applied to an electromagnetic inverse problem, IEEE Trans. Antennas and Prop. AP-29 (1981), 232-238.

19. B. D. Sleeman, The inverse problem of acoustic scattering, IMA J. Applied Math., to appear.

20. B. D. Sleeman, Two-dimensional inverse scattering and conformal mappings, IMA J. Applied Math. 27 (1981), 19-31.

21. R. Smith, Ph.D. Thesis, University of Delaware, in preparation.

22. F. Ursell, On the exterior problems of acoustics, Proc. Camb. Phil. Soc. 74 (1973), 117-125.

ON THE HUKUHARA-KNESER PROPERTY FOR SOME CAUCHY PROBLEMS IN LOCALLY CONVEX TOPOLOGICAL VECTOR SPACES

Jacques Dubois and Pedro Morales

1. Introduction

In this note we introduce a net version of the important Palais-Smale condition [14] (see [13] for other application of this assumption). This allows us to deduce, from a recent result of Szufla [21], the basic Lemma 2.1, which gives a topological characterization of the set of fixed points of a non-linear operator in a function topological vector space. In the Banach space case, a more delicate characterization is given by Lemma 2.2 in terms of the notion of R_δ-set introduced by Aronszajn [1]. This Lemma contains as Corollary an interesting result of Vidossich [22].

As illustration, we apply these Lemmas to describe topological properties of the set of solutions of some Cauchy problems in locally convex topological vector spaces. This permits us to unify several results due to Aronszajn [1], Astala [2], Deimling [5], Pulvirenti [15] and Szufla ([18], [19], [20]), and to show also that the Hukuhara-Kneser property holds for the Cauchy problem under the hypotheses, for the existence case, established by Kato [11].

2. Basic lemmas

Consider the class $\mathcal{M}$ of all metrizable spaces. A topological space S is called an absolute retract for metrizable spaces, in symbols $S \in AR(\mathcal{M})$, if $S \in \mathcal{M}$ and, for each closed subset F of a metrizable space M, every continuous function $h:F \to S$ admits a continuous extension $\bar{h}:M \to S$ [3, p. 87]. For example, if S is a convex subset of a normed space, then $S \in AR(\mathcal{M})$ [8, Corollary 4.2]. A topological space S is called an absolute retract, in symbols $S \in AR$, if S is compact and $S \in AR(\mathcal{M})$. We note that if $S \in AR$, then S is non-empty, compact and connected. A topological space S is called an R_δ-set if $S \in \mathcal{M}$ and S is homeomorphic to the intersection of a decreasing sequence in AR. We note that every r-image of an AR space is an AR space [3, p. 101], and, in particular, if $S \in AR$ and $h:S \to S'$ is a homeomorphism, then $S' \in AR$.

We denote by $H^* = (H^q)_{q=1}^{\infty}$ the Čech cohomological functor with coefficients in $\mathbb{Z}$, defined on the category of topological pairs (Y, A), and we denote by $\tilde{H}^q$ the reduced cohomology (see [17] for more details). A topological space Y is said to be acyclic if $\tilde{H}^q(Y) = \{0\}$ for all $q = 1, 2, 3, \ldots$ We note the following fundamental property: Every R_δ-set is acyclic [12, p. 110].

Let K and X be topological spaces. The set of all continuous functions from K to X will be denoted by $C(K, X)$. If X is a uniform space, the symbol τ_u denotes the topology of uniform convergence on every non-empty subset of

$C(K, X)$. If X is a Hausdorff topological vector space, the symbol $C_b(K, X)$ denotes the vector space of all bounded continuous functions from K to X. In this case, it is well-known that $C_u(K, X) = (C_b(K, X), \tau_u)$ is a Hausdorff topological vector space whose topology has as local base the collection of all sets of the form $N(U) = \{x \in C_b(K, X): x(t) \in U$ for all $t \in K\}$, where U is a neighbourhood of 0 in X.

If u_0 is an element of a normed space and $r > 0$, the symbol $\bar{B}_r(u_0)$ denotes the closed ball of center u_0 and radius r.

2.1 Lemma Let K be a bounded and convex subset of a normed space with norm $||\cdot||$, let X be a Hausdorff topological vector space, let $Y = C_u(K, X)$ and let $F:Y \to Y$. Assume that F satisfies the following conditions:

i) F is continuous.

ii) There exist $t_0 \in K$ and $x_0 \in X$ such that $F(y)(t_0) = x_0$ for all $y \in Y$.

iii) For every $\epsilon > 0$, the relations $x, y \in Y$ and $x|K_\epsilon = y|K_\epsilon$ imply $F(x)|K_\epsilon = F(y)|K_\epsilon$, where $K_\epsilon = K \cap \bar{B}_\epsilon(t_0)$.

iv) For every neighbourhood U of 0 in X, there exists $\epsilon > 0$ such that $t, s \in K$ and $||t - s|| < \epsilon$ imply $F(x)(t) - F(x)(s) \in U$ for all $x \in Y$.

v) For every net $(x_\alpha)_{\alpha\in D}$ in Y such that $x_\alpha - F(x_\alpha) \to 0$, there exists a convergent subnet $(y_\beta)_{\beta\in D'}$ of $(x_\alpha)_{\alpha\in D}$.

Then the set $Fix(F)$ of all fixed points of F is non-empty, compact and connected.

Proof. To prove that $S = Fix(F)$ is non-empty and connected, it suffices, by Theorem 2 of [21], to show that, for every closed subset V of Y, $0 \in \overline{(I - F)(V)}$ implies $0 \in (I - F)(V)$, where I denotes the identity mapping on Y. Suppose $0 \in \overline{(I - F)(V)}$. Then there exists a net $(x_\alpha)_{\alpha\in D}$ in V such that $(I - F)(x_\alpha) \to 0$, so $x_\alpha - F(x_\alpha) \to 0$. It follows from (v) that there exist a subnet $(y_\beta)_{\beta\in D'}$, of $(x_\alpha)_{\alpha\in D}$ and a function $y \in C_b(K, X)$ such that $y_\beta \to y$. Since $y_\beta \in V$ for all $\beta \in D'$ and V is closed in Y, $y \in V$. Moreover, since F is continuous, $F(y_\beta) \to F(y)$. But $y_\beta - F(y_\beta) \to 0$. So $y - F(y) = 0$, and therefore $0 \in (I - F)(V)$.

The compactness of S follows immediately from (v).

2.2 Lemma Let K be a bounded and convex subset of a normed space with norm $||\cdot||$, let X be a Banach space, let $Y = C_u(K, X)$ and let $F:Y \to Y$. Assume that F satisfies the conditions i), ii), iii) and iv) of Lemma 2.1 and the Palais-Smale condition: If (x_n) is a sequence in Y such that $x_n - F(x_n) \to 0$, then (x_n) has a convergent subsequence.

Then the set $Fix(F)$ of all fixed points of F is an R_δ-set.

Proof. In this case, Y is a Banach space with the supremum norm $||\cdot||_\infty$. By Lemma 1 of [21] (see also proof of Theorem 1.1 of [22]) there exists a sequence

(F_n) of continuous mappings from Y to Y such that each $I - F_n$ is a homeomorphism of Y and $\lim_{n\to\infty} F_n(x) = F(x)$ uniformly in $x \in Y$. Put $T = I - F$, $T_n = I - F_n$ for all $n = 1, 2, \ldots$ and $S = \mathrm{Fix}(F)$. Then we may suppose (by choosing a subsequence of (F_n) if necessary) that $||T_n(x) - T(x)||_\infty \le \frac{1}{n}$ for all $x \in Y$ and all $n = 1, 2, 3, \ldots$, and therefore, for every n, the compact set $T_n(S)$ is contained in the closed ball of center 0 and radius $\frac{1}{n}$. Let $Q_n = \overline{co}(T_n(S))$. Then $Q_n \in AR(M)$ and, since Q_n is compact according to the Mazur theorem [9, p. 416], $Q_n \in AR$. Since every T_n is a homeomorphism, $T_n^{-1}(Q_n) = R_n \in AR$.

It is clear that $S \subseteq R_n$ for all $n = 1, 2, 3, \ldots$ So $S \subseteq \bigcup_{n=1}^{\infty} \bigcap_{k=n}^{\infty} R_k = \lim_n \inf R_n$. Let now $x \in \lim_n \sup R_n = \bigcap_{n=1}^{\infty} \bigcup_{k=n}^{\infty} R_k$. Then there exists a subsequence (R_{k_n}) of (R_n) such that $x \in R_{k_n}$ for all $n = 1, 2, 3, \ldots$ Thus $T_{k_n}(x) \in Q_{k_n}$ and therefore $||T_{k_n}(x)||_\infty \le \frac{1}{k_n}$. Since also $||T_{k_n}(x) - T(x)|| \le \frac{1}{k_n}$ it follows that $T(x) = 0$. So $x \in S$, and this shows that $S = \lim_n R_n$. Let V be a neighbourhood of S. Suppose that there exists a positive integer k_0 such that $n > k_0$ implies $R_n \not\subseteq V$. For every $n = k_0 + 1, k_0 + 2, k_0 + 3, \ldots$ choose $x_n \in R_n - V$. Let $y_n = x_{n+k_0}$ for all $n = 1, 2, 3, \ldots$ Since $||T(y_n)||_\infty \le \frac{2}{n+k_0}$, it follows that $y_n - F(y_n) \to 0$. Then, by the Palais-Smale condition, there exist a subsequence (y_{k_n}) of (y_n) and an element $y \in Y$ such that $\lim_{n\to\infty} y_{k_n} = y$. So $\lim_{n\to\infty} T(y_{k_n}) = T(y)$, and therefore $T(y) = 0$. Then $y \in S \subseteq V$. This a contradiction because $y_{k_n} \notin V$ for all $n = 1, 2, 3, \ldots$ Consequently, for each neighbourhood V of S there exists a subsequence (R_{k_n}) of (R_n) such that $R_{k_n} \subseteq V$ for every $n = 1, 2, 3, \ldots$ By Lemma 5 of [4] (which is a correction of the Théorème B of [1]) it follows that S is an R_δ-set.

A function h from a topological space X to a topological space Y is said to be <u>compact</u> if h is continuous and $\overline{h(X)}$ is compact.

2.3 <u>Corollary</u> ([22, Theorem 1.2]). Let K be a compact convex subset of a normed space, let X_0 be a closed convex subset of a Banach space X, let $Y_0 = (C(K, X_0), \tau_u)$ and let $F:Y_0 \to Y_0$. If F is a compact mapping satisfying the conditions ii) and iii) of Lemma 2.1, then the set $\mathrm{Fix}(F)$ of all fixed points of F is an R_δ-set in $C_u(K, X)$.

<u>Proof</u>. Since K is compact, $C_b(K, X) = C(K, X)$. Let $Y = C_u(K, X)$. Since $X_0 \in AR(M)$ and X_0 is a closed subset of the Banach space X, there exists a continuous function $r:X \to X_0$ such that $r(x) = x$ for all $x \in X_0$. Let $G:Y \to Y$ be the mapping defined by the formula $G(u) = F(r \circ u)$. By the argument used in the proof of Theorem 1.2 of [22], we conclude that G is a compact mapping and $\mathrm{Fix}(F) = \mathrm{Fix}(G)$. Since K is compact and $\overline{G(Y)}$ is τ_u-compact, $G(Y)$ is uniformly equicontinuous, and therefore the condition iv) of Lemma 2.1 is verified. To

verify the condition of Palais-Smale, let (u_n) be a sequence in Y such that $u_n - G(u_n) \to 0$. Since $G(u_n) \in G(Y)$ and $\overline{G(Y)}$ is τ_u-compact, there exist a subsequence (u_{k_n}) of (u_n) and a function $u \in C(K, X)$ such that $\lim_{n\to\infty} G(u_{k_n}) = u$. But $u_{k_n} - G(u_{k_n}) \to 0$. So $\lim_{n\to\infty} u_{k_n} = u$. By Lemma 2.2, $Fix(G)$ is an R_δ-set.

2.4 Lemma Let K be a compact subset of a normed space, let X_0 be a non-empty subset of a Hausdorff locally convex topological vector space X and let $h:K \times X_0 \to X$ be a continuous function. Then for every $u \in C(K, X_0)$ and every closed, balanced and convex neighbourhood U of 0 in X, there exists a neighbourhood V of 0 in X such that $h(t, v(t)) - h(t, u(t)) \in U$ for all $t \in K$ whenever $v \in C(K, X_0)$ and $v(t) - u(t) \in V$ for every $t \in K$.

Proof. Let $s \in K$. Since h is continuous, there exist $\delta_s > 0$ and a neighbourhood W_s of 0 in X such that $t \in K \cap B_{\delta_s}(s)$ and $y \in X_0 \cap (u(s) + W_s)$ imply $h(t, y) - h(s, u(s)) \in \frac{1}{2} U$.

Let V_s be a closed, balanced and convex neighbourhood of 0 in X contained in W_s. Since $u \in C(K, X_0)$ there exists $\eta_s > 0$ such that $t \in K \cap B_{\eta_s}(s)$ implies $u(t) - u(s) \in \frac{1}{2} V_s$. Let $\epsilon_s = \text{Min}\,\{\delta_s, \eta_s\}$. Since $K \subseteq \bigcup_{s\in K} B_{\epsilon_s}(s)$ and K is compact, there exists a finite sequence $(s_i)_{i=1}^n$ in K such that $K \subseteq \bigcup_{i=1}^n B_{\epsilon_{s_i}}(s_i)$. Let $V = \frac{1}{2} \bigcap_{i=1}^n V_{s_i}$.

Assume that $v \in C(K, X_0)$ and $v(t) - u(t) \in V$ for every $t \in K$. Let $t \in K$. Then there exists $1 \le i \le n$ such that $t \in B_{\epsilon_{s_i}}(s_i)$. So $u(t) \in u(s_i) + \frac{1}{2} V_{s_i} \subseteq u(s_i) + V_{s_i}$ and $v(t) = u(t) + (v(t) - u(t)) \in u(s_i) + \frac{1}{2} V_{s_i} + V \subseteq u(s_i) + \frac{1}{2} V_{s_i} + \frac{1}{2} V_{s_i} = u(s_i) + V_{s_i}$. Therefore $h(t, v(t)) - h(t, u(t)) = (h(t, v(t)) - h(s_i, u(s_i)) + (h(s_i, u(s_i)) - h(t, u(t)) \in \frac{1}{2} U + \frac{1}{2} U = U$.

3. Some applications

For $t_0 \in R$ and $a > 0$, let $I = [t_0, t_0 + a]$. Let $X = (X, \tau)$ be a Hausdorff locally convex topological vector space, let $x_0 \in X$ and let X_0 be a closed convex subset of X containing x_0.

Consider the Cauchy problem

(CP) $$x' = f(t, x),\ x(t_0) = x_0$$

where f is an X-valued function defined on $I \times X_0$. An X_0-valued function u defined on some non-degenerate compact subinterval K of I, containing t_0, is a solution of (CP) if u is τ-differentiable on K, $u(t_0) = x_0$ and $u'(t) = f(t, u(t))$ for all $t \in K$.

It is well-known that, under mild conditions on f and with a suitable meaning of the integral, (CP) is equivalent to the following equation for integrable functions from K to X_0:

$$\text{(IE)} \qquad x(t) = x_0 + \int_{t_0}^{t} f(s, x(s))ds$$

which allows us to define the integral operator

$$F(x)(t) = x_0 + \int_{t_0}^{t} f(s, x(s))ds$$

whose set $Fix(F)$ of all fixed points gives exactly the set $Sol(f)$ of all solutions of (CP) in the space $C(K, X)$. We note that F satisfies trivially the conditions ii) and iii) of Lemma 2.1.

By way of illustration we consider the following examples:

I) Suppose $X = R^n$, $X_0 = \overline{B}_r(x_0)$ and $f \in C(I \times X_0, X)$.

Let $M = \sup\{||f(t, x)||: (t, x) \in I \times X_0\}$ and let $b = \min\{a, \frac{r}{M}\}$. Consider the compact subinterval of $I: K = [t_0, t_0 + b]$.

In this case, the integral appearing in (IE) is the Lebesgue integral for R^n-valued functions. Let $x \in C(K, X_0)$. Since $||F(x)(t) - x_0|| \leq (t - t_0)\cdot \sup\{||f(s, x(s))||: s \in [t_0, t]\} \leq Mb \leq r$ and $||F(x)(t) - F(x)(s)|| \leq M|t-s|$ for all $s, t \in K$, we conclude that $F(x) \in C(K, X_0)$. This shows also that $F(C(K, X_0))$ is an equicontinuous subset of $C(K, X_0)$. Since $F(C(K, X_0))[t] = \{F(x)(t): x \in C(K, X_0)\}$ is a bounded subset of R^n for every $t \in K$, the Ascoli theorem implies that $\overline{F(C(K, X_0))}$ is τ_u-compact. Therefore the mapping $F:(C(K, X_0), \tau_u) \to (C(K, X_0), \tau_u)$ is compact.

By Corollary 2.3 it follows that the set $Sol(f)$ is an R_δ-set in $C_u(K, X)$. This result is due to Aronszajn [1] (see also [10]). In particular, $Sol(f) \neq \emptyset$ (Peano's Teorem) and $Sol(f)$ is compact and connected (Hukuhara-Kneser property).

II) Suppose X is a Banach space, $X_0 = \overline{B}_r(x_0)$ and f satisfies the Carathéodory conditions:

a) For every $x \in X$, $f(\cdot, x)$ is strongly measurable.

b) For every $t \in I$, $f(t, \cdot)$ is continuous.

c) There exists a positive function $h \in L^1(I)$ such that $||f(t, x)|| \leq h(t)$ for every $(t, x) \in I \times X_0$.

By the Lemma of [15], for every $u \in C(I, X_0)$, the function $t \to f(t, u(t))$ is Bochner integrable with respect to Lebesgue measure on I. Then the integral appearing in (IE) is the Bochner Integral with respect to Lebesgue measure on I, and therefore any solution of (CP) is an absolutely continuous X_0-valued function defined on some non-degenerate compact subinterval of I, containing t_0.

Choose $0 < b \leq a$ such that $\int_{t_0}^{t_0+b} h(t)dt \leq r$, and consider the compact subinterval of $I:K = [t_0, t_0 + b]$. Let $x \in C(K, X_0)$. Since $t \to \int_{t_0}^{t} f(s, x(s))\,ds$ is absolutely continuous and $||F(x)(t) - x_0|| \le \int_{t_0}^{t} ||f(s, x(s)||ds \le \int_{t_0}^{t} h(s)ds \le r$ for all $t \in K$, it follows that $F(x) \in C(K, X_0)$. To prove that $F:(C(K, X_0), \tau_u) \to (C(K, X_0), \tau_u)$ is continuous, let (x_n) be a sequence in $C(K, X_0)$ which τ_u-converges to x. Then $f(s, x_n(s)) \to f(s, x(s))$ for all $s \in K$. Then, by the dominated convergence theorem for Bochner integrals [6, p. 45], $\lim_{n\to\infty} \int_{t_0}^{t} f(s, x_n(s))ds = \int_{t_0}^{t} f(s, x(s))ds$ for every $t \in K$, and therefore $\lim_{n\to\infty} F(x_n)(t) = F(x)(t)$ for every $t \in K$. But the set $\{F(x_n): n = 1, 2, 3, \ldots\}$ is equicontinuous, because given $\epsilon > 0$ choose $\delta > 0$ such that $t, t' \in K$ and $|t - t'| < \delta$ imply $|\int_{t}^{t'} h(s)ds| < \epsilon$; then $||F(x_n)(t') - F(x_n)(t)|| \le |\int_{t}^{t'} ||f(s, x_n(s))||ds| \le |\int_{t}^{t'} h(s)ds| < \epsilon$. So the sequence $(F(x_n))$ τ_u-converges to $F(x)$.

With a suitable additional hypothesis we can describe the topological properties of Sol(f). Consider two examples:

α) Suppose that $F(C(K, X_0))$ is τ_u-compact. Then, by Corollary 2.3, Sol(f) is an R_δ-set in $C_u(K, X)$. This result improves the Theorem II of Pulvirenti [15].

β) Let $g:X \to X_0$ be a continuous function such that $g(x) = x$ for every $x \in X_0$, and define $G(x)(t) = x_0 + \int_{t_0}^{t} f(s, g(x(s))ds$ for every $x \in C(K, X)$ and every $t \in K$. It is clear that $G:C_u(K, X) \to C_u(K, X)$ is a continuous mapping satisfying the conditions ii), iii) and iv) of Lemma 2.1. For every $n = 1, 2, 3, \ldots$ Let $S_n = \{x \in C(K, X): x(t_0) = x_0$ and $||x - G(x)||_\infty \le \frac{1}{n}\}$.

Suppose that $\lim_{n\to\infty} \alpha(S_n) = 0$, where α denotes the Kuratowski measure of non-compactness. It can be shown that G satisfies the Palais-Smale condition [7], and therefore Lemma 2.2 implies that Fix(G) = Sol(f) is an R_δ-set. This result improves two theorems of Szufla ([18, Theorem 6] and [19, Theorem]) (which contain several earlier results) and the Theorem 2.3 of Deimling [5].

III) Suppose $(E, ||\cdot||)$ is a Banach space, $X = (E, \tau_w)$ where τ_w denotes the weak topology, $X_0 = X$ and $f \in C(I \times X, X)$ such that $\overline{f(I \times X)}$ is t_w-compact

We note that, for every $x \in C(I, X)$, the function $t \to f(t, x(t))$ is τ_w-continuous. Also there exists a real constant $M > 0$ such that $||f(t, x)|| \le M$ for every $(t, x) \in I \times X$.

Let $K = I$ and let $x \in C(K, X)$. Since, by the Krein-Šmulian theorem [9, p. 434], $\overline{co}\,(f(K \times X))$ is τ_w-compact, the function $t \to f(t, x(t))$ is

Pettis integrable with respect to Lebesgue measure on K [16, p. 74-75]. Then the integral appearing in (IE) is the Pettis integral with respect to Lebesgue measure on K.

Let $u \in C(K, X)$. Let x^* be any element of E^* such that $||x^*|| = 1$. Since $|x^*(F(u)(t) - F(u)(t'))| = |x^* \int_{t'}^{t} f(s, u(s))ds| = |\int_{t'}^{t} x^*(f(s, u(s)))ds| \leq M|t' - t|$ for all $t, t' \in K$, it follows that $F(u) \in C(K, X)$.

To show that $F:C_u(K, X) \to C_u(K, X)$ is continuous, let $u \in C(K, X)$ and let W be a neighbourhood of 0 in X. Choose a closed, balanced and convex neighbourhood U of 0 in X such that $U \subseteq \frac{1}{a} W$. By Lemma 2.4 there exists a neighbourhood V of 0 in X such that $f(t, x(t)) - f(t, u(t)) \in U$ for all $t \in K$ whenever $x \in C(K, X)$ and $x(t) - u(t) \in V$ for every $t \in K$. Let $x \in C(K, X)$ be such that $x - u \in N(V)$. For $s \in K$ put $h(s) = f(s, x(s)) - f(s, u(s))$. Since $h(K) \subseteq U$ and $F(x)(t) - F(u)(t) = \int_{t_0}^{t} h(s)ds \in (t - t)$.

$\overline{co}\,(h(K)) \subseteq (t - t_0)U \subseteq a\,U \subseteq W$ for all $t \in K$, we conclude that $F(x) - F(u) \in N(W)$.

To verify the condition iv) of Lemma 2.1, let U be a closed, balanced and convex neighbourhood of 0 in X. Since $f(K \times X)$ is (strongly) bounded, there exists a real number $\epsilon > 0$ such that $\overline{co}\,(f(K \times X)) \subseteq \frac{1}{\epsilon} U$. Let $t, t' \in K$ be such that $0 \leq t - t' < \epsilon$. Then, for every $x \in C(K, X)$, we have $F(x)(t) - F(x)(t') = \int_{t'}^{t} f(s, x(s))ds \in (t - t')\,\overline{co}\,(f(K \times X)) \subseteq \frac{t - t'}{\epsilon} U \subseteq U$.

To verify the condition v) of Lemma 2.1, let $(x_\alpha)_{\alpha\in D}$ be a net in $C_u(K, X)$ such that $x_\alpha - F(x_\alpha) \to 0$. Let $s \in K$. Since $f(s, x_\alpha(s)) \in f(K \times X)$ and $f(K \times X)$ is τ_w-compact, there exists a subnet $(y_\beta)_{\beta\in D'}$ of $(x_\alpha)_{\alpha\in D}$ such that $(f(s, y_\beta(s)))_{\beta\in D'}$ τ_w-converges to an element $y(s) \in E$. Since $|x^* f(s, y_\beta(s))| \leq M\,||x^*||$ for every $x^* \in E^*$, then, by the dominated convergence for Pettis integrals, y is Pettis integrable and $\int_{t_0}^{t} f(s, y_\beta(s)ds \to \int_{t_0}^{t} y(s)ds = u(t)$ for every $t \in K$. So $F(y_\beta)(t) \to x_0 + u(t)$ for every $t \in K$. Since $|x^*(F(y_\beta)(t) - f(y_\beta)(t'))| \leq M|t - t'|$ for all $x^* \in E^*$ such that $||x^*|| = 1$ and all t, $t \in K$, the set $\{F(y_\beta): \beta \in D'\}$ is equicontinuous, so $u \in C(K, X)$ and $F(y_\beta) \to x_0 + u$. But $y_\beta - F(y_\beta) \to 0$. Therefore $y_\beta \to x_0 + u$.

By Lemma 2.1 it follows that the set $Sol(f) \neq \phi$(Kato's Theorem [11]) and it is compact and connected. This result generalizes a result of Szufla [20] established when X is reflexive.

IV) Suppose E is a barreled normed space, τ is a Hausdorff locally convex topology on E^* which is stronger than the w^*-topology and weaker than the topology of compact convergence, $X = (E^*, \tau)$ and $f \in C(R \times X, X)$.

By the Banach-Mackey theorem there exists a real number $M \geq 1$ such that $|t - t_0| \leq 1$ and $||x - x_0|| \leq 1$ imply $||f(t, x)|| \leq M$, where $||\cdot||$ denotes the norm of E^*. It follows from proposition 6 and Lemma 10 of [2] that the function $t \to f(t, x(t))$ is Riemann integrable for every $x \in C(K, X)$, where $K = [t_0, t_0 + \frac{1}{M}]$. Then the integral appearing in (IE) is the Riemann integral.

Let $x \in C(K, X)$. By Proposition 4 of [2], $F(x) \in C(K, X)$. Using Lemma 2.4 and some results of [2] it can be shown that $F: C_u(K, X) \to C_u(K, X)$ is a continuous mapping satisfying the conditions iv) and v) of Lemma 2.1. Then $Fix(F) = Sol(f)$ is a non-empty, compact and connected subset of $C_u(K, X)$. This result is due to Astala [2, Theorem 13].

REFERENCES

1. N. ARONSZAJN, Le correspondant topologique de l'unicité dans la théorie des équations différentielles, Ann. of Math. 43 (1942), 730-738.

2. K. ASTALA, On Peano's Theorem in Locally Convex Spaces, Reports of the Department of Mathematics, University of Helsinki (1980), 1-14.

3. K. BORSUK, Theory of Retracts, Polish Scientific Publishers, Warszawa (1967).

4. F. E. BROWDER and C.P. GUPTA, Topological Degree and Nonlinear Mappings of Analytical Type in Banach Spaces, J. Math. Anal. Appl. 26 (1969), 390-402.

5. K. DEIMLING, Ordinary Differential Equations in Banach Spaces, Lect. Notes in Math. 596, Springer-Verlag, New York (1977).

6. J. DIESTEL and J. J. UHL, Vector Measures, Math. Surveys 15, Amer. Math. Soc., Providence (1977).

7. J. DUBOIS and P. MORALES, Structure de l'ensemble des solutions du problème de Cauchy sous les conditions de Carathéodory, Ann. Sc. Math. Québec (to appear).

8. J. DUGUNDJI, An Extension of Tietze's Theorem, Pacific J. Math. 1 (1951), 353-367.

9. N. DUNFORD and J. SCHWARTZ, Linear Operators, Part I, Interscience Publishers, Inc., New York (1958).

10. C. J. HIMMELBERG and F.S. VAN VLECK, On the Topological Triviality of Solution Sets, Rocky Mountain J. Math. 10 (1980), 247-252.

11. S. KATO, On Existence and Uniqueness Conditions for Nonlinear Ordinary Differential Equations in Banach Spaces, Funkcial. Ekvac. 19 (1976), 239-245.

12. J. M. LASRY and R. ROBERT, Analyse non linéaire multivoque, Cahier de Math. de la décision no. 7611, Paris (1978).

13. L. NIRENBERG, Variational and Topological Methods in Nonlinear Problems, Bull. Amer. Math. Soc., Vol. 4, Number 3 (1981), 267-302.

14. R. S. PALAIS, Critical Point Theory and the Minimax Principle, Proc. Sympos. Pure Math., Vol. 15, Amer. Math. Soc., Providence (1970), 185-212.

15. G. PULVIRENTI, Equazioni Differenziali in uno spazio di Banach. Teorema di esistenza e struttura del pennello delle soluzioni in ipotesi di Carathéodory, Ann. Mat. Pura Appl. 56 (1961), 281-300.

16. W. RUDIN, Functional Analysis, McGraw-Hill Book Company, New York (1973).

17. E. SPANIER, Algebraic Topology, Springer-Verlag, New York (1966).

18. S. SZUFLA, Solutions Sets of Nonlinear Equations, Bull. Acad. Polon. Sci., Sér. Sci. Math. Astronom. Phys. 21 (1973), 971-976.

19. S. SZUFLA, Some Properties of the Solutions Set of Ordinary Differential Equations, Bull. Acad. Polon. Sci., Sér. Sci. Math. Astronom. Phys. 22(1974), 675-678.

20. S. SZUFLA, Kneser's Theorem for weak Solutions of Ordinary Differential Equations in Reflexive Banach Spaces, Bull. Acad. Polon. Sci., Sér. Sci. Math. Astronom. Phys. 26 (1978), 407-413.

21. S. SZUFLA, Sets of Fixed Points of Nonlinear Mappings in Function Spaces, Funkcial. Ekvac. 22(1979), 121-126.

22. G. VIDOSSICH, A Fixed Point Theorem for Function Spaces, J. Math. Anal. Appl. 36 (1971), 581-587.

Essential Self-Adjointness and Self-Adjointness for Generalized Schrödinger Operators.

by NGUYEN XUAN DUNG

By "generalized Schrödinger Operators", we mean an elliptic operator of the form $T = \sum_{0<|\alpha|,|\beta|\leq m} (-1)^{|\alpha|} D^{\alpha} a_{\alpha\beta}(x) D^{\beta} + q(x)$ on $L^2(R^n)$, where $D^{\alpha} = \left(\frac{\partial}{\partial x_1}\right)^{\alpha_1} \ldots \left(\frac{\partial}{\partial x_n}\right)^{\alpha_n}$. In this paper, we wish to present two results. One concerns the essential self-adjointness of T on $C_0^{\infty}(R^n)$, and the other concerns the self-adjointness of T on $H^{2m}(R^n) \cap D(q)$, where $D(q) = \{u \varepsilon L^2(R^u) \mid qu \varepsilon L^2(R^n)\}$. We will try to motivate our results, and will only sketch the proofs.

I. General Assumptions and Statements of Theorems.

1/ $a_{\alpha\beta}$ complex-valued with $a_{\alpha\beta} = \bar{a}_{\beta\alpha}$,

2/ $q \varepsilon L^{\infty}_{loc}(R^n)$, real-valued,

3/ Uniform Strong Ellipticity

4/ $a_{\alpha\beta} \varepsilon C^{|\alpha|}(R^n)$ and bounded for $|\alpha|+|\beta| \leq 2m$

5/ $a_{\alpha\beta}$ uniformly continuous for $|\alpha|=|\beta|=m$

Theorem 1.

If $q(x) \geq -q^*(|x|)$, where $q^*(r)$ is non-negative, monotone nondecreasing in $r>0$, and $q^*(r) = 0\left(r^{2m/(2m-1)}\right)$ as $r \to +\infty$, then under the above general assumptions, T is essentially self-adjoint on $C_0^{\infty}(R^n)$.

Comments. The growth rate $q^*(r) = 0\left(r^{2m/(2m-1)}\right)$ was also obtained by R.G. Keller, using different assumptions [3]. It is well-known in the theory of ODE (case n=1) that this growth rate is the best possible. From the point of view of classical mechanics, when the negative part of the potential $q(x)$ drops as $0\left(|x|^{2m/(2m-1)}\right)$, the classical particle will "escape to infinity in finite time."

Before we sketch the proof, we shall state the self-adjointness result. We need to change the general assumptions 2/,4/,5/ to the following: 2'/ $q\varepsilon C^m(R^n)$, $q\geq 1$ (or bounded from below),

4'/ $a_{\alpha\beta}\varepsilon C^{|\alpha|}(R^n)$ and has bounded derivatives.

Theorem 2.

There exists a constant $E>0$, depending on $m,n,a_{\alpha\beta}$ such that if $|D^\alpha q|\leq Eq^{1+\frac{|\alpha|}{2m}}$ with $0<|\alpha|\leq m$, then T is self-adjoint on $H^{2m}(R^N)\cap D(q)$.

Comments. This condition on q is a generalization of a condition for second order operators given by Everitt and Giertz with $G=R^n$ [2]. For $T=-\Delta+q$, the condition becomes $|\nabla q|\leq(2-\varepsilon)q^{3/2}$, where ε is any small positive number. This nonlinear condition on q admits potentials of exponential growth, for example $\exp|x|$, $\exp(\exp|x|)$, etc..., but does not handle oscillatory potentials.

From the point of view of Quantum Mechanics, $-\Delta$ and $q(x)$ represent, respectively, the kinetic and the potential energy. Consider now $T=(-\Delta)^m+q(x)$; if $(-\Delta)^m$ and $q(x)$ are to represent any physical quantity, they must have the same physical unit. Heuristically speaking, if we assign to $(-\Delta)^m$ the unit ℓ^{-2m} (ℓ for length), then $q(x)$ must also have unit ℓ^{-2m}. Consequently, $D^\alpha q$ and $q^{1+\frac{|\alpha|}{2m}}$ will have the same unit $\ell^{-2m-|\alpha|}$. The constant E in theorem 2 is therefore unit free.

We now sketch the proofs of theorems 1 and 2..

II. Proof of theorem 1.

Theorem 1 will be proved through a series of propositions and lemmas. There are three main ingredients in the proof: Gårding's inequality, local elliptic regularity, and an interpolation estimate. We first state the interpolation estimate.

Interpolation lemma.

Given m,n, there exist positive constants E,C with the following properties:

1/ If $0\leq\Psi(x)\varepsilon C_0^\infty(R^n)$, and for $|\nu|=1$, $|D^\nu\Psi(x)|\leq E\Psi^{1-\frac{1}{2m}}(x)$, then for any $|\gamma|\leq m$, and any $u\in H^m_{loc}(R^n)\cap L^2(R^n)$, one has

$$||\Psi^{\frac{|\gamma|}{2m}} D^\gamma u||^2 \leq C\left(||u||^2 + \sum_{|\alpha|=m} ||\Psi^{1/2}D^\alpha u||^2\right) \quad (I.1).$$

2/ If, in addition, for some $K\geq 0, 0<\varepsilon\leq 1$, $\sum_{|\alpha|=m} ||\Psi^{\frac{1}{2}}D^\alpha u||^2 \leq K\varepsilon^{-2m}||u||^2$,

then there exists $K'\geq 0$, depending only on m,u, and K such that for all $\ell\leq m$,

$$\sum_{|\gamma|=\ell} ||\Psi^{\frac{|\gamma|}{2m}} D^\gamma u||^2 \leq K'\ell^{-2\ell}||u||^2 \quad (I.2).$$

Here, $||.||$ denotes the L^2-norm.

Proof.

First, we prove for the case n=1. Suppose $|D\Psi|\leq\varepsilon\Psi^{1-\frac{1}{2m}}$. Let $A_k=||\Psi^{\frac{k}{2m}}D^k u||^2$. Then, using integration by parts, Schwarz's inequality and the inequality $|2ab|\leq a^2+b^2$, we get

$$A_k \leq \frac{\varepsilon}{2}(A_{k-1}+A_k) + \frac{1}{2}(A_{k+1}+A_{k-1}), \text{ for } |\leq|\leq m-1.$$

We rewrite this system in matrix form as follows:

$$M(\varepsilon)\underline{A} = \begin{bmatrix} 1-\frac{\varepsilon}{2} & -\frac{1}{2} & 0 \\ -\frac{1+\varepsilon}{2} & 1-\frac{\varepsilon}{2} \;\diagdown & -\frac{1}{2} \;\diagdown \\ 0 & & \end{bmatrix} \begin{bmatrix} A_1 \\ A_2 \\ \vdots \\ A_{m-1} \end{bmatrix} \leq \begin{bmatrix} \frac{1+\varepsilon}{2} A_0 \\ \vdots \\ 0 \\ \vdots \\ 0 \\ A_m \end{bmatrix} = \underline{B}$$

The matrix $M = \begin{bmatrix} 1 & -\frac{1}{2} & 0 \\ -\frac{1}{2} & 1 \;\diagdown & \diagdown \\ 0 & \diagdown & \end{bmatrix}$ can be easily shown to have a positive inverse (an inverse with positive coefficients). Therefore, the matrix $M(\varepsilon)$ will have a positive for small enough ε. (I.1) follows immediately.

For general n, we integrate by parts one variable at a time, and

and the proof is similar.

(I.2) is proved by making the change of variable $\Phi(x)=\Psi(\varepsilon x)$ and $v(x)=u(\varepsilon x)$, and by applying (I.1) to $||\Phi^{|\gamma|/2m}\, D^{\gamma}v||^2$.

Application of the interpolation lemma.

Fix $\phi\varepsilon C_0^{\infty}(R^n)$ with $0\leq\phi(x)\leq 1$, $\phi(x)=1$ if $|x|\leq \frac{1}{2}$, and $\phi(x)=0$ of $|x|\geq 1$. For $0<\varepsilon\leq 1$, let $\Psi_{\varepsilon}(x)=\phi^{2m}(\varepsilon x)$. For the sake of clarity, we will write Ψ_{ε} as Ψ. It can be shown by induction that for $1\leq\gamma\leq m$,

$$\text{(i)}\quad |D^{\gamma}\Psi| \leq c(\varepsilon)\Psi^{1-\frac{|\gamma|}{2m}}$$

and (ii) $|D^{\gamma}\Psi^{\frac{1}{2}}| \leq c(\varepsilon)\Psi^{\frac{1}{2}-\frac{|\gamma|}{2m}}$, where $c(\varepsilon)=0(\varepsilon)$ as $\varepsilon\to 0$.

Now let $u\varepsilon H^m_{loc}(R^n)\cap L^2(R^n)$, $||\cdot||_m$ denote the Sobolev H^mnorm. Let (,) denote the usual L^2 inner product. Applying (i), (ii) together with the interpolation estimate (I.1), we can prove the following estimates.

Proposition 2.1.

There exist positive constants E, C depending only on n,m, such that whenever $\varepsilon\leq E$, $|\alpha|+|\beta|\leq 2m$,

$$\sum_{|\alpha|=m} ||\Psi^{\frac{1}{2}}D^{\alpha}u||^2 \leq C(||\Psi^{\frac{1}{2}}u||_m^2 + c(\varepsilon)||u||^2) \tag{1}$$

$$|(D^{\beta}u\ ,\ (D^{\gamma-\gamma}\Psi)D^{\gamma}u)| \leq c(\varepsilon)C(||u||^2 + ||\Psi^{\frac{1}{2}}u||_m^2),\ \text{if } \gamma<\alpha \tag{2}$$

$$||(D^{\alpha-\gamma}\Psi^{\frac{1}{2}})D^{\gamma}u||^2\leq c^2(\varepsilon)C(||u||^2 + ||\Psi^{\frac{1}{2}}u||_m^2),\ \text{if } \gamma<\alpha \tag{3}$$

$$||(D^{\alpha-\gamma}\Psi^{\frac{1}{2}})D^{\gamma}u||^2\leq C(||u||^2 + ||\Psi^{\frac{1}{2}}u||_m^2),\ \text{if } \gamma\leq\alpha \tag{4}$$

Local elliptic regularity.

Let T_{max} be the operator T with domain $D(T_{max})=\{u\varepsilon L^2|Tu\varepsilon L^2(R^n)$ as a distribution$\}$.

Let T_{min} be the operator T with domain $D(T_{min})=C_0^{\infty}(R^n)$.

Let T^*_{min} be the adjoint of T_{min}.

Proposition 2.2.

$$D(T_{max}) = D(T^*_{min}) \subset H^m_{loc}(R^n) \cap L^2(R^n).$$

Proof. $D(T_{max}) = D(T^*_{min})$ is obvious from the definitions. The gene- assumption 3/ and 4/ enable us to apply a (more or less) well-known local regularity result ([1], theorem 6.3), and obtain $D(T^*_{min}) \subset H^m_{loc}(R^n)$

Gårding's inequality.

Consider the quadratic form associated with P=T-q defined by

$$B(f) = \sum_{0<|\alpha|,|\beta|\leq m} (a_{\alpha\beta}D^\beta f, D^\alpha f), \qquad \text{for } f\in C_0^\infty(R^n).$$

Because $a_{\alpha\beta}$ are bounded, B can be extended to $H^m(R^n)$. The general assumptions 3/,4/ and 5/ guarantee that Gårding's inequality holds for B, i.e.

$$||f||_m^2 \leq C(B(f) + ||f||^2), \text{ for } f\in H^m(R^n), \text{ (G)},$$

where C only depends on m,n, the ellipticity constant, and $a_{\alpha\beta}$.

Now, let $u\in D(T^*_{min})$, then $u\in H^m_{loc}(R^n)\cap L^2(R^n)$ by proposition 2.2, so that $\Psi^{1/2}u\in H^m(R^n)$. Proposition 2.1 and Gårding's inequality immediately yield

Proposition 2.3.

There exist positive constant E,C, such that, whenever $\varepsilon\leq E, |\alpha|, |\beta|\leq m$,

$$|(D^\beta u, (D^{\alpha-\gamma}\Psi)D^\gamma u)| \leq c(\varepsilon)C(||u||^2 + B(\Psi_u^{1/2})), \text{ if } \gamma<\alpha \qquad (2')$$

$$||(D^{\alpha-\gamma}\Psi^{1/2})D^\gamma u||^2 \leq c^2(\varepsilon)C(||u||^2 + B(\Psi^{1/2}u)), \text{ if } \gamma<\alpha \qquad (3')$$

$$||(D^{\alpha-\gamma}\Psi^{1/2})D^\gamma u||^2 \leq C(||u||^2 + B(\Psi^{1/2}u)), \text{ if } \gamma\leq\alpha \qquad (4').$$

Proof of essential self-adjointness.

It suffices to show $\ker\{T^{*}_{min} \pm i\lambda\}=\{0\}$ for some non zero real λ. Suppose $T^{*}_{min}u = \pm i\lambda u$, we have

$$\mp i\lambda(\Psi u,u) = (\Psi u,\ T^{*}_{min}u)$$

$$= \sum_{\alpha,\beta} (D^{\alpha}(\Psi u), a_{\alpha\beta}\, D^{\beta}u) + (\Psi u, qu) \qquad (*)$$

By comparing the real parts of (*). we deduce

$$B(\Psi^{\frac{1}{2}}u) = -(\Psi u,qu) - \mathrm{Re}\{\text{linear combination of } ((D^{\alpha-\gamma})D^{\gamma}u, a_{\alpha\beta}D^{\beta}u)$$

$$+ \text{linear combination of } ((D^{\alpha-\gamma}\Psi^{\frac{1}{2}})D^{\gamma}u, a_{\alpha\beta}(D^{\beta-\delta}\Psi^{\frac{1}{2}})D^{\delta}u)\} \qquad (L\text{-}1)$$

By comparing the imaginary parts of (*), we get

$$|\lambda|\ ||\Psi^{\frac{1}{2}}u||^{2} = |\mathrm{Im}\ \{\text{linear combination of } ((D^{\alpha-\gamma}\Psi)D^{\gamma}u, a_{\alpha\beta}D^{\beta}u)\}| \qquad (L\text{-}2)$$

To see how proposition 2.3 comes into play, we establish two more estimates: For small enough ε,

$$|B(\Psi^{\frac{1}{2}}u)| \le \text{Constant } (c(\varepsilon)\,||u||^{2} + \int q^{*}\Psi|u|^{2}) \qquad (L\text{-}3)$$

$$|(D^{\beta}u, (D^{\alpha-\gamma}\Psi)D^{\gamma}u)| \le \text{Constant } ||u||^{2} \qquad (L\text{-}4)$$

Here the Constant does not depend on ε nor on λ. (Recall $q^{*}(|x|)$ is the limiting growth rate of $q(x)$). To prove (L-3), we need to use proposition 2.3. To prove (L-4) we need to use the fact that $q^{*}(r)=0(r^{2m/(2m-1)})$, (L-3), Gårding's inequality, proposition 2.1 and the interpolation estimate (I.2).

(L-2) and (L-4) then give

$|\lambda|\ ||\Psi^{\frac{1}{2}}u||^{2} \le$ Constant $||u||^{2}$, where the Constant does <u>not</u> depend on λ <u>nor</u> on ε. So if we choose $|\lambda|$ large enough and let $\varepsilon\to 0$, we must have $||u||=0$.

III. Proof of theorem 2.

The proof of theorem 2 involves Gårding's inequality, global regularity theory, and a criteria for self-adjointness by H. Sohr [4]. The general assumptions 3/ and 4'/ imply that Gårding's inequality holds for P^2 (recall $P=T-q$). Using this fact together with theorem 1, we can prove the following global regularity result.

Proposition 3.1.

Let P_{max} be the operator P with domain $D(P_{max})=\{u\epsilon L^2(R^n)\,|\,Pu\epsilon L^2(R^n)$ as a distribution$\}$. Then P_{max} is self-adjoint and $D(P_{max})=H^{2m}(R^n)$.

Also by Gårding's inequality, we may assume without loss of generality that for $u\epsilon H^{2m}(R^n)$,

$$(P_{max}u,u) \geq \text{Constant } ||u||_m^2 \geq 0.$$

From now on, for clarity, we will denote P_{max} by P.

Criteria for self-adjointness (Sohr).

Let A,B be positive self-adjoint operators on H with scalar product (,). Suppose: 1/ $D(A)\cap D(B)$ dense in H.

2/ $A \geq \lambda > 0$, for some fixed λ

3/ $\mathrm{Re}(Bx,A^{-1}x) \geq -a(x,x)$, for all $x\epsilon D(B)$, where a is a fixed constant with $0\leq a<1$.

Then A + B is self-adjoint on $D(A+B) = D(A)\cap D(B)$.

We apply Söhr's criteria by letting A=q, B=P. We obser that if $|D^\alpha q| \leq \epsilon q^{1+\frac{|\alpha|}{2m}}$, then $p=q^{-1}$ satisfies (i) and (ii) in section II.

So, all the estimates in section II hold with p in place of Ψ and with $u\epsilon C_0^\infty(R^n)$. Computations show that, for $u\epsilon C_0^\infty(R^n)$,

$$(Pu,q^{-1}u) = \{I\} - \{II\} - \{III\},$$

where $\{I\} = B(p^{\frac{1}{2}}u)$ (the quadratic form associated with P)

$\{II\}$ = linear combination of $(a_{\alpha\beta}(D^{\beta-\delta}p^{\frac{1}{2}})D^\delta u,(D^{\alpha-\gamma}p^{\frac{1}{2}})D^\gamma u)$

$\{III\}$ = linear combination of $(a_{\alpha\beta}D^{\beta}u,(D^{\alpha-\gamma}p)D^{\gamma}u)$

Using the estimates in proposition 2.1, we can show that

$$|\{II\}| \leq c(\varepsilon) \text{ Constant } (||u||^2 + ||p^{\frac{1}{2}}u||^2_m)$$

$$|\{III\}| \leq c(\varepsilon) \text{ Constant } (||u||^2 + ||p^{\frac{1}{2}}u||^2_m)$$

So that , if ε is small enough

$$\text{Re}(Pu,q^{-1}u) \geq -a(u,u) \text{ for some } 0 \leq a < 1.$$

Since $C_0^{\infty}(R^n)$ is a core of P and q (by theorem 1), Sohr's criteria is satisfied and T=P+q is self-adjoint on $H^{2m}(R^n) \cap D(q)$.

References

1. Agmon, S. Lectures on Elliptic Boundary Value Problems. New York: D. Van Nostrand, 1965.

2. Everitt, W.N., and M. Giertz. "Inequalities and Separation for Schrödinger-type Operators." Proc. Roy. Soc. Edinburgh 79A (1978), 257-265.

3. Keller, R.G. "The Essential Self-Adjointness of Differential Operators." Proc. Roy. Soc. Edinburgh 82A (1979), 305-344.

4. Sohr, Hermann. "Über die Selbstadjungiertheit von Schrödinger-Operatoren." Math. Z. 160(1978), 255-261.

ASYMPTOTIC THEORY FOR A CRITICAL CLASS OF FOURTH-ORDER DIFFERENTIAL EQUATIONS

M.S.P. EASTHAM

1. In this paper we identify and investigate a critical case in the asymptotic theory of the fourth-order equation

$$(ry'')'' - (py')' + qy = 0 \tag{1.1}$$

on $[X,\infty)$. The critical case is where

$$(pq)'/pq \sim (\text{const.})(q/p)^{\frac{1}{2}} \tag{1.2}$$

as $x \to \infty$, and we show that it represents a borderline between situations where all solutions of (1.1) have a certain exponential character in terms of the coefficients as $x \to \infty$ and where only two solutions have this character. It is not necessary to assume that p, q and r are real-valued.

In the familiar example where

$$r(x) = x^{\alpha}, \quad p(x) = bx^{\beta}, \quad q(x) = cx^{\gamma}, \tag{1.3}$$

(1.2) becomes

$$\beta - \gamma = 2, \tag{1.4}$$

and Devinatz (3, Fig.2; 4, §1) drew special attention to the lack of information in this case in terms of the evaluation of the deficiency index. Subsequently, still referring to (1.3) and (1.4), Eastham (5) considered $b < 0$ and $c > 0$ and obtained an O-estimate of the form $y(x) = O(x^{-\theta})$ for any solution. Anikeeva (1) and Paris and Wood (9) both took $\alpha = 0$ but otherwise gave a full asymptotic analysis of the solutions of (1.1) for all non-zero b and c. The feature of the work of Paris and Wood is that they were able to solve (1.1) explictly in terms of generalized hypergeometric functions when (1.3) and (1.4) hold and $\alpha = 0$. We also mention the recent work of the same authors on a corresponding class of higher-order equations in (8).

The asymptotic methods referred to so far have an ad hoc appearance but, in this paper, we develop the general asymptotic theory formulated recently by Eastham and Grudniewicz (6), and which has already been shown to cover (1.3) and (1.4) in (7), and apply it to (1.1) with general coefficients. Our aim is to draw out the significance of (1.2) and to obtain the asymptotic form of four solutions of (1.1) in this case. Except for one delicate situation, our results cover those of Paris and Wood (9) and, avoiding the use of generalized hypergeometric functions, we are able to put the asymptotic results of (9) into a wider context.

2. We begin by following the general procedure of (6, §§2-3) and write (1.1) in the quasi-derivative form

$$Y' = AY, \tag{2.1}$$

where Y has components y, y', ry", (ry")' - py'. The next step is to diagonalize A by writing $A = T^{-1}\Lambda_1 T$, where

$$T = (v_1 \quad v_2 \quad v_3 \quad v_4) \tag{2.2}$$

and
$$\Lambda_1 = dg(\mu_1, \mu_2, \mu_3, \mu_4). \tag{2.3}$$

Here the v_j and μ_j are the eigenvectors and eigenvalues of A.

The transformation
$$Y = TZ \tag{2.4}$$

then takes (2.1) into

$$Z' = (\Lambda_1 - T^{-1}T')Z. \tag{2.5}$$

The μ_j are the roots of the characteristic equation of A, which is
$$r\mu^4 - p\mu^2 + q = 0.$$

Hence, writing
$$D = (p^2 - 4qr)^{\frac{1}{2}},$$

we have

$$\mu_1 = \{(p + D)/2r\}^{\frac{1}{2}}, \quad \mu_2 = -\mu_1, \tag{2.6}$$

$$\mu_3 = \{(p - D)/2r\}^{\frac{1}{2}}, \quad \mu_4 = -\mu_3. \tag{2.7}$$

Also, v_j can be taken to have components

$$1, \ \mu_j, \ r\mu_j^2, \ q\mu_j^{-1}.$$

The matrix T^{-1}, which occurs in (2.5), has rows $m_j^{-1}r_j$, where r_j is the row vector

$$(q\mu_j^{-1} \quad r\mu_j^2 \quad \mu_j \quad 1)$$

and

$$\left.\begin{aligned} m_1 &= 2\mu_1 D, \quad m_2 = 2\mu_2 D = -m_1, \\ m_3 &= -2\mu_3 D, \quad m_4 = -2\mu_4 D = -m_3. \end{aligned}\right\} \tag{2.8}$$

We then find that, in (2.5), $T^{-1}T' = (t_{jk})$, where

$$m_j t_{jk} = \tfrac{1}{2}m_k' + (\mu_j^2 - \mu_k^2)r\mu_k' + (\mu_j - \mu_k)(r\mu_k^2)'. \tag{2.9}$$

3. We have next to work out t_{jk} in some detail and we impose the following conditions on p, q and r.

(i) p, q and r are nowhere zero on $[X,\infty)$,

(ii) $qr = o(p^2)$ as $x \to \infty$,

(iii) $\delta^{\frac{1}{2}}p'/p$, $\delta^{\frac{1}{2}}q'/q$, $\delta^{\frac{1}{2}}r'/r$ are $L(X,\infty)$, where

$$\delta = qr/p^2 = o(1).$$

Using (i) and (ii) in (2.6) and (2.7), we obtain

$$\mu_1 = (p/r)^{\frac{1}{2}}\{1 + O(\delta)\} \sim (p/r)^{\frac{1}{2}}, \tag{3.1}$$

$$\mu_3 = (q/p)^{\frac{1}{2}}\{1 + O(\delta)\} \sim (q/p)^{\frac{1}{2}}, \tag{3.2}$$

and we note that $\mu_3 = o(\mu_1)$. Thus μ_1 and μ_2 dominate μ_3 and μ_4 as $x \to \infty$.

After a calculation involving (2.8), (ii) and (iii), we obtain from (2.9)

$$-T^{-1}T' = T_1 + R_1,$$

where R_1 is $L(X,\infty)$ and

$$T_1 = \begin{pmatrix} \xi & \eta & 0 & 0 \\ \eta & \xi & 0 & 0 \\ \zeta & \zeta & \theta & -\theta \\ \zeta & \zeta & -\theta & \theta \end{pmatrix} \tag{3.3}$$

$$\xi = \tfrac{3}{4}\frac{p'}{p} - \tfrac{1}{4}\frac{r'}{r}, \quad \eta = -\tfrac{1}{4}\frac{(pr)'}{pr}, \quad \zeta = \tfrac{1}{2}\frac{p'}{p}, \quad \theta = -\tfrac{1}{4}\frac{(pq)'}{pq}. \tag{3.4}$$

4. With (2.5) now in the form

$$Z' = (\Lambda_1 + T_1 + R_1)Z, \tag{4.1}$$

we have to carry out a second diagonalization, this time of the matrix $\Lambda_1 + T_1$. It is here that the critical case (1.2) arises. By (3.2), (3.3) and (3.4), the meaning of (1.2) is that μ_3 and μ_4 do not dominate θ in $\Lambda_1 + T_1$ and, in fact, they have the same order of magnitude as θ when $x \to \infty$.

To diagonalize $\Lambda_1 + T_1$, we require further conditions on p, q, r.

(iv) As $x \to \infty$, let

$$p'/p = O\{(q/p)^{\frac{1}{2}}\}, \quad r'/r = O\{(q/p)^{\frac{1}{2}}\},$$
$$(pq)'/pq \sim 4\sigma(q/p)^{\frac{1}{2}}, \tag{4.2}$$

where σ is a non-zero constant (the factor 4 is introduced only for convenience).

(v) Let

$$\{(pq)'/(p^{\frac{1}{2}}q^{\frac{3}{2}})\}', \tag{4.3}$$
$$p''r^{\frac{1}{2}}/p^{\frac{3}{2}}, \quad r''/(rp)^{\frac{1}{2}}$$

all be $L(X,\infty)$.

If in (4.2) we write

$$(pq)'/pq = 4\sigma(q/p)^{\frac{1}{2}}(1 + \phi), \tag{4.4}$$

then (4.3) implies that ϕ' is $L(X,\infty)$.

We can now write

$$\Lambda_1 + T_1 = S^{-1}\Lambda S,$$

where

$$\Lambda = dg(\lambda_1, \lambda_2, \lambda_3, \lambda_4)$$

and, after a calculation using conditions (i)-(v),

$$\left.\begin{aligned} \lambda_1 &= \mu_1 + \xi + r_1 \\ \lambda_2 &= \mu_2 + \xi - r_1 \end{aligned}\right\} \tag{4.5}$$

$$\left.\begin{aligned} \lambda_3 &= \theta + \surd(\mu_3^2 + \theta^2) \\ \lambda_4 &= \theta - \surd(\mu_3^2 + \theta^2) \end{aligned}\right\} \tag{4.6}$$

$$S \to \mathrm{dg}\left(\begin{pmatrix} 1 & 0 \\ 0 & 1 \end{pmatrix}, \begin{pmatrix} 1 & -a \\ a & 1 \end{pmatrix}\right) \tag{4.7}$$

as $x \to \infty$,

$$S' \text{ is } L(X,\infty), \tag{4.8}$$

$$r_1 \text{ is } L(X,\infty),$$

and

$$a = \sigma/\{1 + \surd(1 + \sigma^2)\} \neq \pm 1. \tag{4.9}$$

The only two λ_j which could coincide are λ_3 and λ_4 but, by (3.2), (3.4) and (4.2), we avoid this possibility by assuming that

$$\sigma^2 + 1 \neq 0. \tag{4.10}$$

5. The transformation

$$Z = SU$$

takes (4.1) into

$$U' = (\Lambda + S^{-1}R_1S - S^{-1}S')U$$

which, by (4.7) and (4.8), has the standard form covered by (2, p.88). Hence there are solutions

$$U_j(x) = \{e_j + o(1)\}\exp\left(\int_X^x \lambda_j(t)\,dt\right) \quad (1 \leq j \leq 4), \tag{5.1}$$

where e_j is the coordinate vector with j-th component unity and other components zero. Transforming back to (1.1), we obtain four solutions y_j with the asymptotic forms

$$y_j(x) \sim \exp\left(\int_X^x \lambda_j(t)\,dt\right) \quad (j = 1, 2) \tag{5.2}$$

$$\left.\begin{aligned} y_3(x) &\sim (1 + a)\exp\left(\int_X^x \lambda_3(t)\,dt\right) \\ y_4(x) &\sim (1 - a)\exp\left(\int_X^x \lambda_4(t)\,dt\right), \end{aligned}\right. \tag{5.3}$$

where a is as in (4.9).

By (4.5) and (3.4), we obtain from (5.2)

$$y_j(x) \sim \{p^3(x)/r(x)\}^{\frac{1}{4}}\exp\left(\int_X^x \mu_j(t)\,dt\right) \qquad (j = 1, 2), \tag{5.4}$$

after adjusting a constant multiple in y_1 and y_2, where μ_1 is given by (3.1) and $\mu_2 = -\mu_1$. To deal with y_3 and y_4, we write (4.6) as

$$\lambda_3, \lambda_4 = -\tfrac{1}{4}\frac{(pq)'}{pq}[1 \pm \sqrt{\{1 + \sigma^{-2}(1+\phi)^{-2}\}}] + r_2$$

$$= -\tfrac{1}{4}\frac{(pq)'}{pq}\left\{1 \pm \left(\sqrt{(1+\sigma^{-2})} - \frac{\sigma^{-2}}{\sqrt{(1+\sigma^{-2})}}\phi + O(\phi^2)\right)\right\} + r_2$$

on using (3.2) and (4.4) and terminating the binomial expansion at the ϕ^2 term for example. Here r_2 is $L(X,\infty)$. Hence, assuming that

$$\phi^2(pq)'/pq \text{ is } L(X,\infty) \tag{5.5}$$

and again adjusting constant multiples in y_3 and y_4, from (5.3) we obtain

$$\begin{aligned} y_3(x) &\sim \{p(x)q(x)\}^{-\frac{1}{4}\{1+\sqrt{(1+\sigma^{-2})}\}}\exp I(x) \\ y_4(x) &\sim \{p(x)q(x)\}^{-\frac{1}{4}\{1-\sqrt{(1+\sigma^{-2})}\}}\exp\{-I(x)\}, \end{aligned} \tag{5.6}$$

where

$$I(x) = \tfrac{1}{4}\sigma^{-2}(1+\sigma^{-2})^{-\frac{1}{2}}\int_X^x \phi(pq)'/pq\,dt. \tag{5.7}$$

If, further to (5.5), $\phi(pq)'/pq$ is $L(X,\infty)$, the exponential factors in (5.6) can be replaced by unity. Thus (5.4) and (5.6) are our main asymptotic results for (1.1). It should be added that the general result (5.1) requires the further condition that $\mathrm{Re}\{\lambda_j(x) - \lambda_k(x)\}$ does not change sign in $[X,\infty)$, and this condition is certainly satisfied if, for example, p, q, and r are real-valued.

6. In the case (1.3), we have noted that (1.2) becomes (1.4). The value of σ in (4.2) is then

$$\sigma = \tfrac{1}{2}(\gamma + 1)(b/c)^{\frac{1}{2}}$$

and (4.10) becomes

$$c/b \neq -\tfrac{1}{4}(\gamma + 1)^2.$$

Also $\phi = 0$ in (4.4) and then (5.6) gives

$$y_3(x),\ y_4(x) \sim x^{-[\gamma + 1 \pm \sqrt{\{(\gamma+1)^2 + 4c/b\}}]/2}$$

after adjustment of constant multiples. These formulae, together with (5.4), agree with the results of Paris and Wood (9, Table 1) in the case $\alpha = 0$.

A quite different situation covered by our analysis is

$$r(x) = x^{\alpha} \exp x^{B}, \quad p(x) = bx^{\beta} \exp x^{C}, \quad q(x) = cx^{\gamma} \exp x^{C},$$

where the constants are real with $B \geqslant 0$ and $B < C$. Then the critical situation (4.2) is given by

$$\gamma - \beta = 2(C - 1).$$

The value of σ in (4.2) is $\sigma = \tfrac{1}{2}C(b/c)^{\frac{1}{2}}$, and

$$\phi(x) = \tfrac{1}{2}(\beta + \gamma)C^{-1}x^{-C}.$$

Then (5.6) and (5.7) give

$$y_3(x) \sim x^{-\frac{1}{4}(\beta+\gamma)\{1 + 1/\sqrt{(1+\sigma^{-2})}\}} \exp[-\tfrac{1}{2}\{1 + \sqrt{(1 + \sigma^{-2})}\}x^{C}]$$

$$y_4(x) \sim \exp[\tfrac{1}{2}\{\sqrt{(1 + \sigma^{-2})} - 1\}x^{C}].$$

As a final general comment, we note that the real difference between (5.4) and (5.6) is that the right-hand side of (5.4) involves p, q and r in basically an exponential manner whereas (5.6), especially in the case where exp I(x) can be replaced by unity, has an algebraic nature in terms of the coefficients. The reason for this algebraic nature is (1.2): had μ_3 and μ_4 dominated $(pq)'/pq$ as $x \to \infty$, y_3 and y_4 would also have had the exponential nature. Thus (1.2) represents a borderline between situations where all solutions of (1.1) have asymptotically an exponential behaviour in terms of p, q, r and where two solutions have an algebraic behaviour.

References

1. L.I. Anikeeva, Vestnik Moscow Univ. 6 (1976) 44-52.

2. W.A. Coppel, Stability and asymptotic behaviour of differential equations (Heath, Boston, 1965).

3. A. Devinatz, Quart. J. Math. (Oxford) 23 (1972) 267-286.

4. A. Devinatz, J. London Math. Soc. 7 (1973) 135-146.

5. M.S.P. Eastham, Proc. Roy. Soc. Edinburgh 79A (1977) 51-59.

6. M.S.P. Eastham and C.G.M. Grudniewicz, Lecture Notes in Mathematics 846 (Springer, Berlin, 1981) 88-99.

7. C.G.M. Grudniewicz, London Univ. Ph.D. thesis, 1980.

8. R.B. Paris and A.D. Wood, Proc. Roy. Soc. Edinburgh 90A (1981) 209-236.

9. R.B. Paris and A.D. Wood, Quart. J. Math. (Oxford) 33 (1982) 97-113.

OSCILLATION AND NONOSCILLATION THEOREMS FOR SOME NON-LINEAR ORDINARY DIFFERENTIAL EQUATIONS

Á. ELBERT

In this paper we consider the half-linear second order differential equations

(1) $(p(x')^{\overset{*}{n}})' + nqx^{\overset{*}{n}} = 0 \ , \quad (' = \frac{d}{dt}) \ ,$

and the nonlinear first order differential equation systems

$$y' = ay + bz^{\overset{*}{\frac{1}{n}}}$$

(2)

$$z' = -cy^{\overset{*}{n}} + dz \ ,$$

where the number n is positive, the functions p, q, a, b, c, d of the independent variable t are continuous on $[t^0,\infty)$ for some $t^0 \in R$ and x, y, z are solutions of these differential equations, and the star above the exponent in $x^{\overset{*}{n}}$ denotes that the power functior $x^{\overset{*}{n}}$ preserves the sign of x, i.e. $x^{\overset{*}{n}} = |x|^n \operatorname{sign} x$. The usual condition on p is

(3) $p(t) > 0 \ .$

We emphasize here that we do not assume $q(t)\geq 0$ in (1). By the substitution $y=x$ and $z=px'^{\overset{*}{n}}$ the differential equation (1) can be turned into a system of the form (2) with $a=d=0$, $b=1/p^{\frac{1}{n}}$, $c=nq$. Henceforth by (3) we assume always $b>0$ in (2) and we allow the possibility $\int^{\infty} b<\infty$, too. A quantitative study of the solutions of (1) or (2) can be found in [1] and with the restrictions $a=d=0$ also in [3]-[6].

In [1] (see Theorem 1 and Theorem 2) we proved that the solutions of (1) (or (2)) have the interlacing property of the zeros, i.e. the zeros of the linearly independent solutions separate each other. Hence if there is a solution $x(t)$ of (1) (or $(y(t),z(t))$ of (2) (with infinitely many zeros t_i , $i=1,2,...$ such that $x(t_i)=0$ (or $y(t_i)=0$ for the system) and $\lim_{i\to\infty} t_i=\infty$, then every solution has also infinitely many zeros and in this case we say that the differential equation (1) (or the system (2)) is *oscillatory*. On the other hand if there exists a solution $x(t)$ (or $(y(t),z(t))$) which has only finitely many zeros or it has none then there is an interval $[t_0,\infty)$, $t_0\geqq t^0$, such that $x(t)$ (or $y(t)$) has no zeros on $[t_0,\infty)$. This interval is called *disconjugacy interval* and the differential equations (1) (or system (2)) is *disconjugate* there. Then by the interlacing property of the zeros every other solution may have at most one zero on this interval. In this case the differential equation (1) (or system (2)) is *nonoscillatory*.

Our main result will concern the nonoscillatory case and the method is a generalization of Hartman's method applied to the linear second order differential equations (see [2]). By negation we obtain oscillation criterion for (1) and (2). A corollary is given in which we generalize the wellknown oscillation criterion due to A. Wintner [7]. In order to get a generalized version of a theorem of M. Zlamal [8] we make use of a method owing to J. D. Mirzov [5] which yields a rather sharp oscillation criterion.

The Example 1 shows how we can deduce the corresponding Mirzov's results from our results. In our treatment an essential role is played by the so-called admissible pairs (λ,μ) of the functions $\lambda(t),\mu(t)$. The admissibility depends on the coefficients of the system (2). To apply our theorems we must have such admissible pairs in advance. In Example 2 we give such admissible pairs. Finally in Example 3 we apply our results to (1) in the special case when the coefficients p, q are periodic.

For the sake of brevity the proofs will be carried out only for the system (2) since the results can be converted easily into the corresponding ones for differential equation (1).

First we generalize the Ricatti equations to the system (2). Let the system (2) be nonoscillatory and the interval $[t_0, \infty)$, $t_0 \geq t^0$ be a disconjugacy interval and

$(y(t),z(t))$ be a solution of (2) such that $y(t)\neq 0$ for $t\geqq t_0$. Let the function $r=r(t)$ be defined by

$$(4)\qquad r = \frac{z}{y^{\overset{*}{n}}} .$$

Then r is continuous and satisfies the generalized Riccati equation

$$(5)\qquad r' + (na-d)r + nb|r|^{1+\frac{1}{n}} + c = 0 .$$

This follows immediately by differentiating (4) and by making use of (2).

Let the function $\lambda^*(t)$ be introduced by

$$(6)\qquad \lambda^* = \exp\left(\int_{t^0}^{t} (na-d)\right) .$$

Then we define the set L of the admissible pairs (λ,μ) of the functions $\lambda(t)$, $\mu(t)$ by the following restrictions:

(i) $\lambda(t)$, $\mu(t)$ are continuous, positive and λ^0 is continuously differentiable on $[t^0,\infty)$;

$$(7)\quad \text{(ii)}\ \int_{t_0}^{\infty} \frac{\lambda}{b^n} \left|\frac{\lambda^{*'}}{\lambda^*} - \frac{\lambda'}{\lambda}\right|^{n+1} < \infty ;$$

$$\text{(iii)}\ \lim_{T\to\infty} \int_{t_0}^{T} \mu = \infty ;$$

$$\text{(iv)}\ \limsup_{T\to\infty} \frac{\int_{t_0}^{T} \mu^{n+1} \frac{\lambda}{b^n}}{\left(\int_{t_0}^{T} \mu\right)^{n+1}} < \infty .$$

The existence of the set L depends heavily on the coefficients a, b, d of the system (2) and we suppose that it is not empty, moreover, for the sake of convenience, that there are functions μ such that $(\lambda^*,\mu)\in L$.

Concerning the fourth coefficient c we shall investigate the behaviour of the function $H(T)$ defined by

$$(8)\qquad H = \frac{\int_{t_o}^{T} \mu(\int_{t_o}^{t} \lambda c)dt}{\int_{t_o}^{T} \mu} .$$

This function can be considered as an average of the function $f(t)=\int_{t_o}^{t} \lambda c$. It is not difficult to show that if the relation $\lim_{t\to\infty} f(t)=\tilde{C}$ holds where $\tilde{C}$ may be finite or infinite then $\lim_{T\to\infty} H(T)=\tilde{C}$. This property will be referred later as an averaging property of $H(T)$.

Finally we shall make use of the inequality

$$(9)\qquad |u^{\overset{*}{n}} v| \leqq \frac{n}{n+1} |u|^{n+1} + \frac{1}{n+1}|v|^{n+1} ,$$

which is a generalization of the wellknown inequality between the geometric and the quadratic means.

Our first observation is formulated in form of a lemma.

LEMMA. Let the system (2) be nonoscillatory and let $(y(t),z(t))$ be a solution such that $y(t)\neq 0$ on $[t_o,\infty)$ with some $t_o \geqq t^o$. Let the function $r(t)$ be given by (4). If for some function λ of a pair $(\lambda,\mu)\in L$ the inequality

$$(10)\qquad \int_{t_0}^{\infty} \lambda b|r|^{1+\frac{1}{n}} < \infty$$

holds, then with the corresponding μ the function $H(T)$ defined by (8) is bounded on $[t_0, \infty)$. If $\lambda=\lambda^*$, then the limit $\lim_{T\to\infty} H(T)=C$ exists and is finite.

Proof. The function r satisfies the Riccati differential equation (5). Multiplying (5) by λ and integrating over $[t_0, t]$ we have

$$(11)\qquad r\lambda+\int_{t_0}^{t} r[\lambda\frac{(\lambda^*)'}{\lambda^*}-\lambda']+n\int_{t_0}^{t} b\lambda|r|^{1+\frac{1}{n}}+\int_{t_0}^{t}\lambda c-r(t_0)\lambda(t_0)=0$$

since by (6) $na-b=(\lambda^*)'/\lambda^*$. Put $u=(\frac{n-\varepsilon}{n}(n+1)b\lambda)^{\frac{1}{n+1}} r^{\frac{1}{n}}$

and $v=[\lambda(\lambda^*)'/\lambda^*-\lambda']\,(\frac{n-\varepsilon}{n}(n+1)\,b\lambda)^{-n/n+1}$ in (9) with $0<\varepsilon<n$ then

$$\left|r[\lambda\frac{(\lambda^*)'}{\lambda^*}-\lambda']\right| \leqq (n-\varepsilon)b\lambda|r|^{\frac{n+1}{n}}+\gamma(\varepsilon)\frac{\left|\lambda\frac{(\lambda^*)'}{\lambda^*}-\lambda'\right|^{n+1}}{(b\lambda)^n}$$

where $\gamma(\varepsilon)=\dfrac{(\frac{n}{n-\varepsilon})^n}{(n+1)^{n+1}}$ hence

$$(12)\qquad \left|r\lambda+n\int_{t_0}^{t} b\lambda|r|^{1+\frac{1}{n}}+\int_{t_0}^{t}\lambda c-r(t_0)\lambda(t_0)\right| \leqq \int_{t_0}^{t}|r|\cdot\left|\lambda\frac{(\lambda^*)'}{\lambda^*}-\lambda'\right| \leqq$$

$$\leqq (n-\varepsilon)\int_{t_0}^{t} b\lambda|r|^{\frac{n+1}{n}}+\gamma(\varepsilon)\int_{t_0}^{t}\frac{\left|\lambda\frac{(\lambda^*)'}{\lambda^*}-\lambda'\right|^{n+1}}{(\lambda b)^n}\ .$$

First we show that $H(T)$ in (8) is bounded from above. From (12) it follows that

$$(13)\quad r\lambda+\varepsilon\int_{t_0}^{t} b\lambda|r|^{1+\frac{1}{n}} + \int_{t_0}^{t}\lambda c - r(t_0)\lambda(t_0) \leqq \gamma(\varepsilon)\int_{t_0}^{t}\frac{\left|\lambda\frac{(\lambda^*)'}{\lambda^*}-\lambda'\right|^{n+1}}{(b\lambda)^n},$$

hence by (7) (ii), (10)

$$(14)\qquad \int_{t_0}^{t}\lambda c - C_1 \leqq \lambda|r|$$

where

$$C_1 = r(t_0)\lambda(t_0)+\gamma(\varepsilon)\int_{t_0}^{\infty}\frac{\left|\lambda\frac{(\lambda^*)'}{\lambda^*} - \lambda'\right|^{n+1}}{(b\lambda)^n}.$$

Multiplying (14) by μ and integrating over $[t_0, T]$ we obtain by (8)

$$(15)\quad H(T) - C_1 \leqq \frac{\int_{t_0}^{T}\lambda\mu|r|}{\int_{t_0}^{T}\mu} = L(T).$$

Concerning the function $L(T)$ we shall need two relations. The first is a simple consequence of the Hölder inequality:

(16) $$0 \leqq L(T) \leqq \left\{\frac{\int_{t_o}^{T} \lambda \frac{\mu^{n+1}}{b^n}}{\left(\int_{t_o}^{T} \mu\right)^{n+1}}\right\}^{\frac{1}{n+1}} \cdot \left\{\int_{t_o}^{T} b\lambda |r|^{1+\frac{1}{n}}\right\}^{\frac{n}{n+1}}$$

To get the second relation we choose the number T_1 arbitrary but $T_1 > t_o$. Then again by the Hölder inequality we get

$$L(T) \leqq \frac{\int_{t_o}^{T_1} \lambda\mu|r|}{\int_{t_o}^{T} \mu} + \left\{\frac{\int_{T_1}^{T} \lambda \frac{\mu^{n+1}}{b^n}}{\left(\int_{t_o}^{T} \mu\right)^{n+1}}\right\}^{\frac{1}{n+1}} \left\{\int_{T_1}^{T} b\lambda |r|^{1+\frac{1}{n}}\right\}^{\frac{n}{n+1}} .$$

Hence by (7) (iii)-(iv)

$$\limsup_{T\to\infty} L(T) \leqq \left\{\int_{T_1}^{\infty} b\lambda |r|^{1+\frac{1}{n}}\right\}^{\frac{n}{n+1}} \cdot \left\{\sup_{T>T_1} \frac{\int_{t_o}^{T} \lambda \frac{\mu^{1+n}}{b^n}}{\left(\int_{t_o}^{T} \mu\right)^{n+1}}\right\}^{\frac{1}{n+1}} ,$$

and the condition (10) implies the wanted second relation by letting $T_1 \to \infty$

(17) $$\lim_{T\to\infty} L(T) = 0 .$$

Hence by (15) we have $\limsup_{T\to\infty} H(T) \leq C_1$.

To get a lower estimate for $H(T)$ we need the second inequality involved in (12)

$$r\lambda+n\int_{t_0}^{t} b\lambda|r|^{1+\frac{1}{n}}+\int_{t_0}^{t}\lambda c-r(t_0)\lambda(t_0) \geqq -(n-\varepsilon)\int_{t_0}^{t} b\lambda|r|^{1+\frac{1}{n}}-\gamma(\varepsilon)\int_{t_0}^{t} \frac{\left|\lambda\frac{(\lambda^*)'}{\lambda^*}-\lambda\right|^{n+1}}{(b\lambda)^n}$$

hence

$$\int_{t_0}^{t} \lambda c - C_2 \geqq |r|\lambda$$

where

$$C_2=r(t_0)\lambda(t_0)-(2n-\varepsilon)\int_{t_0}^{\infty} b\lambda|r|^{1+\frac{1}{n}}-\gamma(\varepsilon)\int_{t_0}^{\infty} \frac{\left|\lambda\frac{(\lambda^*)'}{\lambda^*}-\lambda'\right|^{n+1}}{(b\lambda)^n},$$

and as above it follows

$$H(T) - C_2 \geqq -L(T) ,$$

thus by (17) $\liminf_{T\to\infty} H(T)\geqq C_2$, which, together with the above relation $\limsup_{T\to\infty} H(T)\leqq C_1$, proves the first part of the LEMMA.

In the case $\lambda=\lambda^*$ we have by (11)

$$(18) \qquad 0 = r\lambda^* - n \int_{t}^{\infty} b\lambda^*|r|^{1+\frac{1}{n}} + \int_{t_0}^{t} \lambda^* c - C$$

where

$$C = r(t_0)\lambda^*(t_0) - n\int_{t_0}^{\infty} b\lambda^*|r|^{1+\frac{1}{n}} .$$

Multiplying by μ and integrating over $[t_o , T]$ for $T>t_o$

$$|H(T)-C| \leq \frac{\int_{t_o}^{T} n\mu(\int_t^{\infty} b\lambda^*|r|^{1+\frac{1}{n}})dt}{\int_{t_o}^{T}\mu} + \frac{\int_{t_o}^{T}\mu\lambda^*|r|}{\int_{t_o}^{T}\mu}$$

For the function $\int_t^{\infty} b\lambda^*|r|^{1+\frac{1}{n}}$ tends to zero as $t\to\infty$ hence by the averaging property of the function $H(T)$ the first term on the right hand side tends to zero, while the second term is $L(T)$ by (15) and owing to (17) it also tends to zero, therefore $\lim_{T\to\infty} H(T)=C$, which completes the proof of the LEMMA.

It is an interesting fact that under simple conditions a conversion of this LEMMA is valid.

<u>THEOREM 1.</u> Let the system (2) be nonoscillatory and disconjugate on $[t_o , \infty)$, and the pair of the functions (λ,μ) be admissible for (2). If the function $H(T)$ defined by (8) fulfils the relation

$$(19) \qquad \liminf_{T\to\infty} H(T) > -\infty$$

then the inequality (10) in the Lemma is valid and the function $H(T)$ is bounded on $[t_o , \infty)$. Moreover in the special case $\lambda=\lambda^*$

$$\lim_{T\to\infty} H(T) = C$$

exists (as a finite number).

Proof. As in the beginning of the proof of the Lemma we have also a solution (y,z) of (2) with $y \neq 0$ on $[t_0, \infty)$ and then a continuous function r which satisfies the inequality (13). Let the functions $S(T)$ and $M(T)$ be introduced for $T > t_0$ by

$$(20) \qquad S = \int_{t_0}^{T} \mu\Big(\int_{t_0}^{t} b\lambda |r|^{1+\frac{1}{n}}\Big)\, dt, \qquad M = \int_{t_0}^{T} \mu \,.$$

By (7) (iii) $\lim\limits_{T\to\infty} M = \infty$. Our proof will be indirect. We assume that the inequality (10) is not valid, i.e.

$$(21) \qquad \lim_{T\to\infty} \int_{t_0}^{T} b\lambda |r|^{1+\frac{1}{n}} = \infty \,.$$

By the averaging property of the function $H(T)$ we have then

$$(22) \qquad \lim_{T\to\infty} \frac{S(T)}{M(T)} = \infty \,,$$

and

$$(23) \qquad \lim_{T\to\infty} S(T) = \infty \,.$$

By the inequality (7) (ii) we can write (13) as

$$r\lambda + \varepsilon \int_{t_0}^{t} b\lambda |r|^{1+\frac{1}{n}} \leqq C_1 - \int_{t_0}^{t} \lambda c \,,$$

where the constant C_1 is the same as in (14). A multiplication of this inequality by μ and a quadrature gives by (8)

$$\frac{\int_{t_o}^{T} \lambda\mu r}{M(T)} + \varepsilon \frac{S(T)}{M(T)} \leqq C_1 - H(T) .$$

According to the assumption (19) the right hand side is bounded from above, therefore by (22) it will be less than $\frac{1}{2}\varepsilon S(T)/M(T)$ for $T>T_1$ with some T_1 sufficiently large. Thus we get

$$(24) \qquad \frac{1}{2}\varepsilon \frac{S(T)}{M(T)} < \frac{\int_{t_o}^{T} \lambda\mu|r|}{M(T)} = L(T) \qquad \text{for} \quad T>T_1$$

where the function $L(T)$ is already known from (15). Since by (20)

$$S' = \mu(T) \int_{t_o}^{T} b\lambda|r|^{1+\frac{1}{n}} , \quad M'=\mu(T) ,$$

the estimate (16) implies

$$(25) \qquad L^{\frac{n+1}{n}} \leqq \left\{\frac{\int_{t_o}^{T} \frac{\lambda\mu^{n+1}}{b^n}}{M^{n+1}}\right\}^{\frac{1}{n}} \frac{S'}{M'} .$$

By (7) (iv) we have sufficiently large T_1 and N such that

$$(26) \qquad \frac{\int_{t_o}^{T} \frac{\lambda\mu^{n+1}}{b^n}}{\left(\int_{t_o}^{T} \mu\right)^{n+1}} < N \qquad \text{for} \quad T>T_1 .$$

Combining this with (24) and (25) we get

$$(27)\qquad \gamma_1 M' \, M^{-\frac{n+1}{n}} < N^{\frac{1}{n}} S' S^{-\frac{n+1}{n}} \quad \text{for} \quad T > T_1 ,$$

where γ_1 is a positive constant depending only on n and ε. An integration of (27) over $[T,\infty)$ gives for $T > T_1$ by (23) and by (7) (iii)

$$\gamma_1 M^{-\frac{1}{n}} < N^{\frac{1}{n}} S^{-\frac{1}{n}} ,$$

hence

$$\frac{S}{M} < N\gamma_1^{-n} \qquad \text{for} \quad T>T_1 ,$$

which contradicts to (21), and the indirect assumption (20) is not true, i.e. the relation (10) is valid. Thus we can apply the LEMMA and this application completes the proof of Theorem 1.

A criterion sufficient for oscillation of the solutions of the system (2) is formulated in the next theorem.

<u>THEOREM 2.</u> Let (λ,μ) be an admissible pair for the system (2). If for some $t_0 \geqq t^0$ the relation $\lim_{T\to\infty} H(T)=\infty$ holds then the system (2) is oscillatory. If for an admissible pair (λ^*,μ) the relations $\limsup_{T\to\infty} H(T) > \liminf_{T\to\infty} H(T) > -\infty$ hold then the system (2) is oscillatory.

<u>Remark.</u> It is not difficult to show that the limits here are independent of the choice of the value t_0.

Proof. Suppose the contrary, i.e. the system (2) be nonoscillatory. By the assumptions on $H(T)$ the condition (19) is fulfilled, hence all the conditions of Theorem 1 are satisfied. Then by virtue of Theorem 1 the limit of the function $H(T)$, if any, had to be finite, which is the desired contradiction.

The criterion for oscillation in Theorem 2 may be simplified in special cases to one which contains only single integral.

Corollary. Let λ be a function such that there exists at least one function μ satisfying $(\lambda,\mu)\in L$. If the relation

$$\lim_{T\to\infty} \int_{t_0}^{T} \lambda c = \infty \qquad \text{for some } t_0 \geqq t^0 \tag{28}$$

holds then the system (2) is oscillatory.

Proof. Let us consider the function $H(T)$ for $T>t_0$. By the averaging property of $H(T)$ the limit in (28) yields the same limit for $H(T)$, i.e. $\lim H(T)=\infty$. Then by Theorem 2 the system(2)can be only oscillatory.

Following a remark in [2] (see on page 365) a more stringent criterion for nonoscillation can be established under somewhat stronger restrictions on the pair (λ,μ), because we have not exploited all the properties of the method used in the Lemma.

THEOREM 3. Let the system (2) be nonoscillatory and be disconjugate on $[t_0,\infty)$. Let (λ^*,μ) be a pair of the functions λ^* and μ satisfying the conditions (7) (i), (iii) and

(7) (iv)' $$\limsup_{t\to\infty} \frac{\frac{(\lambda^*)^{\frac{1}{n}}\mu}{b}}{\int_{t_0}^{t}\mu} < \infty .$$

Moreover let the relation (19) be valid. Then the function $H(T)$ is convergent, $\lim_{T\to\infty} H(T)=C$, where C is finite and

(29) $$\lim_{T\to\infty} \frac{\int_{t_0}^{T}\mu\left|C-\int_{t_0}^{t}\lambda^* c\right|^{\frac{n+1}{n}}}{\int_{t_0}^{T}\mu} = 0 .$$

<u>Proof.</u> We want to apply Theorem 1 to this case. First we show that the pair (λ^*,μ) with the restrictions of Theorem 3 is admissible, i.e. it fulfils (7) (iv), too. By condition (7) (iv)' we have for sufficiently large N and T_1 that

(30) $$\frac{\frac{\lambda^*\mu^n}{b^n}}{M^n(t)} < N \qquad \text{for all } t>T_1 ,$$

where the function M is the same as in (20). Since $M'(t)=$ $=\mu$, therefore

$$\frac{\lambda^*\mu^{n+1}}{b^n} < N M^n M' \quad \text{for all } t>T_1 ,$$

hence by integration we have for the function

$$K(T) = \int_{t_o}^{T} \lambda^* \frac{\mu^{n+1}}{b^n}$$

that

$$K(T)-K(T_1) < N \frac{M^{n+1}(T)-M^{n+1}(T_1)}{n+1} \quad \text{for} \quad T>T_1 ,$$

hence

$$\limsup_{T\to\infty} \frac{K(T)}{M^{n+1}(T)} \leqq \frac{N}{n+1} ,$$

i.e. the relation (7) (iv) holds.

Thus the pair (λ^*,μ) is admissible and the conditions of Theorem 1 are satisfied, therefore the relation (10) holds and there exists a finite number C with $\lim_{T\to\infty} H(T)=C$. We may repeat again the proof of the LEMMA and now we consider the relation (18) which we rewrite in the form

$$\left|C-\int_{t_o}^{t} \lambda^* c\right|^{\frac{n+1}{n}} = \left|r\lambda^* - n\int_t^{\infty} b\lambda^* |r|^{1+\frac{1}{n}}\right|^{\frac{n+1}{n}} .$$

Since the function $|x|^{\frac{n+1}{n}}$ is convex

$$\left|C-\int_{t_o}^{t} \lambda^* c\right|^{\frac{n+1}{n}} \leqq 2^{\frac{1}{n}}\{|r|^{\frac{n+1}{n}}(\lambda^*)^{\frac{n+1}{n}} + n^{\frac{n+1}{n}}[\int_t^{\infty} b\lambda^* |r|^{1+\frac{1}{n}}]^{\frac{n+1}{n}}\} ,$$

and then

$$(31)\qquad 0 \leq \frac{\int_{t_0}^{T} \mu \left| C - \int_{t_0}^{t} \lambda^* c \right|^{\frac{n+1}{n}}}{M(T)} \leqq 2^{\frac{1}{n}} \frac{\int_{t_0}^{T} \mu |r|^{\frac{n+1}{n}} (\lambda^*)^{\frac{n+1}{n}}}{M(T)} +$$

$$+ 2^{\frac{1}{n}} n^{\frac{n+1}{n}} \frac{\int_{t_0}^{T} \mu \left[\int_t^{\infty} b\lambda^* |r|^{1+\frac{1}{n}} \right]^{\frac{n+1}{n}} dt}{M(T)} = M_1 + M_2 .$$

By the averaging property of the function $H(T)$ the quantity M_2 tends to zero as $T \to \infty$. Let T_1 be as large as in (30), then we have for all $T \geqq T_2 > T_1$

$$2^{-\frac{1}{n}} M_1 = \frac{\int_{t_0}^{T_2} \mu |r|^{\frac{n+1}{n}} (\lambda^*)^{\frac{n+1}{n}} + \int_{T_2}^{T} b\lambda^* |r|^{1+\frac{1}{n}} \cdot (\lambda^*)^{\frac{1}{n}} \frac{\mu}{b}}{M(T)} <$$

$$< \frac{\int_{t_0}^{T_2} \mu |r|^{\frac{n+1}{n}} (\lambda^*)^{\frac{n+1}{n}} + N^{\frac{1}{n}} M(T) \int_{T_2}^{T} b\lambda^* |r|^{1+\frac{1}{n}}}{M(T)} ,$$

therefore

$$\limsup_{T\to\infty} 2^{-\frac{1}{n}} M_1 \leqq N^{\frac{1}{n}} \int_{T_2}^{\infty} b\lambda^* |r|^{1+\frac{1}{n}} \qquad \text{for all } T_2 > T_1 ,$$

hence by (10) $\lim_{T\to\infty} M_1 = 0 = \lim_{T\to\infty} M_2$, thus (31) implies (29), which was to be proved.

The companion criterion for oscillation is formulated

in the next theorem.

THEOREM 4. Let (λ^*,μ) be an admissible pair for the system (2) satisfying the relation (7) (iv)'. If the function $H(T)$ defined by (8) has a finite limit C as T tends to ∞ and

$$\limsup_{T\to\infty} \frac{\int_{t_0}^{T} \mu \left| C - \int_{t_0}^{t} \lambda^* c \right|^{\frac{n+1}{n}}}{\int_{t_0}^{T} \mu} > 0 ,$$

then the system (2) is oscillatory.

Another nonoscillation criterion can be established if we do not assume the relation (7) (ii). To be precise we define the set of $\tilde{L}$ of the pairs (λ,μ) by the conditions (7) (i),(iii) and (iv). Hence the requirement (7) (ii) is dropped, and therefore $L \subset \tilde{L}$.

Let us introduce

$$\text{(32)}\quad \tilde{H}(T) = \frac{\int_{t_0}^{T} \mu \left[\int_{t_0}^{t} (\lambda c - \gamma(\varepsilon) \frac{\left| \lambda \frac{(\lambda^*)'}{\lambda^*} - \lambda' \right|^{n+1}}{(b\lambda)^n} \right) dt}{\int_{t_0}^{T} \mu} .$$

with $\gamma(\varepsilon) = (\frac{n}{n-\varepsilon})^n (n+1)^{-(n+1)}$ for $0<\varepsilon<n$ as above. Now we extend the results formulated in the Lemma and in Theorem 1 in the following way.

THEOREM 5. Let the system (2) be nonoscillatory and let (y,z) be a solution such that $y(t) \neq 0$ on $[t_0, \infty)$. Let the function r be given by (4). If for the function λ of a pair $(\lambda,\mu)\in\tilde{L}$ the inequality (10) holds then with the corresponding μ the function $\tilde{H}(T)$ is bounded from above. On the other hand if $\tilde{H}(T)$ in (32) is bounded from below and the system (2) is nonoscillatory then the inequality (10) holds again and, consequently, $\tilde{H}(T)$ is bounded from above.

Proof. We start with proving the first part of this Theorem. Similarly as in the proof of the LEMMA we have again the relation (13) which will be rewritten here as

$$(33)\quad r\lambda+\varepsilon\int_{t_0}^{t} b\lambda|r|^{1+\frac{1}{n}}+\int_{t_0}^{t}(\lambda c-\gamma(\varepsilon)\frac{\left|\lambda\frac{(\lambda^*)'}{\lambda^*}-\lambda'\right|^{n+1}}{(b\lambda)^n} - r(t_0)\lambda(t_0) \leqq 0 ,$$

hence by (32)

$$\tilde{H}(T) - r(t_0)\lambda(t_0) \leqq \frac{\int_{t_0}^{T}\lambda\mu|r|}{\int_{t_0}^{T}\mu} = L(T)$$

where the function $L(T)$ is the same as in (14). By (7) (iii) (iv) and (10) the relation (17) is true, therefore the function $\tilde{H}(T)$ is bounded from above, as stated.

To prove the second part of Theorem 5 we follow the indirect argumentation of Theorem 1. If the relation (10) is not true then the functions $S(T)$, $M(T)$ given by (20) satisfy the relations in (22), (23). Then by (33)

$$\frac{\int_{t_o}^{T} \lambda\mu r}{M(T)} + \varepsilon \frac{S(T)}{M(T)} \leqq r(t_o)\lambda(t_o) - \tilde{H}(T) .$$

By our assumption on the function $\tilde{H}$ the right hand side is bounded from above, hence we have for sufficiently large T_1 the relation (24) and by the same way we would have the boundedness of the quotient S/M for $T \geq T_1$, which contradicts to (22). Thus we have again the inequality (10) and according to the first part of Theorem 5 the function $\tilde{H}(T)$ is bounded above, hereby the proof is complete.

As a consequence of Theorem 5 we get the next oscillation criterion.

THEOREM 6. Let the pair $(\lambda,\mu) \in \tilde{L}$. If for some $t_o \geqq t^o$ and $0<\varepsilon<n$ the relation

$$\lim_{T\to\infty} \frac{\int_{t_o}^{T} \mu\left(\int_{t_o}^{t} \lambda c - \gamma(\varepsilon) \dfrac{\left|\lambda \dfrac{(\lambda^*)'}{\lambda^*} - \lambda'\right|^{n+1}}{(b\lambda)^n}\right) dt}{\int_{t_o}^{T} \mu} = \infty$$

holds, then the system (2) is oscillatory.

We must remark here that in the above proofs the inequality $0<\varepsilon<n$ plays an essential role. However, the author believes that Theorem 5 and Theorem 6 remain still valid for $\varepsilon=0$, too. This conjecture is supported by the fact that - using an idea due to Mirzov [5] - we can prove a rather strong oscillation criterion.

THEOREM 7. Let λ be a function such that there exists a function μ satisfying $(\lambda,\mu)\in\tilde{L}\setminus L$. If

$$(34)\qquad \lambda' - \lambda\frac{(\lambda^*)'}{\lambda^*} \geqq 0 \qquad \text{for all}\quad t>t_o\geqq t^o$$

$$(35)\qquad \lim_{t\to\infty}\int_{t_o}^{t}\frac{b}{\lambda^{1/n}} = \infty$$

$$(36)\qquad \lim_{t\to\infty}\int_{t_o}^{t}\Bigl(\lambda c-\frac{1}{(n+1)^{n+1}}\,\frac{\bigl(\lambda'-\frac{(\lambda^*)'}{\lambda^*}\bigr)^{n+1}}{(b\lambda)^n}\Bigr) = \infty ,$$

then the system (2) is oscillatory. The constant $1/(n+1)^{n+1}$ cannot be replaced by a smaller one.

Proof. Suppose the contrary, i.e. the system is non-oscillatory. As in the proof of the LEMMA we have a solution $(y(t),z(t))$ such that $y(t)\neq 0$ for $t\geqq t_o$ and the relations (11), (13) hold. Without the loss of the generality we may assume $y(t)>0$. Since now $\varepsilon=0$ the relation (13) is

$$r\lambda \leqq r(t_o)\lambda(t_o)-\int_{t_o}^{t}\Bigl(\lambda c-\gamma(o)\,\frac{\bigl(\lambda'-\frac{(\lambda^*)'}{\lambda^*}\bigr)^{n+1}}{(b\lambda)^n}\Bigr) ,$$

hence by (36) $r<0$, i.e. $z(t)<0$ for all $t\geq t_1$ with some $t_1>t_o$. Thus we have from (11) by (34)

$$(37)\qquad r\lambda + n\int_{t_o}^{t} b\lambda|r|^{1+\frac{1}{n}} < C_3 - \int_{t_o}^{t}\lambda c \qquad \text{for}\quad t>t_1 ,$$

where

$$C_3 = r(t_o)\lambda(t_o) + \int_{t_o}^{t_1} r[\lambda' - \lambda\frac{(\lambda^*)'}{\lambda^*}] \quad .$$

Since by (36) $\lim_{t\to\infty} \int_{t_o}^{t} \lambda c = \infty$, the right hand side of (37) tends to $-\infty$ thus for some $t_2 > t_1$ we obtain

$$r\lambda + n \int_{t_o}^{t} b\lambda |r|^{1+\frac{1}{n}} \leqq -1 \quad \text{for all} \quad t \geqq t_2$$

or

$$(38) \qquad \frac{|r|\lambda}{1+n\int_{t_o}^{t} b\lambda |r|^{1+1/n}} \geqq 1 \ .$$

Multiplying this by $nb|r|^{1/n}$ we have then by a quadrature over $[t_2, t]$ and taking into account (2), (4)

$$\log \frac{1+n\int_{t_o}^{t} b\lambda |r|^{1+\frac{1}{n}}}{K} \geqq n\int_{t_2}^{t} b|r|^{\frac{1}{n}} = -n\int_{t_2}^{t} br^{\overset{*}{\frac{1}{n}}} = -n\int_{t_2}^{t} \frac{y'-ay}{y} = n\int_{t_2}^{t} a - n \log\frac{y(t)}{y(t_2)}$$

where $K = 1+n\int_{t_o}^{t_2} b\lambda |r|^{1+\frac{1}{n}}$, hence by (38)

$$|r|\lambda \geqq 1+n\int_{t_o}^{t} b\lambda |r|^{1+\frac{1}{n}} \geqq K \frac{y^n(t_2)}{y^n(t)} \exp(n\int_{t_2}^{t} a) \quad \text{for} \quad t \geqq t_2$$

thus

$$-z = |r| y^n \geqq K \quad y^n(t_2) \frac{1}{\lambda} \exp(n\int_{t_2}^{t} a)$$

hence

$$z^{\frac{\overset{*}{1}}{n}} \leqq -K^{\frac{1}{n}} y(t_2)\frac{1}{\lambda^{1/n}} \exp\left(\int_{t_2}^{t} a\right) .$$

Multiplying this inequality by $b>0$ we have by (2)

$$y'-ay \leqq -K^{\frac{1}{n}} y(t_2)\frac{b}{\lambda^{1/n}} \exp\left(\int_{t_2}^{t} a\right) ,$$

hence

$$\left(y \exp\left(-\int_{t_2}^{t} a\right)\right)' \leqq -K^{\frac{1}{n}} y(t_2)\frac{b}{\lambda^{1/n}}$$

and then

$$y(t) \exp\left(-\int_{t_2}^{t} a\right)-y(t_2) \leqq -K^{\frac{1}{n}} y(t_2)\int_{t_2}^{t} \frac{b}{\lambda^{1/n}} \text{ for } t>t_2$$

By our assumptions $y(t)>0$ for $t>t_0$ therefore

$$K^{\frac{1}{n}} \int_{t_2}^{t} \frac{b}{\lambda^{1/n}} < 1 \quad \text{for all} \quad t>t_2$$

which contradicts to (35).

In order to show the sharpness of the constant $1/(n+1)^{n+1}$ we consider the differential equation

$$(x'^{\overset{*}{n}})' + \left(\frac{n}{n+1}\right)^{n+1} \frac{1}{t^{n+1}} x^{\overset{*}{n}} = 0$$

to which the function $x=t^{\frac{n}{n+1}}$ is a solution, hence this differential equation is nonoscillatory. The corresponding system is

$$(39)\qquad y' = z^{\frac{1}{n}^{*}}$$

$$z' = -\left(\frac{n}{n+1}\right)^{n+1} \frac{1}{t^{n+1}}\, y^{\overset{*}{n}} .$$

Since $\tau^*=1$ and $b=1$, the function $\lambda=t^n$ satisfies all the requirements of Theorem 7. If in (36) the constant $1/(n+1)^{n+1}$ would be replaceable by a smaller number Γ then we would have in (36) for some $t_0>0$ that

$$\lim_{t\to\infty} n^{n+1}\left[\frac{1}{(n+1)^{n+1}} - \Gamma\right] \int_{t_0}^{t} \frac{dt}{t} = \infty$$

i.e. the system (38) would be oscillatory.

Finally we give some applications.

Example 1. Let us assume that $na=d$ in the system (2). Then by (6) $\lambda^*(t)\equiv 1$. We choose $\lambda=(\int_{t_0}^{t} b)^{\alpha}$ for some constant $\alpha<n$ and $\mu=b$. It is not difficult to show that they form an admissible pair (λ,μ) for (2) if $\int_{t_0}^{\infty} b=\infty$. According to the Corollary a sufficient criterion for oscillation is

$$\lim_{T\to\infty} \int_{t_0}^{T} c\left(\int_{t_0}^{t} b\right)^{\alpha} dt = \infty ,$$

which is comparable with Theorem 2 in [5]. Let us remark that here it is not assumed the relations $a=d\equiv 0$ and $\alpha\geqq 0$ as there. Moreover, the pair $((\int_{t_0}^{t} b)^n, b)$ belongs to $\tilde{L}$ but not to L .

Example 2. Let us consider the half-linear differential

equation

$$(40) \qquad y''|y'|^{n-1} + q(t)\, y^{\overset{*}{n}} = 0 .$$

As in the beginning was told this differential equation can be turned into a system $y'=z^{\frac{1}{n}}$, $z'=-nq\, y^{\overset{*}{n}}$. Now we have again $\lambda^*\equiv 1$ (see Example 1). Let $\lambda=t^\alpha$, $\mu=t^\beta$ with $\alpha<n$, $\beta\geqq-1$. Then the pair (t^α,t^β) is admissible to this system. Moreover the pairs $(1,t^\beta)$ with the restriction $\beta\geqq-1$ satisfy the stronger condition (7) (iv)', too. Another pairs are the following. Let the sequence of the functions $\ell_i=\ell_i(t)$, $i=0,1,2,\ldots$ be defined by the recurrence relation $\ell_0=t$, $\ell_{i+1}=\log(\ell_i)$. If $\lambda_i=(\ell_0\ell_1\ldots\ell_i)^n$ then the pair $(\lambda_i\, ,1)\in$ $\in\tilde{L}\setminus L$.

Example 3. Let the coefficients p , q in (1) be periodic with the period ω , $q\not\equiv 0$, and let

$$Q = \frac{1}{\omega}\int_0^\omega q(t)\, dt .$$

If $Q\geqq 0$ then the differential equation (1) is oscillatory. Let us choose the pair $(1,p)$ which is admissible pair to the equivalent system $y'=\frac{1}{p}\, z^{\frac{\overset{*}{1}}{n}}$, $z'=-nq\, y^{\overset{*}{n}}$. As the Example 1 shows, this system is oscillatory if $Q>0$. To the case $Q=0$ we apply Theorem 4. The relation (7) (iv)' is satisfied automatically. Let

$$Q^* = \frac{\int_0^\omega p\left|\int_0^t q\right|^{\frac{n+1}{n}} dt}{\int_0^\omega p} .$$

By the assumption $q \not\equiv 0$ we have $Q^* > 0$. It is not difficult to check that

$$\lim_{T\to\infty} \frac{\int_0^T p \left| \int_0^t q \right|^{\frac{n+1}{n}} dt}{\int_0^T p} = Q^* ,$$

hence (1) is oscillatory. In the case $Q<0$ our method does not work because $\lim_{T\to\infty} H(T) = -\infty$.

R e f e r e n c e s

[1] Á. Elbert, A half-linear second order differential equation, Colloquia Mathematica Societatis Jǎnos Bolyai, 30 Qualitative theory of differential equations, Szeged (Hungary) (1979) 153-180.

[2] P. Hartman, Ordinary differential equations, John Wiley et Sons, Inc., N.Y.-London-Sydney, (Chapt. XI)

[3] J. D. Mirzov, On some analogs of Sturm's and Kneser's theorems for non-linear systems, J. Math. Anal. Appl. 53 (1976) 418-426.

[4] J. D. Mirzov, Sturm-Liouville boundary value problem to a non-linear system (in Russian), Izv. Vys. Učeb. Zav. 203 (1979) 28-32.

[5] J. D. Mirzov, On the oscillation of the solutions of a differential equation system (in Russian), Mat. Zam. 23 (1978) 401-404.

[6] J. D. Mirzov, Oscillation of the solutions of a differential equation system (in Russian), Diff. Uravn., 17 (1981) 1504-1508.

[7] A. Wintner, On Laplace-Fourier transcendents occuring in mathematical physics, Amer. J. Maths, 69 (1947) 87-97.

[8] M. Zlamal, Oscillation criterions, Časop. pro pest. Mat. Fys. 75 (1950) 213-218.

Entropy numbers, s-numbers and eigenvalues

D.E. Edmunds

That compact linear operators acting in Banach spaces are useful objects will hardly strike anyone these days as a novel observation; but what may well arouse more than a flicker of interest are the connections established in recent years between <u>geometric</u> quantities related to such operators (such as entropy numbers, approximation numbers and n-widths) and more <u>analytical</u> entities, such as eigenvalues. My impression is that these results are not very widely known in this country; and it is because of this, and with the hope of stimulating some interest in the application of the work to differential equations, that I venture to give a brief account of those results of this nature which seem interesting to me. I shall also attempt to indicate some trends in the work, especially those directed to the study of non-compact maps.

As we shall to some extent be concerned with spectral properties we shall take all spaces to be complex. Moreover, if X and Y are Banach spaces we shall denote the set of all bounded linear maps from X to Y by $L(X,Y)$, although $L(X,X)$ will be shortened to $L(X)$; and B_X will stand for $\{x \in X : ||x|| \le 1\}$.

<u>Entropy numbers</u>

Given any $T \in L(X,Y)$ and any $n \in \mathbb{N}$, the nth entropy number of T, $e_n(T)$, is defined by

$$e_n(T) = \inf\{\varepsilon > 0 : T(B_X) \subset \bigcup_{i=1}^{2^{n-1}} \{y_i + \varepsilon B_Y\} \text{ for some } y_1,\dots,y_{2^{n-1}} \text{ in } Y\}.$$

These numbers were introduced by Pietsch in his book [20], although certain functions inverse to them had been used earlier by Mitjagin and Pelczýnski [17] and Triebel [24]. Since the $e_n(T)$ are monotonic decreasing as n increases, their limit,

$$\beta(T) := \lim_{n\to\infty} e_n(T),$$

exists, and clearly

$$\beta(T) = \inf\{\varepsilon > 0 : T(B_X) \text{ can be covered by finitely many balls of radius } \varepsilon\}.$$

We shall refer to $\beta(T)$ as the <u>measure of non-compactness</u> of T: note that $\beta(T) = 0$ if and only if T is compact. Certain properties of the entropy numbers follow directly: if S and T belong to $L(X,Y)$ and $R \in L(Y,Z)$, where Z is a Banach space, then

(1) $$||T|| = e_1(T) \geq e_2(T) \geq \ldots \geq 0,$$

and for all $m,n \in \mathbb{N}$,

(2) $$e_{m+n-1}(S + T) \leq e_m(S) + e_n(T),$$

(3) $$e_{m+n-1}(R \circ S) \leq e_m(R)\, e_n(S).$$

A connection between these rather geometrical quantities and an object connected to operators in a more analytical way comes via the notion of the <u>essential spectrum</u> of a map. We recall that if $T \in L(X)$, then the essential spectrum of T, $\sigma_e(T)$, may be defined as the set of all those points λ in the spectrum $\sigma(T)$ of T such that at least one of the following conditions holds:

(i) the range $\mathcal{R}(\lambda I-T)$ of $\lambda I-T$ is not closed;

(ii) λ is a limit point of $\sigma(T)$;

(iii) $\dim \bigcup_{n=1}^{\infty} \ker\{(\lambda I-T)^n\} = \infty$.

This is Browder's definition [1] of $\sigma_e(T)$: he proves that with this definition, $\lambda_0 \in \mathbb{C} \setminus \sigma_e(T)$ if and only if there exists $\delta > 0$ such that $(\lambda I-T)^{-1} \in L(X)$ for $0 < |\lambda-\lambda_0| < \delta$ and the Laurent expansion of $(\lambda I-T)^{-1}$ about λ_0 has only a finite number of non-zero coefficients with negative indices. Note that if X is a Hilbert space and T is self-adjoint, (i) and (ii) are equivalent. Numerous other definitions of $\sigma_e(T)$ may be found in the literature: Browder's gives rise to the largest set, while Kato's [12] produces the smallest, his definition being

$$\mathbb{C} \setminus \{\lambda \in \mathbb{C} : \lambda I-T \text{ has closed range and either } \dim\ker(\lambda I-T) < \infty \text{ or } \dim X/(\lambda I-T) < \infty\text{;}$$

in other words, $\lambda \in \mathbb{C}$ belongs to the essential spectrum according to Kato if and only if $\lambda I-T$ is not semi-Fredholm.

Dreadful though this multiplicity of definitions may be, there is a ray of hope, for whichever definition of $\sigma_e(T)$ is selected the radius of the essential spectrum, $r_e(T) = \sup\{|\lambda| : \lambda \in \sigma_e(T)\}$, is the same; and moreover, there is a striking connection between $r_e(T)$ and the measure of non-compactness $\beta(T)$:

$$(4) \qquad r_e(T) = \lim_{n\to\infty} (\beta(T^n))^{1/n} \leq \beta(T).$$

This was established independently by Nussbaum [18] and Lebow and Schechter [14] For the case of a bounded self-adjoint map T acting in a Hilbert space H a simple proof of (4) can be given, and in fact we then have

$$(5) \qquad r_e(T) = \beta(T).$$

To prove (5), let $E(\lambda)$ be the resolution of the identity corresponding to T, put $r = r_e(T)$, let $\varepsilon > 0$ and write

$$T = \int \lambda dE_\lambda = \int_{|\lambda|<r+\varepsilon} \lambda dE_\lambda + \int_{|\lambda|\geq r+\varepsilon} \lambda dE_\lambda = S + C, \text{ say.}$$

Then $||S|| \leq r + \varepsilon$, and as C is of finite rank it is compact. Hence $\beta(T) = \beta(S + C) = \beta(S) \leq ||S|| \leq r + \varepsilon$; and as this holds for all $\varepsilon > 0$ it follows that $\beta(T) \leq r$. Assuming, without loss of generality, that $r \in \sigma_e(T)$, put $H_\varepsilon = (E(r+\varepsilon) - E(r-\varepsilon))H$: then $\dim H_\varepsilon = \infty$ (unless $\dim H < \infty$, in which case (5) is obvious), and $||Tx|| \geq (r-\varepsilon)||x||$ for all $x \in H_\varepsilon$, since

$E_\lambda(E(r+\varepsilon) - E(r-\varepsilon)) = 0$ if $\lambda < -\varepsilon$. Thus $\dim T(H_\varepsilon) = \infty$, and since the unit sphere in an infinite-dimensional space cannot be covered by finitely many balls of radius less than 1, it follows that $\beta(T) \geq r-\varepsilon$, and the proof is complete.

We remark that the notion of the measure of non-compactness of a map, and in particular the relationships between it and the essential spectrum, have proved to be very useful: see, for example, Stuart's work [22] on bifurcation theory, and the work of Evans and others (cf., e.g., [5], [9]) on the location of the essential spectra of operators arising from elliptic boundary-value problems.

Now let T be a compact linear map of a Banach space X into itself. The spectrum of T, apart from the point 0, consists solely of eigenvalues of finite (algebraic and geometric) multiplicity: let $(\lambda_n(T))$ be the sequence of all eigenvalues of T, repeated according to their algebraic multiplicities and ordered so that

$$|\lambda_1(T)| \geq |\lambda_2(T)| \geq \ldots \geq 0.$$

If T has only n $(< \infty)$ distinct eigenvalues we put $\lambda_m(T) = 0$ for all $m \in \mathbb{N}$, $m > n$. Recently Carl and Triebel [4] established the following striking connection between eigenvalues and entropy numbers.

<u>Theorem 1</u> <u>Let</u> $T \in L(X)$ <u>be compact and let</u> $\lambda_n(T)$, $e_n(T)$ <u>be as above.</u> <u>Then for all</u> $m, n \in \mathbb{N}$,

$$(6) \qquad |\lambda_n(T)| \leq \Big(\prod_{j=1}^{n} |\lambda_j(T)|\Big)^{1/n} \leq (\sqrt{2})^{(m-1)/n} e_m(T).$$

We shall indicate later how to prove this, and even a more general result. Notice that by taking $m = n + 1$ it follows that for all $n \in \mathbb{N}$,

$$|\lambda_n(T)| \leq \sqrt{2}\, e_{n+1}(T),$$

an inequality which is perhaps rather surprising. More recently, Carl [2] has been able to show that for all $k, n \in \mathbb{N}$,

(7) $$\left(\prod_{j=1}^{n} |\lambda_j(T)|\right)^{1/n} \leq (\sqrt{k})^{1/n} e_k(T).$$

This last inequality can be used to obtain a result which shows that in a sense the entropy numbers may be thought of as a deformation of the norm (cf. [27]):

(8) $$\lim_{k\to\infty} (e_n(T^k))^{1/k} = r(T)$$

for all $n \in \mathbb{N}$, where $r(T)$ is the spectral radius of T. The proof of (8) is simple: given any $n \in \mathbb{N}$, since $e_n(T) \leq ||T||$, $\limsup_{k\to\infty} (e_n(T^k))^{1/k} \leq \lim_{k\to\infty} ||T^k||^{1/k} = r(T)$. Also, (7) implies that $|\lambda_1(T)| \leq \sqrt{n}\, e_n(T)$, so that $|\lambda_1(T)|^k = |\lambda_1(T^k)| \leq \sqrt{n}\, e_n(T^k)$ and $r(T) = |\lambda_1(T)| \leq \liminf_{k\to\infty} (e_n(T^k))^{1/k}$, which completes the proof. Carl [3] has also obtained estimates for the entropy numbers of embedding maps between Besov spaces, and has used these, together with Theorem 1, to establish results concerning the distribution of eigenvalues of certain maps.

Before leaving the subject of entropy numbers for a moment, we mention a related notion, that of <u>entropy.</u> Let $T \in L(X)$ be compact, so that given any $\varepsilon > 0$, $T(B_X)$ can be covered by a finite number of balls of radius ε: let $K(\varepsilon,T)$ be the least number of such balls. The <u>ε-entropy</u> of $T(B_X)$ is defined to be

$$H(\varepsilon,T) = \log_2 K(\varepsilon,T).$$

In a certain sense, $(e_n(T))$ is inverse to $H(\varepsilon,T)$. Our purpose in mentioning H is to point out that it has a connection with quantities which are of interest from the standpoint of the distribution of the eigenvalues $\lambda_n(T)$ of T. For any $\varepsilon > 0$ put

$$N(\varepsilon,T) = \sum_{|\lambda_j(T)|\geq\varepsilon} 1, \quad M(\varepsilon,T) = \sum_{|\lambda_j(T)|\geq\varepsilon} \log_2 (|\lambda_j(T)|/\varepsilon).$$

The connection referred to is given by the following inequalities, due to Carl and Triebel [4]: for any $\varepsilon > 0$,

$$N(4\varepsilon,T) \leq M(2\varepsilon,T) \leq \tfrac{1}{2}H(\varepsilon,T). \tag{9}$$

Quite apart from the intrinsic interest of (9), it can be used to give a simple proof of (6).

Approximation numbers

Given any $T \in L(X,Y)$ put, for all $n \in \mathbb{N}$,

$$a_n(T) = \inf\{||T-L||: L \in L(X,Y),\ \text{rank } L < n\}.$$

These numbers $a_n(T)$ are the so-called approximation numbers of T. They enjoy various properties similar to those of the entropy numbers, notably (1), (2) and (3); moreover, $a_n(T) = 0$ if and only if rank $T < n$; and if $\dim X \geq n$ and $I : X \to X$ is the identity map, then $a_n(I) = 1$.

Let $T \in L(X,Y)$. The sequence $(a_n(T))$ is monotonic decreasing and bounded below by 0; it thus converges, to $\alpha(T)$, say. Then (cf. [27]),

$$\beta(T) \leq \alpha(T). \tag{10}$$

To prove this, let $\mathfrak{F} = \{L \in L(X,Y): \text{rank } L < \infty\}$, $\mathfrak{K} = \{L \in L(X,Y): L \text{ compact}\}$ and note that

$$\alpha(T) = \text{dist }(T,\overline{\mathfrak{F}}) \geq \text{dist }(T,\mathfrak{K}) := \gamma(T). \tag{11}$$

Also, $\gamma(T) \geq \beta(T)$: for if there is a compact map K with $||T-K|| < \delta < \beta(T)$, then there exists $\varepsilon > 0$ so small that $\delta + \varepsilon < \beta(T)$, and since $K(B_X)$ is relatively compact it can be covered by finitely many balls of radius ε, from which we see that $T(B_X)$ can be covered by finitely many balls of radius $\delta + \varepsilon$. Hence $\beta(T) \leq \delta + \varepsilon < \beta(T)$: contradiction.

We can summarise the position as follows for the case $T \in L(X)$:

$$r_e(T) \leq \beta(T) \leq \gamma(T) \leq \alpha(T). \tag{12}$$

Note that it may well be that $\gamma(T) < \alpha(T)$: this is as a consequence of Per Enflo's celebrated work on the basis problem, as a result of which it follows that there exist Banach spaces and compact operators acting in these spaces which cannot be approximated arbitrarily closely by maps of finite rank. This kind of pathology is impossible in Hilbert spaces, when $\alpha(T) = \gamma(T)$. Note also that since $\gamma(T)$ is the norm in $L(X)/\mathcal{K}(X)$ (where $\mathcal{K}(X) = \mathcal{K}(X,X)$) of the equivalence class containing T, then since $\sigma_e(T)$ may also be regarded as that part of the spectrum of T unchanged by compact perturbations, we have

(13) $$r_e(T) = \lim_{n\to\infty} (\gamma(T^n))^{1/n}.$$

The approximation numbers have important connections with eigenvalues, connections which are significant from the point of view of applications to differential equations. The crucial relation is that if X is a Hilbert space and $T \in L(X)$ is compact, then T^*T has a positive compact square root $|T|$, and for all $n \in \mathbb{N}$,

(14) $$a_n(T) = \lambda_n(|T|).$$

The eigenvalues of $|T|$ are often called the <u>singular values</u> of T; they seem to have been first considered by E. Schmidt. These singular values, or s-numbers, have various interesting properties, as we have seen in our discussion of approximation numbers; more generally, Pietsch [20] refers to s-numbers as being any sequence of non-negative real numbers associated with bounded linear operators and having certain properties similar to those possessed by the a_n. If T is, in addition, positive and self-adjoint, then (14) shows that $a_n(T) = \lambda_n(T)$ for all $n \in \mathbb{N}$: this relation is at the heart of much work on the asymptotic distribution of eigenvalues of elliptic operators in the self-adjoint case (cf. [10], [16]), while (14) itself is important for the non-self-adjoint problem.

All this is, of course, what happens in a <u>Hilbert</u> space. In a general Banach space, with less structure, one could hardly expect to obtain results of such sharpness. Nevertheless, something can be said, and one of the most startling developments is that contained in the following theorem due to König [13]:

<u>Theorem 2</u> <u>Let</u> X <u>be a Banach space and let</u> $T \in L(X)$ <u>be compact.</u> <u>Then for all</u> $n \in \mathbb{N}$,

$$(15) \qquad |\lambda_n(T)| = \lim_{k\to\infty} (a_n(T^k))^{1/k}.$$

<u>Remarks</u>

1) In general, no inequality of the form $a_n(T) \geq (\leq)\ \lambda_n(T)$ for all $n \in \mathbb{N}$ can be expected: for if $T \in L(\mathbb{C}^2)$ is represented by the matrix $\begin{pmatrix} 2 & 0 \\ 1 & 1 \end{pmatrix}$, then $\lambda_1(T) = 2$, $\lambda_2(T) = 1$; but since T^*T is represented by $\begin{pmatrix} 5 & 1 \\ 1 & 1 \end{pmatrix}$, which has eigenvalues $3 \pm \sqrt{5}$, it follows that $a_1(T) = (3+\sqrt{5})^{1/2} > \lambda_1(T)$, while $a_2(T) = (3-\sqrt{5})^{1/2} < \lambda_2(T)$. However, Carl and Triebel [4] have shown that if $T \in L(X)$ is compact and $||T|| \leq 1$, then for all $n \in \mathbb{N}$,

$$(16) \qquad \left(\prod_{j=1}^{n} \lambda_j(T)\right)^{1/n} \leq 16\, a_{m+1}^{(n-m)/n}(T) \qquad (m = 0,1,2,\ldots,n-1).$$

With $m = n-1$ this gives

$$\lambda_n(T) \leq 16\, a_n^{1/n}(T) \qquad (n \in \mathbb{N})$$

2) Other connections between eigenvalues and approximation numbers are known in the Hilbert space case. Thus if H is a complex Hilbert space and $T \in L(H)$ is compact, a celebrated inequality due to Weyl [26] states that for all $n \in \mathbb{N}$,

$$(17) \qquad \prod_{j=1}^{n} |\lambda_j(T)| \leq \prod_{j=1}^{n} |a_j(T)|,$$

from which it may be deduced that for all $n \in \mathbb{N}$ and all $p \in [1,\infty)$,

$$(18) \qquad \sum_{j=1}^{n} |\lambda_j(T)|^p \leq \sum_{j=1}^{n} |a_j(T)|^p.$$

This implies that if $(a_j(T)) \in \ell^p$ then $(\lambda_j(T)) \in \ell^p$, a result which can be used in the study of non-self-adjoint elliptic eigenvalue problems to obtain information about the distribution of eigenvalues (cf. [6]). For inequalities similar to (18) but in a Banach space setting see H. König, Studia Math. 67 (1980), 157-172.

Non-compact maps

Turning now to not-necessarily compact maps, let X be a Banach space, suppose that $T \in L(X)$ and put $\Lambda(T) = \{\lambda \in \sigma(T): |\lambda| > r_e(T)\}$. Then $\Lambda(T)$ is at most countable, and any point in it is an eigenvalue of T of finite multiplicity. Order these points, denoted by $\lambda_n(T)$, in such a way that $|\lambda_1(T)| \geq |\lambda_2(T)| \geq \ldots \geq 0$, each eigenvalue being repeated according to its algebraic multiplicity; if there are only n such points $(n = 0,1,2,\ldots)$, including multiplicities, we put $|\lambda_{n+1}(T)| = |\lambda_{n+2}(T)| = \ldots = r_e(T)$. It turns out that it is this part $\Lambda(T)$ of $\sigma(T)$ which can be regarded as the appropriate analogue of the non-zero part of the spectrum of a compact map, for various of the results about eigenvalues of a compact map which we mentioned earlier hold for the eigenvalues in $\Lambda(T)$, where T is merely bounded and not necessarily compact. For example, the analogue of Theorem 1 holds: if $T \in L(X)$ then for all $m,n \in \mathbb{N}$,

$$\left(\prod_{j=1}^{n} |\lambda_j(T)|\right)^{1/n} \leq (\sqrt{2})^{(m-1)/n} e_m(T).$$

Some indication of the proof in the case when X is a Hilbert space and T has an infinite set of eigenvalues which are larger than $r_e(T)$ in modulus may help to give the flavour of these ideas. Let $B_1, B_2, \ldots, B_{2^{m-1}}$ be balls in X of radius $\varepsilon > 0$ which cover $T(B_X)$, and let C_n be the n-dimensional (complex) subspace of X spanned by the eigenvectors belonging to the eigenvalues $\lambda_1(T), \ldots, \lambda_n(T)$. Interpret C_n as a 2n-dimensional real Euclidean space R_{2n}; thus T maps $B_X \cap R_{2n}$ onto an ellipsoid with half-axes $|\lambda_1(T)|$, $|\lambda_1(T)|$, $|\lambda_2(T)|$, $|\lambda_2(T)|, \ldots,$ $|\lambda_n(T)|$, $|\lambda_n(T)|$ and volume

$\prod_{j=1}^{n} |\lambda_j(T)|^2 V_{2n}$, where V_{2n} is the volume of the unit ball in R_{2n}. But the 2n-dimensional volume of $B_j \cap R_{2n}$ is at most $\varepsilon^{2n} V_{2n}$, and since the sets $B_j \cap R_{2n}$ cover $T(B_X) \cap R_{2n}$,

$$2^{m-1} \varepsilon^{2n} V_{2n} \geq \prod_{j=1}^{n} |\lambda_j(T)|^2 V_{2n},$$

so that

$$\prod_{j=1}^{n} |\lambda_j(T)| \leq 2^{(m-1)/2} \varepsilon^n.$$

This completes the proof.

Carl's inequality (7) also holds for non-compact continuous maps, as does Theorem 2, the proofs being natural adaptations of those for compact maps: see [27]. Another interesting inequality is that proved recently by Zemanek [27] following earlier work by Carl [2]. With $g_n(T) = \inf\{k^{1/(2n)} e_k(T) : k \in \mathbb{N}\}$ and $T \in L(X)$, he shows that

$$\Big(\prod_{j=1}^{n} |\lambda_j(T)| \Big)^{1/n} \leq \lim_{k\to\infty} (g_n(T^k))^{1/k} \qquad (n \in \mathbb{N}) .$$

For compact maps T, much interest has been shown in the manner in which $\lambda_n(T) \to 0$ as $n \to \infty$, a famous example of such matters being the question raised by Lorentz and solved, in a form, by Weyl, about the asymptotic behaviour of eigenvalues of elliptic equations on a bounded open set with zero Dirichlet boundary conditions. One question to have received a good deal of attention is that of whether $(\lambda_n(T)) \in \ell^p$ for some p if it is known that $(e_n(T)) \in \ell^p$: this question was settled by Carl's discovery, mentioned earlier, that $|\lambda_n(T)| \leq \sqrt{2}\, e_{n+1}(T)$. For non-compact maps T it is natural to enquire into the speed at which $|\lambda_n(T)| \to r_e(T)$, and to do this more delicacy is required, although a preliminary result in this direction is given by the result in [8] that if T is a bounded self-adjoint map of a Hilbert space into itself and

$0 < p < \infty$, then there exists $c > 0$ such that

$$\sum_{n=1}^{\infty} \frac{1}{n} (|\lambda_n(T)| - r_e(T))^p \leq c \sum_{n=1}^{\infty} n^{-1}(e_n(T) - r_e(T))^p.$$

Various s-numbers

Other numbers may be attached to a map $T \in L(X,Y)$, the most commonly used being the Kolmogorov numbers (or n-widths) and the Gelfand numbers. Given any $n \in \mathbb{N}$, the nth Kolmogorov number $d_n(T)$ is defined by

$$d_n(T) = \inf\{||Q_N^Y \circ T|| : \dim N < n\},$$

where Q_N^Y is the canonical map of Y onto Y/N. Equivalently,

$$d_n(T) = \inf_{N \subset Y, \dim N < n} \ \sup_{x \in B_X} \ \inf_{y \in N} ||Tx-y||.$$

These numbers have been used extensively in the study of the asymptotic distribution of eigenvalues of elliptic operators (cf. [10], [25]). The nth Gelfand number $c_n(T)$ is defined by

$$c_n(T) = \inf\{||T \circ J_M^X|| : \operatorname{codim} M < n\},$$

where J_M^X is the embedding map from M to X.

Letting s_n stand for c_n or d_n it turns out that the s_n have properties similar to those possessed by the approximation numbers, namely,

$||T|| = s_1(T) \geq s_2(T) \geq \ldots \geq 0,$

$s_{m+n-1}(S + T) \leq s_m(S) + s_n(T), \quad s_{m+n-1}(S \circ T) \leq s_m(S)\, s_n(T),$

$s_n(T) = 0$ if rank $T < n$,

$s_n(I_n) = 1$ if I_n is the identity map on a space X with $\dim X \geq n$.

These properties have been taken by Pietsch [20] as the basis for an axiomatic approach to s-numbers: he calls s-numbers any sequence of numbers which satisfy these conditions. All s-numbers coincide when the spaces involved are Hilbert spaces. Moreover,

$$c_n(T) = d_n(T^*), \quad c_n(T^*) \le d_n(T).$$

If we write $c(T) = \lim_{n\to\infty} c_n(T)$, $d(T) = \lim_{n\to\infty} d_n(T)$, then from the above inequalities we may obtain inequalities for $c(T)$ and $d(T)$. More importantly, Carl has shown that there is a number $\rho > 0$, independent of X,Y and $T \in L(X,Y)$, such that

$$\rho^{-1}\beta(T) \le c(T) \le \rho\beta(T), \quad \rho^{-1}\beta(T) \le d(T) \le \rho\beta(T).$$

Also, $c(T) = 0$ if and only if $d(T) = 0$, and if and only if T is compact; $c(S + T) = c(S)$ and $d(S + T) = d(S)$ if T is compact; and

$$r_e(T) = \lim_{n\to\infty} (c(T^n))^{1/n} = \lim_{n\to\infty} (d(T^n))^{1/n}.$$

Quite recently, Pietsch [21] has introduced the <u>Weyl numbers</u> $x_n(T)$: if $T \in L(X,Y)$, then

$$x_n(T) := \sup\{a_n(T \circ S) : S \in L(\ell^2,X), \ ||S|| \le 1\}.$$

These are s-numbers in the sense of Pietsch; it can be shown that

$$x_n(T) = \sup\{c_n(T \circ S) : S \in L(\ell^2,X), \ ||S|| \le 1\},$$

this result following because $a_n(R) = c_n(R)$ for all $R \in L(\ell^2,Y)$. Interesting results can be proved by means of the Weyl numbers: for example, Pietsch [21] has used them to give sufficient conditions for the eigenvalues of a compact map T to belong to a Lorentz space; and he identifies a class of maps $T \in L(X)$, the so-called absolutely $(p,2)$-summing maps $(2 < p < \infty)$, which have the happy property that $\lambda_n(T) = O(n^{-1/p})$ as $n \to \infty$.

<u>Interpolation theory</u>

To conclude, we discuss briefly the interaction between interpolation theory and the ideas we have mentioned. The manner in which the entropy numbers and the various s-numbers behave under interpolation has been a subject of much discussion (see, for example, [20] and [25]), but there have recently been some interesting developments which we describe briefly.

We recall that a pair of Banach spaces $\{A_0,A_1\}$ is an <u>interpolation pair</u> if A_0 and A_1 are continuously embedded in a Hausdorff topological vector space $\mathcal{A}$. Given such a pair, $A_0 \cap A_1$ is a Banach space when given the norm

$$||a||_{A_0\cap A_1} = \max\,(||a||_{A_0},\ ||a||_{A_1}),$$

and so is $A_0 + A_1 = \{a \in \mathcal{A} :\ a = a_0 + a_1,\ a_i \in A_i \text{ for } i = 0,1\}$ when normed by

$$||a||_{A_0+A_1} = \inf\{||a_0||_{A_0} + ||a_1||_{A_1} : a = a_0 + a_1,\ a_i \in A_i \text{ for } i = 0,1\};$$

clearly $A_0 \cap A_1 \subset A_j \subset A_0 + A_1$ $(j = 0,1)$. A Banach space A such that $A_0 \cap A_1 \subset A \subset A_0 + A_1$ algebraically and topologically is called an <u>interpolation space</u> with respect to $\{A_0,A_1\}$. Given two interpolation pairs $\{A_0,A_1\}$, $\{B_0,B_1\}$ we denote by $L(\{A_0,A_1\}, \{B_0,B_1\})$ the set of all linear operators $T : A_0 + A_1 \to B_0 + B_1$ such that the restriction of T to A_j is in $L(A_j,B_j)$ for $j = 0,1$. If A and B are interpolation spaces with respect to $\{A_0,A_1\}$, $\{B_0,B_1\}$ respectively they are called <u>interpolation spaces of exponent</u> θ $(0 < \theta < 1)$ with respect to $\{A_0,A_1\}$, $\{B_0,B_1\}$ if, and only if, given any $T \in L(\{A_0,A_1\}, \{B_0,B_1\})$, the restriction of T to A is in $L(A,B)$ and

$$||T||_{L(A,B)} \le ||T||^{1-\theta}_{L(A_0,B_0)}\ ||T||^{\theta}_{L(A_1,B_1)}\ .$$

We shall refer to the restriction of T to A, viewed as an element of $L(A,B)$, by $T_{A,B}$; T_{A_j,B_j} will have a similar meaning.

Several methods of constructing interpolation spaces of exponent θ with respect to given pairs $\{A_0,A_1\}$ and $\{B_0,B_1\}$ are known, notably the K-method (cf. [25], p.23) which leads to the spaces $(A_0,A_1)_{\theta,p}$, $(B_0,B_1)_{\theta,p}$ $(1 \le p \le \infty)$

defined by

$$A = (A_0,A_1)_{\theta,p} = \{a \in A_0 + A_1 : \|a\|_A = (\int_0^\infty [t^{-\theta}K(t,a)]^p t^{-1}dt)^{1/p} < \infty\}$$

if $p < \infty$;

$$A = (A_0,A_1)_{\theta,\infty} = \{a \in A_0 + A_1 : \|a\|_A = \sup_{0<t<\infty} t^{-\theta}K(t,a) < \infty\};$$

where

$$K(t,a) = \inf\{\|a_0\|_{A_0} + t\|a_1\|_{A_1} : a = a_0 + a_1,\ a_0 \in A_0,\ a_1 \in A_1\}.$$

Given any interpolation pair $\{A_0,A_1\}$ and any $\theta \in (0,1)$, an interpolation space A is said to belong to $J(\theta;\ A_0,A_1)$ if there is a positive constant C such that $\|a\|_A \le C\|a\|_{A_0}^{1-\theta}\|a\|_{A_1}^{\theta}$ for all $a \in A_0 \cap A_1$; it belongs to $K(\theta;\ A_0,A_1)$ if there is a positive constant c such that for all $a \in A$ and all $t > 0$,

$$\inf\{\|a_0\|_{A_0} + t\|a_1\|_{A_1} : a = a_0 + a_1,\ a_0 \in A_0,\ a_1 \in A_1\} \le ct^{\theta}\|a\|_A.$$

The spaces $(A_0,A_1)_{\theta,p}$ belong to $J(\theta;\ A_0,A_1)$ and $K(\theta;\ A_0,A_1)$.

Now let A and B be interpolation spaces with respect to interpolation pairs $\{A_0,A_1\}$ and $\{B_0,B_1\}$ respectively, and let $T \in L(\{A_0,A_1\}, \{B_0,B_1\})$. A question which has attracted much attention is whether T, viewed as a map from A to B, inherits any compactness properties which it may possess as an element of $L(A_i,B_i)$. Lions and Peetre [15] gave a partial answer, showing that if $B_0 = B_1$ and $A \in K(\theta;\ A_0,A_1)$ for some $\theta \in (0,1)$, then T_{A,B_0} is compact if either T_{A_0,B_0} or T_{A_1,B_0} is compact; a similar result was established if $A_0 = A_1$ and $B_0 \neq B_1$. Shortly afterwards Persson [19] proved that in the general case, with $A_0 \neq A_1$ and $B_0 \neq B_1$, $T_{A,B}$ is compact if T_{A_0,B_0} is compact. To do this he had to assume that B_0 and B_1 had a certain approximation property, an assumption removed in the later work of Hayakawa [11], although at the expense of the additional assumption that T_{A_1,B_1} is also compact. A good deal of the subsequent work in this area has been to do with the interpolation properties of entropy ideals and width ideals of compact operators, an excellent account being given in Triebel's book [25]: a feature of this work is that, like that of Lions and Peetre mentioned above, either $A_0 = A_1$ or $B_0 = B_1$. Our object here is to report on recent developments which have led to the relaxation of these restrictions.

We begin with a discussion of the behaviour of the measure of non-compactness under interpolation. It is comparatively easy (cf. [7]) to prove that if A is a Banach space, $\{B_0,B_1\}$ is an interpolation pair, $0 < \theta < 1$, $B \in J(\theta;\ B_0,B_1)$ and $T \in L(\{A,A\},\ \{B_0,B_1\})$, then

$$\beta(T_{A,B}) \leq C(\beta(T_{A,B_0}))^{1-\theta}\ (\beta(T_{A,B_1}))^{\theta}.$$

A similar result holds if $A \in K(\theta;\ A_0,A_1)$ and $T \in L(\{A_0,A_1\},\ \{B,B\})$. Since $\beta(T) = 0$ if and only if T is compact, these results include as a special case those of Lions and Peetre mentioned above. To deal with maps $T \in L(\{A_0,A_1\},\ \{B_0,B_1\})$, where $A_0 \neq A_1$ and $B_0 \neq B_1$, it is convenient to impose a hypothesis on the spaces, just as Persson did.

Given an interpolation pair $\{B_0,B_1\}$, we say that it has the approximation property H_1 if there is a positive constant c such that given any compact subset K of B_0 and $\varepsilon > 0$, there is a map $P \in L(\{B_0,B_1\},\ \{B_0,B_1\})$ satisfying:

(i) $P(B_i) \subset B_0 \cap B_1$ $(i = 0,1)$;

(ii) $||P_{B_i,B_i}|| \leq c$ $(i = 0,1)$;

(iii) $||x-Px||_{B_0} < \varepsilon$ for all $x \in K$.

This is exactly the condition imposed by Persson in his work on the interpolation of compact maps. With this condition the following result can be established (cf. [23]):

<u>Theorem 3</u> <u>Let</u> $\{A_0,A_1\}$, $\{B_0,B_1\}$ <u>be interpolation pairs, suppose that</u> $\{B_0,B_1\}$ <u>has property</u> H_1, <u>let</u> $A = (A_0,A_1)_{\theta,p}$, $B = (B_0,B_1)_{\theta,p}$ <u>for some</u>

$\theta \in (0,1)$ <u>and</u> $p \in [1,\infty]$, <u>and suppose that</u> $T \in L(\{A_0,A_1\}, \{B_0,B_1\})$ <u>is such that</u> T_{A_0,B_0} <u>is compact.</u> <u>Then for all</u> $m,n \in \mathbb{N}$,

$$(19) \qquad e_{m+n-1}(T_{A,B}) \leq 2\,c\,e_m(T_{A_0,B_0})^{1-\theta}\, e_n(T_{A_1,B_1})^{\theta}.$$

To dispose of the hypothesis that T_{A_0,B_0} is compact, H_1 has to be strengthened to H_2 : there is a positive constant c such that given any $\varepsilon > 0$ and any finite sets $F_i \subset B_i$ $(i = 0,1)$, there exists $P \in L(\{B_0,B_1\}, \{B_0,B_1\})$ such that (i) and (ii) of H_1 hold, together with:
(iii') $||x-Px||_{B_i} < \varepsilon$ for all $x \in F_i$ $(i = 0,1)$.

The L^p spaces $(1 \leq p < \infty)$ satisfy both H_1 and H_2. If $\{B_0,B_1\}$ has property H_2, the hypothesis that T_{A_0,B_0} is compact may be dropped in Theorem 3, the final inequality then being (cf. [23]):

$$(20) \qquad e_{m+n-1}(T_{A,B}) \leq 4 \max(1,c)\, (e_m(T_{A_0,B_0}))^{1-\theta}\, (e_n(T_{A_1,B_1}))^{\theta}.$$

These results have various implications. For example, let $0 < p < \infty$ and denote by E_p the operator ideal of all linear maps T between arbitrary Banach spaces such that $\sum_{n=1}^{\infty} e_n(T)^p < \infty$: E_p is an <u>entropy ideal</u>, in Triebel's terminology [25]. Then in the notation and with the assumptions of Theorem 3, (19) implies that if $T_{A_i,B_i} \in E_{p_i}$ $(i = 0,1)$, where $p_0,p_1 \in (1,\infty)$, then $T_{A,B} \in E_p$, where $p^{-1} = (1-\theta)\, p_0^{-1} + \theta p_1^{-1}$.

Another result comes on letting $m,n \to \infty$ in (20):

$$(21) \qquad \beta(T_{A,B}) \leq 4 \max(1,c)\, (\beta(T_{A_0,B_0}))^{1-\theta}\, (\beta(T_{A_1,B_1}))^{\theta}.$$

This leads to information about the essential spectrum (cf. [7]).

<u>Theorem 4</u> <u>Let</u> $\{A_0,A_1\}$ <u>be an interpolation pair which has property</u> H_2, <u>let</u> $A = (A_0,A_1)_{\theta,p}$ <u>for some</u> $\theta \in (0,1)$ <u>and</u> $p \in [1,\infty]$, <u>and let</u>

$T \in L(\{A_0,A_1\}, \{A_0,A_1\})$. *Then*

$$r_e(T_{A,A}) \leq (r_e(T_{A_0,A_0}))^{1-\theta} (r_e(T_{A_1,A_1}))^{\theta}.$$

Proof. From (21) we have for all $n \in \mathbb{N}$,

$$\beta(T^n_{A,A}) \leq 4 \max(1,c) (\beta(T^n_{A_0,A_0}))^{1-\theta} (\beta(T^n_{A_1,B_1}))^{\theta}.$$

The Theorem follows immediately on using the identity $r_e(T) = \lim_{n\to\infty} (\beta(T^n))^{1/n}$.

Knowledge of how eigenvalues behave under interpolation would be available if results of a nature similar to Theorem 3 could be obtained for the Kolmogorov or Gelfand numbers. There seem to be some promising developments in this direction.

References

1. F.E. Browder, On the spectral theory of elliptic differential operators, I, Math. Ann. 142 (1960/61), 22-130.
2. B. Carl, Entropy numbers, entropy moduli, s-numbers and eigenvalues of operators in Banach spaces (preprint).
3. B. Carl, Entropy numbers of embedding maps between Besov spaces with an application to eigenvalue problems, Proc. Roy. Soc. Edinburgh 90A (1981), 63-70.
4. B. Carl and H. Triebel, Inequalities between eigenvalues, entropy numbers and related quantities of compact operators in Banach spaces, Math. Ann. 251 (1980), 129-133.
5. D.E. Edmunds and W.D. Evans, Elliptic and degenerate elliptic operators in unbounded domains, Ann. Scuola Norm. Sup. Pisa 27 (1973), 591-640
6. D.E. Edmunds, W.D. Evans and J. Fleckinger, to appear.
7. D.E. Edmunds and M.F. Teixeira, Interpolation theory and measures of non-compactness, Math. Nachrichten (to appear).

8. D.E. Edmunds and H. Triebel, Entropy numbers for non-compact self-adjoint operators in Hilbert spaces, Math. Nachrichten 100 (1981), 213-219.

9. W.D. Evans, Semi-bounded Dirichlet integrals and the invariance of the essential spectra of self-adjoint operators, Proc. Roy. Soc. Edinburgh A75 (1975), 41-66.

10. J. Fleckinger, Estimation des valeurs propres d'opérateurs elliptiques sur des ouverts non bornés, Ann. Fac. Sci. Toulouse 12 (1980), 157-180.

11. K. Hayakawa, Interpolation by the real method preserves compactness of operators, J. Math. Soc. Japan 21 (1969), 189-199.

12. T. Kato, Perturbation theory for linear operators (Berlin-Heidelberg-New York: Springer-Verlag, 1966).

13. H. König, A formula for the eigenvalues of a compact operator, Studia Math. 65 (1979), 141-146.

14. A. Lebow and M. Schechter, Semigroups of operators and measures of non-compactness, J. Functional Anal. 7 (1971), 1-26.

15. J.-L. Lions and J. Peetre, Sur une classe d'espaces d'interpolation, Inst. Hautes Études Sci. Publ. Math. 19 (1964), 5-68.

16. G. Métivier, Valeurs propres de problèmes aux limites elliptiques irréguliers, Bull. Soc. Math. France Mem. 51-52 (1977), 125-219.

17. B.S. Mitjagin and A. Pelczýnski, Nuclear operators and approximative dimension, Proc. I.C.M. Moscow (1966), 366-372.

18. R.D. Nussbaum, The radius of the essential spectrum, Duke Math. J. 38 (1970), 473-478.

19. A. Persson, Compact linear mappings between interpolation spaces, Arkiv Math. 5 (1964), 215-219.

20. A. Pietsch, Operator ideals (Berlin: Verlag der Wissenschaften and North-Holland, 1978/80).

21. A. Pietsch, Weyl numbers and eigenvalues of operators in Banach spaces, Math. Ann. 247 (1980), 149-168.

22. C.A. Stuart, Some bifurcation theory for k-set contractions, Proc. Lond. Math. Soc. 27 (1973), 531-550.

23. M.F. Teixeira, Entropy numbers and interpolation, Math. Nachrichten (to appear).

24. H. Triebel, Interpolationseigenschaften von Entropie- und Durchmesseridealen kompakter Operatoren, Studia Math. 34 (1970), 89-107.

25. H. Triebel, Interpolation theory, function spaces, differential operators (Berlin: Verlag der Wissenschaften and North-Holland, 1978).

26. H. Weyl, Inequalities between the two kinds of eigenvalues of a linear transformation, Proc. Nat. Acad. Sci. U.S.A. 35 (1949), 408-11.

27. J. Zemanek, The essential radius and the Riesz part of spectrum (1980 preprint).

The singular perturbation approach to flame theory with chain and competing reactions

by

Paul C. Fife and Basil Nicolaenko

Combustion theory provides an extremely fertile ground for asymptotic analysis. This is due in part to the fact that the theory combines fluid dynamics and chemical reaction theory, both in themselves rich in difficult problems, and in part to the prevalence of small and large parameters appearing in the underlying equations.

One of the most notable large parameters expresses the fact that in combustion problems, the rates of some important chemical reactions typically depend very strongly on the temperature. In the context of the simplest cases of steady plane laminar flame theory, this fact results in a problem of the form [2]

$$LU = f\left(\frac{U-U_+}{\varepsilon}, \varepsilon\right)$$

This research supported by NSF Grant 79-04443 and by the Center for Nonlinear Studies, Los Alamos National Lab.

where ε is a small parameter. Here L is a (vector valued) second order differential operator in the space variable x which does not depend on ε, and U_+ is some reference vector. The solution of this problem has the following "layer" feature: the derivative U' undergoes an abrupt change with magnitude order of unity (as $\varepsilon \to 0$) in a small interval with length order of ε. This interval is called a flame layer because it is here, and only here, that the source (reaction) term f has a significant effect. The function U, therefore, can be approximated for small ε by a function with derivative discontinuous at one point.

This latter piecewise smooth function itself satisfies a boundary value problem (the "reduced problem") in which the parameter ε no longer appears: the nonlinearity has been replaced by a condition specifying the jump in U' at its point of discontinuity (i.e., across the flame).

The above description holds when there is only one chemical reaction in the flame process, and it is exothermic with high activation energy. See [1],[3] for a comprehensive treatment. When there is more than one reaction, the situation outlined above must be modified. Depending on the specific reaction network and the other given chemical data, a wide variety of flame configurations become possible. In this talk we shall explore these possibilities in the case of sequential and competing reaction pairs. In the former case there is some overlap of our results with those of Kapila and Ludford [4]. Pairs of reactions have also been considered in [5-8]. Cases were treated in [6-8] which are in some sense limiting cases of those described here. A much fuller account of our results will be given in [3]. In particular, the latter paper contains an extension of our methods to flames with competing-fuel reaction pairs. They admit immediate extension to most reaction networks with two or three exothermic reactions with high activation energy.

Our analysis is largely formal, although great care is taken to justify the approximations made. They are based on the smallness of certain parameters ε_i. Often, other parameters will be assumed to be "order of unity", or "O(1)". This concept means these parameters are neither very large nor very small; otherwise it is left without strict definition. Effectively, it means that the error introduced by an approximation should be estimable in terms of some ε_i, or a positive power of it, multiplied by a coefficient which depends on these other parameters. The other parameters, therefore, should not be so large or small as to vitiate the accuracy of the approximation.

We begin with a description of the asymptotics and reduced problem for the standard case of a single reaction.

1. Simple flames

Without giving the physical derivation, we simply write the basic problem for laminar flames with a single reaction $A \to P$. We call them simple flames. Here $U = \begin{bmatrix} T \\ Y \end{bmatrix}$, T and Y are dimensionless temperature and concentration of reactant A, respectively. The problem is to determine the "profile" $U(x)$ and the mass flux M from the conditions

$$DU'' - MU' = -\omega(U)K, \tag{1a}$$

$$U(-\infty) = U_-, \; Y(\infty) = 0. \tag{1b}$$

Here D is a positive definite transport matrix which is taken for simplicity to be constant (this is not necessary [3]), K is the vector $\begin{bmatrix} Q \\ -1 \end{bmatrix}$, and ω is a reaction rate, which we take to be of Arrhenius type:

$$\omega = BYe^{-E/T}, \tag{2}$$

with B and E (proportional to the "activation energy") positive constants. Throughout this paper, we assume that all elements of matrices such as D, D^{-1}, and vectors K are $O(1)$ (see [3] for more general circumstances).

By integrating (1a) with use of (1b), it is seen immediately that $U(\infty)$ satisfies

$$U(\infty) = U_+ \equiv \begin{bmatrix} QY_- \\ 0 \end{bmatrix}. \tag{3}$$

Thus $T_+ = QY_-$. We assume

$$\varepsilon \equiv T_+^2/E \ll 1 \tag{4}$$

and on the basis of this relation, approximate ω by

$$\omega = \frac{\eta}{\varepsilon}\left(\frac{Y}{\varepsilon}\exp\left[(T - T_+)/\varepsilon\right]\right), \tag{5}$$

where

$$\eta = B\,\varepsilon^2 \exp(-E/T_+) . \tag{6}$$

This approximation is obtained by writing the exponent of (2) as a function of $t = (T-T_+)/\varepsilon$, and taking the first two terms in its Taylor series expansion in t about $t = 0$. The smallness of ε implies (see [3]) that it can be neglected for temperatures outside a neighborhood of the maximum temperature T_+.

We now perform a scale transformation $\bar{x} = x\eta^{1/2}$, $\bar{M} = M\eta^{-1/2}$; (1a) becomes (differentiation is now with respect to $\bar{x}$)

$$D\ddot{U} - \bar{M}\dot{U} = -\frac{1}{\varepsilon}\,\omega^*(u)K, \tag{7}$$

where $u = \begin{bmatrix} t \\ y \end{bmatrix} = (U-U_+)/\varepsilon$, and

$$\omega^*(u) = \frac{\varepsilon}{\eta}\,\omega(U) = ye^t . \tag{8}$$

We postulate a "reduced problem" of the following form, for $U(\bar{x})$ (assumed continuous) and $\bar{M}$:

(DE) $\quad D\ddot{U} - \bar{M}\dot{U} = 0, \quad \bar{x} \neq 0,$ (9a)

(BC) $\quad U(\pm\infty) = U_\pm$ (9b)

(CC) $\quad [D\dot{U}]_{0-}^{0+} = -\,\beta K.$ (9c)

Here the scalar β in the "connection condition" (9c) is to be determined a priori by a layer analysis from the function ω^*.

By stretching variables $\xi = x/\varepsilon$, $u = \dfrac{U-U_+}{\varepsilon}$, and then neglecting the term formally of order ε, we write (7) as

$$Du'' + \omega^*(u)K = 0 \tag{10}$$

This is to be solved for $u = u(\xi)$ under the boundary condition

$$u(\infty) = 0 \tag{11}$$

Since, from (10), u'' is a multiple of $D^{-1}K$ for each ξ , it follows from (11) that u itself must be also:

$$u(\xi) = \sigma(\xi)\, D^{-1} K. \tag{12}$$

Using (12) in (10) and applying (11), we find a problem for σ alone:

$$\sigma'' + \omega^*(\sigma D^{-1} K) = 0, \; \sigma(\infty) = 0. \tag{13}$$

This has a first integral

$$(\sigma')^2 - \Omega(\sigma) = 0,$$

where $\Omega(\sigma) = 2 \int_\sigma^0 \omega^*(sD^{-1}K)ds.$

Assuming now that $D^{-1}K$ has first component positive and the second negative, we see from (8) that $\Omega(-\infty)$ exists, can be calculated, and represents $(\sigma'(-\infty))^2$. The matching condition between the flame layer and the "outer" solution is

$$u'(-\infty) = \dot{U}(0-). \tag{14}$$

From this, (12), and the fact that necessarily $U \equiv U_+$ for $\bar{x} > 0$, we may identify the constant β in (9c) as

$$\beta = \sigma'(-\infty) = (2 \int_{-\infty}^0 \omega^*(sD^{-1}K)ds)^{1/2}. \tag{15}$$

The specification of β completes the definition of the reduced problem (9). What remains is to solve it. To do so, we note, by integrating (9a), that

$$[D\dot{U} - \bar{M}U]_{-\infty}^{0-} = 0 = D\dot{U}(0-) - \bar{M}U_+ + \bar{M}U_-.$$

$$= \beta K - \bar{M}Y_- K,$$

hence

$$\bar{M} = \beta / Y_- . \tag{16}$$

With $\bar{M}$ known, (9a) may now be solved for $\bar{x} < 0$, with known initial conditions at $\bar{x} = 0$. There is a unique solution, and it approaches the correct limit value U_- as $\bar{x} \to -\infty$. This completes the solution of the reduced problem.

2. The effects of other reactions on simple flames

Consider a simple flame profile, as described in sec. 1, involving a reaction term $\omega_1(U)K_1$ as in (2), (5), in which the subscript "1" is appended to the quantities ε, K, η, E, B, and Y. Thus Y_1 is the concentration of the species A_1 which is depleted at the flame. We now perturb the problem by adding another reaction, with a concomitant second term $\omega_2 K_2$ on the right of (1a). Our main result will be a simple criterion for judging whether the second reaction constitutes a negligible perturbation of the given flame. Rescaling with use of η_1, we obtain the following analog of (7):

$$D\ddot{U} - \overline{M}\dot{U} = -\frac{1}{\varepsilon_1}\overset{*}{\omega}_1(u)\,K_1 - \frac{1}{\eta_1}\omega_2(U)K_2. \tag{17}$$

We assume that

$$\varepsilon_2 = T_+^2/E_2 \ll 1$$

is also small, so that the effect of ω_2 for $\bar{x} < 0$ is negligible, in comparison with its effect in the flame and for $\bar{x} > 0$.

There are two important cases to be considered:

(I) No reactant for the second reaction is depleted in the original flame;

(II) A reactant for the second reaction is so depleted.

Case I. This case will apply, for example, to our study of sequential reactions in sec. 3. In this case, ω_2 remains strictly positive behind the flame, and may be represented by

$$\omega_2 = B_2 Y_2 e^{-E_2/T} .$$

We are assuming, for simplicity only, that the second reaction, as well as the first, is first order: $A_2 \to P$. As long as T deviates by amount $o(\varepsilon_2)$ from T_+, we have

$$\omega_2 \leq CB_2 e^{-E_2/T_+}$$

for some constant $C = O(1)$, and the last term in (17) is bounded in magnitude by

$$q \equiv CB_2 e^{-E_2/T_+} \Big/ B_1 \varepsilon_1^2 e^{-E_1/T_+} . \tag{18}$$

Since the first term, due to depletion of A_1, vanishes behind the flame, the effect of the second reaction can be gauged by examining the solution of the initial value problem

$$D\ddot{U} - \overline{M}\dot{U} = f(\overline{x}),\ U(0) = U_+,\ \dot{U}(0) = 0,$$

for vectors f with $|f| \leq q$. The representation

$$U(\overline{x}) = \int_0^{\overline{x}} \exp[\overline{M}D^{-1}(\overline{x}-s)]D^{-1}f(s)ds + U_+$$

yields

$$|U(\overline{x})-U_+| \leq Cqe^{\lambda\overline{x}},$$

for some constants C and $\lambda = O(1)$, since $\overline{M}$ and D^{-1} are $O(1)$ quantities. If, therefore,

$$q = O(\varepsilon_2),$$

it follows that $|U(\overline{x}) - U_+| \ll 1$ for $\overline{x} \geq 0$ restricted to any interval of length $O(1)$.

We now define functions

$$H_i(T) = \overline{\varepsilon}(\ell n\ B_i - E_i/T), \tag{19}$$

where

$$\overline{\varepsilon} = T_+^{\ 2}/\mathrm{Min}(E_1,E_2). \tag{20}$$

It is easily seen that $H_i'(T)$ is at most $O(1)$ for each i, and $H_k'(T_+) = 1$, where k is such that $E_k = \mathrm{Max}[E_1,E_2]$. Suppose, now, that

$$H_2(T_+) - H_1(T_+) = -\nu < 0. \tag{21}$$

then from (18),

$$q = \frac{c}{\varepsilon_1^{\ 2}} e^{-\nu/\overline{\varepsilon}} \leq C\overline{\varepsilon}^{\ -2}\ e^{-\nu/\overline{\varepsilon}} = o(\varepsilon_2),$$

if (for example) ν is an $O(1)$ quantity.

Our conclusion is that (21) is a sufficient condition for ω_2 to have a negligible effect on the original simple flame in regions a distance $O(1)$ fro the flame (actually, it is clear that much larger regions could be allowed).

Case II. In this case, we assume for simplicity that $Y_2 = Y_1 = Y$. A prime example will be the branching network in Sec. 4. The last reaction rate in (17) becomes

$$\frac{1}{\eta_1}\omega_2 = \frac{1}{\eta_1}B_2 Y e^{-E_2/T}$$
$$\cong \frac{\varepsilon_1}{\eta_1} B_2 e^{-E_2/T_+} y e^{\gamma t}$$
$$= \frac{1}{\varepsilon_1} r\, \omega_2^*(u), \qquad (22)$$

where we have set

$$\gamma = {}^{E_2}/E_1 \, , \; \omega_2^* = y e^{\gamma t}, \qquad (23)$$

and

$$r = B_2 e^{-E_2/T_+}/B_1 e^{-E_1/T_+} . \qquad (24)$$

We assume $\gamma = O(1)$ for this exposition. Since Y is depleted, $U = \text{const} = U_+$ for $\overline{x} > 0$, , so the stretched variable $u = (U - U_+)/\varepsilon_1$ satisfies the limit condition $u(\infty) = 0$. The layer problem is found from (17), (22), and assumes the form

$$Du'' + \omega_1^*(u)K_1 + r\omega_2^*(u)K_2 = 0, \qquad (25a)$$

$$u(\infty) = 0. \quad (25b)$$

Reasoning in the same manner as before, we see any solution must be of the form

$$u(\xi) = -\sigma_1(\xi)D^{-1}K_1 - \sigma_2(\xi)D^{-1}K_2 \qquad (26)$$

for some scalars σ_i. Substituting into (25), we obtain the problem

$$\sigma_1'' - \omega_1^*(-\Sigma\, \sigma_i D^{-1}K_i) = 0, \qquad (27a)$$

$$\sigma_2'' - r\omega_2^*(-\Sigma\sigma_i D^{-1}K_i) = 0, \qquad (27b)$$

$$\sigma_i(\infty) = 0. \qquad (27c)$$

The following result will be useful to us latter, as well as in the present context.

Lemma. Assume the vectors $D^{-1}K_i$ have positive first and negative second components. Then the problem (27) has, for each $r > 0$, a unique (except for translation of the independent variable ξ) positive solution $\sigma_i(\xi)$. It satisfies $-\infty < \sigma_i'(-\infty) < 0$. The quantities $-a_i = \sigma_i'(-\infty)$ depend continuously on r, and satisfy

$$\lim_{r\downarrow 0} \frac{a_2}{a_1} = 0 \; ; \quad \lim_{r\uparrow\infty} \frac{a_2}{a_1} = \infty . \tag{28}$$

Proof: The equation (27 a,b) may be written as a first order system

$$\begin{aligned} \sigma_1' &= -p_1 , \\ p_1' &= -\omega_1^* , \\ \sigma_2' &= -p_2 , \\ p_2' &= -r\omega_2^*. \end{aligned} \tag{29}$$

The object is to find a trajectory approaching the critical point 0 as $\xi \to \infty$ from the positive quadrant $P = \{\sigma_i > 0, \; p_i > 0\}$. It may be checked that the stable manifold is one dimensional and one branch of it indeed approaches 0 from P. We follow that branch backward: in terms of $s = -\xi$ with "$\cdot$" = "$\frac{d}{ds}$", (29) becomes

$$\begin{aligned} \dot\sigma_1 &= p_1 \\ \dot p_1 &= \omega_1^* \\ \dot\sigma_2 &= p_2 \\ \dot p_2 &= r\omega_2^* . \end{aligned} \tag{30}$$

Let the positive constants β be defined by

$$D^{-1}K_i = \begin{bmatrix} \beta_{1i} \\ -\beta_{2i} \end{bmatrix} .$$

Then for $w = \begin{bmatrix} \sigma_1 \\ p_1 \\ \sigma_2 \\ p_2 \end{bmatrix} \in P,$

$$y = \sum_i \beta_{2i}\sigma_i > 0 , \qquad (31)$$

so $\omega_i^* > 0$. Therefore the right sides of (30) are positive in P, and it follows that the trajectory entering P stays there.

In fact, $w(\xi)$ is monotone increasing, and from (30), (31), we have, for some constants C_i, c_i

$$\sigma_i \geq c_1(s-1)$$

$$y \leq C_2(\sigma_1 + \sigma_2) ,$$

$$t \leq -c_3(\sigma_1 + \sigma_2) ,$$

$$\omega_1^* \leq C_2(\sigma_1 + \sigma_2)e^{-c_3(\sigma_1+\sigma_2)}$$

$$\leq C_4 e^{-c_5(\sigma_1+\sigma_2)}$$

$$\leq C_5\, e^{-c_6 s} .$$

Hence from (30) again,

$$p_1 \leq C_6 + \int_0^\infty C_5 e^{-c_6 s}\, ds \leq C_7 .$$

Similar reasoning shows

$$p_2 \leq C_7 .$$

Therefore $p_i(s)$ approach finite limits $a_i = p_i(\infty)$. This proves the first parts of the lemma. We skip the proof of continuity in r. When $r = 0$, the stable manifold we are following backwards lies entirely on the plane

$\sigma_2 = p_2 = 0$, so we clearly have $a_2 = 0$, $a_1 > 0$. This establishes the first limit relation (28). For the second, we transform (30) by setting $s = \bar{s}r^{-1/2}$, $p_i = \bar{p}_i\, r^{1/2}$, $\sigma_i = \bar{\sigma}_i$, then

$$\dot{\bar{\sigma}}_1 = \bar{p}_1 ,$$

$$\dot{\bar{p}}_1 = \frac{1}{r}\, \omega_1^* (\bar{\sigma}_1, \bar{\sigma}_2),$$

$$\dot{\bar{\sigma}}_2 = \bar{p}_2 ,$$

$$\dot{\bar{p}}_2 = \omega_2^* .$$

As $r \uparrow \infty$, $\frac{1}{r} \downarrow 0$ and we use the argument symmetric to the above: it follows that $\bar{p}_1(\infty) \downarrow 0$, hence $p_1(\infty) = r^{1/2}\, \bar{p}_1(\infty) \downarrow 0$. Thus $a_1 \downarrow 0$, $a_2 > 0$. This gives us the second of (28), and completes the proof.

From (26), we now have

$$Du'(-\infty) = a_1 K_1 + a_2 K_2 .$$

The matching condition is $\dot{U}(0-) = u'(-\infty)$; therefore

$$D\dot{U}(0-) = a_1(r) K_1 + a_2(r) K_2 . \tag{32}$$

This enables us to formulate a reduced problem for Case (II), analogous to (9):

$$D\ddot{U} - \bar{M}\dot{U} = 0, \quad \bar{x} \neq 0; \tag{33a}$$

$$U(\pm\infty) = U_\pm ; \tag{33b}$$

and (32). The solution is obtained as follows: from (33a) we have

$D\dot{U}(0-) - \overline{M}(U_+ - U_-) = 0$; then from (32) and this,

$$U_+ - U_- = \frac{a_1(r)}{\overline{M}} K_1 + \frac{a_2(r)}{\overline{M}} K_2 \equiv \alpha_1 K_1 + \alpha_2 K_2 .$$

There are actually two conceivable formulations of the problem here:

(a) Given U_+, determine $\overline{M}$, r, and $U(\overline{x})$;

(b) Given r, determine $\overline{M}$, U_+, and $U(\overline{x})$.

Of course, either case ignores the fact that r and T_+ are also related through (24), and so (24) will have to serve as a consistency check.

Both problems are solvable. For example in (a), $\alpha_i > 0$ will be given, so we first determine r from the condition

$$\frac{a_2(r)}{a_1(r)} = \frac{\alpha_2}{\alpha_1} . \qquad (34)$$

By the lemma (in particular (28)), there exists a value of r which satisfies this relation. It is not clear at this point whether r is unique. Knowing r, we then determine $\overline{M} = a_1/\alpha_1$, and hence $U(\overline{x})$ from (33a).

We now return to the functions H_i (19), and suppose that (21) holds. It follows from (24) that $\overline{\varepsilon} \ln r = -\nu$, so that

$$r = e^{-\nu/\overline{\varepsilon}} << 1.$$

Therefore by the lemma again, $a_2(r) << 1$. But when $a_2 = 0$, the reduced problem (33), (32) is identical to (9), and it is easy to see that the solution depends continuously on a_2. Hence it is justified to neglect reaction (ii).

Our conclusion, then, is the same as in Case I: (21) is the criterion by which the second reaction does not significantly alter the original flame. Having developed a general criterion, we proceed to look at specific networks.

3. Reactions in series

We now consider a combustion process involving a pair of reactions in series:

$$A \xrightarrow{(i)} B \xrightarrow{(ii)} P ,$$

both exothermic with high activation energy. The unburned state, in the form $U_- = \begin{bmatrix} T_- \\ Y_{A_-} \\ Y_{B_-} \end{bmatrix}$ is prescribed. To conceptualize the possible flame structures, we first follow the approach of sec. 2: we imagine a simple flame resulting from only one of the reactions, the other being artificially suppressed. We then ask whether the criterion is met, whereby the other reaction, if no longer suppressed, would have a negligible effect. If so, then that simple flame is a realizable structure. After this is done, we investigate possible nonsimple flames.

There are four basic simple flames:

F1, produced by reaction (i) only, converting the given unburned state to a partially burned one:

$$U_- \to U_1 \equiv \begin{bmatrix} T_1 \\ 0 \\ Y_{A_-}+Y_{B_-} \end{bmatrix} .$$

From (3) it is seen that $T_1 = Q_1 Y_{A-} + T_-$.

F2, produced by (ii) only:

$$U_- \to U_2 \equiv \begin{bmatrix} T_2 \\ Y_{A_-} \\ 0 \end{bmatrix},$$

with $T_2 = T_- + Q_2 Y_{B_-}$.

F12, produced by (ii) acting on the product (final state) of F1:

$$U_1 \to U_+ \equiv \begin{bmatrix} T_+ \\ 0 \\ 0 \end{bmatrix} .$$

where $T_+ = T_- + Q_1 Y_{A-} + Q_2(Y_{A-} + Y_{B-})$.

F21, acting on the product of F2. Here the reaction is of the type $A \to P$ with rate identical to that of (i), but with heat release equal to the sum $Q_1 + Q_2$. One can think of this flame as converting A to B by reaction (i), and B thereupon being almost immediately converted to P by a faster reaction (ii).

We again define functions $H_i(T)$ according to (19), with $\bar{\varepsilon}$ given by (20). We assume that T_-, T_1, T_2, and T_+ are all of the same order of magnitude, and $T_1 - T_-$, $T_2 - T_-$, $T_+ - T_1$, $T_+ - T_2$, are all $O(1)$ quantities. The temperature scale is adjusted so that $T_+ - T_- = 1$.

We consider the consequences of various inequalities involving the functions H_i. When such an inequality is written, it will be assumed that the difference of the two quantities concerned is $O(1)$. This is the typical case, because the range of at least one of the functions H_i is of order unity.

Immediate consequences of sec. 2 are:

1) If $H_2(T_1) < H_1(T_1)$, then F1 is realizable.

2) If $H_1(T_2) < H_2(T_2)$, then F2 is.

The next structures to be considered are possible combinations of 2 flames. Note, from (16) and the fact that $M = \overline{M}\eta^{1/2}$, that the mass flux for F1 is

$$M_1 = C_1 B_1 \varepsilon_1^2 \exp(-E_1/T_1) \,, \tag{35}$$

where

$$\varepsilon_1 = T_1{}^2/E_1 \,,$$

and C_1 is some constant, $C_1 = O(1)$.

Hence

$$\overline{\varepsilon}\,\ell n M_1 = \overline{\varepsilon}(\ell n\, B_1 - E_1/T_1 + 2\,\ell n\,\varepsilon_1)) + O(\overline{\varepsilon})$$

But $|\ell n\,\varepsilon_1| = \ell n \frac{1}{\varepsilon_1} \leq \ell n\, 1/\overline{\varepsilon}$;

$$\overline{\varepsilon}\,\ell n\, M_1 = H_1(T_1) + O(\overline{\varepsilon}\,\ell n\, 1/\overline{\varepsilon}).$$

In a similar manner,

$$\overline{\varepsilon}\,\ell n\, M_2 = H_2(T_2) + o(1) \,,$$

$$\overline{\varepsilon}\,\ell n\, M_{12} = H_2(T_+) + o(1) \,,$$

$$\overline{\varepsilon}\,\ell n\, M_{21} = H_1\,(T_+) + o(1) \,.$$

It turns out that the mass fluxes of these four simple flames are ordered in the same way as their velocities. Thus, for example, if $H_2(T_+) < H_1(T_1)$, then F12, built on the products of F1, is slower than F1 itself, and so one can conceive of the two flames both existing, but the distance between them ever increasing. The flame F1 precedes, and is faster than, F12. We call this total configuration

F1,12:

3) If $H_2(T_+) < H_1(T_1)$, then F1,12 is realizable.

Similarly,

4) If $H_1(T_+) < H_2(T_2)$, then F2,21 is.

The most interesting flames, however, are seen when neither 3) nor 4) hold. This last case is split into two subcases: $H_1(T_1) < H_2(T_+) < H_1(T_+)$, and $H_2(T_2) < H_1(T_+) < H_2(T_+)$. It can be easily checked that these two are exclusive and, excluding cases when equality holds somewhere, exhaust the complement of 3) and 4). Since 3) and 4) do not hold, the diverging configurations F1,12 and F2,21 are not possible. Indeed if such a combination of flames like F1,12 existed at one time, then the rear flame F12 would approach the forward flame F1. As it does so, its effect is to heat up the forward one. As this happens, the speed of the forward increases (this is feasible because our expression for M, as in (35), is an increasing function of T). The question now is whether this heating from the rear is sufficient to raise the velocity of the forward part to equal that of the trailing F12. Of course, the temperature of F1 cannot by this mechanism be raised above T_+, which is the temperature of F12. And it seems reasonable that the equality of speeds will be attained at a temperature T_o such that $H_1(T_o) = H_2(T_+)$. Therefore our conjecture (to be verified) is that such a tandem configuration, with F1 (ahead) and F12 (behind) traveling at a fixed distance apart, is realizable if there is a T_o satisfying the above equation. But this is true precisely in the first subcase. We call this tandem flame F1-12.

5) If $H_1(T_1) < H_2(T_+) < H_1(T_+)$, then F1-12 is realizable.

Similarly,

6) If $H_2(T_2) < H_1(T_+) < H_2(T_+)$, then F2-21 is realizable.

It should again be emphasized that at this point, (5) and (6) are just conjectures. However, they can be verified in the following manner. If a tandem structure exists, it should correspond to the solution of a reduced problem in the spirit of (9), and this reduced problem should be uniquely solvable.

This is, in fact, the case. For F1-12, for example, the reduced problem takes the form

$$D\ddot{U} - \overline{M}\dot{U} = 0, \ \overline{x} \neq 0, \ \overline{x} \neq \ell;$$

$$U(\pm\infty) = U_{\pm} ,$$

$$[D\dot{U}]_{0-}^{0+} = -\beta_1 K_1, \quad [D\dot{U}]_{\ell-}^{\ell+} = -\beta_2 K .$$

Here β_2, $U_\pm$, and $T(0) = T_o$ constitute the given data; the problem is then to determine β_1, ℓ, and $U(\bar{x})$. The problem has a unique solution. The constant β_2 is known by a layer analysis very like that in sec. 2. However, the coefficient β_1 should also be involved with a layer analysis. And indeed such an analysis can be performed; it yields a functional relation

$$\beta_1 = F(r, \dot{U}(0+)),$$

where r is given by (24) with T_+ in the denominator replaced by T_o. This relation can be inverted to yield

$$r = R(\beta_1, \dot{U}(0+)). \tag{36}$$

The proper consistency check on the above reduced problem, then, is (1) solve it, and find β_1 and $\dot{U}(0+)$; (2) determine r from (36); see if this r satisfies (24) (T_+ in denominator replaced by T_o). We rewrite (24) in the form

$$\bar{\varepsilon} \ln r = H_2(T_+) - H_1(T_o). \tag{37}$$

Now T_o was chosen to make the right side vanish; but if we now reason that this is only the first approximation to the true value of T_o, then (37) can be used to obtain a corrected value for T_o. Since $H_1'(T_o) = O(1)$, and $r = O(1)$, it turns out that the correction will only be $O(\bar{\varepsilon})$, so the original value of T_o will be correct within our general range of accuracy. The details of the tandem analysis can be found in [3].

In summary, the possibility of the various flame types 1,2, etc., can be found by testing each of the cases 1) - 6) in turn. Cases 3) - 6) are mutually exclusive, but any of the latter could hold in conjunction with 1) or 2). There result nonunique flame structures; which one is realized in a given observation would depend on how and to what extent the gas mixture is ignited. An example of nonuniqueness is the case when $T_1 < T_2$, and

$$H_2(T_1) < H_1(T_1) < H_1(T_2) < H_2(T_2) < H_2(T_+) < H_1(T_+) .$$

In this case, types 1, 2, and 1-12 are all realizable.

4. Reactions in parallel

The combustion process now allows the reactant A to be consumed in two competing fashions:

$$A \quad \begin{array}{l} \text{(i)} \quad\longrightarrow\quad P_1 \\ \\ \text{(ii)} \quad\longrightarrow\quad P_2 \end{array}$$

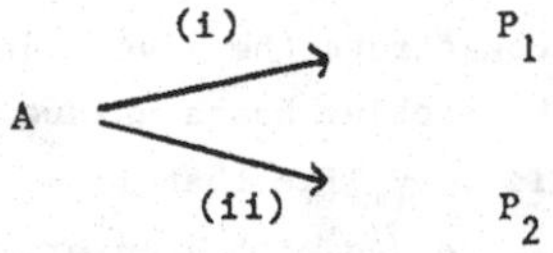

Again, each reaction is exothermic with high activation energy, and the

unburned state $U(-\infty) = U_- = \begin{bmatrix} T_- \\ Y_- \end{bmatrix}$ is prescribed (here Y represents the

dimensionless concentration of A).

There are only two conceivable simple flames: F1 and F2, with final
temperatures $T_1 = T_- + 0_1 Y_-$ and $T_2 = T_- + 0_2 Y_-$ respectively. For the sake
of definiteness, we assume $0_1 < 0_2$, hence

$$T_1 < T_2 .$$

Temperature is scaled so that $T_2 - T_- = 1$. Assume
$T_2 - T_1$, $T_1 - T_- = O(1)$. We define H_i again by (19) and (20) with T_+
replaced by T_2. According to the criterion in sec. 2, F1 is realizable if
$H_1(T_1) > H_2(T_1)$, and F2 is if $H_2(T_2) > H_1(T_2)$. Four cases therefore arise:

    (a) $H_1(T_i) < H_2(T_i)$, $i = 1,2$

    (b) $H_1(T_i) > H_2(T_i)$, $i = 1,2$,

    (c) $H_1(T_1) > H_2(T_1)$, $H_1(T_2) < H_2(T_2)$,

    (d) $H_1(T_1) < H_2(T_1)$, $H_1(T_2) > H_2(T_2)$.

The cases are depicted by these four sample graphs of functions H_1 (dotted lines
represent H_1 (T), dashed lines $H_2(T)$).

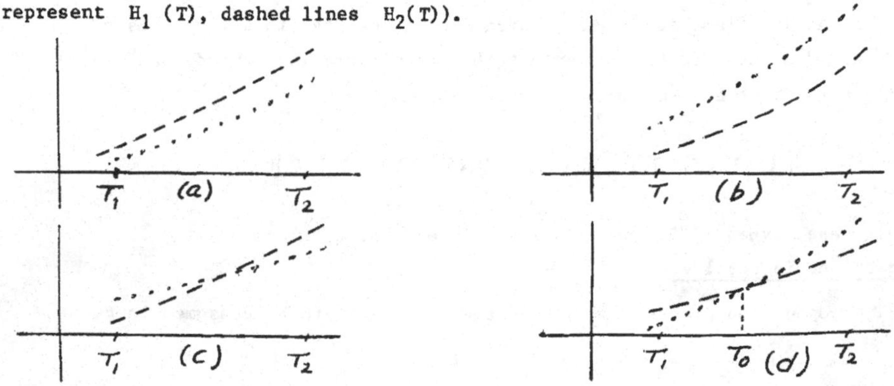

In case (a), only F2 is possible; in case (b), only Fl; in case (c), either Fl or F2, and in case (d), neither Fl nor F2. This suggests there may be an interactive, nonsimple, flame type in case (d), since neither simple flame can exist. The only way both reactions can be significant in the burning process is for their H's to be the same. It is clear from the graph that this occurs at a value $T_o \in (T_1, T_2)$. We therefore postulate a flame burning at temperature T_o in which both reactions take part. This situation can be fit into the framework of case II in sec. 3. such a flame would be governed by (17), with $U = \begin{bmatrix} T \\ Y \end{bmatrix}$. The reduced problem would be (33), (32). More specifically, variant (A) of that problem applies, since

$U_+ = \begin{bmatrix} T_o \\ 0 \end{bmatrix}$ is given.

We now indicate the solution of that problem. First, it is clear that for some $\theta \in (0,1)$, T_o may be represented as $T_o = \theta T_1 + (1 - \theta) T_2$. Hence $T_o - T_- = \theta Y_- Q_1 + (1-\theta) Y_- Q_2$ and since $K_1 = \begin{bmatrix} Q_1 \\ -1 \end{bmatrix}$, we have

$$U_+ - U_- = \alpha_1 K_1 + \alpha_2 K_2 ,$$

where $\alpha_1 = \theta Y_-$, $\alpha_2 = (1-\theta) Y_-$. The ratio r is now determined from (34), and hence $\overline{M} = a_1/\alpha_1$, and the profile $U(\overline{x})$ from (33a).

The consistency check now requires (24) to hold with T_+ replaced by T_o. This equation can be written

$$\overline{\varepsilon}\, \ell n r = H_2(T_o) - H_1(T_o) . \tag{38}$$

Since T_o was chosen so the right side vanishes, there appears to be an inconsistency. But the inconsistency is only superficial; the equation (38) simply seems that the previous value of T_o should be adjusted (by an amount $O(\overline{\varepsilon})$) to obtain a more accurate value.

It should be remarked that the "mixed" flame constructed in case (d) apparently also exists in case (c), because again, H_i have a common value at T_o. But in that case, the mixed flame is unstable in a certain sense: if its temperature is perturbed by a small amount, it will evolve into one of the two simple flames. In case (d), this instability does not occur.

Bibliography

1. J. Buckmaster and G. S. S. Ludford 1982, Theory of Laminar Flames, Cambridge University Press, New York.

2. P. C. Fife 1982, Propagating fronts in reactive media, in: Nonlinear Problems, Present and Future, A. Bishop, D. Campbell, B. Nicolaenko, eds., North-Holland.

3. P. Fife and B. Nicolaenko 1982, Two-reaction flame propagation, in preparation.

4. A. Kapila and G. S. S. Ludford 1977, Two-step sequential reactions for large activation energies, Combustion and Flame, 167-176.

5. A. Liñan 1971, A theoretical analysis of premixed flame propagation with an isothermal chain reaction, Tech. Report, Inst. Nac. Tec. Aerospacial "Esteban Terradas", Madrid.

6. S. B. Margolis 1982, Lean flame propagation with competing chemical reactions, Combust. Sci. and Technol., to appear.

7. S. B. Margolis and M. J. Matkowsky 1981, Flame propagation with multiple fuels, SIAM J. Appl. Math., to appear.

8. S. B. Margolis and B. J. Matkowsky 1981, Flame propagation with a sequential reaction mechanism, SIAM J. Appl. Math., to appear.

On the singular values of non-self-adjoint operators of Schrödinger type

J. Fleckinger

In recent times non-self-adjoint elliptic operators, and especially non-self-adjoint Schrödinger operators, have attracted considerable attention. Here we study the asymptotic behaviour of the singular values of operators of this latter type. In particular, we obtain results for the Schrödinger operator, with a complex potential, on an unbounded domain Ω in $\mathbb{R}^n$ and with homogeneous Dirichlet boundary conditions.

In [2] estimates were established for the spectra of Schrödinger operators with complex potential, even with Neumann boundary conditions, by using tesselations by cubes. Indeed, [2] is an extension of [1] to deal with complex potentials. The present paper deals with the spectrum and the asymptotic behaviour of the singular values of much more general operators but with Dirichlet boundary conditions only.

We are concerned with operators T of 'Schrödinger type': $T = A + q + R$, where A is a 'nice' elliptic operator of order $2m$, q is a smooth, positive potential tending to ∞ at infinity, and the perturbation R is a differential operator of order $k < 2m$. Under suitable assumptions we prove that the spectrum of T, with Dirichlet boundary conditions, is discrete and we obtain an asymptotic estimate for the singular values s_j of T, that is, the eigenvalues of the positive square root of T^*T. These results are obtained by comparison of the s_j with the eigenvalues λ_j of the self-adjoint operator $S = A + q$, using the known asymptotic estimates for the λ_j (cf. [1], [4], [5], [6], [8], [11]) and the max-min principle. In particular we adapt to our operators Ky Fan's inequality and various results concerning compact operators

([3], [7], [10]). A consequence of the theorems established here is that we can extend the estimates valid for a 'smooth' potential ([4], [5], [6], [8], [11]) to some non-smooth ones. Since the asymptotic results proved in [1] and [4] do not assume that the boundary $\partial\Omega$ of the domain Ω is compact, our results are valid for non-compact $\partial\Omega$. Moreover, when $\Omega = \mathbb{R}^n$ results with a remainder term are obtained.

I. Facts about the self-adjoint case

Here we recall the results of [4] and [5] that we shall need. Let Ω be an unbounded domain in $\mathbb{R}^n$ and let m be a positive integer. Denote by $H^m(\Omega)$ and $H^m_0(\Omega)$ the usual Sobolev spaces of order m defined on Ω and with norm $||\cdot||_{H^m(\Omega)}$; $||\cdot||$ will stand for the $L^2(\Omega)$ norm.

(1) Let a be an integrodifferential form defined on $H^m(\Omega)$ by

$$a(u,v) = \int_\Omega \sum_{|\alpha|,|\beta|\le m} a_{\alpha\beta}(x)\, D^\alpha u(x)\, \overline{D^\beta v(x)}\, dx;$$

we suppose that a is hermitian, continuous and coercive on $H^m_0(\Omega)$, and that $a_{\alpha\beta} = \bar{a}_{\beta\alpha} \in L^\infty(\Omega)$ when $|\alpha| + |\beta| < 2m$; $a_{\alpha\beta} \in C(\bar{\Omega})$ if $|\alpha| = |\beta| = m$.

(2) Let q be a real-valued function defined on Ω, tending to $+\infty$ at infinity, bounded below by a positive number and in $L^1_{loc}(\Omega)$.

Denote by $V(\Omega)$ the completion of $C^\infty_0(\Omega)$ with respect to the norm

$$||u||_{V(\Omega)} = \{\int_\Omega \sum_{|\alpha|\le m} (|D^\alpha u(x)|^2 + q(x)|u(x)|^2)\, dx\}^{1/2}.$$

It is easy to prove that equipped with this norm $V(\Omega)$ is a Hilbert space and that the embedding of $V(\Omega)$ in $L^2(\Omega)$ is compact. The integrodifferential form $a_q = a + q$ defined on $V(\Omega)$ by

$$a_q(u,v) = a(u,v) + \int_\Omega q(x)u(x)\overline{v(x)}\, dx$$

is hermitian, continuous and coercive on $V(\Omega)$.

Let us denote by A (resp. L) the positive, self-adjoint, unbounded operator in $L^2(\Omega)$, associated with the variational problem $(H_0^m(\Omega), L^2(\Omega), a)$ (resp. $(V(\Omega), L^2(\Omega), a_q)$). We deduce from the compactness of the embedding of $V(\Omega)$ in $L^2(\Omega)$ that the spectrum of L is wholly discrete; it consists of an infinite sequence of eigenvalues of finite multiplicity:

$$0 < \lambda_1 \leq \lambda_2 \leq \ldots \leq \lambda_j \leq \ldots ,$$

where $\lambda_j \to \infty$ as $j \to \infty$; and we study the asymptotic behaviour, as $\lambda \to \infty$, of $N(\lambda, L, \Omega) = \sum_{\lambda_j \leq \lambda} 1$, the number of eigenvalues less than or equal to λ. When $A = -\Delta$ and $\Omega = \mathbb{R}^n$ the result is well-known; and in [1] an estimate for N is obtained when $A = -\Delta$ and Ω is unbounded, for homogeneous Dirichlet or Neumann boundary conditions.

The following hypotheses are made:

H1 : There exists $\varepsilon_0 > 0$ such that q and all the $a_{\alpha\beta}$ with $|\alpha| = |\beta| = m$ can be extended to $\tilde{\Omega} = \{x \in \mathbb{R}^n : \text{dist}(x, \Omega) < \varepsilon_0\}$ so that the extensions (still denoted by q and $a_{\alpha\beta}$) are such that for all $\varepsilon \in (0, \varepsilon_0)$, there exists $\eta > 0$ such that for all $(x,y) \in \tilde{\Omega} \times \tilde{\Omega}$, $|x-y| < \eta$ implies that $|q(x) - q(y)| \leq \varepsilon q(x)$ and $|a_{\alpha\beta}(x) - a_{\alpha\beta}(y)| \leq \varepsilon |a_{\alpha\beta}(x)|$ for all α, β with $|\alpha| = |\beta| = m$.

H2 : There is a constant $k > 0$ such that for all domains $\omega \subset \tilde{\Omega}$ and all u in $V_q(\omega)$, the set of restrictions to ω of elements of $V(\tilde{\Omega})$ endowed with the norm of $V(\omega)$, we have

$$a_q'(u,u) := (a' + q)(u,u) := \int_\omega \Big(\sum_{|\alpha|=|\beta|=m} a_{\alpha\beta}(x) D^\alpha u(x) \overline{D^\beta u(x)} + q(x)|u(x)|^2 \Big) dx$$

$$\geq k \|u\|^2_{V_q(\omega)}.$$

H3 : For every $\lambda > 0$, $\Omega_\lambda := \{x \in \Omega : q(x) < \lambda\}$ is Lebesgue measurable with measure $|\Omega_\lambda|$, and there exist $\lambda_1 \geq 0$ and $k' > 0$ such that for all $\lambda \geq \lambda_1$, $|\Omega_\lambda| \leq k' |\Omega_{\lambda/2}|$.

H4 : We consider tesselations of $\mathbb{R}^n$, $(Q_r)_{r\in\mathbb{Z}^n}$, by disjoint cubes with centres x_r and sides η; we suppose that $\lim_{\eta\to 0} \text{card}\,(J\backslash I)/\text{card}\,I = 0$ for all $\lambda \ge \lambda_1$, where $I = \{r \in \mathbb{Z}^n : \overline{Q}_r \subset \Omega_\lambda\}$ and $J = \{r \in \mathbb{Z}^n : Q_r \cap \overline{\Omega}_\lambda \neq \phi\}$.

These hypotheses are satisfied for the Schrödinger operator $-\Delta + (1 + |x|^2)^r$ $(r > 1)$, but $-\Delta + \log |x|$ $(|x| > 2)$ does not satisfy H3.

We have the following result:

Theorem 0. Suppose that (1), (2), H1 to H4 are satisfied. Then as $\lambda \to \infty$,

$$N(\lambda,L,\Omega) \sim \int_{\Omega_\lambda} \mu(x)\ (\lambda - q(x))^{n/2m}\ dx$$

where

$$\mu(x) = (2\pi)^{-n}\ \text{meas}\{\xi\ \ \mathbb{R}^n : \sum_{|\alpha|=|\beta|=m} a_{\alpha\beta}(x)\ \xi^{\alpha+\beta} < 1\}.$$

II. The perturbed operator

Let R be a differential operator of order $k \le m$, defined on Ω and with complex coefficients:

$$R = \sum_{|\alpha|\le k} r_\alpha D^\alpha, \quad \text{with } D(R) \supset D(L).$$

We suppose that

(3) the coefficients $r_\alpha (|\alpha| \neq 0)$ and all their derivatives are bounded on Ω;

(4) there is a positive constant $\delta \in (0,\frac{1}{2})$ such that $r_0 q^{-\delta}$ is bounded.

Then we have:

Lemma 1 Suppose that (1) - (4) hold. Then there exist two numbers $\gamma > 0$ and $a \in (0,1)$ such that for all $u \in D(L)$,

$$||Ru|| \le \gamma ||Lu||^{1-a}\ ||u||^a.$$

<u>Proof.</u> From (3) and (4) we see that if $u \in D(L)$,

$$||Ru||^2 \le c_1 \sum_{0<|\alpha|\le k} ||D^\alpha u||^2 + c_2 \int_\Omega |r_0 u|^2\, dx.$$

For all $u \in C_0^\infty(\Omega)$ and all $\alpha \in \mathbb{N}^n$ with $|\alpha| = j \le k \le m$,

$$\begin{aligned}
||D^\alpha u||^2 &= ||\widehat{D^\alpha u}||^2 \le c_3^j \int_{\mathbb{R}^n} (1+|\xi|^2)^j\, |\hat{u}(\xi)|^2\, d\xi \\
&\le c_3^j \left(\int_{\mathbb{R}^n} (1+|\xi|^2)^m\, |\hat{u}(\xi)|^2 d\xi\right)^{j/m} \left(\int_{\mathbb{R}^n} |\hat{u}(\xi)|^2\, d\xi\right)^{1-(j/m)} \\
&\le c_3^j \left(\sum_{|\beta|=m} ||D^\beta u||^2\right)^{j/m} ||u||^{2(1-j/m)} \\
&\le c_3^j\, ||u||_{V(\Omega)}^{2j/m}\, ||u||^{2(1-j/m)}.
\end{aligned}$$

Thus by the coerciveness of a_q,

$$(5) \qquad ||D^\alpha u||^2 \le c_4\, (a_q(u,u))^{j/m}\, ||u||^{2(1-j/m)}.$$

In addition,

$$(6) \qquad \int_\Omega |r_0 u|^2 dx \le c_5 \int_\Omega q^{2\delta} |u|^2\, dx \le c_6\left(\int_\Omega q\, |u|^2\, dx\right)^{2\delta} \left(\int_\Omega |u|^2\, dx\right)^{1-2\delta}$$

$$c_7\, (a_q(u,u))^{2\delta}\, ||u||^{2(1-2\delta)}.$$

But $C_0^\infty(\Omega)$ is dense in $V(\Omega)$ and so (5) and (6) hold on $V(\Omega)$. In particular they hold for all u in $D(L)$, and since for such u,

$$a_q(u,u) = (Lu,u),$$

it follows that

$$||Ru||^2 \le C||Lu||^\sigma\, ||u||^{2-\sigma},$$

which is equivalent to

$$||Ru||^2/||u||^2 \le C(||Lu||/||u||)^\sigma,$$

with $\sigma \in (0,1)$. Lemma 1 follows with $1 - a = \sigma/2$.

<u>Lemma 2</u> <u>The operator</u> R <u>is</u> L-<u>bounded with</u> L-<u>bound zero</u>.

<u>Proof</u>. From Lemma 1 and Young's inequality we see that for all $\varepsilon > 0$,

$$||Ru|| \leq c||Lu||^{1-a}\,||u||^{a} \leq c(1-a)\varepsilon||Lu|| + c\,a\,\varepsilon^{(a-1)/a}||u||.$$

Since ε may be made arbitrarily small the result follows.

Standard arguments (cf. [9]) enable us to deduce from Lemma 2 that $T = L + R$ is closed and m-sectorial. We also have the following

<u>Theorem 1.</u> <u>Under the hypotheses</u> (1) <u>to</u> (4), <u>the spectrum of</u> T <u>is discrete</u>.

<u>Proof.</u> For real $\lambda \neq 0$,

$$(7) \qquad T + i\lambda = L + i\lambda + R = [I + R\,(L + i\lambda)^{-1}]\,(L + i\lambda).$$

By Lemma 2, $||Ru|| \leq k||Lu|| + k'||u||$ with $k < 1$, and thus

$$||R(L + i\lambda)^{-1}|| \leq k||L(L + i\lambda)^{-1}u|| + k'||(L + i\lambda)^{-1}u||$$
$$\leq k||u|| + k'|\lambda|^{-1}\,||u||$$

since $||(L + i\lambda)u||^2 = ||Lu||^2 + |\lambda|^2\,||u||^2$. As $k < 1$ we may choose $|\lambda|$ large enough to show that there exists $k'' \in (k,1)$ such that $||R(L + i\lambda)^{-1}|| < 1$. Hence $(I + R(L + i\lambda)^{-1})^{-1}$ exists and is a bounded map of $L^2(\Omega)$ to itself. Since $(L + i\lambda)^{-1}$ is compact, so is $(T + i\lambda)^{-1} = (L + i\lambda)^{-1}\,[I + R(L + i\lambda)^{-1}]^{-1}$. Thus the spectrum of T is discrete.

III. Estimates for the singular values

A singular value s_j of T is an eigenvalue of $(T^*T)^{1/2}$. We now study the asymptotic behaviour of $M(\lambda,T,\Omega) = \sum_{s_j \leq \lambda} 1$ as $\lambda \to \infty$.

<u>Theorem 2</u> <u>Suppose that the hypotheses</u> (1) <u>to</u> (4) <u>and</u> (H1) <u>to</u> (H4) <u>are satisfied and that</u> $N(\lambda,L,\Omega) \sim c\lambda^{x}$ <u>for some</u> x, <u>as</u> $\lambda \to \infty$. <u>Then</u>

$M(\lambda,T,\Omega) \sim N(\lambda,L,\Omega)$, that is,

$$M(\lambda,T,\Omega) \sim \int_{\Omega_\lambda} \mu(x)\ (\lambda-q(x))^{n/2m}\, dx \quad \text{as} \quad \lambda \to \infty.$$

Examples $T_1 = -\Delta + (1+|x|^2)^k + i\frac{\partial}{\partial x}$ defined on $\Omega_1 = \{(x_1,x_2,\ldots,x_n) \in \mathbb{R}^n : x_i < x_n^2$ for $i = 1,2,\ldots,n-1\}$ satisfies the hypotheses above, and the behaviour of $M(\lambda,T_1,\Omega_1)$ is given by Theorem 2. The same estimate holds for $M(\lambda,T_2,\Omega_1)$, with $T_2 = -\Delta + (1+|x|^2)^k + i(1+|x|^2)^{k/p}$, $p \geq 2$.

As a consequence of Theorem 2 we can improve Theorem 0 in that the hypothesis on the smoothness of the potential can be weakened. In fact we deduce from Theorem 2 the following

Corollary 1. Let q_0 be a potential satisfying (2) and suppose that there exists a positive function q such that

(H5) $L = A + q$ satisfies the hypothesis of Theorem 0;

(H6) there exists $\delta \in (0,\frac{1}{2})$ such that $(q-q_0)\, q_0^{-\delta}$ is bounded.

Then if $N(\lambda,A+q,\Omega) \sim c\,\lambda^x$ for some x as $\lambda \to \infty$, we have $N(\lambda,A+q_0,\Omega) \sim N(\lambda,A+q,\Omega)$ as $\lambda \to \infty$.

Proof of Theorem 2

Here we adapt the results for compact operators of [3], [7] and principally [11]. Notice that $T = L + R = (I + RL^{-1})L$. Denote by E_j the j-dimensional space spanned by the eigenfunctions of L associated with the first j eigenvalues. Using the max-min principle we have a 'Ky Fan inequality' :

$$(8)\quad s_{j+k+1}^2 \geq \inf_{u \perp E_{j+k}} \frac{||Tu||^2}{||Lu||^2}\,\frac{||Lu||^2}{||u||^2} \geq \inf_{u \perp E_k} \frac{||(I+RL^{-1})Lu||^2}{||Lu||^2}\ \inf_{u \perp E_j} \frac{||Lu||^2}{||u||^2}$$

$$= \lambda_{j+1}^2 \inf_{Lu \perp E_k} \frac{||(I+RL^{-1})Lu||^2}{||Lu||^2}\,.$$

By Lemma 1, for all $v \in L^2(\Omega)$,

$$\frac{||RL^{-1}v||}{||L^{-1}v||} \le \gamma\left(\frac{||v||}{||L^{-1}v||}\right)^{1-a};$$

hence

$$(9) \qquad \frac{||RL^{-1}v||}{||v||} \le \gamma\left(\frac{||L^{-1}v||}{||v||}\right)^{a}.$$

Choose k_0 so large that for all $k \ge k_0$,

$$(10) \qquad \gamma\,\lambda_k^{-a} < \tfrac{1}{2};$$

that is, $\sup_{v\perp E_k} \gamma\left(\frac{||L^{-1}v||}{||v||}\right)^{a} < \frac{1}{2}$. For $Lu \perp E_k$, (9) implies that

$$\frac{||RL^{-1}(Lu)||}{||Lu||} \le \gamma\left(\frac{||L^{-1}(Lu)||}{||Lu||}\right)^{a} \le \gamma\,\lambda_k^{-a},$$

and so

$$\frac{||(I+RL^{-1})Lu||^2}{||Lu||^2} \ge (1-\gamma\,\lambda_k^{-a})^2.$$

Hence by (8),

$$(11) \qquad s_{j+k+1}^2 \ge \lambda_{j+1}^2\,(1+O(\lambda_k^{-a})).$$

Now denote by F_j the j-dimensional subspace spanned by the j eigenfunctions of T^*T associated with the eigenvalues $s_1^2,\dots,s_j^2$:

$$s_{j+1}^2 = \inf_{u\perp F_j} \frac{||Tu||^2}{||u||^2}.$$

Choose k so large that

$$(12) \qquad 2^{1-a}\,\gamma\,s_k^{-a} < \tfrac{1}{2}, \quad \sup_{v\perp G_k} \frac{||RL^{-1}v||}{||v||} < \tfrac{1}{2}, \quad G_k = (I+RL^{-1})^{*}F_k$$

Then

$$\inf_{v \perp G_k} \frac{||(I+RL^{-1})v||}{||v||} \geq \tfrac{1}{2},$$

and hence

$$(13) \quad \sup_{v \perp G_k} \frac{||v||}{||(I+RL^{-1})v||} = \sup\left\{\frac{||(I+RL^{-1})^{-1}w||}{||w||} : w = (I+RL^{-1})v,\ v \perp F_k\right\} \leq 2.$$

Proceeding as before we see that

$$(14) \quad \lambda^2_{j+2k+1} \geq \inf\left\{\frac{||Lu||^2}{||u||^2} : u \perp F_{j+k},\ u \perp T^*G_k\right\}$$

$$= \inf_{u \perp T^*G_k} \frac{||(I+K)Tu||^2}{||Tu||^2} \quad \inf_{u \perp F_{j+k}} \frac{||Tu||^2}{||u||^2},$$

where $K = -RL^{-1}(I + RL^{-1})^{-1}$, which is defined on the orthogonal complement of G_k. We actually have $L = (I + K)T$.

From (9) we deduce that for $w \perp F_k$,

$$\frac{||Kw||}{||w||} = \frac{||RL^{-1}(I+RL^{-1})^{-1}w||}{||(I+RL^{-1})^{-1}w||} \cdot \frac{||(I+RL^{-1})^{-1}w||}{||w||}$$

$$\leq \gamma\left(\frac{||L^{-1}(I+RL^{-1})^{-1}w||}{||(I+RL^{-1})^{-1}w||}\right)^a \frac{||(I+RL^{-1})^{-1}w||}{||w||},$$

and hence

$$(15) \quad \frac{||Kw||}{||w||} \leq \gamma\left(\frac{||T^{-1}w||}{||w||}\right)^a \left(\frac{||(I+RL^{-1})^{-1}w||}{||w||}\right)^{1-a}.$$

Thus from (14), (12) and (15) we have

$$(16) \quad \lambda^2_{j+2k+1} \geq s^2_{j+k+1} \inf\{||(I+K)w||^2/||w||^2 : w = Hu \perp G_k\} \geq s^2_{j+k+1}(1 + O(s_k^{-a}))$$

Theorem 1 now follows easily from (11) and (16) on letting j tend to ∞ with k/j tending to 0.

IV. Estimates with a remainder term

Theorem 3 Suppose that hypotheses (1) - (4) are satisfied and that

$$\lambda_j = cj^r (1 + O(j^{-t})) \text{ as } j \to \infty, \tag{17}$$

where $t < ar/(1 + ar)$ and a is defined in Lemma 1. Then

$$s_j = cj^r (1 + O(j^{-t})) \text{ as } j \to \infty.$$

Estimates like (17) are obtained in [6] and [8]; we can extend them to some non-self-adjoint operators and to some Schrödinger operators with non-smooth or complex potentials.

Example Consider $T_2 = -\Delta + (1 + |x|^2)^k + i(1 + |x|^2)^{k/p}$, with $k > n/(n-2)$ and $p > 2$, defined on $\mathbb{R}^n$. With the same notation as above, $L_2 = -\Delta + (1 + |x|^2)^k$ and $N(\lambda, L_2, \mathbb{R}^n) = c\,\lambda^{n(k+1)/2k} (1 + O(\lambda^{-(k+1)/2k}))$ as $\lambda \to \infty$, so that [6]

$$\lambda_j = cj^{2k/(n(k+1))} (1 + O(j^{-1/n})) \quad \text{as} \quad j \to \infty. \tag{18}$$

We notice that Lemma 1 holds with $a = 1 - p^{-1}$ and, if $p > 1 + (1 + k^{-1})(1 - k^{-1} - 2n^{-1})^{-1}$, the estimate (18) holds for the s_j also

Proof of Theorem 3

The proof is the same as in [10]. If λ_j satisfies (10), then

$$\begin{aligned}\frac{\lambda_{j+\ell}}{\lambda_j} &= \left(\frac{j+\ell}{j}\right)^r (1 + O(j^{-t}))^{-1} (1 + O((j+\ell)^{-t})) \\ &= \left(\frac{j+\ell}{j}\right)^r (1 + O(j^{-t})) (1 + O(j^{-t} (1 + \tfrac{\ell}{j})^{-t})).\end{aligned}$$

Hence

$$\frac{\lambda_{j+\ell}}{\lambda_j} = (1 + \tfrac{\ell}{j})^r (1 + O(j^{-t})) = 1 + O(\tfrac{\ell}{j}) + O(j^{-t}). \tag{19}$$

We deduce from (16) that

$$s^2_{j+\ell+k+1} \le \lambda^2_{j+2k+2\ell+1} (1 + O(s^{-a}_{k+\ell})).$$

From (11) we have

$$s_{k+\ell}^{-2} \le \lambda_\ell^{-2} \quad (1 + O(\lambda_k^{-a})),$$

and thus

$$s_{j+\ell+k+1}^{2} \le \lambda_{j+2k+2\ell+1}^{2} \quad (1 + O(\lambda_\ell^{-a}) + O(\lambda_k^{-a})).$$

By (11) again,

$$s_{j+\ell+k+1}^{2} \ge \lambda_{j+\ell+1}^{2} \quad (1 + O(\lambda_k^{-a}));$$

taking $\ell = k$ we obtain

$$\lambda_{j+k+1}^{2} \quad (1 + O(\lambda_k^{-a})) \le s_{j+2k+1}^{2} \le \lambda_{j+4k+1}^{2} \; (1 + O(\lambda_k^{-a})). \tag{20}$$

Using (17), (19) and (20) we can write

$$1 + O(k^{-ar}) + O(k/j) + O(j^{-t}) + O(j^{-1}) \le s_{j+2k+1}^{2}/\lambda_j^{2} \le 1 + O(k^{-ar})$$
$$+ O(k/j) + O(j^{-t}) + O(j^{-1}).$$

Choosing $k/j = j^{-z}$ we therefore have

$$s_{j+2k+1}/\lambda_j = 1 + O(j^{-\nu}),$$

where

$$\nu = \min\left(t, \frac{ar}{1+ar}\right).$$

References

1. D.E. Edmunds and W.D. Evans, On the distribution of eigenvalues of Schrödinger operators, to appear.

2. D.E. Edmunds, W.D. Evans and J. Fleckinger, On the spectrum and the distribution of singular values of Schrödinger operators with a complex potential, to appear.

3. K. Fan, Maximum properties and inequalities for the eigenvalues of completely continuous operators, Proc. Nat. Acad. Sci. USA 37 (1951), 760.

4. J. Fleckinger, Estimate of the number of eigenvalues for an operator of Schrödinger type, Proc. Roy. Soc. Edinburgh 89A (1981), 355.

5. J. Fleckinger, Répartition des valeurs propres d'opérateurs de type Schrödinger, Comptes rendus Acad. Sci. Paris 292 (1981), 359.

6. B. Helffer and D. Robert, Comportement asymptotique précise du spectre d'opérateurs globalement elliptiques dans $\mathbb{R}^n$, Seminaire EDP, École Polytechnique, 1980-81, exposé 2.

7. I.C. Gohberg and M.G. Krein, Introduction to the theory of linear non selfadjoint operators, Transl. Amer. Math. Soc. 18 (1969).

8. L. Hörmander, On the asymptotic distribution of the eigenvalues of pseudodifferential operators in $\mathbb{R}^n$, Ark. För Math. 17 (1979), 296.

9. T. Kato, Perturbation theory for linear operators (Springer-Verlag, 1966)

10. A.G. Ramm, Spectral properties of some non-self-adjoint operators and some applications, in Spectral theory of differential operators, Math. Studies 55 (North Holland 1981).

11. M. Reed and B. Simon, Modern methods of mathematical physics (Academic Press, 1978).

Optimal Control of Systems Governed by Elliptic Operator of Infinite Order

I.M. Gali*

Abstract

In the present paper, using the theory of J.L. Lions [6,7] we find the set of inequalities defining an optimal control of systems governed by elliptic operator of infinite order. The questions treated in this paper are related to a previous result by I.M. Gali; et al. [5], but in different direction, by taking the case of operators of infinite order with finite dimension.

§1. Necessary Spaces and Differential Operators of Infinite Order

Ju.A. Dubinskii [1,2,3,4] initiated the investigation of the theory of problems of infinite order linear differential equations.

$$\text{Let } W^{\infty}\{a_\alpha,2\} = \{u(x)\,\varepsilon\, C^{\infty}(R^n) : \sum_{|\alpha|=0}^{\infty} a_\alpha \|D^\alpha u\|_2^2 < \infty\}$$

be a Sobolev space of infinite order of periodic function defined on all of R^n. Here $\alpha = (\alpha_1,\ldots,\alpha_n)$ is a multi-index

----

Key Words: Optimal Control, Elliptic Operator of Infinite Order.

*Current Address : Department of Mathematics, Qatar University, Doha, P.O. Box 2713.

for differentiation, $|\alpha| = \alpha_1 + \ldots + \alpha_n$, $D^\alpha \equiv \partial^{|\alpha|}/\partial x_1^{\alpha_1} \ldots \partial x_n^{\alpha_n}$, $a_\alpha \geq 0$ is a numerical sequence and $||.||_2$ is the canonical norm in the space $L_2(R^n)$, (all functions are assumed to be real valued).

The imbedding problem for nontrivial Sobolev spaces of infinite order are investigated in [3]. Imbedding conditions expressed in "algebraic" terms, in particular include conditions in terms of the characteristic functions of these spaces.

In this case $W^\infty\{a_\alpha,2\}$ is everywhere dense in $L_2(R^n)$ with topological inclusion, and $W^{-\infty}\{a_\alpha,2\}$ denotes the topological dual space with respect to $L_2(R^n)$ and then we have the following inclusion:

$$W^\infty\{a_\alpha,2\} \subseteq L_2(R^n) \subseteq W^{-\infty}\{a_\alpha,2\}$$

In this article our model A is the elliptic operator of infinite order [4],

$$Au = \sum_{|\alpha|=0}^{\infty} (-1)^{|\alpha|} a_\alpha D^{2\alpha}u \; ; \; A\varepsilon \mathbf{L}(W_o^\infty\{a_\alpha,2\}, W_o^{-\infty}\{a_\alpha,2\}) \qquad (1)$$

for which has a self-adjoint closure where $W_o^\infty\{a_\alpha,2\}$ is the set of all functions of $W^\infty\{a_\alpha,2\}$ which vanish on the boundary Γ of R^n.

§2. Our Main Results

To set our problem, we need to prove the following two lemmas which enable us to formulate our main theorem.

We introduce

$$\pi(u,v) = \sum_{|\alpha|=1}^{\infty} ((-1)^{|\alpha|} a_\alpha D^{2\alpha} u(x), v(x))_{L_2(R^n)}$$

$$+ (q(x)u(x), v(x))_{L_2(R^n)} \qquad (2)$$

where q(x) is a real valued function from $L_2(R^n)$ such that $q(x) \geq \nu$, $1 \geq \nu > 0$. Then,

$$\pi(u,v) = \sum_{|\alpha|=1}^{\infty} \int_{R^n} a_\alpha (D^\alpha u)(x)(D^\alpha v)(x)dx + \int_{R^n} q(x)u(x)v(x)dx \qquad (3)$$

Lemma 1

Consider the continuous bilinear form (3) on $W_o^\infty\{a_\alpha, 2\}$, then $\pi(u,v)$ is coercive, that is

$$\pi(u,u) \geq \nu \, ||u||^2 \qquad ; \qquad u \varepsilon W_o^\infty\{a_\alpha, 2\} \qquad (4)$$

Proof

It is well known that the ellipticity of A is sufficient for the coerciveness of $\pi(u,v)$ on $W_o^\infty\{a_\alpha, 2\}$ (see [7]).
In fact,

$$\pi(u,u) = \sum_{|\alpha|=1}^{\infty} \int_{R^n} a_\alpha (D^\alpha u)(x)(D^\alpha u)(x)dx + \int_{R^n} q(x)u(x)u(x)dx$$

$$\geq \sum_{|\alpha|=1}^{\infty} a_\alpha((D^\alpha u)(x),(D^\alpha u)(x))_{L_2(R^n)} + \nu(u(x),u(x))_{L_2(R^n)}$$

$$= \sum_{|\alpha|=1}^{\infty} a_\alpha ||D^\alpha u||^2_{L_2(R^n)} + \nu||u||^2_{L_2(R^n)}$$

$$= \sum_{|\alpha|=1}^{\infty} a_\alpha ||D^\alpha u||^2_{L_2(R^n)} + \nu \sum_{|\alpha|=1}^{\infty} a_\alpha ||D^\alpha u||^2_{L_2(R^n)} -$$

$$- \nu \sum_{|\alpha|=1}^{\infty} a_\alpha ||D^\alpha u||^2_{L_2(R^n)} + \nu||u||^2_{L_2(R^n)}$$

$$= \nu||u||^2_{W_o^\infty\{a_\alpha,2\}} + (1-\nu) \sum_{|\alpha|=1}^{\infty} a_\alpha ||D^\alpha u||_{L_2(R^n)}$$

Then,

$$\pi(u,u) \geq \nu||u||^2_{W_o^\infty\{a_\alpha,2\}}$$

which proves the coerciveness of the bilinear form.

<u>Lemma 2</u>

Assume that (4) is satisfied. For a given f in $W_o^{-\infty}\{a_\alpha,2\}$, there exists a unique u in $W_o^\infty\{a_\alpha,2\}$ such that

$$\pi(u,v) = (f,v)$$

Then, we have the following operator equation

$$Au = f \ , \quad f \in W_o^{-\infty}\{a_\alpha,2\}$$

where A has the form (1).

Proof

If L is a continuous linear form on $W_o^\infty\{a_\alpha,2\}$, then

$$L(\psi) = (f,\psi)_{L_2(R^n)} \quad ; \quad f\varepsilon W_o^\infty\{a_\alpha,2\}$$

From the coerciveness condition, for given f, there exists a unique element $u\varepsilon W_o^\infty\{a_\alpha,2\}$ such that

$$\pi(u,\psi) = (f,\psi) \text{ for all } \psi\varepsilon W_o^\infty\{a_\alpha,2\} \tag{5}$$

Hence, from (2), equation (5) is equivalent to an operator equation

$$Au = f$$

We may now formulate our theorem. For this purpose, we have the statement of the control problem. The space $L_2(R^n)$ being the space of control, is given. A system which is governed by the model A is given.

For a control u the state of the system y(u) is given by the solution of

$$Ay(u) = f + u$$

We are also given an observation equation

$$Z(u) = y(u)$$

Finally, we are given $N\varepsilon L(L_2(R^n),L_2(R^n))$. N is Hermitian positive definite; i.e.,

$$(Nu,u)_{L_2(R^n)} \geq \gamma||u||^2_{L_2(R^n)} \qquad \gamma > 0 \tag{6}$$

With every control $u \in U$ we associate the cost:

$$J(u) = ||y(u) - z_d||^2_{L_2(R^n)} + (Nu,u)_{L_2(R^n)} \tag{7}$$

where z_d is a given element in $L_2(R^n)$.

Let U_{ad} be a closed convex set in $L_2(R^n)$, in the sequel, it will be the set of admissible controls.

The control problem is to minimize the quadratic functional $J(u)$.

Theorem

Assume that (4) holds. The cost function being given by (7), a necessary and sufficient condition for u to be an optimal control is that the following equation and inequalities be satisfied:

$$Ay(u) = F + u \tag{8}$$

$$AP(u) = y(u) - z_d \tag{9}$$

$$u \in U_{ad}, \quad (P(u) + Nu, v-u)_{L_2(R^n)} \geq 0 \text{ for every } v \in U_{ad} \tag{10}$$

where $P(u)$ is the adjoint state. If N satisfies the positive definiteness condition (6), then the optimal control is unique. If $N = 0$ and U_{ad} is bounded then the system of equations (8), (9) and (10) admit at least one solution.

Proof

Let $J(u)$ be written in the form:

$$J(u) = ||y(u) - y(0) + y(0) - z_d||^2_{L_2(R^n)} + (Nu,u)_{L_2(R^n)}$$

[f we set

$$\pi(u,v) = (y(u) - y(0), y(v) - y(0))_{L_2(R^n)} + (Nu,v)_{L_2(R^n)}$$

$$L(v) = (z_d - y(0) , y(v) - y(0))_{L_2(R^n)}$$

'he form $\pi(u,v)$ is continuous bilinear form on $L_2(R^n)$ and we ıave

$$J(v) = \pi(v,v) - 2L(v) + ||z_d - y(0)||^2_{L_2(R^n)}$$

.nd from (6) we have

$$\pi(v,v) \geq \nu||v||^2_{L_2(R^n)} \quad \text{for every} \quad v \in L_2(R^n)$$

herefore, we have reduced the problem to a form where theorems f Lions [6] can be applied, i.e., there exists a unique element εU_{ad} such that

$$J(u) = \inf_{v \in U_{ad}} J(v)$$

nd this element is characterized by

$$J'(u)(v-u) \geq 0 \quad \text{for all } v \in U_{ad} \tag{11}$$

ince A is a canonical isomorphism from $W_o^{\infty}\{a_\alpha,2\}$ onto $_o^{-\infty}\{a_\alpha,2\}$ we may write

$$y(u) = A^{-1}(f + u)$$

ıence

$$y'(u).\psi = A^{-1}\psi$$

and hence

$$y'(u)(v-u) = A^{-1}(v-u) = y(v) - y(u)$$

Therefore, (after division by 2), (11) is equivalent to

$$(y(u) - z_d, y(v) - y(u))_{L_2(R^n)} + (Nu, v-u)_{L_2(R^n)} \geq 0 \qquad (12)$$

For a control $v \varepsilon L_2(R^n)$ the adjoint state $P(v) \varepsilon W_o^\infty\{a_\alpha, 2\}$ is defined by

$$AP(v) = y(v) - z_d \qquad (13)$$

Let us now transform (12); from (13) we deduce

$$(AP(u), y(v)-y(u))_{L_2(R^n)} = (y(u)-z_d, y(v)-y(u))_{L_2(R^n)}$$

$$= (P(u), Ay(v)-Ay(u))_{L_2(R^n)} = (P(u), v-u)_{L_2(R^n)}$$

and hence (12) is equivalent to

$$(P(u) + Nu, v-u)_{L_2(R^n)} \geq 0 \quad \text{for all } v \varepsilon U_{ad}$$

which completes the proof.

Acknowledgement

The author would like to express his gratitude to Professor J.L. Lions for suggesting the problem and critically reading the manuscript, also to Professor W.N. Everitt for valuable discussion.

References

[1] Ju.A. Dubinskii, Sobolev spaces of infinite order and the behavior of solutions of some boundary value problem with unbounded increase of the order of the equation. Math. USSR Sb. 27 (1975) 143-162.

[2] ______________, Non triviality of Sobolev spaces of infinite order for a full Euclidean space and atours. Math. USSR Sb. 29 (1976).

[3] ______________, Some imbedding theorems for Sobolev spaces of infinite order.Soviet Math. Dokl. Vol. 19 (1978) 1271-1274.

[4] ______________, About one method for solving partial differential equation, Dokl. Akad Nauk SSSR 258 (1981) 780-784.

[5] I.M. Gali and H.A. El-Saify, Optimal control of systems governed by a self-adjoint elliptic operator with an infinite number of variables. International Conference Functional-Differential Systems and Related Topics II, 3-10 May 1981 Poland.

[6] J.L. Lions, Optimal control of systems governed by partial differential equations, Springer-Verlag Band 170 (1971).

[7] J.L. Lions and E. Magenes, Non-homogeneous boundary value problem and applications, Springer-Verlag, Vol. III (1972).

COMPARISON PRINCIPLES FOR SOME FOURTH ORDER ELLIPTIC PROBLEMS

Vinod B Goyal and Philip W Schaefer

1. INTRODUCTION

Many authors have contributed to the development of Sturmian type comparison and oscillation theorems for elliptic partial differential equations and systems (see [1], [7], [8] and references cited therein). These theorems are obtained as a consequence of a variational principle for a suitable quadratic functional. Our concern here is with the global-type comparison theorem ($u \leq v$ in Ω) which one deduces immediiately from a maximum principle in the case of second order elliptic partial differential equations (see [2], [5]). We extend this latter kind of comparison principle to fourth order elliptic problems (where there is no maximum principle on the solution) by introducing a suitable functional of the solution to the fourth order problem and obtaining a maximum principle for this functional. This approach has been used recently to deduce other results of interest for higher order elliptic problems such as uniqueness of the solution [6], nonexistence of a solution [3], and integral and pointwise estimates [4]. The comparison theorems we obtain are useful in approximating the solution to boundary value problems since they allow for the determination of a bound on the error in the approximation (e.g., see [5]).

In section 2 we prove a simple lemma for the class of differential inequalities

$$\Delta(p\Delta u) - (c + pr)\Delta u + cru \geq 0 , \tag{1.1}$$

where c is a nonnegative constant, p and r are sufficiently smooth functions, and Δ is the n-dimensional Laplacian, whereby we conclude that $u \geq 0$ in a bounded domain Ω. This leads directly to the comparison result. Although p and r could be constants, the form of (1.1) dictates a certain relationship between the constant coefficients Thus we modify the argument in order to obtain similar results for

$$\Delta^2 u - a\Delta u + bu \geq 0 , \tag{1.2}$$

where a and b are constants such that $0 \leq b \leq \frac{1}{4} a^2$. We then briefly indicate some extensions of this approach and theorem to more general operators and higher order operators in section 3.

2. RESULTS

In the following calculations we use the comma notation to denote partial differentiation, i.e., $u_{,i} = \partial u/\partial x_i$, and adopt the summation convention for repeated indices, i.e., $u_{,ii} = \Delta u = \Sigma_{i=1}^{n} \partial^2 u/\partial x_i^2$.

Our basic lemma may be formulated as

LEMMA: Let c be a nonnegative constant and p and r be functions defined in a bounded domain $\Omega \subset R^n$ such that $p \in C^2(\Omega)$, $p(x) \geq p_0 > 0$ and $r \in C(\Omega)$. If there exists a positive function q such that

$$q^{-1} q_{,i} \ , \quad q^{-1}\Delta q - r \text{ , are bounded on every closed ball in } \Omega , \tag{2.1}$$

$$\Delta q - rq \leq 0 \ , \quad \text{in } \Omega \ , \tag{2.2}$$

then any function $u \in C^4(\Omega) \cap C^2(\bar{\Omega})$ which satisfies

$$\Delta(p\Delta u) - (c + pr)\Delta u + cru \geq 0 \ , \quad \text{in } \Omega \tag{2.3}$$

$$u \geq 0 \quad , \quad \Delta u \leq 0 \quad , \qquad \text{on } \partial\Omega \ , \tag{2.4}$$

is such that $u \geq 0$ in $\bar{\Omega}$.

Proof: For any u satisfying (2.3), let

$$w = \frac{p\Delta u - cu}{q} \ , \tag{2.5}$$

where q is as yet unspecified, and compute

$$w_{,i} = \frac{(p\Delta u - cu)_{,i}}{q} - \frac{q_{,i}(p\Delta u - cu)}{q^2} \ ,$$

$$w_{,ii} = \frac{\Delta(p\Delta u - cu)}{q} - \frac{2q_{,i}(p\Delta u - cu)_{,i}}{q^2}$$

$$- \frac{(\Delta q)(p\Delta u - cu)}{q^2} + \frac{2q_{,i}q_{,i}(p\Delta u - cu)}{q^3} \ .$$

It follows that

$$\Delta w + \frac{2q_{,i} w_{,i}}{q} + \frac{(\Delta q)w}{q} = \frac{\Delta(p\Delta u - cu)}{q}$$

and by (2.3) that

$$\Delta w + \frac{2q_{,i}\, w_{,i}}{q} + \frac{(\Delta q - rq)w}{q} \geq 0 \, .$$

Thus if q satisfies (2.1) and (2.2) and u satisfies (2.4), we have $w \leq 0$ in $\bar{\Omega}$ by the Hopf maximum principle [5]. By (2.5) and the minimum principle, we conclude that $u \geq 0$ in $\bar{\Omega}$.

From this theorem one easily deduces a comparison result for operators of the form

$$Lu \equiv \Delta(p\Delta u) - (c + pr)\Delta u + cru \, .$$

Thus we have

THEOREM 1: Let c, p, and r be as in the lemma and assume that a positive function q satisfying (2.1) and (2.2) exists. If $u, v \in C^4(\Omega) \cap C^2(\bar{\Omega})$ satisfy

$$Lu \geq Lv \, , \text{ in } \Omega \, ; \quad u \geq v \, , \quad \Delta u \leq \Delta v \, , \quad \text{on } \partial\Omega \, ,$$

then $u \geq v$ in $\bar{\Omega}$.

We observe that (2.3) includes the special cases

i) $(\Delta - c)(\Delta - c)u \geq 0$, $p = 1, r = c \, , q = 1$,

ii) $(\Delta - c)(\Delta - d)u \geq 0$, $p = 1 \, , r = d \geq 0 \, , d \neq c \, , q = 1$,

iii) $(\Delta - c)(\Delta + c)u \geq 0$, $p = 1 \, , r = -c \, , \Delta q + cq \leq 0$,

iv) $\Delta(p\Delta u) - \frac{c^2}{p} u \geq 0$, $r = -\frac{c}{p}$, $\Delta q + \frac{c}{p} q \leq 0$, where the existence of a function q is trivial in cases i) and ii). The need for some conditions on q relative to Ω is exemplified by the eigenvalue situation (case iii))

$$\Delta^2 u - 4u = 0 \quad , \quad \text{in } \Omega = (0, 2\pi) \times (0, 2\pi)$$

$$u = \Delta u = 0 \quad , \quad \text{on } \partial\Omega .$$

This boundary value problem has the nontrivial solution $u = \sin x \sin y$ which changes sign in this Ω. Hence here the theorem implies the non-existence of a positive function q which satisfies (2.1) and (2.2).

In order to obtain a comparison result for a constant coefficient operator which is not a special case of (2.3), we modify the functional used in the proof of the lemma. Suppose that u satisfies

$$Mu \equiv \Delta^2 u - a\Delta u + bu \geq 0 \quad , \quad \text{in } \Omega , \tag{2.6}$$

and (2.4) on $\partial\Omega$. We shall determine conditions on the constants a and b under which a comparison result is possible.

Let

$$w = \Delta u - ku , \tag{2.7}$$

where k is an undetermined constant. Then by (2.6)

$$\Delta w = \Delta^2 u - k\Delta u$$

$$\geq (a - k)(w + ku) - bu .$$

If we choose $k = (a \pm \sqrt{a^2 - 4b})/2$, then

$$\Delta w - (a - k)w \geq 0 .$$

Now if $k \leq a$, then w takes its maximum on $\partial\Omega$, and if $k \geq 0$, as well, then $w \leq 0$ on $\partial\Omega$ and hence in $\overline{\Omega}$. Consequently, since $k \geq 0$, we conclude that $u \geq 0$ in $\overline{\Omega}$ by the minimum principle applied to (2.7).

From the requirements on k , namely,

$$0 \leq k = \frac{1}{2}(a \pm \sqrt{a^2 - 4b}) \leq a ,$$

we see that the constants a and b can both be zero or such that $a > 0$ and $0 \leq b \leq a^2/4$. Formally, we have

THEOREM 2: If $u, v \in C^4(\Omega) \cap C^2(\overline{\Omega})$ satisfy

$$Mu \geq Mv , \quad \text{in } \Omega \quad ; \quad u \geq v , \quad \Delta u \leq \Delta v , \quad \text{on } \partial\Omega,$$

for constants $a = 0$, $b = 0$ or $a > 0$, $0 \leq b \leq a^2/4$ in the operator M , then $u \geq v$ in $\overline{\Omega}$.

3. REMARKS

We note that one could replace the Laplacian in either (2.3) or (2.6) by a general uniformly elliptic operator of the form $Au \equiv (a_{ij}(x)\, u_{,i})_{,j}$ where $a_{ij} = a_{ji}$ and $a_{ij}(x)\, \xi_i \xi_j \geq \gamma |\xi|^2$, for $\gamma > 0$ and all $x \in \Omega$ and real n-tuples ξ.

Finally, one can extend the approach used here to certain higher order problems such as

$$\Delta^3 v - a\Delta^2 v - b\Delta v + abv \geq 0 \quad , \quad \text{in } \Omega ,$$

$$v \leq 0 \ , \ \Delta v \geq 0 \ , \ \Delta^2 v \leq 0 \ , \qquad \text{on } \partial\Omega ,$$

where a and b are nonnegative constants. Since the differential inequality may be rewritten as

$$\Delta^2(\Delta v - av) - b(\Delta v - av) \geq 0 ,$$

one can apply special case iii) with $u = \Delta v - av$, $p = 1$, and $c = \sqrt{b}$ to conclude that $\Delta v - av \geq 0$ in Ω and hence that $v \leq 0$ in $\bar{\Omega}$. This inequality then leads one to a comparison principle for this higher order problem.

REFERENCES

[1] G.J. Etgen and R.T. Lewis, The Oscillation of Elliptic Systems, Math. Nachr. Bd 94, 43-50 (1980).

[2] D.Gilbarg and N.S. Trudinger, Elliptic Partial Differential Equations of Second Order, Springer-Verlag, Berlin (1977).

[3] V.B. Goyal and P.W. Schaefer, On a Subharmonic Functional in Fourth Order Nonlinear Elliptic Problems, J. Math. Anal. Appl. 83, 20-25 (1981).

[4] L.E. Payne, Some Remarks on Maximum Principles, J. Analyse Math. 30, 421-433 (1976).

[5] M.H. Protter and H.F. Weinberger, Maximum Principles in Differential Equations, Prentice-Hall Inc., Englewood Cliffs (1967).

[6] P.W. Schaefer, Uniqueness in Some Higher Order Elliptic Boundary Value Problems, ZAMP, 28, 693-697 (1978).

[7] C.A. Swanson, A Generalization of Sturm's Comparison Theorem, J. Math. Anal. Appl. 15, 512-519 (1966).

[8] C.A. Swanson, Comparison and Oscillation Theory of Linear Differentia Equations, Mathematics in Science and Engineering vol. 48, Academic Press Inc., New York (1968).

WEAK SOLUTIONS OF INTEGRODIFFERENTIAL EQUATIONS AND APPLICATIONS

R. C. Grimmer W. Schappacher

1. Introduction

In this paper we consider the initial value problem for the Volterra integrodifferential equation

$$\text{(VE)} \qquad x'(t) = Ax(t) + \int_0^t B(t-s)x(s)ds + f(t), \ t \geq 0,$$
$$x(0) = x_0$$

in a Banach space X. It is assumed throughout this paper that A is the generator of a C_0 semigroup which has domain D(A) and that B(t) is closed with domain at least D(A), a.e. $t \geq 0$. Also it shall be assumed that $f \in C([0,\infty),X)$ or $f \in LL^1((0,\infty),X)$.

With the aid of various additional assumptions concerning the operators A and B(t) one can show the existence of a unique solution to this initial value problem. If in addition there is continuity with respect to initial data one can define a resolvent operator R(t) on X by $R(t)x_0$ is the solution of

$$x'(t) = Ax(t) + \int_0^t B(t-s)x(s)ds, \ t \geq 0,$$
$$x(0) = x_0,$$

(cf. [3], [6], [7], [8], [9], [10], [12], [13], [14], [17], [18],

[23].) This resolvent operator is then used to obtain a variation of parameters formula

(VP) $$x(t) = R(t)x_0 + \int_0^t R(t - s)f(s)ds$$

which gives the solution of (VE) when it exists.

When $X = \mathbb{R}^n$, Grossman and Miller, [15], [16], showed the existence of a resolvent $R(t,s)$ for the non convolution type equation

$$x'(t) = A(t)x(t) + \int_0^t B(t,s)x(s)ds + f(t)$$
$$x(0) = x_0$$

under quite mild assumptions on $A(t)$ and $B(t,s)$. The associated variation of parameters formula then enables one to obtain qualitative information about the solutions of (VE) if enough is known about the resolvent. When X is infinite dimensional the non-convolution case is examined in [5], [12], and [13].

When $B(t) \equiv 0$ so that (VE) reduces to the differential equation

(1.1) $$x'(t) = Ax(t) + f(t)$$
$$x(0) = x_0,$$

$R(t)$ will be the semigroup generated by A. In general, however, $R(t + s) \neq R(t)R(s)$ so that $R(t)$ is not a semigroup although the family $\{R(t)\}$, $t \geq 0$, will have many of the same properties as the semigroup generated by A.

Differential equations in a Banach space have been studied extensively and a great deal is known (cf. [19], [21], [25]). One of the most useful concepts concerning the differential equation

$$(1.2) \qquad x'(t) = Ax(t)$$
$$x(0) = x_0$$

is the circle of ideas relating the wellposedness of (1.2) with C_0 semigroups. That is, (DE) is uniformly wellposed if and only if A generates a C_0 semigroup if and only if the resolvent $R(\lambda,A) = (\lambda I - A)^{-1}$ satisfies the condition $\|R^n(\lambda,A)\| \leq M/(\mathrm{Re}\lambda - \omega)^n$, $\mathrm{Re}\lambda > \omega$, $n = 1,2,\ldots$ for some ω and M, [21]. This last characterization, known as the Hille-Yosida-Phillips Theorem, is useful as it gives an algebraic characterization of when (DE) is wellposed. In particular, if M = 1 this condition may be reduced to $\|R(\lambda,A)\| \leq 1/(\mathrm{Re}\lambda - \omega)$.

Unfortunately, the extension of the Hille-Yosida-Phillips Theorem to (VE) involves the nth derivative of $(\lambda I - A - \hat{B}(\lambda))^{-1}$ which cannot be expressed as a power of $(\lambda I - A - \hat{B}(\lambda))^{-1}$ (here ^ indicates Laplace transform). This makes the implementation of the extension of the Hille-Yosida-Phillips Theorem often very difficult, [8, p. 218]. A further difficulty in dealing with (VE) is the already noted fact that $R(t + s) \neq R(t)R(s)$. That is, R(t) is not a semigroup.

It thus seems necessary to seek a different characterization of wellposedness of (1.1) or (1.2) which can be extended to a useful characterization of wellposedness of (VE). Such a characterization of wellposedness of (1) or (2) was given by Ball [2]. In particular, denote the adjoint of A by A* and the pairing between X and its dual space X* by $\langle\ ,\ \rangle$. A function $x \in C([0,T],X)$ is a weak solution of (1) if and only if for every $v \in D(A^*)$ the function $\langle x(t),v\rangle$ is absolutely continuous on $[0,T]$ and

$$\frac{d}{dt}\langle x(t),v\rangle = \langle x(t),A^*v\rangle + \langle f(t),v\rangle$$

for almost all $t \in [0,T]$. With $f \in L^1((0,T),X)$, Ball [2] proved that A generates a C_0 semigroup if and only if for each $x_0 \in X$ there exists a unique weak solution x(t) of (1) with $x(0) = x_0$. (In Hilbert space the "only if" part of this theorem was obtained in Balakrishnan [1].) It is this characterization which we will extend to (VE). We shall then apply this characterization of wellposedness of (VE) to a number of problems involving resolvent operators.

2. Preliminaries

If A is the generator of a C_0 semigroup on X we know that D(A) is dense in X, A is closed, and $(A - \omega I)^{-1}$ exists if $\omega > \beta$ for some β. Hence, we may consider the equation

$$y'(t) = (A - \omega I)y(t) + \int_0^t e^{-\omega(t-s)}B(t - s)y(s)ds + g(t)$$

where $y(t) = e^{-\omega t}x(t)$ and $g(t) = e^{-\omega t}f(t)$ rather than (VE) when considering questions of wellposedness. Thus, without loss of generality, we may assume that A has a bounded inverse A^{-1}. This implies also that A* also has a bounded inverse and $(A^*)^{-1} = (A^{-1})^*$

Throughout this paper we shall assume the hypotheses (H1)-(H3).

(H1) A is the generator of a C_0 semigroup on X and $0 \in \rho(A)$, the resolvent set of A.

As A and A^* are closed we denote by Y the Banach space formed from $D(A)$ with the graph norm and by Z the Banach space formed by $D(A^*)$ with the graph norm.

(H2) $D(B(t)) \supset D(A)$ and $D(B^*(t)) \supset D(A^*)$ a.e. $t \geq 0$ with $B(t)x$ and $B^*(t)v$ strongly measurable for $x \in D(A)$ and $v \in D(A^*)$. Further, there exists $b \in LL^1(0,\infty)$ so that $\|B(t)y\| \leq b(t)\|y\|_Y$ and $\|B^*(t)z\| \leq b(t)\|z\|_Z$ for $y \in Y$ and $z \in Z$.

(H3) There exists a set $D \subset D(A^{*2}) \cap D(A^*B^*(t)) \cap D(B^*(t)A^*) \cap D(B^*(t)B^*(s))$ a.e. in $t,s \geq 0$ with the property that A^*D is weak*dense in X^*.

We remark that if $B(t) = a(t)A$ with $a \in LL^1(0,\infty)$ (H2) is automatically satisfied and, as $D(A^*)$ is weak*dense in X^* ([19, p. 43]), (H3) is satisfied by $D = D(A^{*2})$. Frequently D may be somewhat more complicated, however. If $X = L^2(0,1)$, $A = \frac{d^2}{dx^2}$ with Dirichlet boundary conditions, $B(t)$ could be a first order differential operator. Say $B(t) = a(t)\frac{d}{dx}$ with $D(B(t)) = \{u \in X : \frac{du}{dx} \in X,\ u(0) = 0\}$. Then (H2) and (H3) are easily seen to be satisfied with $D = C_0^\infty$.

Two lemmas will be critical in our work.

<u>Lemma 2.1</u>. ([2], [11, p. 127]) Let $x,z \in X$ satisfy $\langle z,v\rangle = \langle x,A^*v\rangle$ for all $v \in D(A^*)$. Then $x \in D(A)$ and $z = Ax$.

<u>Lemma 2.2</u>. The operator $A^{-1}B(t)$ has a bounded extension, $\widetilde{A^{-1}B(t)}$, to X a.e. $t \geq 0$ such that $\widetilde{A^{-1}B(t)}x$ is locally Bochner integrable for each $x \in X$.

Proof. First note that $B^*(t)(A^*)^{-1}$ is a bounded operator on X^* a.e. $t \geq 0$. Also, considering $A^{-1}B(t)$ as an operator on X^{**} we see that $A^{-1}B(t)$ and $B^*(t)(A^*)^{-1}$ are adjoint to each other Thus $[B^*(t)(A^*)^{-1}]^*$ is a bounded extension of $A^{-1}B(t)$ to X^{**}, [20, p. 167]. Restricting this operator to X we obtain the desired extension of $A^{-1}B(t)$ as an operator on X.

As D(A) is dense in X and $A^{-1}B(t)x$ is measurable for each $x \in D(A)$ it follows that $\widetilde{A^{-1}B(t)}x$ is measurable for each $x \in X$. Also, $\|[B^*(t)(A^*)^{-1}]^*\| = \|B^*(t)(A^*)^{-1}\|$ and, for $x \in X^*$,

$$\begin{aligned}\|B^*(t)(A^*)^{-1}x\| &\leq b(t)\{\|x\| + \|(A^*)^{-1}x\|\}\\ &\leq Mb(t)\|x\|.\end{aligned}$$

Thus, $\|\widetilde{A^{-1}B(t)}x\|$ is locally integrable for each $x \in X$ and it follows that $\widetilde{A^{-1}B(t)}x$ is locally Bochner integrable for each $x \in X$.

3. Weak Solutions and Resolvents

A number of possible definitions of a solution of (VE) have been given. We are motivated here by the concept of a solution of a differential equation in semigroup theory when $f \in C([0,\infty),X)$.

Definition 3.1. A solution of (VE) on $[0,\infty)$ is a function $x \in C([0,\infty),Y) \cap C^1([0,\infty),X)$ with $x(0) = x_0 \in D(A)$ which satisfies (VE) on $[0,\infty)$.

In a similar manner our concept of resolvent operator is modelled after a C_0 semigroup.

Definition 3.2. A resolvent operator for (VE) is a bounded operator valued function $R(t) \in B(X)$, $0 \leq t < \infty$, having the following properties:

(a) $R(t)$ is strongly continuous for $t \geq 0$ with $R(0) = I$.

(b) $R(t) \in B(Y)$ and $R(t)$ is strongly continuous, $t \geq 0$, on Y.

(c) For each $x \in Y$, $R(t)x$ is continuously differentiable, $t \geq 0$, with

$$R'(t)x = A\,R(t)x + \int_0^t B(t-u)R(u)x\,du$$

and

$$R'(t)x = R(t)Ax + \int_0^t R(t-u)B(u)x\,du\,.$$

From the previous definitions it is clear that if $R(t)$ is a resolvent operator and $x_0 \in Y$ and $f \equiv 0$ then $R(t)x_0$ is a solution of (VE). In addition, one can easily prove the following result (cf. [13] or [14]).

Theorem 3.3. Suppose $R(t)$ is a resolvent operator for (VE), $x_0 \in Y$ and $f \in C([0,\infty),X)$. If $x(t)$ is a solution of (VE) then it is given by

$$\text{(VPR)} \qquad x(t) = R(t)x_0 + \int_0^t R(t-s)f(s)ds.$$

Conversely, if $x_0 \in Y$ and $f \in C([0,\infty),Y)$ then $x(t)$ given by (VPR) is a solution of (VE).

It is clear that the concept of solution could be slightly changed in Definition 3.1 to accomodate $f \in LL^1((0,\infty),X)$ but we shall be interested in a more general weakening of the concept of solution.

Definition 3.4. A function $x \in C([0,\infty),X)$ is a weak solution of (VE) if and only if for every $v \in D(A^*)$, $\langle x(t),v\rangle$ is absolutely continuous on $[0,\infty)$ and

$$\frac{d}{dt}\langle x(t),v\rangle = \langle x(t),A^*v\rangle + \int_0^t \langle x(s),B^*(t-s)v\rangle ds + \langle f(t),v\rangle$$

for almost all $t \geq 0$.

We are now in a position to state our main theorem.

Theorem 3.5. Let $f \in LL^1((0,\infty),X)$. The integrodifferential equation (VE) has a resolvent operator $R(t)$ if and only if for each $x_0 \in X$ (VE) has a unique weak solution. The weak solution is given by (VPR).

Proof. Assume that (VE) has a resolvent operator $R(t)$. Then for $x \in D(A)$, $R(t)x$ is a solution of (VE) if $f \equiv 0$ and thus a weak solution. If $x_0 \in X$, as $D(A)$ is dense in X we may choose $x_n \in D(A)$, $n \geq 1$, with $x_n \to x_0$. For each $n \geq 1$, an integration yields for $v \in D(A^*)$

$$\langle R(t)x_n,v\rangle = \langle x_n,v\rangle + \int_0^t \langle R(s)x_n,A^*v\rangle ds + \int_0^t\int_0^s \langle R(u)x_n,B^*(s-u)v\rangle du\, ds .$$

It follows from Dominated Convergence that

$$\langle R(t)x_0,v\rangle = \langle x_0,v\rangle + \int_0^t \langle R(s)x_0,A^*v\rangle ds + \int_0^t\int_0^s \langle R(u)x_0,B^*(s-u)v\rangle du\, ds$$

so that $R(t)x_0$ is a weak solution of (VE) when $f \equiv 0$ for all

$x_0 \in X$. Now suppose $f \in C([0,\infty),X)$ and $x(t)$ is given by (VPR). As $R(t)x$ is a weak solution for every $x \in X$,

$$\frac{d}{dt}\int_0^t \langle R(t-s)f(s),v\rangle ds = \langle f(t),v\rangle + \int_0^t \langle R(t-s)f(s),A^*v\rangle ds$$
$$+ \int_0^t\int_0^{t-s} \langle R(u)f(s),B^*(t-s-u)v\rangle du\, ds, \text{ a.e}$$

It now follows from Fubini's Theorem that

$$\frac{d}{dt}\langle x(t),v\rangle = \frac{d}{dt}\langle R(t)x_0,v\rangle + \frac{d}{dt}\langle\int_0^t R(t-s)f(s)ds,v\rangle$$
$$= \langle R(t)x_0,A^*v\rangle + \int_0^t \langle R(s)x_0,B^*(t-s)v\rangle ds$$
$$+ \langle f(t),v\rangle + \langle\int_0^t R(t-s)f(s)ds,A^*v\rangle$$
$$+ \int_0^t \langle\int_0^s R(s-u)f(u)du,\ B^*(t-s)v\rangle ds$$

so that $x(t)$ is a weak solution for each $x_0 \in X$. If $f \in LL^1((0,\infty),X)$, let $T > 0$ and $t \in [0,T]$. On $[0,T]$ $\|R(u)\|$ is bounded so if $f_n \in C([0,T],X)$ $f_n \to f$ in $L^1((0,T),X)$ and

$$x_n(t) = R(t)x_0 + \int_0^t R(t-s)f_n(s)ds$$
$$x(t) = R(t)x_0 + \int_0^t R(t-s)f(s)ds$$

it follows that $x_n \to x$ in $C([0,T],X)$. As

$$\langle x_n(t),v\rangle = \langle x_0,v\rangle + \int_0^t \langle x_n(s),A^*v\rangle ds + \int_0^t \langle f_n(s),v\rangle ds$$
$$+ \int_0^t\int_0^s \langle x_n(u),B^*(s-u)v\rangle du\, ds$$

it follows that $x(t)$ is a weak solution of (VE).

To show uniqueness, suppose that $x_1(t)$ and $x_2(t)$ are weak solutions of (VE) with $x_1(0) = x_2(0)$. We then see, for $v \in D(A^*)$,

$$\langle x_1(t) - x_2(t), v\rangle = \langle \int_0^t (x_1(s) - x_2(s)ds, A^*v\rangle$$
$$+ \int_0^t \langle x_1(s) - x_2(s), \int_s^t B^*(u - s)v\, du\rangle\, ds.$$

Define $z(t) = \int_0^t (x_1(s) - x_2(s))ds$ and integrate by parts to get

$$\langle z'(t), v\rangle = \langle z(t), A^*v\rangle + \int_0^t \langle z(s), B^*(t - s)v\rangle ds.$$

On $D(A^*)$, $B^*(t) = B^*(t)(A^*)^{-1}A^*$ and $B^*(t)(A^*)^{-1}$ is a bounded operator a.e. Further, by Lemma 2.2 $A^{-1}B(t)$ has a bounded extension to X which is the restriction of $[B^*(t)(A^*)^{-1}]^*$ to X so z satisfies

$$\langle z'(t), v\rangle = \langle z(t) + \int_0^t \widetilde{A^{-1}B(t - s)}z(s)ds, A^*v\rangle.$$

From Lemma 2.1 we see that $\omega(t) = z(t) + \int_0^t \widetilde{A^{-1}B(t - s)}z(s)ds$ is in $D(A)$ and $z'(t) = A\omega(t)$. Now

$$\omega(t) = z(t) + \int_0^t \widetilde{A^{-1}B(t - s)}z(s)ds \tag{3.1}$$

has a unique solution $z(t) \in C([0,\infty), X)$ by Lemma 2.2. However, $A^{-1}B(t)$ maps Y into itself and for $x \in Y$

$$\|A^{-1}B(t)x\|_Y = \|B(t)x\| + \|A^{-1}B(t)x\|$$
$$\leq b(t)\|x\|_Y + M\|x\|_X$$

and so $A^{-1}B(t) \in \mathcal{B}(Y)$ and as $B(t)x$ is measurable in X, $A^{-1}B(t)x$ is measurable on Y. Thus $A^{-1}B(t)x$ is locally Bochner integrable in Y so (3.1) has a unique solution in $Y \subset X$. That is, $z(t) \in Y$ so that

$$z'(t) = Az(t) + \int_0^t B(t - s)z(s)ds.$$

Thus, $z(t) = R(t)z(0) = 0$ and $x_1(t) \equiv x_2(t)$.

The proof that the existence of a unique weak solution of (VE) for each $x_0 \in X$ implies the existence of a resolvent operato is technically more difficult. We first note that the difference of two weak solutions of (VE) is a weak solution of the homogeneous equation. There is thus no loss of generality in assuming $f \equiv 0$. Denote by $x(t,x_0)$ the unique weak solution of (VE) with $x(0) = x_0$ and define $R(t)$ by $R(t)x_0 = x(t,x_0)$. It then follows that $R(t)$ is linear, $R(t)x_0$ is continuous in t and $R(0) = I$.

Now let $x_0 \in D(A)$ and $z(t) = \int_0^t R(s)x_0 ds$. As $R(t)x_0$ is a weak solution of (VE) one obtains for $v \in D(A^*)$

$$\langle z'(t) - x_0, v\rangle = \langle z(t), A^*v\rangle + \int_0^t \langle z(s), B^*(t - s)v\rangle ds.$$

As before, $z'(t) - x_0 = A\omega(t)$ where $\omega(t)$ is given by (3.1). Arguing as before, using Lemmas 2.1 and 2.2 yields that $z(t)$ is continuous into Y so that $Az(t)$ and $z(t)$ are continuous. Hence, $\int_0^t R(s)x_0 ds \in D(A)$ and $A\int_0^t R(s)x_0 ds$ is continuous.

Now define $z(t)$ by

$$z(t) = \int_0^t R(s)Ax_0 ds - A\int_0^t R(s)x_0 ds + \int_0^t \int_0^s R(s - u)B(u)x_0 du\, ds - \int_0^t B(t - s)\int_0^s R(u)x_0 du\, ds.$$

We wish to show $z(t) \equiv 0$. Noting that $z(0) = 0$ we see that if it can be shown that $z(t)$ is a weak solution of the homogeneous equation (VE) it will follow from the uniqueness of the weak solution that $z(t) \equiv 0$. This can be shown by choosing $v \in D$ where D is given by (H3) and differentiating $\langle z(t),v\rangle$. After a number of calculations involving triple integrals and terms of the form $A^*B(t)v$ and $B^*(\omega)B^*(u)v$, etc., one shows

$$\frac{d}{dt}\langle z(t),v\rangle = \langle z(t),A^*v\rangle + \int_0^t \langle z(s),B^*(t - s)v\rangle ds$$

for $v \in D$. Now integrate both sides of this equation and rearrange to get

$$(3.2) \quad \langle A^{-1}z(t) - \int_0^t z(s)ds - \int_0^t\int_0^s A^{-1}B(s - u)z(u)du\,ds\ ,A^*v\rangle = 0$$

for $v \in D$. Given $t \geq 0$ and $v \in D(A^*)$ one may approximate A^*v by elements of the form $A^*\omega$ where $\omega \in D$ as A^*D is weak*dense in X^*. One sees that (3.2) is valid for all $v \in D(A^*)$ so $z(t)$ is a weak solution and $z(t) \equiv 0$. Thus, for $v \in D(A)$,

$$0 = \frac{d}{dt}\langle z(t),v\rangle$$

$$= \langle R(t)Ax_0,v\rangle - \langle R(t)x_0 A^*v\rangle$$

$$+ \langle\int_0^t R(t - u)B(u)x_0du,v\rangle - \int_0^t \langle R(t - s)x_0,B^*(s)v\rangle ds$$

or

$$\langle R(t)Ax_0 + \int_0^t R(t - u)B(u)x_0du,v\rangle$$

$$= \langle R(t)x_0 + \int_0^t \widetilde{A^{-1}B(u)}R(t - u)x_0du,A^*v\rangle.$$

This says

$$\omega(t) = R(t)x_0 + \int_0^t \widetilde{A^{-1}B(u)}R(t - u)x_0du \in D(A)$$

and

$$A\omega(t) = R(t)Ax_0 + \int_0^t R(t - u)B(u)x_0du.$$

Arguing as above, this implies $R(t)x_0$ is continuous in Y (and hence $R(t) : Y \to Y$). This implies

$$AR(t)x_0 + \int_0^t B(u)R(t - u)x_0 du$$
$$= R(t)Ax_0 + \int_0^t R(t - u)B(u)x_0 du.$$

Now consider $x_0 \in D(A)$ and $v \in D(A^*)$. Then as $R(t)x_0$ is continuous in Y and $R(t)x_0$ is a weak solution of (VE),

$$\langle R(t)x_0 - x_0, v\rangle = \int_0^t \langle R(s)x_0, A^*v\rangle ds$$
$$+ \int_0^t \int_0^s R(u)x_0, B^*(s - u)v\rangle du\, ds$$
$$= \langle \int_0^t \{AR(s)x_0 + \int_0^s B(s - u)R(u)x_0 du\} ds, v\rangle,$$

for all $v \in D(A^*)$. It now follows from Lemma 2.1 with $x = 0$ that

$$R(t)x_0 - x_0 = \int_0^t \{AR(s)x_0 + \int_0^s B(s - u)R(u)x_0 du\} ds$$

so that $R(t)x_0$ is strongly differentiable with

$$R'(t)x_0 = AR(t)x_0 + \int_0^t B(t - u)R(u)x_0 du$$
$$= R(t)Ax_0 + \int_0^t R(t - u)B(u)x_0 du.$$

To show $R(t)$ is a bounded operator for each $t \geq 0$, fix $T > 0$ and consider the map $\Theta : X \to C([0,T],X)$ defined by $\Theta(x) = R(\cdot)x$. Given $x \in X, R(t)x$ satisfies

$$\langle R(t)x - x, v\rangle = \int_0^t \langle R(s)x, A^*v\rangle ds$$
$$+ \int_0^t \int_0^s \langle R(u)x, B^*(s - u)v\rangle du\, ds$$

for each $v \in D(A^*)$. Suppose now that $x_n \to x$ and $\Theta x_n \to y$. Then

$$\langle y(t),v\rangle - \langle x,v\rangle = \int_0^t \langle y(s),A^*v\rangle ds + \int_0^t\int_0^s \langle y(u),B^*(s-u)v\rangle du\, ds$$

and $y(t)$ is a weak solution with $y(0) = x$. By uniqueness, $y(t) = R(t)x$ and Θ is closed. This implies Θ and, thus, $R(t)$ is bounded. This completes the proof.

As an easy corollary we obtain the following result which can be used to unify much of the work on existence and uniqueness of solutions of (VE).

<u>Corollary 3.6</u>. Suppose there exists a dense subspace $X_0 \subset X$ such that for $x_0 \in X_0$ (VE) has a unique solution $x(t,x_0)$. Further, assume for each $T > 0$ there is a constant $M(T) > 0$ so that $x_0,x_1 \in X_0$ implies $\|x(t,x_0) - x(t,x_1)\| \leq M(T)\|x_0 - x_1\|$ for $0 \leq t \leq T$. Then (VE) has a resolvent operator.

Proof. Without loss of generality we may assume $f \equiv 0$ so that $\|x(t,x_0)\| \leq M(T)\|x_0\|$, $x_0 \in X$, $0 \leq t \leq T$. For $x_0 \in X$ and $v \in D(A^*)$ we have

$$(3.3) \qquad \langle x(t,x_0) - x_0,v\rangle = \int_0^t \langle x(s,x_0),A^*v\rangle ds + \int_0^t\int_0^s \langle x(u,x_0),B^*(s-u)v\rangle du\, ds.$$

It is easy to see that for $x_0 \in X$ there exists $\{x_n\} \subset X_0$ with $x_n \to x_0$ and $x(t,x_n)$ converges uniformly on $[0,T]$, $T > 0$, to a function which we label $x(t,x_0)$ and that this function must satisfy (3.3). It is clear $x(t,x_0)$ is a weak solution. Arguing as in Theorem 3.5 one can show there is at most one weak solution satisfying (VE). The result now follows from the previous theorem.

Another interesting application is the following theorem which characterizes when (VPR) yields a solution of (VE). For differential equations in a Banach space the corresponding result is proved using semigroup theory, [25, p. 111].

<u>Theorem 3.7</u>. Let f be continuous and suppose (VE) has a resolvent operator. The initial value problem (VE) has a solution for every $x_0 \in D(A)$ if and only if

$$\omega(t) = \int_0^t R(t - s)f(s)ds$$

is continuously differentiable.

Proof. We know $\omega(t)$ is a weak solution of (VE) with $x_0 = 0$ by Theorem 3.5. Thus, for $v \in D(A^*)$,

$$\frac{d}{dt}\langle \omega(t), v\rangle = \langle \omega(t), A^*v\rangle + \int_0^t \langle \omega(s), B^*(t - s)v\rangle ds + \langle f(t), v\rangle.$$

However, as $\omega(t)$ is differentiable, $\langle \omega'(t), v\rangle = \frac{d}{dt}\langle \omega(t), v\rangle$ and

$$\langle \omega'(t) - f(t), v\rangle = \langle \omega(t) + \int_0^t \widetilde{A^{-1}B}(t - s)\omega(s)ds, A^*v\rangle.$$

This says

$$g(t) = \omega(t) + \int_0^t \widetilde{A^{-1}B}(t - s)\omega(s)ds \tag{3.4}$$

is in D(A) and $Ag(t) = \omega'(t) - f(t)$ is continuous. As in the proof of Theorem 3.5 this implies $\omega(t)$ is continuous into Y and so

$$\omega'(t) - f(t) = A\omega(t) + \int_0^t B(t - s)\omega(s)ds.$$

Thus the solution of (VE) with $x_0 \in D(A)$ exists and is given by $R(t)x_0 + \omega(t)$. This is just (VPR). The converse is clear.

A number of other applications are possible and will appear in a forthcoming paper.

ACKNOWLEDGMENT

The results in this paper were obtained while the first author was visiting the Institute for Mathematics at the University of Graz. The first author gratefully acknowledges the support of the University of Graz.

REFERENCES

1. A. V. Balakrishnan, Applied functional analysis, Applications of Mathematics, vol. 3, Springer, New York, 1976.

2. J. M. Ball, Stongly continuous semigroups, weak solutions, and the variation of constants formula. Proc. Amer. Math. Soc., 63(1977), 370-373.

3. R. W. Carr and K. B. Hannsgen, A non homogeneous integrodifferential equation in Hilbert space, SIAM J. Math. Anal. 10(1979), 961-984.

4. G. Chen and R. Grimmer, Semigroups and integral equations, J. Integral Equations 2(1980), 133-154.

5. G. Chen and R. Grimmer, Integral equations as evolution equations, J. Diff. Equations (to appear).

6. Ph. Clement and J. A. Nohel, Abstract linear and nonlinear Volterra equations with completely positive kernels, SIAM J. Math. Anal. 12(1981), 514-535.

7. G. DaPrato and M. Iannelli, Linear abstract integrodifferentia equations of hyperbolic type in Hilbert spaces, Rend. Sem. Mat. Padova, 62(1980), 191-206.

8. G. DaPrato and M. Iannelli, Linear integrodifferential equations in Banach spaces, Rend. Sem. Mat. Padova, 62(1980), 207-219.

9. A. Friedman, Monotonicity of solutions of Volterra integral equations in Banach space, Trans. Amer. Math. Soc., 138(1969), 129-148.

10. A. Friedman and M. Shinbrot, Volterra integral equation in Banach space, Trans. Amer. Math. Soc., 126(1967), 131-179.

11. S. Goldberg, Unbounded Linear Operators: Theory and Applications, McGraw-Hill, New York, 1966.

12. R. C. Grimmer, Resolvent operators for integral equations in a Banach space, Trans. Amer. Math. Soc. (to appear).

13. R. C. Grimmer, Resolvents for integral equations in abstract spaces, Proceedings of Conference on Differential Equations and Applications, Graz, Austria, June, 1981.

14. R. C. Grimmer and A. J. Pritchard, Analytic resolvent operators for integral equations in Banach space, (to appear).

15. S. I. Grossman and R. K. Miller, Perturbation theory for Volterra integrodifferential systems, J. Diff. Equations, 8(1970), 457-474.

16. S. I. Grossman and R. K. Miller, Nonlinear integrodifferential systems with L^1-kernels, J. Diff. Equations, 13(1973), 551-566.

17. K. B. Hannsgen, The resolvent kernel of an integrodifferential equation in Hilbert space, SIAM J. Math. Anal., 7(1976), 481-490.

18. K. B. Hannsgen, Uniform L^1 behavior for an integrodifferential equation with parameters, SIAM J. Math. Anal., 8(1977), 626-639

19. E. Hille and R. S. Phillips, Functional Analysis and Semi-groups, Rev. ed., Amer. Math. Soc., Providence, RI, 1957.

20. T. Kato, Perturbation Theory for Linear Operators, Springer, New York, 1976.

21. S. G. Krein, Linear Differential Equations in Banach Space, Amer. Math. Soc., Providence, RI, 1971.

22. R. K. Miller, Nonlinear Volterra Integral Equations, Benjamin, Menlo Park, Calif., 1971.

23. R. K. Miller, Volterra integral equations in a Banach space, Funkcial, Ekvac., 18(1975), 163-193.

24. R. K. Miller and R. L. Wheeler, Well-posedness and stability of linear Volterra integrodifferential equations in abstract spaces, Funkcial. Ekvac., 21(1978), 279-305.

25. A. Pazy, Semi-groups of Linear Operators and Applications to Partial Differential Equations, University of Maryland, 1974

TITCHMARSH'S λ-DEPENDENT BOUNDARY CONDITIONS FOR HAMILTONIAN SYSTEMS

Don Hinton[1] and Ken Shaw[2]

1. INTRODUCTION. Associated with the differential equation,

$$(1.0)\qquad -y'' + q(x)y = \lambda y\ ,\ 0 \le x < \infty\ ,$$

is a certain function $m(\lambda)$ called the Titchmarsh-Weyl coefficient. The function $m(\lambda)$ is defined by introducing a fundamental system $\theta(\cdot, \lambda)$, $\phi(\cdot, \lambda)$ for (1.0) by the initial conditions

$$\begin{pmatrix} \theta(0, \lambda) & \phi(0, \lambda) \\ \theta'(0, \lambda) & \phi'(0, \lambda) \end{pmatrix} = \begin{pmatrix} \cos\alpha & \sin\alpha \\ -\sin\alpha & \cos\alpha \end{pmatrix}$$

for some $\alpha \in [0, \pi)$. The function $m(\lambda)$ is a limit (sequential in the limit-circle case) of functions (cf. [13, p. 24])

$$\ell = \ell(\lambda, b, \beta) = -\frac{\theta(b, \lambda)\cot\beta + \theta'(b, \lambda)}{\phi(b, \lambda)\cot\beta + \phi'(b, \lambda)}\ .$$

In his treatise on eigenfunction expansions, Titchmarsh imposes a boundary condition of the form

$$(1.1)\qquad \lim_{x\to\infty} W(\theta(x, \lambda) + m(\lambda)\,\phi(x, \lambda), f(x)) = 0 \quad \text{for all } \lambda,\ \mathrm{Im}\,\lambda \ne 0\ .$$

[1] Research supported by NSF Grant No. MCS-8101712.

[2] Research supported by NSF Grant No. MCS-8101536.

where $W(g, f) = g'f - f'g$. In the development of the expansion theory, an important role is played by the relation [13, p. 26],

$$\lim_{x\to\infty} W(\theta(x, \lambda) + m(\lambda)\, \Phi(x, \lambda), \theta(x, \lambda') + m(\lambda')\, \Phi(x, \lambda')) = 0 , \tag{1.2}$$

for all λ, λ' with $\mathrm{Im}\,\lambda \neq 0$, $\mathrm{Im}\,\lambda' \neq 0$.

We develop here a theory of boundary conditions for Hamiltonian systems which parallels the Titchmarsh theory for second-order scalar equations. We do not consider the question of eigenfunction expansions, however. The system considered is

$$J\, \vec{y}\,' = [\lambda A(x) + B(x)]\vec{y} , \quad a \leq x < b \leq \infty \tag{1.3}$$

where $\vec{y}$ is a 2n-vector and λ is a complex parameter.

The coefficients A, B, and J satisfy

(1.4) $A(x)$ and $B(x)$ are $2n \times 2n$ Hermitian matrices of locally Lebesgue integrable functions, $A(x) \geq 0$, and

$$J = \begin{pmatrix} 0 & -I_n \\ I_n & 0 \end{pmatrix} ,$$

where I_n is the $n \times n$ identity matrix. A solution of (1.3) is said to be of integrable square if $\int_a^b \vec{y}^* A\vec{y} < \infty$, and we denote this by $y \in \mathcal{L}_A^2(a, b)$ or simply $\vec{y} \in \mathcal{L}_A^2$. We also assume Atkinson's definiteness condition [1, p. 253], i.e., if $\vec{y}$ is a nontrivial solution of (1.3), then

$$\int_c^d \vec{y}^* A\vec{y} > 0 \quad \text{for all} \quad a < c < d < b \ . \tag{1.5}$$

To consider the singular equation (1.3) we first recall some facts from Atkinson [1, Chap. 9] and Kogan and Rofe-Beketov [10]. The regular boundary value problems associated with (1.3) are of the form

$$\text{(R)} \quad \begin{cases} J\,\vec{y}\,' = [\lambda\, A(x) + B(x)]\vec{y} + \vec{f} \quad , \quad a \leq x \leq d \\ \\ \vec{y}(a) = N_1\,\vec{v}\ ,\ \vec{y}(d) = N_2\,\vec{v} \end{cases}$$

where $\vec{v}$ is a 2n-vector and N_1 and N_2 are $2n \times 2n$ matrices such that

$$N_1^* J N_1 = N_2^* J N_2 \ ,\ N_1\,\vec{v} = N_2\,\vec{v} = 0 \Rightarrow \vec{v} = 0 \ . \tag{1.6}$$

The number λ is called an eigenvalue of (R) if for $\vec{f} = 0$ there is a nontrivial $\vec{y}$ satisfying (R) . The symmetry condition (1.6) implies that all eigenvalues are real. We consider here only boundary conditions in separated form. Let $\alpha_1, \alpha_2, \beta_1, \beta_2$ be $n \times n$ matrices such that

$$\text{rank}[\alpha_1, \alpha_2] = \text{rank}[\beta_1, \beta_2] = n \ ,\ \alpha_1\alpha_2^* = \alpha_2\alpha_1^* \ ,\ \beta_1\beta_2^* = \beta_2\beta_1^* \ . \tag{1.7}$$

If we define

$$N_1 = \begin{pmatrix} 0 & \alpha_2^* \\ 0 & -\alpha_1^* \end{pmatrix} , \quad N_2 = \begin{pmatrix} \beta_2^* & 0 \\ -\beta_1^* & 0 \end{pmatrix} ,$$

then the conditions (1.6) are equivalent to (1.7), and (R) can be written as

$$(R^*) \quad \begin{cases} J \vec{y}\,' = [\lambda A(x) + B(x)]\vec{y} + f , \quad a \leq x \leq d , \\ [\alpha_1, \alpha_2]\vec{y}(a) = 0 , \quad [\beta_1, \beta_2]\vec{y}(d) = 0 . \end{cases}$$

For λ not an eigenvalue and $\vec{f}$ Lebesgue integrable, (R) has a unique solution $\vec{y}$ given by

$$\vec{y}(x) = \int_a^d K(x, t, \lambda) \vec{f}(t) \, dt$$

where

$$(1.8) \qquad K(x, t, \lambda) = \begin{cases} Y(x, \lambda)[F(\lambda) J + (1/2)I] \, Y(t, \lambda)^{-1} J^{-1} , & x \leq t , \\ Y(x, \lambda)[F(\lambda) J - (1/2)I] Y(t, \lambda)^{-1} J^{-1} , & x > t , \end{cases}$$

Y is the fundamental matrix for (1.3) with $Y(a, \lambda) = I$,

and

$$(1.9)\qquad F = F(\lambda) = Y(d,\lambda)^{-1} N_2[N_1 - Y(d,\lambda)^{-1} N_2]^{-1} J^{-1} + (1/2) J^{-1} .$$

F is the characteristic function of Atkinson. The matrix function F satisfies the inequality (with equality for F given by (1.9))

$$(1.10)\qquad [F + (1/2) J^{-1}]^* (J/i)[F + (1/2) J^{-1}]$$

$$\geq [F - (1/2) J^{-1}]^* (Y(d,\lambda)^* J\, Y(d,\lambda)/i)\, [F - (1/2) J^{-1}]$$

The set $\mathcal{F}_d$ of matrices F satisfying (1.10) is nested, i.e., if $d < d'$, then $\mathcal{F}_{d'} \subset \mathcal{F}_d$. Further, the set $\mathcal{F}_d$ is bounded independently of d and λ, when λ is restricted to a compact set not intersecting the real axis. This yields that there are functions F_∞, defined and analytic for $\operatorname{Im}\lambda \neq 0$, which are sequential limits of functions $F = F(d, N_1, N_2, \lambda)$.

If we replace F in (1.8) by F_∞ and K by K_∞, then by Lemma 2.1 of [10],

(i) $\int_a^b K_\infty^*(x,t,\lambda)\, A(t)\, K_\infty(x,t,\lambda)\, dt < \infty$;

(ii) if $\vec{f} \in \mathcal{L}_A^2(a,b)$ and

$$(1.11)\qquad \vec{y}(x) = \int_a^b K_\infty(x,t,\lambda) A(t)\, \vec{f}(t)\, dt ,$$

then $\vec{y} \in \mathcal{L}_A^2(a,b)$, $\vec{y}$ satisfies $J\,\vec{y}\,' = (\lambda A + B)\vec{y} + A\vec{f}$, and

$$(1.12)\qquad \int_a^b \vec{y}^* A\vec{y} \le (\operatorname{Im}\lambda)^{-2} \int_a^b \vec{f}^* A\vec{f} .$$

We will show that $\vec{y}$ defined by (1.11) satisfies boundary conditions of the Titchmarsh type (1.1). However, we are concerned primarily with the limit-point or limit-circle case. To define these set

$$N(\lambda) = \dim\{\vec{y} \in \mathcal{L}_A^2 : \ \vec{y} \text{ satisfies } (1.1)\} .$$

Then $N(\lambda)$ is constant in $\operatorname{Im}\lambda > 0$ and in $\operatorname{Im}\lambda < 0$ [10, 12]. In analogy to the classical case considered by Weyl [14], we call $N(i) = N(-i) = n$ the <u>limit-point</u> case and $N(i) = N(-i) = 2n$ the <u>limit-circle</u> case.

In section 2 below we develop the necessary theory to establish the Titchmarsh boundary conditions. In section 3 we state a theorem defining the boundary value problem which (1.11) solves. In section 4 we consider this problem when both endpoints a and b are singular. An important corollary to this development is the derivation of a formula which links the limiting characteristic function F_∞ to the matrix Titchmarsh-Weyl coefficients

2. <u>PROPERTIES OF THE TITCHMARSH-WEYL COEFFICIENT</u>. We will use frequently the following identity for (1.3). If

$$J\,\vec{y}' = [\lambda\, A(x) + B(x)]\vec{y} + \vec{f}$$

and

$$J\,\vec{z}' = [\mu\, A(x) + B(x)]\vec{z} + \vec{g} ,$$

then

$$(2.1) \qquad (\vec{y}^* J \vec{z})' = (\mu - \bar{\lambda})\vec{y}^* A\vec{z} + \vec{y}^* \vec{g} - \vec{f}^* \vec{z} .$$

In the problem (R*) we assume without loss of generality that $\alpha_1\alpha_1^* + \alpha_2\alpha_2^* = I_n$ (since $\alpha_1\alpha_1^* + \alpha_2\alpha_2^* > 0$) and define Y_α to be the fundamental matrix of (1.3) satisfying $Y_\alpha(a, \lambda) = E_\alpha$ where

$$E_\alpha = \begin{pmatrix} \alpha_1^* & -\alpha_2^* \\ \alpha_2^* & \alpha_1^* \end{pmatrix} .$$

Note that $E_\alpha^{-1} = E_\alpha^*$, $E_\alpha^* J E_\alpha = E_\alpha J E_\alpha^* = J$, and $Y(x, \lambda) = Y_\alpha(x, \lambda) E_\alpha^{-1}$.

We decompose Y_α into $n \times n$ blocks by writing

$$Y_\alpha = (\vec{\theta}_\alpha, \vec{\Phi}_\alpha) = \begin{pmatrix} \theta_\alpha & \Phi_\alpha \\ \hat{\theta}_\alpha & \hat{\Phi}_\alpha \end{pmatrix}$$

Then some calculation (cf. [7]) yields that F given by (1.9) satisfies

$$(2.2) \qquad E_\alpha^{-1}[FJ + (1/2)I]E_\alpha = \begin{pmatrix} 0 & 0 \\ -M_\beta & I_n \end{pmatrix}, \quad E_\alpha^{-1}[FJ - (1/2)I]E_\alpha = \begin{pmatrix} -I_n & 0 \\ -M_\beta & 0 \end{pmatrix},$$

where

$$(2.3) \qquad M_\beta = M_\beta(d, \alpha, \lambda) = -[\beta_1\Phi_\alpha(d, \lambda) + \beta_2\hat{\Phi}_\alpha(d, \lambda)]^{-1}[\beta_1\theta_\alpha(d, \lambda) + \beta_2\hat{\theta}_\alpha(d, \lambda)]$$

If we define for $\operatorname{Im} \lambda \neq 0$,

$$(2.4)\qquad \begin{pmatrix} \mathcal{A} & \mathcal{B}^* \\ \mathcal{B} & \mathcal{D} \end{pmatrix} = \begin{pmatrix} \mathcal{A} & \mathcal{B}^* \\ \mathcal{B} & \mathcal{D} \end{pmatrix}(d, \lambda) = \begin{cases} -i\ Y_\alpha^*(d, \lambda) J Y_\alpha(d, \lambda), & \operatorname{Im}\lambda > 0 \\ \\ i\ Y_\alpha^*(d, \lambda)\ J Y_\alpha(d, \lambda)\ , & \operatorname{Im}\lambda < 0\ , \end{cases}$$

then M_β given by (2.3) satisfies $E(M_\beta) = 0$ where E is defined by

$$(2.5)\qquad E(M) = E_{d,\lambda}(M) = [I, M^*] \begin{pmatrix} \mathcal{A} & \mathcal{B}^* \\ \mathcal{B} & \mathcal{D} \end{pmatrix} \begin{bmatrix} I \\ M \end{bmatrix} .$$

Further, $E(M) = 0$ implies M is given by (2.4) for some β_1, β_2 . We refer to [2, 8] for the details of these calculations. It is convenient to write $E(M)$ in a matrix circle from which is analogous to the Weyl circle. From (2.5) we have that

$$\begin{aligned} (2.6)\qquad E(M) &= M^* \mathcal{D} M + M^* \mathcal{B} + \mathcal{B}^* M + \mathcal{A} \\ &= (M + \mathcal{D}^{-1} \mathcal{B})^* \mathcal{D}(M + \mathcal{D}^{-1} \mathcal{B}) + \mathcal{A} - \mathcal{B}^* \mathcal{D}^{-1} \mathcal{B} \\ &= (M - C)^* R_1^{-2}(M - C) - R_2^2 \end{aligned}$$

where

$$C = C(d, \lambda) = -\mathcal{D}^{-1} \mathcal{B}$$

$$R_1 = R_1(d, \lambda) = \mathcal{D}^{-1/2}$$

$$R_2 = R_2(d, \lambda) = [\mathcal{B}^* \mathcal{D}^{-1} \mathcal{B} - \mathcal{A}]^{1/2} .$$

To see that $\mathcal{D} > 0$, we have from (2.1) and (2.4) that

$$(2.7) \qquad \mathscr{D} = -i(\mathrm{sgn}(\mathrm{Im}\,\lambda))\, \vec{\Phi}_\alpha(d,\lambda)^* \, J \, \vec{\Phi}_\alpha(d,\lambda)$$
$$= 2|\mathrm{Im}\,\lambda| \int_c^d \vec{\Phi}_\alpha^* \, A \, \vec{\Phi}_\alpha \; .$$

By following the argument of McIntosh, Hehenberger and Reyes-Sanchez [11] for the discrete case and [2] for the second-order matrix case, it may be shown that [8] $R_2(d, \lambda) = R_1(d, \bar{\lambda})$. In the case where $A(x)$ is real, then $Y_\alpha(d, \bar{\lambda}) = \overline{Y_\alpha(d, \lambda)}$ and hence $R_2(d, \lambda) = \overline{R_1(d, \lambda)}$.

Equation (2.7) shows that $\mathscr{D}$ increases as d increases; hence as $d \to b$, $R_1(d, \lambda)$ and $R_2(d, \lambda)$ decrease to nonnegative limits. Further, it can be shown $C(d, \lambda)$ also has a limit [2, 12]. The equation $E(M) = 0$ can be written as

$$[R_1^{-1}(M - C)\, R_2^{-1}]^* \, [R_1^{-1}(M - C)\, R_2^{-1}] = I_n$$

so that

$$M = C + R_1 \, U \, R_2$$

for some unitary matrix U .

Finally, we note that for $\vec{\Psi} = Y_\alpha \begin{bmatrix} I \\ M \end{bmatrix}$, it follows from (2.1) that

$$(2.8) \qquad E(M) = -i(\mathrm{sgn}(\mathrm{Im}\,\lambda))\, \vec{\Psi}(d)^* \, J \, \vec{\Psi}(d)$$
$$= -i(\mathrm{sgn}(\mathrm{Im}\,\lambda))\, \vec{\Psi}(a)^* \, J \, \vec{\Psi}(a) + 2|\mathrm{Im}\,\lambda| \int_a^d \vec{\Psi}^* \, A \, \vec{\Psi} \; .$$

This relation yields that the sets

$$\mathcal{S}(d, \lambda) = \{M: \ E(M) \leq 0\}$$

are nested, i.e., $\mathcal{S}(d_2, \lambda) \subset \mathcal{S}(d_1, \lambda)$ if $d_2 > d_1$. Members M of $\mathcal{S}(d, \lambda)$ have the representation $M = C + R_1 V R_2$ with $V^* V \leq I_n$. This shows $\mathcal{S}(d, \lambda)$ is compact. If $E(M) \leq 0$, then (2.8) yields

$$\int_a^d \vec{\Psi}^* A \vec{\Psi} \leq i[M^* - M]/2 \operatorname{Im} \lambda$$

since $\vec{\Psi}(a)^* J \vec{\Psi}(a) = M^* - M$. This inequality can be used to establish the existence of $\mathcal{L}_A^2(a, b)$ solutions of (1.3). The number of linearly independent such solutions is related to the rank of the limit as $d \to b$ of $R_1(d, \lambda)$ and is discussed in [2].

Suppose now $\tilde{M}$ is a limit of $\{M_n\}$ where $E(M_n) = 0$, i.e., M_n is given by (2.3) for some d_n and $\beta_n = (\beta_{1n}, \beta_{2n})$ and that $d_n \to b$ as $n \to \infty$. The nesting property ensures that $E(\tilde{M}) \leq 0$ and hence $\tilde{M}$ has the representation

$$\tilde{M} = C(d_n, \lambda) + R_1(d_n, \lambda) V_n R_2(d_n, \lambda)$$

with $V_n^* V_n \leq I$. From (2.6),

$$E_{d_n,\lambda}(\tilde{M}) = R_2(d_n,\lambda) V_n^* V_n R_2(d_n, \lambda) - R_2(d_n, \lambda) R_2(d_n, \lambda) .$$

Assuming now $V_n \to V$ as $n \to \infty$, (note that $\{V_n\}$ is bounded), we have that $E_{d_n,\lambda}(\tilde{M}) \to 0$ as $n \to \infty$ iff

$$(2.9) \qquad R_2(b, \lambda)[\tilde{V}^* V - I] R_2(b, \lambda) = 0 \quad .$$

Thus whenever (2.9) holds,

$$\int_a^b \vec{\Psi}^* A \vec{\Psi} = \frac{i[\tilde{M}^* - \tilde{M}]}{2 \operatorname{Im} \lambda}$$

where $\vec{\Psi} = Y_\alpha \begin{bmatrix} I \\ \tilde{M} \end{bmatrix}$.

We conclude this section with a theorem from [8] which is a direct generalization of Lemma 2.3 of [13]. We refer to [8] for the proof.

Theorem 2.1. Suppose (1.3) is in either the limit-point or limit-circle case at b . Let $M_\infty(\lambda)$ be an analytic function on $\operatorname{Im} \lambda \neq 0$ determined by a sequential limit, i.e.,

$$M_\infty(\lambda) = \lim_{n\to\infty} M_{\beta(n)}(d_n, \lambda)$$

for some $d_n \to b$. Then for all λ, μ not real,

$$\lim_{n\to\infty} [I, M_\infty(\lambda)^*]\, Y_\alpha(d_n, \lambda)^* J\, Y_\alpha(d_n, \mu) \begin{bmatrix} I \\ M_\infty(\mu) \end{bmatrix} = 0 \; .$$

Corollary 2.1. Let M be as in Theorem 2.1 and set $\vec{\Psi}_\alpha = Y_\alpha \begin{bmatrix} I \\ M_\infty \end{bmatrix}$. Then

$$\lim_{n\to\infty} \vec{\Psi}_\alpha(d_n)^* J\, \vec{\Psi}_\alpha(d_n) = 0 \; .$$

Proof. Set $\mu = \lambda$ in Theorem 2.1.

Corollary 2.2. Let M_∞ and $\vec{\Psi}_\alpha$ be as in Corollary 2.1. Then

(2.10) $$\int_a^b \vec{\Psi}_\alpha^* A \vec{\Psi}_\alpha = i[M_\infty(\lambda)^* - M_\infty(\lambda)]/2 \operatorname{Im} \lambda .$$

Proof. This relation follows by application of Corollary 2.1 to (2.1) with $\vec{y} = \vec{z} = \vec{\Psi}_\alpha$.

Equation (2.10) plays an important role in relating the singular structure of $M_\infty(\lambda)$ to the spectrum of differential operators [4, 6, 7].

3. SINGULAR BOUNDARY VALUE PROBLEMS. Let M_∞ be as in Theorem 2.1, but without assuming (1.3) is limit-point or limit-circle at b . By (2.2) we see that K_∞ may be written as

$$K_\infty(x, t, \lambda) = \begin{cases} Y_\alpha(x, \lambda)\begin{pmatrix} 0 & 0 \\ -M_\infty & I_n \end{pmatrix} Y_\alpha(t, \lambda)^{-1} J^{-1} , & a \le x \le t \\ Y_\alpha(x, \lambda)\begin{pmatrix} -I_n & 0 \\ -M_\infty & 0 \end{pmatrix} Y_\alpha(t, \lambda)^{-1} J^{-1} , & x > t . \end{cases}$$

Since $Y_\alpha(t, \lambda)^{-1} = J^{-1} Y_\alpha(t, \overline{\lambda})^* J$ [5, 7], we have then

(3.1) $$K_\infty(x, t, \lambda) = \begin{cases} \vec{\Phi}_\alpha(x, \lambda) \vec{\Psi}_\alpha(t, \overline{\lambda})^* & , a \le x \le t \\ \vec{\Psi}_\alpha(x, \lambda) \vec{\Phi}_\alpha(t, \overline{\lambda})^* & , x > t \end{cases}$$

where $\vec{\Psi}_\alpha = \vec{\theta}_\alpha + \vec{\Phi}_\alpha M_\infty$.

Lemma 3.1. Let $\operatorname{Im} \lambda \neq 0$, $\vec{f} \in \mathcal{L}_A^2$, and

(3.2) $$\vec{y}(x, \lambda) = \int_a^b K_\infty(x, t, \lambda) A(t) \vec{f}(t)\, dt .$$

Suppose μ is such that $\operatorname{Im} \mu \neq 0$ and

$$(3.3) \qquad \lim_{x\to b} \vec{\Psi}^*_\alpha(x, \lambda)\ J\ \vec{\Psi}_\alpha(x, \mu) = 0 .$$

Then

$$(3.4) \qquad \lim_{x\to b} \vec{y}(x, \lambda)^*\ J\ \vec{\Psi}_\alpha(x, \mu) = 0 .$$

Proof. Let f_d be f restricted to $[a, d]$, and let $\vec{y}_d$ be given by (3.2) for f replaced by f_d. Then for $x > d$,

$$\vec{y}_d(x, \lambda) = \vec{\Psi}_\alpha(x, \lambda) \int_a^d \vec{\Phi}^*_\alpha(t, \overline{\lambda})\ A(t)\ \vec{f}(t)\ dt .$$

Hence (3.3) implies (3.4) for $\vec{y}_d$. From (2.1) we have that

$$(3.5) \qquad \vec{y}^*_d\ J\ \vec{\Psi}_\alpha(\cdot, \mu)\Big|_a^x = (\mu - \overline{\lambda}) \int_a^x \vec{y}^*_d\ A\ \Psi_\alpha(\cdot, \mu) - \int_a^x \vec{f}^*_d\ A\ \vec{\Psi}_\alpha(\cdot, \mu) .$$

Now by (1.12), (note that $\vec{y} \in \mathscr{L}^2_A$ by (1.11)),

$$(3.6) \qquad \int_a^b (\vec{y} - \vec{y}_d)^*\ A(\vec{y} - \vec{y}_d) \le (\operatorname{Im} \lambda)^{-2} \int_d^b \vec{f}^*\ A\ \vec{f} ,$$

since $\vec{f} - \vec{f}_d = \vec{f}$ on $[d, b)$. Further, (3.1) shows $\vec{y}_d$ converges uniformly to $\vec{y}$ on compact sets as $d \to b$. Now let $x \to b$ in (3.5). By (3.4) for $\vec{y}_d$ we obtain

$$(3.7) \qquad -\vec{y}_d(a, \lambda)^*\ J\ \vec{\Psi}_\alpha(a, \mu) = (\mu - \overline{\lambda}) \int_a^b \vec{y}^*_d\ A\ \vec{\Psi}_\alpha(\cdot, \mu) - \int_a^d \vec{f}^*\ A\ \vec{\Psi}_\alpha(\cdot, \mu) .$$

Application of (3.6) to (3.7) yields

$$(3.8) \qquad -\vec{y}(a, \lambda)^*\ J\ \vec{\Psi}_\alpha(a, \mu) = (\mu - \overline{\lambda}) \int_a^b \vec{y}^*\ A\ \vec{\Psi}_\alpha(\cdot, \mu) - \int_a^b \vec{f}^*\ A\ \vec{\Psi}_\alpha(\cdot, \mu) .$$

On the other hand, if $d \to b$ in (3.5), we have

$$(3.9) \qquad \vec{y}^* J \vec{\Psi}_\alpha(\cdot, \mu)|_a^x = (\mu - \overline{\lambda}) \int_a^x \vec{y}^* A \vec{\Psi}_\alpha(\cdot, \mu) - \int_a^x \vec{f}^* A \vec{\Psi}_\alpha(\cdot, \mu) .$$

Letting $x \to b$ in (3.9) and comparing with (3.8) completes the proof.

We note that (3.3) holds for $\mu = \overline{\lambda}$ since by (2.1),

$$\vec{\Psi}_\alpha^*(x, \lambda) J \vec{\Psi}_\alpha(x, \overline{\lambda}) \equiv \vec{\Psi}_\alpha^*(a, \lambda) J \vec{\Psi}_\alpha(a, \overline{\lambda})$$

$$= [I, M_\infty^*(\lambda)] J \begin{bmatrix} I \\ M_\infty(\overline{\lambda}) \end{bmatrix}$$

$$= M_\infty(\overline{\lambda}) - M_\infty^*(\lambda) = 0$$

since $M_\infty(\lambda)^* = M_\infty(\overline{\lambda})$ [5]. In addition if (1.3) is in either the limit-point or limit-circle case at b, then by Theorem 2.1, (3.3) and hence (3.4) hold for all μ with $\operatorname{Im} \mu \neq 0$ (sequential in the limit-circle case). In this case we have the generalization of Titchmarsh's boundary condition (1.1). In any case however, we have the following theorem.

<u>Theorem 3.1</u>. <u>Suppose</u> (1.4)-(1.5), (1.7), <u>and</u> (3.1) <u>hold</u>, $\operatorname{Im} \lambda \neq 0$, L_a <u>is an</u> n-<u>vector</u>, <u>and</u> $\vec{f} \in \mathcal{L}_A^2$. <u>Then there is a unique</u> $\vec{y}$ <u>locally absolutely continuous such that</u>

$$(3.10) \qquad J \vec{y}' = [\lambda A(x) + B(x)]\vec{y} + A(x) \vec{f}$$

$$(3.11) \qquad [\alpha_1, \alpha_2]\vec{y}(a) = 0$$

$$(3.12) \qquad \lim_{x \to b} \vec{y}(x)^* J \vec{\Psi}_\alpha(x, \overline{\lambda}) = \vec{L}_a^*$$

<u>Moreover</u>, $\vec{y}$ <u>is given by</u>

$$(3.13) \qquad \vec{y}(x, \lambda) = \vec{\Phi}_\alpha(x, \lambda) L_a + \int_a^b K_\infty(x, t, \lambda) A(t) \vec{f}(t)\, dt .$$

<u>Proof</u>. If $\vec{y}$ is given by (3.13), then (3.10) follows by differentiation. Since $\vec{\Phi}_\alpha(\cdot, \lambda)$ satisfies the condition (3.11), then $\vec{y}$ given by (3.13) also satisfies (3.11). Now by (2.1),

$$(3.14) \qquad \vec{\Phi}_\alpha(x, \lambda)^* J \vec{\Psi}_\alpha(x, \overline{\lambda}) \equiv \vec{\Phi}_\alpha(a,\lambda)^* J \vec{\Psi}_\alpha(a, \overline{\lambda})$$

$$= [-\alpha_2, \alpha_1] J E_\alpha \begin{pmatrix} I \\ M_\infty(\overline{\lambda}) \end{pmatrix}$$

$$= [-\alpha_2, \alpha_1] \begin{pmatrix} -\alpha_2^* & -\alpha_1^* \\ \alpha_1^* & -\alpha_2^* \end{pmatrix} \begin{bmatrix} I \\ M_\infty(\overline{\lambda}) \end{bmatrix}$$

$$= I .$$

Hence by Lemma 3.1, $\vec{y}$ given by (3.13) satisfies (3.12). If the solution to (3.10)-(3.12) is not unique, then there is a solution $\vec{z}$ satisfying (1.3), (3.11), and

$$(3.15) \qquad \lim_{x \to b} \vec{z}(x)^* J \vec{\Psi}_\alpha(x, \overline{\lambda}) = 0 .$$

However since $\vec{z}$ satisfies (1.3), $\vec{z} = Y_\alpha(\cdot, \lambda) \vec{c}$ for some c. Thus $\vec{z}(a) = E_\alpha \vec{c}$ and (3.11) implies $\vec{c} = \begin{bmatrix} 0 \\ c_1 \end{bmatrix}$ or $\vec{z} = \vec{\Phi}_\alpha(\cdot, \lambda)c_1$. Substitution of this relation into (3.15) yields $c_1 = 0$ by (3.14). Hence $\vec{z} = 0$ and uniqueness is established. Note that $y \in \mathcal{L}_A^2$ if $L_a = 0$ or b is limit-circle

4. <u>TWO SINGULAR ENDPOINTS</u>. Assume now (1.3) holds on $-\infty \leq a < x < b \leq \infty$. To consider two singular endpoints, it is convenient to introduce a fundamental matrix Z of (1.1) where $Z(e, \lambda) = I$ with e fixed as $c \to a$ and $d \to b$. If we define

$$(4.1) \qquad \tilde{F} = \tilde{F}(\lambda) = Z(c, \lambda)^{-1} F(\lambda) J Z(c, \lambda) J^{-1} ,$$

then the formula for K in (1.6) becomes (with a replaced by c in (R))

$$(4.2) \qquad K(x, t, \lambda) = \begin{cases} Z(x, \lambda) [\tilde{F} J + (1/2)I] Z(t, \lambda)^{-1} J^{-1} , & x \le t \\ Z(x, \lambda) [\tilde{F} J - (1/2)I] Z(t, \lambda)^{-1} J^{-1} , & x > t . \end{cases}$$

If we let $Z_c = Z(c, \lambda)$ and $Z_d = Z(d, \lambda)$, then from (1.9) and (4.1) we have (note that $Y = Z Z_c^{-1}$)

$$\begin{aligned} \tilde{F} J - (1/2)I &= Z_c^{-1} F J Z_c - (1/2) I \\ &= Z_d^{-1} N_2[N_1 - Z_c Z_d^{-1} N_2]^{-1} Z_c \\ &= Z_d^{-1} N_2[E_\alpha^{-1} N_1 - E_\alpha^{-1} Z_c Z_d^{-1} N_2]^{-1} E_\alpha Z_c . \end{aligned}$$

If we write

$$Z = \begin{pmatrix} \theta & \Phi \\ \hat{\theta} & \hat{\Phi} \end{pmatrix} ,$$

then it can be shown [8] that

$$(4.3) \qquad \tilde{F}J - (1/2)I = \begin{pmatrix} [M_\beta - M_\alpha]^{-1} M_\alpha & -[M_\beta - M_\alpha]^{-1} \\ M_\beta[M_\beta - M_\alpha]^{-1} M_\alpha & -M_\beta[M_\beta - M_\alpha]^{-1} \end{pmatrix}$$

$$= \begin{pmatrix} I \\ M_\beta \end{pmatrix} (M_\beta - M_\alpha)^{-1} (M_\alpha, -I ,$$

where

$$M_\beta(\lambda) = -[\beta_1 \ \Phi(d,\lambda) + \beta_2 \ \hat{\Phi}(d, \lambda)]^{-1}[\beta_1 \ \theta(d, \lambda) + \beta_2 \hat{\theta}(d, \lambda)]$$

$$M_\alpha(\lambda) = -[\alpha_1 \ \Phi(c, \lambda) + \alpha_2 \ \hat{\Phi}(c, \lambda)]^{-1}[\alpha_1 \theta(c, \lambda) + \alpha_2 \ \hat{\theta}(c, \lambda)] \ .$$

From (4.3) it follows readily that

$$(4.4) \qquad \tilde{F}J + (1/2)I = \begin{pmatrix} I \\ M_\alpha \end{pmatrix} (M_\beta - M_\alpha)^{-1} \ (M_\beta, -I) \ .$$

We may now consider sequential limits as in the case of one singular point. Suppose $c_n, d_n, N_1(n), N_2(n)$ are such that $c_n \to a$, $d_n \to b$ as $n \to \infty$, and for $\text{Im } \lambda \neq 0$,

$$\tilde{F}_\infty(\lambda) = \lim_{n \to \infty} F(c_n, d_n, N_1(n), N_2(n), \lambda)$$

is analytic on $\text{Im } \lambda \neq 0$. Suppose

$$M^+(\lambda) = \lim_{n \to \infty} \ M_\beta(d_n, \lambda)$$

and

$$M^-(\lambda) = \lim_{n \to \infty} M_\alpha(c_n, \lambda) \ .$$

Then by (4.3) and (4.4),

$$(4.5) \qquad \tilde{F}_\infty J - (1/2)I = \begin{pmatrix} I \\ M^+ \end{pmatrix} (M^+ - M^-)^{-1} \ (M^-, -I)$$

and

$$(4.6) \qquad \tilde{F}_\infty J + (1/2)I = \begin{pmatrix} I \\ M^- \end{pmatrix} (M^+ - M^-)^{-1} (M^+, I) .$$

The boundedness of $\tilde{F}$, independent of c and d, may be established as that for F in the case of one singular endpoint. This ensures that $(M^+ - M^-)^{-1}$ is analytic on $\text{Im}\,\lambda \neq 0$. From the above formulas, we see that from (4.2), (4.5), and (4.6) that the limiting Green's function K_∞ may be written as

$$(4.7) \qquad K_\infty(x, t, \lambda) = \begin{cases} \vec{\Phi}_a(x, \lambda)[M^+(\lambda) - M^-(\lambda)]^{-1} \vec{\Phi}_b(t, \bar{\lambda})^* , & x \leq t \\ \vec{\Phi}_b(x, \lambda)[M^+(\lambda) - M^-(\lambda)]^{-1} \vec{\Phi}_a(t, \bar{\lambda})^* , & x > t , \end{cases}$$

where

$$\vec{\Phi}_a(x, \lambda) = Z(x, \lambda) \begin{pmatrix} I \\ M^-(\lambda) \end{pmatrix} , \quad \vec{\Phi}_b(x, \lambda) = Z(x, \lambda) \begin{pmatrix} I \\ M^+(\lambda) \end{pmatrix} .$$

Equation (4.7) yields a representation of the Green's function which is the same as the second order scalar case (compare with [13, p. 29]). Further solving (4.5) for $\tilde{F}_\infty$ yields (compare with [3, p. 251] and [9, p. 926] for the second-order scalar case),

$$(4.8) \qquad \tilde{F}_\infty(\lambda) = \begin{pmatrix} (M^+ - M^-)^{-1} & \frac{1}{2}(M^+ - M^-)^{-1}(M^+ + M^-) \\ \frac{1}{2}(M^+ + M^-)(M^+ - M^-)^{-1} & M^+(M^+ - M^-)^{-1} M^- \end{pmatrix}$$

In [8] we also establish theorems similar to Theorem 3.1 for two singular endpoints. When both endpoints are either limit-point or limit-circle, we further establish the analog of (2.10). It is

$$\frac{\tilde{F}_\infty(\lambda) - \tilde{F}_\infty(\lambda)^*}{2 \text{ Im } \lambda} = \int_a^e \{Z(\tilde{F}_\infty + (1/2)\ J^{-1})\}^* \ A\{Z(\tilde{F}_\infty + (1/2)\ J^{-1})\}$$

$$+ \int_e^b \{Z(\tilde{F}_\infty - (1/2)\ J^{-1})\}^* A\{Z(\tilde{F}_\infty - (1/2)\ J^{-1})\} .$$

REFERENCES

1. F.V. Atkinson, Discrete and Continuous Boundary Problems, Academic Press, New York, 1964.

2. E.S. Birger and G.A. Kalyabin, The theory of Weil limit-circles in the case of non-self-adjoint second-order differential-equation systems, Differentsial'nye Uravneniya, 12(1976), 1531-1540.

3. E.A. Coddington and N. Levinson, Theory of Ordinary Differential Equations, McGraw-Hill, New York, 1955.

4. J. Chandhuri and W.N. Everitt, On the spectrum of ordinary second order differential operators, Proc. Royal Soc. Edin., 68A (1967-68), 95-119.

5. D.B. Hinton and J.K. Shaw, On Titchmarsh-Weyl M(λ)-functions for linear Hamiltonian systems, J. Differential Eqs., 40 (1981), 316-342.

6. __________, On the spectrum of a singular Hamiltonian system, to appear in Quaestiones Mathematicae.

7. __________, Titchmarsh-Weyl theory for Hamiltonian systems, Spectral Theory of Differential Operators, I.W. Knowles and R.T. Lewiis, eds., North Holland Amsterdam, 1981, 219-231.

8. _________, On boundary value problems for Hamiltonian systems with two singular points, submitted.

9. K. Kodaira, The eigenvalue problem for ordinary differential equations of the second order and Heisenberg's theory of S-matrices, Amer. J. Math. 71(1949), 921-945.

10. V.I. Kogan and F.S. Rofe-Beketov, On square-integrable solutions symmetric systems of differential equations of arbitrary order, Proc. Royal Soc. Edin., 74A (1974), 5-40.

11. H.V. McIntosh, M. Hehenberger, and R. Reyes-Sanchez, Lattice dynamics with second-neighbor interactions,III. International J. Quan. Chem., 11(1977), 189-211.

12. S.A. Orlov, Nested matrix disks analytically depending on a parameter, and theorems on the invariance of ranks of radii of limiting disks, Izv. Akad. Nauk SSSR, 40 (1970), 593-644. Math. USSR Izvestija, 10 (1976), 565-613.

13. E.C. Titchmarsh, Eigenfunction Expansions Associated with Second-order Equations, I, Oxford Univ. Press, London, 1962.

14. H. Weyl, Über genöhnliche Differentialqleichungen mit Singularitäten und die Zugehörigen Entwicklungen, Math. Ann., 68 (1910), 220-269.

EXTERIOR BOUNDARY VALUE PROBLEMS
FOR PERTURBED EQUATIONS OF ELLIPTIC TYPE

F. A. Howes

1. INTRODUCTION

Many phenomena in mathematical physics can be formulated in terms of solutions to certain boundary value problems for partial differential equations in unbounded regions of Euclidean space. Indeed, several important areas of classical analysis were developed to treat such problems in the special, but important case of linear equations. Many transform techniques derive their efficacy from the unboundedness of the region which permits otherwise formidable integrals to be calculated, often in closed form. However most, if not all, of these classical methods are applicable only because the problem is linear. There are simply no general solution strategies for nonlinear boundary value problems which give even a qualitative description of the solution and its properties.

In this paper we treat briefly a class of nonlinear elliptic boundary value problems in unbounded regions, for which some general existence and comparison results are available. These problems can be viewed as simplified models of various types of flow phenomena described by the Navier-Stokes equations in the case of large Reynolds number. Our detailed analysis of this specific application will be published separately.

2. STATEMENT OF THE PROBLEM

Consider the following singularly perturbed Dirichlet problem for small positive values of ε

$$(P_\varepsilon)\qquad \begin{aligned} &\varepsilon\nabla^2 u = A(x,u)\cdot\nabla u + h(x,u), \quad x \text{ in } \mathcal{E},\\ &u = \varphi(x), \quad x \text{ on } \Gamma = \partial\mathcal{E}, \end{aligned}$$

where $x = (x_1,\ldots,x_N)$ is a vector in $\mathbb{R}^N$, $\cdot$ is the usual Euclidean inner product, $\nabla = (\partial_1,\ldots,\partial_N)$ is the N-dimensional gradient, $\nabla^2 = \nabla\cdot\nabla$ is the Laplacian, and $A = (a_1(x,u),\ldots,a_N(x,u))$. The region $\mathcal{E}$ is assumed to be an open subset of $\mathbb{R}^N$ which contains the point at infinity and whose boundary Γ is a smooth (N-1)-dimensional manifold. For ease of exposition, we assume, in fact, that $\mathcal{E}$ can be described in terms of a smooth scalar function F in the sense that

$$\mathcal{E} = \{x\colon F(x) < 0\};$$

whence, $\Gamma = \partial\mathcal{E} = F^{-1}(0)$ and $\nabla F(x)$ is the outer normal to Γ at x. Finally we assume that the functions a_i $(1 \leq i \leq N)$ and h are sufficiently smooth in appropriate subdomains of $\bar{\mathcal{E}} \times \mathbb{R}$, so as to guarantee the existence of a classical solution of (P_ε).

The task before us is then to prove the existence of a strong solution $u = u(x,\varepsilon)$ of (P_ε) within a class of functions having a prescribed growth at infinity, and to study the limiting behavior of u as $\varepsilon \to 0^+$. It turns out that the smallness of ε allows us to use the techniques of perturbation analysis coupled with the theory of differential inequalities to give sufficient conditions on A, h and F for the existence of such solutions. At the same time, our approach provides rather sharp estimates on these solutions which improve as ε decreases to zero.

Before presenting some general results, let us examine several representative linear problems where the asymptotic phenomena reveal themselves in a straightforward way.

3. LINEAR EXAMPLES

For ease of exposition, let us assume that the region $\mathcal{E}$ in the following four examples is the exterior of the unit circle S^1 in $\mathbb{R}^2$ and let us write the coordinates (x_1,x_2) as (x,y). Then $\mathcal{E}$ is the set $\{(x,y)\colon F(x,y) = \frac{1}{2}(1-x^2-y^2) < 0\}$, and the boundary Γ of $\mathcal{E}$ is simply S^1. Moreover, in all of the problems in this section, we look for solutions of class $C^{(2)}(\mathcal{E}) \cap C(\bar{\mathcal{E}})$ which decay to zero at infinity. The given boundary data is always assumed to be continuous on Γ.

Our first example is

$$\text{(E1)}\qquad \begin{aligned} \varepsilon(u_{xx}+u_{yy}) &= m^2u, \qquad (x,y) \text{ in } \mathcal{E},\\ u &= \varphi(x,y), \qquad (x,y) \text{ on } \Gamma, \end{aligned}$$

where m is a positive constant. It may be thought of as a simple model of a reaction-diffusion system with no convection present. An easy calculation shows that the problem (E1) has a unique solution $u = u(x,y;\varepsilon)$ for each sufficiently small $\varepsilon > 0$ which satisfies the estimate

$$(3.1)\qquad |u(x,y;\varepsilon)| \leq \|\varphi\|_\infty \exp[m_1F(x,y)/\varepsilon^{\frac{1}{2}}]$$

for (x,y) in $\bar{\mathcal{E}}$ and $0 < m_1 < m$. (Here and throughout the paper $\|\cdot\|_\infty$ denotes the supremum norm restricted to the boundary Γ.) Thus u converges to zero as $\varepsilon \to 0^+$ in any closed subset of $\mathcal{E}$, while near Γ there is a thin layer (boundary layer) of width $\mathcal{O}(\varepsilon^{\frac{1}{2}})$ in which u differs from zero by the amount $\|\varphi\|_\infty$. The identically zero function is the solution of the so-called reduced equation $m^2u = 0$ obtained by formally setting ε equal to zero in the differential equation.

The second example shows the influence that convective terms may have on the behavior of solutions, namely

$$\text{(E2)}\qquad \begin{aligned} &\varepsilon(u_{xx}+u_{yy}) = -(x,y)\cdot\nabla u + u, \qquad (x,y) \text{ in } \mathcal{E}, \\ &u = \varphi(x,y), \qquad (x,y) \text{ on } \Gamma. \end{aligned}$$

Let us proceed a little differently than in Example (E1), and first set ε equal to zero so as to obtain the reduced equation

$$(3.2) \qquad (x,y)\cdot\nabla u = u,$$

which is Euler's relation for homogeneous functions of degree one. Now we are seeking a solution u of (E2) which is zero at infinity. Since ε is small, we anticipate that such a solution must be close to a solution of (3.2) throughout most of $\mathcal{E}$, and so we select the solution $u_0 \equiv 0$ of (3.2) as our candidate for the limiting value of u in $\mathcal{E}$ as $\varepsilon \to 0^+$. In fact, a straightforward application of the theory of differential inequalities for elliptic boundary value problems in unbounded regions (cf. for example [9] or [6]) shows that the problem (E2) has a unique solution $u = u(x,y;\varepsilon)$ for each sufficiently small $\varepsilon > 0$. Moreover, we obtain simultaneously the estimate

$$(3.3) \qquad |u(x,y;\varepsilon)| \le \|\varphi\|_\infty \exp[(1-\delta)F(x,y)/\varepsilon]$$

for (x,y) in $\overline{\mathcal{E}}$ and $0 < \delta < 1$, since the function on the right is a barrier function for (E2) as $\varepsilon \to 0^+$. As in the last example, the solution converges to zero in $\mathcal{E}$ away from the boundary Γ, but let us note an important difference between the estimates (3.1) and (3.3). In the former estimate the boundary layer (region of nonuniform convergence) has thickness of order $\mathcal{O}(\varepsilon^{\frac{1}{2}})$, while in the latter the boundary layer is thinner, with thickness of order $\mathcal{O}(\varepsilon)$. The thinning of the layer is of course due to the presence of the gradient terms in the differential equation, which implies that the reduced equation (3.2) has the family of straight lines $y = (\text{const.})x$ as characteristic curves. These characteristics <u>exit</u> $\mathcal{E}$ through Γ nontangentially, that is,

$$-(x,y)\cdot\nabla F(x,y) = x^2 + y^2 = 1 > 0$$

for (x,y) on Γ.

It is essential that the characteristics exit $\mathcal{E}$ for there to be such a boundary layer along Γ. Consider, as an illustration, the Dirichlet problem (E2') for the related differential equation $\varepsilon(u_{xx}+u_{yy}) = (x,y)\cdot\nabla u + u$ in $\mathcal{E}$. The characteristic curves of the reduced equation $(x,y)\cdot\nabla u + u = 0$ are again the family of straight lines $y = (\text{const.})x$, but now these curves <u>enter</u> $\mathcal{E}$ through Γ. There is <u>no</u> boundary layer along Γ; indeed, since Γ is not itself a characteristic curve, the solution $u = u_0(x,y)$ of the reduced equation is determined by the requirement that $u_0(x,y) = \varphi(x,y)$ for (x,y) on Γ. If the data φ is such that u_0 decays to zero at infinity, then one can show that u_0 is a uniformly valid approximation of the solution of (E2') in any subset of $\overline{\mathcal{E}}$.

We consider next an example in which the differential equation contains gradient terms, and yet the solution behaves like the solution of Example (E1), namely

$$\text{(E3)} \qquad \begin{aligned} &\varepsilon(u_{xx}+u_{yy}) = (y,-x)\cdot\nabla u + u, \qquad (x,y) \text{ in } \mathcal{E}, \\ &u = \varphi(x,y), \qquad (x,y) \text{ on } \Gamma. \end{aligned}$$

As in the last example, we look first at the reduced equation

$$\text{(3.4)} \qquad (y,-x)\cdot\nabla u + u = 0.$$

Its characteristic curves are the family of concentric circles $x^2 + y^2 = (\text{const.})^2$, and so the boundary Γ is itself a characteristic. This means that it is not possible, in general, to find a solution of (3.4) which satisfies the given boundary data anywhere along Γ. Since we are looking for a solution of (E3) which decays to zero at infinity, we select $u_0 \equiv 0$ as the solution of (3.4) which we think will serve as the limiting value in closed subsets of $\mathcal{E}$. In addition, we see that

$$(y,-x)\cdot\nabla F(x,y) \equiv 0$$

(that is, Γ is a characteristic curve), and so an estimate like (3.3) cannot hold for the solution of (E3). Instead we can show that the function $w(x,y;\varepsilon) = \|\varphi\|_\infty \exp[m_1 F(x,y)/\varepsilon^{\frac{1}{2}}]$ $(0 < m_1 < 1)$ is a barrier function for (E3), and so the solution $u = u(x,y;\varepsilon)$ satisfies

$$|u(x,y;\varepsilon)| \leq w(x,y;\varepsilon)$$

for (x,y) in $\overline{\mathcal{E}}$ as $\varepsilon \to 0^+$.

These four examples reveal that the behavior of solutions of the general problem (P_ε) for small ε is governed by the interaction of the characteristic curves of the corresponding reduced solution with the boundary of $\mathcal{E}$. If the characteristics either are nonexistent (cf. (E1)) or are tangent to a portion Γ' of the boundary (cf. (E3)), then there is a boundary layer along Γ' with thickness of order $\mathcal{O}(\varepsilon^{\frac{1}{2}})$. While if the characteristics exit $\mathcal{E}$ through Γ' nontangentially, then there is a boundary layer with thickness of order $\mathcal{O}(\varepsilon)$. Finally if the characteristics enter $\mathcal{E}$ through Γ', then there is no boundary layer along Γ'; indeed, the reduced solution u_0 is chosen by the requirement that $u_0 = \varphi$ on Γ'. These results have been established for the problem (P_ε) in bounded regions; the interested reader can consult [7], [2] and [4] for details and further references. In the next section we prove some similar statements about the exterior problem.

4. GENERAL RESULTS

Guided by the examples and observations in the last section, let us turn now to a consideration of two fairly representative results for the general problem

$$(P_\varepsilon) \qquad \begin{aligned} \varepsilon\nabla^2 u &= A(x,u)\cdot\nabla u + h(x,u), \quad x \text{ in } \mathcal{E} \subset \mathbb{R}^N, \\ u &= \varphi(x), \quad x \text{ on } \Gamma, \end{aligned}$$

where the various functions and sets have the properties described in Section 2. In order to study the behavior of solutions of (P_ε) as $\varepsilon \to 0^+$, we first examine the solutions $u = u_0(x)$ of the corresponding reduced problem

$$(P_0) \qquad \begin{aligned} A(x,u)\cdot\nabla u + h(x,u) &= 0, \quad x \text{ in } \mathcal{E}, \\ u &= \varphi(x), \quad x \text{ on } \Gamma_- \subset \Gamma. \end{aligned}$$

Here $\Gamma_- = \{x \text{ on } \Gamma\colon \gamma(x) = A(x,u(x))\cdot\nabla F(x) < 0\}$ is the (possibly empty) set of boundary points at which the characteristic curves of (P_0) enter $\mathcal{E}$ (cf. Example (E2')) or the theory for bounded regions in [7] and [4]). Since the region $\mathcal{E}$ is unbounded, we must select solutions of (P_0) from a class $\mathcal{K}$ of smooth functions which have a restricted growth at infinity. For instance, in the examples of Section 3 the reduced solutions belonged to the class of functions which approach zero as $\|x\|$ tends to infinity. In general, we may define $\mathcal{K}$ as the class of smooth functions $v = v(x)$ such that

$$\lim_{R\to\infty}\left[\sup_{\|x\| = R} \frac{v(x)}{U(x)}\right] = 0, \quad x \text{ in } \mathcal{E},$$

where U is a smooth positive function in $\bar{\mathcal{E}}$ having the property that $\varepsilon\nabla^2 U \leq A(x,U)\cdot\nabla U + h(x,U)$ in $\mathcal{E}$. Such a function U is called an anti-barrier at infinity (cf. [9], [6]), and these references show that the problem (P_ε) may be studied by using barrier functions which themselves belong to the class $\mathcal{K}$. Before discussing these results, let us indicate briefly some of the earlier work in this area.

The exterior linear Dirichlet problem in two dimensions has been studied by Mauss [8] and Eckhaus [1] for the particular equation $\varepsilon(u_{xx}+u_{yy}) = -u_y$ in regions $\mathcal{E}$ contained in the half-plane $y > y_0$ (fixed) > 0. They used as an anti-barrier at infinity the function $U(r,\theta) = I_0(\frac{r}{2\varepsilon})\exp[\frac{r\sin\theta}{2\varepsilon}]$, where I_0 is the modified Bessel function of the first kind, of order zero, and (r,θ) are polar coordinates with respect to an arbitrary reference point (x_0,y_0) defined by $x - x_0 = r\cos\theta$, $y - y_0 = r\sin\theta$. For large positive values of its argument, $I_0(z) \sim e^z/(2\pi z)^{\frac{1}{2}}$, and so U is exponentially unbounded at infinity. Thus the solutions constructed by Mauss and

Eckhaus were allowed to be large at infinity, provided they grew slower than U as $r \to \infty$. The particular regions $\mathcal{E}$ included the upper half-plane and nonconvex sets like $\mathcal{E}_1 = \{(x,y): y > 0 \text{ for } x \geq 0, y > 1 \text{ for } x < 0\}$ whose boundary has a "step" at $x = 0$, and the exterior $\mathcal{E}_2$ of the unit circle in $\mathbb{R}^2$. The nonconvexity of such sets leads naturally to the occurrence of detached boundary layers called free boundary layers, as well as the usual boundary layers illustrated in Section 3. In the case of $\mathcal{E}_1$, there is a boundary layer along the segment $\{(x,y): x = 0, 0 \leq y \leq 1\}$ which becomes a free boundary layer along the vertical line $x = 0$, $y \geq 1$. These layers have thickness of order $\mathcal{O}(\varepsilon^{\frac{1}{2}})$. The free layer arises from the fact that, in general, the corresponding reduced solution $u = u_0(x)$ is discontinuous along the positive y-axis. Consequently the solution $u = u(x,y;\varepsilon)$ of the Dirichlet problem in $\mathcal{E}_1$ satisfies an estimate of the form

$$|u(x,y;\varepsilon)-u_0(x)| \leq L \exp[-x^2/(2\varepsilon)]$$

for (x,y) in $\overline{\mathcal{E}}_1$, where $L = |\varphi(0^+,0) - \varphi(0^-,1)|$ is the magnitude of the difference in the boundary data at $x = 0$; cf. [5] for some related results in bounded regions. The line $x = 0$ is of course a characteristic curve of the reduced equation $u_y = 0$. In the case of the region $\mathcal{E}_2$, it turns out that there is a boundary layer on the lower semicircle which becomes a free boundary layer along each of the lines $\{x = \pm 1, y > 0\}$. The boundary layer on the semicircle has thickness of order $\mathcal{O}(\varepsilon)$ away from the points $(\pm 1,0)$. In neighborhoods of these points and along the lines $x = \pm 1$, $y \geq 0$, this layer and its continuations as the free layers are fatter with thickness of order $\mathcal{O}(\varepsilon^{\frac{1}{2}})$, owing to the fact the lines are characteristic curves of the reduced equation which are tangent to Γ at $(\pm 1,0)$; cf. Example (E3).

In order to deal now with the general problem (P_ε), let us introduce the functions

$$\gamma(x,u) = A(x,u) \cdot \nabla F(x)$$

and

$$H(x,u) = A(x,u) \cdot \nabla u_0(x) + h(x,u),$$

where $u = u_0(x)$ is a smooth solution of the reduced problem (P_0). Let us also define the domain

$$\mathcal{D}(u_0) = \overline{\mathcal{E}} \times \{u: |u-u_0(x)| \leq d(x)\},$$

where $\|\varphi-u_0\|_\infty \leq d(x) \leq \|\varphi-u_0\|_\infty + \delta$ for $\text{dist}(x,\Gamma) < \delta/2$ and $d(x) \leq \delta$ for $\text{dist}(x,\Gamma) \geq \delta$ (with δ a small positive constant). It is in this domain $\mathcal{D}(u_0)$ we will look for solutions of (P_ε) for small ε, in the case that the characteristic curves are exiting $\mathcal{E}$ everywhere along Γ. The first theorem treats curves which exit nontangentially (cf. Example (E2)).

Theorem 4.1. Suppose that the reduced problem (P_o) has a solution $u = u_0(x)$ in the class $\mathcal{K}$ such that

(1) there exists a positive constant k for which $\gamma(x,u) \geq k(\nabla F\cdot\nabla F)(x)$ in $\mathcal{D}_\delta(u_0)$ $(= \mathcal{D}(u_0)$ with $\mathrm{dist}(x,\Gamma) < \delta)$;

(2) there exists a positive constant m for which $H_u(x,u) \geq m^2 > 0$ in $\mathcal{D}(u_0)$.

Then there exists an $\varepsilon_0 > 0$ such that the problem (P_ε) has a solution $u = u(x,\varepsilon)$ in $\mathcal{K}$ of class $C^{(2)}(\mathcal{E}) \cap C(\overline{\mathcal{E}})$ whenever $0 < \varepsilon \leq \varepsilon_0$. Moreover, for x in $\overline{\mathcal{E}}$ we have that

$$|u(x,\varepsilon)-u_0(x)| \leq \|\varphi-u_0\|_\infty \exp[k_1F(x)/\varepsilon] + c\varepsilon,$$

for $0 < k_1 < k$ and c a positive constant depending on u_0, k and m.

The next theorem deals with characteristic curves which may exit $\mathcal{E}$ tangentially.

Theorem 4.2. Suppose that the reduced solution u_0 is such that the assumptions of Theorem 4.1 hold with assumption (1) replaced by

$$\gamma(x,u) \geq 0 \quad \text{in} \quad \mathcal{D}_\delta(u_0).$$

Then the conclusion of Theorem 4.1 is valid with the term $\exp[k_1F(x)/\varepsilon]$ replaced by $\exp[m_1F(x)/\varepsilon^{\frac{1}{2}}]$ for $0 < m_1 < m$.

Thus the boundary layer is thicker wherever the characteristics exit $\mathcal{E}$ tangentially, as we have noted already for our simple linear examples.

The idea behind the proof of these two results is the observation that the function $w(x,\varepsilon) = u_0(x) + \|\varphi-u_0\|_\infty\, \rho(x,\varepsilon) + \varepsilon K m^{-1}$ (with $\rho(x,\varepsilon) = \exp[k_1F(x)/\varepsilon]$ or $\exp[m_1F(x)/\varepsilon^{\frac{1}{2}}]$) is a barrier function for the problem (P_ε) in the sense that

$$|\varphi(x)| \leq w(x,\varepsilon) \quad \text{on} \quad \Gamma,$$

and in $\mathcal{E}$,

$$\varepsilon\nabla^2 w \leq A(x,w)\cdot\nabla w + h(x,w),$$

$$\varepsilon\nabla^2(-w) \geq A(x,-w)\cdot\nabla(-w) + h(x,-w),$$

for appropriately chosen positive constants k_1, m_1 and K, and for ε sufficiently small, say $0 < \varepsilon \leq \varepsilon_0$. Then a theorem of Kusano [6] allows us to conclude that for this range of ε, the problem (P_ε) has a smooth solution $u = u(x,\varepsilon)$ in $\mathcal{K}$ such that $|u(x,\varepsilon)-u_0(x)| \leq \|\varphi-u_0\|_\infty\, \rho(x,\varepsilon) + c\varepsilon$. Similar calculations for the case of a bounded region were performed in [4] and [5], where the interested reader can find complete details.

We close with two remarks. The first one is that our results in [5] on the existence and properties of free boundary layers for problems in bounded regions can be applied mutatis mutandis to the problems in unbounded regions discussed here. If the reduced solution u_0 under consideration is either discontinuous or nondifferentiable along certain (N-1)-dimensional manifolds in $\mathcal{E}$, then this irregularity of u_0 gives rise to free boundary layers in neighborhoods of such manifolds which serve to smooth out the function u_0 there. Nonsmooth reduced solutions are present in many of these problems in unbounded regions, as we have seen above in the two examples of Mauss and Eckhaus.

The second remark is that we can apply the theory in [3] to problems like (P_ε) in which the Dirichlet boundary condition is replaced by a boundary condition of Neumann or Robin type on all or part of Γ. Such problems occur frequently in the study of transport phenomena.

ACKNOWLEDGMENTS

The author gratefully acknowledges the support of the National Science Foundation under grant no. MCS 80-01615. He also wishes to thank Mrs. Ida Zalac for her usual superb typing job.

REFERENCES

1. W. Eckhaus, Boundary Layers in Linear Elliptic Singular Perturbation Problems, SIAM Rev. 14(1972), 225-270.

2. W. Eckhaus and E. M. deJager, Asymptotic Solutions of Singular Perturbation Problems for Linear Differential Equations of Elliptic Type, Arch. Rational Mech. Anal. 23(1966), 26-86.

3. F. A. Howes, Robin and Neumann Problems for a Class of Singularly Perturbed Semilinear Elliptic Equations, J. Differential Equations 34(1979), 55-73.

4. F. A. Howes, Some Singularly Perturbed Nonlinear Boundary Value Problems of Elliptic Type, Proc. Conf. Nonlinear P.D.E.'s in Engrg. and Applied Sci., ed. by R. L. Sternberg, Marcel Dekker, New York, 1980, pp. 151-166.

5. F. A. Howes, Perturbed Boundary Value Problems Whose Reduced Solutions are Nonsmooth, Indiana U. Math. J. 30(1981), 267-280.

6. T. Kusano, On Bounded Solutions of Exterior Boundary Value Problems for Linear and Quasilinear Elliptic Differential Equations, Japan J. Math. 35(1965), 31-59.

7. N. Levinson, The First Boundary Value Problem for $\varepsilon\Delta u + A(x,y)u_x + B(x,y)u_y + C(x,y)u = D(x,y)$ for Small ε, Ann. Math. 51(1950), 428-445.

8. J. Mauss, Etude des Solutions Asymptotiques de Problèmes aux Limites Elliptiques pour des Domaines non Bornés, Compte Rendus Acad. Sci., Ser. A 269(1969), 25-28.

9. N. Meyers and J. Serrin, The Exterior Dirichlet Problem for Second Order Elliptic Partial Differential Equations, J. Math. Mech. 9(1960), 513-538.

LINEAR TRANSPORT THEORY
AND
AN INDEFINITE STURM-LIOUVILLE PROBLEM*

Hans G. Kaper, C. Gerrit Lekkerkerker,** Anton Zettl***

Abstract

Linear transport processes occur whenever particles move in a host medium, carrying mass, momentum, and energy from one point of the medium to another. Mathematical models of such transport processes involve two operators, one accounting for free streaming of the particles, the other for interactions between the particles and the atoms or molecules of the surrounding host medium. We investigate a time-independent electron transport problem, where the free streaming operator is the multiplicative coordinate operator in $L^2(-1,1)$ and the interaction operator is the Legendre differential operator.

I. Introduction

Transport theory. Linear transport theory is the study of equations that describe linear transport phenomena in matter. The phenomena may relate to neutron transport in a nuclear reactor or a nuclear scattering experiment, radiative transfer in a stellar or planetary atmosphere, electron transport in a metal, penetration of x-rays or γ-rays through scattering media, and similar processes. In each case, the transport mechanism involves the migration of particles (neutrons, photons, electrons, etc.) through a host medium. As long as a particle is not subject to any force, it moves with a constant velocity -- that is, in a straight line with constant speed. The particle motion may

*This work was supported by the Applied Mathematical Sciences Research Program (KC-04-02) of the Office of Energy Research of the U.S. Department of Energy under Contract W-31-109-ENG-38.

**Permanent address: Department of Mathematics, University of Amsterdam, Amsterdam, Netherlands.

***Permanent address: Department of Mathematical Sciences, Northern Illinois University, DeKalb, IL 60115.

be affected along the way by external accelerative or decelerative forces (gravity, electric field, etc.). More important, however, are the collisions that take place between the particles and the atoms and molecules of the host medium. A collision is an interactive process governed by the laws of classical or quantum mechanics. The study of individual collisions belongs to the field of physics; transport theory begins with given laws governing the results of individual collisions, and considers the statistical problem of determining the result of a large number of collisions governed by these laws.

The object of investigation in transport theory is the distribution of particles in phase space. Phase space is the direct product of configuration space and velocity space; in general, it is six-dimensional. A transport equation is a balance equation for the (expected) number of particles in an infinitesimal volume element of phase space. Let (x,ξ) denote the coordinates of a point in phase space, x being the position in configuration space and ξ the position in velocity space, and let $f(x,\xi,t)dxd\xi$ denote the (expected) number of particles in an infinitesimal volume element $dxd\xi$ centered at (x,ξ) at time t. Let Ω be the region in configuration space where the transport processes take place, and let S be the velocity domain of the particles involved in the transport processes. Then a transport equation is an equation for the phase space distribution function f of the form

$$\frac{\partial f}{\partial t} = -\frac{\partial}{\partial x} \cdot \xi f - \frac{\partial}{\partial \xi} \cdot af + \left(\frac{\partial f}{\partial t}\right)_{coll} . \tag{1.1}$$

The first term in the right member represents the effect of free streaming; it is the spatial divergence of the particle flux vector. The second term represents the effect of external forces; it is the divergence of the particle flux vector in velocity space, a is the force per unit mass. The last term represents the rate of change of f due to collisions; in linear transport theory, it depends linearly on f. Eqn. 1.1 is supposed to hold for all $(x,\xi) \in \Omega\times S$, and for all $t > 0$ say.

Simplifying assumptions. Having stated the general form of a transport equation, we shall now introduce various simplifications. To begin with, we shall ignore the effects of external forces; this eliminates the second term in the right member of 1.1. Next, we shall assume that the transport system is in a state of equilibrium, so the rate of change of f, i.e., the left member of 1.1, is identically zero. Furthermore, if all particles involved in the transport process have the same speed (i.e., the same magnitude of the velocity vector, $|\xi|$) and this speed is not changed in a collision, then $|\xi|$ becomes a parameter of the problem; consequently, the velocity variable ξ is effectively replaced by the unit vector $\omega = \xi/|\xi|$ as independent variable. There are, in fact, many systems for which a one-speed transport equation is entirely realistic.

A difficulty with the transport equation in arbitrary domains is the coupling between the spatial and the velocity variable through the operator $(\partial/\partial x)\cdot\xi$ in the streaming term. This coupling can only be broken in one-dimensional plane slab configurations, where $\Omega = \Delta_1 \times \mathbb{R}^2$, Δ_1 some open interval on the real axis, and the data (material properties of the host medium, boundary data, etc.) are independent of x_2 and x_3. The solution of the transport equation is then also independent of x_2 and x_3. Under these circumstances, assuming a one-speed approximation, we find that the operator $(\partial/\partial x)\cdot\xi$ reduces to $(\partial/\partial x_1)(\omega\cdot e_1)$, where e_1 is the unit vector in the direction of increasing x_1. Now, if we let e_1 define the polar axis in a polar coordinate system, and ω has the polar angle θ $(0 \le \theta \le \pi)$, then $\omega\cdot e_1 = \cos\theta$. It is customary to denote the cosine of the polar angle of ω by μ: $\mu = \cos\theta$; this variable ranges over the interval $[-1,1]$. Note that a positive value of μ represents movement in the direction of increasing x_1, a negative value of μ movement in the direction of decreasing x_1.

Of course, ω is determined not only by its polar angle, but also by its azimuth, so, in general, the distribution function f will depend on this variable as well. However, we shall ignore this dependence and assume that our transport system has azimuthal symmetry, i.e., that the system is invariant under a rotation about the polar axis e_1.

With these assumptions we have achieved that the phase space is essentially two-dimensional, one coordinate to represent the position inside the slab (measured perpendicularly from some fixed plane of reference), another coordinate to represent the direction of motion (measured by the cosine of the angle between the velocity vector and the direction of increasing depth inside the slab). We shall use the letter x to denote the position coordinate (normalized in some convenient way); the direction coordinate is μ, as we observed earlier. In terms of these variables, the free streaming operator is of the form $(\partial/\partial x)\mu$.

So far we have not said anything about the term $(\partial f/\partial t)_{coll}$ in Eqn. 1.1. In stationary transport problems, t does not enter at all, of course, but f still changes because of collisions. The exact expression for the collision term depends on the particular physical problem under discussion. In general, collisions are regarded as instantaneous and localized events, so they affect only the dependence of the distribution function on the velocity variable. With all the simplifying assumptions introduced earlier, collisions affect only the dependence of f on μ, not its dependence on x; the position variable is simply a parameter.

Boundary value problem. The particular problem that we shall discuss in this article comes from electron transport theory. It goes back to an article by Bethe, Rose, and Smith [1938], although the differential equation can already be found in an earlier article by Bothe [1929]. The equation is

$$\frac{\partial}{\partial x}\mu\phi(x,\mu) - \frac{\partial}{\partial\mu}(1-\mu^2)\frac{\partial\phi}{\partial\mu}(x,\mu) = 0 , \qquad (x,\mu) \in \Delta\times J , \qquad (1.2)$$

where $\Delta = (0,\tau)$, $J = (-1,1)$. The unknown function ϕ represents the electron distribution in phase space. One recognizes the first term as the free streaming term, the second term as the collision term.

With Eqn. 1.2 are prescribed boundary conditions at the endpoints $x = 0$ and $x = \tau$. The nature of these boundary conditions is somewhat unusual from the mathematical point of view, but easy to understand if we keep the physics behind the equation in mind. Eqn. 1.2 describes what happens inside the plate; the boundary conditions embody what we can prescribe at the free surfaces. Physically, we expect that we can prescribe the flux of those electrons that move into the plate; the flux of electrons moving out of the plate should follow as part of the solution. The electron flux used to be given by the vector ξf in Eqn. 1.1; in the simplified model 1.2 it corresponds to the quantity $\mu\phi$. Recall that positive μ-values represent motion towards increasing values of x, negative μ-values motion towards decreasing values of x. The following specification of the boundary conditions should now be understandable:

$$\lim_{x \downarrow 0} \mu\phi(x,\mu) = g_+(\mu) , \quad 0 \leq \mu \leq 1 , \tag{1.3-1}$$

$$\lim_{x \uparrow \tau} \mu\phi(x,\mu) = g_-(\mu) , \quad -1 \leq \mu \leq 0 , \tag{1.3-2}$$

where g_+ and g_- are given functions, which are defined on [0,1] and [-1,0], respectively.

The boundary value problem 1.2,1.3 was considered by Bethe, Rose, and Smith, who used formal expansion techniques. A few years ago, Beals [1977] proved existence and uniqueness of solutions for this problem in a suitable weak formulation. In the last decade, a considerable effort has been spent on the development of spectral methods for linear transport equations. A monograph on the subject, written by Kaper, Lekkerkerker, and Hejtmanek [1982], will appear shortly.

Functional formulation. The approach is based on a representation of Eqn. 1.2 in terms of vector-valued functions and operators on a Hilbert space.

Let $J = (-1,1)$, and let H be the usual Hilbert space $L^2(J)$. In H we define the multiplicative coordinate operator T,

$$Tf(\mu) = \mu f(\mu) , \qquad \mu \in J, \quad f \in H . \tag{1.4}$$

Note that T is injective, bounded and selfadjoint; its inverse T^{-1} is unbounded and defined on im T.

Let $p = [1-\mu^2 : \mu \in J]$, and let M be the maximal operator associated with the expression $-(pf')'$, i.e.,

$$Mf = -(pf')' , \qquad f \in \text{dom } M , \tag{1.5}$$

where dom M = $\{f \in H$: pf' absolutely continuous on compact subintervals of J, $(pf')' \in H\}$. We define the collision operator (or: scattering operator) A in H as the restriction of M to dom A,

$$Af = Mf , \qquad f \in \text{dom } A , \tag{1.6}$$

where dom A = $\{f \in \text{dom } M: \lim_{\mu\uparrow 1} f(\mu)$ and $\lim_{\mu\downarrow -1} f(\mu)$ exist and are finite$\}$. That is, A is the well-known Legendre differential operator.

Furthermore, let $H_\pm = L^2(J_\pm)$, where $J_+ = (0,1)$ and $J_- = (-1,0)$; H_+ and H_- can be identified with proper subspaces of H, and $H = H_+ \oplus H_-$. Let $P_\pm$ be the (orthogonal) projection which maps H onto $H_\pm$ along $H_\mp$.

Then the boundary value problem 1.2,1.3 can be formulated as an equation in H for the vector-valued function $\phi: [0,\tau] \to H$,

$$(T\phi)'(x) + A\phi(x) = 0 , \qquad x \in (0,\tau) , \tag{1.7}$$

$$P_+ \lim_{x\downarrow 0} T\phi(x) = g_+ , \quad P_- \lim_{x\uparrow\tau} T\phi(x) = g_- . \tag{1.8}$$

In Eqn. 1.7, ' denotes differentiation with respect to x. Because T is injective, we can also formulate the equation in terms of the function ψ, where $\psi(x) = T\phi(x)$ for all $x \in (0,\tau)$,

$$\psi'(x) + AT^{-1}\psi(x) = 0 \ , \qquad x \in (0,\tau) \ , \tag{1.9}$$

$$P_+ \lim_{x\downarrow 0} \psi(x) = g_+ \ , \quad P_- \lim_{x\uparrow \tau} \psi(x) = g_- \ . \tag{1.10}$$

Thus, the rate of change of ψ is determined by the operator AT^{-1}, to which we shall refer as the transport operator. Note that dom $AT^{-1} = \{f \in H: f \in \text{im } T, T^{-1}f \in \text{dom } A\}$. The function ϕ is commonly known as the angular density, the function ψ as the current density. The (macroscopically observable) total electron density and electron current density are obtained by integrating ϕ and ψ, respectively, over all directions μ.

In the following Section II we investigate the operator A in some detail. In Section III we study the operator AT^{-1} and present what is known in transport theory as the full-range theory. The objective here is to give a spectral representation of AT^{-1}. In Section IV we use the results of the full-range theory to construct the general solution of the differential equation. Section V contains further details about an indefinite Sturm-Liouville boundary value problem that is equivalent with the eigenvalue problem for $T^{-1}A$. In Section VI we discuss what is known in transport theory as the half-range theory. The objective here is to show, roughly speaking, that one half of the eigenfunctions are sufficient to expand a function defined on one half of the range of the independent variable. The discussion is very speculative and meant to stimulate research on this challenging and open problem.

II. Scattering Operator

In this section we summarize some relevant properties of the scattering operator A. This operator was defined in Eqn. 1.6 as a restriction of the maximal operator M. Note that the equation $-(pf')' = 0$ is singular at both endpoints of J, as p^{-1} is not integrable in a neighborhood of either of these points. The equation $-(pf')' = 0$ has two fundamental solutions, namely $f_1 = 1$ and $f_2 = [\ell n((1+t)/(1-t)): t \in (-1,1)]$, which are both in dom M. Hence, the maximal operator M is limit-circle at both endpoints and M is not selfadjoint. To obtain a selfadjoint realization of the expression $-(pf')'$, a boundary condition is needed at each endpoint. The boundedness condition in the definition of dom A serves this purpose. Notice that it eliminates the solution f_2 from the domain, i.e., $f_2 \in$ dom M, but $f_2 \notin$ dom A. It turns out that there are several equivalent definitions of the operator A.

THEOREM 2.1. Suppose $f \in$ dom M. Then the following conditions are equivalent:

(i) f is bounded on (-1,1);

(ii) $\lim_{\mu\uparrow 1} f(\mu)$ and $\lim_{\mu\downarrow -1} f(\mu)$ exist and are finite;

(iii) $\lim_{\mu\uparrow 1} p(\mu)f'(\mu) = \lim_{\mu\downarrow -1} p(\mu)f'(\mu) = 0$;

(iv) $p^{1/2} f' \in H$.

PROOF. The equivalence of (ii), (iii), and (iv) is established in Akhiezer and Glazman [1981, Vol. 2, Appendix 2]. The equivalence of (i) and (ii) follows similarly. ///

The next theorem shows that the boundedness condition imposed on the elements of dom A is the "right" restriction, i.e., it determines a selfadjoint realization of the expression $-(pf')'$ in H.

THEOREM 2.2. *The (unbounded) operator* A *is selfadjoint in* H; *its spectrum* $\sigma(A)$ *is discrete and consists of simple eigenvalues:* $\sigma(A) = \{n(n+1): n=0,1,...\}$; *the eigenfunction corresponding to the eigenvalue* $n(n+1)$ *is the Legendre polynomial of degree* n,

$$P_n(\mu) = \frac{1}{2^n n!} \frac{d^n}{d\mu^n} (\mu^2-1)^n, \quad n=0,1,... \tag{2.1}$$

PROOF. See, for example, Akhiezer and Glazman [1981, Vol. 2, Appendix 2] or Kamke [1971, Part C, Section 2.240]. ///

For a comprehensive discussion of the Legendre operator, using both the "right definite" and the "left definite" approach of the Sturm-Liouville theory, we refer the reader to the recent article by Everitt [1978]. This article also contains a discussion of half-range expansions, i.e., expansions of elements of H_+ or H_- in terms of Legendre polynomials.

Notice that A has a nontrivial kernel, $\ker(A) = sp(P_0)$. Because P_0 has the constant value 1 on J, we prefer to write 1, instead of P_0, so $\ker(A) = sp(1)$.

THEOREM 2.3. *The equation* $Au = f$ *is solvable in* H *if and only if* $(f,1) = 0$; *if the solvability condition is met, then the solution* u *is of the form* $u = K_0f+c1$, *where* $c \in \mathbb{C}$ *is arbitrary and* K_0 *is the integral operator*

$$K_0f(\mu) = \int_{-1}^{1} k(\mu,\mu')f(\mu')d\mu' , \quad \mu \in J, \quad f \in H ; \tag{2.2}$$

the kernel k *is given by the expression*

$$k(\mu,\mu') = -\tfrac{1}{2} \ln((1+\mu_>)(1-\mu_<)) , \quad (\mu,\mu') \in J^2 , \tag{2.3}$$

with $\mu_> = \max(\mu,\mu')$, $\mu_< = \min(\mu,\mu')$.

PROOF. Any element f in the range of A satisfies the identity $(f,1) = 0$. The converse statement, as well as the second assertion of the theorem, follows from a direct computation. ///

Notice that $k(\cdot,\mu') \in L^2(J)$ for each fixed $\mu' \in J$, and $k(\mu,\cdot) \in L^2(J)$ for each fixed $\mu \in J$, so k is a Hilbert-Schmidt kernel. Hence, K_0 is compact and selfadjoint in H. Obviously, $\sigma(K_0) = \{(n(n+1))^{-1}: n=1,2,\ldots\}$ with $K_0P_n = (n(n+1))^{-1}P_n$ for $n=1,2,\ldots$, which shows that K_0 is (strictly) positive on the orthogonal complement of 1 in H. The vector $K_0 1$ is a negative multiple of 1: $K_0 1 = -2(\ell n2-\frac{1}{2})1$. Consequently,

$$(K_0 f,1) = (f,K_0 1) = -2(\ell n2 - \tfrac{1}{2})(f,1) , \quad f \in H , \tag{2.4}$$

which shows that K_0 maps the orthogonal complement of 1 into itself.

We use the operator K_0 to define a new operator K in H,

$$Kf = K_0 f + 2(\ell n2 - \tfrac{1}{2})(f,1)1 , \quad f \in H . \tag{2.5}$$

The operator K, being a perturbation of K_0 by a one-dimensional symmetric operator, is also compact and selfadjoint in H. Furthermore, $\sigma(K) = \{(n(n+1))^{-1}: n=1,2,\ldots\}$, with $KP_n = (n(n+1))^{-1}P_n$ for $n=1,2,\ldots$, so K is also (strictly) positive on the orthogonal complement of 1 in H. Observe that K1 is a positive multiple of 1: $K1 = 2(\ell n2-\frac{1}{2})1$. Hence, K is (strictly) positive on the entire space H.

For future reference we summarize the relationship between A and K:

$$KAf = f - \tfrac{1}{2}(f,1)1 , \quad f \in \text{dom } A , \tag{2.6}$$

$$AKf = f , \quad f \in H , \tag{2.7}$$

and note the identities

$$(Kf,1) = 2(\ell n2-\tfrac{1}{2})(f,1) , \quad f \in H , \tag{2.8}$$

$$(Kf,T1) = \tfrac{1}{2}(f,T1)\,, \qquad f \in H\,. \tag{2.9}$$

III. Full-Range Theory

Reduction of AT^{-1}. Full-range theory is the study of the transport operator AT^{-1}. The first step consists of a reduction of AT^{-1} to isolate the generalized eigenspace associated with the eigenvalue at the origin.

THEOREM 3.1. *The Hilbert space H admits a decomposition $H = H_0 \oplus H_1$, such that the pair $\{H_0,H_1\}$ reduces the operator AT^{-1}. In particular, $H_0 = sp(T1,T^2 1)$ and $H_1 = \{f \in H: (f,1) = (f,T1) = 0\}$.*

PROOF. Because $\ker(A) = sp(1)$ and T is injective, we have $\ker(AT^{-1}) = sp(T1)$. A straightforward calculation shows that $\ker((AT^{-1})^2) = sp(T1,T^2 1)$ and $\ker((AT^{-1})^n) = \ker((AT^{-1})^2)$ for n=3,4,... . Thus, H_0 is the generalized eigenspace associated with the eigenvalue 0 of AT^{-1}. Note that $H_0 \subset \mathrm{dom}\, AT^{-1}$. Now consider the subspace H_1, as defined in the theorem. If $f \in H_1 \cap \mathrm{dom}\, AT^{-1}$, then $(AT^{-1}f,1) = (T^{-1}f,A1) = 0$ and $(AT^{-1}f,T1) = (T^{-1}f,AT1) = 2(T^{-1}f,T1) = 2(f,1) = 0$, so $AT^{-1}f \in H_1$. That is, H_1 is an invariant subspace. Clearly, $H_0 \cap H_1 = \{0\}$. ///

The projection operator P_0 which maps H onto H_0 along H_1 is easily found,

$$P_0 f = \frac{3}{2}(f,T1)T1 + \frac{3}{2}(f,1)T^2 1\,, \qquad f \in H\,. \tag{3.1}$$

We denote the projection operator which maps H onto H_1 along H_0 by P,

$$Pf = f - \frac{3}{2}(f,T1)T1 - \frac{3}{2}(f,1)T^2 1\,, \qquad f \in H\,. \tag{3.2}$$

THEOREM 3.2. (i) <u>The restriction</u> $AT^{-1}|H_0$ <u>is defined on</u> H_0 <u>and is represented by the matrix</u> $\begin{pmatrix}0&2\\0&0\end{pmatrix}$ <u>relative to the basis</u> $(T1,T^2 1)$ <u>of</u> H_0.
(ii) <u>The restriction</u> $AT^{-1}|H_1$ <u>is injective and</u> $(AT^{-1}|H_1)^{-1} = PTK|H_1$.

PROOF. (i) Immediate consequence of the identities $A1 = 0$ and $AT1 = 2T1$.
(ii) Take any $f \in H_1$. Then $TKf \in \text{dom}\, AT^{-1}$ and

$$AT^{-1}PTKf = AT^{-1}[TKf-P_0TKf] = AKf-3(TKf,1)T1 .$$

Because AK is the identity on H, we have $AKf = f$. Furthermore, $Kf \in H_1$, so $(TKf,1) = (Kf,T1) = 0$. Hence, $AT^{-1}PTKf = f$ for all $f \in H_1$.

Next, take any $f \in H_1 \cap \text{dom}\, AT^{-1}$. Then

$$PTKAT^{-1}f = PT[T^{-1}f - \tfrac{1}{2}(T^{-1}f,1)1] = Pf - \tfrac{1}{2}(T^{-1}f,1)PT1 .$$

But $Pf = f$ and $PT1 = 0$, so $PTKAT^{-1}f = f$ for all $f \in H_1 \cap \text{dom}\, AT^{-1}$. ///

Let the operator B be defined on H_1 by the expression

$$Bf = PTKf , \quad f \in H_1 . \tag{3.3}$$

According to Theorem 3.2(ii), B coincides with the inverse of AT^{-1} on H_1. As K is compact and P and T are bounded on H, B is compact on H_1. The study of AT^{-1} has thus been reduced to the study of the compact operator B on H_1.

<u>**Structure of B.**</u> The operator B is clearly not symmetric with respect to the inner product of H. However, B can be made into a symmetric operator if we introduce a new inner product $(\cdot,\cdot)_A$ on H,

$$(f,g)_A = (K^{1/2} f, K^{1/2} g) , \qquad f,g \in H . \tag{3.4}$$

The A-inner product and the corresponding norm $\|\cdot\|_A$ define a new topology on H. On the finite-dimensional subspace H_0 this topology is equivalent with the topology induced by the ordinary inner product, but on H_1 it is weaker. Hence, the linear space H endowed with the A-inner product, which we denote by $(H,\|\cdot\|_A)$, is an inner product space, but not a Hilbert space. We obtain a Hilbert space by completing $(H,\|\cdot\|_A)$ with respect to the A-inner product; we denote this Hilbert space by H_A. Similarly, the linear space $(H_1,\|\cdot\|_A)$ is an inner product space which, upon completion with respect to the A-inner product, yields the Hilbert space $H_{1,A}$. Clearly, $H_A = H_0 \oplus H_{1,A}$.

The projection operators P_0 and $P = I-P_0$ are continuous in the A-norm. They can therefore be extended by continuity to H_A; we denote the extensions by the same symbols, P_0 and P. The formula $P = I-P_0$ remains valid in H_A, but P_0 and P are, of course, no longer given by the expressions 3.1 and 3.2.

The following lemma shows that T can also be extended by continuity to H_A.

LEMMA 3.3. The operator T is bounded on $(H,\|\cdot\|_A)$.

PROOF. We already know that the Legendre polynomials P_n, n=0,1,..., are eigenfunctions of $K^{1/2}$. Specifically, $K^{1/2}P_n = \rho_n^{1/2}P_n$, where $\rho_0 = 2(\ell n 2 - \frac{1}{2})$ and $\rho_n = (n(n+1))^{-1}$ for n=1,2,... . Furthermore,

$$TP_n = (2n+1)^{-1}[(n+1)P_{n+1}+nP_{n-1}] , \qquad n=0,1,\ldots,$$

where $P_{-1} = 0$.

Let $f \in H$ be arbitrary. Then $f = \sum_{n=0}^{\infty} (n+\frac{1}{2})^{1/2} a_n P_n$, where $a_n = (f,P_n)$, and

$$\|f\|_A^2 = \|K^{1/2} f\|^2 = \sum_{n=0}^{\infty} \rho_n |a_n|^2 ,$$

$$\|Tf\|_A^2 = \|K^{1/2} Tf\|^2 = \sum_{n=0}^{\infty} \theta_n \rho_n |a_n|^2 .$$

The constants θ_n are uniformly bounded and positive. In the derivation of the expression for $\|Tf\|_A^2$ we have used the fact that the ratio ρ_{n+1}/ρ_n is uniformly bounded above and below, and the elementary inequality $2|\mathrm{Re}(ab)| \leq |a|^2+|b|^2$ for any $a,b \in \mathbb{C}$. The lemma follows. ///

We extend the operator T to H_A by continuity, denoting the extension by the same symbol T. The extension of the operator K to H_A is trivial, as $\|Kf\|_A = \|K(K^{1/2} f)\| \leq |K| \|K^{1/2} f\| = |K| \|f\|_A$ for every $f \in H$. We denote the extended operator also by K. We now extend the operator B to $H_{1,A}$ by the expression

$$Bf = PTKf , \qquad f \in H_{1,A} . \tag{3.5}$$

Note that B maps $H_{1,A}$ into itself. Actually, we have a stronger result.

LEMMA 3.4. The operator B maps $H_{1,A}$ into H_1.

PROOF. Let $\{f_n: n=1,2,\ldots\}$ be a Cauchy sequence in the inner product space $(H_1, \|\cdot\|_A)$. Then $\{K^{1/2} f_n: n=1,2,\ldots\}$ is a Cauchy sequence in the Hilbert space H_1, so there exists an element $g \in H_1$, such that $\|K^{1/2} f_n - g\| \to 0$ as $n \to \infty$. Because $PTK^{1/2}$ is bounded in $\mathbf{L}(H)$, it follows that the sequence $\{Bf_n: n=1,2,\ldots\}$ converges to the element $PTK^{1/2} g$. This element belongs to H_1, so the (extended) operator B maps $H_{1,A}$ into H_1. ///

The argument used in the proof of the lemma also shows that the operator B is injective. As B^{-1} coincides with AT^{-1} on $\mathrm{dom}\, AT^{-1} \cap H_1$, we see that B^{-1} extends $AT^{-1}|H_1$, with $\mathrm{dom}\, B^{-1} = \mathrm{im}\, B$. Note that $\mathrm{dom}\, B^{-1} \subset H_1$, but $\mathrm{im}\, B^{-1}$ may contain elements of $H_{1,A}$ which are not in H_1. It is not clear whether $\mathrm{dom}\, B^{-1} \subset \mathrm{im}\, PT$.

THEOREM 3.5 (i) The operator B is compact and selfadjoint on $H_{1,A}$.
(ii) The function $P1$ is a cyclic vector for B.

PROOF. (i) Because $\|Kf\|_A = \|K(K^{1/2}f)\|$ for every $f \in H$, the compactness of $K^{1/2}$ on H implies the compactness of K on H_A. Furthermore, PT is bounded on H_A, so B is compact on $H_{1,A}$. Using the expression 3.1 for P_0 one readily verifies that the product operator P_0T is selfadjoint on H. As $PT = T-P_0T$ and T is selfadjoint on H, the product PT is also selfadjoint on H. Hence, $(Bf,g)_A = (f,Bg)_A$ for all $f,g \in H_1$. The identity extends to all $f,g \in H_{1,A}$.
(ii) To show that $P1$ is a cyclic vector for B, it suffices to consider the action of B on the Legendre polynomials. Notice that $P1 = -2P_2$; furthermore, $BP_2 = (1/10)P_3$, and

$$BP_n = (2n+1)^{-1}[n^{-1}P_{n+1} + (n+1)^{-1}P_{n-1}] , \qquad n = 3,4,\ldots .$$

Hence, P_1 is a cyclic vector for B in H_1. Because the A-norm is weaker than the usual norm, any cyclic vector in H_1 is _a fortiori_ a cyclic vector in $H_{1,A}$. ///

Because B is compact and selfadjoint on $H_{1,A}$ its spectrum consists of a countably infinite sequence of real eigenvalues with an accumulation point at the origin. As B is injective, the origin itself does not correspond to an eigenvalue. The existence of a cyclic element implies that the spectrum of B is simple.

Eigenfunction expansions. Because B is selfadjoint on $H_{1,A}$, we can use the Spectral Theorem to obtain eigenfunction expansions. We write the eigenvalue equation in the form

$$B\chi_n = \lambda_n^{-1}\chi_n , \qquad n = \pm 1, \pm 2, \ldots \tag{3.6}$$

THEOREM 3.6. The eigenvectors $\{\chi_n: n = \pm1,\pm2,\ldots\}$ form an orthogonal basis in $H_{1,A}$; the eigenfunction expansion

$$f = \sum_{\substack{n=-\infty \\ (n\neq 0)}}^{\infty} \frac{(f,\chi_n)_A}{\|\chi_n\|_A^2} \chi_n , \qquad f \in H_{1,A} , \tag{3.7}$$

converges in the topology of H_A.

PROOF. Immediate consequence of the Spectral Theorem. ///

The eigenfunction expansion 3.7 can be interpreted in the framework of H_1. The operator B maps $H_{1,A}$ into H_1, so the eigenvectors χ_n belong, in fact, to H_1. Let the vectors ϕ_n be defined by

$$\phi_n = K\chi_n , \qquad n = \pm1,\pm2,\ldots . \tag{3.8}$$

From Eqns. 2.8 and 2.9 we see that K maps H_1 into itself; hence, each ϕ_n belongs to H_1. On H_1, A and K are each other's inverses. Therefore, the definition 3.8 is equivalent with

$$\chi_n = A\phi_n , \qquad n = \pm1,\pm2,\ldots . \tag{3.9}$$

The ϕ_n's satisfy the equation $KPT\phi_n = \lambda_n^{-1}\phi_n$. We observe that KPT is the adjoint of the restriction of B to H_1 considered as an operator in $\mathbf{L}(H_1)$. Thus, if we denote this adjoint by B', we have the identities

$$B'\phi_n = \lambda_n^{-1}\phi_n , \qquad n = \pm1,\pm2,\ldots . \tag{3.10}$$

That is, the ϕ_n's are eigenvectors of B' in H_1.

Because P1 is a cyclic vector for B in H_1, the inner product $(P1,\phi_n)$ is nonzero for each n. We can therefore normalize the ϕ_n's by the condition

$$(P1,\phi_n) = 0 , \qquad n = \pm1,\pm2,\ldots . \tag{3.11}$$

Note that $\chi_n \in \text{dom } AT^{-1}$ and $\phi_n \in \text{dom } T^{-1}A$ for each n, with

$$AT^{-1}\chi_n = \lambda_n\chi_n , \quad T^{-1}A\phi_n = \lambda_n T^{-1}PT\phi_n , \quad n = \pm1,\pm2,\ldots . \tag{3.12}$$

THEOREM 3.7. _The eigenvectors_ $\{\chi_n: n=\pm1,\pm2,\ldots\}$ _and_ $\{\phi_n: n = \pm1,\pm2,\ldots\}$ _form a biorthogonal system in_ H_1, _in the sense that_ $(\chi_m,\phi_n) = 0$ _if_ $m \neq n$ _and_ $(\chi_n,\phi_n) \neq 0$ _for each_ n. _The eigenfunction expansion_ 3.7 _can be written in the form_

$$f = \sum_{\substack{n=-\infty \\ (n\neq 0)}}^{\infty} \frac{(f,\phi_n)}{(\chi_n,\phi_n)} \chi_n , \qquad f \in H_{1,A} . \tag{3.13}$$

PROOF. The theorem follows from Theorem 3.6 and Eqn. 3.8. ///

The eigenvalues λ_n have a certain symmetry which is most easily detected when one uses the so-called switch operator S,

$$Sf(\mu) = f(-\mu) , \qquad \mu \in J , \quad f \in H . \tag{3.14}$$

Clearly, S is a unitary operator on H. It maps H_1 onto itself, commutes with K, and anticommutes with PT. Hence, S anticommutes with B and B' on H_1, i.e., $BSf = -SBf$ and $B'Sf = -SB'f$ for all $f \in H_1$. It follows that, if ϕ_n is an eigenvector of B' at the eigenvalue λ_n^{-1}, then $S\phi_n$ is an eigenvector of B' at the eigenvalue $-\lambda_n^{-1}$. The eigenvalues and eigenvectors can therefore be indexed such that

$$\lambda_{-n} = -\lambda_n , \quad \chi_{-n} = S\chi_n, \quad \phi_{-n} = S\phi_n, \qquad n = \pm1,\pm2,\ldots , \tag{3.15}$$

with $\lambda_n > 0$ for n=1,2,... . We shall assume that the positive eigenvalues are ordered such that $\lambda_n < \lambda_{n+1}$ for n=1,2,... . Note that $(\chi_{-n},\phi_{-n}) = (\chi_n,\phi_n)$ for each n.

Diagonalization. Theorem 3.6 and 3.7 imply that the space $H_{1,A}$ is topologically isomorphic with the sequence space ℓ^2_σ of all square summable (with respect to the weight σ) sequences $c = [c_n: n = \pm1,\pm2,\ldots]$, $c_n \in \mathbb{C}$, where the weight σ is given by

$$\sigma_{\pm n} = (\chi_n,\phi_n)^{-1} , \qquad n = 1,2,\ldots . \tag{3.16}$$

The isomorphism F which maps $H_{1,A}$ onto ℓ^2_σ and its inverse are given by

$$Ff = [(f,\phi_n): n = \pm1,\pm2,\ldots] , \qquad f \in H_{1,A} , \tag{3.17}$$

$$F^{-1}c = \sum_{\substack{n=-\infty \\ (n\neq 0)}}^{\infty} \sigma_n c_n \chi_n , \qquad c \in \ell^2_\sigma . \tag{3.18}$$

The transformation F diagonalizes the operator B on $H_{1,A}$,

$$FBf = [\lambda_n^{-1}(f,\phi_n): n = \pm1,\pm2,\ldots] , \qquad f \in H_{1,A} . \tag{3.19}$$

We can write this result more succinctly if we introduce the (unbounded) operator Λ on ℓ^2_σ:

$$\Lambda c = [\lambda_n c_n: n = \pm1,\pm2,\ldots] , \qquad c \in \operatorname{dom}\Lambda , \tag{3.20}$$

where $\operatorname{dom}\Lambda = \{c \in \ell^2_\sigma: \sum_{\substack{n=-\infty \\ (n\neq 0)}}^{\infty} \sigma_n |\lambda_n c_n|^2 < \infty\}$. Then

$$FBf = \Lambda^{-1}Ff , \qquad f \in H_{1,A} . \tag{3.21}$$

The normalization condition 3.11 implies the identity $FP1 = 1_N$, where $1_N \in \ell^2_\sigma$ is the vector all components of which are equal to 1.

IV. Solution of the Transport Equation

Reduction of the equation. In this section we consider the problem of constructing the general solution of the transport equation 1.9. In view of the results of the full-range theory we extend the equation to an equation in the space $H_A = H_0 \oplus H_{1,A}$. We recall that the operator AT^{-1} is defined on all of H_0, and that B^{-1} is the extension of $AT^{-1}|H_1$ in $H_{1,A}$. Thus, the extension of Eqn. 1.9 consists of the following pair of differential equations:

$$(P_0\psi)'(x) + AT^{-1}P_0\psi(x) = 0 , \tag{4.1}$$

$$(P\psi)'(x) + B^{-1}P\psi(x) = 0 . \tag{4.2}$$

The solution of Eqn. 4.1 is of the form

$$P_0\psi(x) = m_1(x)T1 + m_0(x)T^2 1 , \tag{4.3}$$

where the scalar quantities $m_0(x)$ and $m_1(x)$ satisfy the pair of differential equations

$$m_1'(x) + 2m_0(x) = 0 , \qquad m_0'(x) = 0 . \tag{4.4}$$

The solution can be written down immediately,

$$m_1(x) = \alpha+2\beta(\tfrac{1}{2}\tau-x) , \qquad m_0(x) = \beta , \tag{4.5}$$

where $\alpha,\beta \in \mathbb{C}$ are arbitrary. (The constant $\frac{1}{2}\tau$ has been inserted for later convenience.) Thus, the problem has been reduced to the construction of the general solution of the differential equation 4.2 in $H_{1,A}$.

Decomposition of $H_{1,A}$. The construction of the general solution of the differential equation 4.2 is rather delicate, because of the presence of an infinite number of positive and an infinite number of negative eigenvalues which accumulate at $+\infty$ and $-\infty$. We are only interested in solutions that remain bounded as $x \to \pm\infty$. For that reason we shall treat vectors in the space of the eigenvectors χ_n with positive index (for which λ_n is positive) separately from those in the span of the eigenvectors χ_n with negative index (for which λ_n is negative). We shall do so by decomposing the space $H_{1,A}$,

$$H_{1,A} = H_{1,p} \oplus H_{1,m} , \tag{4.6}$$

where $H_{1,p}$ = cl sp(χ_n: n = 1,2,...) and $H_{1,m}$ = cl sp(χ_{-n}: n = 1,2,...), the closure being taken in the A-norm. Clearly, the pair $\{H_{1,p},H_{1,m}\}$ reduces B, with $\sigma(B|H_{1,p}) = \{\lambda_n^{-1}: n = 1,2,\dots\}$ and $\sigma(B|H_{1,m}) = \{-\lambda_n^{-1}: n = 1,2,\dots\}$. That is, the decomposition 4.6 reduces B to an accretive operator in $H_{1,p}$ and a dissipative operator in $H_{1,m}$. This reduction will enable us to write down the general solution of Eqn. 4.2 in a form suitable for our purpose.

Let $P_{1,p}$ denote the projection operator which maps H_A onto $H_{1,p}$ along $H_0 \oplus H_{1,m}$, and $P_{1,m}$ the projection operator which maps H_A onto $H_{1,m}$ along $H_0 \oplus H_{1,p}$. The representations of $P_{1,p}$ and $P_{1,m}$ are

$$P_{1,p}f = \sum_{n=1}^{\infty} \sigma_n(Pf,\phi_n)\chi_n , \qquad f \in H_A , \tag{4.7}$$

$$P_{1,m}f = \sum_{n=1}^{\infty} \sigma_n(Pf,\phi_{-n})\chi_{-n} , \qquad f \in H_A . \tag{4.8}$$

The differential equation 4.2 for $P\psi(x)$ is equivalent with the following pair of differential equations for the components $P_{1,p}\psi(x)$ and $P_{1,m}\psi(x)$:

$$(P_{1,p}\psi)'(x) + B^{-1}P_{1,p}\psi(x) = 0 , \tag{4.9}$$

$$(P_{1,m}\psi)'(x) + B^{-1}P_{1,m}\psi(x) = 0 . \tag{4.10}$$

We solve these equations by semigroup methods.

__Solution operator.__ The accretive operator $B|H_{1,p}$ defines the exponential operator $\exp(-zB^{-1})$ on $H_{1,p}$ for $z \in \mathbb{C}$, Rez > 0:

$$\exp(-zB^{-1})f = F^{-1}e^{-z\Lambda}F\, f , \qquad f \in H_{1,p} , \quad \text{Re}z > 0 . \tag{4.11}$$

This operator is holomorphic in $\mathrm{Re}z > 0$, uniformly bounded for $|\arg z| \leq \frac{1}{2}\pi-\varepsilon$ ($\varepsilon > 0$), and strongly continuous (within the ε-sector) at $z = 0$, where its limit coincides with the identity on $H_{1,p}$. The general solution of Eqn. 4.9 can be expressed in terms of this exponential operator,

$$P_{1,p}\psi(x) = \exp(-xB^{-1})P_{1,p}h \ . \tag{4.12}$$

Here $P_{1,p}h \in H_{1,p}$ is arbitrary. The solution is such that $P_{1,p}\psi(x)$ remains bounded (in fact, decays exponentially) as $x \to \infty$, and its limit as $x \downarrow 0$ exists in $H_{1,p}$ and is equal to $P_{1,p}h$.

The switch operator S defined in Eqn. 3.14 and extended continuously to H_A maps $H_{1,p}$ onto $H_{1,m}$ and vice versa. We use it to define the exponential operator $\exp(-zB^{-1})$ on $H_{1,m}$ for $z \in \mathbb{C}$, $\mathrm{Re}z < 0$:

$$\exp(-zB^{-1})f = S\exp(zB^{-1})Sf \ , \qquad f \in H_{1,m} \ , \quad \mathrm{Re}z < 0 \ . \tag{4.13}$$

This operator is holomorphic in $\mathrm{Re}z < 0$, uniformly bounded for $|\arg z-\pi| < \frac{1}{2}\pi-\varepsilon$ ($\varepsilon > 0$), and strongly continuous (within the ε-sector) at $z = 0$, where its limit coincides with the identity on $H_{1,m}$. In terms of this exponential operator, the general solution of Eqn. 4.8 is

$$P_{1,m}\psi(x) = \exp((\tau-x)B^{-1})P_{1,m}h \ , \tag{4.14}$$

where $P_{1,m}h \in H_{1,m}$ is arbitrary. It is such that $P_{1,m}\psi(x)$ remains bounded as $x \to -\infty$, and its limit as $x \uparrow \tau$ exists in $H_{1,m}$ and is equal to $P_{1,m}h$.

The general solution of Eqn. 4.2 is obtained by adding the expressions 4.12 and 4.14, and the general solution of the full transport equation by adding the expression for $P_0\psi(x)$ obtained earlier to it. The final result can be formulated succinctly if we introduce the family of operators $\{U(z): 0 < \mathrm{Re}z < \tau\}$ in H_A thus:

$$U(z) = \exp((\tfrac{1}{2}\tau - z)AT^{-1})P_0 + \exp(-zB^{-1})P_{1,p} + \exp((\tau-z)B^{-1})P_{1,m} . \tag{4.15}$$

THEOREM 4.1. (i) U(z) is holomorphic in the open strip $S = \{z \in \mathbb{C}: 0 < \mathrm{Re}\, z < \tau\}$, uniformly bounded for $z \in S_\varepsilon = \{z \in \mathbb{C}: |\arg z| \leq \tfrac{1}{2}\pi - \varepsilon,\ |\arg(\tau - z) - \pi| \leq \tfrac{1}{2}\pi - \varepsilon\}$, $\varepsilon > 0$, and strongly continuous (within S_ε) at $z = 0$ and $z = \tau$.
(ii) The function $\psi(x) = U(x)h$ satisfies the pair of equations 4.1,4.2 for $0 < x < \tau$ for any $h \in H_A$; conversely, each solution of the pair of equations 4.1,4.2 has the form $\psi(x) = U(x)h$ for some $h \in H_A$.

PROOF. On H_0, U(z) is represented by the matrix

$$\begin{pmatrix} 1 & 2(\tfrac{1}{2}\tau - z) \\ 0 & 1 \end{pmatrix}$$

relative to the basis $(T1, T^2 1)$; this matrix is clearly holomorphic for all $z \in \mathbb{C}$. The sum $\exp(-zB^{-1})P_{1,p} + \exp((\tau-z)B^{-1})P_{1,m}$ is holomorphic in S, uniformly bounded in S_ε, and strongly continuous at $z = 0$ and $z = \tau$.
(ii) The image of U(z) is contained in $H_0 \oplus \mathrm{dom}\, B^{-1}$. If $\psi(x) = U(x)h$, $h \in H_A$, then $m_1(x)$ and $m_0(x)$ have the form 4.5, where $P_0 h = \alpha T1 + \beta T^2 1$ and $P_{1,p}\psi(x)$ and $P_{1,m}\psi(x)$ are given by 4.12 and 4.14, respectively. ///

V. An Indefinite Sturm-Liouville Boundary Value Problem

General remarks. The eigenvalue problem $T^{-1}Ay = \lambda y$ for the operator $T^{-1}A$ is equivalent with an indefinite Sturm-Liouville boundary value problem. The boundary value problem consists of the differential equation

$$My = -(py')' = \lambda r y , \tag{5.1}$$

where $p = [1-t^2: t \in J]$, $r = [t: t \in J]$, and the boundary condition

$$y \text{ bounded on } (-1,1) \ . \tag{5.2}$$

This problem exhibits a number of interesting features. Firstly, the differential equation is singular at both endpoints of the interval (-1,1), as p^{-1} is not integrable in a right neighborhood of -1 or in a left neighborhood of +1. Each endpoint is a limit circle type of singularity: all solutions of $My = \lambda y$ are in H for any $\lambda \in \mathbb{C}$, the same is true for all solutions of Eqn. 5.1 -- see Bradley [1972]. Secondly, the "weight" function r is positive on a subinterval of J and negative on another subinterval. Thirdly, when $\lambda = 0$, two fundamental solutions of Eqn. 5.1 are $y_1(t) = 1$ and $y_2(t) = \ell n((1+t)/(1-t))$. Both solutions are in H, but only y_1 satisfies the boundary condition 5.2. Thus, y_1 is an eigenfunction belonging to the eigenvalue $\lambda = 0$.

In the literature on Sturm-Liouville problems associated with the general equation

$$-(py')' + qy = \lambda ry \ , \tag{5.3}$$

such problems are called "right definite" when the weight function r is positive, "left definite" when $p > 0$ and $q \geq 0$ but q not identically zero. For the left definite case, the additional assumption

$$|r(t)| \leq c|q(t)| \ , \quad c \text{ constant} \ , \tag{5.4}$$

is often made, e.g. in Everitt [1974], Daho and Langer [1977].

The fact that $\lambda = 0$ is an eigenvalue of the boundary value problem is a nontrivial complication. In the right definite case the eigenvaue $\lambda = 0$ can be "removed" by a substitution of the type $\lambda \mapsto \lambda+\alpha$. In the present case, such a substitution would yield an equation of the type 5.3, where both q and r change sign.

Orthogonality. If the vectors ϕ_n and ϕ_m are eigenvectors of $T^{-1}A$ in H_1 at the eigenvalues λ_n and λ_m, respectively, then $A\phi_n = \lambda_n T\phi_n$ and $A\phi_m = \lambda_m T\phi_m$. Taking the inner product of the first equation on the left with ϕ_m, and of the second equation on the right with ϕ_n, and subtracting the two resulting equations, we obtain the identity $(\lambda_n-\lambda_m)(T\phi_n,\phi_m) = 0$. The eigenvalues λ_n and λ_m are distinct if $n \neq m$. Hence, the eigenvectors ϕ_n and ϕ_m satisfy the "orthogonality relation"

$$\int_{-1}^{1} \mu\phi_n(\mu)\phi_m(\mu)d\mu = 0 , \qquad n \neq m , \tag{5.5}$$

Obviously, this is not a true orthogonality relation, because the factor μ does not define a weight function. Nevertheless, Eqn. 5.5 is what used to be called a full-range orthogonality relation in transport theory. We now show how it is related to the biorthogonality relation,

$$(\chi_n,\phi_m) = 0 , \qquad n \neq m , \tag{5.6}$$

which we established earlier.

From the eigenvalue equation $\chi_n = \lambda_n B\chi_n$ and the relation $\chi_n = K\phi_n$ we obtain the identity

$$\chi_n = \lambda_n PT\phi_n . \tag{5.7}$$

Combined with the biorthogonality relation 5.6, this identity yields the result $(PT\phi_n,\phi_m) = 0$ if $n \neq m$. But ϕ_n and ϕ_m both belong to H_1, so $(PT\phi_n,\phi_m)$ $= (T\phi_n,\phi_m)$. Hence, $(T\phi_n,\phi_m) = 0$ if $n \neq m$, which is precisely Eqn. 5.5. It is thus evident that the occurrence of the indefinite weight factor μ in Eqn. 5.5 is really due to the elimination of the eigenvector χ_n from the biorthogonality relation 5.6. The occurrence of a biorthogonality relation which involves the eigenvectors of both B^{-1} and its adjoint should not come as a surprise, as we are dealing with operators which are not selfadjoint. The existence of a simple relation like 5.7, which enabled us to eliminate χ_n, is entirely fortuitous and has no obvious analogue in more complicated models.

Completeness. We now show that the set of eigenvectors of the boundary value problem 5.1,5.2 is complete in the entire space H in the usual topology. The proof is a modification of a proof of Kamke [1939,1942], who derived a similar result for left-definite Sturm-Liouville problems. The details are rather straightforward and will be omitted.

Kamke considered the general boundary value problem

$$Ny = -(py')' + qy = \lambda ry \,, \tag{5.8}$$
$$U(y) = 0 \,, \tag{5.9}$$

on a compact interval [a,b]. The coefficients p, q and r are assumed continuous on [a,b], with p strictly positive; r changes sign on [a,b] at least once, but no more than a finite number of times. The (two-point) boundary condition 5.9 is such that the boundary value problem described by Eqns. 5.8 and 5.9 is selfadjoint in $L^2(a,b)$. Furthermore, it is assumed that $\lambda = 0$ is not an eigenvalue and that $\int_a^b yNy \geq 0$ for every function $y \in C^2([a,b])$ which satisfies the boundary condition 5.9.

Kamke showed that the boundary value problem 5.8,5.9 has an infinite number of positive eigenvalues, as well as an infinite number of negative eigenvalues, which can be ordered such that $\lambda_{-n-1} \leq \lambda_{-n} < 0$ and $0 < \lambda_n < \lambda_{n+1}$ for n = 1,2,... . Each eigenvalue has a finite multiplicity, and $\sum_n |\lambda_n|^{-1} = \infty$. Furthermore, there exists an infinite system of eigenvectors $\{\psi_n : n = \pm 1, \pm 2, \ldots\}$, which can be normalized such that $\int_a^b r\psi_m\psi_n = \delta_{mn}\,\mathrm{sgn}(\lambda_n)$. The series $\sum_n |\lambda_n|^{-2}\psi_n^2(t)$ converges for all $t \in [a,b]$, and the sum S(t) satisfies the inequality $S(t) \leq G(t,t)$, where G is the Green's function of the boundary value problem 5.8,5.9 when $r(t) = 1$. Finally, each function $f \in C^2([a,b])$ which satisfies the boundary condition 5.9 has an eigenfunction expansion of the form

$$f = \sum_{\substack{n=-\infty \\ (n\neq 0)}}^{\infty} c_n\psi_n \,, \quad \text{where} \quad c_n = \mathrm{sgn}(\lambda_n)\int_a^b f\psi_n r\,dt \,,$$

and this series expansion converges absolutely and uniformly on [a,b].

Kamke presented two proofs of the above results; one proof based on Hilbert's theorem on polar integral equations can be found in Kamke [1939, Section 3], another proof based on the calculus of variations in Kamke [1942].

The boundary value problem 5.1,5.2 does not fit directly in the framework of Kamke's results, so some modifications and extensions of Kamke's proofs are necessary.

Firstly, the boundary condition 5.2 is not of the type 5.9; U(y) represents a linear combination of the values of y and py' at a and b. The specific form of 4.9 allows Kamke to get the boundary terms to vanish upon integration by parts. However, as we have seen in Theorem 2.1, the boundary conditions 5.2 are equivalent with several other boundary conditions; using these equivalences, we can apply Kamke's proof almost verbatim.

Secondly, the boundary value problem 5.1,5.2 is singular, whereas the boundary value problem 5.8,5.9 is regular. This difference is not as significant as it might seem, however, because the generalized Green's function of 5.1,5.2 is still a Hilbert-Schmidt kernel. The generalized Green's function is given by

$$\ell(t,s) = k(t,s)|t|^{1/2}|s|^{1/2}, \qquad t,s \in J ,$$

where k is the generalized Green's function of the Legendre differential operator, which was given in Eqn. 2.3. Kamke's proof applies, but instead of uniform convergence on (-1,1), we only obtain uniform convergence on compact subintervals of (-1,1).

Thirdly, $\lambda = 0$ is an eigenvalue for the boundary value problem 5.1,5.2, with the eigenfunction $\phi_0 = 1$. This case is excluded by assumption in Kamke's proof. Nevertheless, by considering the boundary value problem 5.1,5.2 in the space $G = \{1\}^{\perp}$, rather than in the full space $H = L^2(J)$, and using the generalized Green's function k, we recover all of Kamke's results in the closed subspace G, including the completeness of the eigenvectors $\{\psi_n: n=\pm1,\pm2,\ldots\}$, and all that needs to be done is supplement this system by the single vector $\psi_0 = 1$ to achieve completeness in the entire space H.

The above three points refer to both methods of proof of Kamke's. To see how Hilbert's theorem on polar integral equations is relevant to the boundary value problem 5.1,5.2, we observe that the nonzero eigenvalues and corresponding eigenfunctions can be characterized as eigenvalues and eigenvectors of a polar integral equation.

LEMMA 5.1. _The function_ $y \in G = \{1\}^{\perp}$ _is a solution of the boundary value problem 4.1,4.2 at the eigenvalue_ λ _if and only if the function_ z _defined by_ $z(t) = |t|^{1/2} y(t)$, $t \in J$, _is a solution of the polar integral equation_

$$z(t) = \lambda \int_{-1}^{1} k(t,s)V(s)|t|^{1/2}|s|^{1/2} z(s)ds , \qquad t \in J , \tag{5.10}$$

where $V(s) = 1$ _if_ $0 < s < 1$, $V(s) = -1$ _if_ $-1 < s < 0$.

PROOF. The proof is straightforward and will be omitted. ///

Hilbert's theorem on polar integral equations yields that Eqn. 5.10 admits a system of solutions $\{z_n: n = \pm1,\pm2,\ldots\}$ which is complete in G. The eigenvectors of 5.1,5.2 which correspond to the nonzero eigenvalues are then given by $y_n(t) = |t|^{-1/2} z_n(t)$, $n = \pm1,\pm2,\ldots$.

THEOREM 5.2. _The system of functions_ $\{y_n: n = \pm1,\pm2\}$, _supplemented by the function_ $y_0 = 1$, _is complete in_ H.

PROOF. Suppose $h \in G = \{1\}^{\perp}$ and $(h,y_n) = 0$ for $n = \pm1,\pm2,\ldots$. Then

$$0 = (h,y_n) = \int_{-1}^{1} h(t)|t|^{-1/2}|t|^{1/2} y_n(t)dt$$

$$= \int_{-1}^{1} h(t)|t|^{-1/2} z_n(t)dt .$$

Hence, $h(t)|t|^{-1/2} = 0$, a.e., consequently, $h = 0$. ///

Asymptotics. Asymptotic estimates of the eigenvalues λ_n and the normalization constant (χ_n,ϕ_n) were obtained recently by Veling [1982]. His results can be summarized as follows:

$$\lambda_n = A(n+1/2)^2 + B + O(n^{-1}), \qquad n \to \infty, \tag{5.11}$$

$$(\chi_n,\phi_n) = (L/2\pi + o(1))\lambda_n^{3/2}, \qquad n \to \infty, \tag{5.12}$$

where the constants A, B, and L involve certain elliptic integrals; numerical values are $A = 6.87518590$, $B = -0.91184985$, $L = 1.19814023$. It follows from Eqns. 5.11 and 5.12 that the spectral measure σ_n decays like n^{-3} as $n \to \infty$.

Numerical results. Using the code "SLEIGN" developed by P. Bailey [1978] at Sandia Laboratory we computed the first ten positive eigenvalues λ_n of the boundary value problem 5.1,5.2. Although this code was developed and justified only for Sturm-Liouville problems with a positive weight function, we believe that it is also effective in our case. Veling [1982] independently computed λ_1 and λ_{10} and his results agree with ours. The computed eigenvalues are listed below.

$$\begin{aligned} \lambda_1 &= 14.527, & \lambda_2 &= 42.050, \\ \lambda_3 &= 83.304, & \lambda_4 &= 138.31, \\ \lambda_5 &= 207.06, & \lambda_6 &= 289.56, \\ \lambda_7 &= 385.82, & \lambda_8 &= 495.81, \\ \lambda_9 &= 619.57, & \lambda_{10} &= 757.07. \end{aligned}$$

Properties of the eigenfunctions. In general the eigenfunctions of a Sturm-Liouville problem on an interval (a,b) have the property that the eigenfunction belonging to the smallest eigenvalue has no zero on (a,b), the eigenfunction belonging to the next smallest eigenvalue has exactly one zero on (a,b), and so on. The corresponding property of the eigenfunctions of the boundary value problem 5.1,5.2 was established recently by Kwong [1982].

THEOREM 5.3 (Kwong). For each n (n=1,2,...) the eigenfunction ϕ_n has exactly n zeros on (-1,1), which are all positive.

As a consequence of the symmetry relation 3.11, Theorem 5.3 implies that, for each n (n=1,2,...), the eigenfunction ϕ_{-n} has also exactly n zeros on (-1,1); these zeros are all negative.

VI. Half-Range-Theory

Heart of the matter. As we have seen in the Introduction, boundary conditions for transport equations are of the so-called half-range type. They involve only one half of the range of the independent variable μ at each endpoint. Although as a result of the full-range theory the structure of the general solution of the differential equation is known, this knowledge cannot be transferred immediately to the half-range subspaces H_+ and H_-, where the boundary data are specified. The basic reason is, of course, that the transport operator AT^{-1} does not commute with the projection operators P_+ and P_-.

The half-range theory is meant to deal with this difficulty. Here, the objective is to extend vectors $g_+ \in H_+$ and $g_- \in H_-$ with components in H_- and H_+, respectively, in such a way that the extended vectors, E_+g_+ and E_-g_- say, belong to subspaces of H where the structure of AT^{-1} is known. Whether this objective can be achieved for the electron transport operator under consideration is still an open problem. The physics suggest that it must be possible, and that the extension operators E_+ and E_- are well-defined at least for some classes of vectors g_+ and g_-. But this type of evidence is suggestive at best. Also, there is an important class of transport problems for which a half-range theory has been developed successfully. These transport problems involve collision operators A that are close to the identity, e.g., compact perturbations of the identity; see the recent monograph by Kaper, Lekkerkerker and Hejtmanek [1982]. It would be interesting to see whether these theories could be extended to include unbounded collision operators A.

At this point we could stop our exposition, but we prefer to take a somewhat unorthodox approach. The development of a half-range theory for the electron transport operator (and similar operators) poses a challenge that we don't intend to monopolize, and we believe that we have some ideas worth presenting. However, we emphasize that most of the following is purely speculative and that further research is certainly called for.

__Eigenvectors revisited.__ We begin with the observation of an interesting fact, namely that the set of eigenvectors $\{\chi_n: n = \pm1,\pm2,...\}$ of the operator B constitutes a basis not only in the topology of the A-inner product, as shown in Theorem 3.6, but also in the stronger topology of the ordinary inner product.

LEMMA 6.1. __The vectors__ $\{\chi_n: n = \pm1,\pm2,...\}$ __form a basis in the space__ H_1 __in the topology induced by the usual inner product on__ H.

PROOF. We already know that each χ_n belongs to H_1. (In fact, $\chi_n \in \mathrm{dom}\, AT^{-1}|H_1$.) Suppose $(f,\chi_n) = 0$ for $n = \pm1,\pm2,...$ for some $f \in H_1$. Then $(PTf,\chi_n)_A = (f,PTK\chi_n) = (f,B\chi_n) = \lambda_n^{-1}(f,\chi_n) = 0$ for each n. The completeness of the set of eigenvectors $\{\chi_n: n = \pm1,\pm2,...\}$ in $H_{1,A}$ implies that $PTf = 0$. Because $f \in H_1$, it follows that Tf is a multiple of T1, i.e., f is a multiple of 1. But now $f \in H_1$ implies that $f = 0$. This proves the lemma. ///

Decomposition of H_1. The lemma shows that we can represent vectors in H_1 as limits (in the ordinary topology) of linear combinations of the vectors χ_n. In particular, we can use the two sets of vectors $\{\chi_n: n = 1,2,...\}$ and $\{\chi_n: n = -1,-2,...\}$ to define the subspaces $H_{1,p}$ and $H_{1,m}$ of H_1,

$$H_{1,p} = \text{cl sp}(\chi_n: n=1,2,...), \quad H_{1,m} = \text{cl sp}(\chi_n: n=-1,-2,...) \tag{6.1}$$

where the closure is taken in the ordinary topology. Thus we obtain the following decomposition of H_1:

$$H_1 = H_{1,p} \oplus H_{1,m} . \tag{6.2}$$

Note that $H_{1,p}$ and $H_{1,m}$ are mapped onto each other by the switch operator S defined in Eqn. 3.14.

The decomposition 6.2, although not orthogonal with respect to the usual inner product, nevertheless induces an interesting "orthogonality property" on dom $AT^{-1}|H_1$. The orthogonality is with respect to the indefinite inner product $\{\cdot,\cdot\}$, which can be defined on dom T^{-1}

$$\{f,g\} = (f,T^{-1}g) , \qquad f,g \in \text{dom } T^{-1} . \tag{6.3}$$

One verifies that $\{f,g\} = 0$ for any pair $f \in H_{1,p} \cap \text{dom } AT^{-1}, g \in H_{1,m} \cap \text{dom } AT^{-1}$. Moreover, $\{f,f\} \geq 0$ for all $f \in H_{1,p} \cap \text{dom } AT^{-1}$ and $\{f,f\} \leq 0$ for all $f \in H_{1,m} \cap \text{dom } AT^{-1}$, with equality if and only if $f = 0$, so $\{\cdot,\cdot\}$ is positive on $H_{1,p} \cap \text{dom } AT^{-1}$ and negative on $H_{1,m} \cap \text{dom } AT^{-1}$. These considerations will guide us in the decomposition of the two-dimensional subspace H_0.

Decomposition of H_0. We now decompose H_0 into two subspaces $H_{0,p}$ and $H_{0,m}$ such that $H_{0,p}$ and $H_{0,m}$ enjoy the same properties as $H_{1,p}$ and $H_{1,m}$: they have a trivial intersection, they are mapped onto each other by the switch operator S, they are orthogonal with respect to the indefinite inner product $\{\cdot,\cdot\}$ (note that $H_0 \subset \text{dom } AT^{-1}$), and $\{\cdot,\cdot\}$ is positive on $H_{0,p}$ and negative on $H_{0,m}$. Such a decomposition is indeed possible:

$$H_0 = H_{0,p} \oplus H_{0,m} , \tag{6.4}$$

where

$$H_{0,p} = sp(\theta T1+T^2 1) , \quad H_{0,m} = sp(\theta T1-T^2 1) . \tag{6.5}$$

Here, θ can be any positive number. (This result shows that H_0 is a Krein space; cf. Bognar [1974, Section V.1].) The constant θ is determined uniquely, for example, by requiring that

$$H_{0,p} = H_0 \cap (H_+ \oplus H_{1,m}) . \tag{6.6-1}$$

Then one also has

$$H_{0,m} = H_0 \cap (H_- \oplus H_{1,p}) . \tag{6.6-2}$$

This choice of θ is, in fact, most convenient for the following analysis; cf. the discussion in Kaper, Lekkerkerker and Hejtmanek [1982, Section 9.1].

<u>**Decomposition of H.**</u> We now combine the decompositions 6.2 and 6.4 to obtain a decomposition of H. We define

$$H_p = H_{0,p} \oplus H_{1,p} , \quad H_m = H_{0,m} \oplus H_{1,m} . \tag{6.7}$$

Then, because $H = H_0 \oplus H_1$,

$$H = H_p \oplus H_m . \tag{6.8}$$

It is this decomposition 6.8 that we want to connect with the direct sum decomposition

$$H = H_+ \oplus H_- . \tag{6.9}$$

Notice that $\{f,g\} = 0$ for any pair $f \in H_p \cap \text{dom}\, AT^{-1}$, $g \in H_m \cap \text{dom}\, AT^{-1}$, as well as for any pair $f \in H_+ \cap \text{dom}\, T^{-1}$, $g \in H_- \cap \text{dom}\, T^{-1}$. Moreover, $\{\cdot,\cdot\}$ is positive on $H_p \cap \text{dom } AT^{-1}$ and $H_+ \cap \text{dom } T^{-1}$, and negative on $H_m \cap \text{dom } AT^{-1}$ and $H_- \cap \text{dom } T^{-1}$. Another property that the decompositions 6.8 and 6.9 have in common is that the component subspaces are mapped onto each other by the switch operator S.

The idea is to prove that the projections P_+ and P_- define surjective bijections of H_p onto H_+ and of H_m onto H_-, respectively. Such a result would imply that we also had the direct sum decompositions $H = H_+ \oplus H_m$ and $H = H_- \oplus H_p$. We could then define E_+ on H_+ as the inverse of $P_+|H_p$ and E_- on H_- as the inverse of $P_-|H_m$. Trivially extended to the entire space H, these operators would then be the projection operators which mapped H onto H_p along H_- and onto H_m along H_+, respectively. Or, viewed differently, E_+ would extend an element $g_+ \in H_+$ to the element $E_+g_+ \in H_p$ and E_- would extend an element $g_- \in H_-$ to the element $E_-g_- \in H_m$. On H_p and H_m we know the structure of the operator AT^{-1}, so this would solve the problem of the half-range theory.

Of course, it may be that P_+ and P_- are injective only on certain subspaces of H_p and H_m. In that case the definitions of E_+ and E_- have to be restricted to the corresponding ranges of P_+ and P_-. The identification of these ranges may be a major problem. Also note that P_+ and P_- do not extend continuously to H_A.

Connecting transformations. Let P_p and P_m denote the projection operators associated with the decomposition 6.8. The projection operators associated with the decomposition 6.9 are P_+ and P_-. The pairwise connections of the subspaces in the decompositions 6.8 and 6.9 are established by means of the so-called connecting transformations,

$$V = P_+P_p + P_-P_m \,, \quad V^\# = P_pP_+ + P_mP_- \,, \tag{6.9}$$

$$W = P_+P_m + P_-P_p \,, \quad W^\# = P_pP_- + P_mP_+ \,. \tag{6.10}$$

They satisfy several interesting identities with regard to the indefinite inner product $\{\cdot,\cdot\}$. Using the abbreviations

$$(f,g)_B = \{(P_p-P_m)f,g\} , \tag{6.11}$$

$$(f,g)_T = \{(P_+-P_-)f,g\} , \tag{6.12}$$

whenever these expressions are meaningful, one verifies that

$$(Vf,g)_T = (f,V^{\#}g)_B , \quad (Wf,g)_T = -(f,W^{\#}g)_B , \tag{6.13}$$

which suggests that V and $V^{\#}$ are, in a sense, each other's adjoints, W and $W^{\#}$ each other's anti-adjoints. Furthermore,

$$(Vf,Vf)_T - (Wf,Wf)_T = (f,f)_B , \tag{6.14}$$

$$(V^{\#}f,V^{\#}f)_B - (W^{\#}f,W^{\#}f)_B = (f,f)_T . \tag{6.15}$$

In all these identities, f and g must be restricted to the appropriate domains.

Shadow operators. We conclude this speculative discussion of the half-range theory with another suggestive remark about the role of the extension operators E_+ and E_-. Basically, half-range theory is concerned with the analysis of the (full-range) operator AT^{-1} on the half-range subspaces H_+ and H_-. The way this is achieved is by extending elements of these half-range spaces to elements of the full-range space. That is, by means of E_+ an element of H_+ is extended to an element of H_p; on H_p, the action of AT^{-1} is known; having applied AT^{-1} to the extended element, we then project back into H_+ by means of the projection operator P_+. A similar procedure is followed for H_-. As a result we have the decomposition

$$AT^{-1} = P_+AT^{-1}E_+ + P_-AT^{-1}E_- . \tag{6.16}$$

The operators $P_+AT^{-1}E_+$ and $P_-AT^{-1}E_-$ might be called the shadows of AT^{-1} on H_+ and H_-.

References

Akhiezer, N. I. and I. M. Glazman [1981], Theory of Linear Operators in Hilbert Space, Pitman, London.

Bailey, P. B. [1978], SLEIGN, An Eigenvalue-Eigenfunction Code for Sturm-Liouville Problems, Sand 77-2044, Sandia National Laboratory, Albuquerque, New Mexico.

Beals, R. [1977], On an Equation of Mixed Type from Electron Scattering, J. Math. Anal. and Applic. 58, 32-45.

Bethe, H. A., M. E. Rose and L. P. Smith [1938], The Multiple Scattering of Electrons, Proc. Am. Phil. Soc. 78, 573-585.

Bognar, J. [1974], Indefinite Inner Product Spaces, Springer-Verlag, New York.

Bothe [1929], Zeitschr. f. Physik 54, 161.

Bradley, J. S. [1972], Comparison Theorems for the Square Integrability of Solutions of $(r(t)y')' + q(t)y = f(t,y)$, Glasgow Math. J. 13, 75-79.

Daho, K. and H. Langer [1977], Sturm-Liouville Operators with an Indefinite Weight Function, Proc. Roy. Soc. Edinburgh 78A, 161-191.

Everitt, W. N. [1974], Some Remarks on a Differential Expression with an Indefinite Weight Function, in: Spectral Theory and Asymptotics of Differential Equations, E. M. de Jager (Ed.), Mathematics Studies, Vol. 13, North-Holland Publ. Co., Amsterdam.

Everitt, W. N. [1978], Legendre Polynomials and Singular Differential Operators, in: Ordinary and Partial Differential Equations, W. N. Everitt (Ed.), Lecture Notes in Mathematics, Vol. 827, Springer-Verlag, New York.

Kamke, E. [1939], Zum Entwicklungssatz bei polaren Eigenwertaufgaben, Math. Zeitschrift 45, 706-718.

Kamke, E. [1942], Über die definiten selbstadjungierten Eigenwertaufgaben bei gewöhnlichen linearen Differentialgleichungen, I,II,III,IV, Math. Zeitschrift, 45 (1939), 759-787; 46 (1940), 231-250 and 251-286; 48 [1942), 67-100.

Kamke, E. [1971], Differentialgleichungen, Lösungsmethoden und Lösungen , Chelsea Publ. Co., New York.

Kaper, H. G., C. G. Lekkerkerker and J. Hejtmanek [1982], Spectral Methods in Linear Transport Theory, Birkhäuser, Basel.

Kato, T. [1966], Perturbation Theory for Linear Operators, Springer-Verlag, New York.

Kwong, M. K. [1982], personal communication.

Veling, E. J. M. [1982], Asymptotic Solution of the Eigenfunctins of a Linear Transport Equation Arising in the Theory of Electron Scattering (to appear).

Non-normalizable eigenfunction expansions for ordinary differential operators

Robert M. Kauffman

0. Introduction

In physics, one frequently wishes to represent arbitrary waves in terms of standing waves. This is especially true in quantum mechanics, where the standing waves are eigenstates of the Schroedinger operator. It is important that this representation be a pointwise representation, rather than a representation in some abstract space, because, as will be illustrated in this note, facts of mathematical and physical significance may be easily deduced from such a concrete representation. At a point of the continuous spectrum of an ordinary differential operator in $L_2(R)$, where R denotes the real numbers, there need be no square-integrable eigenfunctions; hence one must in general deal with eigenfunctions which are not square-integrable.

What should be the asymptotic behavior at infinity for the eigenfunctions of the representation? If we ask our eigenfunctions to be too well-behaved at infinity, we will not have enough eigenfunctions to complete the representation. However, some restriction on the rate of growth of the eigenfunctions at infinity seems reasonable, because it is well-known in physics that not all solutions to the differential equation are acceptable eigenfunctions, and physicists in many examples use requirements of "smallness at infinity" to separate the acceptable from the unacceptable eigenfunctions.

In this note, we state such a representation theorem for self-adjoint ordinary differential operators in $L_2(-\infty,\infty)$ of arbitrary order. We then apply the representation to prove new results, some of which seem to have qualitative physical significance, about such ordinary differential operators and their associated unitary groups. The growth rate at infinity we shall use for an eigenfunction f is the condition that $f = (M\phi)'$, where $M = (x^2+1)^{3/4+\varepsilon}$ and $\phi \in L_1(R)$. We do not prove the representation theorem, as the proof is technical and will appear elsewhere.

Our starting point is the work of Gelfand and Kostyuchenko, which appears in Gelfand-Shilov [2]. This work does not give a pointwise representation, but does guarantee that eigenfunctions with the growth rate given above may be used to calculate the Fourier coefficients. There are some points in the proof given in [2] which are difficult to follow, particularly the assertions involving the differentiation of a Hilbert-space valued measure with respect to another measure. For this reason, we outline briefly another approach to their theory, built around the Radon-Nikodym theorem. (The details will appear elsewhere.)

Another pointwise representation theorem has been recently given by Bennewitz [1]. This interesting theorem, which also applies to the more general situation of pairs of differential operators, is based on a completely different construction from the one we use. His method does not give a rate of growth at infinity for the eigenfunctions, but may give more information than ours does about other aspects of the problem. In applications (such as the ones given below) which do not involve the rate of growth of the eigenfunctions at infinity, Bennewitz's representation may be used in place of ours. The relationship between the two representations deserves further study.

While this note deals only with the case of an ordinary differential operator, the author plans to discuss the case of an elliptic partial differential operator in a later paper.

1. The representation.

Notation 1.1. Let $M = (x^2 + 1)^{3/4+\varepsilon}$, for any positive real number ε which will remain fixed throughout this note. Let Φ denote the closure of $C_0^\infty(R)$ in the norm given by $||\theta||_\Phi = ||M\theta'||_\infty$, where R denotes the real numbers. Let $L = \Sigma_0^N a_i D^i$ be an Nth order formally symmetric ordinary differential expression, where $D = d/dx$, and each a_i is in $C^\infty(R)$, with $a_N(x) \neq 0$ for any x in R. (L is said to be <u>formally</u>

symmetric if $L = L^+$, where L^+ is the formal (Lagrange) adjoint of L.) Let H denote a self-adjoint extension of the minimal operator $T_0(L)$ in $L_2(R)$. Let S(e) denote the closed linear span in $L_2(R)$ of $\{U_t e\}$, where e is in $L_2(R)$ and U_t is exp(iHt). Let $E(\Delta)$ denote the spectral projection for H corresponding to the Borel set Δ. Let $\sigma_e(\Delta) = (E(\Delta)e,e)$, for e in $L_2(R)$. Let $E_\lambda = E(-\infty,\lambda]$. Let $P_{S(e)}$ denote the orthogonal projection in $L_2(R)$ on S(e). If ϕ is in $L_2(R)$, let $C_{\phi,e}(\lambda)$ denote the Fourier coefficient of $P_{S(e)}\phi$, which is defined by the relation

$$P_{S(e)}\phi = \int C_{\phi,e}(\lambda)\, dE_\lambda e.$$

$C_{\phi,e}$ exists and is well-defined by standard functional-analytic arguments using the spectral theorem.

Theorem 1.2. (Gelfand-Kostyuchenko). For any e in $L_2(R)$, there exists a family $\{f_{\lambda,e}\}$ of C^∞ complex-valued functions on R such that, for almost every real λ with respect to σ_e, the following hold:

a) $f_{\lambda,e} = (M\phi_{\lambda,e})'$ for some C^∞ function $\phi_{\lambda,e}$ in $L_1(R)$;

b) $Lf_{\lambda,e} = \lambda f_{\lambda,e}$;

c) for any ϕ in Φ, $C_{\phi,e}(\lambda) = \int_{-\infty}^{\infty} \bar{f}_{\lambda,e}(x)\phi(x)\, dx$.

Outline of proof: For any ϕ in Φ and any e in $L_2(R)$ define $u_{\phi,e}(\Delta)$ to be $(E(\Delta)\phi,e)$ for any Borel set Δ. It follows from a theorem of Kostyuchenko (see Gelfand-Shilov [2], Theorem IV.6.1) that there exists a positive constant K such that for any partition $\{\Delta_i\}_1^k$ of $(-\infty,\infty)$,

$$\Sigma_1^k \, ||E(\Delta_i)e||_{\Phi'} \le K. \tag{1}$$

where we identify $E(\Delta)e$ with the element F of Φ' given by $F(\phi) = (\phi,E(\Delta)e)$.

Since $u_{\phi,e}$ is absolutely continuous with respect to σ_e, it follows that $du_{\phi,e} = h_{\phi,e}d\sigma_e$ for some element $h_{\phi,e}$ of $L_1(\sigma_e)$. We define $f_{\lambda,e}(\phi)$ to essentially be $h_{\phi,e}(\lambda)$, although it takes some work, using (1), to show that this definition makes sense for almost every λ with respect to σ_e and gives an element of Φ'. It is then possible to show that $f_{\lambda,e}(H\phi) = \lambda f_{\lambda,e}(\phi)$ for all ϕ in $C_0^\infty(R)$. By distribution theory, it follows that $f_{\lambda,e}$ is a C^∞ function and $Lf_{\lambda,e} = \lambda f_{\lambda,e}$. Since $f_{\lambda,e}$ is in Φ', one may fairly easily obtain the desired asymptotic behavior of $f_{\lambda,e}$ at ∞. That $f_{\lambda,e}$ may be used in the desired way to calculate the Fourier coefficients follows directly from the relation $P_{S(e)}\phi = \int C_{\phi,e}(\lambda)dE_\lambda e$, since $(P_{S(e)}\phi, E(\Delta)e) = \int_\Delta h_{\phi,e}d\sigma_e$ for all Δ.

The details of the proof will appear elsewhere.

Remark: We now state the main theorem of the paper. The proof will appear elsewhere.

Theorem 1.3. Suppose that, for some positive real number ε, $(Hf,f) \geq \varepsilon(f,f)$ for all f in domain H. Suppose ϕ is in the domain of $H^{1/N}$, and ϕ is in S(e). Then $C_{\phi,e}(\lambda)\ f_{\lambda,e}(x)$ is in $L_1(\sigma_e \times R)$, and, for almost every x in R,

$$\phi(x) = \int C_{\phi,e}(\lambda) f_{\lambda,e}(x)\ d\sigma_e(\lambda).$$

Also, for each real t,

$$U_t\phi(x) = \int C_{\phi,e}(\lambda)e^{i\lambda t}\ f_{\lambda,e}(x)\ d\sigma_e(\lambda)$$

for almost every real x.

Remark: The above theorem leads fairly easily to the following corollary, which gives a decomposition of an arbitrary member of the domain of $H^{1/N}$.

Corollary 1.4. Suppose that $(Hf,f) \geq \varepsilon(f,f)$ for all f in domain H, where ε is a positive real number. Then there exists a set $\{e_i\}_1^k$ of members of $L_2(R)$ such that $k \leq N$ and

a) $S(e_i) \perp S(e_j)$ if $i \neq j$;
b) $\Sigma_1^k\ S(e_i) = L_2(R)$;
c) $\sigma_{e_1}(\Delta) = 0$ if and only if $E(\Delta) = 0$;
d) σ_{e_j} is absolutely continuous with respect to σ_{e_i} for $i < j$;
e) $\sigma_{e_j}(\Delta) = 0$ iff $E(\Delta)P_{S(e_m)} = 0$ for all $m \geq j$;
f) if $\Delta = \{\lambda | \{f_{\lambda,e_i}\}$ is linearly dependent$\}$, then $E(\Delta) = 0$;
g) any ϕ in domain $H^{1/N}$ is given by

$\phi(x) = \Sigma_{i=1}^k \int C_{\phi,e_i}(\lambda)\ f_{\lambda,e_i}(x)\ d\sigma_{e_i}(\lambda)$ for almost every real x;

h) if ϕ is in the domain of $H^{1/N}$, and t is any real number, $U_t\phi(x) = \Sigma_{i=1}^{k} \int e^{i\lambda t}\, C_{\phi,e_i}(\lambda)\, f_{\lambda,e_i}(x)\, d\sigma_{e_i}(\lambda)$ for almost every real x.

Remark: We now apply the representation to obtain information about $U_t\phi$.

Corollary 1.5. Let H be as in Theorem 1.3. Suppose that ϕ is in the domain of $H^{1/N}$, and that, for all t, $U_t\phi(x) = 0$ for almost every x in (a,b). Then $\phi(x) = 0$ almost everywhere.

Proof: For any θ in $C_0^\infty(a,b)$, since $C_{\phi,\phi}(\lambda) = 1$, Theorem 1.3, together with the Fubini theorem, guarantees that

$$(U_t\phi,\theta) = \int e^{i\lambda t} \left[\int_a^b f_{\lambda,\phi}(x)\, \bar{\theta}(x)dx\right] d\sigma_\phi(\lambda).$$

Hence, if $U_t\phi$ vanishes on (a,b) for all t, it follows that the Fourier transform of the measure $p(\lambda)d\sigma_\phi(\lambda)$ is zero, where $p(\lambda) = \int_a^b f_{\lambda,\phi}(x)\, \bar{\theta}(x)dx$. However, this implies that $p(\lambda) = 0$ for almost every λ with respect to σ_ϕ. Hence, for almost every λ with respect to σ_ϕ, $\int_a^b f_{\lambda,\phi}(x)\, \bar{\theta}(x)dx = 0$ for all θ in a countable dense subset of $C_0^\infty(a,b)$ which is dense in $L_2(a,b)$. Thus, for almost every λ with respect to σ_ϕ, $f_{\lambda,\phi}(x) = 0$ for all x in (a,b). Since $Lf_{\lambda,\phi} = \lambda f_{\lambda,\phi}$, it follows that, for almost every λ with respect to σ_ϕ, $f_{\lambda,\phi}(x) = 0$ for all x. By Theorem 1.3, this implies that $\phi(x) = 0$ for almost every x (with respect to Lebesgue measure).

Corollary 1.6. Let H be as in Theorem 1.3. Suppose that, for some bounded Borel set Δ, $E(\Delta)\phi = \phi$ for some ϕ in $L_2(R)$. Suppose that $\phi(x) = 0$ for almost every x in (a,b). Then $\phi(x) = 0$ for all x.

Proof: First note that the restriction of H to $E(\Delta)L_2(R)$ is a bounded linear operator, since Δ is bounded. Hence such a ϕ is necessarily in domain H, and is therefore continuous. Therefore, $\phi(x) = 0$ for every x in (a,b). But $U_t\phi = \Sigma_0^\infty (itH)^n\phi/n!$, since the restriction of H to $E(\Delta)L_2(R)$ is bounded. Since H is a differential operator, it is clear that $H^n\phi(x) = 0$ for all x in (a,b). Hence $U_t\phi(x) = 0$ for all such x. Since ϕ is in the domain of $H^{1/N}$, the previous corollary guarantees that $\phi(x) = 0$ for all x.

Remark: The last two corollaries, in the case where H is a one-dimensional Schroedinger operator, provide a qualitative explanation of a physical phenomenon known as tunnelling. Corollary 1.5 shows that, if the wave function ϕ is in the domain of $H^{1/N}$, $U_t\phi$ spreads throughout one-dimensional space at some past or future time. No potential barrier except a singularity is high enough to contain $U_t\phi$ for all t. However, Corollary 1.6 tells us that this is not surprising, since either the wave function is spread over all space at time t = 0, or else the wave function has a positive probability associated with arbitrarily high energy values, because $E(\Delta)\phi$ is the component of ϕ with energies lying in the Borel set Δ.

References

1. C. Bennewitz, "Spectral theory for pairs of differential operators," Ark. Matematik. 15(1977) 33-61.

2. I. Gelfand and G. Shilov, "Generalized functions, vol. 3," Academic Press, New York, 1977.

SOME ASPECTS AND RECENT DEVELOPMENTS IN LINEAR CONTROL THEORY.

H.W.Knobloch

Introduction.

A typical situation which underlies many considerations in modern control theory is depicted in the diagram of Fig. 1. The "plant" represents a given system which reacts in two ways with its environment namely through a pair of input variables (u,v) and a pair of output variables (y,z). u is the control, i.e. the portion of the input which we have at our disposal. v represents the disturbances acting on the system which cannot be controlled directly and which in most cases are not accessible to precise measurement. The output consists of the observed variable y (information which is available about the instantaneous situation of the plant) and the variable which is to be controlled, namely z.

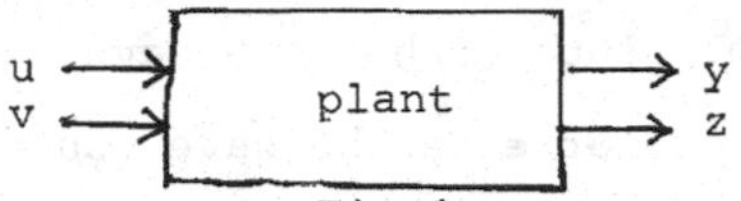

Fig.1

In general terms the problem one is concerned with can then be phrased as follows: To find a strategy for u which is based on past observations (i.e. the choice of u in a particular moment t depends upon evaluation of y(s) for $s \leq t$) and which reduces the influence of v upon z.

Our considerations will be based upon a state-space-model for the configuration of Fig. 1. We assume that this model is finite dimensional, time invariant and linear. In other words, we establish a link between input and output by introducing a state variable x and assuming relations of the form

$$\dot{x} = Ax + Bu + Gv, \quad y = Cx, \quad z = Dx . \tag{1.1}$$

A,B,C,D,G are constant matrices of appropriate dimensions. Note that the state variable plays the role of an auxiliary variable and does not have the same unquestioned physical meaning as input and output. It will sometimes be convenient to incorporate in x elements which appear in the disturbance model and have nothing to do with the usual interpretation of the state of a plant.

If one works with a state space model of the form (1.1) then the aim of a control action is mostly to bring the controlled variable close to the value 0 (note that 0 is the output in the absence of disturbance and for the initial state $x = 0$). The action itself should be generated by a feedback/observer structure. In other words: We wish to describe the strategy in the form of a control law $u = -F\hat{x}$, where $\hat{x}$ represents the state of a suitable dynamical observer, that is a linear system with input u,y. This more special type of an admissible strategy is usually called the synthesis of the regulator problem. We will restrict ourselves in this paper to the consideration of this class of strategies only and we also will regard as design objective to bring the output z close to the value 0. We call this the problem of disturbance rejection.

Let us first consider the case that v is a momentary disturbance, that is v is equal to zero for $t \geq t_o$ (or decays exponentially to zero for $t \to \infty$). The natural thing to do is then to wait until the disturbance has disappeared and then the problem of disturbance rejection becomes a problem of stabilization. The equilibrium will be restored automatically in the course of the time if the feedback/observer structure creates a stable closed loop system. A new type of problem arises if v is a persistent disturbance and we here are mainly concerned with this situation.

To be more specific we will assume that all disturbance signals belong to a finite-dimensional space of functions and that this space can be identified with the output of a linear system. In other words, we assume that all disturbance signals belong to a class of functions with a known dynamics. The objective of regulator design is then to find a feedback/observer strategy which is simultaneously effective against all disturbances in the given class. This is a generalization of the classical idea of "proportional plus integral" control and has been extensively studied by Wonham ([1], Chapter 8); the solution of the corresponding synthesis problem is often called the "internal model prinicple." The philosophy behind this principle is that the whole system (plant plus controller) should "learn" in the course of the time from its own reaction as much as it needs about the nature of disturbance. How much is needed will of course depend upon the design objective (i.e. upon the choice of z). It makes therefore a difference whether one assumes - as it is done in [1]-that the whole controlled variable z can be observed or whether one considers the problem of acquiring information about the disturbance as a problem in its own right as we do it here. Hence the controlled variable will play no role in our further considerations. We start with a system of the form

$$\dot{x} = Ax + Bu + Gv, \quad y = Cx \tag{1.2}$$

and assume in addition that v is solution of a linear differential-equation $\dot{v} = A'v$ with unknown initial data. Our aim is to find linear functionals $k^T v$ on the state of the disturbance model which can be asympotically evaluated using the state of some linear system with input (u,y). This is done by a method which presents a new angle on the classical concept of state reconstruction. The novelty of this approach lies in the structure of the observers which we will introduce in Sec. 2 and 3. Compared with the standard dynamical observer (see e.g. [2]) this structure seems more suitable to answer questions of robustness (i.e. the question: How much does the performance of the observer depend upon good modelling?). As an example we will discuss at the end of Sec. 4 three specific questions, namely: What happens to the system (plant plus controller) as a whole if the design of the observer was based on a wrong disturbance model? Is it possible to recognize wrong modelling on the output? Is it possible to adjust or refine the disturbance model without interruption of the control action?

The material in the following sections is almost self contained; the three simple lemmas at the end of this section will be all what we need beyond textbook knowledge. We will adopt standard matrix notation with small (big) Latin letters a,p,x etc. $(A,B,C$ etc.) signifying column vectors (matrices). B^T denotes the transpose and $\mathrm{Im}(B)$ the range of a matrix B(i.e. the linear space spanned by the columns of B). Small greek letters denote complex numbers; $\mathrm{Re}(\ldots)$ and $\mathfrak{Im}(\ldots)$ respectively stands for the real and imaginary part respectively of a number or a vector.

We use from now on the abbreviation "d.e." für the word "differential equation " and we will write for shortness

$$e(t) \approx 0 \tag{1.3}$$

if the funktion $e(t)$ (scalar or vector-valued) decays exponentially to zero for $t \to \infty$.

Lemma 1.1. If $y(t)$ is a nontrivial solution of the differential equation $\dot{y} = -A^T y$ then $y(t)^T x = \text{const.}$ is a first integral of the differential equation $\dot{x} = Ax$.
Proof. Straightforward. □

Lemma 1.2. If $Dx(t) \approx 0$ for each solution of a d.e. $\dot{x} = Ax$ and if $e(t) \approx 0$ then we have also $Dx(t) \approx 0$ for each solution of the inhomogeneous d.e. $\dot{x} = Ax + e(t)$.

Proof. We make a transformation of the state variable such that the system and the matrix D assumes the form

$$\dot{x} = \begin{pmatrix} A^- & 0 \\ 0 & A^+ \end{pmatrix} x, \quad D = (D^-, D^+) ,$$

and the eigenvalues of $A^-(A^+)$ have negative (nonnegative) real parts. $Dx(t) \approx 0$ for any solution of the homogeneous d.e. implies $D^+ = 0$. Hence $Dx(t) = D^- x'(t)$, where $x'(t)$ is solution of $\dot{x}' = A^- x'$. But $x'(t) \approx 0$ and this property remains unchanged if an inhomogeneous terme $e(t) \approx 0$ is added to the d.e. □

Lemma 1.3. Let a be an eigenvector of A^T and let the corresponding eigenvalue α have negative real part. Let $p = \text{Re}(a)$ or $p = \mathcal{I}m(a)$. Then $p^T x(t) \approx 0$ for any solution of the d.e. $\dot{x} = Ax$.
Proof. $e^{-\alpha t} a^T x = \text{const.}$ is a first integral of the d.e. $\dot{x} = Ax$, according to Lemma 1.1. □

2. Reconstructing differential equations.

Given a linear system

$$\dot{x} = Ax + Bu, \quad y = Cx \tag{2.1}$$

We consider a sequence of subspaces

$$[0] = \mathcal{G}_0 \subseteq \mathcal{G}_1 \subseteq \cdots \subseteq \mathcal{G}_p \tag{2.2}$$

of the state space which has the following property. For each $i=1,\dots,p$ there exists a set of linearly independent vectors $k_{i,j}$, $j = 1,\dots,r_i$, and complex numbers $\alpha_{i,j}$ such that

$$\begin{aligned} \mathcal{G}_i &= \mathcal{G}_{i-1} + \text{span}\{k_{i,1},\dots,k_{i,r_i}\}, \\ (A^T - \alpha_{i,j} I) k_{i,j} &\in \mathcal{G}_{i-1} + \text{Im}(C^T), \text{Re}(\alpha_{i,j}) < 0 \quad . \end{aligned} \tag{2.3}$$

<u>Theorem 2.1.</u> Given a sequence of subspaces having the properties (2.2), (2.3). Then one can find p coupled linear systems of the form

$$\dot{x}_i = Ax_i - \sum_{j=1}^{i-1} L_{i,j}K_{i,j}^T(x_i-x_j) - L_{i,o}(Cx_i-y)+Bu \tag{2.4}$$

such that the following is true.

Let, for a given control function $u(t)$, $x(t)$ be a solution of (2.1) and let $x_i(t)$, $i=1,\dots,p$ be the solution of (2.4) (with $u \to u(t), y \to y(t)$, where $y(t)$ is the output of (2.1)) with initial value

$$x_i(0) = 0. \tag{2.5}$$

Then

$$k_{i,j}^T(x(t)-x_i(t)) \approx 0 \quad \text{for} \quad i=1,\dots,p, j=1,\dots,r_i \tag{2.6}$$

<u>Proof.</u> We construct the d.e. (2.4) recursively. That is we assume that, given a certain i, the equations (2.4) have been defined for $i'<i$ (i' instead of i) such that the relations (2.6) hold true (with $i \to i'$). It follows then that

$$(A^T-\alpha_{i,j}I)k_{i,j},\ j=1,\dots,r_i, \tag{2.7}$$

can be represented as a linear combination of certain $k_{i',j}$, $i'<i$ (we denote by $k_{o,j}$ the rows of the matrix C^T), i.e. we have representations of the form

$$(A^T - \alpha_{i,j}I)k_{i,j} = \tilde{K}u_j,\ j=1,\dots,r_i, \tag{2.8}$$

$\tilde{K}$ is a matrix made out of columns $k_{i',j}, i'<i$.

Since the $k_{i,j}$ are linearly independent one can find a matrix L such that

$$u_j = Lk_{i,j},\ j=1,\dots,r_i .$$

It follows then from (2.7), (2.8) that $k_{i,j}$ is eigenvector of the matrix $A^T-\tilde{K}L$ and the corresponding eigenvalue is $\alpha_{i,j}$. Hence we infer from Lemma 1.3 this statement:

$$k_{i,j}^T e(t) \approx 0,\ j=1,\dots,r_i, \text{ for each solution of the d.e. } \dot{e} = (A-L^T\tilde{K}^T)e. \tag{2.9}$$

On the other hand it follows from what was said above about the columns of $\tilde{K}$ that we can write

$$L^T\tilde{K}^T = \sum_{h=1}^{i-1} L_{i,h}K^T_{i,h} + L_{i,o}C, \tag{2.10}$$

where for fixed h the rows of $K^T_{i,h}$ can be identified with vectors of the form $k^T_{h,j}$. Since, by hypothesis of induction, the relations (2.6) are assumed to be true if $i \to i' < i$ we have

$$K^T_{i,h}(x(t)-x_h(t)) \approx 0, \; h \le i-1. \tag{2.11}$$

Consider now the d.e. (2.4) for the given i and substitute for x_j, u, y respectively the functions $x_j(t), u(t), y(t) = Cx(t), j<i$. Here $x(t)$ is solution of (2.1) for the input function $u(t), x_j(t)$ is solution of (2.4)(with $i \to j$). Let $x_i(t)$ be a solution of the d.e. thus obtained. It follows then from (2.11) that $x_i(t)$ is then solution of a d.e. of the form

$$\dot{x}_i = Ax_i - \sum_{j=1}^{i-1} L_{i,j}K^T_{i,j}(x_i - x(t)) - L_{i,o}(Cx_i - Cx(t)) + Bu(t) + \tilde{e}(t)$$

where $\tilde{e}(t) \approx 0$. Because of (2.10) this equation can now be rewritten in the form

$$\dot{x}_i = Ax_i - L^T\tilde{K}^T(x_i - x(t)) + Bu(t) + \tilde{e}(t).$$

Hence $e(t) := x(t) - x_i(t)$ is solution of the d.e

$$\dot{e} = (A - L^T\tilde{K}^T)e + \tilde{e}(t),$$

and the desired result $k_{i,j}e(t) \approx 0$ follows from (2.9) and Lemma 1.2. □

<u>Remarks.</u> (i) We call (2.4) a <u>reconstructing d.e.</u> for the factor space $\mathcal{C}_i/\mathcal{C}_{i-1}$.

(ii) Let us rewrite the d.e. (2.4) as follows

$$\dot{x}_i = (A - L^T\tilde{K}^T)x_i + \sum_{h=1}^{i-1} L_{i,h}K^T_{i,h}x_h + L_{i,o}y + Bu. \tag{2.4'}$$

From what was said about the rows of the matrix $K^T_{i,h}$ it is clear that the row vector $K^T_{i,h}x_h$ can be expressed in terms of the linear functionals $k^T_{h,j}x_h$, $h<i$. Furthermore $k_{i,j}$ is eigenvector of the matrix $A^T - \tilde{K}L$ to the eigenvalue $\alpha_{i,j}$. Hence one can derive from (2.4') a system of d.e.'s for the linear functionals

$$\xi_{i,j} := k^T_{i,j}x_i \tag{2.12}$$

simply by multiplication from the left with $k^T_{i,j}$:

$$\dot{\xi}_{i,j} = \alpha_{i,j}\xi_{i,j} + \sum_{l=1}^{i-1} \sum_k \lambda_{i,j,l,k}\xi_{l,k} + \lambda^T_{o,i,j}y + b^T_{i,j}u \tag{2.13}$$

Hence the reconstructing differential equations (2.4) can be replaced by a system of differential equations for the reconstructed linear functionals. The system matrix is triangular and has all eigenvalues in the left half plane.

(iii) Let $\mathcal{C}_1$ be the subspace of $\mathbb{R}^n$ which is spanned by the columns of the matrix $(C^T, A^TC^T, (A^T)^2C^T, \ldots)$. Then

$$[0] = \mathcal{C}_o \subset \mathcal{C}_1$$

is a sequence having the properties (2.2), (2.3). The numbers $\alpha_{1,j}$, $j=1,\ldots,n_1 = \text{Dim}(\mathcal{C}_1)$ can be prescribed arbitrarily. This can be inferred by standard arguments (pole placement), if one assumes that the system is given in normal form (cf.[2],Th. 1.35).

For later purposes we wish to describe the reconstructing differential equation (2.13) in a special case in more explicit terms. Here we do not require linear independence of the $k_{i,j}$.

<u>Corollary 2.1.</u> Let $\text{Re}(\alpha)<0$ and let k be such that

$$(A^T-\alpha I)k = \sum_{i=1}^{p} \sum_{j=1}^{r_i} \rho_{i,j}k_{i,j} + C^T l \in \mathcal{C}_p + \text{Im}(C^T). \tag{2.14}$$

$\rho_{i,j}$ are scalars and l a vector of appropriate dimension.

Then the linear functional

$$\xi := k^T x$$

can be reconstructed from the scalar differential equation

$$\dot{\xi} = \alpha\xi + \sum_{i=1}^{p} \sum_{j=1}^{r_i} \rho_{i,j}\,\xi_{i,j} + l^T y + k^T Bu\,. \tag{2.15}$$

(for the definition of $\xi_{i,j}$ cf. (2.12)). If in particular

$$(A^T-\alpha I)k = k' = \sum_{j=1}^{r_i} \rho_{i,j}k_{i,j}$$

then ξ can be reconstructed from the equation

$$\dot{\xi} = \alpha\xi + \xi', \quad \xi' = (k')^T x_i$$

("can be reconstructed" means: $\xi(t)-k^T x(t) \approx 0$, $\xi(t)$ being solution of (2.15)).

<u>Proof.</u> Let $x(t)$ be a solution of (2.1). Then

$$k^T\dot{x}(t) = k^T(Ax(t) + Bu(t))$$

and hence, in view of (2.14),

$$\frac{d}{dt}(k^T x(t)) = \sum_{i=1}^{p}\sum_{j=1}^{r_i} \rho_{i,j}k_{i,j}^T\, x(t) + 1^T Cx(t) + k^T Bu(t) + \alpha k^T x(t).$$

Since $Cx(t) = y(t)$ and $k_{i,j}^T x(t) - k_{i,j}^T x_i(t) \approx 0$ it is clear that $k^T x(t)$ satisfies the d.e. (2.15) up to an error term ≈ 0. Therefore the statement of the corollary follows from Lemma 1.2. □

We now turn to the consideration of linear systems which include a model for possible disturbances, i.e. systems which are given in the form (1.2) plus the additional d.e. $\dot{v} = A'v$. It is convenient to treat x,v as parts of the same state variable and to work with one d.e. of the form

$$\dot{x} = \begin{pmatrix} A_{11} & A_{12} \\ 0 & A_{22} \end{pmatrix} x + \begin{pmatrix} B_1 \\ 0 \end{pmatrix} u, \quad y = (C_1, 0)x. \qquad (2.16)$$

Again n is the dimension of x. We write x in the form $(x', x'')^T$ where the partitioning is obvious from the form of the d.e.(2.16). The dimension of x', x'' is denoted by n', n'' $(n=n'+n'')$. Furthermore we introduce the reduced system

$$\dot{x}' = A_{11}x' + B_1 u, \quad y = C_1 x' \qquad (2.16_r)$$

The remaining portion of this section will be devoted to the problem of finding subspaces $\mathcal{C}_i$ for the system (2.16) under the assumption that a sequence of such spaces is known for the reduced system (2.16_r). Here we make use of the fact that the transpose of the system matrix can be written in this form

$$A^T = \begin{pmatrix} A_{11}^T & 0 \\ A_{12}^T & A_{22}^T \end{pmatrix}$$

Hence let there be given a sequence of subspaces $\mathcal{G}'_i, i=1,\ldots,p$ which are contained in $\mathbb{R}^n$ and have the properties (2.3) with respect to the equation (2.16_r):

$$\mathcal{G}'_i = \mathcal{G}'_{i-1} + [k'_{i,1},\ldots,k'_{i,r_i}],$$
$$(A^T_{11}-\alpha_{i,j}I)k'_{i,j} \in \mathcal{G}'_{i-1} + \mathrm{Im}(C^T_1), \; \mathrm{Re}(\alpha_{i,j}) < 0. \tag{2.17}$$

We wish to construct for the original system (2.16) a corresponding sequence

$$\mathcal{G}_o = [0] \subseteq \mathcal{G}_1 \subseteq \ldots \subseteq \mathcal{G}_p$$

which we call the <u>canonical extension</u> of the given sequence $\mathcal{G}'_i$. This extension is possible, according to the scheme described below if all eigenvalues of A_{22} lie in the closed right half plane. The canonical extension should satisfy the relations (2.3) with $p, r_i, \alpha_{i,j}$ being the same numbers as those appearing in (2.17). Furthermore we require that the first component of $k_{i,j}$ equals the vector $k'_{i,j}$ appearing in (2.17). Finally the $k_{i,j}$ should be chosen such that the following two relations hold true with suitable real numbers $\lambda_{l,\nu}$ (depending upon i,j) :

$$(A^T_{11}-\alpha_{i,j}I)k'_{i,j} - \sum_{l=1}^{i-1}\sum_{\nu} \lambda_{l,\nu}k'_{l,\nu} \in \mathrm{Im}(C^T_1),$$
$$(A^T-\alpha_{i,j}I)k_{i,j} - \sum_{l=1}^{i-1}\sum_{\nu} \lambda_{l,\nu}k_{l,\nu} \in \mathrm{Im}\begin{pmatrix} C^T_1 \\ 0 \end{pmatrix}. \tag{2.18}$$

Note that this requirement is in accordance with the second condition (2.3). That it can be fulfilled follows from the corresponding part of condition (2.17). If one takes the special form of the system matrix into account (cf. (2.16)) one sees that (2.18) is equivalent with a condition of the form

$$(A^T_{22}-\alpha_{i,j}I)k''_{i,j} = l_{i,j}, \tag{2.18'}$$

where the expression on the right hand side depends upon $k''_{i',j}$ for $i'<i$. This condition can always be met by a proper choice of $k''_{i,j}$, since the matrix $(A^T_{22}-\alpha_{i,j}I)$ is non-singular.

Let i be fixed for the moment. As we have seen in the proof of Theorem 2.1 there exist matrices K',L, where K' consists of column vectors $k_{i',j}$, $i'<i$, such that

$$(A_{11}^T - \alpha_{i,j}I)k'_{i,j} = K'u_j, \quad u_j = Lk'_{i,j}, \quad j=1,\dots,r_i .$$

Using (2.18) this relation can be extended to an analogous relation involving the $k_{i,j}$:

$$(A^T-\alpha_{i,j}I)k_{i,j} = \begin{pmatrix} K' \\ K'' \end{pmatrix} u_j = \begin{pmatrix} K' \\ K'' \end{pmatrix} (L,0)k_{i,j} .$$

The reconstructing equation for the functionals $k_{i,j}^T x$ assumes then the form

$$\dot{x}_i = \begin{pmatrix} A_{11} & A_{12} \\ 0 & A_{22} \end{pmatrix} x_i - \begin{pmatrix} \sum_{j=1}^{i-1} L_{ij}K_{ij}(x'_i-x'_j) + L_{i,o}(Cx'_i-y)+B_1u+P_i \\ 0 \end{pmatrix} \tag{2.19}$$

where P_i consists of homogeneous linear functions of x''_j, $j<i$.

From the form of this equation the following statement can be immediately inferred.

If $x_h(t) = (x'_h(t),x''_h(t)),h=1,\dots,i$ is a solution of the system (2.19)(with $i\to h$) satisfying the initial condition $x''_h(0) = 0$, then $x''_h(t) = 0$ for all t and $h=1,\dots,i$, and $x'_i(t)$ is solution of the reconstructing equation

$$\dot{x}'_i = A_{11}x'_i - \sum_{j=1}^{i-1} L_{ij}K_{ij}(x'_i-x'_j) - L_{io}(Cx'_i-y)+B_1u \tag{2.20}$$

belonging to the reduced system and the space $\mathcal{G}'_i/\mathcal{G}'_{i-1}$.

In other words: Canonical extension preserves the reconstructing differential equations, provided one normalizes the states of reconstructing d.e.'s according to (2.5).

We summarize what we found so far as

<u>Theorem 2.2.</u> General hypothesis: The eigenvalues of A_{22} have non-negative real parts. Assume furthermore that there is given a sequence of subspaces $\mathcal{G}'_i$ with properties (2.17) and let

$$\dot{x}_i' = A_{11}x_i' - \sum_{j=1}^{i-1} L_{ij}K_{ij}(x_i'-x_j') - L_{io}(Cx_i'-y) + B_1u \qquad (2.21)$$

be the reconstructing differential equation for $\mathcal{G}_i'/\mathcal{G}_{i-1}'$. Note that these spaces and d.e.'s are based on the reduced system (2.16_r). Let $\mathcal{G}_i$ then be the canonical extension of the $\mathcal{G}_i'$ and let $k_{i,j} = (k_{i,j}', k_{i,j}'')$ be a set of generators satisfying the condition (2.18).

<u>Claim.</u> If $x(t) = (x'(t),x''(t))$ is a solution of (2.16) and if $(x_1'(t),\dots,x_i'(t))^T$ is a solution of the d.e.'s (2.21) then these asymptotic relations hold true:

$$(k_{i,j}')^T(x'(t) - x_i'(t)) + (k_{i,j}'')^T x''(t) \approx 0 \qquad (2.22)$$

or equivalently

$$k_{i,j}^T x(t) - \xi_{i,j}'(t) \approx 0, \quad \xi_{i,j}'(t) := (k_{i,j}')^T x_i'(t), \qquad (2.22')$$

$j=1,\dots,r_i$.

<u>Remark.</u> If $k \in \mathcal{G}_p$ and $x(t)$ a solution of (2.16) then $k^T x(t)$ can be asymptotically evaluated with the help of the $\xi_{i,j}'(t)$. Indeed $k \in \mathcal{G}_p$ means that we have a representation of the form

$$k = \sum_{i,j} \lambda_{i,j} k_{i,j} \quad .$$

It follows then from (2.22') that

$$k^T x(t) - \sum_{i,j} \lambda_{i,j}\, \xi_{i,j}'(t) = \sum_{i,j} \lambda_{i,j} k_{i,j}^T (x(t) - x_i'(t)) \approx 0. \quad \square$$

We finally wish to have a closer look on those elements $k \in \mathcal{G}_p$ which are of the form

$$k = (0, k'')^T. \qquad (2.23)$$

The linear functional $k^T x = (k'')^T x''$ depends x'' only. As we have just remarked one can reconstruct the functional $(k'')^T x''$ from the $\xi_{i,j}'$ (cf. (2.22')). A natural problem is then the following one: Find further functionals on x'' which can be reconstructed from $(k'')^T x''$ using observer equations for x'' only. This problem has a rather obvious solution which roughly speaking can be described as follows: Construct observers for the system $\dot{x}'' = A_{22}x''$ regarding $y'' = (k'')^T x''$ formally as observed variable. A more precise version is given in the following theorem.

Theorem 2.3. Hypothesis: $k_\nu = (0, k''_\nu)^T$, $\nu=1,\dots,s$, are elements of $\mathcal{C}_p$ and $\mathcal{S}$ is the linear subspace of the x''-space spanned by the vectors

$$(A_{22}^T)^\rho k''_\nu \ , \quad \nu=1,\dots,s, \quad \rho=0,1,\dots,n''-1 \ .$$

Furthermore let C_2^T be the matrix $(k''_1, \dots, k''_s)$.

Claim. (i) There exist row vectors $l_{i,j}^T$ such that

$$C_2^T = \sum_{i,j} l_{i,j}^T k''_{i,j} \quad \text{and} \quad 0 = \sum_{i,j} l_{i,j}^T k'_{i,j} \, . \tag{2.24}$$

(ii) Let K_2 be a matrix (of appropriate dimension) with this property: $\mathcal{S}$ is spanned by eigenvectors of $(A_{22}-K_2C_2)^T$ and the corresponding eigenvalues have negative real parts. Then for any $\tilde{k}'' \in \mathcal{S}$ the following statement holds true

$$(\tilde{k}'')^T (x''(t) - \hat{x}''(t)) \approx 0,$$

where $x''(t)$ is the second component of any solution $x(t)$ of the d.e. (2.16) and $\hat{x}''(t)$ is any solution of the d.e.

$$\dot{\hat{x}}'' = (A_{22}-K_2C_2)\hat{x}'' + K_2 \sum_{i,j} l_{i,j} \xi'_{i,j}(t) \, . \tag{2.25}$$

Remark. A matrix K_2 with the property mentioned in the statement of the theorem does always exist (cf. remark (iii) following the proof of Theorem 2.1).

Proof. (i) is clear, since $k_\nu \in \mathcal{C}_p$ and hence can be written as a linear combination of the $k_{i,j}$. In order to prove (ii) note that we have, in view of (2.24), (2.22'),

$$C_2 x''(t) - \sum_{i,j} l_{i,j} \xi'_{i,j}(t) = \sum_{i,j} l_{i,j} (k''^T_{i,j} x''(t) - \xi'_{i,j}(t)) + (\sum_{i,j} l_{i,j} k'_{i,j}) x'(t)$$

$$= \sum_{i,j} l_{i,j} (k_{i,j}^T x(t) - \xi'_{i,j}(t)) \approx 0.$$

The difference $e''(t) := x''(t) - \hat{x}''(t)$ is hence solution of a d.e. of the form

$$\dot{e}'' = (A_{22}-K_2C_2)e'' + \tilde{e}(t), \ \tilde{e}(t) \approx 0.$$

It follows then from Lemma 1.2, 1.3 that $k^T e''(t) \approx 0$, whenever k is eigenvector of $(A_{22}-K_2C_2)^T$ and the corresponding eigenvalue has negative real part. □

3. Auxiliary observers.

We return to a system of the form (2.1) and consider a sequence $\mathcal{C}_i$ and reconstructing differential equations as explained before (see (2.2)-(2.4)). It should be noted that we allow $\mathcal{C}_{i-1} = \mathcal{C}_i$, i.e. we do not exclude the possibility that

$$k_{i,j} \in \mathcal{C}_{i-1}, \; j=1,\dots,r_i. \tag{3.1}$$

The reason is that a relation $\mathcal{C}'_{i-1} = \mathcal{C}'_i$ does not necessarily imply that the canonical extensions $\mathcal{C}_{i-1}, \mathcal{C}_i$ are also equal. The system-theoretic interpretation of this fact is rather obvious. If (3.1) holds, then $k^T_{i,j}x_i(t)$ is asymptotically equal to a linear combination of $k^T_{i',j}\,x_{i'}(t)$, $i'<i$, provided the input y in equations (2.4) is the output of the unperturbed equation (2.1). This statement however is not longer true if a perturbing signal is added to the right hand side of the differential equation $\dot{x} = Ax + Bu$.

In this section we go a step further and will introduce what one may call a canonical extension of the vector $k' = 0$ to a vector $(0,k'')$ which of course is in general not zero. We adopt formally the previous definition (see (2.18)). $\{\mathcal{C}_i\}$ denotes the canonical extension of $\{\mathcal{C}'_i\}$.

Definition 3.1 Let $k_{l,\nu} = (k'_{l,\nu}, k''_{l,\nu}) \in \mathcal{C}_p$. Assume that

$$\tilde{k} := \sum_{l=1}^{p} \sum_{\nu} \lambda_{l,\nu} k'_{l,\nu} \in \mathrm{Im}(C_1^T). \tag{3.2}$$

Furthermore let α be a number such that $\mathrm{Re}(\alpha)<0$.

Define then k'' as the solution of the linear equation

$$(A_{22}^T - I)k'' - \sum_{l=1}^{p} \sum_{\nu} \lambda_{l,\nu} k''_{l,\nu} = 0. \tag{3.3}$$

We call then $k := (0,k'')^T$ an extension of 0 and the differential equation

$$\dot{\xi} = \alpha\xi + \sum_{l=1}^{p} \sum_{\nu} \lambda_{l,\nu}\xi'_{l,\nu}(t) - p^T y \tag{3.4}$$

a reconstructing differential equation for the linear functional

$$k^T x = (k'')^T x.$$

The column vector p is defined in terms of the relation

$$\sum_{l=1}^{p} \sum_{\nu} \lambda_{l,\nu} k'_{l,\nu} = C_1^T p \tag{3.2'}$$

which can be derived from (3.2).

The following theorem states the analogue of Theorem 2.2 for extensions of 0. Note that we cannot include this theorem into the results of the previous section since the $k'_{i,j}$ need not longer be linearly independent vectors.

Theorem 3.1 Let $k := (0,k'')$ be an extension of 0 and let (3.4) be the corresponding reconstructing differential equation. Let $\xi'_{i,j}(t)$ have the same meaning as in Theorem 2.2 (cf. (2.22')). Then

$$\xi(t) - (k'')^T x''(t) \approx 0$$

whenever $x(t) = (x'(t),x''(t))$ is a solution of the d.e. (2.16) and $\xi(t)$ a solution of the d.e. (3.4) (with $y \to y(t)$ = output of (2.16)).

Proof. We first note that the two relations (3.3), (3.2') are equivalent with the single relation

$$(A^T-\alpha I)k = (A^T-\alpha I)(0,k'')^T = \sum_{i=1}^{p} \sum_{\nu} \lambda_{i,\nu} k_{i,\nu} - \begin{pmatrix} C_1^T \\ 0 \end{pmatrix} p .$$

It follows therefore from Corollary 2.1 that $\xi = k^T x$ can be reconstructed from the differential equation

$$\dot{\xi} = \alpha\xi + \sum_{i=1}^{p} \sum_{\nu} \lambda_{i,\nu}(k'_{i,\nu} x'_i(t) + k''_{i,\nu} x''(t)) - p^T y . \qquad (3.5)$$

Here $(x'_i(t), x''_i(t))$ is solution of the reconstructing d.e for $\mathcal{C}_i / \mathcal{C}_{i-1}$. The sequence $\{\mathcal{C}_i\}$ however is the canonical extension of the sequence $\{\mathcal{C}'_i\}$ and we have therefore $x''_i(t) = 0$, $i=1,\ldots p$, as has been noted in the proof of Theorem 2.2. Hence the above differential equation is in fact identical with (3.4) and the statement of the theorem follows from Corollary 2.1. □

In the sequel we refer to a linear functional of the type as described in Definition 3.1 as an auxiliary observer. More precisely we will speak of an auxiliary observer belonging to the pair $(\alpha,\tilde{k})$, where $\tilde{k}$ is defined as in (3.2). Note that the differential equation of an auxiliary observer does not depend upon the disturbance model, but can be constructed according to (3.2') from any sequence of spaces $\mathcal{C}'_\nu$ for the unperturbed equation (2.16_r). Nothing can be gained from auxiliary observers for the reconstruction of states for the unperturbed equation. In the absence of $x_2(t)$ every solution (3.4) is asymptotically 0 (this is a consequence of the relation (3.2')!).

4. Systems with disturbances: State reconstruction.

In this section we consider a system of the form

$$\dot{x} = \begin{pmatrix} A_{11} & A_{12} & \cdots & A_{1N} \\ 0 & A_{22} & & 0 \\ \vdots & & \ddots & \vdots \\ \vdots & & & 0 \\ 0 & \cdots & 0 & A_{NN} \end{pmatrix} x + \begin{pmatrix} B_1 \\ 0 \\ \vdots \\ 0 \end{pmatrix} u, \quad y = (C',0,\dots,0)x \tag{4.1}$$

According to the partitioning of the matrix A we write

$x = (x_1,\dots,x_N)^T$, x_i being of dimension n_i, $n = \sum_i n_i$.

We assume that the following general assumptions hold:

(i) All eigenvalues of A_{ii} are in the closed right half plane, $i>1$. A_{ii} and A_{jj} have no eigenvalues in common, $i \neq j$, $i,j > 1$.

ii) There exists a sequence $\mathcal{C}_0' = [0] \subseteq \mathcal{C}_1' \subseteq \dots \subseteq \mathcal{C}_p'$ of subspaces for the reduced system

$$\dot{x}_1 = A_{11}x_1 + B_1u, \; y = C'x_1 \tag{4.1$_r$}$$

having the properties (2.17). For simplicity we write k_ν' instead of $k_{i,j}'$, $\nu=1,\dots$ and α_ν instead of $\alpha_{i,j}$.

iii) There exist M linear combinations of the k_ν which belong to $\mathrm{Im}(C')$:

$$l_1^{(\mu)} := \sum_{\nu=1}^{N} \lambda_\nu^{(\mu)} k_\nu' \in \mathrm{Im}(C'), \; \mu=1,\dots,M. \tag{4.2}$$

Note that we do not assume that the k_ν' are independent or that the linear combinations (4.2) are non-zero.

We denote by $\mathcal{C}_0 \subseteq \mathcal{C}_1 \subseteq \dots \subseteq \mathcal{C}_p$ the canonical extension of the spaces $\mathcal{C}_i'$, according to the definition given in Sec. 2 (the role of A_{22} is now played by the matrix diag $(A_{22},\dots,A_{NN})$. As before we denote by k_ν the canonical extension of each k_ν' and put

$$l^{(\mu)} := \sum_{\nu=1}^{N} \lambda_\nu^{(\mu)} k_\nu \; , \; \mu=1,\dots,M. \tag{4.3}$$

Let $l^{(\mu)}$ be partitioned as

$$l^{(\mu)} = (l_1^{(\mu)},\dots,l_N^{(\mu)})^T \tag{4.4}$$

and let α be a complex number with $\mathrm{Re}(\alpha) < 0$. It follows then from (4.2) and Definition 3.1 that the vector

$$\hat{1}^{(\mu)}(\alpha) := (0,(A_{22}^T-\alpha I)^{-1} 1_2^{(\mu)},\ldots\ldots,(A_{NN}^T-\alpha I)^{-1} 1_N^{(\mu)})^T \tag{4.5}$$

is an extension of 0 belonging to the pair $(\alpha, 1_1^{(\mu)})$.

Finally we introduce the following matrices and spaces:

$$\text{(i)} \quad C_i := (1_i^{(1)},\ldots,1_i^{(M)})^T, \; i=1,\ldots,N$$

$$\text{(ii)} \quad \mathcal{S}_i = \mathrm{Im}((C_i^T, A_{ii}^T C_i^T, (A_{ii}^T)^2 C_i^T,\ldots,)), \; i=1,\ldots N. \tag{4.6}$$

$1_i^{(\mu)}$ is the i-th component of $1^{(\mu)}$ (cf. (4.3),(4.4)) and an n_i-dimensional column vector. C_i is hence a matrix of type (M,n_i), $\mathcal{S}_i$ a subspace of $\mathbb{R}^{n_i}$

<u>Theorem 4.1</u> Any linear functional $k^T x$, $k=(k_1,\ldots,k_N)^T$, $k_i \in \mathcal{S}_i$, can be reconstructed from the output y using a dynamical observer with the following structure. It consists of

(i) reconstructing d.e.'s (of the type (2.4) of (2.13)) for the reduced system (4.1_r), based on the sequence $\mathcal{C}_1',\ldots, \mathcal{C}_p'$.

(ii) M auxiliary observers for (4.1_r), belonging to pairs $(\alpha_\mu, 1_1^{(\mu)})$ (cf. (4.2)). α_μ is an arbitrary number with $\mathrm{Re}(\alpha_\mu)<0$.

(iii) A sufficient number of additional "integrating actions", provided by d.e.'s of the form

$$\dot{\xi}_{\mu,\rho} = \alpha_\mu \xi_{\mu,\rho} + \xi_{\mu,\rho-1}, \quad \mu=1,\ldots,M, \rho=1,2,\ldots. \tag{4.7}$$

$\xi_{\mu,o}$ satisfies the d.e. for the auxiliary observer belonging to the pair $(\alpha_\mu, 1_1^{(\mu)})$.

(iv) a standard observer of the form

$$\dot{x}_i' = (A_{ii}-K_i C_i)x_i' + q_i(\xi_{1,1},\xi_{1,2},\ldots,\ldots,\xi_{M,1},\ldots..) \tag{4.8}$$

for each $i=2,\ldots,N$.

<u>Remarks.</u> (i) The reconstructing d.e.'s for the reduced system and the auxiliary observers do not depend upon the disturbance model. (ii) The <u>number</u> of "integrating actions" does depend upon the disturbances. (iii) q_i (cf. (4.8)) depends upon A_{jj} for $j \leq i$ only.

In particular, if a disturbance model

$$\dot{x}_{N+1,N+1} = A_{N+1,N+1}\, x_{N+1,N+1}$$

is added to the existing ones, and if C_{N+1} is defined according to (4.6) then the reconstructing d.e.'s for the "enriched" disturbance structure are obtained from the existing ones by increasing (if necessary) the number of integrating actions and by adding a standard observer (for i=N+1) of the form (4.8). (iv) The reconstructing d.e.'s for the reduced system as well as the standard observers of type (4.8) can be replaced by a triangular system of the form (2.13) with $Re(\alpha_{i,j})<0$.

Proof. Let μ be a fixed integer with $1\le\mu\le M$. For each pair i,j with $i,j\ge 2, i \ne j$, the two matrices $(A_{ii}^T-\alpha_\mu I)^{-1}$ and $(A_{jj}^T-\alpha_\mu I)^{-1}$ have no eigenvalues in common and their respective characteristic polynomials are hence relatively prime. This is a consequence of our general hypothesis (i). There exists hence for every $j=2,\dots,N$ a polynomial $\psi_j(s)$ with these properties

$$\psi_j((A_{jj}^T-\alpha_\mu I)^{-1}) = A_{jj}^T - \alpha_\mu I,\ \psi_j((A_{ii}^T-\alpha_\mu I)^{-1}) = 0,\ i\ne j,\ i,j\ge 2 \quad (4.9)$$

One simply has to choose ψ_j according to these conditions:

ψ_j is divisible by the characteristic polynomial of $(A_{ii}^T-\alpha_\mu I)^{-1}$.

$s\psi_j(s)-1$ is divisible by the characteristic polynomial of $(A_{jj}^T-\alpha_\mu I)^{-1}$.

This is always possible according to the Chinese remainder rule. Note that 0 is not an eigenvalue of any of the matrices $(A_{ii}^T-\alpha_\mu I)^{-1}$, hence none of the characterisitc polynomials is divisible by s. We denote by $\tilde{A}_\mu$ the matrix (of type (n,n))

$$\mathrm{diag}\ (0,(A_{22}^T-\alpha_\mu I)^{-1},\dots,(A_{NN}^T-\alpha_\mu I)^{-1}). \quad (4.10)$$

It follows then from (4.5) and (4.9) that

$$\psi_j(\tilde{A}_\mu)\hat{1}^{(\mu)}(\alpha_\mu) = (0,\dots,0,1_j^{(\mu)},0,\dots,0).$$

On the other hand, taking into account the special form of the matrix $\tilde{A}_\mu$ (cf. (4.10)) and of the coefficient matrix A appearing in the given equation (4.1) one sees immediately that

$$(A^T-\alpha_\mu I)(\tilde{A}_\mu^\rho) = \tilde{A}_\mu^{\rho-1}$$

It follows then from Corollary 2.1 that the functionals $((\tilde{A}_\mu)^\rho\hat{1}^{(\mu)}(\alpha_\mu))^T x, \rho=0,1,\dots,$ can be reconstructed recursively from

$\hat{1}^{(\mu)}(\alpha_\mu F)x$; the reconstructing differential equations are just given by (4.7).

Hence we are arrived at this result. One can find a sequence $\mathcal{C}_o = [0] \subseteq \mathcal{C}_1 \subseteq \ldots \subseteq \mathcal{C}_{p'} \subseteq \mathbb{R}^n$ for the system (4.1) having all properties stated in Sec. such that $\mathcal{C}_{p'}$ contains all vectors of the form

$$(1_1^{(\mu)}, *,\ldots,*), \mu=1,\ldots.M ,$$

and

$$(0,\ldots,0,1_j^{(\mu)},0,\ldots,0), j=2,\ldots,N, \mu=1,\ldots,M.$$

Furthermore the corresponding reconstructing differential equations are of the type as described in the statements (i)-(iii) of the theorem. It is then clear, in view of Theorem 2.3 how one can find an observer of the form (4.8) in order to reconstruct the functionals $k^T x$, where

$$k = (0,\ldots,0,c_j,0,\ldots,0), c_j \in \mathcal{S}_j. \qquad \square$$

We conclude this section with some remarks about the range and possible applications of the theorem.

(i) The theorem emphasizes explicit reconstruction of special functionals on the state rather than reconstruction of the state itself. This aspect of our method fits well into the general frame of regulator synthesis. What is needed for the design of a controller is an observer which provides asymptotic evaluation of certain functionals (in case of a scalar control only one funtional, namely $u=-Fx$) not necessarily an observer which reconstructs the full state (and which in general will not exist).

(ii) The observer structure is basically a recursive scheme, as already remarked. Hence additional disturbances can be taken care of simply by adding further observers without changing the existing structure.

(iii) The reconstructing d.e.'s can assumed to be stable (see remark (ii) following the proof of Theorem 2.1) and they preserve their stability under certain changes of the data of the disturbance model. To make this remark more transparent let us consider a special case. Take a system which is subject to (scalar) disturbances of quasi-periodic character with (unknown) basis frequencies $\omega_2', \ldots, \omega_N'$. We guess that the basic frequencies are $\omega_2, \ldots, \omega_N$ and accordingly design the observer part of a controller based on a model of type

(4.1) where

$$A_{ii} = \begin{pmatrix} 0 & -\omega_i^2 \\ 1 & 0 \end{pmatrix} . \qquad (4.11).$$

In order to meet the general requirements stated at the beginning of this section we assume that

$$\omega_j > 0 \ , \ \omega_i \neq \omega_j \quad \text{if} \quad i \neq j. \quad i,j > 1. \qquad (4.12)$$

Let there be given then as before M elements $1_1^{(\mu)}$ of the state space having the property (iii). These elements of course do not depend upon the choice of the ω_i. Let then C_i be the matrices defined according to (4.3)-(4.6). These matrices depend upon the choice of ω_j. It follows now by standard arguments and in view of the special type of matrices A_{ii} (cf. (4.11)) that $\mathcal{S}_i = \mathbb{R}^{n_i} (=\mathbb{R}^2)$ provided these conditions hold

$$C_i \ \text{real}, \quad C_i \neq 0, \ i = 2,\dots,N. \qquad (4.13).$$

(4.12) and (4.13) represent a kind of robustness specification for the observer structure described in this section. If these conditions are satisfied then the corresponding feedback/observer structure is a linear system driven by quasi-periodic motions with frequencies $\omega'_1,\dots,\omega'_N$ and a stable coefficient matrix. The eigenvalues of this matrix are independent from ω_j and can be placed arbitrarily in the left half plane. This follows from the construction given in the proof of the last theorem. So regardless what choice of the ω_j has been made - provided (4.12) and (4.13) are satisfied - the output of the system will be a motion of quasi-periodic type and the basic frequencies are numbers from the set $\{\omega'_2,\dots, \omega'_N\}$. A certain frequency ω'_j will disappear in the output if the observer is adjusted such that $\omega_j = \omega'_j$.

(iv) Certain changes in the environment in which the system operates can be taken care of by the observer. Assume for example that the disturbance in the first environment has basic frequency ω_2, in the second the basic frequency ω_3 etc., and that these data are known in advance. Then a feedback/observer structure will operate equally well in all environments if the observer part has been chosen according to Theorem 4.1 and if for every possible forequency ω_i a disturbance model with a matrix of the form (4.11) has been included in the system representation (4.1). In the first environment e.g. the substructure of the observer which is associated with A_{22} automatically will take over responsibility, the others will operate also but in such a way that they do not impede the first one; their contribution to state reconstruction will actually die out in the course of the time.

References

[1] M.W.WONHAM: Linear Multivariable Control. Lecture Notes in Economics and Mathematical Systems, Vol. 101. Springer Verlag 1974.

[2] H.KWAKERNAAK, R.SIVAN: Linear Optimal Control Systems. Wiley-Interscience 1972.

EIGENVALUE PROBLEMS AND THE RIEMANN ZETA-FUNCTION

IAN KNOWLES

1. Introduction

For complex numbers $s = \sigma + it$ with $\sigma > 1$, the Riemann zeta-function, $\zeta(s)$, is defined by the Dirichlet series

$$\zeta(s) = \sum_{n=1}^{\infty} \frac{1}{n^s}. \tag{1.1}$$

It is known (see e.g. [6]) that the analytic function defined by the above series has an analytic continuation (also dènoted by ζ) that is regular for all values of s except $s = 1$, where there is a pole of order 1. Furthermore, ζ satisfies the functional equation

$$\zeta(s) = 2^s \pi^{s-1} \sin\left(\tfrac{1}{2}\pi s\right) \Gamma(1-s) \zeta(1-s) \tag{1.2}$$

where Γ denotes the gamma function. Finally, we have the so-called Euler product formula for ζ:

$$\zeta(s) = \prod_{p \text{ prime}} \left(1 - \frac{1}{p^s}\right)^{-1}, \quad \sigma > 1. \tag{1.3}$$

The latter identity, relating on the one hand the prime numbers and on the other, complex function theory, explains much of the interest in the zeta function. In particular, the precise location of the zeros of $\zeta(s)$ has been the subject of much study (see e.g. [3, 6]). It is known that $\zeta(s) = 0$ when $s = -2, -4, -6, \ldots$ (the "trivial zeros") and that $\zeta(s)$ has infinitely many zeros in the vertical strip $0 < \sigma < 1$ symmetrically placed with respect to

$\sigma = \frac{1}{2}$ and $t = 0$, and no zeros outside this region, other than the trivial zeros. It was conjectured by Riemann [5; see also 3, p. 299] that these complex zeros of ζ all lie on the line $\sigma = \frac{1}{2}$. Although much is known, including the fact that the first 81,000,001 zeros above the real axis are simple and on the line $\sigma = \frac{1}{2}$, [2], the conjecture remains unproven despite the best efforts of many mathematicians during the intervening 123 years.

The objective of this lecture is to outline a proof of the following theorem on the zero-free region for ζ:

<u>Theorem</u>. Let $\theta > 1$ be a fixed real number. Then there exists a kernel $k(r,t,s)$ such that

(i) the integral equation

$$u(r,s) = 1 + \int_r^\infty k(r,t,s)u(t,s)dt \tag{1.4}$$

has a unique bounded solution $u(r,s)$ analytic in s for $\mathrm{Re}(s+\theta) > 1$;

(ii) if (1.4) has a unique bounded solution in any sub-region of the strip $\frac{1}{2} < \mathrm{Re}(s+\theta) \le 1$ contiguous to $\mathrm{Re}(s+\theta) > 1$, then $\zeta(s+\theta) \ne 0$ in that sub-region.

Thus, in particular, the Riemann hypothesis is true if (1.4) has a unique bounded solution for all s with $\mathrm{Re}(s+\theta) > \frac{1}{2}$.

Although the details of the proof are somewhat complicated (see [4]), the basic idea is to first produce a suitable explicit construction of a boundary value problem, E_N, on $[0,\infty)$ with eigenvalue equation of the form $\xi_N(s) = 0$, where ξ_N is the N-th partial product of the function

$$\frac{\zeta(s+\theta)}{\zeta\left(2(s+\theta)\right)} = \prod_{n=1}^{\infty} \left(1 + \frac{1}{p_n^{s+\theta}}\right) , \tag{1.5}$$

$\theta > 1$ is an arbitrary, but fixed, real number, and $\{p_n : n = 1,2,...\}$ denotes the sequence of prime numbers, beginning with $p_1 = 2$. The integral equation (1.4) is then derived from the differential equation obtained by a formal limiting procedure involving the differential equations of these approximating boundary value problems.

2. The Eigenvalue Problem E_n

As in §1, let $\{p_n\}$ denote the sequence of prime numbers, and let θ be a fixed real number greater than one. Let also δ be an arbitrary positive number. We define the sequence $\{\varepsilon(n) : n = 1,2,...\}$ by

$$\varepsilon(1) = 1 \quad , \quad \varepsilon(2) = \frac{1}{p_1^\theta} , \tag{2.1}$$

and for $n \geq 1$

$$\varepsilon(2^n+k) = \frac{1}{p_{n+1}^\theta} \varepsilon(k) \quad , \quad 1 \leq k \leq 2^{n-1} , \tag{2.2a}$$

$$= \frac{1+\delta}{p_{n+1}^\theta} \varepsilon(k) \quad , \quad 1 + 2^{n-1} \leq k \leq 2^n . \tag{2.2b}$$

One can then deduce

<u>Lemma 2.1</u>. For integers $n \geq 1$ and $q \geq 0$

(i) $\varepsilon(1 + 2^{n-1} + q\cdot 2^{n+1}) = \frac{1}{p_n^\theta} \varepsilon(1 + q\cdot 2^{n+1})$

(ii) $\varepsilon(1 + 3\cdot 2^{n-1} + q\cdot 2^{n+1}) = \frac{1+\delta}{p_n^\theta} \varepsilon(1 + 2\cdot 2^{n-1} + q\cdot 2^{n+1})$.

Next, for $n \geq 0$ and $0 \leq r \leq n$ define

$$w_n^{(r)} = \sum_{k=0}^{2^{n-r}-1} \varepsilon(1 + k\cdot 2^r) \tag{2.3}$$

and set $w_n^{(0)} = w_n$. Using Lemma 2.1, it is then not difficult to prove

<u>Lemma 2.2</u>. For $3 \le n < \infty$ and $r = 0, 1$,

$$w_n^{(r)} = \left(1 + \frac{1}{p_n^\theta}\right) w_{n-1}^{(r)} + \frac{\delta}{p_n^\theta}\left(w_{n-1}^{(r)} - w_{n-2}^{(r)}\right). \tag{2.4}$$

Also, for $n \ge 1$,

$$w_n \le \prod_{r=1}^{n} \left(1 + \frac{1+\delta}{p_r^\theta}\right). \tag{2.5}$$

For $n \ge 1$ and $1 \le r \le n+1$ define $\alpha_r^{(n)}$ and $\beta_r^{(n)}$ by

$$\alpha_1^{(n)} = 0, \tag{2.6a}$$

$$\beta_1^{(n)} = -w_n, \tag{2.6b}$$

$$\beta_r^{(n)} = -\frac{w_{r-1}^{(1)} w_n}{w_{r-1}}, \quad 2 \le r \le n+1, \tag{2.6c}$$

$$\alpha_r^{(n)} = -\frac{\left(w_r^{(1)} - w_{r-1}^{(1)}\right) w_n}{w_r - w_{r-1}}, \quad 2 \le r \le n+1. \tag{2.6d}$$

Using (2.4) one can deduce

<u>Lemma 2.3</u>. For $n \ge 1$

(i) $$\alpha_r^{(n)} - \beta_r^{(n)} = \frac{\delta^{r-1}}{(p_1 p_2 \cdots p_{r-1})^\theta} \cdot \frac{w_n}{w_{r-1}[w_{r-1} + \delta(w_{r-1} - w_{r-2})]},$$

for $1 \le r \le n+1$, while for $1 \le r \le n$,

(ii) $$\beta_{r+1}^{(n)} - \beta_r^{(n)} = \frac{\delta^{r-1}}{(p_1 p_2 \cdots p_r)^\theta} \cdot \frac{w_n}{w_{r-1} w_r}.$$

We also have

Lemma 2.4. For $n \geq 1$ and $1 \leq r \leq n$,

(i) $$\frac{\alpha_r^{(n)} - \beta_{r+1}^{(n)}}{\alpha_r^{(n)} - \beta_r^{(n)}} = \frac{w_{r-1}}{w_r} ;$$

(ii) $$\frac{\beta_{r+1}^{(n)} - \beta_r^{(n)}}{\alpha_r^{(n)} - \beta_r^{(n)}} = \frac{w_r - w_{r-1}}{w_r} ;$$

(iii) $$\frac{\alpha_r^{(n)} - \alpha_{r+1}^{(n)}}{\alpha_r^{(n)} - \beta_r^{(n)}} = \frac{1}{p_{r+1}^{\theta}} \frac{w_{r-1}}{w_{r+1} - w_r} ;$$

(iv) $$\frac{\alpha_{r+1}^{(n)} - \beta_r^{(n)}}{\alpha_r^{(n)} - \beta_r^{(n)}} = \frac{1+\delta}{p_{r+1}^{\theta}} \frac{w_r - w_{r-1}}{w_{r+1} - w_r} .$$

We are now in a position to state the eigenvalue problem E_n. Let $b_r^{(n)}$ and $c_r^{(n)}$ be defined by

$$b_r^{(n)} = \alpha_r^{(n)} + \beta_r^{(n)} \tag{2.7}$$

$$c_r^{(n)} = \alpha_r^{(n)} \cdot \beta_r^{(n)} \tag{2.8}$$

Also, noting that $\alpha_r^{(n)} - \beta_r^{(n)} > 0$, $n \geq 1$, $1 \leq r \leq n+1$, by Lemma 2.3 (i), we define $a_r^{(n)}$, $n \geq 1$, $0 \leq r \leq n+1$ by $a_0^{(n)} = 0$, $a_{n+1}^{(n)} = \infty$ and, for $1 \leq r \leq n$,

$$a_r^{(n)} = a_{r-1}^{(n)} + \frac{\ell n(p_r)}{\alpha_r^{(n)} - \beta_r^{(n)}} . \tag{2.9}$$

Finally, for later use we note

Lemma 2.5. For $n \geq 1$, $1 \leq k \leq n$,

$$a_k^{(n)} = O\left(\frac{(p_1 p_2 \cdots p_{k-1})^{\theta}}{\delta^{k-1}}\right) ,$$

where the order constant does not depend on n.

For fixed n, the set $\{a_r^{(n)} : 0 \le r \le n\}$ clearly forms a partition of the interval $[0,\infty)$. Define functions $b_n(x)$ and $c_n(x)$ on $[0,\infty)$ by

$$b_n(x) = b_r^{(n)} \tag{2.10a}$$

$$c_n(x) = c_r^{(n)} \tag{2.10b}$$

for $a_{r-1}^{(n)} \le x < a_r^{(n)}$, $n \ge 1$, $1 \le r \le n+1$. The eigenvalue problem E_n consists of the differential equation

$$y'' - s b_n(x) y' + s^2 c_n(x) y = 0 \tag{2.11}$$

together with the boundary conditions

$$y(0) = 0 \tag{2.12a}$$

$$y(x) \sim \exp\left(\beta_{n+1}^{(n)} sx\right) \quad \text{as} \quad x \to \infty . \tag{2.12b}$$

Consider now the solution, $\phi_n(x,s)$, of (2.11) satisfying (2.12b). On $[a_{r-1}^{(n)}, a_r^{(n)}]$ it is clear that

$$\phi_n(x,s) = A_r^{(n)}(s)\exp\left(\alpha_r^{(n)} sx\right) + B_r^{(n)}(s)\exp\left(\beta_r^{(n)} sx\right) \tag{2.13}$$

where the functions $A_r^{(n)}(s)$ and $B_r^{(n)}(s)$, $1 \le r \le n+1$ are uniquely determined by (2.12b). We set $A_{n+1}^{(n)} = 0$ and $B_{n+1}^{(n)} = 1$, and let

$$A_{n+2-k}^{(n)}(s) = \sum_{i=1}^{2^{k-1}} C_{k,i}^{(n)} \exp\left(\eta_{k,i}^{(n)} s\right) \tag{2.14a}$$

$$B_{n+2-k}^{(n)}(s) = \sum_{i=1}^{2^{k-1}} D_{k,i}^{(n)} \exp\left(\nu_{k,i}^{(n)} s\right) \tag{2.14b}$$

for certain constants $C_{k,i}^{(n)}$, $D_{k,i}^{(n)}$, $\eta_{k,i}^{(n)}$, and $\nu_{k,i}^{(n)}$, where $1 \le k \le n+1$, $C_{1,1}^{(n)} = 0$, $D_{1,1}^{(n)} = 1$, and $\eta_{1,1}^{(n)} = \nu_{1,1}^{(n)} = 0$. The condition that s be an eigenvalue of E_n is now $\phi_n(0,s) = 0$, or $A_1^{(n)}(s) + B_1^{(n)}(s) = 0$. Our next task is to determine formulae for the coefficients in (2.14a-b) in terms of the members of the sequence $\{\varepsilon(n)\}$ defined earlier.

Consider then the solution $\phi_n(x,s)$ and its derivative at a boundary point $a_r^{(n)}$, $1 \le r \le n$. We have

$$A_{r+1}^{(n)} \exp\left(\alpha_{r+1}^{(n)} s\, a_r^{(n)}\right) + B_{r+1}^{(n)} \exp\left(\beta_{r+1}^{(n)} s\, a_r^{(n)}\right) = A_r^{(n)} \exp\left(\alpha_r^{(n)} s\, a_r^{(n)}\right) + B_r^{(n)} \exp\left(\beta_r^{(n)} s\, a_r^{(n)}\right) \tag{2.15}$$

$$\alpha_{r+1}^{(n)} A_{r+1}^{(n)} \exp\left(\alpha_{r+1}^{(n)} s\, a_r^{(n)}\right) + \beta_{r+1}^{(n)} B_{r+1}^{(n)} \exp\left(\beta_{r+1}^{(n)} s\, a_r^{(n)}\right) = \alpha_r^{(n)} A_r^{(n)} \exp\left(\alpha_r^{(n)} s\, a_r^{(n)}\right) + \beta_r^{(n)} B_r^{(n)} \exp\left(\beta_r^{(n)} s\, a_r^{(n)}\right). \tag{2.16}$$

From these equations it follows that

$$A_r^{(n)} = \frac{\alpha_{r+1}^{(n)} - \beta_r^{(n)}}{\alpha_r^{(n)} - \beta_r^{(n)}} \exp\left[\left(\alpha_{r+1}^{(n)} - \alpha_r^{(n)}\right) s\, a_r^{(n)}\right] A_{r+1}^{(n)} + \frac{\beta_{r+1}^{(n)} - \beta_r^{(n)}}{\alpha_r^{(n)} - \beta_r^{(n)}} \exp\left[\left(\beta_{r+1}^{(n)} - \alpha_r^{(n)}\right) s\, a_r^{(n)}\right] B_{r+1}^{(n)} \tag{2.17a}$$

$$B_r^{(n)} = \frac{\alpha_r^{(n)} - \alpha_{r+1}^{(n)}}{\alpha_r^{(n)} - \beta_r^{(n)}} \exp\left[\left(\alpha_{r+1}^{(n)} - \beta_r^{(n)}\right) s\, a_r^{(n)}\right] A_{r+1}^{(n)} + \frac{\alpha_r^{(n)} - \beta_{r+1}^{(n)}}{\alpha_r^{(n)} - \beta_r^{(n)}} \exp\left[\left(\beta_{r+1}^{(n)} - \beta_r^{(n)}\right) s\, a_r^{(n)}\right] B_{r+1}^{(n)}. \tag{2.17b}$$

Thus if

$$A_{n+1-k}^{(n)}(s) = \sum_{i=1}^{2^k} C_{k+1,i}^{(n)} \exp\left(\eta_{k+1,i}^{(n)} s\right)$$

$$B_{n+1-k}^{(n)}(s) = \sum_{i=1}^{2^k} D_{k+1,i}^{(n)} \exp\left(\nu_{k+1,i}^{(n)} s\right)$$

then it follows from (2.17a-b) and Lemma 2.4 that

$$\begin{aligned} C_{k+1,i}^{(n)} &= \frac{1+\delta}{p_{n+2-k}^{\theta}} \cdot \frac{w_{n+1-k} - w_{n-k}}{w_{n+2-k} - w_{n+1-k}} \cdot C_{k,i}^{(n)}, \qquad 1 \le i \le 2^{k-1} \\ &= \frac{w_{n+1-k} - w_{n-k}}{w_{n+1-k}} D_{k,i-2^{k-1}}^{(n)}, \qquad 2^{k-1}+1 \le i \le 2^k \end{aligned} \tag{2.18}$$

$$\begin{aligned} D_{k+1,i}^{(n)} &= \frac{1}{p_{n+2-k}^{\theta}} \cdot \frac{w_{n-k}}{w_{n+2-k} - w_{n+1-k}} \cdot C_{k,i}^{(n)}, \qquad 1 \le i \le 2^{k-1} \\ &= \frac{w_{n-k}}{w_{n-k+1}} D_{k,i-2^{k-1}}^{(n)}, \qquad 2^{k-1}+1 \le i \le 2^k \end{aligned} \tag{2.19}$$

$$\begin{aligned} \eta_{k+1,i}^{(n)} &= \eta_{k,i}^{(n)} + \left(\alpha_{n+2-k}^{(n)} - \alpha_{n+1-k}^{(n)}\right) a_{n+1-k}^{(n)}, \qquad 1 \le i \le 2^{k-1}, \\ &= \nu_{k,i-2^{k-1}}^{(n)} + \left(\beta_{n+2-k}^{(n)} - \alpha_{n+1-k}^{(n)}\right) a_{n+1-k}^{(n)}, \qquad 2^{k-1}+1 \le i \le 2^k, \end{aligned} \tag{2.20}$$

$$\begin{aligned} \nu_{k+1,i}^{(n)} &= \eta_{k,i}^{(n)} + \left(\alpha_{n+1-k}^{(n)} - \beta_{n+1-k}^{(n)}\right) a_{n+1-k}^{(n)}, \qquad 1 \le i \le 2^{k-1} \\ &= \nu_{k,i-2^{k-1}}^{(n)} + \left(\beta_{n+2-k}^{(n)} - \beta_{n+1-k}^{(n)}\right) a_{n+1-k}^{(n)}, \qquad 2^{k-1}+1 \le i \le 2^k. \end{aligned} \tag{2.21}$$

Define $E_{k,i}^{(n)}$ and $F_{k,i}^{(n)}$ by

$$C_{k,i}^{(n)} = \frac{w_{n-k+2} - w_{n-k+1}}{w_n} E_{k,i}^{(n)} \tag{2.22}$$

$$D_{k,i}^{(n)} = \frac{w_{n-k+1}}{w_n} F_{k,i}^{(n)}. \tag{2.23}$$

Then, from (2.18) and (2.19) we have,

$$E_{k+1,i}^{(n)} = \frac{1+\delta}{p_{n+1-k}^{\theta}} E_{k,i}^{(n)}, \qquad 1 \le i \le 2^{k-1},$$

$$= F_{k,i}^{(n)}, \qquad 2^{k-1}+1 \le i \le 2^k, \tag{2.24}$$

$$F_{k+1,i}^{(n)} = \frac{1}{p_{n+2-k}^{\theta}} E_{k,i}^{(n)}, \qquad 1 \le i \le 2^{k-1},$$

$$= F_{k,i}^{(n)}, \qquad 2^{k-1}+1 \le i \le 2^k. \tag{2.25}$$

We now define $L_r^{(n)}(s)$ and $M_r^{(n)}(s)$ by

$$A_r^{(n)}(s) = \exp\left[s\eta_{n+2-r,2^{n+1-r}}^{(n)}\right]\frac{w_r - w_{r-1}}{w_n} L_r^{(n)}(s) \tag{2.26}$$

$$B_r^{(n)}(s) = \exp\left[s\nu_{n+2-r,2^{n+1-r}}^{(n)}\right]\frac{w_{r-1}}{w_n} M_r^{(n)}(s). \tag{2.27}$$

Equations (2.14a-b) now become

$$L_r^{(n)}(s) = \sum_{i=1}^{2^{n+1-r}} E_{n+2-r,i}^{(n)} \exp\left[s\left(\eta_{n+2-r,i}^{(n)} - \eta_{n+2-r,2^{n+1-r}}^{(n)}\right)\right] \tag{2.28}$$

$$M_r^{(n)}(s) = \sum_{i=1}^{2^{n+1-r}} F_{n+2-r,i}^{(n)} \exp\left[s\left(\nu_{n+2-r,i}^{(n)} - \nu_{n+2-r,2^{n+1-r}}^{(n)}\right)\right]. \tag{2.29}$$

Using the properties of the constants $E_{k,i}^{(n)}$, $F_{k,i}^{(n)}$, $\eta_{k,i}^{(n)}$, and $\nu_{k,i}^{(n)}$ one can finally deduce

Lemma 2.6.

(i) $$L_r^{(r)}(s) = 1, \tag{2.30a}$$

$$L_r^{(r+1)}(s) = 1 + \frac{1+\delta}{p_{r+1}^{\theta+s}}, \tag{2.30b}$$

and for $n \ge r+2$

$$L_r^{(n)}(s) = \left(1 + \frac{1}{p_n^{\theta+s}}\right)L_r^{(n-1)}(s) + \frac{\delta}{p_n^{\theta+s}}\left[L_r^{(n-1)}(s) - L_r^{(n-2)}(s)\right]; \tag{2.30c}$$

(ii) $M_r^{(r)}(s) = 1$, (2.31a)

$$M_r^{r+1}(s) = 1 + \frac{1}{p_{r+1}^{\theta+s}} \tag{2.31b}$$

and for $n \geq r+2$

$$M_r^{(n)}(s) = \left[1 + \frac{1}{p_n^{\theta+s}}\right] M_r^{(n-1)}(s) + \frac{\delta}{p_n^{\theta+s}}\left[M_r^{(n-1)}(s) - M_r^{(n-2)}(s)\right]. \tag{2.31c}$$

The recursion formulae (2.30c) and (2.31c) indicate that, for δ small enough, the functions $L_r^{(n)}(s)$ and $M_r^{(n)}(s)$ are essentially partial products of the type we are seeking. More precisely, define $H_r^{(n)}(s)$, $n \geq r+1$, by $H_r^{(r+1)}(s) = 1$ and

$$H_r^{(n)}(s) = \frac{p_n^{\theta+s}\left[L_r^{(n)}(s) - L_r^{(n-1)}(s)\right]}{\left[1 + \frac{1+\delta}{p_{r+1}^{\theta+s}}\right]\left[1 + \frac{1}{p_{r+2}^{\theta+s}}\right] \cdots \left[1 + \frac{1}{p_{n-1}^{\theta+s}}\right]}, \quad n \geq r+2 . \tag{2.32}$$

Then, using (2.30c), one can show by induction that

$$L_r^{(n)}(s) = U_r^{(n)}(s)\left[1 + \frac{\delta}{\left[1 + \frac{p_{r+1}^{\theta+s}}{1+\delta}\right]\left(1 + p_{r+2}^{\theta+s}\right)} + \sum_{j=r+2}^{n-1} \frac{\delta\, H_r^{(j)}(s)}{\left(1 + p_j^{\theta+s}\right)\left(1 + p_{j+1}^{\theta+s}\right)}\right] \tag{2.33}$$

where

$$U_r^{(n)}(s) = \left[1 + \frac{1+\delta}{p_{r+1}^{\theta+s}}\right]\left[1 + \frac{1}{p_{r+2}^{\theta+s}}\right] \cdots \left[1 + \frac{1}{p_n^{\theta+s}}\right] . \tag{2.34}$$

Moreover, from (2.30c) again,

$$H_r^{(n)}(s) = \frac{L_r^{(n-1)}(s)}{U_r^{(n-1)}(s)} + \frac{\delta\, H_r^{(n-1)}(s)}{1 + p_{n-1}^{\theta+s}}$$

$$= \left[1 + \frac{\delta}{\left(1 + \frac{p_{r+1}^{\theta+s}}{1+\delta}\right)\left(1 + p_{r+2}^{\theta+s}\right)} + \sum_{j=r+2}^{n-2} \frac{\delta\, H_r^{(j)}(s)}{\left(1 + p_j^{\theta+s}\right)\left(1 + p_{j+1}^{\theta+s}\right)}\right] + \frac{\delta\, H_r^{(n-1)}(s)}{1 + p_{n-1}^{\theta+s}} \tag{2.35}$$

on using (2.33). Similarly, relative to $M_r^{(n)}(s)$, define $K_r^{(n)}(s)$, $n \geq r+1$, by $K_r^{(r+1)}(s) = 1$ and

$$K_r^{(n)}(s) = \frac{p_n^{\theta+s}\left[M_r^{(n)}(s) - M_r^{(n-1)}(s)\right]}{\left(1 + \frac{1}{p_{r+1}^{\theta+s}}\right)\left(1 + \frac{1}{p_{r+2}^{\theta+s}}\right) \cdots \left(1 + \frac{1}{p_{n-1}^{\theta+s}}\right)}, \quad n \geq r+2. \tag{2.36}$$

Then, from (2.31c) we have that

$$M_r^{(n)}(s) = V_r^{(n)}(s)\left[1 + \sum_{j=r+1}^{n-1} \frac{\delta\, K_r^{(j)}(s)}{\left(1 + p_j^{\theta+s}\right)\left(1 + p_{j+1}^{\theta+s}\right)}\right] \tag{2.37}$$

where

$$V_r^{(n)}(s) = \prod_{j=r+1}^{n} \left(1 + \frac{1}{p_j^{\theta+s}}\right), \tag{2.38}$$

and, from (2.31c) and (2.37),

$$K_r^{(n)}(s) = \frac{M_r^{(n-1)}(s)}{V_r^{(n-1)}(s)} + \frac{\delta\, K_r^{(n-1)}(s)}{\left(1 + p_{n-1}^{\theta+s}\right)}$$

$$= \left[1 + \sum_{j=r+1}^{n-2} \frac{\delta\, K_r^{(j)}(s)}{\left(1 + p_j^{\theta+s}\right)\left(1 + p_{j+1}^{\theta+s}\right)}\right] + \frac{\delta\, K_r^{(n-1)}(s)}{1 + p_{n-1}^{\theta+s}}. \tag{2.39}$$

From (2.35) and (2.39) one can deduce, by a straightforward induction argument,

<u>Lemma 2.7</u>. For each real number σ_0 with $\sigma_0 + \theta > \frac{1}{2}$ there is a value $\delta = \delta(\sigma_0)$ and a constant C such that $|H_r^{(n)}(s)| \leq C$ and

$|K_r^{(n)}(s)| \le C$ for all n and r, $n \ge r+1$, and all complex numbers s with $\mathrm{Re}(s+\theta) \ge \sigma_0 + \theta$.

Thus, the solution $\phi_n(x,s)$ of (2.11) satisfying (2.12b) is now given explicitly by (2.13), (2.26), (2.27), (2.33), (2.34), (2.37), and (2.38).

3. The Differential Equation E

As $\theta > 1$, it follows from (2.5) that for fixed $r \ge 0$ the partial sums (2.3) converge as $n \to \infty$. Define

$$w_\infty^{(r)} = \sum_{k=0}^{\infty} \varepsilon(1 + k\cdot 2^r) . \tag{3.1}$$

It is clear from (2.6a-d) and (2.7-9) that for fixed $r \ge 1$ the sequences $\{\alpha_r^{(n)}\}$, $\{\beta_r^{(n)}\}$, $\{b_r^{(n)}\}$, $\{c_r^{(n)}\}$, $\{a_r^{(n)}\}$ all converge to finite limits, which we denote by α_r, β_r, b_r, c_r, and a_r, respectively. Observe also that $b_r = \alpha_r + \beta_r$, $c_r = \alpha_r\beta_r$, and that $a_0 = 0$ and

$$a_r = a_{r-1} + \frac{\ln(p_r)}{\alpha_r - \beta_r}, \quad r \ge 1, \tag{3.2}$$

from (2.9). Furthermore, it follows from Lemma 2.3 that for $r \ge 1$

$$\alpha_r - \beta_r = \frac{\delta^{r-1}}{(p_1 p_2 \cdots p_{r-1})^\theta} \cdot \frac{w_\infty}{w_{r-1}\left[w_{r-1} + \delta(w_{r-1} - w_{r-2})\right]} \tag{3.3}$$

and

$$\beta_{r+1} - \beta_r = \frac{\delta^{r-1}}{(p_1 p_2 \cdots p_r)^\theta} \cdot \frac{w_\infty}{w_{r-1} w_r} . \tag{3.4}$$

One can also prove that

$$\beta_r + w_\infty^{(1)} = -\frac{\delta^{r-1}}{(p_1 p_2 \cdots p_r)^\theta} \cdot \frac{w_\infty^{(r)} + \delta\left(w_\infty^{(r)} - w_\infty^{(r+1)}\right)}{w_{r-1}} . \tag{3.5}$$

Finally, from Lemma 2.5 we have

$$a_r = O\left(\frac{(p_1 p_2 \cdots p_{r-1})^{\theta}}{\delta^{r-1}}\right) . \tag{3.6}$$

Setting $\gamma = -w_{\infty}^{(1)}$, it follows from (3.3) and (3.5) that $\alpha_r \to \gamma$ and $\beta_r \to \gamma$ as $r \to \infty$.

Define functions $b(x)$ and $c(x)$ on $(0,\infty)$ by $b(x) = b_r$ and $c(x) = c_r$ on $(a_{r-1}, a_r]$, $r \geq 1$. Formally letting $n \to \infty$ in the differential equation (2.11) we obtain the equation

$$y'' - s\,b(x)y' + s^2 c(x)y = 0 , \tag{3.7}$$

while the boundary condition (2.12b) becomes

$$y(x,s) \sim \exp(\gamma s x) \quad \text{as} \quad x \to \infty . \tag{3.8}$$

It is convenient at this stage to transform (3.7-8) by means of the change of dependent variable $y(x,s) = e^{\gamma s x} z(x,s)$:

$$z'' - s\,B(x)z' + s^2 C(x)z = 0 \tag{3.9}$$

$$z(x,s) \sim 1 , \quad \text{as} \quad x \to \infty \tag{3.10}$$

where $B(x) = b(x) - 2\gamma$ and $C(x) = c(x) - \gamma b(x) + \gamma^2$. Equation (3.9) may be rewritten as

$$\frac{d}{dx}\left[\exp\left(-s \int_0^x B\right)\frac{dz}{dx}\right] + s^2 C(x)\exp\left(-s \int_0^x B\right) \cdot z = 0 . \tag{3.11}$$

Set

$$r(x) = \int_0^x B(u)\,du \tag{3.12}$$

and

$$u(r,s) = z\big(x(r), s\big) . \tag{3.13}$$

Then, on changing the independent variable from x to r in (3.11) we obtain the equation

$$\frac{d}{dr}\left[e^{-sr}B(x(r))\frac{du}{dr}\right] + s^2\,\frac{C(x(r))}{B(x(r))}\,e^{-sr}u(r,s) = 0\,. \tag{3.14}$$

Consider now

<u>Lemma 4.1</u> [1]. If $u(r,s)$ is a solution of (3.14) satisfying $u(r,s) \to 1$ and $e^{-sr}B(x(r))\frac{du}{dr} \to 0$ as $r \to \infty$, then $u(r,s)$ satisfies the integral equation

$$u(r,s) = 1 - s^2\int_r^\infty\int_t^\infty e^{-s(v-t)}\,\frac{C(x(v))}{B(x(t))B(x(v))}\,u(v,s)\,dv\,dt\,. \tag{3.15}$$

Furthermore, (3.15) has a unique bounded solution for values $s = \sigma + i\tau$ such that

$$\int_0^\infty\int_t^\infty e^{-\sigma(v-t)}\left|\frac{C(x(v))}{B(x(t))B(x(v))}\right|dv\,dt < \infty\,. \tag{3.16}$$

The latter assertion follows easily from the relevant Neumann series expansion in the space of continuous functions on $[r_0\,,\infty)$, r_0 suitably large, equipped with the usual supremum norm. The former assertion may be established by twice integrating equation (3.14).

Observe that, on $(a_{n-1}\,,a_n]$, $C(x) = C_n$, where

$$\begin{aligned} C_n &= \alpha_n\beta_n - \gamma(\alpha_n+\beta_n) + \gamma^2 \\ &= (\beta_n-\gamma)^2 + (\beta_n-\gamma)(\alpha_n-\beta_n) \\ &= (\beta_n-\gamma)(\alpha_n-\beta_n)\left[1+\frac{\beta_n-\gamma}{\alpha_n-\beta_n}\right] \\ &\sim (\beta_n-\gamma)(\alpha_n-\beta_n) \quad \text{as} \quad n\to\infty \end{aligned} \tag{3.17}$$

by (3.3) and (3.5). Also, on the same interval $B(x) = B_n$, where

$$\begin{aligned} B_n &= 2(\beta_n - \gamma) + (\alpha_n - \beta_n) \\ &= (\alpha_n - \beta_n)\left[1 + \frac{2(\beta_n - \gamma)}{\alpha_n - \beta_n}\right] \\ &\sim (\alpha_n - \beta_n) \quad \text{as} \quad n \to \infty, \end{aligned} \tag{3.18}$$

by a similar argument. By using (3.17-8) directly in (3.16) one can deduce

<u>Lemma 4.2</u>. The integral equation (3.15) has a unique bounded solution $u(r,s)$ whenever $\mathrm{Re}(s+\theta) > 1$.

In addition, as the double integral on the right of (3.15) is now abaolutely convergent for $\mathrm{Re}(s+\theta) > 1$, it follows by Fubini's theorem that for these values of s, (3.15) may be replaced by (1.4) with

$$k(r,t,s) = -s^2 e^{-st}\,\frac{C(x(t))}{B(x(t))}\int_r^t \frac{e^{sv}}{B(x(v))}\,dv \quad (r \le t). \tag{3.19}$$

We now return to a detailed consideration of the solutions $\phi_n(x,s)$ constructed in §2. Set

$$\Psi_r^{(n)}(s) = \frac{L_r^{(n)}(s)}{U_r^{(n)}(s)}, \tag{3.20}$$

and

$$\Phi_r^{(n)}(s) = \frac{M_r^{(n)}(s)}{V_r^{(n)}(s)}. \tag{3.21}$$

It follows from (2.33), (2.37), Lemma 2.7, and the fact that $p_n \sim n\,\ell n\, n$ as $n \to \infty$, that the function sequences $\left\{\Psi_r^{(n)}(s)\right\}$ and $\left\{\Phi_r^{(n)}(s)\right\}$ are convergent as $n \to \infty$ in any fixed half-plane $\mathrm{Re}(s+\theta) > \sigma_0 + \theta > \frac{1}{2}$, provided that the constant $\delta = \delta(\sigma_0)$ is chosen small enough. Set

$$\Psi_r(s) = \lim_{n\to\infty} \Psi_r^{(n)}(s) \tag{3.22}$$

$$\Phi_r(s) = \lim_{n\to\infty} \Phi_r^{(n)}(s) \tag{3.23}$$

Also, for fixed r and $\mathrm{Re}(s+\theta) > 1$, by (1.5) we have $\lim_{n\to\infty} U_r^{(n)}(s) = U_r(s)$ and $\lim_{n\to\infty} V_r^{(n)}(s) = V_r(s)$ where

$$U_r(s) = \frac{\zeta(s+\theta)}{\zeta(2s+2\theta)} \cdot \frac{\left(1+\dfrac{1+\delta}{p_{r+1}^{s+\theta}}\right)}{\left(1+\dfrac{1}{p_1^{\theta+s}}\right)\left(1+\dfrac{1}{p_2^{\theta+s}}\right)\cdots\left(1+\dfrac{1}{p_{r+1}^{\theta+s}}\right)} \tag{3.24}$$

$$V_r(s) = \frac{\zeta(s+\theta)}{\zeta(2s+2\theta)} \cdot \frac{1}{\left(1+\dfrac{1}{p_1^{\theta+s}}\right)\left(1+\dfrac{1}{p_2^{\theta+s}}\right)\cdots\left(1+\dfrac{1}{p_r^{\theta+s}}\right)} . \tag{3.25}$$

From (2.21) we have

$$\begin{aligned} \nu^{(n)}_{n+2-r,2^{n+1-r}} &= \sum_{k=1}^{n+1-r} \left(\nu^{(n)}_{k+1,2^k} - \nu^{(n)}_{k,2^{k-1}}\right) \\ &= \sum_{k=1}^{n+1-r} \left(\beta^{(n)}_{n+2-k} - \beta^{(n)}_{n+1-k}\right) a_k^{(n)} \\ &= \sum_{k=r}^{n} \left(\beta^{(n)}_{k+1} - \beta^{(n)}_{k}\right) a_k^{(n)} , \end{aligned} \tag{3.26}$$

and from (2.20), (3.26), and (2.9),

$$\begin{aligned} \eta^{(n)}_{n+2-r,2^{n+1-r}} &= \nu^{(n)}_{n+1-r,2^{n-r}} + (\beta^{(n)}_{r+1} - \alpha_r^{(n)}) a_r^{(n)} \\ &= \sum_{k=r+1}^{n} \left(\beta^{(n)}_{k+1} - \beta^{(n)}_{k}\right) a_k^{(n)} + \left(\beta^{(n)}_{r+1} - \alpha_r^{(n)}\right) a_r^{(n)} \\ &= \sum_{k=r}^{n} \left(\beta^{(n)}_{k+1} - \beta_j^{(n)}\ a_k^{(n)}\right) - \ell n(p_r) . \end{aligned} \tag{3.27}$$

Now, by Lemmas 2.3 and 2.5, $\left(\beta^{(n)}_{k+1} - \beta^{(n)}_{k}\right) a_k^{(n)} = O(k^{-\theta})$ and therefore

$$\sum_{k=r}^{n} \left(\beta_{k+1}^{(n)} - \beta_k^{(n)}\right) a_k^{(n)} = O(1) , \tag{3.28}$$

where the order constant is independent of n and r. A similar argument using (3.6) shows that we also have

$$\sum_{k=1}^{\infty} (\beta_{k+1} - \beta_k) a_k < \infty . \tag{3.29}$$

We are now in a position to let $n \to \infty$ in the formula (2.13). First note that for fixed r, the sequence $\left\{a_r^{(n)}\right\}$ is decreasing in n. Consequently, if $x \in (a_{r-1}, a_r]$ is fixed, then there is an integer N such that $x \in \left[a_{r-1}^{(n)}, a_r^{(n)}\right]$ for all $n \geq N$. From (2.13), (2.26), and (2.27) it is clear that for $\mathrm{Re}(s+\theta) > 1$, $\lim_{n\to\infty} \phi_n(x,s) = \phi(x,s)$ where

$$\phi(x,s) = A_r(s)\exp(\alpha_r sx) + B_r(s)\exp(\beta_r sx) ; \tag{3.30}$$

here

$$A_r(s) = \frac{1}{p_r^{\theta+s}} \exp\left[s \sum_{k=r}^{\infty} (\beta_{k+1} - \beta_k) a_k\right] \cdot \frac{w_{r-1} + \delta(w_{r-1} - w_{r-2})}{w_\infty} U_r(s)\Psi_r(s) \tag{3.31}$$

and

$$B_r(s) = \exp\left[s \sum_{k=r}^{\infty} (\beta_{k+1} - \beta_k) a_k\right] \cdot \frac{w_{r-1}}{w_\infty} \cdot V_r(s)\Phi_r(s), \tag{3.32}$$

and the functions $\Psi_r(s)$ and $\Phi_r(s)$ (and therefore also $\phi(x,s)$, $s \neq 1$) are analytic in any region $\mathrm{Re}(s+\theta) > \sigma_0 + \theta > \frac{1}{2}$, provided always that the constant $\delta = \delta(\sigma_0)$ is chosen small enough. We define the function $\psi(x,s)$ by

$$\psi(x,s) = e^{-\gamma sx} \phi(x,s) . \tag{3.33}$$

Observe that the function $\psi(x,s)$ is a solution to the differential equation (3.9), because on each interval $(a_{r-1}, a_r]$ ϕ is a linear combination of exponential solutions, and the compatibility at the end-points is established by letting n tend to infinity in the identities (2.15) and (2.16). One can also show by a direct computation that $\psi(x,s) \to 1$ and $\exp\left(-s\int_0^x B\right)\frac{d\psi}{dx} \to 0$ as $x \to \infty$. Thus, from Lemmas 4.1 and 4.2, we see that $\psi(x(r), s)$ is the unique bounded solution of the integral equation (1.4) with kernel (3.19) when $\mathrm{Re}(s+\theta) > 1$. To prove part (ii) of the Theorem let the solution of equ. (1.4) be denoted by $u(r,s)$. Then from the above argument

$$u(r,s) = \psi(x(r), s) \tag{3.34}$$

for $\mathrm{Re}(s+\theta) > 1$. Assume that $u(r,s)$ can be analytically continued into a sub-region, G, of the strip $\frac{1}{2} < \mathrm{Re}(s+\theta) \le 1$. It is clear from (3.30-2) that (provided the constant δ is chosen suitably small a priori) the function $\psi(x(r), s)$ is analytic in a half-plane $\mathrm{Re}(s+\theta) > \sigma_0 + \theta > \frac{1}{2}$ containing G. Consequently, by analytic continuation, $\psi(x(r), s)$ satisfies (1.4) for $s \in G$. In particular, for these values of s, $\psi(x(r), s)$ cannot be identically zero. It then follows immediately from (3.24-5) and (3.30-2) that $\zeta(s+\theta) \ne 0$ for $s \in G$, as required.

References

[1] F.V. Atkinson, private communication.

[2] R. Brent, On the zeros of the Riemann zeta function in the critical strip, Math. Comp. 33 (1979), 1361-1372; see also MR 80g: 10033.

[3] H.M. Edwards, Riemann's Zeta Function, Academic Press, New York, 1974.

[4] I. Knowles, On the zeros of the Riemann zeta-function, in preparation.

[5] B. Riemann, Ueber die Anzahl der Primzahlen unter einer gegebenen Grösse, Monatsberichte der Berliner Akademie, November 1859.

[6] E.C. Titchmarsh, The Theory of the Riemann Zeta-Function, Oxford University Press, London, 1951.

Superposition Principles and Pointwise Evaluation of Sturm-Liouville Eigenfunction Expansions

Mark A. Kon*

Louise A. Raphael†

§1. Introduction

The problem of correctly summing eigenfunction expansions in pointwise and function space topologies has been considered classically [10,11], as well as in relatively recent papers (see, e.g, [3,9]). Current interest in this subject in the Soviet literature has arisen from its relation to the theory of stable summation of expansions, which falls into the area of ill-posed problems [2,6]. The case of expansions in a continuous spectral parameter (arising, for example, from singular Sturm-Liouville problems) has been studied less extensively [7], and techniques which are useful in the case of discrete spectrum often do not directly translate to the continuous case.

In this paper we present an approach which is useful in both situations, and two applications which subsume a broad class of summability methods, including the Abel, resolvent, and Gauss-Weierstrass methods. We begin with some definitions.

Let $\phi(\lambda)$ be a function of a real (or complex) variable with $\phi(0) = 1$ and

$$f(x) \sim \sum_n a_n u_n(x) \tag{1}$$

be an expansion in eigenfunctions of a self-adjoint operator A. The $\sum$ is generic and may represent summation or integration. The expansion (1) is *ϕ-summable* in a given topology if

$$\lim_{\epsilon \to 0} \phi(\epsilon A) f(x) \equiv \lim_{\epsilon \to 0} \sum_n a_n \phi(\epsilon\lambda_n) u_n(x) = f(x),$$

*Partially supported by the NSF under grant MCS800-3407

†Research supported by the ARO, grant DAA-G29-81-G-0011

where λ_n in the spectrum $\sigma(A)$ of A corresponds to the (generalized) eigenfunction $u_n(x)$. Abel, Gauss-Weierstrass, and resolvent summability correspond to $\phi(\lambda) = e^{-\lambda}$, $e^{-\lambda^2}$, and $\frac{1}{1+\lambda}$, respectively.

We consider the following Sturm-Liouville problem on $[0, \infty)$:

$$Au(x,\lambda) \equiv \left(-\frac{d^2}{dx^2} + q(x)\right)u(x,\lambda) = \lambda u(x,\lambda);$$

$$u(0,\lambda)\cos\beta + u'(0,\lambda)\sin\beta = 0, \qquad u(\infty,\lambda) < \infty$$

with $\beta \in [0, 2\pi)$ fixed. The eigenfunctions $u(x,\lambda)$ are normalized by $u(0,\lambda) = \sin\beta$, $u'(0,\lambda) = -\cos\beta$. We assume the real potential $q(x) \in L^1(\mathbf{R})$ to be continuous and bounded (see [7]). The operator A is then self-adjoint and bounded from below, its negative spectrum is discrete and finite and its positive spectrum is continuous. If $f(x) \in L^2(\mathbf{R}^+)$, the expansion

$$f(x) \sim \int_{-b}^{\infty} F(\lambda)u(x,\lambda)d\rho(\lambda)$$

holds, where

$$F(\lambda) \sim \int_0^{\infty} f(x)u(x,\lambda)dx.$$

Above, b is a lower bound of $\sigma(\mathrm{A})$ and $\sim$ denotes convergence in L^2.

If $\phi(\lambda)$ is continuous at 0, then formally

$$\phi(\epsilon A)f \equiv \int_{-b}^{\infty} \phi(\epsilon\lambda)F(\lambda)u(x,\lambda)d\rho(\lambda) \underset{\epsilon\to 0}{\to} f(x). \tag{3}$$

The pointwise convergence of (3) in the cases of Riesz and resolvent summability has been studied in [7] and [5], and the discrete case has been considered much more extensively (see, e.g., [10,11]).

All of the following results hold also for expansions on finite intervals, and the proofs simplify in some cases. For brevity, however, most proofs here will be sketched or omitted.

§ 2. Use of the Superposition Principle

Our aim in this and the following section is to show that the above-mentioned class of results on ϕ-summability in the case of continuous (as well as discrete) spectrum follows from a superposition principle for summability: if an expansion is ϕ_1-summable and ϕ_2-summable, and α_1 and α_2 are complex constants whose sum is 1, then it is $\alpha_1\phi_1 + \alpha_2\phi_2$ summable. This fact extends to infinite or continuous sums through regularity arguments.

Recall x is in the *Lebesgue set* $L(f)$ of f if

$$\lim_{\eta\to 0} \int_{-\eta}^{\eta} |f(x+\eta') - f(x)| d\eta' = 0.$$

THEOREM 1 (Analytic multipliers): *Let* $\gamma > 0$ *and* $\phi(z)$ *be analytic in a neighborhood* D *of the sector* $B_\gamma = \{z \in \mathbf{C} : |\arg z| \le \gamma\}$. *For* $z \in B_\gamma$, *let* $\phi(z) = O(|z|^{-\delta})$ *as* $z \to \infty$ *in* D, *for some* $\delta > 0$. *Then for* $f \in L^2(\mathbf{R}^+)$ *and* $|\arg \epsilon| \le \gamma_1 < \gamma$:

(i) $\phi(\epsilon A)f(x) \underset{\epsilon\to 0}{\to} f(x)$ *for all* x *in the Lebesgue set of* f *(and hence almost everywhere)*

(ii) $\phi(\epsilon A)f \underset{\epsilon\to 0}{\to} f$ *in* $L^p(\mathbf{R}^+)$ *for* $f \in L^2 \cap L^p$, $1 \le p < \infty$.

We sketch a proof of Theorem 1. Initially, let $\phi(\lambda) = \frac{1}{1+\lambda}$. Then

$$\phi(\alpha A)f(x) = (1+\alpha A)^{-1} f(x) = \frac{1}{\alpha}\int_0^\infty G_\alpha(x,x')f(x')dx', \tag{4}$$

where G_α is the Green function of $A + \frac{1}{\alpha}$. We define the mixed p,q norm

$$\|g(x,x')\|_{p,q} = \left(\int_0^\infty \left(\int_0^\infty |g(x,x')|^p dx\right)^{\frac{q}{p}} dx'\right)^{\frac{1}{q}},$$

with the natural definition when p or q is infinite. Let G_α^* denote the Green function of $-\frac{d^2}{dx^2} + \frac{1}{\alpha}$. The identity

$$R_\alpha(x,x') \equiv G_\alpha(x,x') - G_\alpha^*(x,x') = -\int_0^\infty G_\alpha^*(x,x'')q(x'')G_\alpha(x'',x')dx''$$

yields

$$||R_\alpha||_{p,q} \leq \sup_{x''\geq 0} \{||G^*_\alpha(x,x'')||_{p,x}||G_\alpha(x'',x')||_{q,x'}||q||_1\}, \tag{5}$$

where subscripts on the right denote the variables with respect to which norms are taken. A series of estimates of terms on the right side shows $||R_\alpha||_{p,q} \underset{\alpha\to 0}{\to} 0$ uniformly in $|\arg\alpha| \leq \gamma,\ \ \gamma < \pi$, for $1 \leq p,q \leq \infty$.

This together with Hölder's inequality shows the remainder R_α to be negligible in the relevant asymptotic L^p and pointwise estimates of (4). The $\alpha \to 0$ limit is thus unchanged when G_α is replaced by G^*_α, and the latter can be computed:

$$G^*_\alpha(x,x') = \sqrt{\alpha}\frac{e^{\frac{-|x-x'|}{\sqrt{\alpha}}}}{2} + \sqrt{\alpha}\frac{\sin\beta+\sqrt{\alpha}\cos\beta}{2(\sin\beta-\sqrt{\alpha}\cos\beta)}e^{\frac{-|x+x'|}{\sqrt{\alpha}}} \equiv G^{(1)}_\alpha(x,x') + G^{(2)}_\alpha(x,x'). \tag{6}$$

This is essentially a sum of convolution kernels, and the proof in the case $\phi(\lambda) = \frac{1}{1+\lambda}$ is completed by use of a standard theorem in harmonic analysis [8]:

THEOREM: *Let $\psi(x) \in L^1(\mathbf{R}^n)$, $\int_{\mathbf{R}^n}\psi(x)dx = 1$, and define $\beta(x) = \sup_{|t|\geq|x|}|\psi(t)|$. For $\epsilon > 0$, let $\psi_\epsilon(x) = \epsilon^{-n}\psi(\frac{x}{\epsilon})$. If $\beta(x) \in L^1(\mathbf{R}^n)$ and $f \in L^p(\mathbf{R}^n)$, then $\lim_{\epsilon\to 0}(f*\psi_\epsilon) = f(x)$ whenever x belongs to the Lebesgue set of f, and $f*\psi_\epsilon \underset{\epsilon\to 0}{\to} f$ in L^p $(1 \leq p < \infty)$.*

The convolution kernels in (6) are essentially L^1-norm preserving dilations (in the parameter α) of functions with L^1 radially decreasing bounds $\beta(x)$ as defined above, so that the Theorem applies to them. We now only require arguments proving uniformity of convergence in $\arg\alpha$ as $\alpha \to 0$, which will be omitted.

For general $\phi(z)$, we use the Cauchy representation of $\phi(\lambda)$ as a superposition of dilations of $\phi(\lambda) = \frac{1}{1+\lambda}$:

$$\phi(\epsilon\lambda) = \frac{1}{2\pi i}\int_\Gamma \frac{\phi(\epsilon z)}{z}\left(1-\frac{\lambda}{z}\right)^{-1}dz \tag{7}$$

where the contour Γ contains $\sigma(A)$, and coincides with the boundary of the set $\frac{1}{\epsilon}B_\gamma$ for $|z|$ large. Let the operator $\phi(\epsilon A)$ be defined by (3). Then (3), (4) and (7) together with some regularity arguments show that the integral kernel of $\phi(\epsilon A)$ is

$$K_\epsilon(x,x') = \frac{1}{2\pi i}\int_\Gamma \phi(\epsilon z) G_{-\frac{1}{z}}(x,x')dz, \tag{8}$$

and hence

$$\phi(\epsilon A)f = \frac{1}{2\pi i}\int_\Gamma \frac{\phi(z)}{z}\int_0^\infty \left(-\frac{z}{\epsilon}\right) G_{-\frac{z}{\epsilon}}(x,x')f(x')dx'dz. \tag{9}$$

We now proceed formally: as $\epsilon \to 0$, the inner integral converges to $f(x)$; since $\phi(0) = 1$, the proof is completed by use of Cauchy's theorem.

Remark: Although the above method also works for discrete expansions on finite intervals, these results in the discrete case follow more directly from estimates of kernels such as those in [3].

§ 3. Non-Analytic Multipliers

Proof of pointwise summability with non-analytic multipliers ϕ can be accomplished by use of Fourier transforms and the superposition of exponentials.

Let $\phi(\lambda) \in L^1(\mathbb{R})$ be differentiable and satisfy

(i) $\xi\hat{\phi}'(\xi) \in L^1(\mathbb{R})$

(ii) $\xi\hat{\phi}(\xi) \underset{|\xi|\to\infty}{\longrightarrow} 0$

(iii) $\phi(0) = 1$

where $\hat{\phi}$ is the Fourier transform of ϕ. We denote the above conditions by [D]; note that they are essentially differentiability conditions. We continue with our assumptions on the Sturm-Liouville operator A.

THEOREM 2: *If $f(x) \in L^2(\mathbb{R}^+)$, then the expansion of f in eigenfunctions of A is ϕ-summable on the Lebesgue set of f.*

Proof: Let x be a Lebesgue point of f. We have

$$\phi(\epsilon A)f \equiv \int_{-b}^{\infty} \phi(\epsilon\lambda)F(\lambda)u(x,\lambda)d\rho(\lambda) = \int_{-b}^{\infty} \frac{1}{\sqrt{2\pi}} \int_{-\infty}^{\infty} \hat{\phi}(\xi)e^{i\xi(\epsilon\lambda)}d\xi F(\lambda)u(x,\lambda)d\rho(\lambda)$$

$$= \frac{1}{\sqrt{2\pi}} \int_{-\infty}^{\infty} \frac{1}{\epsilon}\hat{\phi}\left(\frac{\xi}{\epsilon}\right) \int_{-b}^{\infty} e^{i\xi\lambda}F(\lambda)u(x,\lambda)d\rho(\lambda)d\xi$$

The interchange of integrals can be justified if the inner integral is interpreted as an L^2 limit. Let

$$h(\xi) \equiv \frac{1}{\sqrt{2\pi}} \int_{-b}^{\infty} e^{i\xi\lambda}F(\lambda)u(x,\lambda)d\rho(\lambda). \tag{10}$$

We define ξ_0 to be a *generalized Lebesgue point* of $h(\xi)$ if

$$\frac{1}{\eta}\int_0^{\eta} h(\xi_0+\xi)d\xi \underset{n\to 0}{\to} h(\xi_0).$$

The following lemma is due to M. Freedman.

LEMMA: *Let $f(\xi) \in L^2(\mathbf{R})$, with ξ_0 a generalized Lebesgue point of f. Let $\phi(\lambda)$ satisfy [D]. Then*

$$\left.\frac{1}{\epsilon}\hat{\phi}\left(\frac{\xi}{\epsilon}\right) * f(\xi)\right|_{\xi=\xi_0} \underset{\epsilon\to 0}{\to} f(\xi_0)$$

Thus to prove Theorem 2 it suffices to show that

(i) 0 is a generalized Lebesgue point of $h(\xi)$

(ii) $h(0) = f(x)$.

We note that

$$\frac{1}{\eta}\int_0^{\eta} h(\xi)d\xi = \int_{-b}^{\infty} \frac{\sin\left(\frac{\lambda\eta}{2}\right)}{\left(\frac{\lambda\eta}{2}\right)} e^{i\frac{\lambda\eta}{2}}F(\lambda)u(x,\lambda)d\rho(\lambda). \tag{11}$$

Let Γ be a contour which encloses $\mathbf{R}^+$, and for $|z|$ large coincides with $\{z : |\text{Im } z| = \frac{1}{3}\ln(\text{Re } z)\}$. For $z \in \Gamma$, $|z|$ large, we have $\left|\frac{\sin(\frac{z}{2})}{(\frac{z}{2})}e^{i\frac{z}{2}}\right| \leq 4|z|^{-\frac{2}{3}}$; hence

$$e^{i\frac{\lambda\eta}{2}}\frac{\sin(\frac{\lambda\eta}{2})}{(\frac{\lambda\eta}{2})} \equiv g(\lambda\eta) = \frac{1}{2\pi i}\int_{\Gamma}\frac{g(z)}{z-\lambda\eta}dz.$$

Through interchange of integrals and use of (4), we have

$$\int_{-b}^{\infty} g(\lambda\eta)F(\lambda)u(x,\lambda)d\rho(\lambda) = \frac{1}{2\pi i}\int_{\Gamma}\frac{g(z)}{z}\left\{\int_{0}^{\infty}\frac{-z}{\eta}G_{-\frac{\eta}{z}}(x,x')f(x')dx'\right\}dz. \quad (12)$$

We have $G_\alpha = G^*_\alpha + R_\alpha$, with R_α defined by (5). If R_α can be appropriately bounded, we are reduced to dealing with G^*_α and the $q \equiv 0$ problem. To this end, according to (7), we calculate $\|G^*\|_{p,x} \equiv \left(\int |G^*(x,x')|^p dx\right)^{\frac{1}{p}}$ explicitly. The L^1 and L^∞ norms are

$$\|G^*_\alpha(x,x')\|_{1,x} = O\left(\frac{|\alpha|^{\frac{1}{2}}}{|\mathrm{Re}\,\alpha^{-\frac{1}{2}}|}\right) \quad (|\alpha| \to 0; |\arg\alpha| \le \gamma_1 < \gamma); \quad \|G^*_\alpha\|_{\infty,x} = O(\sqrt{\alpha}).$$

By interpolation,

$$\|G^*_\alpha(x,x')\|_{p,x} = O\left(|\alpha|^{\frac{1}{2}}\left|\mathrm{Re}\,\alpha^{-\frac{1}{2}}\right|^{-\frac{1}{p}}\right) \quad (13)$$

uniformly in $|\arg\alpha| \le \gamma_1$. On the other hand,

$$\|G_\alpha\|_{1,x} - \|G^*_\alpha\|_{1,x} \le \|G_\alpha - G^*_\alpha\|_{1,x} \le \|q\|_\infty C\|G_\alpha\|_{1,x'}$$

where C is chosen such that

$$\|G^*_\alpha(x,x')\|_{1,x} \le C \qquad (x' \in \mathbf{R}^+). \quad (14)$$

Hence $\|G_\alpha\|_{1,x} \le \frac{C}{1-C\|q\|_\infty}$. A similar argument shows

$$\|G^*_\alpha - G_\alpha\|_\infty = O(|\alpha|), \quad (15)$$

and thus $\|G_\alpha\|_\infty = O(\sqrt{|\alpha|})$, so that (13) holds as well for G_α.

Some calculation shows that along the contour Γ, for $|z|$ large,

$$\left\|G_{-\frac{\eta}{z}}\right\|_{p,x} = O\left(\eta^{\frac{1}{2}+\frac{1}{p}}|z|^{-\frac{1}{2}+\frac{1}{2p}}(\ln|z|)^{-\frac{1}{p}}\right). \quad (16)$$

By (5) and (16)

$$\left|\frac{z}{\eta}\right|\left\|R_{-\frac{\eta}{z}}\right\|_{p,q} = O\left(\eta^{\frac{1}{2}}|z|^{\frac{1}{2}}(\ln|z|)^{-1}\right).$$

Thus by Hölder's inequality

$$\left\| \frac{1}{2\pi i} \int_\Gamma \frac{g(z)}{z} \int_0^\infty \frac{-z}{\eta} R_{-\frac{\eta}{z}}(x,x') f(x') dx' dz \right\|_p \leq O\left(\eta^{\frac{1}{2}}\right) \int_\Gamma \left| \frac{g(z)}{z} \right| 0\left(z^{\frac{1}{2}} (ln|z|)^{-1} \right) dz \underset{\eta \to 0}{\longrightarrow} 0. \tag{17}$$

To prove (i) it now remains to consider (12) with G^* replacing G. In this case $u(x,\lambda) = C_1(\lambda)e^{i\sqrt{\lambda}x} + C_2(\lambda)e^{-i\sqrt{\lambda}x}$, and $b = 0$. Some substitutions show that the left side of (12) is a convolution:

$$\int_0^\infty g(\lambda\eta) F(\lambda) u(x,\lambda) d\rho(\lambda) = \chi_\eta(x) * \int_0^\infty F(\lambda) u(x,\lambda) d\rho(\lambda)$$

where

$$\chi_\eta(x) = \frac{1}{2\pi} \int_{-\infty}^\infty g(k^2\eta) e^{ikx} dk.$$

By the harmonic analysis theorem in §2 it now suffices to prove that $\chi_\eta(x)$ has an L^1 radially decreasing bound, which can be done by explicit computations. Thus

$$\int_0^\infty g(\lambda\eta) F(\lambda) u(x,\lambda) d\rho(\lambda) \underset{\eta \to 0}{\longrightarrow} f(x), \tag{18}$$

from which (i), (ii) and Theorem 2 follow.

§4. Some Remarks on Stable Summability

Much of the motivation for study of summability theory in recent years has arisen in the context of stable summability [4,5,6,9]. This is used in the study of stability of pointwise values of eigenfunction expansions under perturbations of coefficients. Let $\phi(\epsilon A)$ denote the operator in (3). Then ϕ-summation is *pointwise stable* if there exists a non-trivial scaling $\gamma(\epsilon)$ such that for any net $\{f^\gamma\}_{\gamma>0}$ of L^2 functions such that $\|f^\gamma - f\|_2 \leq \gamma$, $\phi(\epsilon A) f^{\gamma(\epsilon)}(x) \underset{\epsilon \to 0}{\longrightarrow} f(x)$ on the Lebesgue set of f.

If error causes the generalized Fourier coefficients of f to be measured as those of f^γ (the L^2 error in Fourier transform space is still $\|f^\gamma - f\|_2$), then pointwise error can be made to vary continuously with L^2 error, (and in particular be small when L^2 error is small) through use of a summability method ϕ and an appropriate scaling $\gamma(\epsilon)$. With probabilistic estimation of L^2 error γ [1], this procedure can be applied, for example, in the recovery of local properties of signals from global properties. The following theorem translates statements on summability into ones on stable summability; it is proved in [4].

THEOREM : *Let $\phi(\lambda)$ be a real-valued function and $f, f^\gamma \in L^2(\mathbb{R}^+)$ (or $L^2[a,b]$). Let*

(i) $\|f - f^\gamma\|_2 \leq \gamma$

(ii) $\int_0^\infty \frac{\phi^2(\lambda)}{\sqrt{\lambda}} d\lambda < \infty$

(iii) $\phi(\epsilon A)f(x) \underset{\epsilon \to 0}{\to} f(x)$ *uniformly on a bounded subset E of $\mathbb{R}^+$ (or $[a,b]$).*

(iv) α be a function of γ such that as $\gamma \to 0, \alpha \to 0$ and $\frac{\gamma}{\alpha^{\frac{1}{4}}} \to 0$.

Then $\phi(\epsilon A)f^{\gamma(\epsilon)}(x) \underset{\epsilon \to 0}{\to} f(x)$ uniformly on E; in particular if E can be an arbitrary point in the Lebesgue set of f, $\phi(\epsilon A)$ is a stable summation method. The scaling in (iv) is the best possible.

In Theorems 1 and 2 hypothesis (iii), a priori summability, is verified, giving the following:

COROLLARY : *If $\phi(\lambda)$ satisfies the hypotheses of Theorems 1 and 2, and condition (ii) above, then ϕ is a stable summation method with optimal scaling* $\frac{\gamma}{\alpha^{\frac{1}{4}}} \to 0$.

This result was proved for resolvent summability on intervals in [9].

REFERENCES

[1] Arsenin, V.I., Optimal Summation of Fourier Series with Approximate Coefficients, Soviet Math. Dokl. 9(1968), p. 1345.

[2] Arsenin,V.I., and A.N. Tikhonov, Solutions of Ill Posed Problems, V.H. Winston & Sons, 1977

[3] Benzinger, H.E., Perturbation of the Heat Equation, Jnl. Diff. Eq. 32(1979), p. 398.

[4] Diamond, H., M.A. Kon, and L. Raphael, Stable Summation Methods for a Class of Singular Sturm-Liouville Expansions, PAMS 81(1981), p. 279.

[5] Diamond, H., M.A. Kon, and L. Raphael, A Regularization of the Pointwise Summation of Singular Sturm-Liouville Expansions, Proceedings of the International Conference on Ill Posed Problems, to appear.

[6] Krukovskii, N.M., On the Tikhonov-Stable Summation of Fourier Series with Perturbed Coefficients by Some Regular Methods, Vestnik Mosk. Univ. Mat., 28, No. 3, p. 17.

[7] Levitan, B.M., and I.S. Sargsjan, Introduction to Spectral Theory: Self-Adjoint Ordinary Differential Operators, AMS, 1975.

[8] Stein, E. and G. Weiss, Introduction to Fourier Analysis on Euclidean Spaces, Princeton University Press, 1971.

[9] Tikhonov, A.N., Stable Methods for the Summation of Fourier Series, Soviet Math. Dokl., 5(1964), p. 641.

[10] Titchmarsh, E.C., Eigenfunction Expansions, Part 1, Oxford, 1962.

[11] Zygmund, A., Trigonometric Series, Cambridge, 1959.

QUALITATIVE THEORY OF HYPERBOLIC CHARACTERISTIC INITIAL VALUE PROBLEMS[1)]

Kurt Kreith

For ordinary differential equations there is an extensive literature dealing with the oscillation and asymptotic behavior of solutions of $u'' + f(t,u) = 0$. Despite physical interpretations which suggest their importance, corresponding questions for hyperbolic equations have received surprisingly scant attention.

The nonlinear telegraph equation

(1) $$\frac{\partial^2 u}{\partial x \partial y} + f(x,y,u) = 0 \quad ; \quad x \geq 0,\ y \geq 0$$

describes the small displacements of an infinite string of unit tension and density. A unique solution of (1) is determined by characteristic initial data of the form

(2) $$u(x,0) = p(x) \quad ; \quad u(0,y) = q(y),$$

where $p(x)$ and $q(y)$ are assumed continuous on $[0,\infty)$ and satisfying $p(0) = q(0)$. If $f(x,y,u)$ is continuous in $R^+ \times R^+ \times R$ and satisfies $uf(x,y,u) > 0$ for $u \neq 0$, then f can be regarded as a nonlinear restoring force. When this force is "sufficiently large" then solutions of (1), (2) can be expected to exhibit an oscillatory form of behavior, even though $p(x)$ and $q(y)$ are of constant sign for $x > 0$ and $y > 0$.

Alternatively, if $|f(x,y,u)|$ is "sufficiently small", then (1) can be regarded as a perturbation of the wave equation

(3) $$u_{xy} = 0,$$

whose solution $p(x) + q(y)$ retains certain of its qualitative features when (3) is replaced by (1).

In the linear case, oscillation properties of

(4) $$u_{xy} + f(x,y)u = 0$$

have been established by the author [2], [3] and Pagan [7], [8] using Sturm-type

---

1) Research supported by a grant of the National Science Foundation, NSF MCS 80-02130.

comparison theorems. Specifically, if $f(x,y) \geq F(x,y)$ in a domain D and

(5) $$v_{xy} + F(x,y)v = 0$$

has a solution satisfying $v = 0$ on the spacelike part of ∂D, then solutions of (4) satisfying prescribed conditions on the characteristic or timelike part of ∂D can be shown to change sign in D. Pagan [7] applied this technique to characteristic initial value problems with $f(x,y) \geq F(x,y) \equiv k^2 > 0$, in which case the solution $v = \sin(Lx + k^2y/L)$ is zero on the spacelike boundary of

$$D_n(L) = \{(x,y) \mid x > 0,\ y > 0,\ n\pi < Lx + k^2y/L < (n+1)\pi\}$$

for arbitrary $L > 0$ and $n = 1,2,\cdots$. It follows that if $f(x,y) \geq k^2 > 0$ and $p'(x) \leq 0$, $q'(y) \leq 0$ in $R^+ \times R^+$, then every solution of (4) must change sign in each $D_n(L)$.

To address the corresponding question for functions $f(x,y)$ which are <u>not</u> bounded away from zero, it is natural to consider $F(x,y) = G(xy)$, in which case $s = xy$ transforms (5) into

(5') $$\frac{d}{ds}(s\,\frac{dV}{ds}) + G(s)V = 0.$$

Assuming (5') to be oscillatory, the zeros $s_1 < s_2 < \cdots$ of $V(s)$ define a sequence of unbounded nodal domains for $v(x,y) = V(xy)$ such that $v(x,y)$ vanishes on the spacelike boundary of

$$D_n = \{(x,y) \mid x > 0,\ y > 0,\ s_{n-1} < xy < s_n\}.$$

However, the unboundedness of the D_n precludes a direct application of the technique of [7].

This difficulty can be overcome by noting that for arbitrary constants $\xi > 0$, $\eta > 0$, the function $w(x,y) = v(x+\xi, y+\eta)$ satisfies $w_{xy} + G((x+\xi)(y+\eta))w = 0$ and that the corresponding nodal domains

(6) $$\tilde{D}_n = \{(x,y) \mid x > 0,\ y > 0;\ s_{n-1} < (x+\xi)(y+\eta) < s_n\}$$

<u>are</u> bounded. This observation enabled Pagan and the author to establish the following.

<u>Theorem 1. Suppose there exist constants</u> $\xi > 0$ <u>and</u> $\eta > 0$ <u>and a continuous function</u> $g(s)$ <u>such that</u>

(i) $\frac{d}{ds}(s\,\frac{dV}{ds}) + g(s)V = 0$ <u>is oscillatory,</u>

<u>and</u>

(ii) $f(x,y) \geq g((x+\xi)(y+\eta))$ *in* $R^+ \times R^+$.

If $u(x,y)$ *is a solution of* (2), (4) *with positive characteristic initial data satisfying* $p'(x) \leq 0$, $q'(y) \leq 0$ *in* R^+, *then* $u(x,y)$ *changes sign in each* $\tilde{D}_n$.

A different approach to nonlinear hyperbolic oscillations was introduced by Yoshida in [10]. Defining

$$U(t) = \int_0^t u(t-y,y)dy \tag{7}$$

for a solution $u(x,y)$ of (1), where $f(x,y,u)$ is assumed to satisfy

$$f(x,y,u) \geq h(x+y)\varphi(u) > 0 \quad \text{in } R^+ \times R^+ \times R \tag{8}$$

and φ is convex in u, it is possible to impose hypotheses on h which assure that $U(t)$, and therefore $u(x,y)$, changes sign.

Such "averaging techniques" have recently been applied in more general situations by Kusano, Yoshida, and the author. For example, if (7) is applied to

$$u_{xy} + f(x,y,u) = F(x,y), \tag{9}$$

subject to characteristic initial data (2), one obtains the ordinary differential inequality

$$(tU)'' + th(t)\varphi(U) \leq p(t) + q(t) + \int_0^t F(t-\tau,\tau)d\tau$$

to which recent oscillation criteria of Kusano and Naito [5] can be applied, yielding results such as the following.

Theorem 2. If for large T

$$\liminf_{t \to \infty} \int_T^t (1 - \frac{s}{t})[p(s) + q(s) + \int_0^s F(s-\sigma,\sigma)d\sigma]ds = -\infty$$

and

$$\limsup_{t \to \infty} \int_T^t (1 - \frac{s}{t})[p(s) + q(s) + \int_0^s F(s-\sigma,\sigma)d\sigma]ds = +\infty$$

then every solution of (9), (2) *has a zero in each*

$$Q_\rho \equiv \{(x,y) \in R^2 | \rho < x+y < \infty\}.$$

Similar techniques can also be applied to boundary value problems associated with

$$U_{tt} - \Delta u + f(t,x,u) = F(t,x) \tag{10}$$

in a cylinder $(0,\infty)\times G \subset R^{n+1}$. Imposing the boundary condition

$$\frac{\partial u}{\partial \nu} = g(t,x) \quad \text{on } \partial G \tag{11}$$

the inequality (8) and the transformation

$$U(t) = \frac{1}{|G|}\int_G u(t,x)dx \quad ; \quad |G| = \text{volume of } G \tag{12}$$

yields

$$U'' + h(t)\varphi(U) \le \frac{1}{|G|}\int_{\partial G} g(t,x)d\bar{x} + \frac{1}{|G|}\int_G F(t,x)dx,$$

to which the previously cited theorem of Kusano and Naito [5] can also be applied, yielding analogous oscillation criteria for (10), (11).

In case $(0,\infty)\times G$ is replaced by a smooth non-cylindrical domain $H \subset R^+ \times R^n$, similar techniques apply in case (11) is replaced by

$$u = 0 \quad \text{on } \partial H \tag{13}$$

and ∂H is timelike for all $t > 0$. Defining $G_\tau = \{(t,x) \in H | t = \tau\}$, the transformation

$$U(t) = \frac{1}{|G_t|}\int_{G_t} u(t,x)dx \tag{12'}$$

and an n-dimensional version of Leibniz's rule yield analogous oscillation criteria for (10), (13) in H.

Turning to questions of asymptotic behavior, we recall [9] that a disconjugate n^{th} order homogeneous ordinary differential equation $\ell n\ u = 0$ has an asymptotically ordered fundamental system of solutions $u_1(x),\dots,u_n(x)$ with the property that $\lim_{x\to\infty} \frac{u_i(x)}{u_{i+1}(x)} = 0$ for $i = 1,\dots,n-1$. For $n = 2$, the question of when the perturbed equation $\ell_n u + f(x,u) = 0$ has solutions with corresponding asymptotic properties goes back to Atkinson [1] and Nehari [6]; very general results of this sort (for functional equations of order n) have recently been given by Kitamura and Kusano [4].

A natural hyperbolic analogue (currently being studied by C. A. Swanson and the author) includes (1) when the initial data $p(x)$ and $q(y)$ are bounded on $[0,\infty)$. In this instance the wave equation $u_{xy} = 0$ has the bounded solution $u(x,y) = p(x) + q(y)$ and it is natural to seek conditions on $f(x,y,u)$ which assure that the solution of (1), (2) is also bounded for large values of x and y.

Writing (1), (2) as an integral equation

$$u(x,y) = p(x) + q(y) - \int_0^x \int_0^y f(\xi,\eta,u(\xi,\eta))d\xi\, d\eta$$

suggests one form of boundedness criterion analogous to those which exist for ordinary differential equations.

Theorem 3. *If $f(x,y,u)$ is super- or sublinear in u and if there exists a constant $c > 0$ such that $f(x,y,c) > 0$ in $R^+ \times R^+$ and*

$$\int^\infty \int^\infty f(x,y,c)d\xi\, d\eta < \infty,$$

then there exist constants $X > 0$, $Y > 0$ and a solution $u(x,y)$ of (1) which is bounded in $\{(x,y)|x \geq X,\ y \geq Y\}$.

The proof consists of showing that

$$\Phi[u] = 2c - \int_X^x \int_Y^y f(\xi,\eta,u(\xi,\eta))d\xi\, d\eta$$

has a fixed point in a Banach space of functions which are continuous in $\{(x,y)|x \geq X,\ y \geq Y\}$ and sufficiently close to $u_0(x,y) \equiv c$ in the sup norm. The existence of such a fixed point follows from the Schauder theorem when X and Y are chosen sufficiently large.

These techniques allow a natural generalization to $\frac{\partial^{m+n}u}{\partial x^m \partial y^n} + f(x,y,u) = 0$, in which case more varied forms of asymptotic behavior must be considered. However, the hyperbolic case also involves some new difficulties insofar as the necessity conditions, which are readily established for ODEs, seem to be considerably more difficult. Furthermore, the existence of bounded solutions "for large x and y" does not insure the existence of a solution bounded for <u>all</u> x and y. Such questions suggest that a hyperbolic generalization of the asymptotic theory of [4] may lead to a number of new problems which will be interesting in their own right.

References

1. F. Atkinson, On second-order nonlinear oscillations, Pacific J. Math. 5(1955), 643-647.
2. K. Kreith, Sturmian theorems for hyperbolic equations, Proc. Amer. Math. Soc. 22(1969), 277-281.
3. ________, Sturm theory for partial differential equations of mixed type, Proc. Amer. Math. Soc. 81(1981), 75-78.
4. Y. Kitamura and T. Kusano, Nonlinear oscillation of higher order functional differential equations with deviating arguments, J. Math. Anal. and Appl., to appear.

5. T. Kusano and M. Naito, Oscillation criteria for a class of perturbed Schrodinger equations, to appear.

6. Z. Nehari, On a class of nonlinear second order differential equations, Trans. Amer. Math. Soc. 95(1960), 101-123.

7. G. Pagan, Oscillation theorems for characteristic initial value problems for linear hyperbolic equations, Rend. Accad. Naz. Lincei 55(1973), 301-313.

8. ______, An oscillation theorem for characteristic initial value problems in linear hyperbolic equations, Proc. Royal Soc. Edinburgh 77A(1977), 265-271.

9. W. Trench, Canonical forms and principal systems for general disconjugate equations, Trans. Amer. Math. Soc. 189(1974), 319-327.

10. N. Yoshida, An oscillation theorem for characteristic initial value problems for nonlinear hyperbolic equations, Proc. Amer. Math. Soc. 76(1979), 95-100.

APPLICATIONS OF A COMPARISON THEOREM FOR QUASI-ACCRETIVE OPERATORS IN A HILBERT SPACE

Roger T. Lewis *

1. Introduction

A considerable amount of activity has been devoted to the study of the qualitative properties of the spectra of differential operators in a Hilbert space. The study is included in what has been said to be the three basic mathematical problems in quantum mechanics: selfadjointness, spectral analysis, and scattering theory (Reed and Simon [17]). In defiance of an impressive history, many important aspects of the study remain open.

A typical problem, which includes the work discussed in this lecture, is to determine characteristics of the spectrum—purely discrete, purely continuous, length of gaps, bounds, etc.—in terms of the coefficients of the differential operator. When the differential operator, $T: H \to H$, is selfadjoint, $T = T^*$, the spectrum consists of only real numbers that belong to either the essential spectrum $\sigma_e(T)$ or the set of isolated eigenvalues of finite multiplicity [19, p. 36]. This particular case has received the most attention. We refer to the books of Dunford and Schwartz [2], Glazman [7], Müller-Pfeiffer [14], Naimark [15], Reed and Simon [18], Schechter [19], and the references contained therein for many selfadjoint results. The nonselfadjoint problem, which has received a limited amount of attention in the past (see the references above), has recently been investigated by Evans [3, 4, 5], Evans, Kwong, and Zettl [6], Knowles and Race [10], and Race [16].

In the study of the nonselfadjoint problem, it is natural to hope to use the extensive literature that already exists for selfadjoint problems. In this lecture we present a comparison theorem, which is proved in [13], that establishes lower bounds on the real or imaginary parts of λ in the essential spectrum of a closed quasi-accretive operator T in a Hilbert space. By comparing Re(Tu,u) or Im(Tu,u) with quadratic forms associated with selfadjoint operators, the real or imaginary parts of $\lambda \in \sigma_e(T)$ are shown to be bounded below by the least limit point of the essential spectrum of the selfadjoint operators. For the sake of completeness, we outline the proofs here. In [13], the result is illustrated by applying it to certain second-order partial differential operators with Dirichlet boundary conditions. There, the application depended on a lemma of Wolf [23, Lemma 2.4] that requires Dirichlet boundary conditions, but it resulted in sharp criteria that are quite accessible We pursue here the problem with more general boundary conditions. The applications in [13] and here are related to recent work of Evans [5] and the author [11,12].

2. A Comparison Theorem

Let $T: H \to H$ be a linear operator in the Hilbert space H with a domain $\mathcal{D}(T)$ that is dense in H. If the numerical range, $\Theta(T) = \{(Tu,u): u \in \mathcal{D}(T), \|u\| = 1\}$, of T is a subset of a half-plane $\{\xi \in \mathbb{C}: \operatorname{Re} \xi \geq \gamma\}$ for some real number γ, then T is said to be *quasi-accretive*. (If $\gamma = 0$, T is said to be *accretive* and -T is *dissipative*.) As a consequence, T is closable [9, p. 268]. We denote the closure of T by $\tilde{T}$.

If a sequence $\{u_n\} \in \mathcal{D}(T)$ that is normal in H has the property that $\{u_n\}$ has no convergent subsequence and $\lim_{n\to\infty} Tu_n = 0$, then $\{u_n\}$ is said to be a *singular sequence* of T. We take the *essential spectrum* $\sigma_e(T)$ of a closed, densely defined operator T to be the set of all complex numbers λ for which $T - \lambda I$ has a singular sequence. Kato [9, p. 243] defines the essential spectrum to be the complement of the semi-Fredholm domain of T, which implies that $T - \lambda I$ has a singular sequence for λ in that set [9, p. 233].

If T is selfadjoint, $T = T^*$, then the spectrum, which contains only real numbers, consists of the union of isolated eigenvalues of finite multiplicity and $\sigma_e(T)$ — see Schechter [19, p. 36]. For a good discussion of the different definitions of the essential spectrum and how they compare, the reader should consult Gustafson and Weidmann [8].

THEOREM 1. *Let* T *be a closed linear operator in the Hilbert space* H *with domain* $\mathcal{D}(T)$ *that is dense in* H. *Let* A *be a selfadjoint operator in* H *that is bounded below with* $\mathcal{D}(T) \subseteq \mathcal{D}(A)$. *If*

$$\mathrm{Re}(Tu,u) \geq (Au,u) \ , \qquad u \in \mathcal{D}(T), \qquad (1)$$

then

$$\sigma_e(T) \subseteq \{\lambda \in \mathbb{C}: \mathrm{Re}\ \lambda \geq \inf \sigma_e(A)\}.$$

If $\sigma_e(A) = \emptyset$ *then* $\sigma_e(T) = \emptyset$.

REMARK. Let B be a bounded-below, selfadjoint operator in H such that $\mathcal{D}(T) \subseteq \mathcal{D}(B)$ and $\mathrm{Im}(Tu,u) \geq (Bu,u)$ for all $u \in \mathcal{D}(T)$. The proof of Theorem 1 given below can be adapted to show that

$$\sigma_e(T) \subseteq \{\lambda \in \mathbb{C}: \mathrm{Im}\ \lambda \geq \inf \sigma_e(B)\}.$$

We refer the reader to [9 , p. 330 and 336] for a discussion of results that are related to Theorem 1 (and Theorem 2 below). When T is selfadjoint, Theorem 1 follows easily from Theorem 12 of Glazman [7 , p. 15] or via form theory from Lemma 3.5 of Schechter [20].

PROOF OF THEOREM 1. If $\lambda \in \sigma_e(T)$ then there is a singular sequence $\{u_j\} \subseteq \mathcal{D}(T)$ of $T - \lambda I$. Let $\alpha = \inf \sigma_e(A)$. (In the case of $\sigma_e(A) = \emptyset$, let α be an arbitrarily large number and the fact that $\sigma_e(\tilde{T}) = \emptyset$ will follow.) For a given $\varepsilon > 0$, let G(A) be the finite-dimensional subspace in H that is spanned by the eigenfunctions corresponding to the finite number of eigenvalues of A in the interval $(-\infty,\alpha-\varepsilon)$ [7, p. 15].

For each u_j, let $u_j = u_j^{(1)} + u_j^{(2)}$ where $u_j^{(2)} \in G(A)$ and $u_j^{(1)} \in G(A)^{\perp}$, the orthogonal complement of $G(A)$ in H. Since $1 = \|u_j\|^2 = \|u_j^{(1)}\|^2 + \|u_j^{(2)}\|^2$, then $\{u_j^{(2)}\}$ is a bounded infinite sequence in a finite dimensional subspace $G(A)$ of H, which implies that $\{u_j^{(2)}\}$ has a convergent subsequence. Let $\{u_j'\}$ be a subsequence of $\{u_j\}$ such that $\{u_j^{(2)'}\}$ converges. Since $\{u_j'\}$ cannot be Cauchy in H then for some $\delta > 0$ there is a subsequence, which we still name $\{u_j'\}$, such that $\|u_j' - u_{j+1}'\| \geq \delta$ for all j. Using $\{u_j'\}$ we form the normalized sequence $\{v_j\}$ by letting

$$v_j = (u_j' - u_{j+1}')/\|u_j' - u_{j+1}'\|$$

with $v_j^{(1)} = \left(u_j^{(1)'} - u_{j+1}^{(1)'}\right)/\|u_j' - u_{j+1}'\| \in G(A)^{\perp}$ and $v_j^{(2)} = \left(u_j^{(2)'} - u_{j+1}^{(2)'}\right)/\|u_j' - u_{j+1}'\| \in G(A)$. Since $\{u_j^{(2)'}\}$ converges then $\|v_j^{(2)}\| \leq \|u_j^{(2)'} - u_{j+1}^{(2)'}\|/\delta \to 0$ as $j \to \infty$. Also, either $(T - \lambda I)v_j \to 0$ or $(T^* - \overline{\lambda} I)v_j \to 0$ as $j \to \infty$.

Let μ be a lower bound for A, then by the spectral theorem for A and (1)

$$\begin{aligned}
\|(T - \lambda)v_j\| + \operatorname{Re}\lambda &\geq \left|\left((T - \lambda)v_j, v_j\right)\right| + \operatorname{Re}\lambda \\
&\geq \operatorname{Re}(Tv_j, v_j) \\
&\geq (Av_j, v_j) \\
&= \left(A(v_j^{(1)} + v_j^{(2)}), v_j^{(1)} + v_j^{(2)}\right) \\
&= (Av_j^{(1)}, v_j^{(1)}) + 2(Av_j^{(2)}, v_j^{(1)}) + (Av_j^{(2)}, v_j^{(2)}) \\
&\geq (\alpha - \varepsilon)(v_j^{(1)}, v_j^{(1)}) + \mu(v_j^{(2)}, v_j^{(2)}) + 2(Av_j^{(2)}, v_j^{(1)}) \\
&= (\alpha - \varepsilon) + (\mu - \alpha + \varepsilon)(v_j^{(2)}, v_j^{(2)}) + 2(Av_j^{(2)}, v_j^{(1)}).
\end{aligned}$$

Since $v_j^{(2)} \in G(A)$, then for some k, $v_j^{(2)} = \sum_{i=1}^{k} c_i \phi_i$ where each ϕ_i is an eigenfunction in $G(A)$ corresponding to an eigenvalue $\lambda_i < \alpha - \varepsilon$ of A. Hence,

$$(Av_j^{(2)}, v_j^{(1)}) = \sum_{i=1}^{k} c_i \lambda_i(\phi_i, v_j^{(1)}) = 0.$$

Now, $v_j^{(2)} \to 0$ and $(T - \lambda)v_j \to 0$ as $j \to \infty$.

Therefore $\text{Re } \lambda \geq \alpha$ since ε is arbitrary, which completes the proof.

In applying Theorem 1 to differential operators below, it is sometimes easier to establish the required inequality (1) for a closable operator T. The next theorem shows that with proper adjustments we can still make the conclusion concerning the location of $\sigma_e(\tilde{T})$.

We let T_{t} denote the m-sectorial operator associated with the sectorial sesquilinear form t as guaranteed by the first representation theorem [9, p. 322].

LEMMA 1. [13] Let T be a closable operator with closure $\tilde{T}$ and let s be a closed symmetric form that is bounded below. If $\mathcal{D}(T) \subseteq \mathcal{D}(s)$ and $\text{Re}(Tu,u) \geq s[u]$ $\left(\text{Im}(Tu,u) \geq s[u]\right)$ for all $u \in \mathcal{D}(T)$ then $\mathcal{D}(\tilde{T}) \subseteq \mathcal{D}(s)$ and $\text{Re}(\tilde{T}u,u) \geq s[u]$ $\left(\text{Im}(\tilde{T}u,u) \geq s[u]\right)$ for all $u \in \mathcal{D}(\tilde{T})$.

THEOREM 2. *Let* T *be a closable linear operator in* H *with closure* $\tilde{T}$ *and domain* $\mathcal{D}(T)$ *which is dense in* H. *Let* s *be a closed, symmetric form that is bounded below with domain* $\mathcal{D}(s)$. *If* $\mathcal{D}(T) \subseteq \mathcal{D}(s)$ *and*

$$\text{Re}(Tu,u) \geq s[u], \qquad u \in \mathcal{D}(T),$$

then

$$\sigma_e(\tilde{T}) \subseteq \{\lambda \in \mathbb{C}: \text{Re } \lambda \geq \inf \sigma_e(T_s)\}.$$

If $\sigma_e(T_s) = \emptyset$ *then* $\sigma_e(\tilde{T}) = \emptyset$.

The Remark following Theorem 1 holds in a similar way here.

The proof of Theorem 2 follows from Lemma 1 and the fact that $\mathcal{D}(T_s)$ is a core of s [9 , p. 322]. Some of the ideas used in the above proofs were used in work of Evans [3] for certain special cases.

3. An Application

Let Ω be an open, connected set in $\mathbb{R}^n$ that is the union of bounded, open, connected sets $\{\Omega_k\}$ with $\Omega_k \subset \Omega_{k+1}$. Let Γ denote the boundary of Ω. Assume that each Ω_k lies on one side of its boundary Γ_k, which is C^1 (as a submanifold of $\mathbb{R}^n$). (This requirement allows an application of the Rellich compact imbedding theorem, which we need—see Showalter [21, p. 49] As an alternative, Agmon [1 , p. 30] only requires that Ω_k be bounded and have the segment property.)

The Sobolev space $H^m(\Omega)$ is defined to be the completion of $C^m(\bar{\Omega}) = \{f|_{\bar{\Omega}}: f \in C_0^m(\mathbb{R}^n)\}$ with respect to the norm $\|\cdot\|_{H^m(\Omega)}$ induced by the inner product

$$(f,g)_{H^m(\Omega)} = \sum \{\int_\Omega D^\alpha f \cdot \overline{D^\alpha g} : |\alpha| \le m\}.$$

Here, $f|_{\bar{\Omega}}$ denotes the restriction of f to the closure of the set Ω. Let m be the usual measure on $\mathbb{R}^n$ and let the real-valued weight function $w(x)$ be measurable on the measure space (Ω,m) and positive for almost every $x \in \Omega$. In addition, assume that $w(x)$ is bounded on each Ω_k. In this section we let $H = L_w^2(\Omega)$, the Hilbert space of all complex-valued functions $u(x)$ satisfying

$$\int_\Omega w(x)\ |u(x)|^2\ dx < \infty.$$

The norm $\|\cdot\|$ and inner product $(\cdot,\cdot)$ will refer to $L_w^2(\Omega)$.

If f is a real-valued, measurable function on the measure space (Ω,m) the <u>essential range of</u> f is the set of all real numbers t such that

$$m\{x \in \Omega: |f(x) - t| < \varepsilon\} > 0$$

for all $\varepsilon > 0$. We denote the essential range of f by $R_e(f)$. Define the multiplication operator $M_f: L_w^2(\Omega) \to L_w^2(\Omega)$ by $M_f u = fu$ with $\mathcal{D}(M_f) = \{u \in L_w^2(\Omega): fu \in L_w^2(\Omega)\}$. The next lemma can be found in the book by Reed and Simon [17, p. 229] .

LEMMA 2. The spectrum of M_f is the essential range of f; i.e. $\sigma(M_f) = \mathcal{R}_e(f)$.

The proof of Lemma 2 follows easily by showing that $\lambda \notin \mathcal{R}_e(f)$ if and only if λ is in the resolvent set of M_f.

A sesquilinear form $\tau(\cdot,\cdot)$ on a Hilbert space H is said to be H-coercive if there is a number $c > 0$ such that $|\tau[v]| \geq c\|v\|_H^2$ for all $v \in H$. A related notion is the idea of relative compactness of operators on H. Let T and A be operators on H with $\mathcal{D}(T) \subset \mathcal{D}(A)$. Suppose that for any sequence $\{u_n\} \subseteq \mathcal{D}(T)$ such that $\{u_n\}$ and $\{Tu_n\}$ are bounded, $\{Au_n\}$ contains a convergent subsequence. Then, A is said to be T-compact. It is well-known [9 , p. 244] for closed operators T and A that $\sigma_e(T) = \sigma_e(T + A)$ when A is T-compact. We use this fact in the next lemma.

LEMMA 3. Let $p(x)$ be a real-valued, measurable function that is bounded on each Ω_k and bounded below on Ω. Define

$$p_k(x) = \begin{cases} p(x) - \inf \mathcal{R}_e(p\restriction_{\Omega\sim\Omega_k}), & x \in \Omega_k \\ 0 & x \notin \Omega_k. \end{cases}$$

Define $M_k: L_w^2(\Omega) \to L_w^2(\Omega)$ by $M_k u = -p_k u$ with $\mathcal{D}(M_k) = L_w^2(\Omega)$. Let $T: L_w^2(\Omega) \to L_w^2(\Omega)$ have a domain $\mathcal{D}(T)$, which is dense in $L_w^2(\Omega)$, such that $u \in \mathcal{D}(T)$ implies that $u \in H^1(\Omega_k)$ for each k. If the form (Tu,u) is $H^1(\Omega_k)$-coercive then each M_k is T-compact.

PROOF. If $\{u_n\} \subseteq \mathcal{D}(T)$ then for each k there is a constant $c_k > 0$ such that

$$c_k\|u_n\|^2_{H^1(\Omega_k)} \leq |(Tu_n,u_n)| \leq \|Tu_n\|\,\|u_n\|.$$

If $\{u_n\}$ and $\{Tu_n\}$ are bounded, then $\{u_n\restriction_{\Omega_k}\}$ is a bounded sequence in $H^1(\Omega_k)$. By the Rellich compact imbedding theorem there is a subsequence $\{u_n'\}$ such that $\{u_n'\restriction_{\Omega_k}\}$ converges in $L^2(\Omega)$. Since $w(x)$ and $p(x)$

are bounded on Ω_k, then $\{M_k u'_n\}$ converges in $L^2_w(\Omega)$. Therefore, each M_k is T-compact.

A consequence of Lemma 3 is the fact that $\sigma_e(T) = \sigma_e(T + M_k)$ for each k assuming that T is closed and satisfies the hypothesis of Lemma 3. This fact produces the following corollary which is central in our applications.

COROLLARY TO THEOREM 2. *Let* $T: L^2_w(\Omega) \to L^2_w(\Omega)$ *be a closable linear operator with closure* $\tilde{T}$ *and domain* $\mathcal{D}(T)$ *that is dense in* $L^2_w(\Omega)$. *Assume that* $\mathcal{D}(\tilde{T}) \subseteq H'(\Omega_k)$ *and that* $\tilde{T}$ *is* $H^1(\Omega_k)$*-coercive for each* k. *Let* p(x) *and* q(x) *be real-valued measurable functions that are bounded on each* Ω_k *and bounded below on* Ω. *If*

$$\operatorname{Re}(Tu,u) \geq \int_\Omega w(x)p(x)\,|u(x)|^2\,dx, \qquad u \in \mathcal{D}(T), \tag{2}$$

then $\lambda \in \sigma_e(\tilde{T})$ *implies that* $\operatorname{Re}\lambda \geq \liminf_{k\to\infty} R_e(p\rceil_{\Omega\sim\Omega_k})$. *If*

$$\operatorname{Im}(Tu,u) \geq \int_\Omega w(x)q(x)\,|u(x)|^2\,dx, \qquad u \in \mathcal{D}(T) \tag{3}$$

then $\lambda \in \sigma_e(\tilde{T})$ *implies that* $\operatorname{Im}\lambda \geq \liminf_{k\to\infty} R_e(q\rceil_{\Omega\sim\Omega_k})$.

PROOF. Let M_k be defined as in Lemma 3. By (2)

$$\operatorname{Re}\big((T+M_k)u,u\big) \geq \inf R_e(p\rceil_{\Omega\sim\Omega_k}) \int_{\Omega_k} w(x)|u(x)|^2dx + \int_{\Omega\sim\Omega_k} w(x)p(x)|u(x)|^2dx,$$

i.e., if $\hat{p}_k(x)$ is defined to be p(x) for $x \in \Omega \sim \Omega_k$ and $\inf R_e(p\rceil_{\Omega\sim\Omega_k})$ for $x \in \Omega_k$ then

$$\operatorname{Re}\big((T + M_k)u,u\big) \geq \big(\hat{p}_k(x)u(x),u(x)\big), \qquad u \in \mathcal{D}(T).$$

By Theorem 2 and Lemma 2, $\sigma_e(\tilde{T} + M_k) \subseteq \{\lambda \in \mathbb{C}: \operatorname{Re}\lambda \geq \inf R_e(\hat{p}_k)\}$. Since M_k is $\tilde{T}$-compact then by Lemma 3 $\sigma_e(\tilde{T}) = \sigma_e(\tilde{T} + M_k)$ and the conclusion follows. The proof in the case of (3) is similar.

As an example of an application of the above corollary we consider a nonselfadjoint, second-order differential operator T. The boundary conditions required for domain elements of T include the Dirichlet, Neumann, and mixed

Dirichlet-Neumann boundary value problems as well as a certain weighted boundary condition. Let

$$T = \frac{1}{w(x)}\{-\sum_{i,j=1}^{n} \frac{\partial}{\partial x_i} a_{ij}(x)\frac{\partial}{\partial x_j} + \sum_{i=1}^{n} b_i(x)\frac{\partial}{\partial x_i} + c(x)\} \tag{4}$$

where the coefficients are complex-valued in Ω. Each $a_{ij}(x)$ has partial derivatives that are bounded on Ω_k for each k. Also, $c(x)$, each $a_{ij}(x)$, and each $b_i(x)$ are bounded on Ω_k for each k. Let $A(x) = (a_{ij}(x))$ and let $A^*(x)$ be the conjugate-transpose of $A(x)$. Let $\mu_R(x)$ be the minimum eigenvalue of $(A(x) + A^*(x))/2$ and let $\mu_I(x)$ be the minimum eigenvalue of $(A(x) - A^*(x))/2i$. The coefficients and the weight $w(x)$ may become unbounded for some x on the boundary Γ of Ω (Γ includes ∞ if Ω is unbounded). Similarly, $w(x)$ may vanish for some $x \in \Gamma$. All such singularities we denote by S and we require that $S \subset \Gamma \sim \Gamma_k$ for each k. For each k let $N_k \subseteq \mathbb{R}^n \sim \overline{\Omega}_k$ be a neighborhood of S.

Let $\gamma(s)$ be a piecewise smooth, nonnegative function on Γ. Let $\delta(s)$ assume only the value 1 or 0 on Γ and assume that $\gamma(s)$ and $\delta(s)$ are not both zero for any $s \in \Gamma$. For $\nu(s) = (\nu_1(s),\cdots,\nu_n(s))$ the unit outward normal at $s \in \Gamma$ define

$$\frac{\partial u(s)}{\partial \nu_A} = \sum_{i=1}^{n} (\sum_{j=1}^{n} a_{ij} \nu_j) \frac{\partial u}{\partial x_i} = (\nabla u)^t A \cdot \nu$$

for $s \in \Gamma$. We now define the domain of T to be

$$\mathcal{D}(T) = \{u: u = \phi|_\Omega \text{ for some } \phi \in C_0^\infty(\mathbb{R}^n \sim \overline{N}_k) \text{ and some } k;$$

$$Tu \in L_w^2(\Omega); \text{ and } \delta(s)\frac{\partial u(s)}{\partial \nu_A} + \gamma(s)u(s) = 0\}. \tag{5}$$

Note that each $u \in \mathcal{D}(T)$ vanishes in a neighborhood of the singularities of T. Below, we let $(\cdot,\cdot)_E$ and $|\cdot|_E$ denote the usual inner product and norm for vectors in Euclidean n-space.

LEMMA 4. If $u \in \mathcal{D}(T)$, then for any $\varepsilon > 0$

$$\operatorname{Re}(Tu,u) \geq \int_\Omega [\left(\mu_R(x) - \varepsilon \max_i |b_i(x)|\right)|\nabla u|_E^2 + \left(\operatorname{Re} c(x) - \varepsilon^{-1}\sum_{i=1}^n |b_i(x)|\right)|u|^2]dx \quad (6)$$

and

$$\operatorname{Im}(Tu,u) \geq \int_\Omega [\left(\mu_I(x) - \varepsilon \max_i |b_i(x)|\right)|\nabla u|_E^2 + \left(\operatorname{Im} c(x) - \varepsilon^{-1}\sum_{i=1}^n |b_i(x)|\right)|u|^2]dx. \quad (7)$$

PROOF. By the usual Stokes-type formula (see Treves [22, p. 78 and p. 360])

$$(Tu,v) = \int_\Omega [\left(A(x)\nabla u, \nabla v\right)_E + \sum_{i=1}^n b_i(x)\frac{\partial}{\partial x_i} u\,\overline{v} + c(x)u\,\overline{v}]dx - \int_\Gamma \frac{\partial u(s)}{\partial \nu_A}\overline{v}(s)ds$$

for $u, v \in \mathcal{D}(T)$. Since $\operatorname{Re}(A\nabla u,\nabla u)_E = \frac{1}{2}\{(A\nabla u,\nabla u)_E + \overline{(A\nabla u,\nabla u)_E}\} = \frac{1}{2}\left((A + A^*)\nabla u,\nabla u\right)_E$, then

$$\operatorname{Re}(Tu,u) \geq \int_\Omega [\mu_R(x)|\nabla u|_E^2 + \operatorname{Re}\sum_{i=1}^n b_i(x)\frac{\partial}{\partial x_i} u\,\overline{u} + \operatorname{Re} c(x)|u|^2]\,dx$$

for all $u \in \mathcal{D}(T)$. If we let $u_1 = \operatorname{Re} u$ and $u_2 = \operatorname{Im} u$ then calculations show that

$$\operatorname{Re}(b_i \frac{\partial}{\partial x_i} u\,\overline{u}) = \operatorname{Re} b_i(u_1 \frac{\partial}{\partial x_i} u_1 + u_2 \frac{\partial}{\partial x_i} u_2) - \operatorname{Im} b_i(u_1 \frac{\partial}{\partial x_i} u_2 - u_2 \frac{\partial}{\partial x_i} u_1).$$

Since $2ab \leq \varepsilon a^2 + \varepsilon^{-1}b^2$ for any $\varepsilon > 0$ and arbitrary real numbers a and b, then

$$\operatorname{Re} b_i \frac{\partial}{\partial x_i} u\,\overline{u} \geq -(|\operatorname{Re} b_i| + |\operatorname{Im} b_i|)(\varepsilon|\frac{\partial}{\partial x_i} u|^2 + \varepsilon^{-1}|u|^2)/2.$$

Since $2^{-1}(|\operatorname{Re} b_i| + |\operatorname{Im} b_i|) \leq |b_i|$, then inequality (6) now follows. Inequality (7) can be established by an identical procedure.

LEMMA 5. [12] Let $g \in C^2(\Omega)$ be a real-valued function with a nonzero Laplacian on Ω. Then

$$\int_\Omega |\Delta g(x)|\ |\phi(x)|^2\,dx \leq 2\int_\Omega |\phi(x)|\ |\nabla g(x)|_E\ |\nabla\phi(x)|_E\,dx$$

$$\leq 4\int_\Omega |\Delta g(x)|^{-1}\ |\nabla g(x)|_E^2\ |\nabla\phi(x)|_E^2\,dx \quad (8)$$

for all $\phi(x) \in C_o^1(\mathbb{R}^n)$ that satisfy

$$(-1)^\eta \int_\Gamma \nabla g(s)\cdot\nu(s)\ |\phi(s)|^2\,ds \leq 0 \quad (9)$$

where $\eta = 0$ if $\Delta g > 0$ on Ω and $\eta = 1$ if $\Delta g < 0$ on Ω.

For Dirichlet boundary conditions $(\delta(s) \equiv 0,\ \gamma(s) \equiv 1)$, (9) is satisfied for any $g(x) \in C^2(\Omega)$ with $\Delta g \neq 0$ on Ω. More generally, we need either $\phi(s) = 0$ or

$$(-1)^n \nabla g(s)\cdot\nu(s) \leq 0$$

for $s \in \Gamma$. For any $g \in C^2(\Omega)$ with nonzero Laplacian on Ω we partition the boundary Γ into subsets Γ_g^+ and Γ_g^- according to whether the dot product $(-1)^n \nabla g(s)\cdot\nu(s)$ is positive or nonpositive, respectively, at $s \in \Gamma$. For example, if $g(x) = \alpha(|x|)$ for some positive-valued function $\alpha \in C^1(\mathbb{R})$ with positive-valued first and second derivatives, then

$$(-1)^n \nabla g \cdot \nu(s) = |s|^{-1}\alpha'(|s|)(s_1,\cdots,s_n)\cdot\nu(s).$$

Hence, $s \in \Gamma_g^+$ if, and only if the angle θ between the point vector $(s_1,\cdots,s_n)$ and the unit outward normal $\nu(s)$ at $s \in \Gamma$ satisfies the inequality $\pi > \theta > \pi/2$.

THEOREM 3. *Let* T *and* $\mathcal{D}(T)$ *be given by* (4) *and* (5). *Let* $g \in C^2(\Omega)$ *be a real-valued function with a nonzero Laplacian on* Ω. *Assume that for all* $u \in \mathcal{D}(T)$ $u(s) = 0$ *for all* $s \in \Gamma_g^+$ (*i.e.* $\delta(s) \equiv 0$ *on* Γ_g^+). *Suppose that there is an* $\varepsilon > 0$ *such that*

$$\left(\mu_R(x) - \varepsilon \max_i |b_i(x)|\right) \geq |\nabla g(x)|^2/|\Delta g(x)|, \qquad x \in \Omega,$$

and that $\operatorname{Re} c(x) - \varepsilon^{-1}\sum_{i=1}^{n} |b_i(x)|$ *is bounded below on* Ω. *If we define*

$$p(x) \equiv [|\Delta g(x)|/4 + \operatorname{Re} c(x) - \varepsilon^{-1}\sum_{i=1}^{n} |b_i(x)|]/w(x)$$

then $\lambda \in \sigma_e(\tilde{T})$ *implies that* $\operatorname{Re}\lambda \geq \liminf_{k\to\infty} R_e(p\rceil_{\Omega\sim\Omega_k})$.

PROOF. The proof follows from the corollary to Theorem 2 and Lemmas 4 and 5. Note that since $g \in C^2(\Omega)$ then $\mu_R(x) - \varepsilon \max_i |b_i(x)|$ must have a positive lower bound on each Ω_k. If $\operatorname{Re} c(x) - \varepsilon^{-1}\sum_{i=1}^{n} |b_i(x)|$ has a positive lower bound on each Ω_k then the coercivity of $\tilde{T}$ follows from (6) While this may not be true, our assumptions do require that

$\mathrm{Re}\ c(x) - \varepsilon^{-1} \sum_{i=1}^{n} |b_i(x)|$ be bounded below--say by β. The theorem would then follow for p replaced by $p + \beta/w$ and T replaced by $T + \beta/w$ from which the conclusion above could be drawn.

Obviously, results similar to Theorem 3 related to the imaginary parts of the coefficients of T hold as well. We refer the reader to [11,12,13] for choices of $g(x)$ which lead to more specific applications of Theorem 3.

REFERENCES

1. S. Agmon, *Elliptic Boundary Value Problems*, D. Van Nostrand Co., Inc., Princeton, N. J., 1965.

2. N. Dunford and J. Schwartz, *Linear Operators, Part II*, Interscience, New York, 1963.

3. W. D. Evans, *On the essential spectrum of second order degenerate elliptic operators*, J. London Math. Soc. (2), 8(1974), 463-482.

4. _____, *On the spectra of Schrödinger operators with a complex potential*, Math. Ann., 255(1981), 57-76.

5. _____, *On the spectra of non-self-adjoint realizations of second-order elliptic operators*, Proc. Royal Soc. Edinburgh, 90A(1981), 71-105.

6. W. D. Evans, M. K. Kwong, and T. Zettl, *On the spectra of $2n^{th}$-order differential operators*. Manuscript.

7. I. M. Glazman, *Direct Methods of Qualitative Spectral Analysis of Singular Differential Operators*, Israel Program for Scientific Translations, Jerusalem, 1965.

8. K. Gustafson and J. Weidmann, *On the essential spectrum*, J. Math. Analysis and Applications, 25(1969), 121-127.

9. T. Kato, *Perturbation Theory for Linear Operators*, 2nd ed., Springer-Verlag, Berlin, Heidelburg and New York, 1976.

10. I. Knowles and D. Race, *On the point spectra of complex Sturm-Liouville operators*, Proc. Royal Soc. Edinburgh, 85A(1979), 263-289.

11. R. T. Lewis, *The spectra of some elliptic operators of second order*, Spectral Theory of Differential Operators, (Math. Studies Series No. 55), North-Holland Pub. Co., Amsterdam, New York, Oxford, 1981.

12. _____, *Singular elliptic operators of second order with purely discrete spectra*, Transactions A. M. S., in press.

13. _____, *The essential spectrum of quasi-accretive operators*, Manuscript (submitted for publication).

14. E. Müller-Pfeiffer, *Spectral Theory of Ordinary Differential Operators*, Ellis Horwood Limited, Chichester, England, 1981.

15. M. A. Naimark, *Linear Differential Operators*, Part II, Frederick Ungar Pub. Co., New York, 1968.

16. D. Race, *On the essential spectra of linear* $2n^{th}$ *order differential operators with complex coefficients*, Manuscript.

17. M. Reed and B. Simon, *Methods of Mathematical Physics, Part I: Functional Analysis*, Academic Press, New York, San Francisco, London, 1972.

18. _____, *Methods of Mathematical Physics, Part IV: Analysis of Operators*, Academic Press, New York, San Francisco, London, 1978.

19. M. Schechter, *Spectra of Partial Differential Operators*, North-Holland Pub. Co., Amsterdam, New York, Oxford, 1971.

20. _____, *Hamiltonians for singular potentials*, Indiana University Math. J., 22(1972), 483-503.

21. R. E. Showalter, *Hilbert Space Methods for Partial Differential Equations*, Pitman Pub. Co., New York, 1977.

22. F. Treves, *Basic Linear Partial Differential Equations*, Academic Press, New York, San Francisco, London, 1975.

23. F. Wolf, *On the essential spectrum of partial differential boundary problems*, Comm. Pure Appl. Math., 12(1959), 211-228.

*The author was partly supported by NSF grant MCS-8005811.

A Singular Sixth Order Differential Equation with Orthogonal Polynomial Eigenfunctions

by

Lance L. Littlejohn[1]

and

Allan M. Krall[2]

Abstract

In this paper, we develop the eigenfunction expansion theory of a selfadjoint operator generated by a symmetric sixth order differential equation $L_6(y) = \lambda y$. This differential equation has regular singular points at $x = 1$, and is in the limit-5 case at each end. This means that two boundary conditions are needed at ± 1 to ensure a well-posed boundary value problem. Not many examples are known of such higher order singular differential equations. The example that we give is interesting because the eigenvalue problem $L_6(y) = \lambda y$ has a sequence of polynomial solutions that are orthogonal on $[-1,1]$ with respect to the weight distribution $W(x) = \frac{1}{A}\delta(x+1) + \frac{1}{B}\delta(x-1) + C$.

1) Department of Mathematics, Computer Science and Systems Design, The University of Texas at San Antonio, San Antonio, Texas, 78285.

2) 845 North Thomas Street, State College, PA 16801

A Singular Sixth Order Differential Equation with Orthogonal Polynomial Eigenfunctions

§1 Introduction

In [8], we found a set of polynomial solutions to the sixth order equation

$$
\begin{aligned}
L_6(y) = {} & (x^2-1)^3 y^{(vi)}(x) + 18x(x^2-1)^2 y^{(v)}(x) \\
& + (x^2-1)\{(3AC+3BC+96)x^2 - (3AC+3BC+36)\}y^{(iv)}(x) \\
& + x(x^2-1)(24AC+24BC+168)y'''(x) \\
& + \{(12ABC^2+42AC+42BC+72)x^2 + (12BC-12AC)x \\
& \quad - (12ABC^2+30AC+30BC+72)\}y''(x) \\
& + \{(24ABC^2+12AC+12BC)x + (12BC-12AC)\}y'(x) \\
& = \lambda_n y(x)
\end{aligned} \tag{1}
$$

where

$$
\begin{aligned}
\lambda_n = {} & (24ABC^2+12AC+12BC)n + (12ABC^2+42AC+42BC+72)n(n-1) \\
& + (24AC+24BC+168)n(n-1)(n-2) + (3AC+3BC+96)n(n-1)(n-2)(n-3) \\
& + 18n(n-1)(n-2)(n-3)(n-4) + n(n-1)(n-2)(n-3)(n-4)(n-5)
\end{aligned}
$$

To see how we derived (1), the reader is referred to [8] or [9].

The polynomial solution of degree n is given by

$$
K_n(x) = \sum_{j=0}^{n} \frac{(-1)^{[j/2]}(2n-j)!\,Q(n,j)x^{n-j}}{2^{n+1}\left(n-\left[\frac{j+1}{2}\right]\right)!\left[\frac{j}{2}\right]!\,(n^2+n+AC+BC)}
$$

where
$$
\begin{aligned}
Q(n,j) = {} & \frac{1+(-1)^j}{2}\{(n^4+(2AC+2BC-1)n^2+4ABC^2)+2j(n^2+n+AC+BC)\} \\
& + \frac{1-(-1)^j}{2}(4BC-4AC)
\end{aligned}
$$

and [·] denotes the greatest integer function. These polynomials, named the Krall polynomials after the many contributions by H. L. Krall to the theory of orthogonal polynomials, satisfy the orthogonality relation:

$$\int_{-1}^{1} K_n(x)K_m(x)w(x)\,dx =$$

(2)
$$\frac{\{n^4 + (2AC+2BC-1)n^2 + 4ABC^2\}\{(n+1)^4 + (2AC+2BC-1)(n+1)^2 + 4ABC^2\}}{2(2n+1)(n^2+n+AC+BC)^2}\, C\, \delta_{nm}$$

where

$$w(x) = \frac{1}{A}\,\delta(x+1) + \frac{1}{B}\,\delta(x-1) + C, \quad -1 \leqq x \leqq 1$$

and $0 < A,\ B \leqq \infty,\ C > 0$. Various other properties and their connection with the classical orthogonal polynomials are discussed in [8].

When $A = B = 2,\ C = \alpha/2,$ the polynomials were previously known, [6], and called the Legendre type polynomials. In this case, H. L. Krall [7] showed that these polynomials satisfy the fourth order differential equation

(3)
$$\begin{aligned}(x^2-1)^2y^{(iv)}(x) + 8x(x^2-1)y'''(x) + (4\ \ +12)(x^2-1)y''(x) + 8\alpha xy'(x)\\ = [8\alpha n + (4\alpha+12)n(n-1) + 8n(n-1)(n-2) + n(n-1)(n-2)(n-3]y(x)\end{aligned}$$

However, when $A \neq B$, $K_n(x)$ does not satisfy a fourth order equation like (3).

§2 The Appropriate Boundary Value Problem

Since $x = \pm 1$ are regular singular points of (1), Frobenius' method can be applied. At either point, the indicial equation is

$$\rho(\rho-1)^2(\rho-2)(\rho-3)(\rho+1) = 0.$$

This implies that there are five solutions to (1) that are square integrable on [-1,1] with respect to the weight $w(x)$. Therefore, two boundary conditions are necessary at $x = \pm 1$ [3].

Note that if y is a polynomial, then

$$L_6(y)(1) = 24BCy''(1) + (24ABC^2 + 24BC)y'(1)$$

and

$$L_6(y)(-1) = 24ACy''(-1) + (-24ABC^2 - 24AC)y'(-1).$$

Thus, it is apparent that one boundary condition at $x = -1$ is the λ-dependent boundary condition

$$24ACy''(-1) + (-24ABC^2 - 24AC)y'(-1) = \lambda y(-1).$$

Similarly, one of the boundary conditions at $x = 1$ is

$$24BCy''(1) + (24ABC^2 + 24BC)y'(1) = \lambda y(1)$$

We now discuss the other boundary condition at $x = \pm 1$, By introducing the notation $W_{ij} = z^{(i)}y^{(j)} - z^{(j)}y^{(i)}$, we see that:

$$\begin{aligned} zL_6(y) - yL_6(z) = {} & (x^2-1)^3 W_{06} + 18x(x^2-1)^2 W_{05} \\ & + (x^2-1)\{(3AC+3BC+96)x^2 - (3AC+3BC+36)\}W_{04} \\ & + x(x^2-1)(24AC+24BC+168)W_{03} \\ & + \{(42AC+42BC+12ABC^2+72)x^2 + (12BC-12AC)x \\ & \quad - (30AC+30BC+12ABC^2+72)\}W_{02} \\ & + \{(12AC+12BC+24ABC^2)x + (12BC-12AC)\}W_{01} \end{aligned}$$

Denote the right side of this equation by $[y,z]_1$. Also, Lagrange's identity [1 yields the identity:

$$zL_6(y) - yL_6(z) = \frac{d[y,z]_2}{dx}$$

where

$$\begin{aligned}[y,z]_2 = &\{(12ABC^2 + 6AC + 6BC)x^2 + (12BC - 12AC)x \\ &+ (-12ABC^2 - 18AC - 18BC - 24)\}W_{01} + \{(12AC + 12BC + 24)x^3 \\ &+ (-12AC - 12BC - 48)x\}W_{02} + \{(3AC + 3BC + 36)x^4 \\ &+ (-6AC - 6BC - 60)x^2 + (3AC + 3BC + 24)\}W_{03} + \\ &+ 12x(x^2 - 1)^2 W_{04} + (x^2 - 1)^3 W_{05} + \{(-3AC - 3BC - 6)x^4 \\ &+ (6AC + 6BC + 24)x^2 + (-3AC - 3BC - 18)\}W_{12} - 6x(x^2 - 1)^2 W_{13} \\ &- (x^2 - 1)^3 W_{14} + (x^2 - 1)^3 W_{23}.\end{aligned}$$

Hence

$$C\int_{-1}^{1} (zL_6(y) - yL_6(z))dx = C \lim_{x\to 1} [y,z]_2 - C \lim_{x\to -1} [y,z]_2$$

if these limits exist. Similarly,

$$\int_{-1}^{1} (zL_6(y) - yL_6(z))[\tfrac{1}{A}\,\delta(x+1) + \tfrac{1}{B}\,\delta(x-1)]dx$$

$$= \frac{1}{A} \lim_{x\to -1} [y,z]_1 + \frac{1}{B} \lim_{x\to -1} [y,z]_1$$

if these limits exist. Combining these latter two equations yields

$$\int_{-1}^{1} (zL_6(y) - yL_6(z))w(x)dx =$$

$$= \lim_{x\to 1} \{y,z\}_1 + \lim_{x\to -1} \{y,z\}_2$$

where

$$\{y,z\}_1 = C[y,z]_2 + \frac{1}{B}[y,z]_1$$

and

$$\{y,z\}_2 = -C[y,z]_2 + \frac{1}{A}[y,z]_1$$

It appears then that the other boundary condition at -1 is $\lim_{x\to -1} \{y,K_0\}_2 = 0$ and the other boundary conditions at 1 is $\lim_{x\to 1} \{y,K_0\}_1 = 0$, where $K_0 = K_0(x)$. These conditions are certainly necessary. We believe that the conditions $\lim_{x\to -1} \{y,K_0\}_2 = \lim_{x\to -1} \{z,K_0\}_2 = \lim_{x\to 1} \{y,K_0\}_1 = \lim_{x\to 1} \{z,K_0\}_1 = 0$ are sufficient to guarantee that

$$\int_{-1}^{1} (zL_6(y) - yL_6(z))W(x)dx = 0.$$

In summary, then, the appropriate boundary value problem is:

a) $L_6(y) = \lambda y \quad -1 < x < 1$

b) i) $24ACy''(-1) + (-24ABC^2 - 24AC)y'(-1) = \lambda y(-1)$

ii) $\lim_{x\to -1} \{y,K_0\}_2 = 0$

c) i) $24BCy''(1) + (24ABC^2 + 24BC)y'(1) = \lambda y(1)$

ii) $\lim_{x\to 1} \{y,K_0\}_1 = 0$

Alternatively, we can describe the boundary value problem so that the operator has a domain not depending on λ [5]. Let $H = L_w^2[-1,1] \otimes \mathbb{R} \otimes \mathbb{R}$ where $L_w^2[-1,1]$ indicates the real Hilbert space generated by $w(x)$. If $Y \in H$, we write $Y = (y(x), y_{-1}, y_1)^T$. Define the inner product of $Y = (y(x), y_{-1}, y_1)^T$ and

$Z = (z(x), z_{-1}, z_1)^T$ as:

$$(Y,Z) = C\int_{-1}^{1} y(x)z(x)dx + \frac{1}{A} z_{-1}y_{-1} + \frac{1}{B} z_1 y_1$$

Clearly, with this inner product, H is a Hilbert space. We define the operator A as follows: let D_A consist of all elements $Y = (y(x), y_{-1}, y_1)^T$ satisfying the conditions:

1. $y(x) \in L_w^2[-1,1]$
2. $y', y'', y''', y^{(iv)}, y^{(v)}$ exist and $y^{(v)}$ is absolutely continuous on $(-1,1)$
3. $L_6(y)$ exists a.e. and is in $L_w^2[-1,1]$
4. $y(-1) = y_{-1}$
5. $y(1) = y_1$
6. $\lim_{x\to 1} \{y, K_0\}_1 = 0$
7. $\lim_{x\to -1} \{y, K_0\}_2 = 0$

For $Y \in D_A$, define the operator A by:

$$AY = \begin{pmatrix} L_6(Y) \\ L_6(y)(-1) \\ L_6(y)(1) \end{pmatrix} = \begin{pmatrix} L_6(y) \\ 24ACy''(-1) + (-24ABC^2 - 24AC)y'(-1) \\ 24BCy''(1) + (24ABC^2 + 24BC)y'(1) \end{pmatrix}$$

Theorem 1 The operator A is self adjoint in H.

The proof is by standard techniques.

We now turn our attention to the eigenfunction expansion theory of the self adjoint operator A. Becuase of the weights assigned to ± 1, the completeness of the Krall polynomials, represented by $(K_n(x), K_n(-1), K_n(1))^T$, in H is not immediately obvious. Let H_0 be the subspace spanned by the set $\{(K_n(x), K_n(-1), K_n(1))^T\}_{n=0}^{\infty}$. We show that H_0 is dense in H.

Lemma 1 Let $F = (f(x), f_{-1}, f_1)$ be orthogonal to H_0. Let $X^n = (x^n, (-1)^n, 1)$. Then $(F, X^n) = 0$, $n = 0,1,\ldots$.

Proof. This follows by taking appropriate linear combinations of

$$0 = (F, K_j) = C\int_{-1}^{1} f(x)K_j(x)dx + \frac{1}{A} f_{-1}K_j(-1) + \frac{1}{B} f_1 K_j(1) \quad j = 0,1,\ldots n. \quad \blacksquare$$

Lemma 2 The measure ϕ generated by $\phi(E) = \int_E f(x)w(x)dx$ is of bounded variation.

Proof. By Schwarz' inequality, the total variation $|\phi|$ satisfies

$$|\phi|(-1,1) = \int_{-1}^{1} |f(x)|w(x)dx \le \left(\int_{-1}^{1} |f(x)|^2 w(x)dx\right)^{1/2} \left(\int_{-1}^{1} w(x)dx\right)^{1/2} < \infty. \quad \blacksquare$$

Lemma 3 The Fourier transform of ϕ, $\hat{\phi}(\lambda)$, is 0.

Proof

$$\hat{\phi}(\lambda) = \frac{1}{\sqrt{2\pi}} \int_{-1}^{1} e^{-i\lambda x} f(x)w(x)dx = \frac{1}{\sqrt{2\pi}} \int_{-1}^{1} \sum_{n=0}^{\infty} \frac{(-i\lambda x)^n}{n!} f(x)w(x)dx$$

$$= \frac{1}{\sqrt{2\pi}} \int_{-1}^{1} \lim_{N\to\infty} \sum_{n=0}^{N} \frac{(-i\lambda x)^n}{n!} f(x)w(x)dx = \frac{1}{\sqrt{2\pi}} \lim_{N\to\infty} \int_{-1}^{1} \sum_{n=0}^{N} \frac{(-i\lambda x)^n}{n!} f(x)w(x)dx$$

$$= \frac{1}{\sqrt{2\pi}} \lim_{N\to\infty} \sum_{n=0}^{N} \frac{(-i\lambda)^n}{n!} \int_{-1}^{1} x^n f(x)w(x)dx = 0,$$

from Lebesgue's dominated convergence theorem and Lemma 2. ∎

<u>Theorem 2</u> $\{K_n(x), K_n(-1), K_n(1)\}_{n=0}^{\infty}$ <u>forms a complete set of vectors in H.</u>

<u>Proof</u>. Recall that H_0 is the span of this set of vectors. We show that if $F = (f(x), f_{-1}, f_1)^T \in H$ and F is orthogonal to H_0, then $F = 0$. By inverting the Fourier transform ϕ, we see that $\phi = 0$. (see [4]) Hence $f(x) = 0$ a.e. in $(-1,1)$ and $f_{-1} = f_1 = 0$. We now have our expansion theorem:

<u>Theorem 3</u> <u>If</u> $F = (f(x), f_{-1}, f_1)^T \in H$, <u>then</u> $F = \sum_{n=0}^{\infty} c_n (K_n(x), K_n(-1), K_n(1))^T$

<u>where the generalized Fourier coefficients</u> c_n <u>are given by</u>

$$c_n = \frac{[C\int_{-1}^{1} f(x)K_n(x)dx + \frac{1}{A} f_{-1}K_n(-1) + \frac{1}{B} f_1 K_n(1)]}{(\int_{-1}^{1} K_n^2(x)w(x)dx)^2}$$

The proof is standard. We note that the formula may be further simplified by using

$$K_n(1) = \frac{BC(n^2+n+2AC)}{n^2+n+AC+BC} \quad \text{and} \quad K_n(-1) = \frac{(-1)^2 AC(n^2+n+2BC)}{n^2+n+AC+BC},$$

and by using (2). ∎

<u>Corollary 1</u> If $f \in L_w^2(-1,1)$, <u>then</u> $f(x) = \sum_{n=0}^{\infty} c_n K_n(x)$

<u>where</u> $c_n = \dfrac{C\int_{-1}^{1} f(x)K_n(x)dx}{\int_{-1}^{1} K_n^2(x)w(x)dx}$

References

[1] R. H. Cole, Theory of Ordinary Differential Equations, Appleton-Century-Crofts, New York, 1968.

[2] W. N. Everitt, "Fourth Order Singular Differential Equations", Math. Ann., 49(1963), 320-340.

[3] ____________ and V. Krishna Kumar, "On the Titchmarsh-Weyl Theory of Ordinary Symmetric Expressions I: The General Theory", N. Arch. V. Wisk. 24(1976), 109-145.

[4] W. Feller, An Introduction to Probability Theory and Its Applications, Vol. II, John Wiley and Sons Inc., New York, 1966.

[5] C. Fulton, "Two Point Boundary value Problems with the eigenvalue parameter contained in the Boundary Conditions", Proc. Roy. Soc. Edin., 77(1977), 293-308.

[6] A. M. Krall, "Orthogonal Polynomials Satisfying Fourth Order Differential Equations", Proc. Roy. Soc. Edin., 87(A), 271-288.

[7] H. L. Krall, "Certain Differential Equations for Tchebycheff Polynomials", Duke Math. Jour. 4(1938), 705-718.

[8] L. L. Littlejohn, "The Krall Polynomials: A New Class of Orthogonal Polynomials," Quaestiones Mathematicae, to appear.

[9] _______________ and S. D. Shore, "Nonclassical Orthogonal Polynomials as Solutions to Second Order Differential Equations", Can. Bull. (to appear).

10] S. D. Shore, "On the Second Order Differential Equation which has Orthogona Polynomial Solutions", Bull. Calcutta Math. Soc. 56(1964), 195-198.

On the asymptotics of some Volterra equations with locally finite measures and large perturbations

Stig-Olof Londen

1. Introduction

In this lecture we discuss the asymptotics of the scalar, real nonlinear Volterra differential equation of convolution type

$$(1.1) \qquad x'(t) + \int_{[0,t]} g(x(t-s))d\mu(s) = f(t), \; t \in R^+ = [0,\infty), \; x(0) = x_0.$$

Here g and f are given real functions, μ is a given real Borel measure on R^+ and $x(t)$ is the unknown solution. We concentrate on the case when both μ and f are large; that is on the case when μ is only locally finite and f vanishes at infinity but $f \in L^p(R^+)$, for some $p < \infty$, does not necessarily hold. Our analysis is based on a transformation of (1.1) to an equation of the same type but with a finite measure and, on showing that the limit equation corresponding to this transformed version of (1.1) has sufficiently small ($\in L^2(R)$) solutions.

Throughout the lecture the following basic assumptions will be made:

$$(1.2) \quad \begin{cases} \text{(i)} & g \in C(R), \\ \text{(ii)} & \mu \text{ is a real, locally finite, positive definite measure} \\ & \text{on } R^+, \\ \text{(iii)} & f \in L^1_{loc}(R^+), \\ \text{(iv)} & x \in (L^\infty \cap LAC)(R^+), \; x \text{ satisfies (1.1) a.e. on } R^+. \end{cases}$$

Observe that we do assume the existence of a uniformly bounded, locally absolutely continuous solution of (1.1).

For u,v functions, ν a measure, all supported on R^+, we define the convolutions $(u*v)(t)$, $(u*\nu)(t)$ by

$$(u*v)(t) = \int_0^t u(t-\tau)v(\tau)d\tau, \quad (u*\nu)(t) = \int_{[0,t]} u(t-\tau)d\nu(\tau), \quad t \in R^+.$$

Also let

$$\tilde{u}(z) = \int_{R^+} e^{-zt}u(t)dt, \quad \tilde{\nu}(z) = \int_{R^+} e^{-zt}d\nu(t),$$

and write $\hat{u}(\omega) = \tilde{u}(i\omega)$, $\hat{\nu}(\omega) = \tilde{\nu}(i\omega)$, $\omega \in R$.

The classical way to analyze the asymptotics of solutions of (1.1) is to define the quadratic form $Q(\varphi,\mu,T)$ for $\varphi \in L^2_{loc}(R^+)$, μ satisfying (1.2ii), $T > 0$, by

$$Q(\varphi,\mu,T) = \int_0^T \varphi(t)(\varphi*\mu)(t)dt$$

and then to try to somehow obtain

$$\sup_{T>0} Q(g(x(t)),\mu,T) < \infty. \tag{1.3}$$

It is well-known [6, 7] that a detailed knowledge of the asymptotics of $g(x(t))$ can be obtained from (1.3). But to get (1.3) one needs to take f very small, i.e. $f \in L^1(R^+)$, which of course excludes many interesting applications where f (in addition to (1.2iii)) only satisfies

$$\lim_{t\to\infty} f(t) = 0. \tag{1.4}$$

However, by taking μ small enough, in particular by assuming (at least) $\int_{R^+} d|\mu|(t) < \infty$ and by working with the limit equation corresponding to (1.1),

$$y'(t) + \int_{R^+} g(y(t-s))d\mu(s) = 0, \quad t \in R, \tag{1.5}$$

one may obtain asymptotic results on bounded solutions of (1.1) even if f satisfies only (1.4), [2, 3, 8].

But the question as to what happens when μ is only locally finite (with f satisfying only (1.4)) remains and is much harder to answer. An obvious difficulty that one now encounters is that (1.5) is meaningless.

A first partial answer to this question is the following Theorem 1 which essentially is due to Gripenberg [1], although he formulates it somewhat differently, in particular note that μ is taken finite in [1].

Theorem 1. *Let* (1.2) *hold and assume the solution* r *of*

(1.6) $$r'(t) + (r*\mu)(t) = 0 \text{ a.e. on } R^+,\ r(0) = 1,$$

satisfies

(1.7) $$r' \in (L^1 \cap NBV)(R^+).$$

Define ν *to be the measure corresponding to* $-r'$, *thus* $\nu([0,t]) = -r'(t)$, $t \geq 0$, *and let*

(1.8) $$\int_{R^+} t\,d|\nu|(t) < \infty.$$

Suppose

(1.9) $$|\hat{\mu}(\omega)| < \infty,\ \omega \neq 0,$$

and let the set Z *defined by* $Z = \{\omega |\ \omega \neq 0, \operatorname{Re} \hat{\mu}(\omega) = 0\}$ *be at most denumerable and such that* $\operatorname{Im} \hat{\mu}(\omega) = 0$, *for* $\omega \in Z$. *Finally let* z *be the locally absolutely continuous solution of*

(1.10) $$z'(t) + (z*\mu)(t) = f(t),\ t \in R^+,\ z(0) = x_0.$$

Then if $\lim_{t\to\infty} z(t) = z(\infty)$ *exists (and is finite) one has*

$$\lim_{t\to\infty} [x(t+d) - x(t)] = 0, \quad \forall d > 0,$$

$$\lim_{t\to\infty} [r(\infty)x(t) + (1-r(\infty))g(x(t))] = z(\infty).$$

If in addition $z' \in L^{\infty}(R^+)$, $\lim_{t\to\infty} \operatorname{ess\,sup}_{s\geq t} |z'(s)| = 0$ *then* $\lim_{t\to\infty} \operatorname{ess\,sup}_{s\geq t} |x'(s)| = 0.$

Thus, if the solution $z = x_0 r + r*f$ of the linear equation (1.10) tends to a limit then so does, under certain hypotheses, a bounded solution of the nonlinear equation (1.1) having the same data.

The transform $\hat{\mu}(\omega)$, $\omega \neq 0$, is defined as $\lim_{\substack{s\to i\omega \\ \mathrm{Re}\, s>0}} \tilde{\mu}(s)$ where $\tilde{\mu}(s) = \int_{R^+} e^{-st} d\mu(t)$. One can show that from (1.6), (1.7) follows that this is welldefined, possibly infinite, for $\omega \neq 0$. The assumption (1.9) does however exclude this possibility.

Theorem 1 has the advantage of having a short and lucid proof. Also observe that nothing but continuity is imposed on g. The assumptions do however include a moment condition (1.8) on the second derivative of the differential resolvent of μ which in general is difficult to check. It is satisfied if $d\mu(t) = a(t)dt$ with $(-1)^k a^{(k)}(t) \geq 0$ for $k = 0,1,2,3$; but apart from this example it is hard to find classes of only locally finite measures which do satisfy this assumption. One might also note that Theorem 1 requires $\hat{\mu}(\omega)$ to be finite for $\omega \neq 0$, thus excluding interesting cases like $d\mu(t) = a(t)dt$ with $a(t) = t^{-1/2}\cos t$.

One is consequently motivated to try to remove the hypotheses (1.8), (1.9). This was recently done in [4] where the following result was obtained.

Theorem 2. *Let* (1.2) *be satisfied and let*

(1.11) $g(x)$ be locally Lipschitzian, $x \in R$.

Let $r \in LAC(R^+)$ satisfy (1.6), (1.7) and suppose $\lim_{t\to\infty} r(t) = \lim_{t\to\infty} (r*f)(t) = 0$. Also let

(1.12) $\operatorname{Im} \hat{\mu}(\omega) = 0$ for $\omega \in \{\omega |\ \omega \neq 0,\ \operatorname{Re} \hat{\mu}(\omega) = 0\}$,

(1.13) $xg(x) > 0,\ x \neq 0$,

(1.14) $\liminf_{|x|\to 0} x^{-1}g(x) > 0$.

Then $\lim_{t\to\infty} x(t) = 0$. If in addition $\lim_{t\to\infty} \operatorname{ess\,sup}_{s\geq t} |f(s)| = 0$ then $\lim_{t\to\infty} \operatorname{ess\,sup}_{s\geq t} |x'(s)| = 0$.

Note that (1.8), (1.9) are now absent from the assumptions. Certain other conditions have instead been added, in particular on $g(x)$. These additional assumptions (1.11), (1.13), (1.14) on g clearly have the advantage of being easily checked and at least (1.11), (1.14) are not overly restrictive. The relation (1.13) does however limit the class of nonlinearities to which Theorem 2 can be applied. In case μ is a finite measure then nothing like (1.13) is needed and it was suggested that the positive definiteness of μ and (1.7) should allow us to abolish (1.13) even if μ is nonfinite. That this is indeed the case is shown by

Theorem 3, [5]. Let (1.2) be satisfied, define $X = \{x \in R |\ g(x) = 0\}$ and assume that

(1.15) $0 < \liminf_{|x-x_e|\to 0} \left|\frac{g(x)}{x-x_e}\right| \leq \limsup_{|x-x_e|\to 0} \left|\frac{g(x)}{x-x_e}\right| < \infty$ for $x_e \in X$.

Let

(1.16) $\inf_{\omega \in S} \operatorname{Re} \hat{\mu}(\omega) > 0$ *for any compact set* $S \subset R$,

and suppose r, f *satisfy* (1.6), (1.7) *and*

(1.17) $\lim_{t\to\infty} r(t) = \lim_{t\to\infty} (r*f) = 0.$

Then $\lim_{t\to\infty} \text{dist}\,(x(t),X) = 0$. *If in addition* $\lim_{t\to\infty} f(t) = 0$, *then* $\lim_{t\to\infty} x'(t) = 0$.

The sign condition (1.13) on $g(x)$ has obviously disappeared but we still need to assume that $|g(x)|$ is neither too small nor too large for x near X. It is apparent from the proof of Theorem 3 that the relation (1.16) may be significantly weakened without altering the conclusions. How much is still an open question; current work is devoted to this.

2. Outline of proof of Theorem 3

Convolve (1.1) by r and use (1.6). Also define $h(x) = g(x) - x$, $x \in R$; $a(t) = -r'(t)$, $t \in R^+$. This gives

$$x(t) + \int_0^t h(x(t-s))a(s)ds = z(t), \quad t \in R^+,$$

where $z(t) = x_0 r(t) + (r*f)(t)$ by (1.17) satisfies $\lim_{t\to\infty} z(t) = 0$. As $a \in L^1(R^+)$ and as $h \in C(R)$ we therefore have that given any sequence $t_n \to \infty$ there exists a subsequence t_{n_K} and $y \in (L^\infty \cap LAC)(R)$ such that $x(t+t_{n_K}) \to y(t)$ uniformly on compact sets where y satisfies

(2.1) $$y(t) + \int_{R^+} h(y(t-s))a(s)ds = 0, \quad t \in R.$$

Define

$$Y = \{y|\ y \in (LAC \cap L^{\infty})(R),\ \text{there exist}\ t_n \to \infty\ \text{such that}\ x(t+t_n) \to y(t)\ \text{uniformly on compact sets}\}.$$

Note that if $y \in Y$ then y not only satisfies (2.1) but also

(2.2) $$y'(t) + \int_{R^+} h(y(t-s))d\alpha(s) = 0, \quad \text{a.e. on } R,$$

where $\alpha([0,t]) = a(t)$. Straightforward computations give

(2.3) $$\text{Re}\ \hat{\alpha}(\omega) = \omega^2 \text{Re}\ \hat{\mu}(\omega)|i\omega + \hat{\mu}(\omega)|^{-2},\ \omega \neq 0,$$

(2.4) $$\text{Im}\ \hat{\alpha}(\omega) = \omega^2 \text{Im}\ \hat{\mu}(\omega)|\ i\omega + \hat{\mu}(\omega)|^{-2} + \omega|\hat{\mu}(\omega)|^2\ |i\omega + \hat{\mu}(\omega)|^{-2},\ \omega \neq 0$$

(2.5) $$\hat{\alpha}(0) = 0.$$

Thus, by (1.2ii) and (1.7) α is a positive definite finite measure. Note however that although (1.16) is assumed then $\text{Re}\ \hat{\alpha}(\omega) > 0$, $\omega \neq 0$, does not necessarily hold. Indeed, if $\omega_0 \neq 0$ is such that $|\hat{\mu}(\omega_0)| = \infty$ then $\text{Re}\ \hat{\alpha}(\omega_0) = 0$. Furthermore observe that for such ω_0-values we have $\text{Im}\ \hat{\alpha}(\omega_0) = \omega_0 \neq 0$. This fact, together with (2.5), makes (2.2) quite difficult to handle.

We begin by showing that Y does contain constants.

Lemma 1. There does not exist $y \in Y$ such that for some positive δ, T

(2.6) $$\int_{t-T}^{t} |y'(s)|ds \geq \delta,\ t \in R.$$

Proof of Lemma 1. Suppose the Lemma is false and take $y \in Y$ such that (2.6) is satisfied for some δ, T. Define $Z_\infty = \{\omega \neq 0|\ |\hat{\mu}(\omega)| = \infty\}$ and observe that as $\hat{\alpha} = i\omega\hat{a}$, $\omega \neq 0$; $\hat{a}(0) = 1$, then $\hat{a}(\omega) = 1$ for $\omega \in Z_\infty \cup \{0\}$. From (1.7) and (2.4) follows that the set Z_∞ is bounded.

Choose $\hat{\varepsilon} > 0$ arbitrarily, let $\omega_0 > 0$ be such that $|1-\hat{a}(\omega)| \leq \hat{\varepsilon}$, $|\omega| \leq \omega_0$, take ε (depending on $\hat{\varepsilon}$) sufficiently smal and define

$$(2.7) \qquad V_1 = \{\omega \mid \operatorname{dist}(\omega, Z_\infty) \leq \varepsilon,\ |1-\hat{a}(\omega)| \leq \hat{\varepsilon},\ \omega_0 \leq |\omega| \leq \varepsilon^{-1}\}$$

$$(2.8) \qquad V = V_1 \cup [-\omega_0, \omega_0].$$

Write

$$v_t(\tau) = \begin{cases} y(\tau), & 0 \leq \tau \leq t \\ 0 & , \tau < 0;\ \tau > t \end{cases} ;\quad z_t(\tau) = \begin{cases} h(y(\tau)), & 0 \leq \tau \leq t \\ 0 & , \tau < 0;\ \tau > t\ . \end{cases}$$

Then one can show by estimates of [4] that if T is fixed sufficiently large then

$$(2.9) \qquad \int_{R \setminus V} |\hat{v}_t(\omega)|^2 d\omega \leq \varepsilon c_1 \int_V |\hat{z}_t(\omega)|^2 d\omega + c_2, \quad t \geq T,$$

where the constant c_1 depends on $\hat{\varepsilon}$ but not on t, T, ε and where the constant c_2 although depending on ε, $\hat{\varepsilon}$ and T is independent of t.

Define F by $F(t) = y(t) - (a*y)(t)$, $t \in R^+$. Using the second part of (1.15), (2.7)-(2.9) one can show that there exist intervals $I_n \subset R^+$ such that $m(I_n) \to \infty$, $\limsup_{n\to\infty} \sup_{\tau \in I_n} |F(\tau)| = 0$. Consequently some sequence of translates of y converges (uniformly on compact sets) to a solution w of

$$w(t) - \int_{R^+} w(t-s)a(s)ds = 0, \ t \in R.$$

But as y (and therefore by the translation invariance of (2.1) any translate of y) is a solution of (2.1) we also have

$$w(t) + \int_{R^+} h(w(t-s))a(s)ds = 0, \ t \in R.$$

Consequently the spectrum σ of $w(t) + h(w(t))$ satisfies $\sigma(w + h(w)) \subset \{\omega | \hat{a}(\omega) = 0\}$ which is empty. One concludes that $w(t) + h(w(t)) \equiv 0$ and so $g(w(t)) \equiv 0$. But this clearly violates (2.6).

The next lemma is rather technical but essential to the proof of Theorem 3.

Lemma 2. Define

$$Y_0 = \{y \in Y | \ h(y(t)) \notin L^2(R); \text{ there exist } c > 0$$

(may depend on y) such that for some $t_n \to \infty$,

$$\sup_{-\infty < s \leq t_n} \int_{s-2t_n}^{s} |h(y(v))|^2 dv \leq c \int_{-t_n}^{t_n} |h(y(v))|^2 dv\}, \text{ and let } y \in Y_0.$$

Then for any $\varepsilon > 0$ if n sufficiently large,

$$\int_{-t_n}^{t_n} g^2(y(\tau))d\tau \leq \varepsilon \int_{-t_n}^{t_n} h^2(y(\tau))d\tau.$$

The (fairly long) proof of Lemma 2 makes in particular use of some estimates of [3], [4] and of the second part of (1.15) and of (1.16). Once Lemma 2 is proved it is not difficult to obtain the following result on the size of the solutions of the limit equation which behave well at infinity. The proof of Lemma 3 is where the first part of (1.15) is essentially needed.

Lemma 3. Define

$$Y_\infty = \{y \in Y | \ y_1 \overset{\text{def}}{=} \lim_{t\to-\infty} y(t), \ y_2 \overset{\text{def}}{=} \lim_{t\to\infty} y(t) \text{ both exist}\}.$$

Then

$$g(y(t)) \in L^2(R), \quad (y(t) - y_1) \in L^2(R^-), \quad y(t) - y_2 \in L^2(R^+).$$

Suppose for a moment that we have:

Lemma 4. Let $y \in Y_\infty$ with $g(y(t)) \not\equiv 0$ and let

$G(x) \overset{\text{def}}{=} \int_0^x g(u)du$. Then

(2.10) $G(y_2) < G(y_1)$.

The proof of Theorem 3 may then be completed as follows. Suppose $\limsup_{t\to\infty} \text{dist}(x(t),X) \neq 0$. Let Y_c denote the set of constant limit functions, i.e. define $Y_c = \{y \in R \mid y \in Y\}$. By Lemma 1 Y_c is not empty. Let $y_c \in Y_c$ be such that

(2.11) $G(y_c) = \inf_{y \in Y_c} G(y)$.

Take a sequence of translates $x_n(t) = x(t+t_n)$ of $x(t)$ converging to y_c uniformly on compact sets. Define $d = \text{dist}(y_c, Y_c \setminus y_c)$. By (1.15) $d > 0$. Take $\delta \in (0,d)$ sufficiently small so that $T_n \overset{\text{def}}{=} \inf\{\tau \mid \tau \geq 0, |x_n(\tau) - y_c| = \delta\}$ is well-defined for all n. Let $\tilde{x}_n(t) = x_n(t+T_n)$, thus for some $\nu > 0$ $|g_n(\tilde{x}_n(0))| \geq \nu$. By Lemma 1 there exists a subsequence $\{n_k\}$ of $\{n\}$ and $y \in Y_\infty$ such that $\tilde{x}_{n_k}(t) \to y(t)$ uniformly on compact sets, $g(y(t)) \not\equiv 0$ and $y(-\infty) = y_c$. But then from (2.10) $G(y_c) > G(y(\infty))$ which violates (2.11) and completes the proof.

We are thus left with proving (2.10). We sketch the proof of this in what follows.

Let $r_\lambda(t)$, $\lambda > 0$, $t \in R^+$, be the differential resolvent corresponding to $\lambda\mu$, thus

(2.12) $r_\lambda'(t) + \lambda(r_\lambda * \mu)(t) = 0$ a.e. on R^+, $r_\lambda(0) = 1$.

We have the following result concerning the size of r_λ.

Lemma 5. [4]. Let μ satisfy (1.2ii) and (1.16) and let

$r_1' \in L^1(R^+)$. <u>Then</u> $r_\lambda' \in L^1(R^+)$ <u>for</u> $\lambda > 0$. <u>If in addition</u> $r_1' \in NBV(R^+)$ <u>then</u> $r_\lambda' \in NBV(R^+)$ <u>for</u> $\lambda > 0$.

Convolve (1.1) by r_λ, use (2.12). This gives

$$x(t) + \int_{[0,t]} \left[\frac{g(x(t-s))}{\lambda} - x(t-s)\right] a_\lambda(s)ds = z_\lambda(t),$$

where $z = x_0 r_\lambda + r_\lambda * f$; $a_\lambda = -r_\lambda'$. From Lemma 5 and (1.17) follows easily $z_\lambda(t) \to 0$ as $t \to \infty$ for λ fixed. Thus $y \in Y_\infty$ also satisfies

$$\text{(2.13)} \quad y(t) + \int_{R^+} \left[\frac{g(y(t-s))}{\lambda} - y(t-s)\right] a_\lambda(s)ds = 0, \; t \in R.$$

Define $\alpha_\lambda([0,t]) = a_\lambda(t)$, differentiate (2.13), define

$$f_\lambda(t) = \lambda^{-1} g(y(t)) - y(t) + y_1, \; t < 0;$$

$$f_\lambda(t) = \lambda^{-1} g(y(t)) - y(t) + y_2, \; t \geq 0$$

and use the fact that $\alpha_\lambda(R^+) = 0$. This gives

$$\text{(2.14)} \quad y'(t) + \int_{R^+} f_\lambda(t-s)d\alpha_\lambda(s)ds = h_\lambda(t), \; t \in R,$$

where $h_\lambda(t) = 0$, $t < 0$; $h_\lambda(t) = (y_2 - y_1)a_\lambda(t)$, $t \geq 0$. Multiply (2.14) by f_λ, integrate over R (can be done as by Lemma 3 $f_\lambda \in L^2(R)$) and apply Parseval's relation. This gives

$$\text{(2.15)} \quad \begin{cases} G(y_2) - G(y_1) + \lambda\left(\frac{y_2^2}{2} - \frac{y_1^2}{2} - y(0)y_2 + y(0)y_1\right) \\ + \lambda \int_R |\hat{f}_\lambda(\omega)|^2 \mathrm{Re}\, \hat{\alpha}_\lambda(\omega)d\omega = \lambda \int_{R^+} f_\lambda(\tau)h_\lambda(\tau)d\tau. \end{cases}$$

Note that $h_\lambda \in L^1(R^+)$ and that for fixed λ we can make $\sup_{\tau \in R^+} |f_\lambda(\tau)$ arbitrarily small by translating y. (This translation does not affect $|\hat{f}_\lambda(\omega)|$). Thus the right side of (2.15) can be made arbitrarily small. We therefore have, without loss of generality,

$$\lim_{\lambda \downarrow 0} [G(y_2) - G(y_1) + \lambda \int_R |\hat{f}_\lambda(\omega)|^2 \text{Re}\, \hat{\alpha}_\lambda(\omega) d\omega] = 0,$$

and so, if (2.10) does not hold,

$$\text{(2.16)} \quad \lim_{\lambda \downarrow 0} \lambda \int_R |\hat{f}_\lambda(\omega)|^2 \text{Re}\, \hat{\alpha}_\lambda(\omega) d\omega \leq 0.$$

Substitute the expression for f_λ into (2.16). We wish to show that $(\hat{g}(\omega) \overset{\text{def}}{=} \widehat{g(y(s))}(\omega))$

$$\int_R |\hat{g}(\omega)|^2 \lambda^{-1} \text{Re}\, \hat{\alpha}_\lambda(\omega) d\omega$$

is the dominant part in the resulting expression and that this term is bounded away from zero. We have

$$\lambda^{-1} \text{Re}\, \hat{\alpha}_\lambda(\omega) = \omega^2 (\text{Re}\, \hat{\mu}) \, |i\omega \ + \lambda\hat{\mu}|^{-2}$$

and so, if $\omega \neq 0$ and such that $|\hat{\mu}(\omega)| < \infty$, then $\lim_{\lambda \downarrow 0} \lambda^{-1} \text{Re}\, \hat{\alpha}_\lambda(\omega) =$ $\text{Re}\, \hat{\mu}(\omega)$. By Egoroffs theorem we therefore have that for any $\varepsilon > 0$ and for any compact set $K \subset R$ there exists $E \subset K$, $m(E) < \varepsilon$, such that the converge is uniform on $K \setminus E$. (Note that by (1.7) $m(\{\omega | \, |\hat{\mu}(\omega)| = \infty\}) = 0$.) As also $\lambda^{-1} \text{Re}\, \hat{\alpha}_\lambda(\omega) \geq 0$, $\forall \omega$, $\forall \lambda$, we have that for λ sufficiently small

$$\text{(2.17)} \quad \int_R |\hat{g}(\omega)|^2 \lambda^{-1} \text{Re}\, \hat{\alpha}_\lambda(\omega) d\omega \geq \frac{1}{2} \int_{K \setminus E} |\hat{g}(\omega)|^2 \text{Re}\, \hat{\mu}(\omega) d\omega.$$

But as $\text{Re}\, \hat{\mu} > 0$, $\omega \in R$, we may assume that K, E are such that the right side of (2.17) $(\overset{\text{def}}{=} \rho)$ is strictly positive.

Next one shows that

$$\text{(2.18)} \quad \lim_{\lambda \downarrow 0} \lambda \int_R |\hat{y}_{12}(\omega)|^2 \ \text{Re}\, \hat{\alpha}_\lambda(\omega) = 0.$$

Here $y_{12}(t) = y(t) - y_1$, $t < 0$; $y_{12}(t) = y(t) - y_2$, $t \geq 0$. The relation (2.18) follows after some calculations, from (1.7) and (1.16). Finally one observe that (2.17), (2.18), $\rho > 0$ violate (2.16). Thus

Lemma 4 and hence Theorem 3 are proved.

References

1. G. Gripenberg, On nonlinear Volterra equations with nonintegrable kernels, SIAM J. Math. Anal., 11 (1980), 668-682.

2. S-O. Londen, On a Volterra integrodifferential equation with L^∞-perturbation and noncountable zero-set of the transformed kernel, J. Integral Eqs., 1 (1979), 275-280.

3. S-O. Londen, On an integral equation with L^∞-perturbation. To appear, J. Integral Eqs.

4. S-O. Londen, On some integral equations with locally finite measures and L^∞-perturbations. To appear, SIAM J. Math. Anal.

5. S-O. Londen, Manuscript in preparation.

6. O. Staffans, Positive definite measures with applications to a Volterra equation, Trans. Amer. Math. Soc., 218 (1976), 219-237.

7. O. Staffans, Tauberian theorems for a positive definite form, with applications to a Volterra equation, Trans. Amer. Math. Soc., 218 (1976), 239-259.

8. O. Staffans, On a nonlinear integral equation with a nonintegrable perturbation, J. Integral Eqs., 1 (1979), 291-307.

PERIODIC OSCILLATIONS OF FORCED PENDULUM-LIKE EQUATIONS

Jean MAWHIN

Introduction.

The influence of the problem of the pendulum on the developement of mathematics, mechanics and technology has been enormous. Most part of the history deals with the study of the free oscillations of the pendulum with Galilée, Marci, Torricelli, Descartes, Huyghens, Newton, Euler, Poisson, Bessel, Jacobi and others and its link with the theory of elliptic functions is well known. The case of the forced pendulum equation, and in particular the existence of its periodic solutions has started much later, with Duffing [9] in 1918 who started with the equation

$$x'' + a \sin x = e(t) \quad (a > 0) \tag{0.1}$$

and was led to the so-called Duffing equation

$$x'' + ax + cx^3 = e(t)$$

by retaining the two first terms in the Taylor series of sin x. This analysis was therefore only valid for (1) when $|x|$ is sufficiently small. Hamel [11], in 1922 wrote a very interesting paper on (1) with $f(t) = b \sin t$, where he first proves, by a variational method in the line of the work of Hilbert and Caratheodory, that the equation

$$x'' + a \sin x = b \sin t \tag{0.2}$$

has a 2π-periodic solution for every $a \geqslant 0$ and $b \in \mathbb{R}$. This approach works for $\beta \sin t$ replaced by a C^1-function with mean value zero. After some remarks on the Ritz method for approximating the solutions, he considers the existence of odd 2π-periodic solutions of (2) by reducing it to the nonlinear integral equation

$$x(t) = -a\int^{\pi} K(t,\tau) \sin x(\tau)\, d\tau - b \sin t$$

where $K(t,\tau) = t(\frac{\tau}{\pi} - 1)$ if $t < \tau$ and $\tau(\frac{t}{\pi} - 1)$ if $t > \tau$, i.e. to the solution of the first boundary value problem

$$x'' + a \sin x = b \sin t, \quad x(0) = x(\pi) = 0 \tag{0.3}$$

followed by extending the solution on $[-\pi, 0]$ by the relation $x(-t) = -x(t)$. He then proves the existence of a unique solution (up to translations of 2π) when $a < 1$, by the method of successive approximations. Notice that the famous Birkhoff and Kellogg's paper on fixed point theory [2], published the same year, contains an application to rather general boundary value problems for n^{th} order ordinary differential equations which, applied to (0.3) would imply the existence of a solution for every a. Hammerstein [12] obtained in 1930 the existence of a solution for

$$x'' + a \sin x = e(t), \tag{0.4}$$

$$x(0) = x(\pi) = 0, \tag{0.5}$$

for arbitrary a and arbitrary continuous e as an application of his variational method for nonlinear integral equations and the uniqueness when $a < 1$. Another proof of those results was given by Iglisch [13] the same year by a method of continuation. In 1944, Lettenmeyer [16] proves the existence of a solution for (0.1) with the boundary conditions (0.5) or

$$x(0) = x'(\tfrac{\pi}{2}) = 0 \tag{0.6}$$

by a shooting method. He then shows that solutions of (0.4) satisfying the boundary conditions (0.6) can be extended to a 2π-periodic solution. Those results are rediscovered by Marlin [17] as special cases of existence results for symmetric periodic solutions for systems of the form

$$x'' = f(t,x)$$

where f is sufficiently smooth, periodic in t and x and has some symmetries. Those results are extended by topological degree arguments in [19] to systems of the form

$$x'' + \text{grad } F(x') + f(t,x,x') = 0$$

(see also [23]). When specialized to the scalar case, one obtains for the problem

$$x'' + g(x') + a \sin x = e(t), \quad x(0) - x(2\pi) = x'(0) - x'(2\pi) = 0,$$

the existence of a solution satisfying the relation $x(t+\pi) = -x(t)$ when g is an arbitrary odd continuous function and when $e(t+\pi) = -e(t)$, generalizing Lettenmeyer's result to some dissipative cases. Let us mention also some results, more in the direction of numerical analysis, by Borges, Cesari and Sanchez [4], Cesari and Bowman [6], Bononcini [3] and B. Schmitt and Brzezinski [24], which all deal with various types of symmetric periodic solution for (1) or special cases of it.

All the above results deal with the first and second boundary value problems or with some symmetric periodic solutions if the forcing term $e(t)$ and the dissipative term $g(x')$ have some symmetry properties. In the abstract setting, they correspond to nonlinear perturbations of an invertible linear operator whose solutions are a priori bounded so that existence follows easily. The general case of periodic boundary conditions without symmetry is of a different nature because a necessary condition for the solvability of (0.1) with periodic boundary conditions is that

$$\left|\frac{1}{2\pi} \int_a^{2\pi} e(t)\, dt\right| \leq a,$$

as it follows easily by integrating (0.1) over one period. So the range of the operator $\frac{d^2}{dt^2} + a \sin(.)$ on the space of 2π-periodic functions is no more the whole space. The same conclusion holds for the homogeneous Neumann boundary conditions. Maybe the first paper dealing with the general situation is that of Knobloch [14] introducing the use of upper and lower solutions in the study of periodic boundary value problems. As an example, he considers the problem

$$x'' - |x'|x' + \sin x = -\sin t, \quad x(0) - x(2\pi) = x'(0) - x'(2\pi) = 0$$

and proves the existence of a solution; the interest of his method is that it holds when $-\sin t$ is replaced by any continuous $e(t)$ such that $|e(t)| \leq 1$. He also shows that the problem

$$x'' + ax'^2 + \sin x = -\sin t + c, \quad x(0) - x(2\pi) = x'(0) - x'(2\pi) = 0$$

has no solution if $c > 2$ and that for each $c \in]0,2[$, it has a solution for sufficiently large a. We shall see in section 2 and 3 how the use of lower and upper solutions techniques can give further information. In 1968, Marlin and Ullrich [18] proved by the use of the Poincaré operator that the problem

$$x'' + a \sin x = e(t), \quad x(0) - x(2\pi) = x'(0) - x'(2\pi) = 0$$

where

$$\int_0^{2\pi} e(t)\, dt = 0 \tag{0.7}$$

has a solution if

$$a + \max_{t \in [0,2\pi]} |e(t)| < \frac{1}{8\pi}. \tag{0.8}$$

When (0.7) holds, (0.8) was improved to

$$a < 1 \tag{0.9}$$

by Castro [5], using a min-max method and (0.9) was shown to be superflous by Willem [25] and independently by Dancer [7] using a variational approach. We deal with it in section 4. For cases where (0.7) does not necessary hold, (0.8) was successively improved by Koneckny [15], Drabek [8] and Zanolin [26] (who allows some dissipative terms) and we shall describe results of this type in section 1.

The problem of characterizing the range of $\frac{d^2}{dt^2} + a \sin (.)$ was considered by Koneckny [15], Castro [5], Bates [1] and Dancer [7] and we describe extensions of their results to $\frac{d^2}{dt^2} + f(.) \frac{d}{dt} + \alpha \sin (.)$ in section 3. Finally, in the naturally associated cylinder space, the equation

$$x'' + cx' + a \sin x = 0$$

has two singular points of different nature (a center or focus and a saddle point), that means at least two periodic solutions. We present in section 2 a generalization of this result to the forced case, allowing moreover the presence of dissipative terms of rather general nature.

Of course most of the described methods work in situations much more general that ones described here and some of them will be developed in subsequent papers.

1. EXPLICIT CONDITIONS OF PERTURBATIONS TYPE FOR THE EXISTENCE OF MULTIPLE PERIODIC SOLUTIONS

Let us consider the problem

$$x'' + f(x)x' + a \sin x = e(t),$$

$$(1.1) \qquad x(0) - x(2\pi) = x'(0) - x'(2\pi) = 0$$

where $f : \mathbb{R} \to \mathbb{R}$ is continuous, $a > 0$ and $e \in L^1(0,2\pi)$.

The results of this section will unify and extend in various direction previous ones by Marlin and Ullrich [18], Konecny [15], Drabeck [8], Nagle and Singkofer [22] and Zanolin [26]. Except for the multiplicity results, they follow essentially [26]. If $g \in L^1(0,2\pi)$, we shall write

$$\bar{g} = (2\pi)^{-1} \int_0^{2\pi} g(t)\, dt, \quad \tilde{g}(t) = g(t) - \bar{g},$$

if $g \in L^p(0,2\pi)$, $(p \geq 1)$, we write

$$|g|_p = \left(\int_0^{2\pi} |g(t)|^p\, dt\right)^{1/p},$$

and if $g \in L^\infty(0,2\pi)$, then

$$|g|_\infty = \operatorname*{ess\,sup}_{t \in [0,2\pi]} |g(t)|.$$

THEOREM 1. *Assume that*

$$(1.2) \qquad -a < \bar{e} < a, \quad 2\pi a + |\tilde{e}|_1 < 3,$$

and that

$$(1.3) \qquad \arcsin \frac{|\bar{e}|}{a} < \frac{\pi}{2} - \frac{\pi}{6}(2\pi a + |\tilde{e}|_1)$$

Then (1.1) *has at least two distinct solutions* x *such that* $\bar{x} \in [-\frac{\pi}{2}, \frac{3\pi}{2}]$.

Proof. We want to apply the continuation theorem of [10] and we refer to the current literature for the usual reduction of (1.1) to a semi linear operator equation betwen the spaces $C^1(0,2\pi)$ and $L^1(0,2\pi)$ satisfying the regularity conditions of [10]. To check the a priori bounds and degree conditions of the continuation theorem, let $\lambda \in]0,1[$ and x be a possible solution of

$$(1.4) \qquad x'' + \lambda f(x)x' + \lambda a \sin x = \lambda e(t),$$

$$(1.5) \qquad x(0) - x(2\pi) = x'(0) - x'(2\pi) = 0.$$

Multipying both members of (1.4) by $\tilde{x}$ and using the boundary conditions, we get

$$(1.6)\qquad \int_0^{2\pi} \tilde{x}'^2(t)\,dt \leq a|\tilde{x}|_\infty \int_0^{2\pi} |\sin x(t)|\,dt + |\tilde{x}|_\infty |\tilde{e}|_1,$$

so that, by the well known inequality (see e.g. [23]),

$$(1.7)\qquad |\tilde{x}|_\infty \leq \frac{\pi}{\sqrt{6}} |\tilde{x}'|_2,$$

we obtain

$$(1.8)\qquad |\tilde{x}|_\infty < \frac{\pi}{6} (2\pi a + |\tilde{e}|_1).$$

Define

$$\Omega_1^0 = \{x \in C^1(0,2\pi) : x \text{ satisfies } (1.5),\ |\tilde{x}|_\infty < \frac{\pi}{6} (2\pi a + |\tilde{e}|_1) \text{ and } \bar{x} \in]-\frac{\pi}{2},\frac{\pi}{2}[\ \}.$$

If $x \in \Omega_1^0$ and satisfies (1.4)-(1.5), then, using (1.4), we get

$$|x''|_1 \leq C_1 |x'|_2 + C_2 < C_3$$

where the C_i depend only on a and e. Therefore, as x' necessarily vanishes at one point, we have

$$(1.9)\qquad |x'|_\infty \leq |x''|_1 < C_3$$

for every $x \in \Omega_1^0$ satisfying (1.4)-(1.5). Now define

$$\Omega_1^0 = \{x \in C^1(0,2\pi): x \text{ satisfies } (1.5),\ |\tilde{x}|_\infty < \frac{\pi}{6} (2\pi a + |\tilde{e}|_1),\ |\bar{x}| < \frac{\pi}{2} \text{ and } |x'|_\infty < C_3\}$$

It is an open bounded set of $C^1(0,2\pi)$ and if x is a solution of (1.4)-(1.5) such that $x \in \partial\Omega_1$, then necessarily, by (1.8) and (1.9) we must have $\bar{x} = -\frac{\pi}{2}$ or $\bar{x} = \frac{\pi}{2}$. Now, integrating (1.4) on $[0,2\pi]$ and using (1.5), we get

$$(1.10)\qquad \frac{1}{2\pi} \int_0^{2\pi} \sin(\bar{x} + \tilde{x}(t))\,dt = \frac{\bar{e}}{a}.$$

But, if $\bar{x} = \frac{\pi}{2}$, then, by (1.2), (1.3) and (1.8), we have

$$\sin(\frac{\pi}{2} + \tilde{x}(t)) > \frac{\bar{e}}{a}$$

for all $t \in [0,2\pi]$ and hence

$$\frac{1}{2\pi} \int_0^{2\pi} \sin(\frac{\pi}{2} + \tilde{x}(t))\,dt > \frac{\bar{e}}{a}.$$

Similarly, if $\bar{x} = -\frac{\pi}{2}$, then

$$\frac{1}{2\pi} \int_0^{2\pi} \sin(-\frac{\pi}{2} + \tilde{x}(t))\,dt < \frac{\bar{e}}{a}$$

and we conclude that $x \notin \partial\Omega_1$ for every solution of (1.4)-(1.5). On the other hand, by (1.2) the mapping

$$F : \bar{x} \to a \sin \bar{x} - \bar{e}$$

has a Brouwer degree $d_B[F,]-\frac{\pi}{2},\frac{\pi}{2}[, 0]$ with respect to $]-\frac{\pi}{2},\frac{\pi}{2}[$ equal to +1. Thus (1.4)-(1.5) has a least one solution x such that $\bar{x} \in]-\frac{\pi}{2},\frac{\pi}{2}[$. One can proceed similarly with

$$\Omega_2^0 = \{x \in C^1(0,2\pi) : x \text{ satisfies } (1.5),$$
$$|\tilde{x}|_\infty < \frac{\pi}{6}(2\pi a + |\tilde{e}|_1) \text{ and } \bar{x} \in]-\frac{\pi}{2},\frac{3\pi}{2}[\ \}.$$

and obtain the existence of a solution x such that $\bar{x} \in]\frac{\pi}{2},\frac{3\pi}{2}[$, so that the proof is complete.

2. DEGREE THEORY AND THE EXISTENCE OF MULTIPLE SOLUTIONS FOR THE PENDULUM-LIENARD EQUATION.

For $c \neq 0$, if the problem

(2.1) $$x'' + cx' + a \sin x = \bar{e}$$
$$x(0) - x(2\pi) = x'(0) - x'(2\pi) = 0.$$

has a solution, then, multiplying the equation by x' and integrating over $[0,2\pi]$, we get

$$c \int_0^{2\pi} x'^2(t)\, dt = 0$$

and hence x is constant, i.e. $x(t) = \bar{x}$, with

$$a \sin \bar{x} = \bar{e}.$$

Consequently, (2.1) has two distinct solutions in $[0,2\pi]$ if $\bar{e} \in]-a,a[$, one solution in $[0,2\pi]$ if $|\bar{e}| = a$ and no solution if $|\bar{e}| > a$. We shall consider the extension of this elementary result to problems of the form

(2.2) $$x'' + g(t,x,x') + a \sin x = e(t)$$
$$x(0) - x(2\pi) = x'(0) - x'(2\pi) = 0$$

where $g : [0,2\pi] \times \mathbb{R}^2 \to \mathbb{R}$ and $e : [0,2\pi] \to \mathbb{R}$ are continuous,

$$g(t,x,0) = 0$$

for every $(t,x) \in [0,2\pi] \times \mathbb{R}$, g satisfies the following *Nagumo type condition* : there exists a constant $C > 0$ such that every possible solution of (2.2) verifying

$$|x|_\infty \leq \frac{3\pi}{2}$$

is such that

$$|x'|_\infty < C.$$

Example of admissible g are the following ones :

a) g depends only on x' (see [20])

b) $|g(t,x,y)| \leqslant h(|y|)$ for $(t,x,y) \in [0,2\pi] \times [-\frac{3\pi}{2}, \frac{3\pi}{2}] \times \mathbb{R}$ where h is positive continuous and such that

$$\int_0^\infty \frac{s\ ds}{h(s)} = +\infty$$

(see e.g. [10]). In particular, $g(t,x,y) = f(x)|y|y^\delta$ with an arbitrary continuous f satisfies this condition if $\gamma \geqslant 0$, $\delta \geqslant 0$ and $\gamma + \delta \leqslant 2$.

We have now the following result whose existence part for one solution at least was already given Knobloch [14] when $g(t,x,y) = y|y|$.

THEOREM 2. *If the above conditions on* g *are satisfied then the problem (2.1) has at least two solutions not differing by a multiple of* 2π *if*

$$|e|_\infty < a \tag{2.3}$$

and at least one solution if

$$|e|_\infty = a.$$

<u>**Proof**</u>. By assumption, we have, for all $k \in \mathbb{Z}$,

$$e(t) - a\sin(\frac{\pi}{2} + 2k\pi) - g(t, \frac{\pi}{2} + 2k\pi, 0) = e(t) + a \leqslant 0$$

for all $t \in [0,2\pi]$, with a strict inequality if (2.3) holds. Similarly, for all $l \in \mathbb{Z}$,

$$e(t) - a\sin(\frac{3\pi}{2} + 2l\pi) - g(t, \frac{3\pi}{2} + 2l\pi, 0) = e(t) - a \geqslant 0$$

with a strict inequality if (2.3) holds. Taking just $k = l = 0$, the existence of at least one solution such that $\frac{\pi}{2} \leqslant x(t) \leqslant \frac{3\pi}{2}$ for $t \in [0,2\pi]$ when $|e|_\infty \leqslant a$ then follows from classical results (see e.g. [10] and [20]). Now if we define

$$\text{dom } L = \{x \in C^1(0,2\pi) : x \text{ is of class } C^2,$$
$$x(0) - x(2\pi) = x'(0) - x'(2\pi) = 0\},$$

$$L : \text{dom } L \subset C^1(0,2\pi) \to C^0(0,2\pi),\ x \to x''$$

$$N : C^1(0,2\pi) \to C^0(0,2\pi),\ x \to e - a \sin x - g(.,x,x'),$$

$$\Omega_{k,l} = \{x \in C^1(0,2\pi) : \frac{\pi}{2} + 2k\pi < x(t) < \frac{3\pi}{2} + 2l\pi,\ t \in [0,2\pi]$$

$$|x'|_\infty < C\} \qquad (l \geq k)$$

where C is the constant given by the Nagumo condition, it follows from a slight modification of the proof of Theorem V.5 in [10] that the coincidence degree $D[(L,N), \Omega_{0,0}]$ is defined when (2.3) holds, and that

$$D[(L,N), \Omega_{0,0}] = d_B[\bar{e} - a \sin(.),]\frac{\pi}{2}, \frac{3\pi}{2}[, 0] = 1.$$

We find similarly that if (2.3) holds, we have

$$D[(L,N), \Omega_{-1,-1}] = d_B[\bar{e} - a \sin(.),]-\frac{3\pi}{2}, -\frac{\pi}{2}[, 0] = +1$$

and

$$D[(L,N), \Omega_{-1,0}] = d_B[\bar{e} - a \sin(.),]-\frac{3\pi}{2}, \frac{3\pi}{2}[, 0] = +1.$$

But

$$\Omega_{0,0} \subset \Omega_{-1,0},\ \Omega_{-1,-1} \subset \Omega_{-1,0},$$

and hence, by the excision property of degree, we get

$$1 = D[(L,N), \Omega_{-1,0}] = D[(L,N), \Omega_{-1,-1}] + D[(L,N), \Omega_{0,0}] +$$

$$+ D[(L,N), \Omega_{-1,0} \setminus (\bar{\Omega}_{-1,-1} \cup \bar{\Omega}_{0,0})] = -2 +$$

$$+ D[(L,N), \Omega_{-1,0} \setminus (\bar{\Omega}_{-1,-1} \cup \bar{\Omega}_{0,0})],$$

and hence

$$D[(L,N), \Omega_{-1,0} \setminus (\bar{\Omega}_{-1,-1} \cup \bar{\Omega}_{0,0})] = -1.$$

Therefore there exists a solution of (2.2) in $\Omega_{-1,0} \setminus (\bar{\Omega}_{-1,-1} \cup \bar{\Omega}_{0,0})$ i.e. a solution such that

$$-\frac{3\pi}{2} < x(t) < \frac{3\pi}{2}$$

for all $t \in [0,2\pi]$, $x(\tau) > -\frac{\pi}{2}$ for at least one $\tau \in [0,2\pi]$ and $x(\tau') < \frac{\pi}{2}$ for at least one $\tau' \in [0,2\pi]$. Thus this solution cannot differ from the one in $\Omega_{0,0}$ by a multiple of 2π. Notice that in contrast the solutions in $\Omega_{0,0}$ and $\Omega_{-1,-1}$ may just be 2π-translates one from another.

3. UPPER AND LOWER SOLUTIONS AND THE RANGE OF DIFFERENTIAL EQUATIONS OF THE PENDULUM-LIENARD TYPE

We first recall that a lower (resp. upper) solution of the differential equation

(3.1) $$x''(t) = f(t,x(t),x'(t)),$$

where $f : [0,2\pi] \times \mathbb{R} \times \mathbb{R} \to \mathbb{R}$ is continuous, is a C^2-function $\alpha : [0,2\pi] \to \mathbb{R}$ (resp. $\beta : [0,2\pi] \to \mathbb{R}$) such that

$$\alpha''(t) \geqslant f(t,\alpha(t),\alpha'(t))$$

(resp. $\beta''(t) \leqslant f(t,\beta(t),\beta''(t))$)
for all $t \in [0,2\pi]$. (Special cases of constant lower or upper solutions have occured in section 2). The following lemma is elementary but useful.

LEMMA 1. *Assume that there exists some* $T>0$ *such that*

$$f(t,x,y) = f(t,x+T,y)$$

for all $(t,x,y) \in [0,2\pi] \times \mathbb{R} \times \mathbb{R}$. *Then, if (3.1) has a lower solution* α_0 *and an upper solution* β_0, *it also has a lower solution* α *and an upper solution* β *such that* $\alpha(t) \leqslant \beta(t)$ *for all* $t \in [0,2\pi]$.

Proof. Take $\alpha = \alpha_0$ and $\beta = \beta_0 + kT$ with $k \in \mathbb{N}^*$ such that $k \geqslant T^{-1} \max_{t \in [0,2\pi]} [\alpha_0(t) - \beta_0(t)]$.

We now consider the problem

(3.2) $$x'' + f(x)x' + a \sin x = e(t)$$

$$x(0) - x(2\pi) - x'(0) - x'(2\pi) = 0$$

where $a>0$, $f : \mathbb{R} \to \mathbb{R}$, 2π-periodic, and $e : [0,2\pi] \to \mathbb{R}$ are continuous. The results described in this section unify and extend in various directions previous ones of Koneckny [15], Castro [5], Bates [1] and Dancer [7]. We first establish a series of lemmas.

LEMMA 2. *If* $|\bar{e}| > a$, *then (3.2) has no solution and, for* $|\bar{e}| = a$, *only constant solutions* $\pm\frac{\pi}{2}$ *are possible.*

Proof. If (3.2) has a solution, then, integrating over $[0,2\pi]$ and using the boundary conditions we get

$$\frac{a}{2\pi}\int_0^{2\pi} \sin x(t)\,dt = \bar{e}$$

and hence $|\bar{e}| \leq a$. If $x \neq \pm\frac{\pi}{2} + 2k\pi$, then $|\sin x(t)| < 1$ on a set of positive measure and $|\bar{e}| < a$.

LEMMA 3. *If the problems*

$$\text{(3.3)}\qquad \begin{aligned} &x'' + f(x)x' + a\sin x = \bar{e}_i + \tilde{e}(t) \\ &x(0) - x(2\pi) = x'(0) - x'(2\pi) = 0 \end{aligned} \qquad (i = 1,2)$$

have solutions and if, say

$$\text{(3.4)}\qquad -a \leq \bar{e}_1 \leq \bar{e}_2 \leq a,$$

then the problem

$$\text{(3.5)}\qquad \begin{aligned} &x'' + f(x)x' + a\sin x = \bar{e} + \tilde{e}(t) \\ &x(0) - x(2\pi) = x'(0) - x'(2\pi) = 0 \end{aligned}$$

has a solution for every $\bar{e}$ *such that*

$$\text{(3.6)}\qquad \bar{e}_1 \leq \bar{e} \leq \bar{e}_2.$$

Proof. If ξ_i is a solution of (3.3) (i = 1,2), then, by (3.6), we have for all $t \in [0,2\pi]$,

$$\xi_1''(t) \leq -f(\xi_1)\xi_1' - a\sin\xi_1 + \bar{e} + \tilde{e}(t),$$

$$\xi_2''(t) \geq -f(\xi_2)\xi_2' - a\sin\xi_2 + \bar{e} + \tilde{e}(t)$$

so that ξ_1 (resp. ξ_2) is an upper (resp. lower) solution of (3.5) satisfying the periodic boundary conditions. By lemma 1 with T = 2π, we deduce the existence of a lower solution α and an upper solution β such that $\alpha(t) \leq \beta(t)$ $(t \in [0,2\pi])$ which satisfy the boundary conditions. As our equation clearly satisfies a Nagumo conditi on the set $\{t,x,y : \alpha(t)\leqslant x\leqslant\beta(t),\ t\in[0,2\pi],\ y\in\mathbb{R}\}$, the existence of a solution of (3.5) such that

$$\alpha(t)\leqslant x(t)\leqslant\beta(t),\ t\in[0,2\pi]$$

follows from classical results (see e.g. [10]).

For $\tilde{e} : [0,2\pi]\to\mathbb{R}$ continuous and such that

$$\int_0^{2\pi}\tilde{e}(t)\,dt = 0,$$

let $R(\tilde{e}) = \{\bar{e}\in\mathbb{R}$ such that (3.5) has a solution$\}$. By lemmas 2 and 3, $R(\tilde{e})$ is a (possibly empty) interval contained in $[-a,a]$.

LEMMA 4. *For every* $\bar{x}\in\mathbb{R}$, *the problem*

$$\tilde{x}'' + f(\bar{x}+\tilde{x})\tilde{x}' + a\sin(\bar{x}+\tilde{x}) - \frac{a}{2\pi}\int_0^{2\pi}\sin(\bar{x}+\tilde{x}(s))\,ds = \tilde{e}(t)$$

(3.7)

$$\tilde{x}(0)-\tilde{x}(2\pi) = \tilde{x}'(0)-\tilde{x}'(2\pi) = 0,$$

$$\int_0^{2\pi}\tilde{x}(t)\,dt = 0$$

has at least one solution.

Proof. Let, for $k\in\mathbb{N}$,

$$\tilde{C}^k(0,2\pi) = \{\tilde{x}\in C^k(0,2\pi) : \int_0^{2\pi}\tilde{x}(s)\,ds = 0\}$$

with the induced norm and let

$$H : \tilde{C}^0(0,2\pi)\to\tilde{C}^1(0,2\pi)$$

be the mapping defined by

$$(H\tilde{x})(t) = \int_0^t\tilde{x}(s)\,ds - \frac{1}{2\pi}\int_0^{2\pi}\int_0^t\tilde{x}(s)\,ds\,dt.$$

Notice that $\mathrm{Im}\,H\subset\{x\in\tilde{C}^1(0,2\pi) : x(0)-x(2\pi) = 0\}$, and clearly if we define $N_{\bar{x}} : \tilde{C}^1(0,2\pi)\to\tilde{C}^0(0,2\pi)$ by

$$(N_{\bar{x}}\tilde{x})(t) = \tilde{e}(t) - f(\bar{x}+\tilde{x}(t))\tilde{x}'(t) - a[\sin(\bar{x}+\tilde{x}(t)) - \frac{1}{2\pi}\int_0^{2\pi}\sin(\bar{x}+\tilde{x}(s)\,ds],$$

$$(t\in[0,2\pi]),$$

then (3.7) is equivalent to the fixed point problem

(3.8) $$\tilde{x} = H^2 N_{\bar{x}}(\tilde{x})$$

in $\tilde{C}^1(0,2\pi)$. Now $H^2 N_{\bar{x}}$ is easily shown to be a completely continuous mapping on $\tilde{C}^1(0,2\pi)$ so that (3.7) will have a solution, according to Leray-Schauder's theory (see e.g. [21]), if the set of all possible solutions of

(3.9) $$\tilde{x} = \lambda H^2 N_{\bar{x}}(\tilde{x}),\ \lambda \in [0,1]$$

is a priori bounded independently of λ. If $\tilde{x}$ is a fixed point of (3.9) for some λ, then $\tilde{x} \in \tilde{C}^2(0,2\pi)$, satisfies the periodic boundary conditions and the equation

(3.10) $$\tilde{x}'' + \lambda f(\bar{x} + \tilde{x})\tilde{x}' + \lambda a[\sin(\bar{x} + \tilde{x}) - \frac{1}{2\pi}\int_0^{2\pi} \sin(\bar{x} + \tilde{x}(s))\, ds] = \lambda \tilde{e}(t).$$

Multiplying (3.10) by $\tilde{x}$ and using the boundary conditions, we get

$$\frac{1}{2\pi}\int_0^{2\pi} \tilde{x}'^2(t)\, dt = \frac{\lambda a}{2\pi}\int_0^{2\pi} \tilde{x}(t)\sin(\bar{x} + \tilde{x}(t))\, dt - \frac{\lambda}{2\pi}\int_0^{2\pi} \tilde{x}(t)\tilde{e}(t)dt,$$

and hence, by Schwarz and Wirtinger inequalities,

$$|\tilde{x}'|_2 \leqslant a' + |\tilde{e}|_2 \qquad (a' = \sqrt{2\pi}\, a)$$

which implies in turn, by (1.7),

$$|\tilde{x}|_\infty \leqslant \frac{\pi}{\sqrt{6}}(a' + |\tilde{e}|_2).$$

But then, integrating $|\tilde{x}''|$ taken from (3.10), we get

$$|\tilde{x}''|_1 \leqslant C$$

where C depends only on a and $\tilde{e}$ (and not of $\bar{x}$), hence, as $\tilde{x}'$ necessarily vanishes at one point, this gives

$$|\tilde{x}'|_\infty \leqslant 2\pi C$$

and the possible solutions of (3.9) are a priori bounded in $\tilde{C}^1(0,2\pi)$.

LEMMA 5. *For each* $\tilde{e} \in \tilde{C}(0,2\pi)$,

(3.11) $$R(\tilde{e}) = \{\frac{a}{2\pi}\int_0^{2\pi} \sin(\bar{x} + \tilde{x}_{\bar{x}}(t))\, dt : \tilde{x}_{\bar{x}} \text{ satisfies } (3.7)$$

$$\text{and } \bar{x} \in \mathbb{R}\} \neq \phi.$$

Proof. That the second set is nonempty follows from lemma 4. Let $\bar{e} \in R(\tilde{e})$; then there exists a solution x of (3.5) and we can write it $x = \bar{x} + \tilde{x}$.

But (3.5) is equivalent to

$$\tilde{x}'' + f(\tilde{x} + \bar{x})\tilde{x}' + a[\sin(\bar{x} + \tilde{x}) - \frac{1}{2\pi}\int_0^{2\pi} \sin(\bar{x} + \tilde{x}(s))\, ds] = \tilde{e}(t),$$

$$\frac{a}{2\pi}\int_0^{2\pi} \sin(\bar{x} + \tilde{x}(s))\, ds = \bar{e}$$

which shows that

$$\bar{e} \in \{\frac{a}{2\pi}\int_0^{2\pi} \sin(\bar{x} + \tilde{x}_{\bar{x}}(t))\, dt : \tilde{x}_{\bar{x}} \text{ satisfies (3.7) and } \bar{x} \in \mathbb{R}\}. \tag{3.12}$$

New let $\bar{e}$ satisfy (3.12). Then there is $\bar{x} \in \mathbb{R}$ and $\tilde{x}_{\bar{x}}$ which verifies (3.7) and is such that $\bar{e} = \frac{a}{2\pi}\int_0^{2\pi} \sin(\bar{x} + \tilde{x}_{\bar{x}}(t)dt$. Letting $x = \bar{x} + \tilde{x}_{\bar{x}}$ we see that x verifies the periodic boundary conditions and the equation

$$x'' + f(x)x' + a \sin x = \frac{a}{2\pi}\int_0^{2\pi} \sin(\bar{x} + \tilde{x}_{\bar{x}}(s))\, ds + \tilde{e}(t) = \bar{e} + \tilde{e}(t)$$

so that $\bar{e} \in R(\tilde{e})$, and the proof is complete.

Summarizing and completing the above lemmas, we obtain the following

THEOREM 3. *For each* $\tilde{e} \in \tilde{C}(0,2\pi)$, $R(\tilde{e})$ *is a non-empty closed subinterval of* $[-a,a]$ *characterized by the relation (3.11).*

Proof. If remains only to prove that $R(\tilde{e})$ is closed. Let $(\bar{e}_k)$ be a sequence in $R(\tilde{e})$ which converges to $\bar{e}$ and let x_k be a solution of

$$\begin{aligned} &x'' + f(x)x' + a \sin x = \bar{e}_k + \tilde{e}(t) \\ &x(0) - x(2\pi) = x'(0) - x'(2\pi) = 0, \quad (k \in \mathbb{N}^*). \end{aligned} \tag{3.13}$$

As $x_k + 2m\pi$ is solution together with x_k for every $m \in \mathbb{Z}$, we can assume without loss of generality that $\bar{x}_k \in [0,2\pi]$ $(k \in \mathbb{N}^*)$. From (3.13), we see like in the proof of lemma 4, that $\tilde{x}_k$ satisfies the fixed point equation

(3.14) $$\tilde{x}_k = H^2 N_{\bar{x}_k}(\tilde{x}_k)$$

and that

$$|\tilde{x}_k|_\infty \leq \frac{\pi}{\sqrt{6}} (a' + |\tilde{e}|_2), \quad |\tilde{x}_k'|_\infty \leq 2\pi C, \quad |\tilde{x}_k''|_1 \leq C.$$

By the compactness of $[0,2\pi]$ and Ascoli-Arzela theorem, there exists a subsequence (x_{j_k}) such that $(\bar{x}_{j_k})$ converges to some $\bar{x} \in [0,2\pi]$ and $(\tilde{x}_{j_k})$ converges in $\tilde{C}^1(0,2\pi)$ to some $\tilde{x}$. From (3.14) we deduce then that

$$\tilde{x} = H^2 N_{\bar{x}}(\tilde{x}),$$

and from the relation, deduced from (3.13),

$$\frac{a}{2\pi} \int_0^{2\pi} \sin(\bar{x}_{j_k} + \tilde{x}_{j_k}(t))\, dt = \bar{e}_{j_k},$$

we obtain

$$\frac{a}{2\pi} \int_0^{2\pi} \sin(\bar{x} + \tilde{x}(t))\, dt = \bar{e}.$$

Consequently, $x = \bar{x} + \tilde{x}$ is a solution of (3.5) and the proof is complete.

Remark 1. If we apply theorem 2 to problem (3.2) and notice that $|e|_\infty \leq |\bar{e}| + |\tilde{e}|_\infty$, we see that *for each* $\tilde{e}$ *with* $|\tilde{e}|_\infty \leq a$, *we have*

$$R(\tilde{e}) \supset \{\bar{e} \in \mathbb{R} : |\bar{e}| + |\tilde{e}|_\infty \leq a\}.$$

4. VARIATIONAL METHODS AND PERIODIC SOLUTIONS OF THE PENDULUM EQUATIONS WHEN THE FORCING TERM HAS MEAN VALUE ZERO.

We describe here the independant work of Willem [25] and Dancer [7] on the problem

$$\text{(4.1)}\qquad \begin{aligned} &x'' + a\sin x = e(t) \\ &x(0) - x(2\pi) = x'(0) - x'(2\pi) = 0 \end{aligned}$$

when $e \in L^1(0,2\pi)$ and $\bar{e} = 0$. We refer to [25] for abstract corresponding results and for considerations on the stability of solutions.

THEOREM 4. *If* $e \in L^1(0,2\pi)$ *and* $\bar{e} = 0$, *then (4.1) has at least one solution which is a minimum of the functional*

$$\varphi : H^1_{2\pi}(0,2\pi) \to \mathbb{R},\ x \to \int_0^{2\pi} \left[\frac{(x'(t))^2}{2} + a\cos x(t) + e(t)x(t)\right] dt.$$

Proof. Recall that $H^1_{2\pi}(0,2\pi)$ is the space of absolutely continuous and 2π-periodic functions such that $u' \in L^2(0,2\pi)$, with the inner product

$$(u,v)_{H^1} = \int_0^{2\pi} [u(t)v(t) + u'(t)v'(t)]\, dt.$$

It is easily shown that φ is weakly lower semi-continuous. Now, as $\bar{e} = 0$, we have, using will known inequalities,

$$\begin{aligned} \varphi(x) &= \int_0^{2\pi} \frac{(\tilde{x}'(t))^2}{2} + a\cos x(t) + e(t)\tilde{x}(t)\, dt \geq \\ \text{(4.2)}\qquad &\geq \int_0^{2\pi} \frac{(\tilde{x}'(t))^2}{2} - 2\pi a - |e|_1 |\tilde{x}|_\infty \geq \\ &\geq \frac{1}{2}|\tilde{x}|^2_{H^1} - \frac{\pi|e|_1}{\sqrt{6}} |\tilde{x}|_{H^1}, \end{aligned}$$

and hence $\varphi(x) \to +\infty$ if $|\tilde{x}|_{H^1} \to \infty$. Let (x_k) be a minimizing sequence for φ. As $\varphi(x)$ $\varphi(x+2\pi)$ $(x \in H^1_{2\pi})$, we can assume without loss of generality that $\bar{x}_k \in [0,2\pi]$ and, by (4.2) $(\tilde{x}_k)$ is bounded in $H^1_{2\pi}(0,2\pi)$, so that (x_k) has the same property. Thus (x_k) has a subsequence which weakly converges to some x and by the weak lower semi-continuity of φ, x is a minimum of φ. It is easily checked that the Euler equation satisfied by x implies that x is a solution of (4.1).

Remark. The above theorems completes the results of section 2 by showing that, *for* $f \equiv 0$, $0 \in R(\tilde{e})$ *for every* $\tilde{e} \in \tilde{C}(0,2\pi)$. We do not know of any nonvariational proof of this result which would make possible an extension to the not necessarily conservative case.

REFERENCES
++++++++++

[1] BATES P.W. A variational approach to solving semilinear equations at resonance (in Nonlinear Phenomena in Math. Sci., Arlington, 1980, to appear).

[2] BIRKHOFF G.D and KELLOGG O.D. Invariant points in function space (Trans. Amer. Math. Soc. 23 (1922) 96-115).

[3] BONONCINI V.E. Soluzioni periodiche di equazioni differenziali del secondo ordine (Riv. Mat. Univ. Parma (4) 3 (1977) 391-399).

[4] BORGES C.A., CESARI L. and SANCHEZ D.A. Functional analysis and the method of harmonic balance (Quarterly Appl. Math. 32 (1975) 457-464).

[5] CASTRO A. Periodic solutions of the forced pendulum equation (in "Differential Equations", Ahmad and Lazer ed., Academic Press, New York, 1980, 149-160).

[6] CESARI L. and BOWMAN T.T. Some error estimates by the alternative method (Quarterly Appl. Math. 34 (1977) 121-128).

[7] DANCER E.N. On the use of asymptotics in nonlinear boundary value problems (preprint).

[8] DRABEK P. Remarks on multiple solutions of nonlinear ordinary differential equations (Comment. Math. Univ. Carolinae 21 (1980) 155-160).

[9] DUFFING G. Erzwungene Schwingungen bei veränderlicher Eigenfrequenz (Monographie 41/42, Vieweg, Braunschweig, 1918).

[10] GAINES R.E. and MAWHIN J. Coincidence Degree and Nonlinear Differential Equations (Lecture Notes in Math. n° 568, Springer, Berlin, 1977).

[11] HAMEL G. Uber erzwungene Schwingungen bei endlichen Amplituden (Math. Ann. 86 (1922) 1-13).

[12] HAMMERSTEIN A. Nichtlineare Integralgleichungen nebst Anwendungen (Acta Math. 54 (1930) 117-176).

[13] IGLISCH R. Zur Theorie Schwingungen (Monatsh. Math. Phys. 37 (1930) 325-342).

[14] KNOBLOCH H.W. Eine neue Methode zur Approximation periodischer Lösungen nicht-linearer Differentialgleichungen zweiter Ordnung (Math. Z. 82 (1963) 177-197).

[15] KONECNY M. Remarks on periodic solvability of nonlinear ordinary differential equations (Comment. Math. Univ. Carolinae 18 (1977) 547-562).

[16] LETTENMEYER F. Uber die von einem Punkt ausgehenden Integralkurven einer Differentialgleichung zweiter Ordnung (Deutsche Math. 7 (1944) 56-74).

[17] MARLIN J.A. Periodic motions of coupled simple pendulums with periodic disturbances (Intern. J. Nonlinear Mech. 3 (1968) 439-447).

[18] MARLIN J.A. and ULLRICH D.F. Periodic solutions of second order nonlinear differential equations without damping (SIAM J. Appl. Math. 16 (1968) 998-1010).

[19] MAWHIN J. Une généralisation de théorèmes de J.A. Marlin (Int. J. Non-linear Mech. 5 (1970) 335-339).

[20] MAWHIN J. Boundary value problems for nonlinear second order vector differential equations (J. Differential Equations 16 (1974) 257-269).

[21] MAWHIN J. Topological Degree Methods in Nonlinear Boundary value Problems (CBMS Regional Confer. in Math. n° 40, Amer. Math. Soc., Providence, 1979).

[22] NAGLE R.K. and SINGKOFER K. Nonlinear ordinary differential equations at resonance with slowly varying nonlinearities (Applicable Analysis 11 (1980) 137-149).

[23] ROUCHE N. and MAWHIN J. Ordinary Differential Equations. Stability and Periodic solutions (Pitmann, Boston, 1980).

[24] SCHMITT B. and BRZEZINSKI R. Localisation numérique de solutions périodiques (Equadiff 78, Conti, Sestini and Villari ed. Firenze, 1978, 99-108).

[25] WILLEM M. Oscillations forcées de systèmes hamiltoniens (Publications de l'Université de Besançon, 1981).

[26] ZANOLIN F. Remarks on multiple solutions for nonlinear ordinary differential systems of Liénard type (Boll. Un. Mat. Ital., to appear).

Author Adress : Institut Mathématique
Université Catholique de Louvain
Chemin du Cyclotron, 2
B-1348 Louvain-La-Neuve
Belgium.

A SIMPLE LAYER POTENTIAL METHOD FOR THREE-DIMENSIONAL EDDY CURRENT PROBLEMS

R.C. MacCamy - E. Stephan

1. INTRODUCTION

Let S be a closed analytic surface which divides $\mathbb{R}^3$ into simply connected disjoint domains, an interior Ω' (bounded) and an exterior Ω (unbounded). Ω is to represent air and Ω' a metallic conductor. Thus the exterior domain Ω is characterized by constitutive parameters ε_o,μ_o denoting permitivity and permeabilty and is assumed to have zero conductivity. The interior domain Ω' is characterized by constants ε,μ,σ where the conductivity σ may be infinite. The total electromagentic field $(\underset{\sim}{E},\underset{\sim}{H})$ will consist of the sum of incident $(\underset{\sim}{E}^o,\underset{\sim}{H}^o)$ and scattered $(\underset{\sim}{E}^s,\underset{\sim}{H}^s)$ terms where the incident field is assumed to originate in Ω. All fields are assumed to be time harmonic and monochromatic, i.e. to have a time harmonic dependence $e^{-i\omega t}$. The frequency ω and constitutive parameters are related to wave numbers appropriate to Ω and Ω' by

$$k = \omega\sqrt{\varepsilon_o\mu_o} \quad , \qquad k' = \omega\sqrt{(\varepsilon + \frac{i\sigma}{\omega})\,\mu} \tag{1}$$

and the time harmonic Maxwell equations are

$$\text{curl}\,\underset{\sim}{E} = i\omega\mu_o\underset{\sim}{H} \;, \quad \text{curl}\,\underset{\sim}{H} = -i\omega\varepsilon_o\underset{\sim}{E} \qquad \text{in } \Omega \tag{2}$$

$$\text{curl}\,\underset{\sim}{E} = i\omega\mu\,\underset{\sim}{H} \;, \quad \text{curl}\,\underset{\sim}{H} = (-i\omega\varepsilon + \sigma)\underset{\sim}{E} \quad \text{in } \Omega' \tag{3}$$

It is noted that the incident field satisfies (2) almost everywhere in $\mathbb{R}^3$ and the field quantities are infinitely differentiable except at source points in Ω and Ω'. Across the interface $S = \partial\Omega = \partial\Omega'$ the tangential component of the total field $(\underset{\sim}{E},\underset{\sim}{H})$ must be continuous. Therefore after an appropriate scaling ($\underset{\sim}{\mathrm{H}} = i\omega\mu_o\underset{\sim}{H}$, $\underset{\sim}{\mathrm{E}} = \underset{\sim}{E}$ in Ω ; $\underset{\sim}{\mathrm{H}} = i\omega\mu\underset{\sim}{H}$, $\underset{\sim}{\mathrm{E}} = \underset{\sim}{E}$ in Ω') the <u>eddy current</u> problem is given by

$$\begin{aligned} &\text{curl}\,\underset{\sim}{\mathrm{E}} = \underset{\sim}{\mathrm{H}} \;, && \text{curl}\,\underset{\sim}{\mathrm{H}} = \alpha^2\mathrm{E} \;\text{ in } \Omega \\ &\text{curl}\,\underset{\sim}{\mathrm{E}} = \underset{\sim}{\mathrm{H}} \;, && \text{curl}\,\underset{\sim}{\mathrm{H}} = i\beta\underset{\sim}{\mathrm{E}} \;\text{ in } \Omega' \qquad (P_{\alpha\beta}) \\ &\text{and } (\underset{\sim}{n}\times\underset{\sim}{\mathrm{E}})^+ = (\underset{\sim}{n}\times\underset{\sim}{\mathrm{E}})^- \;, && (\underset{\sim}{n}\times\underset{\sim}{\mathrm{H}})^+ = (\underset{\sim}{n}\times\underset{\sim}{\mathrm{H}})^- \;\text{ on } S \end{aligned}$$

Here $\alpha^2 = \omega^2\varepsilon_o\mu_o$, $\beta = (\omega\mu\sigma - i\omega^2\mu\varepsilon)$ are dimensionless parameters and $\beta = \omega\mu\sigma > 0$ if displacement currents are neglected in metal $(\varepsilon=0)$. The superscripts plus and minus denote limits from Ω and Ω' where $\underset{\sim}{n}$ is outward directed normal on the surface S .

At higher conductivity the constant β is usually large and this leads to the <u>perfect conductor</u> approximation. Formally this means solving only the Maxwell equations (2) in Ω for the scattered field and requiring that the tangential component of the total electric field vanishes on S ($\underset{\sim}{n}\times\underset{\sim}{E} = 0$ on S) , i.e.

$$\begin{aligned} \operatorname{curl} \underset{\sim}{E}^s = \underset{\sim}{H}^s , \quad \operatorname{curl} \underset{\sim}{H}^s &= \alpha^2 \underset{\sim}{E}^s && \text{in } \Omega \\ (\underset{\sim}{n}\times\underset{\sim}{E}^s)^+ &= -(\underset{\sim}{n}\times\underset{\sim}{E}^o)^+ && \text{on } S \end{aligned} \qquad (P_{\alpha\infty})$$

If in addition the scattered field satisfies the Sommerfeld radiation condition the following uniqueness theorem holds:

<u>Theorem 1: There exists at most one solution of</u> $(P_{\alpha\beta})$ <u>for any</u> $\alpha>0$ <u>and</u> $0<\beta\leq\infty$.

The proof of uniqueness for $(P_{\alpha\infty})$ can be found in [5] and the corresponding one for $(P_{\alpha\beta})$ (with $\sigma\neq 0$ in Ω') in [3, p. 592].

In order to avoid additional difficulties we need the technical assumption

$$\operatorname{curl} \underset{\sim}{E} = \underset{\sim}{H} , \ \operatorname{curl} \underset{\sim}{H} = \alpha^2 \underset{\sim}{E} \text{ in } \Omega' , \ \underset{\sim}{n}\times\underset{\sim}{E} = 0 \text{ on } S \ \text{ implies } \underset{\sim}{E} = \underset{\sim}{H} = \underset{\sim}{0} \text{ on } \Omega' \qquad (4)$$

i.e. α^2 is not an eigenvalue of the interior Dirichlet problem for Maxwell's equations.

2. Our simple layer potential procedure

In the following we give new boundary integral equation methods for solving both $(P_{\alpha\beta})$ and $(P_{\alpha\infty})$. To this end we introduce the <u>simple layer</u> V_γ with continuous density ψ on the surface S by

$$V_\gamma(\psi)(x) = \frac{1}{4\pi} \int_S \psi(y)\phi_\gamma(|x-y|)\,dS_y , \qquad x \in \mathbb{R}^3 \quad . \qquad (5)$$

Here

$$\phi_\gamma(|x-y|) = \frac{e^{i\gamma|x-y|}}{|x-y|} \qquad (6)$$

is the fundamental solution of the Helmholtz equation $\Delta w = -\gamma^2 w$ satisfying the Sommerfeld radiation condition for $\operatorname{Re}\gamma \neq 0$. For a vector field $\underset{\sim}{v}$ on S we define $V_\gamma(\chi)$ by (5) again with χ replacing ψ. There hold the following well-known properties of the simple layer potential (see [4],[5]).

Lemma 2: For any complex $\gamma, 0 \le \arg\gamma \le \pi/2$ and any continuous ψ, χ on S:

(i) $V_\gamma(\psi)$ is continuous in $\mathbb{R}^3$

(ii) $\Delta V_\gamma(\psi) = -\gamma^2 V_\gamma(\psi)$ in $\Omega \cup \Omega'$

(iii) $V_\gamma(\psi)(x) = O(|x|^{-1} e^{i\gamma|x|})$ as $|x| \to \infty$.

(iv) $(\frac{\partial}{\partial n} V_\gamma(\psi)(x))^{\pm} = \mp \frac{1}{2}\psi(x) + \int_S K_\gamma(x,y)\psi(y)\,dS_y$ on S

(v) $(n \times \operatorname{curl} V_\gamma(\chi)(x))^{\pm} = \pm \frac{1}{2}\chi(x) + \int_S K_\gamma(x,y)\chi(y)\,dS_y$ on S

where the function (matrix function) K_γ (K_γ) is $O(|x-y|^{-1})$ as $y \to x$.

Our methods are based on the Stratton-Chu representation formulas of electromagnetic fields from [7] yielding for the scattered fields (E,H) in $(P_{\alpha\infty})$:

$$\begin{aligned} E &= V_\alpha(n\times H) - \operatorname{curl} V_\alpha(n\times E) + \operatorname{grad} V_\alpha(n\cdot E) \\ H &= \operatorname{curl} V_\alpha(n\times H) - \operatorname{curl}\operatorname{curl} V_\alpha(n\times E) \end{aligned} \quad \text{in } \Omega \tag{7}$$

If $n\times H$, $n\times E$ and $n\cdot E$ were all known on S then (7) would yield a solution of $(P_{\alpha\infty})$ - but only $n\times E$ is known. The standard treatment of $(P_{\alpha\infty})$ starts from (7) but sets $n\times H$ and $n\cdot E$ equal to zero and replaces $-n\times E$ by an unknown tangential field L (see [2]):

$$E = \operatorname{curl} V_\alpha(L)\ , \quad H = \operatorname{curl}\operatorname{curl} V_\alpha(L)\ . \tag{8}$$

Imposition of the boundary condition in $(P_{\alpha\infty})$ then yields an integral equation of second kind for L in the tangent space to S. The method (8) is analogous to solving the Dirichlet problem for the scalar Helmholtz equation with a double layer. It has the drawback, for the eddy current problem, that having found L it is hard to determine $n\times H$ on S. The latter involves finding a second normal derivative of $V_\alpha(L)$ on S.

Our method for $(P_{\alpha\infty})$ is analogous to solving the scalar problems with a simple layer (see [1]). We set $n\times E = 0$ in (7) and replace $n\times H$ and $n\cdot E$ by unknowns J and M. Thus we take

$$\underset{\sim}{E} = V_\alpha(\underset{\sim}{J}) + \text{grad } V_\alpha(M) \ , \quad \underset{\sim}{H} = \text{curl } V_\alpha(\underset{\sim}{J}) \tag{9}$$

Having deterimed $\underset{\sim}{J}$ we can use Lemma 2 (v) to determine $\underset{\sim}{n}\times\underset{\sim}{H}$ on S .

An easy computation shows that the ansatz (9) gives a solution of $(P_{\alpha\infty})$ if $\text{div } \underset{\sim}{E} = 0$ holds in Ω ; because in (9) only $\underset{\sim}{H}$ is automatically divergence free. But from (9) and Lemma 2 follows

$$\Delta \text{ div } \underset{\sim}{E} = -\alpha^2 \text{div } \underset{\sim}{E} \quad \text{in } \Omega .$$

Therefore by uniqueness of the exterior Dirichlet problem (for the Helmholtz equation) the boundary condition $\text{div } \underset{\sim}{E} = 0$ on S guarantees $\text{div } \underset{\sim}{E} \equiv 0$ in the exterior domain Ω . Imposing the boundary conditions of $(P_{\alpha\infty})$ in (9) and the constraint $\text{div } \underset{\sim}{E} = 0$ on S we obtain a coupled system of pseudodifferential equations on the boundary surface S for the unknown densities $\underset{\sim}{J}, M$:

$$\begin{aligned} V_\alpha(\underset{\sim}{J})_T + \text{grad}_T V_\alpha(M) &= -(\underset{\sim}{n}\times\underset{\sim}{E}^0) \\ V_\alpha(\text{div}_T \underset{\sim}{J}) - \alpha^2 V_\alpha(M) &= 0 \end{aligned} \qquad (E_{\alpha\infty})$$

where $V_\alpha(\underset{\sim}{J})_T$ denotes the tangential component of the vector function $V_\alpha(\underset{\sim}{J})$. Here we have used the relations with the surface divergence div_T and the surface gradient grad_T on S for any $\gamma\in\mathbb{C}$ with $0 \le \arg \gamma \le \frac{\pi}{2}$:

$$\text{div } V_\gamma(\underset{\sim}{v}) = V_\gamma(\text{div}_T \underset{\sim}{v}) \ , \quad \text{div}_T \text{grad}_T V_\gamma(M) = -\gamma^2 V_\gamma(M) .$$

Our procedure for $(P_{\alpha\beta})$ proceeds as follows. This time we let $\underset{\sim}{E}$ and $\underset{\sim}{H}$ denote the total fields and use again (9) in Ω and its analog in Ω' . Thus we put,

$$\begin{aligned} \underset{\sim}{E} &= \underset{\sim}{E}^0 + V_\alpha(\underset{\sim}{J}) + \text{grad } V_\alpha(M) \ , \quad \underset{\sim}{H} = \underset{\sim}{H}^0 + \text{curl } V_\alpha(\underset{\sim}{J}) \ \text{ in } \Omega \ , \\ \underset{\sim}{E} &= V_{\sqrt{i\beta}}(\underset{\sim}{j}) + \text{grad } V_{\sqrt{i\beta}}(m) \ , \quad \underset{\sim}{H} = \text{curl } V_{\sqrt{i\beta}}(\underset{\sim}{j}) \quad \text{in } \Omega' . \end{aligned} \tag{10}$$

Again the constraint $\text{div } \underset{\sim}{E} \equiv 0$ in $\Omega\cup\Omega'$ together with the boundary conditions in $(P_{\alpha\beta})$ give a coupled system of pseudodifferential equations for the unknown layers $(\underset{\sim}{J}, M, \underset{\sim}{j}, m)$ on S :

$$
\left.\begin{aligned}
&V_\alpha(\underset{\sim}{J})_T + \mathrm{grad}_T V_\alpha(M) - V_{\sqrt{i\beta}}(\underset{\sim}{j}) - \mathrm{grad}_T V_{\sqrt{i\beta}}(m) = -\underset{\sim}{E}_T^{\,o} \\
&V_\alpha(\mathrm{div}_T \underset{\sim}{J}) - \alpha^2 V_\alpha(M) = 0 \\
&\underset{\sim}{J} + K_\alpha(\underset{\sim}{J}) + \underset{\sim}{j} - K_{\sqrt{i\beta}}(\underset{\sim}{j}) = -2(\underset{\sim}{n}\times\underset{\sim}{H}^o) \\
&V_{\sqrt{i\beta}}(\mathrm{div}_T \underset{\sim}{j}) - i\beta\, V_{\sqrt{i\beta}}(m) = 0
\end{aligned}\right\} \qquad (E_{\alpha\beta})
$$

Equation $(E_{\alpha\beta})_1$ is caused by the interface condition $(\underset{\sim}{n}\times\underset{\sim}{E})^- = (\underset{\sim}{n}\times\underset{\sim}{E})^+$ on S, equations $(E_{\alpha\beta})_2$, $(E_{\alpha\beta})_4$ are consequences of the constraint $\mathrm{div}\,\underset{\sim}{E} = 0$ in $\Omega \cup \Omega'$ whereas equation $(E_{\alpha\beta})_3$ is resulting from the second interface condition $(\underset{\sim}{n}\times\underset{\sim}{H})^+ = (\underset{\sim}{n}\times\underset{\sim}{H})^-$ on S together with Lemma 2 (v) if we define

$$
\underset{\sim}{K}_\gamma(\underset{\sim}{v})(x) = 2\int_S \underset{\sim}{K}_\gamma(x,y)\underset{\sim}{v}(y)\,dS_y \; .
$$

Collecting the above results there holds the following representation theorem.

Theorem 3 [4]: (i) If $(\underset{\sim}{J},M)$ solve $(E_{\alpha\infty})$ with $\underset{\sim}{J}$ differentiable and M continuous then (9) yields a solution of $(P_{\alpha\infty})$.

(ii) If $(\underset{\sim}{J},M,\underset{\sim}{j},m)$ solve $(E_{\alpha\beta})$ with $\underset{\sim}{J},j$ differentiable and M,m continuous then (10) yields a solution of $(P_{\alpha\beta})$.

3. Existence, uniqeness and regularity of our integral equations

In the half-space case $\Omega = \{x\in\mathbb{R}^3 \mid x_3>0\}$ ($\underset{\sim}{n}(x) = \underset{\sim}{e}_3$ on $S = \mathbb{R}^2$) the system $(E_{\alpha\infty})$ is reduced on $x_3 = 0$ to

$$
\left.\begin{aligned}
&\phi_\alpha * \underset{\sim}{J} + \frac{\partial}{\partial x_1}\phi_\alpha * M\underset{\sim}{e}_1 + \frac{\partial}{\partial x_2}\phi_\alpha * M\underset{\sim}{e}_2 = -4\pi(\underset{\sim}{e}_3 \times \underset{\sim}{E}^o) \; , \\
&\phi_\alpha * \mathrm{div}\,\underset{\sim}{J} - \alpha^2\phi_\alpha * M = 0 \; ,
\end{aligned}\right. \qquad (11)
$$

where the star denotes convolution. Via Fourier transformation we find in [4]

$$
\hat{\phi}_\alpha(\xi) = (|\xi|^2 - \alpha)^{-1/2} \qquad (12)
$$

and the explicit solution of (11) reads with $\underset{\sim}{E}_T^o = \underset{\sim}{e}_3 \times \underset{\sim}{E}^o$:

$$
M = 4\,V_\alpha(\mathrm{div}\,\underset{\sim}{E}_T^o)\;,\quad \underset{\sim}{J} = -\mathrm{grad}_T(4\,V_\alpha \mathrm{div}\,\underset{\sim}{E}_T^o) + 2(\Delta+\alpha^2)V_\alpha(\underset{\sim}{E}_T^o)\;. \qquad (13)
$$

Let $H^r(\mathbb{R}^2)$ denote the Sobolev space of order r on $\mathbb{R}^2$, i.e. the completion of $C_0^\infty(\mathbb{R}^2)$ under the norm

$$\|\psi\|_r^2 = \int_{\mathbb{R}^2} (1 + |\xi|^2)^r \, |\hat{\psi}(\xi)|^2 d\xi \quad ,$$

and let $\underset{\sim}{H}^r(\mathbb{R}^2)$ be the space of vector functions with components in $H^r(\mathbb{R}^2)$. Now, V_α is a pseudodifferential operator of order minus one in the sense of [6] with symbol (12) and maps $H^r(\mathbb{R}^2)(\underset{\sim}{H}^r(\mathbb{R}^2))$ continuously into $H^{r+1}(\mathbb{R}^2)(\underset{\sim}{H}^{r+1}(\mathbb{R}^2))$ The operators div_T and grad_T are of order plus one and take $\underset{\sim}{H}^r(\mathbb{R}^2)$, $H^r(\mathbb{R}^2)$ into $H^{r-1}(\mathbb{R}^2)$, $\underset{\sim}{H}^{r-1}(\mathbb{R}^2)$ respectively. Though the result (13) is only formal, it indicates the mapping properties of the system $(E_{\alpha\infty})$ in the general case.

Theorem 4 [4] : For any real r the mapping $A_\alpha : (\underset{\sim}{J}, M) \to (-\underset{\sim}{E}_T^{\,o}, 0)$ defined by the system $(E_{\alpha\infty})$ is bijective from $\underset{\sim}{H}^{r-1}(S) \times H^r(S)$ onto $\underset{\sim}{H}^r(S) \times H^{r-1}(S)$.

The proof of Theorem 4 given in [4] uses partition of unity and local coordinate systems on the analytic surface S together with standard techniques for pseudodifferential operators. The expansion

$$\phi_\alpha(r) = \frac{e^{i\alpha r}}{r} = \frac{1}{r} \sum_{j=0}^{\infty} \frac{\delta^j}{j!} r^j \quad , \qquad \delta \in \mathbb{C} \ , \ r = |x-y| \tag{14}$$

induces the decomposition

$$V_\alpha \psi = V_i \psi + W_\alpha \psi \tag{15}$$

where the Bessel potential V_i is a pseudodifferential operator of order minus one and W_α is a smoothing operator of order -3. A standard Green's formula argument together with Lemma 2 shows that the Bessel potential operator V_i is injective. On the other hand V_i is a Fredholm operator of index zero (in any Sobolev space) since V_i is a self-adjoint elliptic pseudodifferential operator. Due to the decomposition (15) the bijectivity of $(E_{\alpha\infty})$ is proved in [4] by a perturbation argument, since $(E_{\alpha\infty})$ differs by compact operators from

$$\begin{aligned} V_i(\underset{\sim}{J})_T + \mathrm{grad}_T \, V_i(M) &= \underset{\sim}{F} \quad , \\ V_i(\mathrm{div}_T \, \underset{\sim}{J}) + V_i(M) &= G \quad . \end{aligned} \tag{16}$$

The system (16) can be reduced to a Riesz-Schauder system for $(\underset{\sim}{J},M)$ since $V_i,(V_i)_T$ and $(\Delta_T - I)$ are bijective and for any $\gamma\in\mathbb{C}$, $0 \le \arg\gamma \le \frac{\pi}{2}$, there holds

$$\operatorname{div}_T V_\gamma(\underset{\sim}{v})_T = V_\gamma(\operatorname{div}_T \underset{\sim}{v}) + \mathcal{J}_\gamma(\underset{\sim}{v}) \tag{17}$$

with a continuous map $\mathcal{J}_\gamma$ from $\underset{\sim}{H}^r(S)$ into $H^{r+1}(S)$, $r\in\mathbb{R}$.
The obtained Riesz-Schauder system for $(\underset{\sim}{J},M)$ is uniquely solvable which follows by potential theoretic arguments from Theorem 1 (together with jump relations).
In a second step we reduce our original system $(E_{\alpha\infty})$ to a Riesz-Schauder system making use of the bijectivity of (16). This again is uniquely solvable due to Theorem 1 and our technical assumption (4).

In a similar way we treat in [4] the system $(E_{\alpha\beta})$ for the eddy current problem. Using the bijectivity of $V_{\sqrt{i\beta}}$ from $H^r(S)$ onto $H^{r+1}(S)$ $(r\in\mathbb{R})$ and the relation (17) we reduce $(E_{\alpha\beta})$ to a Riesz-Schauder system on $\underset{\sim}{H}^{r-1}(S)\times H^r(S)$:

$$\begin{aligned}
\underset{\sim}{j} &= \underset{\sim}{J} + D_1(M,\underset{\sim}{J},\underset{\sim}{j}) + \underset{\sim}{g}\\
\underset{\sim}{j} + \underset{\sim}{J} &= K_{\sqrt{i\beta}}(\underset{\sim}{j}-\underset{\sim}{J}) - L_1(\underset{\sim}{J}) + \underset{\sim}{H}{}_T^{o}\\
m - M &= L_2(M,\underset{\sim}{J},\underset{\sim}{j}) + f\\
i\beta m + \alpha^2 M &= D_2(M,\underset{\sim}{J},\underset{\sim}{j}) + h
\end{aligned} \tag{18}$$

where $\underset{\sim}{g}\in H^{r-1}(S)$; $f,h\in H^r(S)$ for given data $\underset{\sim}{H}{}_T^{o}\in H^{r+1}(S)$, $\underset{\sim}{E}{}_T^{o}\in H^r(S)$.
Here $D_1,D_2,K_{\sqrt{i\beta}}$ and L_1,L_2 are pseudo-differential operators of order minus one and minus two, respectively. Again (18) turns out to be uniquely solvable guaranteeing the following existence and regularity result for $(E_{\alpha\beta})$.

<u>Theorem 5 [4] :</u> <u>For any real</u> r <u>and given</u> $\underset{\sim}{E}{}_T^{o}\in\underset{\sim}{H}^r(S)$, $\underset{\sim}{H}{}_T^{o}\in\underset{\sim}{H}^{r+1}(S)$ <u>the system</u> $(E_{\alpha\beta})$ <u>has a unique solution</u>

$$\underset{\sim}{J},\underset{\sim}{j}\in\underset{\sim}{H}^{r-1}(S)\ ;\quad M,m\in H^r(S)\ .$$

<u>Remark 6 :</u> Since the field quantities in (2), (3) and hence in $(P_{\alpha\beta})$ and $(P_{\alpha\beta})$ are infinitely differentiable except at source points in Ω and Ω' the regularity of the solution $(\underset{\sim}{J},M)$ of $(E_{\alpha\beta})$ and $(\underset{\sim}{J},M,\underset{\sim}{j},m)$ of $(E_{\alpha\beta})$ hinges only on the smoothness of the boundary surface S.

References :

[1] Hsiao, G. and MacCamy, R.C., Solutions of boundary value problems by integral equations of the first kind, SIAM Review 15 (1973), 687-705.

[2] Knauff, W. and Kress, R., On the exterior boundary value problem for the time harmonic Maxwell equations, Journ. Math. Anal. Appl. 72 (1979), 215-235.

[3] Koshlyakov, N.S., Smirnov, M.M. and Gliner, E.B., Differential Equations of Mathematical Physics, North-Holland Publishing Company, Amsterdam, 1964.

[4] MacCamy, R.C. and Stephan, E., Solution procedures for three-dimensional Eddy Current Problems, to appear.

[5] Müller, C., Foundations of the Mathematical Theory of Electromagnetic Waves, Springer-Verlag, Berlin - Heidelberg - New York, 1969.

[6] Seeley, R., Pseudo-Differential Operators, CIME, Cremonese, Rome (1969) (coordinated by L. Nirenberg).

[7] Stratton, I.A., Electromagnetic Theory, McGraw-Hill, New York, N.Y., 1941.

R.C. MacCamy
Carnegie-Mellon University
Pittsburgh, PA 15213

Ernst Stephan
Technische Hochschule Darmstadt
Federal Republic of Germany

Index Laws for some Ordinary Differential Operators

Adam C. McBride

§1. Much work has been done in recent years on developing general theories of fractional powers of operators. Many of these theories have considered various classes of operators in Banach spaces and have made extensive use of the Dunford integral; see, for instance, [1] and [5]. However, when the operators concerned have a more concrete form, other approaches are possible. For example, let T be a pseudo-differential operator in $L^2(R^n)$ with symbol g. Then, for appropriate functions $\phi \epsilon L^2(R^n)$,

$$(\mathfrak{F}(T\phi))(x)=g(x)(\mathfrak{F}\phi)(x) \qquad (x\epsilon R^n) \tag{1.1}$$

so that, formally,

$$T=\mathfrak{F}^{-1}g\,\mathfrak{F} \tag{1.2}$$

where $\mathfrak{F}$ denotes the Fourier transform. From examination of positive integral powers of T, we are led to define T^α, for suitable complex numbers α, by

$$T^\alpha=\mathfrak{F}^{-1}g^\alpha\,\mathfrak{F} \tag{1.3}$$

provided that this is meaningful. As a particular case, we can define fractional powers of $-\Delta=-\frac{\partial^2}{\partial x_1^2}-\ldots-\frac{\partial^2}{\partial x_n^2}$, for which $g(x)=|x|^2$, and study their properties [3].

Similarly, we can consider multiplier transforms T in weighted versions of the spaces $L^p(0,\infty)$ which take the form

$$(\mathfrak{M}(T\phi))(s)=g(s)\,\mathfrak{M}\phi(s) \tag{1.4}$$

where $\mathfrak{M}$ is the Mellin transform defined formally by

$$\mathfrak{M}\psi(s)=\int_0^\infty x^{s-1}\psi(x)dx \tag{1.5}$$

and s is an appropriately restricted complex variable; see [8], [9]. Again, the formal definition

$$T^\alpha=\mathfrak{M}^{-1}g^\alpha\,\mathfrak{M} \tag{1.6}$$

is suggested. Although (1.3) and (1.6) seem simple enough, there may be problems in defining g^α, with the possibility of branch points arising. In contrast, we propose to study transforms which obey a rule similar to (1.4) but which do not lead to the same difficulties of interpretation. The operators T above may be thought of as operators from a space of functions into the <u>same</u> space, whereas the operators which we shall consider can be thought of as mapping one space into a <u>different</u> space. It is perhaps a little paradoxical that we can sometimes make more progress in the second case than in the first. However, we hope to illustrate this clearly by examining a general class of ordinary differential operators on $(0,\infty)$.

§2. We consider operators T which satisfy the formal relation

$$(\mathfrak{M}(T\phi))(s)=\frac{h(s)}{h(s+\gamma)}\mathfrak{M}\phi(s+\gamma) \tag{2.1}$$

where $\mathfrak{M}$ is defined via (1.5), γ is a fixed <u>non-zero</u> complex number and h is a fixed function defined on some domain in the complex plane. For the moment, we shall argue formally. We can give two examples of operators which are of the form (2.1).

<u>Example 2.1</u>

(i) For any fixed complex number γ, let T_γ be defined by

$$(T_\gamma\phi)(x)=x^\gamma\phi(x) \qquad (x\in(0,\infty)). \tag{2.2}$$

Then, trivially, $(\mathfrak{M}(T_\gamma\phi))(s)=\mathfrak{M}\phi(s+\gamma)$ if both expressions are meaningful and (2.1) is satisfied with $h(s)\equiv 1$.

(ii) For the differentiation operator D, integration by parts gives

$$\int_0^\infty x^{s-1}(D\phi)(x)dx=\left[x^{s-1}\phi(x)\right]_0^\infty-\int_0^\infty (s-1)x^{s-2}\phi(x)dx.$$

If ϕ is such that the integrated terms vanish, then

$$(\mathfrak{M}(-D\phi))(s)=(s-1)\mathfrak{M}\phi(s-1)=\frac{\Gamma(s)}{\Gamma(s-1)}\mathfrak{M}\phi(s-1) \tag{2.3}$$

and (2.1) is satisfied with $\gamma=-1$ and $h(s)=\Gamma(s)$.

From (2.1), it follows easily by induction that, for $r=1,2,\ldots,$

$$(\mathfrak{M}(T^r\phi))(s)=\frac{h(s)}{h(s+r\gamma)}\mathfrak{M}\phi(s+r\gamma). \tag{2.4}$$

If we assume that T^{-1} exists and we replace ϕ and s in (2.1) by $T^{-1}\phi$ and $s-\gamma$ repsectively, we might be tempted to say that (2.4) holds for $r=-1$ also and hence, by induction again, for <u>all</u> integers r (the case $r=0$ being trivial). This suggests the following definition.

Definition 2.2 For suitable complex numbers α and with T as in (2.1), define T^α by

$$(\mathcal{M}(T^\alpha\phi))(s)=\frac{h(s)}{h(s+\alpha\gamma)}\mathcal{M}\phi(s+\alpha\gamma). \tag{2.5}$$

Continuing with our formal manipulations, we can ask whether or not the normal index laws of ordinary algebra carry over. That $T^\alpha T^\beta=T^{\alpha+\beta}=T^\beta T^\alpha$ for complex numbers α and β is no great surprise. However, it is of much more interest to observe that we can form powers of T^α since, by (2.5), T^α satisfies a relation of the form (2.1) with the same function h but with γ replaced by $\alpha\gamma$. In particular, if we replace T, α and γ in (2.5) by T^α, β and $\alpha\gamma$ respectively, we obtain

$$(\mathcal{M}[(T^\alpha)^\beta\phi])(s)=\frac{h(s)}{h(s+\beta\alpha\gamma)}\mathcal{M}\phi(s+\beta\alpha\gamma)=(\mathcal{M}(T^{\alpha\beta}\phi))(s) \tag{2.6}$$

for complex numbers α and β. This suggests that $(T^\alpha)^\beta=T^{\alpha\beta}$, a law which presents considerable problems in other theories such as that based on the Dunford integral [11].

Without further proof, we state the following results for completeness.

Theorem 2.3 (Index Laws) Let α and β be complex numbers. Then

(i) $T^\alpha T^\beta=T^{\alpha+\beta}=T^\beta T^\alpha$; $T^0=I$; $(T^\alpha)^{-1}=T^{-\alpha}$. (2.7)

(ii) $(T^\alpha)^\beta=T^{\alpha\beta}=(T^\beta)^\alpha$. (2.8)

§3. It is natural to ask at this stage if there are circumstances in which all the formal calculations in §2 can be justified. Fortunately, there are and we shall examine one particular instance. Firstly, we describe suitable spaces of functions.

Definition 3.1 Let μ be any complex number.

(i) For $1\le p<\infty$,

$$F_{p,\mu}=\{\phi\in C^\infty(0,\infty) : x^kD^k(x^{-\mu}\phi)\in L^p(0,\infty) \text{ for } k=0,1,2,\ldots\},$$

$$\gamma_k^{p,\mu}(\phi)=\|x^kD^k(x^{-\mu}\phi)\|_p \quad (k=0,1,2,\ldots).$$

(ii) $F_{\infty,\mu}=\{\phi\in C^\infty(0,\infty) : x^kD^k(x^{-\mu}\phi)\to 0 \text{ as } x\to 0+ \text{ and as } x\to\infty \text{ for } k=0,1,2,\ldots\}$,

$$\gamma_k^{\infty,\mu}(\phi)=\|x^kD^k(x^{-\mu}\phi)\|_\infty \quad (k=0,1,2,\ldots).$$

Here, $D\equiv d/dx$ and $\|\ \|_p$ is the usual $L^p(0,\infty)$ norm $(1\le p\le\infty)$. For each p and μ, $F_{p,\mu}$ is a vector space and a Fréchet space with respect to the topology generated by the collection $\{\gamma_k^{p,\mu}\}_{k=0}^\infty$ of seminorms. Analytic and topological properties of these spaces are described in detail in [6, Chapter 2].

Secondly, we describe a suitable class of operators. Let n be a positive integer and let $a_1,\ldots,a_{n+1}$ be complex numbers such that

$$a=\sum_{k=1}^{n+1} a_k \text{ is real} \tag{3.1}$$

$$m=|a-n|>0. \tag{3.2}$$

Let T denote the nth order ordinary differential expression

$$T=x^{a_1}Dx^{a_2}\ldots x^{a_n}Dx^{a_{n+1}} \qquad (D\equiv d/dx). \tag{3.3}$$

Any attempt to get a manageable expression for T^r (r=1,2,3,...) directly from (3.3) seems doomed to failure. However, by writing

$$D=mx^{m-1}d/dx^m$$

with m as in (3.2), progress can be made. Equivalently, although T is the composition of 2n+1 operators of the types in Example 2.1, the lack of commutativity prevents the immediate derivation of a manageable expression for T in the form (2.1) so that transcription in terms of m is again helpful. For convenience, we shall write

$$b_k=\left(\sum_{i=k+1}^{n+1} a_i+k-n\right)/m \qquad (k=1,\ldots,n). \tag{3.4}$$

The following result is proved in [7].

<u>Theorem 3.1</u> For $1\leqslant p\leqslant\infty$ and any complex number μ, T is a continuous linear mapping from $F_{p,\mu}$ into $F_{p,\mu+a-n}$. Further, T is a homeomorphism from $F_{p,\mu}$ onto $F_{p,\mu+a-n}$ iff $\mathrm{Re}(mb_k+\mu)\neq 1/p$ (k=1,...,n).

In attempting to obtain an expression of the form (2.1), we must study the Mellin transform on $F_{p,\mu}$. From the work of Rooney [8], we can say that, if $\phi\varepsilon F_{p,\mu}$, then $\mathcal{M}\phi(s)$ is well-defined provided that $1\leqslant p\leqslant 2$ and $\mathrm{Re}\ s=1/p-\mathrm{Re}\ \mu$. In practice, the restriction $1\leqslant p\leqslant 2$ can often be removed at a later stage, as we shall see below. As the theory unfolds, it is necessary to distinguish between the cases a<n and a>n (a=n being excluded by (3.2)). For simplicity, we consider only the case a<n. From [7], we quote the following result.

<u>Theorem 3.2</u> Let $1\leqslant p\leqslant 2$, $\phi\varepsilon F_{p,\mu}$, m=n-a>0, Re s=1/p-Re μ+m. Then

$$(\mathcal{M}(T\phi))(s)=m^n\prod_{k=1}^{n}\frac{\Gamma(b_k+2-s/m)}{\Gamma(b_k+1-s/m)}\mathcal{M}\phi(s-m) \tag{3.5}$$

provided that $\mathrm{Re}(mb_k+\mu)+m\neq 1/p-m\ell$ (ℓ=0,1,2,...). This is of the form (2.1) where $\gamma=-m$ and

$$h(s)=m^{ns/m}\left[\prod_{k=1}^{n}\Gamma(b_k+1-s/m)\right]^{-1}. \tag{3.6}$$

(We notice in passing that (3.5) includes (2.3) as a special case.) We can now apply Definition 2.2 in conjunction with (3.6).

Definition 3.3 Let T be as in (3.3), $1\leq p\leq 2$, $\phi\in F_{p,\mu}$, $m=n-a$, Re $s=1/p-\mathrm{Re}(\mu+m\alpha)$. We define $T^{\alpha}\phi$ by

$$(\mathcal{M}(T^{\alpha}\phi))(s)=m^{n\alpha}\prod_{k=1}^{n}\frac{\Gamma(b_k+1+\alpha-s/m)}{\Gamma(b_k+1-s/m)}\mathcal{M}\phi(s-m\alpha) \tag{3.7}$$

provided that $\mathrm{Re}(mb_k+\mu)+m\neq 1/p-m\ell$ $(\ell=0,1,2,\ldots)$.

Since T is a continuous linear mapping from $F_{p,\mu}$ into $F_{p,\mu-m}$ (Theorem 3.1), we might expect that T^{α} would be a continuous mapping from $F_{p,\mu}$ into $F_{p,\mu-m\alpha}$ under appropriate conditions and this is indeed so. Under the hypotheses of Definition 3.3, the mapping properties can be deduced from the work of Rooney in [8], [9]. However, as hinted above, the range of possible values of p can be extended from $1\leq p\leq 2$ to $1\leq p\leq\infty$. One way to achieve this is to express T^{α} in terms of the so-called Erdélyi-Kober operators which we will discuss briefly.

§4. For concrete representation of the Erdélyi-Kober operator $I_m^{\eta,\gamma}$, we proceed in easy stages. With m real and positive and $\phi\in F_{p,\mu}$ throughout, we first consider complex numbers η and γ such that $\mathrm{Re}(m\eta+\mu)+m>1/p$ and Re $\gamma>0$. We define $I_m^{\eta,\gamma}\phi$ by

$$(I_m^{\eta,\gamma}\phi)(x)=[\Gamma(\gamma)]^{-1}\int_0^1(1-t^m)^{\gamma-1}t^{m\eta}\phi(xt)d(t^m) \qquad (0<x<\infty) \tag{4.1}$$

where $d(t^m)=mt^{m-1}dt$. The definition is then extended to values of γ with Re $\gamma\leq 0$ by means of the formula

$$I_m^{\eta,\gamma}\phi=(\eta+\gamma+1)I_m^{\eta,\gamma+1}\phi+m^{-1}I_m^{\eta,\gamma+1}\delta\phi \tag{4.2}$$

where $(\delta\phi)(x)=xd\phi/dx$. Finally, if $\mathrm{Re}(m\eta+\mu)+m<1/p$ but $\mathrm{Re}(m\eta+\mu)+m\neq 1/p-m\ell$ for $\ell=1,2,\ldots$, we let k be the unique positive integer such that $1/p-mk<\mathrm{Re}(m\eta+\mu)+m<1/p-m(k-1)$ and define $I_m^{\eta,\gamma}\phi$, via the previous cases, by

$$I_m^{\eta,\gamma}\phi=I_m^{\eta+k,\gamma-k}x^{-m(\eta+k)}(D_m)^{-k}x^{m\eta}\phi \tag{4.3}$$

where $(D_m)^{-k}$ is the kth power of the inverse of the operator $D_m\equiv d/dx^m$. The justification for the various formulae above can be found in [6].

From the original results of Kober [4] and from Definition 3.1 above, we find that the operator $I_m^{\eta,\gamma}$ is a continuous linear mapping from $F_{p,\mu}$ into itself provided that $1\leq p\leq\infty$ and $\mathrm{Re}(m\eta+\mu)+m\neq 1/p-m\ell$ ($\ell=0,1,2,\ldots$). Further, when $1\leq p\leq 2$ we have, from [8],

$$(\mathcal{M}(I_m^{\eta,\gamma}\phi))(s)=\frac{\Gamma(\eta+1-s/m)}{\Gamma(\eta+\gamma+1-s/m)}(\mathcal{M}\phi)(s) \tag{4.4}$$

where $\phi\in F_{p,\mu}$ and $\mathrm{Re}\ s=1/p-\mathrm{Re}\ \mu$. It follows easily from (3.7) and (4.4) that under these conditions,

$$(T^\alpha\phi)(x)=m^{n\alpha}x^{-m\alpha}\prod_{k=1}^{n} I_m^{b_k,-\alpha}\phi(x) \tag{4.5}$$

with α any complex number and b_k ($k=1,\ldots,n$) as in (3.4). Bearing in mind that $C_0^\infty(0,\infty)$ is dense in $F_{p,\mu}$ for every μ and $1\leq p\leq\infty$ [6, p.18], we see that Kober's results enable us to show that (4.5) holds, not merely for $1\leq p\leq 2$ but for $1\leq p\leq\infty$ (and any μ) provided only that $\mathrm{Re}(mb_k+\mu)+m\neq 1/p-m\ell$ ($\ell=0,1,2,\ldots$) for $k=1,\ldots,n$.

In the case $\mathrm{Re}\ \alpha<0$, we would expect that T would be an integral operator rather than a differential operator and this is so, as we can see explicitly in the simplest case which is to be found in [7].

<u>Theorem 4.1</u> If $m=n-a>0$, $\mathrm{Re}(mb_k+\mu)+m>1/p$ ($k=1,\ldots,n$) and $\mathrm{Re}\ \alpha<0$, then

$$(T^\alpha\phi)(x)=m^{n\alpha}x^{-m\alpha-m}\int_0^x G_{n,n}^{n,0}\left(\frac{t^m}{x^m}\,\middle|\,\begin{matrix}b_1-\alpha,\ldots,b_n-\alpha\\ b_1,\ldots,b_n\end{matrix}\right)\phi(t)d(t^m) \quad (\phi\in F_{p,\mu}) \tag{4.6}$$

where $G_{n,n}^{n,0}$ is Meijer's G-function. For the case n=2, (4.6) becomes

$$(T^\alpha\phi)(x)=[\Gamma(-2\alpha)]^{-1}m^{2\alpha}x^{-mb_1}\int_0^x (x^m-t^m)^{-2\alpha-1}{}_2F_1(b_2-b_1-\alpha,-\alpha;-2\alpha;1-x^m/t^m)\times t^{mb_1+m\alpha}\phi(t)d(t^m) \tag{4.7}$$

where ${}_2F_1$ denotes the Gauss hypergeometric function.

We mention that our results for the case n=2 are in agreement with those of Sprinkhuizen-Kuyper [10] who made an extensive study of operators of the form

$$D^2+\nu x^{-1}D \tag{4.8}$$

corresponding to the values $a_1=-1$, $a_2=2-\nu$, $a_3=\nu-1$ in our notation. Operators of the form (4.8) turn up in many places e.g. in GASPT and many relationships exist between these operators and the Erdélyi-Kober operators. For instance, if we write

$$L_\eta=D^2+(2\eta+1)x^{-1}D \tag{4.9}$$

then $I_2^{\eta,\gamma}L_\eta=L_{\eta+\gamma}I_2^{\eta,\gamma}$ under appropriate conditions. When we use (3.2) in the case of the operator L_η we obtain m=2 and hence the appearance of the number 2 in the formula is a natural consequence of our theory. We can extend the result by proving, under conditions of great generality, that

$$I_2^{\eta,\gamma}(L_\eta)^\alpha=(L_{\eta+\gamma})^\alpha I_2^{\eta,\gamma} \tag{4.10}$$

[7], along with other similar results. Whether or not (4.10) is of any practical use remains to be seen.

§5. We have not so far dealt with the index law in Theorem 2.3(ii) in the context of the operator T given by (3.3). The sight of the G-function in (4.6) suggests that this law might be related to some results involving special functions. We shall merely indicate one aspect of this. At this stage we shall return again to formal analysis, bearing in mind that rigour can be supplied by working in the spaces $F_{p,\mu}$ and imposing conditions on the parameters similar to those in §4.

In (2.7), let $\alpha=r$, a positive integer. Then for complex numbers β, we have

$$(T^r)^\beta=T^{r\beta}. \tag{5.1}$$

On the one hand, we can write down an expression for $T^{r\beta}$ directly from (4.5), namely,

$$(T^{r\beta}\phi)(x)=m^{nr\beta}x^{-mr\beta}\prod_{k=1}^{n} I_m^{b_k,-r\beta}\phi(x). \tag{5.2}$$

On the other hand we can examine T^r in a similar manner to that in which we examined T. By stringing out r copies of T, we obtain an expression of the form (3.3) in which the n+1 numbers $a_1,\ldots,a_{n+1}$ are replaced by a set of rn+1 numbers related to them in a simple way. Similarly $b_1,\ldots,b_n$ are replaced by a set of rn numbers and an easy calculation shows that m has to be replaced by rm. After another completely routine calculation, we find that

$$(T^r)^\beta\phi=(mr)^{nr\beta}x^{-mr\beta}\prod_{k=1}^{n}\prod_{\ell=1}^{r} I_{mr}^{(b_k-r+\ell)/r,-\beta}\phi\ . \tag{5.3}$$

Comparison of (5.2) with (5.3) suggests that

$$r^{r\beta}\prod_{\ell=1}^{r} I_{mr}^{(b_k-r+\ell)/r,-\beta}=I_m^{b_k,-r\beta}$$

as operators. Equivalently, if we write $\gamma=-r\beta$, $\eta=b_k$, we might expect that

$$I_m^{\eta,\gamma}=r^{-\gamma}\prod_{\ell=1}^{r} I_{mr}^{(\eta-r+\ell)/r,\gamma/r}. \tag{5.4}$$

We can show that (5.4) is true by using Mellin transforms, whereupon we discover that the result can be thought of as being a version of the classical formula

$$\prod_{\ell=1}^{r} \Gamma(z+(\ell-1)/r)=(2\pi)^{(r-1)/2} r^{1/2-rz}\Gamma(rz) \qquad (r=2,3,4,\ldots) \tag{5.5}$$

for the gamma function due to Gauss and Legendre [2, p.4]. Indeed, from (4.4), the Mellin transforms of the two sides of (5.4) are formally equal if and only if

$$\frac{\Gamma(\eta+1-s/m)}{\Gamma(\eta+\gamma+1-s/m)}=r^{-\gamma}\prod_{\ell=1}^{r}\frac{\Gamma((\eta+\ell-s/m)/r)}{\Gamma((\eta+\gamma+\ell-s/m)/r)}. \tag{5.6}$$

However, by (5.5), the right-hand side can be written as

$$\frac{(2\pi)^{(r-1)/2} r^{1/2-\eta-1+s/m}\Gamma(\eta+1-s/m)r^{-\gamma}}{(2\pi)^{(r-1)/2} r^{1/2-\eta-\gamma-1+s/m}\Gamma(\eta+\gamma+1-s/m)}$$

and (5.6) follows at once, so that (5.4) is established formally. In the special case r=2, an alternative proof of (5.4) can be obtained from scratch via the formula [2, p.112]

$$\begin{aligned} &{}_2F_1(a,b;a+b-1/2;z) \\ &=(1-z)^{-1/2}\left[1/2+(1-z)^{1/2}/2\right]^{1-2a} {}_2F_1\left(2a-1,a-b+1/2;a+b-1/2;\ \frac{(1-z)^{1/2}-1}{(1-z)^{1/2}+1}\right), \end{aligned}$$

which is not surprising in view of (4.7).

(5.4) does not seem to appear in the books on fractional calculus. Whether it is new of not, it would be fair to say that it emerges naturally from our second index law (2.8) without the need for complicated manipulations involving special functions. It seems that our theory could suggest similar results closely related to properties of special functions and these we hope to discuss at some future date.

References

1. A.V. Balakrishnan, Fractional powers of closed operators and semigroups generated by them. Pacific J. Math. 10(1960), 419-437.

2. A. Erdélyi et al., Higher Transcendental Functions, Vol. 1 (McGraw-Hill, New York, 1953).

3. R. Johnson, Weighted estimates for fractional powers of partial differential operators. Trans. Amer. Math. Soc. 265(1981), 511-525.

4. H. Kober, On fractional integrals and derivatives. Quart. J. Math. (Oxford) 11(1940), 193-211.

5. H. Komatsu, Fractional powers of operators. Pacific J. Math. 19(1966), 285-346.

6. A.C. McBride, Fractional Calculus and Integral Transforms of Generalized Functions. (Research Notes in Mathematics No. 31, Pitman, London, 1979).

7. A.C. McBride, Fractional powers of a class of ordinary differential operators. To appear in Proc. London Math. Soc.

8. P.G. Rooney, On the ranges of certain fractional integrals. Canad. J. Math. 24(1972), 1198-1216.

9. P.G. Rooney, A technique for studying the boundedness and extendability of certain types of operators. Canad. J. Math. 25(1973), 1090-1102.

10. I.G. Sprinkhuizen-Kuyper, A fractional integral operator corresponding to negative powers of a certain second-order differential operator. J. Math. Anal. Appl. 72(1979), 674-702.

11. K. Yosida, Functional Analysis. (Springer-Verlag, Berlin, 1971).

CONVERSE INITIAL VALUE PROBLEMS FOR A CLASS OF HEAT EQUATIONS

Peter A. McCoy
Department of Mathematics
United States Naval Academy
Annapolis, Maryland 21402

INTRODUCTION.

In the classical initial value problem for the heat equation

$$(1)\qquad \{\partial_t - \partial_x[(1-x)^{\alpha+1}(1+x)^{\beta+1}\partial_x]\}\Psi(x,t) = 0\ ,\ -1 < x < 1\ ,\ t > 0$$

with parameters $\alpha \geq \beta$ and $\beta \geq -1/2$ or $\alpha + \beta > 0$, "arbitrary" initial data

$$\Psi(x,0) = f(x),\ -1 \leq x \leq +1$$

is specified and one seeks the temperature function $\Psi(x,t)$ solving eqn (1) for which $\Psi \to f$ as $t \to 0^+$. The well-known solution can be found in a number of ways. One is by taking the Weierstrass transform of the Jacobi series expansion of data in L^p ($p \geq 2$); another is representation by function theoretic methods [3-6] as an integral transform of analytic data.

Specific problems are connected with Ψ and its initial values. The connection with symmetric Poisson processes on the segment (-1, +1) was established by S. Bochner [2] when $\alpha = \beta > -1$. With $\alpha = \beta = (k-3)/2$, the solutions represent temperatures of the segment (-1, +1) for k = -1, and in higher-dimensional spaces E^k, correspond under the Chebyshev map to conically symmetric temperatures. Time independent solutions with L^p initial data represent Jacobi series whose Abel means solve boundary value problems related to the GASP and Euler-Poisson-Darboux equations [5]. Moreover, the initial value problem arises in the theory of Weierstrass summability [1].

These problems are direct; the solution is determined from specified initial data. This paper considers the converse problem of identifying the initial data from the solution. It maintains that the solution exists uniquely in the space H^*

of initial data, hyperfunctions on [-1, +1], by constructing isomorphisms between the space H^* and the space H of temperature functions. Thus, solutions of the heat equation are identified with their generalized boundary values. An essential feature characterizes the temperature as a series whose terms depend on a sequence of continuous functions satisfying a growth condition on [-1, +1].

The characterization combines function-theoretic [3-6] and special function methods [1] to extend the converse Dirichlet problem for harmonic functions in the disk developed by G. Johnson [8] to these parabolic equations. For the extensions of his work to elliptic partial differential equations, refer to Saylor in [7], J. Kelingos & P. Staples [9] and A. Zayed [13].

THE INITIAL VALUE PROBLEM.

The normalized Jacobi polynomials $R_n(x) := R_n^{(\alpha,\beta)}(x) = P_n^{(\alpha,\beta)}(x)/P_n^{(\alpha,\beta)}(1)$ for $n \in N := \{0,1,2,\ldots\}$ are orthogonal functions

$$\int_{-1}^{+1} R_n(s)R_m(s)d\mu_{\alpha\beta}(s) = \omega_n^{-1}\delta_n ,$$

$$d\mu_{\alpha\beta}(s) = (1-s)^{\alpha}(1+s)^{\beta}ds$$

$$\omega_n = \frac{(2n+\alpha+\beta+1)\Gamma(n+\alpha+\beta+1)\Gamma(n+\alpha+1)}{\Gamma(n+\beta+1)\Gamma(n+1)\Gamma(\alpha+1)\Gamma(\beta+1)} .$$

They form a set of eigenfunctions $\Delta_x^{(\alpha,\beta)}R_n(x) = \lambda_n R_n(x)$, $-1 < x < 1$ corresponding to the eigenvalues $\lambda_n = n(n+\alpha+\beta+1)$ for the operator

$$\Delta_x^{(\alpha,\beta)} = \partial_x[(1-x)^{\alpha+1}(1+x)^{\beta+1}\partial_x]$$

which is complete in the space $X := L^p(-1, +1) \cup C(-1, +1)$ for $p \geq 2$. Expand $f \in X$ as the Jacobi series

$$f(x) = \sum_{n=0}^{\infty} a_n \omega_n R_n(x), \quad -1 \leq x \leq 1$$

with Fourier-Jacobi coefficients

$$a_n = a_n(f) := \int_{-1}^{+1} f(s)R_n(s)d\mu_{\alpha\beta}(s) , \quad n \in N$$

and form the Weierstrass means [1] on $-1 \le x \le 1$, $t > 0$ as

$$\Psi(x,t) = \sum_{n=0}^{\infty} a_n \omega_n \psi_n(x,t), \quad \psi_n(x,t) := e^{-\lambda_n t} R_n(x), \quad n \in N .$$

These means are Weierstrass summable to f [1] and the functions ψ_n are particular solutions of the heat equation. We summarize now.

THEOREM 1.

Let $f \in X$. The solution of the initial value problem

$$\left\{ \begin{array}{c} [\partial_t - \Delta_x^{(\alpha,\beta)}]\Psi(x,t) = 0 , \ -1 < x < 1, \ t > 0 \\ \Psi(x,0) = f(x) , \ -1 \le x \le 1 \end{array} \right\}$$

is the unique temperature function

$$\Psi(x,t) := \sum_{n=0}^{\infty} a_n(f) \omega_n \psi_n(x,t) .$$

This reformulates in terms of the Weierstrass transform as

$$\Psi(x,t) = \int_{-1}^{+1} K(x,s,t)\, f(s)\, d\mu_{\alpha\beta}(s)$$

with the Weierstrass-Jacobi heat kernel

$$K(x,s,t) = \sum_{n=0}^{\infty} \omega_n e^{-\lambda_n t} R_n(x)\, R_n(s), \ -1 \le x, \ s \le 1, \ t > 0 .$$

THE CONVERSE TO THE INITIAL VALUE PROBLEM.

A structure theorem, analogous to the one found by Johnson for harmonic functions is given. It shows that aspects of the initial value problem have as antecendents, boundary value problems for elliptic equations, beginning with the work of Poisson-Stieltjes [7] and Faber [12]. Apparently, the key roles played by Weierstrass summability and Abel summability correspond. We next view the converse problem in terms of the uniform norm $||h|| := \sup\{|h(x)|: -1 < x < 1\}$.

THEOREM 2.

The function Ψ is a temperature function

$$[\partial_t - \Delta_x^{(\alpha,\beta)}]\Psi(x,t) = 0 \;, \; -1 < x < 1,\; t > 0$$

if, and only if, there exists a sequence $\{g_n\}$ of continuous functions on $[-1, +1]$ such that

$$\lim (2n! \, ||g_n||)^{1/n} = 0$$

and

$$\Psi(x,t) = \sum_{n=0}^{\infty} (\Delta_x^{(\alpha,\beta)})^n \int_{-1}^{+1} K(x,s,t)\; g_n(s)\; d\mu_{\alpha\beta}(s)$$

where K is the Weierstrass-Jacobi heat kernel.

PROOF:

Consider a sequence $\{g_n\} \subset C[-1, +1]$ with the specified growth condition. Expand g_n in a Jacobi series

$$g_n(s) = \sum_{k=0}^{\infty} a_{k,n} R_k(s) \;, \; -1 \leq s \leq 1 \;, \; n \in N$$

and form the Weierstrass transforms

$$\int_{-1}^{+1} K(x,s,t) g_n(s)\; d\mu_{\alpha\beta}(s) = \sum_{k=0}^{\infty} a_{k,n}\omega_k\psi_k(x,t) \;, \; n \in N$$

for $-1 \leq x \leq 1$, $t > 0$. If we now set

$$\Phi_n(x,t) = (\Delta_x^{(\alpha,\beta)})^n \int_{-1}^{+1} K(x,s,t) g_n(s)\; d\mu_{\alpha\beta}(s)$$

it follows that the series

$$\Psi(x,t) = \sum_{n=0}^{\infty} \Phi_n(x,t)$$

is a formal solution of the heat equation as a direct calculation shows that

$$\Phi_n(x,t) = \sum_{k=0}^{\infty} a_{k,n}\omega_k(-\lambda_k)^n \; \psi_k(x,t) \;.$$

It remains to establish uniform convergence of the series Ψ by the method of majorants. The functions Φ_n are bounded by the series

$$|\Phi_n(x,t)| \leq C \; ||g_n|| \sum_{k=0}^{\infty} \omega_k(\lambda_k)^n e^{-\lambda_k t}$$

found by estimating the Fourier-Jacobi coefficients $a_{k,n}$. Applying the fact that $\omega_k \sim O(k^{2\alpha+1})$ to the tail, T_n, of this series gives

$$T_n(t) \leq C \; ||g_n|| \sum_{k=n}^{\infty} k^{2\alpha+1} \; [k(k+\alpha+\beta+1)]^n \; e^{-\lambda_k t} \;.$$

Working with the larger term of this inequality shows that

$$T_n(t) \le B\, 4^n \, ||g_n|| \sum_{k=n}^{\infty} k^{[2\alpha]+2n+1} a^k \le B\, 4^n \, ||g_n|| \, ([2\alpha]+2n+1)!$$

where $0 < a(t) \le 1$. For sufficiently large B, the estimate applies termwise and majorizes the series

$$\Psi(x,t) \le B \sum_{n=0}^{\infty} 4^n \, ||g_n|| \, ([2\alpha]+2n+1)! .$$

Thus, it is (subuniformly) uniformly convergent on compacta of $[-1,1]X(t > 0)$ because

$$\limsup \left(4^n \, ||g_n|| \, ([2\alpha]+2n+1)!\right)^{1/n} = D \limsup \left(2n! ||g_n||\right)^{1/n} = 0 .$$

And, Ψ is Weierstrass summable to the initial data.

Having proved sufficiency, the next step reverses the argument.

Set

$$\Psi(x,t) = \sum_{n=0}^{\infty} a_n \omega_n \psi_n(x,t), \ \limsup |a_n|^{1/n} \le 1 .$$

Given $\varepsilon > 0$, reasoning as in (8,13) quid pro quo finds a constant $B(\varepsilon)$ and sequences $\{a_{k,n}\}$ such that

$$a_k = \sum_{n=0}^{k} a_{k,n}, \ |a_{k,n}| \le \frac{B(\varepsilon)\varepsilon^{[2\alpha]+2n+1} k^{2n}}{([2\alpha]+2n+1)!}, \ 0 \le k \le n .$$

Define $$\Phi_n(x,t) = \omega_n \sum_{k \ge n-[2\alpha]-1} (-1)^n a_{k,n-[2\alpha]-1} \psi_k(x,t) (\lambda_k)^{-n}, \ n \in N .$$

The next estimate shows the series to be subuniformly convergent.

$$|\Phi_n(x,t)| \le \frac{B(\varepsilon)\omega_n \varepsilon^{2n+1}}{2n!} \sum_{k \ge n-[2\alpha]-1} k^{2(n-[2\alpha]-1)} e^{-\lambda_k t} \lambda_k^{-n} \le$$

$$\frac{B(\varepsilon)\omega_n \varepsilon^{2n+1}}{2n!} \sum_{k=1}^{\infty} k^{-2}, \ -1 \le x \le 1, \ t > 0 .$$

Consequently, $\Phi_n(x,0) := g_n(x) \in C[-1, +1]$ for $n \in N$, and $\limsup (2n! ||g_n||)^{1/n} \le \varepsilon$. From the earlier representation and

$$(\Delta_x^{(\alpha,\beta)})^n \Phi_n(x,t) = \omega_n \sum_{k \ge n-[2\alpha]-1} a_{k,n-[2\alpha]-1} \psi_k(x,t)$$

all that remains is to check the sum

$$\sum_{n \ge [2\alpha]+1} (\Delta_x^{(\alpha,\beta)})^n \Phi_n(x,t) = \sum_{n \ge [2\alpha]+1} \omega_n \sum_{k \ge n-[2\alpha]-1} a_{k,n-[2\alpha]-1} \psi_k(x,t) .$$

Finally, we obtain the form that completes the proof,

$$\Psi(x,t) = \sum_{n=0}^{\infty} a_n \omega_n \psi_n(x,t), \ -1 \le x \le 1, \ t > 0 .$$

THE INITIAL VALUE PROBLEM VIA HYPERFUNCTIONS.

We now identify the space of hyperfunctions as the space of initial values corresponding to temperatures. The space H^* of hyperfunctions on [-1, +1] is the strong dual of the space H of analytic functions on [-1, +1] with the topology in [10]. The term generalized function refers only to Schwartz distributions so that each generalized function is a hyperfunction. To characterize hyperfunctions in a way comparable with Weierstrass summability, we derive a representation of elements in H^* generalizing [13].

THEOREM 3.

For integer parameters $\alpha \geq \beta \geq 0$, the Jacobi series $\tilde{f} := \sum_{n=0}^{\infty} a_n \omega_n R_n$ converges to a hyperfunction on [-1, +1] if, and only if, $\limsup |a_n|^{1/n} \leq 1$.

PROOF:

First, consider an analytic function $\phi \in H$. Because of the relation $\omega_n / P_n^{(\alpha,\beta)}(1) \sim O(n^{2\alpha+1})$, the expansion theorem [11] applies, and

$$\phi(x) = \sum_{n=0}^{\infty} b_n \omega_n R_n(x) \quad , \ \limsup |b_n|^{1/n} < 1$$

converges to a function in H whose m-th partial sum is designated as ϕ_m.

Turning to $\tilde{f} \in H^*$, an inner product is defined by

$$(\tilde{f},\ (1-s)^{\alpha}(1+s)^{\beta}\phi) := \int_{-1}^{+1} \tilde{f}(s)\phi(s)\ (1-s)^{\alpha}(1+s)^{\beta} ds, \quad \alpha \geq \beta \geq 0 .$$

If $\tilde{f}$ has a Jacobi series expansion, then

$$(\tilde{f},\ (1-s)^{\alpha}(1+s)^{\beta}\ \phi_m) \to (\tilde{f},\ (1-s)^{\alpha}(1+s)^{\beta}\ \phi), \ m \to \infty .$$

However, the series are orthogonal and we infer from the convergence

$$\sum_{n=0}^{m} a_n b_n \omega_n \to \sum_{n=0}^{\infty} a_n b_n \omega_n \ , \ m \to \infty$$

that $\limsup |a_n b_n \omega_n|^{1/n} \leq 1$. Simple reasoning gives $\limsup |a_n|^{1/n} \leq 1$.

Conversely, given a sequence for which the limit condition is met, define

$$f_m(x) := \sum_{n=0}^{m} a_n \omega_n R_n(x) \ , \ -1 \leq x \leq 1, \ m \in N .$$

To show that f_m converges to a hyperfunction $\tilde{f}$, it suffices that the sequence $\{(f_m, \phi)\}$ converges for each $\phi \in H$. This follows as H and H^* are Frechet-Montel spaces and all Montel spaces are reflexive. Thus, weak and strong convergence coincide.

Consider the function $\phi \in H$ as a Jacobi series. The limit

$$((1-s)^{\alpha}(1+s)^{\beta}f_m,\phi) = \sum_{n=0}^{m} a_n b_n \omega_n \to \sum_{n=0}^{\infty} a_n b_n \omega_n \ , \ m \to \infty$$

exists since $\limsup |a_n b_n \omega_n|^{1/n} \leq 1$. Therefore, $(1-s)^{\alpha}(1+s)^{\beta}\tilde{f}$ is a hyperfunctic as is $\tilde{f}$ because α and β are integers. This completes the proof.

The characterization just derived shows that H^* corresponds to the initial-value set of temperatures. To proceed to the initial value problem, define the convolution mapping $*$: $H^* \times H^* \to H^*$ by the Hadamard product

$$(\tilde{f} * \tilde{g})(x) := \sum_{n=0}^{\infty} a_n b_n \omega_n R_n(x) \ .$$

The space $\{H^*,+,*\}$ is a C-ring with identity element

$$\tilde{j}(x) = \sum_{n=0}^{\infty} \omega_n R_n(x) \ .$$

The Weierstrass means yield elements of H,

$$j_t(x) = K(x,1,t), \ t > 0$$

for which $j_t(x) \to \tilde{j}(x) \in H^*$ as $t \to 0^+$. The identification theorem follows.

<u>THEOREM 4.</u>

The function Ψ is a temperature function

$$[\partial_t - \Delta_x^{(\alpha,\beta)}]\Psi(x,t) = 0 \ , \ -1 < x < 1 \ , \ t > 0$$

if, and only if, a unique hyperfunction $\tilde{f}$ exists on $[-1, +1]$ such that

$$\Psi(x,t) = (j_t * \tilde{f})(x) \ .$$

And, $\Psi(x,t) \to \tilde{f}(x) \in H^*$ as $t \to 0^+$.

<u>PROOF:</u>

Let Ψ be a temperature function and consider its Jacobi expansion with $\limsup |a_n|^{1/n} \leq 1$. Set

$$\tilde{f}(x) = \sum_{n=0}^{\infty} a_n \omega_n R_n(x) \ , \ -1 \leq x \leq 1 \ .$$

Then $\Psi(x,t) = (j_t * \tilde{f})(x)$. The argument of an earlier theorem, together with Weierstrass summability gives $\Psi \to \tilde{f} \in H^*$ as $t \to 0^+$. And, if $\tilde{f} \in H^*$ then

$$\tilde{f}(x) = \sum_{n=0}^{\infty} a_n \omega_n R_n(x), \ \limsup |a_n|^{1/n} \leq 1$$

so that $\Psi = (\tilde{f} * j_t)$ is a temperature function. The proof is complete.

CONCLUDING REMARKS.

The roles of the parameters (α,β) may be reversed because of the symmetries

$$R_n^{(\alpha,\beta)}(-x) = R_n^{(\beta,\alpha)}(x), \ \Delta_{-x}^{(\alpha,\beta)} = \Delta_x^{(\beta,\alpha)} .$$

Several corollaries may be derived for polynomial sequences, or Radon measures, provided the growth conditions are satisfied. Also, the preceeding results extend to a broader class of heat equations by means of generalized Weierstrass summability and fractional differential operators as defined in [1].

REFERENCES

[1] H. Bavinck, Jacobi series and Approximation, Math. Centre Tracts, Math. Centrum, Amsterdam, 1972.

[2] S. Bochner, Strum-Liouville and Heat Equations whose Eigenfunctions are Ultraspherical Polynomials or Associated Bessel Functions, Proc. Conf. on Diff. Eqns., ed. J. B. Diaz & L. E. Payne, Univ. Maryland, 1956.

[3] D. L. Colton, Partial Differential Equations in the Complex Domain, Research Notes in Math., vol. 4, Pitman Publ., San Francisco, 1976.

[4] D. L. Colton, Solutions of Boundary Value Problems by the Method of Integral Operators, Research Notes in Math., Pitman Publ. Co., San Francisco, 1976.

[5] R. P. Gilbert, Function Theoretic Methods in Partial Differential Equations, Math. in Science and Engineering, vol. 54, Academic Press, New York, 1969.

[6] R. P. Gilbert, Constructive Methods for Elliptic Equations, Lecture Notes in Math., vol. 365, Springer-Verlag, New York, 1974.

[7] K. Hoffman, Banach Spaces of Analytic Functions, Prentice-Hall, Englewood Cliffs, N. J., 1962.

[8] G. Johnson, Jr., Harmonic Functions in the Unit Disk, I&II., Illinois J. Math., vol. 12 (1968) 366-396.

[9] J. Kelingos & P. Staples, A Characterization of Solutions to a Perturbed LaPlace Equation, Illinois J. Math., vol. 22 (1978) 208-216.

[10] G. Kothe, Topological Vector Spaces I, Springer-Verlag, New York, 1969.

[11] G. Szego, Orthogonal Polynomials, Amer. Math. Soc. Publ., vol. 23, Providence, 1939.

[12] M. Tsuji, Potential Theory in Modern Function Theory, Maruzen Publ. Co., Tokyo, 1959.

[13] A. I. Zayed, Hyperfunctions as Boundary Values of Generalized Axially Symmetric Potentials, Illinois J. Math., vol. 25 (1981) 306-317

Higher Order Inverse Eigenvalue Problems

Joyce R. McLaughlin

ABSTRACT

The problem to be discussed is as follows. Suppose a mathematica model for a given physical problem results in a self-adjoint eigenvalue problem of the form

$$w^{(4)} + (Aw^{(1)})^{(1)} + Bw - \lambda w = 0$$

$$\sum_{i=1}^{4} \alpha_{ij} w^{(i-1)}(0) = 0 = \sum_{i=1}^{4} \beta_{ij} w^{(i-1)}(1), \ j = 1,2.$$

Suppose A, B and possibly α_{ij}, β_{ij}, $i = 1,\ldots,4$, $j = 1,2$ are unknown but eigenvalues λ_i, $i = 1,2,\ldots$ are known. A constructive technique for finding the unknown coefficients is presented. Additional data which can be required known are two normalization constants for each of the eigenfunctions.

Introduction

Suppose we consider the following inverse problem. We are given a set of positive numbers $\lambda_1 \leq \lambda_2 \leq \lambda_3 \leq \ldots$ (satisfying a particular asymptotic form and where any given number can be repeated at most once). Then we seek a fourth order self-adjoint eigenvalue problem, i.e. coefficients $A(s)$, $B(s)$, $0 \leq s \leq 1$, and real numbers M_{ij}, N_{ij}, $i = 1,2$, $j = 1,2,3,4$ so that the given sequence $\{\lambda_i\}$, $i = 1,2,\ldots$ is the set of eigenvalues for

$$(1) \qquad \begin{aligned} & y^{(4)} + (Ay^{(1)})^{(1)} + By - \lambda y = 0, \ 0 \leq s \leq 1, \\ & \sum_{j=1}^{4} M_{ij} y^{(j-1)}(0) = 0, \ \sum_{j=1}^{4} N_{ij} y^{(j-1)}(1) = 0, \ i = 1,2. \end{aligned}$$

We are interested in developing a method of constructing the unknown coefficients as well as determining what other spectral-type data (normalization constants?) associated with the to be derived self-adjoint problem can (or must) be utilized to determine a solution.

Before continuing with this fourth order problem it may provide some perspective to review what is known for second order self-adjoint inverse eigenvalue problems. The second order problem has been studied more extensively than the fourth order problem. In the second order case the self-adjoint eigenvalue problem to be considered is

$$(2)\qquad \begin{gathered} y^{(2)} + (\lambda - q)y = 0,\ 0 \le s \le 1 \\ \sum_{j=1}^{2} \alpha_j y^{(j-1)}(0) = 0,\quad \sum_{j=1}^{2} \beta_j y^{(j-1)}(1) = 0. \end{gathered}$$

The sequence of distinct numbers $\lambda_1 < \lambda_2 < \ldots$ satisfying a certain asymptotic form is given and the coefficients $q(s)$, α_j, β_j, $j = 1,2$ are sought so that the given sequence of λ_i's contains all the eigenvalues for (2). Historically, extensive work in the problem has been done by Borg [4], Marcenko [15], Krein [9,10], Levinson [12], and Gel'fand and Levitan [6]. It is known that additional information can be utilized to determine the coefficients. In fact, it was first shown by Borg [4] that additional information must be given if the solution, that is the coefficients, are to be uniquely determined. Such additional information could be a second set of eigenvalues [4], [9], [10], [12], [15] where boundary conditions are different (but related) to those in (2). One could assume that q is symmetric about $s = \frac{1}{2}$, [4], [7], [8]. Or, alternatively, unique solutions have been shown to exist, by Gel'fand and Levitan [6], when knowledge of a positive sequence $\rho_1, \rho_2, \ldots$, which are shown to be normalization constants, is also assumed. The existence theorem obtained by a constructive process by Gel'fand and Levitan, [6], can also be applied to show existence when knowledge of two sequences of eigenvalues is assumed. This is done by applying the results of Levitan [13], [14] which state that given the two sequences of eigenvalues in [4], [9], [10], [12], [15], the normalization constants associated with either sequence of eigenvalues may be constructed.

Results for the fourth order inverse eigenvalue problem have been obtained by Barcilon [2], [3], McKenna [16], Leibenzon [11], and the author [17], [18], [19]. Barcilon follows the approach of M. Krein assuming knowledge of three distinct sequences of eigenvalues and associated boundary conditions. A constructive technique is given when it is known, a priori, that the given sequences are eigenvalues for eigenvalue problems which contain the given corresponding sets of boundary conditions. Leibenzon also proves a uniqueness theorem when three given sequences of eigenvalues and three given corresponding sequences of normalization constants are given for eigenvalue problems with related boundary conditions.

The author, in [17], [18], and [19] has developed a method for finding the solution of a well-posed inverse eigenvalue problem. It is assumed that all eigenvalues are simple and that two positive sequences $\lambda_1 < \lambda_2 < \ldots$ and $\rho_1, \rho_2 \ldots$ are given and satisfy certain asymptotic forms. A constructive technique is then developed for finding coefficients in terms of the eigenvalues and L^2 norms of the corresponding eigenfunctions for a known eigenvalue problem, in terms of solutions of the differential equation in the known eigenvalue problem, and in terms of the given positive numbers. The given sequences of λ_i's and ρ_i's are the eigenvalues and L^2 norms of the corresponding eigenfunctions of the derived problem. The derived coefficients are unique in the sense that they define uniquely an eigenvalue problem which has eigenvalues, λ_i, $i = 1,2,\ldots$ and normalization constants, ρ_i, $i = 1,2,\ldots$ and for which an associated non-self-adjoint problem has spectral data identical with that of a known non-self-adjoint eigenvalue problem (same differential equation as in known self-adjoint eigenvalue problem). This technique will be explained in Section 1. It should further be noted that in addition to the results mentioned above there is a continuity result, [19]. This result gives a bound on the derived coefficients in terms of the differences of the eigenvalues and differences of the normalization constants for the derived and known self-adjoint eigenvalue problems.

In this paper we first (Section 1) present the above mentioned ideas in a more general way. This presentation should help motivate the "new" concepts and basically shows how to instruct a multitude of eigenvalue problems which have the same data λ_i, ρ_i, $i = 1,2,\ldots$. This collection of solutions is limited in the sense that all eigenvalues must be simple and because of the property about the spectral data for an associated non-self-adjoint problem mentioned above. The fact that many solutions can be obtained indicates that more "spectral" data possibly could (or should) be used in determining the derived eigenvalue problem.

Thus, we seek a method of constructing solutions of the inverse problem which allows multiple eigenvalues and where an additional sequence of real numbers (another type of normalization constant) associated with the (to be) derived self-adjoint eigenvalue problem may be given. As in previous work, the constructive technique will be a generalization of that of Gel'fand-Levitan.

In order to make the assumptions to be given seem natural, consider the following description of the work of Gel'fand-Levitan. Two sequences are given which satisfy required asymptotic forms. A known (base problem) eigenvalue problem is given. An integral relationship (with unknown kernel K) is shown to exist between solutions of the differential equation in the known eigenvalue problem and solutions of the differential equation in the (to be) derived eigenvalue problem. The kernel K is determined by the two given sequences and the unknown coefficients q, β_j, α_j, $j = 1,2$ are determined in terms of K. One of the given sequences is the set of eigenvalues for the derived problem. If we select corresponding eigenfunctions with L^2 norm equal to 1, then the other sequence, a set of normalization constants, can be interpreted to yield the value of the eigenfunction (or its derivative) at one end point, say $x = 0$. It is important to note that the derived boundary conditions plus the normalization constants completely define a full set of initial conditions for each eigenfunction (the value of the function and its derivative) at one of the endpoints of the interval, say $x = 0$. It should also be emphasized that it is not assumed a priori that the given sequences are eigenvalues and normalization constants for an eigenvalue problem.

In this paper three sequences of real numbers will be given and, as in previous papers by the author, the method of solution will be illustrated by considering the following special problem. It is assumed that the positive sequence $\lambda_1 \leq \lambda_2 \leq \ldots$ satisfies a particular asymptotic form and that a given number can be repeated at most once. The sequences γ_i, ξ_i are given so that they satisfy a given asymptotic form and such that $\gamma_i \geq 0$ and if $\gamma_i = 0$ then $\xi_i > 0$, $i = 1,2,\ldots$. We will seek coefficients $A(s) \in C^3[0,1]$ and $B(s) \in C^1[0,1]$ and real constants a,b,c such that the eigenvalue problem

$$
\begin{aligned}
& y^{(4)} + (Ay^{(1)})^{(1)} + By - \lambda y = 0 \\
& y(1) = y^{(1)}(1) = 0 \\
& y^{(2)}(0) + a\,y^{(1)}(0) - b\,y(0) = 0 \\
& y^{(3)}(0) + (b+A(0))y^{(1)}(0) + c\,y(0) = 0
\end{aligned}
\tag{3}
$$

has eigenvalues λ_i, $i = 1,2,\ldots$. And if y_{λ_i} is the eigenfunction associated with λ_i and having L^2 norm 1, then y_{λ_i} has normalization constants $y_{\lambda_i}(0) = \gamma_i$ and $y'_{\lambda_i}(0) = \xi_i$. It should be observed that with this assigned data for the eigenfunction, once the solution to the inverse eigenvalue problem is determined all of the initial values $[y_{\lambda_i}^{(j)}(0),\ j = 0,1,2,3]$ of each eigenfunction are determined. In this sense the assignment of the given data seems a "natural" generalization of the data required for the solution of the second order inverse eigenvalue problem.

The result which will be stated in Section 2 is an existence and uniqueness result. A method of construction of the solution will be given. It will be assumed that an integral relationship (with kernel K) exists between solutions of the (to be) derived differential equation and solutions of a known differential equation. Under this assumption solutions are unique.

There will be no a priori assumption that the given sequences are normalization constants and eigenvalues for an eigenvalue problem. These sequences will only be required to satisfy asymptotic forms. In addition, it will be shown that some sets of sequences are not allowed in the sense that the constructive technique cannot be applied to find a solution.

Section 1

In this section we will describe an iterative technique for finding solutions as given in [17], [18], [19] of fourth order inverse eigenvalue problems where sequences $0 < \lambda_1 < \lambda_2 < \dots$ and $\rho_i > 0$, $i = 1,2,\dots$ are given. The solution will consist of coefficients $A(s) \in C^3[0,1]$, $B(s) \in C^1[0,1]$ and real constants a,b,c such that the λ_i's are eigenvalues for (3) and each ρ_i is the L^2 norm of the i th eigenfunction, y_{λ_i}, normalized so that $y_{\lambda_i}(0) = 1$. The eigenvalue problem (3) is different from the example chosen in [17], [18], [19] and has been chosen so as to illustrate possible changes which can occur in non-Dirichlet boundary conditions. The solution which will be constructed will be the unique eigenvalue problem which has data (λ_i's and ρ_i's, $i = 1,2,3,\dots$) as given and for which a related non-self-adjoint problem has exactly the same eigenvalues and normalization constants (the L^2 inner product of the eigenfunction and adjoint eigenfunctions) as a given (base) non-self-adjoint eigenvalue problem.

We begin with a known (base) self-adjoint eigenvalue problem

(4) $$z^{(4)} - \lambda z = 0,\ z^{(3)}(0) = z^{(2)}(0) = z(1) = z^{(1)}(1) = 0.$$

The eigenvalues for this problem are $0 < \lambda_1^* < \lambda_2^* < \ldots$ and if $z_{\lambda_i^*}$ is the i th eigenfunction with $z_{\lambda_i^*}(0) = 1$ then the square of the L^2 norm of $z_{\lambda_i^*}$ is ρ_i^*, $i = 1,2,\ldots$. Let $0 \le \beta \le \frac{\pi}{2}$, and let $z_{\lambda,\beta}$ denote the solution of $z^{(4)} - \lambda z = 0$ which also satisfies the initial, boundary value problem

(4A) $z(0) = 1,\ z^{(2)}(0) = 0 = z^{(3)}(0),\ \cos\beta\, z(1) + \sin\beta\, z^{(1)}(1) = 0.$

We will first state a result which will show how (for each fixed β) to construct coefficients, $A_n(s,\beta)$, $B_n(s,\beta)$ and a_n, b_n, and c_n such that the sequences $0 < \lambda_1 < \lambda_2 < \ldots < \lambda_n < \lambda_{n+1}^* < \lambda_{n+2}^* \ldots$ and $\rho_1, \rho_2, \ldots, \rho_n, \rho_{n+1}^*, \rho_{n+2}^*, \ldots$ are the eigenvalues and normalization constants for

(5)
$$\begin{aligned} & y^{(4)} + (A_n y^{(1)})^{(1)} + B_n y - \lambda y = 0 \\ & U_1^n y = y^{(2)}(0) + a_n y^{(1)}(0) - b_n y(0) = 0 \\ & U_2^n y = y^{(3)}(0) + (b_n + A_n(0))y^{(1)}(0) + c_n y(0) = 0 \\ & y(0) = 1 \\ & y(1) = y^{(1)}(1) = 0 \end{aligned}$$

In each case (for a given n, and fixed β) solutions $y_{\lambda,\beta}^n(x)$, for $\lambda > 0$, of the initial, boundary value problem

(5A)
$$\begin{aligned} & y^{(4)} + (A_n y^{(1)})^{(1)} + B_n y - \lambda y = 0, \\ & U_1^n y = 0,\ U_2^n y = 0, \\ & y(0) = 1,\ \cos\beta\, y(1) + \sin\beta\, y'(1) = 0 \end{aligned}$$

are shown to be related to $z_{\lambda,\beta}$ by

$$y_{\lambda,\beta}(x) = z_{\lambda,\beta}(x) + \int_0^x K^n(x,t,\beta) z_{\lambda,\beta}(t)\,dt.$$

where K^n is determined as a solution of an integral equation and A_n, B_n, a_n, b_n, c_n are determined in terms of $K^n(x,t,\beta)$. We will then state a result showing that if the λ_i's and ρ_i's satisfy certain asymptotic forms then $K^n, A^n, B^n, a_n, b_n, c_n$ converge to K, A, B, a, b, c and for each β we will have constructed a solution to the inverse problem.

The sense in which each solution is unique will also be discussed and possible extensions of this result for $\frac{\pi}{2} < \beta < \pi$ will be given.

Before stating the theorems we introduce the following notation. Let

$$(7) \qquad f_{n,\beta}(s,t) = \sum_{i=1}^{n} \left[\frac{z_{\lambda_{i,\beta}}(s) z_{\lambda_{i,\beta}}(t)}{\rho_i} - \frac{z_{\lambda^*_{i,\beta}}(s) z_{\lambda^*_{i,\beta}}(t)}{\rho^*_i} \right]$$

and let

$$(8) \qquad f_{\beta}(s,t) = \sum_{i=1}^{\infty} \left[\frac{z_{\lambda_{i,\beta}}(s) z_{\lambda_{i,\beta}}(t)}{\rho_i} - \frac{z_{\lambda^*_{i,\beta}}(s) z_{\lambda^*_{i,\beta}}(t)}{\rho^*_i} \right].$$

Further observe that $z_{\lambda^*_{i,\beta}}(s)$ is independent of β and is the eigenfunction, corresponding to λ^*_i for (4). Accordingly we will write $z_{\lambda^*_{i,\beta}}(s) = z_{\lambda^*_i}(s)$. Then, we have the following theorems.

Theorem 1: Suppose $K_n(s,t,\beta)$ is continuous for $0 \leq t \leq s \leq 1$ and fixed β, $0 \leq \beta \leq \frac{\pi}{2}$. Then $\{y_{\lambda_{i,\beta}}\}_{i=1}^{n} \cup \{y_{\lambda_{i,\beta}}\}_{i=n+1}^{\infty}$, as defined in (6), is a complete orthogonal set in $L^2(0,1)$ with normalization constants $\rho_1, \rho_2, \ldots, \rho_n, \rho^*_{n+1}\ \rho^*_{n+2}, \ldots$ iff $K^n(s,t,\beta)$ is the unique solution of the integral equation

$$(9) \qquad f_{n,\beta}(s,t) + \int_0^x K_n(s,u,\beta) f_{n,\beta}(t,u)du + K_n(s,u,\beta) = 0.$$

In addition, the resultant $K_n(s,t,\beta)$ is analytic in s and t, $0 \leq t \leq s \leq 1$ Also, $\lambda_1, \lambda_2, \ldots, \lambda_n, \lambda^*_{n+1}, \lambda^*_{n+2}, \ldots$ and $\rho_1, \ldots, \rho_n, \rho^*_{n+1}, \rho^*_{n+2}, \ldots$ are the eigenvalues and normalization constants in (5) with corresponding eigenfunctions $y^n_{\lambda_{i,\beta}}$, $i = 1, \ldots, n$, $y^n_{\lambda^*_{i,\beta}}$, $i = n+1, n+2, \ldots$ iff

$$A^n(s,\beta) = -4 \frac{d}{ds} K_n(s,s,\beta)$$

$$(10) \qquad B^n(s,\beta) = -A^n(s,\beta) \frac{\partial}{\partial s} K_n(s,t,\beta)\Big|_{t=s} + 2 \frac{\partial}{\partial t}\left(\frac{\partial^2}{\partial s^2} K_n - \frac{\partial}{\partial t^2} K_n\right)\Big|_{t=s} - 2 \frac{d^3}{ds^3} K_n(s,s)$$

$$a_n = -K_n(0,0,\beta) = \sum_{i=1}^{n}\left[\frac{1}{\rho_i} - \frac{1}{\rho_i^*}\right]$$

(11) $$b_n = \left[2K_{n,s} + K_{n,t} - (K_n)^2\right]\Big|_{t=s=0} = 3\sum_{i=1}^{n}\left[\frac{z'_{\lambda_i^*}(0)}{\rho_i^*} - \frac{z'_{\lambda_{i,\beta}}(0)}{\rho_i}\right] + (a_n)^2$$

$$c_n = -\left[-3K_{n,s}K_n - 2K_{n,t}K_n + 3K_{n,ss} + 3K_{n,st} + K_{n,tt}\right]\Big|_{t=s=0}$$
$$= +10\,a_n \sum_{i=1}^{n}\left[\frac{z'_{\lambda_i^*}(0)}{\rho_i^*} - \frac{z'_{\lambda_{i,\beta}}(0)}{\rho_i}\right] - 3\sum_{i=1}^{n}\left[\frac{\left(z'_{\lambda_i}(0)\right)^2}{\rho_i^*} - \frac{\left(z'_{\lambda_i^*}(0)\right)^2}{\rho_i}\right]$$

Remark: The kernel in the integral equation has finite rank and thus $K_n(s,t,\beta)$ can be determined simply by solving a set of linear, non-homogeneous equations. Using this notion it can be shown that

(12) $$K_n(s,t,\beta) = \sum_{i=1}^{n}\left[\frac{y^n_{\lambda^*_{i,\beta}}(s)\,z_{\lambda^*_{i,\beta}}(t)}{\rho_i^*} - \frac{y^n_{\lambda_{i,\beta}}(s)\,z^n_{\lambda_{i,\beta}}(t)}{\rho_i}\right].$$

Remark: It can be shown that for fixed sequences $\lambda_i, \rho_i, i = 1,2,\ldots n$ we cannot have both $A_n(s,\beta_1) \equiv A_n(s,\beta_2)$ and $B_n(s,\beta_1) \equiv B_n(s,\beta_2)$ for $\beta_1 \neq \beta_2$. This can be seen from the fact that the initial values at $s = 0$ for the eigenfunctions $y_{\lambda_{i,\beta_1}}$ and $y_{\lambda_{i,\beta_2}}$ are not the same. Further, for each β the eigenvalues and normalization constants [L^2 inner product of an eigenfunction with its corresponding adjoint eigenfunction] are the same for the two problems

(13)
$$z^{(4)} - \lambda z = 0$$
$$z(0) = 0 = z''(0) = z'''(0),\ z'(0) = 1$$
$$\cos\beta\, z(0) + \sin\beta\, z^{(1)}(1) = 0$$

and

(14)
$$y^{(4)} + (A_n y^{(1)})^{(1)} + B_n y - \lambda y = 0$$
$$y(0) = 0,\ y'(0) = 1$$
$$U_1^n y = U_2^n y = 0$$
$$\cos\beta\, y(0) + \sin\beta\, y'(0) = 0$$

where the adjoint eigenfunctions for (13) and (14) satisfy $z'''(0) = 1$ and $U_2^n y = 1$ respectively. Using techniques developed in Leibenzon we can show that the data [normalization constants plus eigenvalues] from the non-self-adjoint problem and from the self-adjoint problem together uniquely determine A_n, B_n, a_n, b_n, c_n. Hence the coefficients are determined uniquely in this sense.

Remark: We can also characterize the uniqueness of the solution by observing that the eigenvalue problem with coefficients A_n, B_n, a_n, b_n, c_n is the only self-adjoint problem of the term (3) with the given eigenvalues and normalization constants, where solutions of (5A) are related to those of (4A) by an integral relationship as described by (6).

Remark: The theorem may be extended to include the values of $\beta, \frac{\pi}{2} < \beta < \pi$ for which $z_{\lambda_{i,\beta}}$ can be defined, $i = 1,\ldots,n$, and for which the set of functions $J = \{z_{\lambda_{i,\beta}}\}_{i=1}^n \cup \{z_{\lambda_i^*}\}_{i=n+1}^\infty$ are defined and complete in $L^2(0,1)$. The exceptional cases then include those values of β for which one of the λ_i's, $i = 1,\ldots,n$ is an eigenvalue for the non-self-adjoint problem (13) for then $z_{\lambda_{i,\beta}}$ cannot be defined for that particular value of λ_i. The remaining exceptional cases are those for which $z''_{\lambda_i^*}(1)\cos\beta + z'''_{\lambda_i^*}(1)\sin\beta = 0$ for some $i = 1,\ldots,n$. In this case the inner product $(z_{\lambda_j}, z_{\lambda_i^*}) = 0$ for $j = 1,\ldots,n$ and the given i and hence the set of functions J is not complete.

The next theorem gives sufficient conditions for the solutions derived in Theorem 1 to converge as $n \to \infty$ to eigenvalue problems which have the given λ_i's and ρ_i's as all the eigenvalues and all the L^2 norms of the corresponding eigenfunctions of the derived problem.

Theorem 2: Let β be fixed with $0 \leq \beta \leq \frac{\pi}{2}$. Let $0 < \lambda_1 < \lambda_2 < \ldots$ and $\rho_i > 0$, $i = 1,2,\ldots$ satisfy $(\lambda_i)^{\frac{1}{4}} + P_i$ and

$$\frac{1}{\rho_i} = \frac{1}{\rho_i^*} + R_i \text{ with } \sum_{i=1}^\infty (\lambda_i)^{\frac{k}{4}} |P_i| < \infty, \quad \sum_{i=1}^\infty (\lambda_i)^{\frac{k}{4}} |R_i| < \infty, \quad k = 0,1,2,3,4.$$

Then, there exists a unique solution $K(s,t,\beta) \in C^4[0 \le t \le s \le 1]$ of

$$(15) \qquad f(s,t,\beta) + \int_0^s f(t,u,\beta)K(s,u,\beta)du + K(s,t,\beta) = 0$$

with the property that

$$\lim_{n\to\infty} \frac{\partial^{j+k}}{\partial s^j \partial t^k} K_n(s,t,\beta) = \frac{\partial^{j+k}}{\partial s^j \partial t^k} K(s,t,\beta)$$

$j,k = 0,1,2,3,4,\ 0 \le j+k \le 4$. Further $y_{\lambda_{i,\beta}}$, $i = 1,2,\ldots$, as defined by

$$(16) \qquad y_{\lambda_{i,\beta}}(s) = z_{\lambda_{i,\beta}}(s) + \int_0^s K(s,t,\beta) z_{\lambda_{i,\beta}}(t)dt$$

forms a complete, orthogonal set in $L^2(0,1)$ with normalization constants $\rho_1, \rho_2, \ldots$ iff $K(s,t)$ satisfies (15).

The sequences $\lambda_1, \lambda_2, \ldots$ and $\rho_1, \rho_2, \ldots$ are eigenvalues and normalization constants for (3) with corresponding eigenfunctions $y_{\lambda_{i,\beta}}$, $i = 1,2,\ldots$ iff $A(s)$, $B(s)$, a,b,c are defined by

$$A(s) = -4 \frac{d}{ds} K(s,s)$$

$$B(s) = -A(s)K_s(s,t,\beta)\Big|_{t=s} + 2(K_{ss}-K_{tt})_t\Big|_{t=s} - 2\frac{d^3}{ds^3} K(s,s,\beta)$$

$$a = -K(0,0,\beta)$$

$$b = 3K_t(0,0,\beta) + a^2$$

$$c = 10\,a\,K_t(0,0,\beta) - 3\sum_{i=1}^{\infty}\left[\frac{(z'_{\lambda_i^*}(0))^2}{\rho_i^*} - \frac{(z'_{\lambda_i^*}(0))^2}{\rho_i^*}\right]$$

Remark: Similar statements, as those following Theorem 1, about uniqueness and about extension of this theorem to the case where $\frac{\pi}{2} < \beta < \pi$ can be made for the limiting case (as $n\to\infty$) as well.

Section 2

In this section we present new existence and uniqueness results where additional data, that is additional normalization constants for the eigenfunctions of the derived problem may be given. What will be given is three sequences. An eigenvalue problem will be derived for which one given sequence is the sequence of eigenvalues. The eigenfunctions will be assumed to have L^2 norm 1. The remaining two sequences will be values of the eigenfunctions and their first derivatives at one of the endpoints of the interval. It will not be assumed a priori that the given sequence arc those from an eigenvalue problem. Asymptotic forms for the sequences will be assumed. In addition a completeness hypothesis will further restrict the relationship between the sequences.

Following the format of the previous section, we will state the existence theorem in two parts. First we present a theorem which shows how to obtain the solution of the inverse problem when only a finite number of elements in the given sequences differ from those in a corresponding known eigenvalue problem. We will then explain (less formally) how to treat the limiting case when all elements of the sequences can differ from those in the corresponding known problem.

Before stating Theorem 3, we establish some additional notation. Consider again the eigenvalue problem (4) and let $w_{\lambda_i^*}(s)$ be the i th eigenfunction with L^2 norm equal to 1. Let $0 < \gamma_i^* = w_{\lambda_i^*}(0)$ and $\xi_i^* = w_{\lambda_i^*}^{(1)}(0)$, $i = 1,2,\dots$. Further, for given real numbers $\gamma_i \geq 0$, ξ_i, $i = 1,\dots,n$ and positive numbers $\lambda_i > 0$, $i = 1,\dots,n$, let $w_{\lambda_i}(s)$ satisfy the initial value problem

$$w^{(4)} - \lambda_i w = 0,$$

$$w(0) = \gamma_i, w'(0) = \xi_i - \gamma_i \Gamma_n, w''(0) = 0, w'''(0) = 0,$$

where $\Gamma_n = \sum_{j=1}^{n} [(\gamma_i^*)^2 - (\gamma_i)^2]$.

Let, for $0 \leq s,t \leq 1$,

$$F_n(s,t) = \sum_{i=1}^{n} [w_{\lambda_i^*}(s)w_{\lambda_i^*}(t) - w_{\lambda_i}(s)w_{\lambda_i}(t)]$$

Theorem 3: Let $\gamma_i \geq 0$, ξ_i, $i = 1,\dots,n$ be given real numbers such that $\xi_i > 0$ if $\gamma_i = 0$. Let $0 < \lambda_1 \leq \lambda_2 \leq \dots \leq \lambda_n < \lambda_{n+1}^* < \dots$. Suppose that $\{w_{\lambda_i}\}_{i=1}^{n} \cup \{w_{\lambda_i^*}\}_{i=n+1}^{\infty}$ is a complete set of functions in $L^2(0,1)$. Let $K_n(s,t)$ be continuous for $0 \leq t \leq s \leq 1$. Define

$$V^n_{\lambda_i} = w_{\lambda_i}(s) + \int_0^s K_n(s,t)w_{\lambda_i}(t)dt, \; i = 1,\dots,n$$

and

$$V^n_{\lambda_i^*} = w_{\lambda_i^*}(s) + \int_0^s K_n(s,t)w_{\lambda_i^*}(t)dt, \; i = 1,2,\dots .$$

Then the set $\{V^n_{\lambda_i}\}_{i=1}^n \cup \{V^n_{\lambda_i^*}\}_{i+n+1}^\infty$ is a complete orthonormal set iff $K_n(s,t)$ satisfies the integral equation

$$F_n(s,t) + \int_0^s K_n(s,u)F_n(t,u)du + K_n(s,t) = 0.$$

The function $K_n(s,t)$ is analytic in s and t for $0 \le t \le s \le 1$. Then the functions $\{V^n_{\lambda_i}\}_{i=1}^n \cup \{V^n_{\lambda_i^*}\}_{i=n+1}^\infty$ have initial values $V^n_{\lambda_i}(0) = \gamma_i$, $V^{n(1)}_{\lambda_i}(0) = \xi_i$, $i = 1,\dots,n$, and $V^n_{\lambda_i^*}(0) = \gamma_i^*$, $V^{n(1)}_{\lambda_i^*}(0) = \xi_i^* + \gamma_i^*\Gamma_n$ $i = n + 1, n + 2,\dots$ and are the eigenfunctions corresponding to the eigenvalues $\lambda_1,\lambda_2,\dots,\lambda_n,\lambda^*_{n+1},\dots$ for (3) iff

$$A_n(s) = - 4 \frac{d}{ds} K_n(s,s)$$

$$B_n = - A_n(s,\beta)K_{n,s}(s,t)\Big|_{t=s} + 2(K_{n,ss}-K_{n,tt})_t\Big|_{t=s} - 2 \frac{d^3}{ds^3} K_n(s,s)$$

$$a_n = - K_n(0,0) = - \Gamma_n$$

$$b_n = 3 \sum_{i=1}^n [\xi_i^*\gamma_i^*-\zeta_i\gamma_i] + [K_n(0,0)]^2$$

$$c_n = - 10 K_n(0,0) \sum_{i=1}^n [\xi_i^*\gamma_i^*-\zeta_i\gamma_i] - 3 \sum_{i=1}^n [(\xi_i^*)^2\gamma_i^*-(\zeta_i)^2\gamma_i]$$

where $\zeta_i = \xi_i - \gamma_i\Gamma_n$, $i = 1,\dots,n$.

Sketch of Proof: We show first that the integral equation is satisfied iff the sequence of functions $\{V^n_{\lambda_i}\}_{i=1}^n \cup \{V^n_{\lambda_i^*}\}_{i=n+1}^\infty$ is complete. We then observe that the integral equation has finite rank. Thus it can be solved simply by solving a set of linear nonhomogeneous equations and $K^n(s,t)$ can be expressed as

$$K_n(s,t) = \sum_{i=1}^n [V^n_{\lambda_i^*}(s)w_{\lambda_i^*}(t)-V^n_{\lambda_i}(s)w_{\lambda_i}(t)].$$

We then verify that $\{V^n_{\lambda_i}\}^n_{i=1} \cup \{V^n_{\lambda_i}\}^\infty_{i=n+1}$ are the eigenfunctions for the stated eigenvalue problem.

Remark: The assumption that $\{w_{\lambda_i}\}^n_{i=1} \cup \{w_{\lambda^*_i}\}^\infty_{i=n+1}$ is a complete set on $L^2(0,1)$ automatically implies that a given number in the sequence of λ_i's can occur at most twice. The completeness assumption also insures that if $\lambda_i = \lambda_{i+1}$ then w_{λ_i} is not a constant multiple of $w_{\lambda_{i+1}}$.

Remark: We can replace the completeness hypothesis by the assumption that a certain determinant is not zero. That is, let M_n be the determinant with components which are the L^2 inner product of w_{λ_j} and $w_{\lambda^*_i}$, $(w_{\lambda_j}, w_{\lambda^*_i})$, i, j = 1,...,n. Then det $M_n \neq 0$ iff $\{w_{\lambda_i}\}^n_{i=1} \cup \{w_{\lambda^*_i}\}^\infty_{i=n+1}$ is complete on $L^2(0,1)$. Further, it can be shown that whether or not det $M_n \neq 0$ depends only on the values of the λ_i and on the ratios of γ_i to ζ_i, $i = 1,\ldots,n$.

A nontrivial example where the completeness assumption is not satisfied can be obtained easily. Simply choose n = 1. Then $(w_{\lambda^*_i}, w_{\lambda_i}) = 0$ iff $(w'''_{\lambda^*_i}(1)w_{\lambda_i}(1) - w''_{\lambda^*_i}(1)w'_{\lambda_i}(1)) = 0$. Hence we choose γ_1, ξ_1 so this last equation is satisfied.

Remark: Without stating a formal theorem, we will explain one way to show existence for the case where $n \to \infty$. This can be done by showing that the iterations, that is the solutions given for each n in Theorem 3 converge to a solution of the desired inverse eigenvalue problem as $n \to \infty$. That is, we assume, in addition to the assumptions of Theorem 3, that $\{w_{\lambda_i}\}^\infty_{i=1}$ is complete and that we have the asymptotic forms $(\lambda_i)^{\frac14} - (\lambda^*_i)^{\frac14} = R_i$, $(\gamma^*_i)^2 - (\gamma_i)^2 = P_i$, and $\zeta_i = \lambda_i^{\frac14}[-\gamma_i + Q_i\{\exp[-(\lambda_i)^{\frac14}]\}]$ where $\sum_{i=1}^\infty |R_i|\lambda_i^{\frac{k}{4}} < \infty$, $\sum_{i=1}^\infty |P_i|\lambda_i^{\frac{k}{4}} < \infty$, $\sum_{i=1}^\infty |Q_i - Q_i|\lambda_i^{\frac{k}{4}} < \infty$, k = 0,1,2,3,4, (and where $\xi^*_i = \lambda_i^{\frac14}[-\gamma^*_i + Q^*_i\{\exp[-(\lambda^*_i)^{\frac14}\}])$. This is certainly sufficient to show that K_n and it's first four derivatives converge to a corresponding function K and it's derivatives. Further, there exist

$A(s) \in C^1$, $B(s) \in C^3$, a,b,c such that $A_n \to A$, $B_n \to B$ uniformly and $a_n \to a$, $b_n \to b$, $c_n \to c$ and the corresponding defined eigenvalue problem is a solution of our inverse eigenvalue problem.

It should be observed that for the limiting case we do not need $\det M_n \neq 0$ for all n. Because of the assumed asymptotic forms we need only select N such that for $i > N$ there is a "small enough" L so that, $|(\lambda_i^*)^{\frac{1}{4}}-(\lambda_i)^{\frac{1}{4}}| \leq \frac{L}{i}$, $|(\gamma_i^*)^2-(\gamma_i)^2| \leq \frac{L}{i}$, and $[|Q_i-Q_i^*|] \leq \frac{L}{i}$.

We then require $\det M_N \neq 0$. The above asymptotic forms then yield $\{w_{\lambda_i}\}_{i=1}^{\infty}$ complete in $L^2(0,1)$ by applying the completeness result in [20].

Remark: The solution derived above is unique in the sense that it is the only solution for which the corresponding eigenfunctions $\{V_{\lambda_i}\}_{i=1}^{\infty}$ are related to the set $\{w_{\lambda_i}\}_{i=1}^{\infty}$ by the integral relationship

$$V_{\lambda_i}(s) = w_{\lambda_i}(s) + \int_0^s K(s,t)w_{\lambda_i}(t)dt.$$

Remark: It is clear that the Dirichlet conditions at s = 1 are preserved but that in general the boundary conditions at s = 0 are not. The necessary and sufficient conditions for preserving the boundary conditions at s = 0 is clearly a = 0, b = 0, c = 0.

References

[1] V. Barcilon, Iterative Solution of the Inverse Sturm-Liouville Problem, J. Math. Phys., 15 (1974), pp. 287-298.

[2] V. Barcilon, On the solution of inverse eigenvalue problems of high orders, Geophys. J. R. Astr. Soc., 39 (1974), pp. 143-154.

[3] V. Barcilon, On the uniqueness of inverse eigenvalue problems, Ibid., 38 (1974), pp. 287-298.

[4] G. Borg, Eine Umkerung der Sturm-Liouvilleschen Eigenvertaufgabe, Acta. Math., 78 (1946), pp. 1-96.

[5] E. A. Coddington and N. Levinson, Theory of Ordinary Differential Equations, McGraw-Hill Book Co., New York, 1955.

[6] I. M. Gel'fand and B. M. Levitan, On the Determination of a Differential Equation from its Spectral Function, Izv. Akad. Nauk SSSR Ser. Mat., 15 (1951), pp. 309-360; English transl., Amer. Math. Soc. Transl., 1 (1955) pp. 253-304.

[7] O. H. Hald, The Inverse Sturm-Liouville Problem with Symmetric Potentials, Acta Math., 141 (1978), pp. 263-291.

[8] H. Hochstadt, The Inverse Sturm-Liouville Problem, Comm. Pure Appl. Math., 26 (1973), pp. 715-729.

[9] M. G. Krein, On a Method of Effective Solution of a Inverse Boundary Problem, Dokl. Akad. Nauk SSSR, 94 (1954), pp. 987-990.

[10] M. G. Krein, Solution of the Inverse Sturm-Liouville Problem, Ibid., 76 (1951), pp. 21-24.

[11] Z. L. Leibenzon, The Inverse Problem of the Spectral Analysis of Ordinary Differential Operators of Higher Order, Trudy Moskov. Mat. Obsc., 15 (1966) pp. 78-163.

[12] N. Levinson, The Inverse Sturm-Liouville Problem, Mat. Tidsskr. B., 25 (1949), pp. 25-30.

[13] B. M. Levitan, Generalized Translation Operators and Some of Their Applications, Fizmatigz, Moscow, 1962; English trans. Israel Program for Scientific Translations, Jerusalem and Davey, New York, 1964.

[14] B. M. Levitan, On the Determination of a Sturm-Liouville Equation by Two Spectra, Izv. Akad., Nauk SSSR Ser. Mat., 38 (1964), pp. 63-78; Amer. Math. Soc. Transl., 68 (1968), pp. 1-20.

[15] V. A. Marcenko, Concerning the Theory of a Differential Operator of the Second Order, Dakl. Adad. Nauk SSSR, 72 (1950), pp. 457-460.

[16] J. McKenna, On the Lateral Vibration of Conical Bars, SIAM J. Appl. Math., 21 (1971), pp. 265-278.

[17] J. R. McLaughlin, An Inverse Eigenvalue Problem of Order Four, SIAM J. Math. Anal., 7 (1976), pp. 646-661.

[18] J. R. McLaughlin, An Inverse Eigenvalue Problem of Order Four - An Infinite Case, SIAM J. Math. Anal., 9 (1978), pp. 395-413.

[19] J. R. McLaughlin, Fourth Order Inverse Eigenvalue Problems, Spectral Theory of Differential Operators, I. W. Knowles and R. T. Lewis (editors), North Holland Publishing Co. (1981), pp. 327-335.

[20] Sz. Nagy, Béla de, Expansion Theorems of Paley-Wiener Type, Duke Math. Journal, 14 (1947), pp. 975-978.

INDEFINITE STURM-LIOUVILLE PROBLEMS

Angelo B. Mingarelli[1]

1. The main stimulus for this paper arose out of an old and somewhat forgotten paper of Ralph Richardson [12]. His paper contained the first actual investigation (cf., [3]) of Indefinite or non-definite Sturm-Liouville problems over a finite interval; that is, problems of the form

$$-(p(t)y')' + q(t)y = \lambda r(t)y \tag{1.1}$$

where $p,q,r \in C[a,b]$ are real-valued and $p(t) > 0$. The indefinite case arises when (1.1), under suitable boundary conditions, is neither "left-"nor "right-definite" in current terminology. The boundary conditions which determine the eigenvalues in (1.1) will be of the form

$$\cos\alpha\, y(a) - \sin\alpha\, y'(a) = 0 \tag{1.2}$$

$$\cos\beta\, y(b) + \sin\beta\, y'(b) = 0 \tag{1.3}$$

where $0 \le \alpha,\beta < \pi$. One of the features of the indefinite case lies in the possible existence of *non-real* eigenvalues. To see this we may appeal to some earlier work of Hilb [7] wherein (1.1) takes the usual form

$$-y'' + q(t)y = \lambda y \tag{1.4}$$

and q is now *complex-valued* and a (non-trivial) solution of (1.4) is required to satisfy

$$y(a) = 0 = y(b). \tag{1.5}$$

Hilb showed that (1.4-5) admits a denumerable infinity of complex eigenvalues with no finite point of accumulation and gave, among other results, their asymptotic representation as $|\lambda| \to \infty$. So if we

1. This research is supported by the Natural Sciences and Engineering Research Council of Canada Grant U0167.

choose $q(t) = q_1(t) + iq_2(t)$ so that $\lambda = \mu + i\nu$, $\nu \neq 0$ is a non-real eigenvalue of (1.4-5), the corresponding non-real eigenfunction will satisfy an equation of the form (1.1) where now $p(t) \equiv 1$, $q \equiv \mu - q_1$, $r \equiv \nu - q_2$ and $\lambda = i$. Hence (1.1) would admit a non-real eigenvalue in this case. An extension of Hilb's work may be found in [1,§12.8].

Some recent investigations have shown that in the *singular* case of (1.1), non-real eigenvalues may arise even when $q(t) \geq 0$, (cf.,[2], [6], [4], [5]).

A point about which $r(t)$ changes its sign will be called a *turning point*, [9]. Problems of the form (1.1) with turning points have a wide variety of applications (e.g., [10]).

The aim of this paper is to study the non-real spectrum of boundary problems of the form (1.1-3) in the indefinite case.

In the sequel $L^2(a,b)$ will denote the space of all (equivalence classes of) complex-valued measurable functions over $[a,b]$ with finite L^2-norm. The usual inner-product on $L^2(a,b)$ will be denoted by (,). As usual $AC_{loc}(a,b)$ will denote the space of all complex valued locally absolutely continuous functions on (a,b).

2. ASSUMPTIONS: (H1) The function in question does not vanish identically on any subinterval of $[a,b]$.

(H2) $\lambda = 0$ is not an eigenvalue of the boundary problem(s) in question.

We should first note that not every $\lambda \in \mathbb{C}$ can be an eigenvalue of (1.1-3). This can be seen by adapting the argument in [1,§8.2] with minor modifications so as to show that a fixed solution $y(t,\lambda)$ of (1.1-2) is an entire function of λ of order not exceeding 1/2, as long as $r(t) \not\equiv 0$ over $[a,b]$. Thus if $y(t,\lambda)$ satisfies (1.3) the resulting equation for the eigenvalues yields a countable number of such with no finite point of accumulation. In fact it was shown by

Richardson [12] that, in the indefinite case, (1.1-2-3) admits infinitely many positive and infinitely many negative eigenvalues. Non-real eigenvalues may or may not exist even if both q and r have turning points in (a,b). To see this let $\alpha = \beta = 0$, say, and for fixed q let $r = Kq$ where $K \neq 0$ is a constant. Then (1.1-2-3) is in the polar case and so the spectrum is real. Sufficient conditions for the existence of non-real eigenvalues are difficult to formulate though some necessary conditions will be derived below.

3. Let $S = \{f \in L^2(a,b)\colon f, f' \in AC_{loc}(a,b),\ f'' \in L^2(a,b)$ and
$\cos\alpha\, f(a) - \sin\alpha\, f'(a) = 0,\ \cos\beta\, f(b) + \sin\beta\, f'(b) = 0$
for some $0 \le \alpha, \beta < \pi\}$.

Then for $f \in S$,

$$\cot\beta\, |f(b)|^2 + \cot\alpha\, |f(a)|^2 + \int_a^b |f'|^2 \geq \lambda_0 \int_a^b |f|^2 \qquad (3.1)$$

if $0 < \alpha, \beta < \pi$. Here λ_0 is the smallest eigenvalue of the problem $-y'' = \lambda y$ where y is subject to (1.2-3). When $\alpha = \beta = 0$, (3.1) is Wirtinger's inequality. (We note that in this case the first two terms in (3.1) disappear). The proof of (3.1) follows from the variational characterization of λ_0, (cf., [8, p 6.4]).

Theorem 1: Assume that (1.1-2-3) admits a non-real eigenvalue.

Then a) r changes sign in (a,b). In addition, if r satisfies (H1) ther $r(t)$ has at least one turning point in (a,b)

b) $q(t)$ has at least one turning point in (a,b) under (H1) if $0 \le \alpha, \beta < \pi/2$ and $q(t_0) > 0$ for some $t_0 \in [a,b]$.

c) There exists a subinterval J of $[a,b]$ over which

$$q(t) < -\lambda_0, \quad t \in J,$$

d) Under (H2) the boundary problem (1.2-3),

$$-y'' + q(t)y = \lambda y,$$

has at least one negative eigenvalue.

Proof. a) Let $y(t,\lambda) \equiv y$ be a corresponding non-real eigenfunction. Multiplying (1.1) by $\overline{y}$ integrating over $[a,b]$ and using (1.2-3) we find (since we may suppose $p(t) \equiv 1$),

$$\cot \beta|y(b)|^2 + \cot \alpha|y(a)|^2 + \int_a^b\{|y'|^2+q|y|^2\} = \lambda\int_a^b r|y|^2. \quad (3.2)$$

Assume, on the contrary, that r does not change sign in (a,b). Since q is real it follows from (3.2) that

$$(\mathrm{Im}\,\lambda) \int_a^b r|y|^2 = 0 \quad (3.3)$$

and since $\mathrm{Im}\,\lambda \neq 0$, the integral must vanish which is impossible unless $r(t) \equiv 0$. The other part also follows from this.

b) If possible let q have no turning point in (a,b). Then $q(t) \geq 0$ in (a,b) by hypothesis. But then the left side of (3.2) is real and positive whereas the right-side of (3.2) equals zero on account of (3.3). This is a contradiction and so q changes sign in (a,b). Under (H1) q must then have a turning point.

c) An application of (3.1) to the left side of (3.2) yields the estimate

$$\int_a^b(\lambda_0+q)|y|^2 \leq 0. \quad (3.4)$$

Assuming $q(t) \geq -\lambda_0$ all t, (3.4) implies $q(t) \equiv -\lambda_0$. (3.2-3) now yields $y = K\phi_0$, $K \in \mathbb{C}$, $\phi_0(t) \neq 0$ an eigenfunction corresponding to λ_0. There now follows $\lambda r(t) \equiv 0$ in (a,b) which is impossible.

d) The smallest eigenvalue ν of (1.4-2-3) is given by (cf., [8]) the minimum of the left side of (3.2) taken over those (normalized) functions $y \in S_0$ where S_0 is similar to S above except that we replace f" by f" - qf in the earlier definition. Now the non-real eigenfunction y makes the left side of (3.2) vanish. Moreover $y \in S_0$. Hence $\nu < 0$ since $\nu = 0$ is not an eigenvalue of (1.4-2-3), i.e., the problem has at least one negative eigenvalue.

Lemma 1. Let λ_i, λ_j be non-real eigenvalues of (1.1-2-3) with $\lambda_i \neq \overline{\lambda}_j$. Let ϕ_i, ϕ_j be two corresponding eigenfunctions. Then $\phi_i \neq \phi_j$ and

$$\cot \beta\, \overline{\phi}_j(b)\phi_i(b) + \cot \alpha\, \overline{\phi}_j(a)\phi_i(a) + \int_a^b (\phi_i'\overline{\phi}_j' + q\phi_i\overline{\phi}_j)dt = 0 \tag{3.5}$$

Proof. This is essentially an orthogonality relation between the non real eigenfunctions. As the proof follows usual lines it will be omitted (cf., also [11, p 213, lemma 4.2.2]). A glance at (3.2) shows that (3.5) is also valid when i = j.

It is evident that non-real eigenvalues of (1.1-2-3) occur in pairs, i.e., if λ is one such then so is $\overline{\lambda}$, as the coefficients are all real. In the following M will denote the number of "pairs" of non-real eigenvalues of (1.1-2-3) while N will denote the number of negative eigenvalues of the corresponding (1.4-2-3). That $N < \infty$ follows by Sturm-Liouville theory.

Theorem 2 Let q, r be real and continuous in [a,b] and assume that (1.1-2-3) gives rise to the indefinite case. Assume that (H2) is satisfied for (1.4-2-3). Then

$$M \leq N \tag{3.6}$$

Proof. We may suppose that $M \geq 1$, else (3.6) is trivial. Since (1.1-2-3) has at least one non-real eigenvalue (1.4-2-3) has at least one negative eigenvalue (theorem 1, (d)), say μ_0 with corresponding eigenfunction ψ_0. If we denote the eigenvalues of (1.4-2-3) and corresponding eigenfunctions by μ_i, ψ_i respectively where $-\infty < \mu_0 < \mu_1 < \ldots < \mu_n \ldots$ we know that

$$\mu_n = \inf \frac{(\tilde{A}f, f)}{(f,f)} \tag{3.7}$$

where $\tilde{A}f \equiv -f'' + qf$ and the "inf" is over all functions $f \not\equiv 0$, $f \in S_0$ for which $(f,\psi_i) = 0$ for $i = 0,1,2,\ldots,n-1$. Let $\lambda_0, \lambda_1, \ldots, \lambda_{M-1}$ denote a collection of non-real eigenvalues labelled in such a way that $\lambda_i \neq \overline{\lambda}_j$ for all i,j, $0 \le i,j \le M-1$. Let $\phi_0, \phi_1, \ldots, \phi_{M-1}$ denote M corresponding eigenfunctions.

If possible assume that $M > N$. This said, let $f(t) = \sum c_i\phi_i$ where the generally complex c_i, $0 \le i \le M-1$, are chosen so that

$$(f,\psi_i) = 0, \quad i = 0,1,2,\ldots,N-1 \tag{3.8}$$

Now (3.8) describes a system of N linear equations in M unknowns. Since $M > N$ this system has a non-trivial solution $c_0, c_1, \ldots c_{M-1}$ which we now fix. We note that $f \not\equiv 0$ and $f \in S_0$. Furthermore use of (3.5) shows that f satisfies

$$\cot\beta|f(b)|^2 + \cot\alpha|f(a)|^2 + \int_a^b(|f'|^2+q|f|^2) = 0. \tag{3.9}$$

There now follows that $\mu_N < 0$ on account of (3.7) and (H1). But this implies that (1.4-2-3) has at least N+1 negative eigenvalues, which is a contradiction. Thus $M \le N$, and so the non-real point spectrum of (1.1-2-3) is necessarily a finite set.

Remarks. It follows from the validity of (3.7) for the class of functions mentioned that the continuity requirements on q,r may be weakened to $q,r \in L$ (a,b) without any essential change in the proof.

The upper bound appearing in (3.6) is sharp in the case of a three-term recurrence relation (cf., [11, p. 206 theorem 4.1.6 and p. 209 example 4.1.1]).

That $M = N$ is, in general, false. For example, for given q(t) let $r(t) = Kq(t)$ for each t and $K \neq 0$ real. Set $\alpha = \beta = 0$ in (1.2-3). The boundary problem (1.1-2-3) in this case has real spectrum only, so that $M = 0$. On the other hand q may be chosen so as to generate any finite number N of negative eigenvalues of (1.4-2-3).

4. The fundamental theorems of Sturm, viz. oscillation, separation, regarding the eigenfunctions of non-indefinite Sturm-Liouville problems do not appear to have any simple counterpart in the indefinite case when the eigenfunctions are non-real. However Richardson [12, p 302] did show that if r(t) has precisely one turning point in (a,b) and $\alpha = \beta = 0$ in (1.2-3), the zeros of the real and imaginary part of a non-real eigenfunction interlace. In the sequel let r satisfy (H1)

Theorem 3. Let λ, $y(t,\lambda)$ be a non-real eigenvalue and associated non-real eigenfunction of (1.1-2-3). If r(t) has precisely n turning points in (a,b) then $y(t,\lambda)$ may vanish at most (n-1)-times in (a,b).

Proof. Let ξ_i, $i = 1,2,\ldots,n$ be the turning points of r arranged so that $\xi_1 < \xi_2 < \ldots < \xi_n$. Let

$$f(t) \equiv \int_a^t r|y|^2 , \quad t \in [a,b]. \tag{4.1}$$

Then $f(a) = f(b) = 0$ (cf., (3.3)). If possible assume, on the contrary, that y vanishes n times at, say, $c_1, c_2, \ldots, c_n$, (arranged in an increasing order), in (a,b). Restricting $y(t,\lambda)$ to $[a,c_i]$ we see that y satisfies (1.2) and $y(c_i) = 0$. Thus y is a non-real eigenfunction corresponding to the non-real eigenvalue λ over $[a,c_i]$. Thus (3.2) holds with b replaced by c_i and so (3.3) implies that $f(c_i) = 0$, $i = 1,2,\ldots,n$. Hence there exists at least one interval among $[a,c_1]$, $[c_1,c_2],\ldots,[c_n,b]$ whose interior is turning point-free. We may assume, without loss of generality, that $[a,c_1]$ is such an interval. Since $f(a) = f(c_1) = 0$, there is a point $t_1 \in (a,c_1)$ such that $0 = f'(t_1) = r(t_1)|y(t_1)|^2$. Hence $y(t_1) = 0$. Since y satisfies (1.2) and $y(t_1) = 0$ there follows $f(t_1) = 0$ by an argument similar to the one above. Since $f(a) = 0 = f(t_1)$ there is a point $t_2 \in (a,t_1)$ so that $0 = f'(t_2) = r(t_2)|y(t_2)|^2$ and so $y(t_2) = 0$. Repeating the previous argument we have $f(t_2) = 0$. Continuing in this way we obtain an infinite sequence $t_n \in [a,c_1]$ on which $y(t_n) = 0$. Since $[a,c_1]$ is compact there is a subsequence $t_{n_k} \to c$ where $c \in [a,c_1]$. By continuity $y(c) = 0$. Writing $y(t) = u(t) + iv(t)$, we have $u(c) = v(c) = 0$. Furthermore $u(c+h) = u(c) + h\,u'(c+\theta h)$, $0 \le \theta < 1$. If we set $h = t_{n_k} - c$ we find that $u'(c+\theta(t_{n_k} - c)) = 0$. Since u' is continuous there follows $u'(c) = 0$. A similar argument shows that $v'(c) = 0$. Thus $y(c) = y'(c) = 0$, and so $y \equiv 0$ over $[a,b]$. This is a contradiction and so y can vanish at most $(n-1)$-times.

<u>Corollary 1</u>. Let $y(t,\lambda)$ be as in the theorem. If $r(t)$ has exactly one turning point in (a,b) then $y(t,\lambda) \neq 0$ in (a,b).

The above corollary can now be used to prove the following result, due to Richardson [12, p 302, theorem $\underline{X}$].

<u>Proposition 1</u>. Let $\alpha = \beta = 0$ in (1.2-3). If, in (a,b), $r(t)$ has precisely one turning point the zeros of the real and imaginary part of any non-real eigenfunction separate one another.

Consolidating some results of [12] with the above we conclude with

<u>Theorem 4</u>. The boundary problem (1.1-2-3) has, in the indefinite case, an infinite number of real eigenvalues $\lambda_n^{\pm}$ with the property that $\lambda_n^{\pm} \to \pm\infty$ respectively and for n sufficiently large corresponding eigenfunctions vanish n times in (a,b). Furthermore there is an at most finite (though possibly empty) set of non-real eigenvalues whose total number (necessarily even) does not exceed twice the number of negative eigenvalues of the corresponding problem (1.4-2-3).

REFERENCES

[1] F.V. Atkinson, *Discrete and Continuous Boundary Problems*, Academic Press, New York, 1964

[2] F.V. Atkinson, W.N. Everitt, K.S. Ong, *On the m-coefficient of Weyl for a differential equation with an indefinite weight-function*, Proc. London Math. Soc 3 (29), (1974), 368-384

[3] M. Bôcher, *Boundary problems in one dimension*, 5th International Congress of Mathematicians, Proceedings, Cambridge U.P. 1912, Vol. 1, 163-195.

[4] K. Daho and H. Langer, *Some remarks on a paper by W.N. Everitt*, Proc. Roy. Soc. Edinburgh, Sect A, 78 (1977) 71-79.

[5] K. Daho and H. Langer, *Sturm-Liouville problems with an indefinite weight-function*, Proc. Roy. Soc. Edinburgh, Sect A, 78 (1977), 161-191.

[6] W.N. Everitt, *Some remarks on a differential expression with an indefinite weight-function*, in Spectral Theory and Asymptotics of Differential Equations, Math. Stud., 13 (Amsterdam: North-Holland 1974).

[7] E. Hilb, *Über Reihenentwicklungen nach den Eigenfunktion linearer Differentialgleichungen 2ter Ordnung*, Mat. Ann. 71 (1912), 76-87.

[8] K. Jörgens, *Spectral Theory of Second-Order Ordinary Differential Equations*, Matematisk Institut, Aarhus University, 1964.

[9] R.E. Langer, *The asymptotic solution of ordinary differential equations of the second order with special reference to a turning point*, Trans. Amer. Math. Soc. 67 (1949), 461-490

[10] R.E. Langer, *Asymptotic solutions of a differential equation in the theory of microwave propagation*, Comm. Pure Appl. Math 3 (1950), 427-438

[11] A.B. Mingarelli, *Volterra-Stieltjes Integral Equations and Generalized Differential Expressions*, Ph.D. Dissertation, University of Toronto (June 1979).

[12] R.G.D. Richardson, *Contributions to the study of oscillation properties of the solutions of linear differential equations of the second order*, Amer. J. Math. 40 (1918), 283-316.

585 King-Edward Av, Ottawa, Ontario, Canada, K1N 9B4.

The Infinitesimal Generator of a Stochastic Functional Differential Equation

S. E. A. Mohammed

§0 Introduction

Trajectories of stochastic retarded functional differential equations are known to possess a Markov property (See §1 below). In this article, we prove a general formula for their weak infinitesimal generator.

§1 Summary of Known Results

Let $r > 0$, $J = [-r, o]$ and $C \equiv C(J, R^n)$ stand for the separable Banach space of all continuous paths $\eta : J \longrightarrow R^n$, given the supremum norm $\|\eta\|_C = \sup |\eta(s)|$, where $|.|$ represents the Enclidean norm on R^n. Let $(\Omega, \mathcal{F}, P)$ be a probability space carrying m-dimensional Brownian motion $W : \Omega \times R^{\geqslant o} \longrightarrow R^m$, with induced filtration $(\mathcal{F}_t)_{t \geqslant o}$, $\mathcal{F}_t = \sigma\{W(.,s) : o \leqslant s \leqslant t\}$, $t \geqslant o$. If $x : \Omega \times [-r, \infty) \longrightarrow R^n$ is any continuous $(\mathcal{F}_t)_{t \geqslant o}$ - adapted process on $(\Omega, \mathcal{F}, P)$, define the *slice* of x at $t \geqslant o$ to be the $\mathcal{F}_t$-measurable map $x_t : \Omega \longrightarrow C$ given by $x_t(w)(s) = x(w, t+s)$ for all $s \in J$ and a.a. $w \in \Omega$. Suppose $L(R^m, R^n)$ is the Banach space of all linear maps $R^m \longrightarrow R^n$ with the uniform operator norm

$$\| M \| = \sup \{ | M(x) | : x \in R^m, |x| \leqslant 1\}, M \in L(R^m, R^n).$$

Then, for Lipschitz "coefficients" $H : C \longrightarrow R^n$ and $G : C \longrightarrow L(R^m, R^n)$, it is known that the *trajectory* $\{ {}^\eta x_t : t \geqslant o\}$ of the stochastic FDE

$$(\text{SFDE}) \quad \begin{cases} d\,{}^\eta x(t) = H({}^\eta x_t)\,dt + G({}^\eta x_t)\,dW(t) & t>o \\ {}^\eta x_o = \eta \end{cases}$$

gives a continuous time-homogenous C-valued Markov process. Indeed, for any Borel subset B of C, one has

$$P({}^\eta x_{t_2} \in B | \mathcal{F}_{t_1}) = P({}^\eta x_{t_2} \in B \mid {}^\eta x_{t_1}) \text{ a.s., whenever } o \leqslant t_1 \leqslant t_2.$$

For further details, examples and proofs, the reader may refer to [6] and [7].

§2 The Semi-group

In this and the next section, we shall assume, in addition to the Lipschitz condition, that the *drift* and *diffusion coefficients* H and G in (SFDE) are globally bounded on C with a common bound $K > 0$. Let C_b be the Banach space of all bounded uniformly continuous functions $\phi : C \longrightarrow R$ given the supremum norm $\|\phi\| = \sup \{|\phi(\eta)| : \eta \in C\}$. For $\phi \in C_b$, $t \geq o$, define $P_t(\phi) : C \longrightarrow R$ by setting $P_t(\phi)(\eta) = E\,\phi({}^{\eta}x_t(.))$ for all $\eta \in C$. From the uniform continuity of ϕ and the map $C \longrightarrow R$, $\eta \longmapsto E\,\|{}^{\eta}x_t(.)\|^2$, it follows that $P_t(\phi)$ is uniformly continuous. Moreover, $\|P_t(\phi)\| \leq \|\phi\|$. By the Markov property and time-homogeneity, it is easy to see that $P_t : C_b \longrightarrow C_b$, $t \geq o$, is a contraction semi-group on C_b; viz $P_o = id_{C_b}$, $P_{t_1+t_2} = P_{t_1} \circ P_{t_2}$, $\|P_t\| \leq 1$, for all $t_1, t_2, t \geq o$ ([4] PP. 19-46).

Theorem 1:

The contraction semi-group $\{P_t\}_{t\geq o}$ is *never* strongly continuous on C_b. In particular, its *domain* of *strong continuity* $C_b^o = \{\phi : \phi \in C_b, \lim_{t \to o+} P_t(\phi) = \phi\}$ is a *proper* closed subalgebra of C_b, *independent* of H, G and W.

Proof:

For each $\eta \in C$ and $t \geq o$, define $\tilde{\eta}_t \in C$ by

$$\tilde{\eta}_t(s) = \begin{cases} \eta(o) & t + s \geq o,\ s \in J \\ \eta(t+s) & -r \leq t + s < o,\ s \in J \end{cases}$$

Define the *shift semi-group* $S_t : C_b \longrightarrow C_b$, $t \geq o$, by $S_t(\phi)(\eta) = \phi(\tilde{\eta}_t)$, $\eta \in C$, $\phi \in C_b$. It is easy to see that $\{S_t\}_{t\geq o}$ is a contraction semi-group on C_b. By the sample-path continuity of the trajectory $\{{}^{\eta}x_t : t \geq o\}$ in C, we have $\lim_{t \to o+} P_t(\phi)(\eta) = \phi(\eta) = \lim_{t \to o+} S_t(\phi)(\eta)$ for every $\phi \in C_b$. Moreover, the Martingale inequality for stochastic integrals ([5] P. 70) yields

$$E\,\|{}^{\eta}x_t(.) - \tilde{\eta}_t\|^2 \leq E \sup \{|\int_o^{t+s} H({}^{\eta}x_u(.))\,du + (.)\int_o^{t+s} G({}^{\eta}x_u(.))\,dW(u)|^2 : t+s\geq o,\ s \in$$

$$\leq 2K^2 t^2 + 8K^2 t,\ t \geq o.$$

Therefore, $\lim_{t \to o+} E\,\|{}^{\eta}x_t(.) - \tilde{\eta}_t\|^2 = 0$ *uniformly* for $\eta \in C$. Using this and the *uniform* continuity of $\phi \in C_b$, it is not hard to see that $\lim_{t \to 0+} [P_t(\phi)(\eta) - S_t(\phi)(\eta)] = 0$ *uniformly* for $\eta \in C$.

Writing

$$P_t(\phi)(\eta) - \phi(\eta) = [P_t(\phi)(\eta) - S_t(\phi)(\eta)] + [S_t(\phi)(\eta) - \phi(\eta)],$$

we see that $\lim_{t \to 0+} P_t(\phi)(\eta) = \phi(\eta)$ *uniformly* for $\eta \in C$ if and only if $\lim_{t \to 0+} S_t(\phi)(\eta) = \phi(\eta)$ *uniformly* for $\eta \in C$. Thus it is sufficient to prove that $\{S_t\}_{t \geq 0}$ is *not* strongly continuous on C_b. Define $\psi : C \longrightarrow R$ by

$$\psi(\eta) = \begin{cases} \eta(s_0) & \|\eta\| \leq 1 \\ \frac{1}{\|\eta\|}\eta(s_0) & \|\eta\| > 1 \end{cases}$$

for a fixed $s_0 \in [-r, 0)$. Clearly ψ is bounded and globally Lipschitz on C; so $\psi \in C_b$. Fix $v \in R^n$ such that $|v| = 1$ and construct a sequence $\{\eta^n\}$ in C with the graph of each η^n looking like

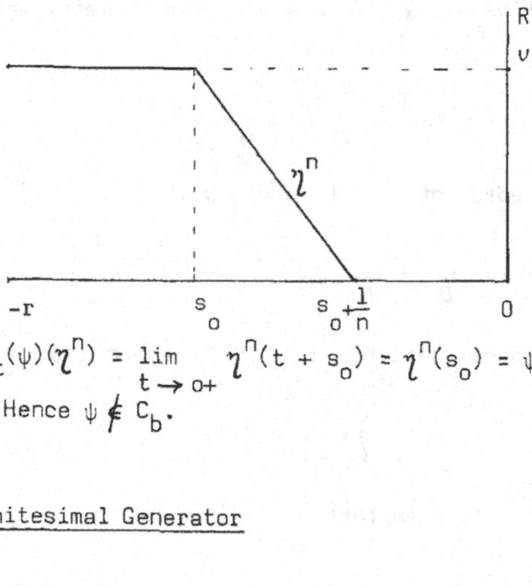

Obviously $\lim_{t \to 0+} S_t(\psi)(\eta^n) = \lim_{t \to 0+} \eta^n(t + s_0) = \eta^n(s_0) = \psi(\eta^n)$, and the convergence is *not* uniform in n. Hence $\psi \notin C_b$.

§3 The Weak Infinitesimal Generator

Give C_b the weak topology induced by the bilinear pairing $<.,.>: C_b \times M \longrightarrow R$, $<\phi,\mu> = \int_C \phi(\eta)\, d\mu(\eta)$, $\phi \in C_b$, $\mu \in M$, the Banach space of all finite regular Borel measures on C with the total variation norm. If $\phi_t, \phi \in C_b$, $t \geq 0$, then say the weak limit $\text{w-lim}_{t \to 0+} \phi_t = \phi$ whenever $\lim_{t \to 0+} <\phi_t,\mu> = <\phi, \mu>$ for all $\mu \in M$. This is equivalent to saying that $\{\|\phi_t\| : t > 0\}$ is bounded and $\phi_t(\eta) \longrightarrow \phi(\eta)$ as $t \to 0+$, for *each* $\eta \in C$. ([4] PP. 36-43). The *weak infinitesimal generator* $A : \mathcal{D}(A) \subset C_b \longrightarrow C_b$ of $\{P_t\}_{t \geq 0}$ is defined by $A(\phi) = \text{w-lim}_{t \to 0+} \frac{1}{t}[P_t(\phi) - \phi]$ whenever the limit exists. Its domain $\mathcal{D}(A)$ is the set of all $\phi \in C_b$ for which the last weak limit exists. Note that

$\text{w-lim}_{t \to o+} P_t(\phi) = \phi$, and $\mathcal{D}(A) \subset C_b^o$.

Let $S : \mathcal{D}(S) \subset C_b^o \longrightarrow C_b$ be the weak infinitesimal generator of $\{S_t\}_{t \geq o}$. To obtain a formula for A, we need to augment the state space C by attaching a canonical n-dimensional direction F_n. Then A equals S plus a second order partial differential operator along F_n. Take $F_n = \{v\chi_{\{o\}} : v \in R^n\}$, where $\chi_{\{o\}} : J \longrightarrow R$ is the characteristic function of $\{o\}$. Form the complete direct sum $C \oplus F_n$ normed by $\|\eta + v\chi_{\{o\}}\| = \|\eta\| + |v|$, $\eta \in C$, $v \in R^n$; i.e. $C \oplus F_n$ is the space of all bounded functions $\xi : J \longrightarrow R^1$ which are continuous on $[-r, o)$ but with at most only a jump discontinuity at o.

In order to establish our formula for the generator, we require the following sequence of lemmas.

Lemma 1:

Each $\alpha \in C^*$ has a unique weakly continuous extension $\bar{\alpha} \in [C \oplus F_n]^*$ i.e. if $\{\xi^k\}_{k=1}^{\infty}$ is bounded in C and $\xi^k(s) \longrightarrow \xi(s)$ as $k \longrightarrow \infty$ for all $s \in J$ and some $\xi \in C \oplus F_n$, then $\alpha(\xi^k) \longrightarrow \bar{\alpha}(\xi)$ as $k \longrightarrow \infty$. The map $\alpha \longmapsto \bar{\alpha}$ is a linear isometry into.

Proof:

Take n=1. By Riesz's representation theorem, write

$$\alpha(\eta) = \int_{-r}^{o} \eta(s)\, d\mu(s) \qquad \eta \in C$$

where μ is a finite regular Borel measure on J. Define $\bar{\alpha} \in [C \oplus F_1]^*$ by

$$\bar{\alpha}(\eta + v\chi_{\{o\}}) = \alpha(\eta) + v\mu\{o\} \qquad \eta \in C,\ v \in R$$

For higher dimensions $n > 1$, note that

$$C(J, R^n) \oplus F_n = [C(J, R) \oplus F_1] \times \cdots\cdots \times [C(J, R) \oplus F_1] \quad \text{(n copies)}$$

Write each $\alpha \in [C(J, R^n)]^*$ in the form

$$\alpha(\eta) = \sum_{i=1}^{n} \alpha^i(\eta^i), \quad \eta = (\eta^1, \ldots, \eta^n) \in C(J, R^n),$$

$\eta^i \in C(J, R)$, $\alpha^i(\tilde{\xi}) = \alpha(o, \ldots, o, \tilde{\xi}, o, \ldots, o)$, $\tilde{\xi} \in C(J, R)$ occupying the ith place, $i=1, \ldots, n$. Finally define $\bar{\alpha} \in [C(J, R^n) \oplus F_n]^*$ by

$$\bar{\alpha}(\eta + v\chi_{\{0\}}) = \sum_{i=1}^{n} \bar{\alpha}^i(\eta^i + v^i\chi_{\{0\}}), \quad \eta \in C(J, R^n), \ v = (v^1,\ldots,v^n) \in R^n.$$

Lemma 2:

Every continuous bilinear form $\beta : C \times C \longrightarrow R$ has a unique weakly continuous bilinear extension $\bar{\beta} : [C \oplus F_n] \times [C \oplus F_n] \longrightarrow R$ i.e. if $\{\xi^k\}_{k=1}^{\infty}$, $\{\eta^k\}_{k=1}^{\infty}$ are bounded in C with $\xi^k(s) \longrightarrow \xi(s)$, $\eta^k(s) \longrightarrow \eta(s)$ as $k \longrightarrow \infty$ for all $s \in J$, $\xi, \eta \in C \oplus F_n$, then $\beta(\xi^k, \eta^k) \longrightarrow \bar{\beta}(\xi, \eta)$ as $k \longrightarrow \infty$.

Proof:

First, take n=1. Regard $\beta : C \times C \longrightarrow R$ as a *continuous linear* map $\beta : C \longrightarrow C^*$. Since C^* is weakly complete, then β is weakly compact; hence there is a unique C^*-valued regular Borel measure λ on J (with finite semi-variation) such that

$$\beta(\xi) = \int_{-r}^{0} \xi(s)\, d\lambda(s) \qquad \xi \in C$$

([3] P. 494). Now extend β to a continuous linear map $\hat{\beta} : C \oplus F_1 \longrightarrow C^*$ as in Lemma 1, by using the dominated convergence theorem for vector-valued measures ([3] P. 328). Define $\bar{\beta} : C \oplus F_1 \longrightarrow [C \oplus F_1]^*$ by

$$\bar{\beta}(\eta + v_1\chi_{\{0\}}) = \overline{\hat{\beta}(\eta + v_1\chi_{\{0\}})}, \quad \eta \in C, \ v_1 \in R$$

It is easy to check that $\bar{\beta}$ is weakly continuous as a bilinear form on $C \oplus F_1$. When n>1, use co-ordinates to reduce to the one-dimensional situation above (cf. Lemma 1).

For uniqueness, observe that $\bar{\beta}$ is uniquely determined on $F_n \times F_n$: approximate $v^1\chi_{\{0\}}, v^2\chi_{\{0\}} \in F_n$ weakly by sequences $\{\xi_0^k\}_{k=1}^{\infty}$, $\{\eta_0^k\}_{k=1}^{\infty}$ in C respectively; note that any weakly continuous extension $\tilde{\beta}$ of β must satisfy

$$\tilde{\beta}(v^1\chi_{\{0\}}, v^2\chi_{\{0\}}) = \lim_{k\to\infty} \beta(\xi_0^k, \eta_0^k) = \bar{\beta}(v^1\chi_{\{0\}}, v^2\chi_{\{0\}}).$$

Lemma 3:

If $\alpha \in C^*$, then $\lim_{t\to 0+} \frac{1}{t} E\alpha({}^{\eta}x_t(.) - \tilde{\eta}_t) = \bar{\alpha}(H(\eta)\chi_{\{0\}})$ for each $\eta \in C$.

Proof:

E denotes expection for C (or R^n) - valued random variables. Since evaluation at each $s \in J$ commutes with expectation in C, then, for each $t>o$,

$$[\frac{1}{t} E({}^{\eta}x_t(.) - \tilde{\eta}_t)](s) = \begin{cases} \frac{1}{t}\int_o^{t+s} E(H({}^{\eta}x_u(.))\,du & t+s \geq o \\ o & -r \leq t+s < o, \; s \in J. \end{cases}$$

Therefore, $\lim_{t \to o+} [\frac{1}{t} E({}^{\eta}x_t(.) - \tilde{\eta}_t)](s) = H(\eta)\chi_{\{o\}}(s)$ for all $s \in J$. Moreover, $\| \frac{1}{t} E({}^{\eta}x_t(.) - \tilde{\eta}_t)\| \leq K$ for all $t > o$; so the result follows from the weak continuity of $\bar{\alpha}$.

Lemma 4:

For each $t \in (o, r]$ and a.a. $w \in \Omega$, define $W_t^*(w) \in C(J, R^m)$ by

$$W_t^*(w)(s) = \begin{cases} \frac{1}{\sqrt{t}}[W(w, t+s) - W(w, o)] & -t \leq s \leq o \\ o & -r \leq s < -t \end{cases}$$

Let β be a continuous bilinear form on C. Then, for each $\eta \in C$,

$$\lim_{t \to o+} [\frac{1}{t} E\, \beta({}^{\eta}x_t(.) - \tilde{\eta}_t, {}^{\eta}x_t(.) - \tilde{\eta}_t) - E\, \beta(G(\eta)oW_t^*(.), G(\eta)oW_t^*(.))] = 0$$

Proof:

Let $o<t<r$ and consider

$$E \sup_{s \in J} |\frac{1}{\sqrt{t}}({}^{\eta}x_t - \tilde{\eta}_t)(s) - [G(\eta)oW_t^*(.)](s)|^2$$

$$< \frac{2}{t} E \sup_{s \in [-t, o]} |\int_o^{t+s} H({}^{\eta}x_u)du|^2 + \frac{2}{t} E \sup_{s \in [-t, o]} |\int_o^{t+s} [G({}^{\eta}x_u) - G(\eta)]dW(u)|^2$$

$$< \int_o^t E|H({}^{\eta}x_u(.))|^2\,du + \frac{8}{t}\int_o^t E\| G({}^{\eta}x_u(.)) - G(\eta)\|^2\,du$$

Letting $t \to o+$ in the above inequality gives

$$\lim_{t \to o+} E\, \|\frac{1}{\sqrt{t}}({}^{\eta}x_t(.) - \tilde{\eta}_t) - G(\eta)oW_t^*(.)\|^2 = 0$$

But $E \| G(\eta)oW_t^*(.)\| \leq \| G(\eta)\|^2$ for all $t \in (o, r]$, so by continuity of β and Hölder's inequality we obtain

$$\left| \frac{1}{t} E\, \beta({}^{\eta}x_t(.) - \tilde{\eta}_t, {}^{\eta}x_t(.) - \tilde{\eta}_t) - E\, \beta(G(\eta)oW_t^*(.), G(\eta)oW_t^*(.)) \right|$$

$$< \|\beta\| E \| \frac{1}{\sqrt{t}} ({}^{\eta}x_t(.) - \tilde{\eta}_t) - G(\eta)oW_t^*(.)) \|^2$$

$$+ 2\|\beta\| [E\| \frac{1}{\sqrt{t}} ({}^{\eta}x_t(.) - \tilde{\eta}_t) - G(\eta)oW_t^*(.)\|^2]^{\frac{1}{2}} [E\| G(\eta)oW_t^*(.)\|^2]^{\frac{1}{2}}$$

We get the required result by letting $t \to o+$ in the above inequality.

Lemma 5:

Let $\{e_j\}_{j=1}^m$ be any basis for R^m. Then

$$(*) \lim_{t \to o+} \frac{1}{t} E\, \beta({}^{\eta}x_t(.) - \tilde{\eta}_t, {}^{\eta}x_t(.) - \tilde{\eta}_t) = \sum_{j=1}^m \bar{\beta}(G(\eta)(e_j)\chi_{\{o\}}, G(\eta)(e_j)\chi_{\{o\}}) \text{ for}$$

each $\eta \in C$.

Proof:

Observe that the right-hand-side of (*) is the trace of the bilinear map $R^m \times R^m \to R$, $(v^1, v^2) \mapsto \bar{\beta}(G(\eta)(v^1)\chi_{\{o\}}, G(\eta)(v^2)\chi_{\{o\}})$, and is therefore independent of the choice of basis in R^m. So we can assume that $e_j = (\delta_{ij})_{i=1}^m$, $j=1,\ldots,m$, is the canonical basis for R^m. Moreover, using co-ordinates and the independence of the real Brownian motions $\{\langle W(.,t), e_j\rangle : t \in R^{\geq 0}\}$, $j=1,\ldots,m$, (*) may be reduced to the one-dimensional case $m=n=1$, viz.

$$(**) \lim_{t \to o+} E\, \beta(W_t^*(.), W_t^*(.)) = \bar{\beta}(\chi_{\{o\}}, \chi_{\{o\}})$$

for one dimensional Brownian motion W, where we have used Lemma 4.

To prove (**), form the complete projective tensor product $C \hat{\otimes}_{\pi} C$ with the norm

$$\| h \|_{\otimes_{\pi}} = \inf \{ \sum_{i=1}^p \| \xi^i \| \, \| \eta^i \| : h = \sum_{i=1}^p \xi^i \otimes \eta^i, \xi^i, \eta^i \in C, i=1,\ldots,p\}$$

where the infimum is taken over all possible representations of h in the algebraic tensor product $C \otimes_{\pi} C$. Let $L^2(\Omega, C)$ denote the Banach space of all measureable maps $\psi : \Omega \to C$ such that the function $w \mapsto \|\psi(w)\|^2$ is P-integerable; the norm on $L^2(\Omega, C)$ is given by $\|\psi\| = [\int_{\Omega} \|\psi(w)\|^2 dP(w)]^{\frac{1}{2}}$. But $C \hat{\otimes}_{\pi} C$ is separable, so for any

$\psi_1, \psi_2 \in L^2(\Omega, C)$ the Bochner integral $E[\psi_1(.)\otimes\psi_2(.)] = \int_\Omega \psi_1(w)\otimes\psi_2(w)\, dP(w)$ is well defined in $C \hat{\otimes}_\pi C$.

Think of β as a member of $[C \hat{\otimes}_\pi C]^*$ ([8] PP. 434-445). Then $E\,\beta(\psi_1(.), \psi_2(.)) = \beta(E\psi_1(.)\otimes\psi_2(.))$ for all $\psi_1, \psi_2 \in L^2(\Omega, C)$. Fix $t \in (o, r]$. Then $W_t^* \in L^2(\Omega, C)$; so define $K_t \in C \hat{\otimes}_\pi C$ by

$$K_t = E[W_t^*(.)\otimes W_t^*(.)]$$

By the covariance property of Brownian motion, one gets

$$K_t(s,s') = [1 + \frac{1}{t}\min(s,s')]\,\chi_{[-t,o]}(s)\,\chi_{[-t,o]}(s'), \quad s,s' \in J$$

Using the last relation, the eigen value problem

$$\int_{-r}^{o} K_t(s,s')\,\xi(s')ds' = \lambda\,\xi(s), \quad s \in J$$

can be solved directly in C, thus obtaining, via Mercer's Theorem ([1] P. 138), the series expansion

$$(***) \qquad K_t = \sum_{k=o}^{\infty} \frac{8}{\pi^2(2k+1)^2}\, \xi^{t,k} \otimes \xi^{t,k}$$

where

$$\xi^{t,k}(s) = \chi_{[-t,o]}(s)\cos\left\{\frac{(2k+1)\pi s}{2t}\right\} \qquad s \in J,\ k=0,1,2,\ldots$$

The series (***) converges in the norm of $C \hat{\otimes}_\pi C$ uniformly with respect to $t \in (o, r]$ due to the fact that $\|\xi^{t,k}\otimes\xi^{t,k}\|^2 \leq \|\xi^{t,k}\|^2 \leq 1$, $k=0,1,2,\ldots$, and the convergence of

$$K_t(o,o) = 1 = \sum_{k=o}^{\infty} \frac{8}{\pi^2(2k+1)^2}$$

Therefore, $\lim_{t\to o+} E\,\beta(W_t^*(.), W_t^*(.)) = \lim_{t\to o+} \beta(K_t)$

$$= \sum_{k=o}^{\infty} \frac{8}{\pi^2(2k+1)^2} \lim_{t\to o+} \beta(\xi^{t,k}, \xi^{t,k}) = \bar{\beta}(\chi_{\{o\}}, \chi_{\{o\}})$$

We may now state the main theorem describing the weak generator A of $\{P_t\}_{t\geq o}$.

Theorem 2:

Let $\phi \in \mathcal{D}(S)$ be C^2 with $D\phi$, $D^2\phi$ globally bounded and $D^2\phi$ Lipschitz on C. Then $\phi \in \mathcal{D}(A)$, and if $\{e_j\}_{j=1}^m$ is any basis for R^m then

$$A(\phi)(\eta) = S(\phi)(\eta) + \overline{D\phi(\eta)}\,(H(\eta)\,\chi_{\{o\}}) + \frac{1}{2}\sum_{j=1}^{m} \overline{D^2\phi(\eta)}\,(G(\eta)(e_j)\chi_{\{o\}},\, G(\eta)(e_j)\,\chi_{\{o\}})$$

for all $\eta \in C$.

Proof:

Apply Taylor's formula ([2] P. 186) to the C^2 function ϕ getting

$$(*)\ \frac{1}{t}\,E[\phi({}^{\eta}x_t) - \phi(\eta)] = \frac{1}{t}\,[S_t(\phi)(\eta) - \phi(\eta)] + \frac{1}{t}\,E\,D\phi\,(\tilde{\eta}_t)({}^{\eta}x_t - \tilde{\eta}_t)$$

$$+ \frac{1}{t}\,E\,R_2(t) \qquad t>o$$

where $R_2(t) = \int_0^1 (1-u)D^2\phi[\tilde{\eta}_t + u({}^{\eta}x_t - \tilde{\eta}_t)]\,({}^{\eta}x_t - \tilde{\eta}_t,\ {}^{\eta}x_t - \tilde{\eta}_t)du$ a.s.

Since $\phi \in \mathcal{D}(S)$, $\lim\limits_{t \to o+} \frac{1}{t}\,[S_t(\phi)(\eta) - \phi(\eta)] = S(\phi)(\eta)$.

For all $t > o$ and $\eta \in C$, we have

$$\left| \frac{1}{t}\,ED\phi(\tilde{\eta}_t)({}^{\eta}x_t - \tilde{\eta}_t) - \frac{1}{t}\,ED\phi(\eta)({}^{\eta}x_t - \tilde{\eta}_t) \right| \leq K\| D\phi(\tilde{\eta}_t) - D\phi(\eta)\|$$

because of Lemma 3. Letting $t \to o+$ in the above inequality yields

$$\lim_{t \to o+} \frac{1}{t}\,ED\phi(\tilde{\eta}_t)({}^{\eta}x_t - \tilde{\eta}_t) = \overline{D\phi(\eta)}(H(\eta)\,\chi_{\{o\}})$$

An easy calculation of fourth-order moments in (SFDE) gives us the estimate

$$E\,\|\,{}^{\eta}x_t - \tilde{\eta}_t\,\|^4 \leq 8K^4\,(t^4 + 144t^2)$$

for all $t \geq o$, $\eta \in C$ ([5] P. 87). Therefore

$$\left| \frac{1}{t}\,ED^2\phi[\tilde{\eta}_t + u({}^{\eta}x_t - \tilde{\eta}_t)]({}^{\eta}x_t - \tilde{\eta}_t,\ {}^{\eta}x_t - \tilde{\eta}_t) - \frac{1}{t}\,ED^2\phi(\eta)({}^{\eta}x_t - \tilde{\eta}_t,\ {}^{\eta}x_t - \tilde{\eta}_t)\right|$$

$$\leq 2\sqrt{2}\,K^2(t^2 + 144)^{\frac{1}{2}}\,\{E\|\,D^2\,\phi[\tilde{\eta}_t + u({}^{\eta}x_t - \tilde{\eta}_t)] - D^2\phi(\eta)\,\|^2\}^{\frac{1}{2}} \to 0$$

as $t \to o+$, uniformly in $u \in [0,1]$, because $D^2\phi$ is Lipschitz. By Lemma 5, we get

$$\lim_{t\to 0+} \frac{1}{t} ER_2(t) = \int_0^1 (1-u) \lim_{t\to 0+} \frac{1}{t} ED^2\phi(\eta)({}^{\eta}x_t - \tilde{\eta}_t, {}^{\eta}x_t - \tilde{\eta}_t)\, du$$

$$= \frac{1}{2}\sum_{j=1}^{m} \overline{D^2\phi(\eta)}(G(\eta)(e_j)\chi_{\{0\}}, G(\eta)(e_j)\chi_{\{0\}}).$$

As $\phi \in \mathcal{D}(S)$ and $D\phi$, $D^2\phi$ are bounded on C, all terms on the right-hand side of (*) are bounded in t and η. Hence $\phi \in \mathcal{D}(A)$ and result holds.

References

1. Courant, R. and Hilbert, D.: Methods of Mathematical Physics I, Interscience, New York (1953).

2. Dieudonne, J.: Foundations of Modern Analysis, Academic Press, New York (1960).

3. Dunford, N. and Schwartz, J.T.: Linear Operators I: General Theory, Interscience, New York (1958).

4. Dynkin, E.B.: Markov Processes I, Springer-Verlag, Berlin (1965).

5. Friedman, A.: Stochastic Differential Equations and Applications I, Academic Press, New York (1976).

6. Mohammed, S.E.A.: Markov Solutions of Stochastic Functional Differential Equations, Proceedings of the Second International Conference on Differential-Delay Systems and Related Topics, Poland (1981), (To appear).

7. Mohammed, S.E.A.: Stochastic Functional Differential Equations, Research Notes in Mathematics, Pitman Publishing Ltd., London, (To appear).

8. Treves, F.: Topological Vector Spaces, Distributions and Kernels, Academic Press New York (1967).

Acknowledgement:

I am grateful to K.D. Elworthy and K.R. Parthasarathy for helpful discussions during the preparation of this paper.

ON SOLUTIONS TO THE INITIAL-BOUNDARY PROBLEM FOR PERTURBED POROUS MEDIUM EQUATION

MITSUHIRO NAKAO

§1 Introduction

In this article we are concerned with the existence, nonexistence and asymptotic behavior of global solutions to the porous medium equation with blowing up term:

$$\begin{cases} \frac{\partial}{\partial t}u - \Delta u^{p+1} - u^{\alpha+1} = 0 \quad \text{on} \quad \Omega \times (0,T) & (1) \\ u(x,0) = u_0(x) \ \text{(given)}, \ u(x,t)|_{\partial\Omega} = 0 \ \text{and} \ u \geqslant 0 \end{cases}$$

where Ω is a bounded domain in R^n with boundary $\partial\Omega$, α and p are nonnegative numbers and T is an arbitrarily fixed positive time.

When $p = 0$ and $\alpha > 0$, as is well known, the problem (1) was investigated by H. Fujita [3]. His result was extended by M. Tsutsumi [6] to the typical nonlinear parabolic equation:

$$\begin{cases} \frac{\partial}{\partial t}u - \sum_{i=1}^{n} \frac{\partial}{\partial x_i}\left(\left|\frac{\partial}{\partial x_i}u\right|^p \frac{\partial}{\partial x_i}u\right) - u^{\alpha+1} = 0 \quad \text{on} \quad \Omega \times (0,T) & (2) \\ u(x,0) = u_0(x), \quad u(x,t)|_{\partial\Omega} = 0 \ \text{and} \ u \geqslant 0. \end{cases}$$

Tsutsumi used the concept of 'potential well' introduced by D. Sattinger [5] to give the criteria on global existence and nonexistence of solutions.

The first object of this article is to discuss the same problem for (1) as in [6]. We note that the nonlinearity $-\Delta u^{p+1}$ causes more delicate difficulty than that in (2). The second object is to give a precise estimate for the rate of decay of $\|\nabla u^{p+1}(t)\|_{L^2}$ as $t \to \infty$ when $\alpha > p$ and a global solution $u(t)$ is admitted. For this a similar technique as in our previous paper [4] is used.

The problem to decide the behavior of solutions for the case $\alpha \leqslant p$ is open. The estimation of $\|u(t)\|_{L^\infty}$ for the case $\alpha > p$ as in porous medium equation (see N. Alikakos [1], D. Aronson & L. Peletier [2]) is also open.

§2 Global Existence for the Case $p > \alpha$

Our result here reads as follows:

Theorem 1

Suppose that $u_0(t) \geqslant 0$ and $u_0^{p+1} \in H_1^0(\Omega)$. Then the problem (1) admits a global solution $u(x,t)$ such that

$$\frac{\partial}{\partial t}(u^{p/2+1}(t)) \in L^2([0,T];\ L^2(\Omega)),\quad u^{p+1}(t) \in L^\infty([0,T];H_1^0(\Omega))$$

and the equation is satisfied in the sense that

$$\int_0^T\int_\Omega \left\{-u(x,t)\phi_t(x,t) + \nabla(u^{p+1})\nabla\phi(x,t) - u^{\alpha+1}\phi(x,t)\right\}dxdt \qquad (3)$$
$$- \int_\Omega u(x,0)\phi(x,0)dx = 0$$

for $\forall\ \phi \in C^1([0,T];\ H_1^0)$ with $\phi(T) = 0$. Moreover, the following estimate holds:

$$\int_t^{t+1}\left\|\frac{\partial}{\partial s}(u^{p/2+1}(s))\right\|_{L^2}^2 ds + \|u^{p+1}(t)\|_{H_1^0}^2 \leqslant C(\|u_0^{p+1}\|_{H_1^0}$$

for $0 \leqslant \forall\ t \leqslant T$.

For the proof of Th.1 and also for later use we rewrite (1) as follows:

$$\begin{cases} \frac{\partial}{\partial t}(|U|^{-p/(p+1)}U) - \Delta U - \Psi(U) = 0 & (4) \\ U(x,0) = U_0,\quad U|_{\partial\Omega} = 0 \quad \text{and} \quad U \geqslant 0 \end{cases}$$

where we set

$$U = u^{p+1},\ U_0 = u_0^{p+1} \quad \text{and} \quad \Psi(U) = U^{(\alpha+1)/(p+1)} \ \text{if}\ U \geqslant 0,$$

0 if $U \leqslant 0$.

To avoid the singularity in (4) we introduce the modified equation

$$\begin{cases} \frac{\partial}{\partial t}(\beta_\varepsilon(U) + \varepsilon U) - \Delta U - \Psi(U) = 0, \quad \varepsilon > 0, & (5) \\ U(x,0) = U_0,\quad U|_{\partial\Omega} = 0,\quad U \geqslant 0 \end{cases}$$

where we set

$$\beta_\varepsilon(U) = \int_0^U \beta_\varepsilon'(\eta)d\eta \quad \text{and} \quad \beta_\varepsilon'(U) = (p+1)^{-1}(|U| + \varepsilon)^{-p/(p+1)}.$$

Now, we begin the proof of Th.1. We employ the Galerkin's method. Let $\{w_j\}_{j=1}^{\infty}$ be a basis of $H_1^0(\Omega)$ and consider the system of ordinary differential equation

$$\begin{cases} ((\beta_\varepsilon'(U_{m,\varepsilon}(t))+\varepsilon)\frac{\partial}{\partial t}U_{m,\varepsilon}(t),w_j) + (\nabla U_{m,\varepsilon}(t),\nabla w_j) \\ \qquad - (\Psi(U_{m,\varepsilon}(t)),w_j) = 0, \quad j = 1,2,\dots,m\ , \\ U_{m,\varepsilon}(0) = U_{0,m} \in [w_1,\dots,w_m] \end{cases} \tag{6}$$

where $U_{m,\varepsilon}(t) = \sum_{j=1}^{m} \lambda_{j,m}(t)w_j$ and $U_{0,m}$ should be chosen in such a way that $U_{0,m} \to U_0$ in H_1^0 as $m \to \infty$. The system (6) admits a unique local solution $U_{m,\varepsilon}(t)$, say, on $[0,T_{m,\varepsilon}]$. Multiplying (6) by $\dot{\lambda}_{J,m}(t)$ and integrating we get

$$\int_0^t\int_\Omega \left[\beta_\varepsilon'(U_{m,\varepsilon})+\varepsilon\right]\left|\frac{\partial}{\partial t}U_{m,\varepsilon}\right|^2 dxds + J(U_{m,\varepsilon}(t)) = J(U_{m,\varepsilon}(0)) \tag{7}$$

for $t \in [0,T_{m,\varepsilon}]$, where we set

$$J(U) \equiv \frac{1}{2}\|\nabla U\|_{L^2}^2 - \frac{p+1}{p+\alpha+2}\int_\Omega \Psi(U)U\,dx. \tag{8}$$

Since $p > \alpha$, $J(U)$ is well defined and it holds that

$$\left|\int_\Omega \Psi(U)U dx\right| \leqslant c + \frac{1}{4}\|\nabla U\|_{H_1^0}^2 . \tag{9}$$

From (7)-(9) we see that $U_{m,\varepsilon}(t)$ exists on $[0,T]$ and

$$\int_0^T\left[\|\frac{\partial}{\partial t}(V_{m,\varepsilon}(t))\|_{L^2}^2 + \varepsilon\|\frac{\partial}{\partial t}U_{m,\varepsilon}(t)\|_{L^2}^2\right]dt + \|U_{m,\varepsilon}(t)\|_{H_1^0}^2 \leqslant c < \infty, \tag{10}$$

$$\|V_{m,\varepsilon}(t)\|_{H_1^0} \leqslant c\,\varepsilon^{-p/2(p+1)}\|\nabla U_{m,\varepsilon}(t)\|_{L^2} \leqslant c_\varepsilon < \infty \tag{11}$$

where we set $V_{m,\varepsilon} = \int_0^{U_{m,\varepsilon}} \sqrt{\beta_\varepsilon'(\eta)}d\eta$.

Applying standard compactness argument we see that $U_{m,\varepsilon}$ and $V_{m,\varepsilon}$ converges as $m \to \infty$ (along a subsequence) to U_ε and V_ε, respectively, in various topologies and it holds that

$$\int_0^T\left[\|\frac{\partial}{\partial t}V_\varepsilon(t)\|_{L^2}^2 + \varepsilon\,\|\frac{\partial}{\partial t}U_\varepsilon(t)\|_{L^2}^2\right]dt + \|\nabla U_\varepsilon(t)\|_{L^2}^2 \leqslant c < \infty \tag{12}$$

where U_ε and V_ε are connected by $V_\varepsilon = \int_0^{U_\varepsilon} \sqrt{\beta_\varepsilon'(\eta)}d\eta$.

U_ε becomes a solution of the modified problem (5). To show the convergence of U_ε as $\varepsilon \to 0$ we observe that

$V_\varepsilon(t) \in Lip(\theta,q)$ with $\theta = (p+2)/2(p+1)$ and $q = 4(p+1)/(p+2)$

for each t. In fact,

$$\int_\Omega |V_\varepsilon(x+h,t) - V_\varepsilon(x,t)|^{4(p+1)/(p+2)} dx$$
$$\leqslant c\int_\Omega |U_\varepsilon(x+h,t) - U_\varepsilon(x,t)|^2 dx \leqslant c\|U_\varepsilon(t)\|^2_{H^0_1} |h|^2$$

for $h = (h_1,\dots,h_n)$ (We define $U_\varepsilon(x+h,t) = 0$ if $x + h \notin \Omega$), and hence

$$\|V_\varepsilon(t)\|_{Lip(\theta,q)} \leqslant c\left\{\|V_\varepsilon\|_q + \|U_\varepsilon\|_{H^0_1}^{(p+2)/2(p+1)}\right\} < c < \infty . \tag{13}$$

Noting that $Lip(\theta,q)$ is compactly imbedded in $L^2(\Omega)$, we can conclude from (12) and (13) that $V_\varepsilon(t) \to {}^\exists V(t)$ strongly in $L^2([0,T];L^2(\Omega))$ and a.e., and $U_\varepsilon(t)$ converges to $U(t)$ in appropriate topologies where

$$V = \frac{1}{\sqrt{p+1}} \int_0^U |\eta|^{-p/2(p+1)} d\eta . \tag{14}$$

$U(x,t)$ becomes a desired solution of (4). It remains to prove the nonnegativity of $u(x,t) \equiv |U|^{-p/(p+1)}U$. Noting $u'(t) \in L^\infty([0,T]:H^{-1})$ and setting $u^- = \min(u,0)$ we have

$$\frac{1}{p+2}\frac{d}{dt}\int_\Omega |u^-(t)|^{p+2}dx = \left\langle \frac{d}{dt}u,\ |u^-|^p u^- \right\rangle_{H^{-1}\times H^0_1}$$
$$= -\|\nabla(|u^-|^p u^-)\|^2_{L^2} \leqslant 0$$

which implies $u^-(t) \equiv 0$, i.e., $u(t) \geqslant 0$.

§3 Global Existence for the Case $p < \alpha$

For the functional $J(U)$ to be well-defined for $U \in H^0_1$ we assume $0 < \alpha \leqslant (p(n+2)+4)/(n-2)$ if $n \geqslant 3$ and $0 < \alpha < \infty$ if $n = 1,2$. Under this assumption we set

$$d \equiv \inf_{\substack{U\in H^0_1 \\ U\neq 0}} \sup_{\lambda \geqslant 0} J(\lambda U).$$

Then it is easy to see

$$\infty > d \geqslant c(\alpha - p)/2(p+\alpha-1) > 0.$$

The potential well associated with (1) or (4) is defined by

$$\mathcal{W} = \{U \in H^0_1 \mid 0 \leqslant J(\lambda U) < d \quad \text{for } \forall\, \lambda \in [0,1]\}.$$

Lemma 1 (Tsutsumi [6]).

$$\mathcal{W} = \mathcal{W}_* \cup \{0\}$$

where

$$\mathcal{W}_* = \{U \in H_1^0 \mid \|\nabla U\|_{L^2}^2 - \int_\Omega \Psi(U)U\,dx > 0 \quad \text{and} \quad J(U) < d\}.$$

The purpose of this section is to prove that (1) has a global solution if $U_0 \equiv u_0^{p+1} \in \mathcal{W}$.

Lemma 2

Let $U_0 \in \mathcal{W}$. Then the approximate solutions $U_{m,\varepsilon}(t)$ constructed through (6) exist on $[0,T]$ (for large m) and $U_{m,\varepsilon}(t) \in \mathcal{W}$ for $\forall\, t \in [0,T]$, and consequently

$$\|U_{m,\varepsilon}(t)\|_{H_1^0}^2 + \int_0^T \left\{\|\sqrt{\beta_\varepsilon'(U_{m,\varepsilon}}\,\frac{\partial}{\partial s} U_{m,\varepsilon}\|_{L^2}^2 + \varepsilon\|\frac{\partial}{\partial s} U_{m,\varepsilon}(s)\|_{L^2}^2\right\}ds \leqslant c \quad (15).$$

Proof. If $U_{m,\varepsilon}(t) \in \mathcal{W}$ for $\forall\, t \in [0,T]$ we see from Lemma 1 that

$$d > J(U_{m,\varepsilon}(t) \geqslant \frac{\alpha - p}{2(p+\alpha-2)}\|\nabla U_{m,\varepsilon}(t)\|_{L^2}^2 \tag{16}$$

which together with (7) yields (15) immediately. To prove the former assertion we assume it was false. Then we can choose $t^* > 0$ such that

$$U_{m,\varepsilon}(t^*) \notin \mathcal{W} \quad \text{and} \quad U_{m,\varepsilon}(t) \in \mathcal{W} \text{ for } 0 \leqslant \forall\, t < t^*.$$

By Lemma 1 either of the following cases must hold:

(i) $J(U_{m,\varepsilon}(t^*) = d$ or (ii) $\|\nabla U_{m,\varepsilon}(t^*)\|_{L^2}^2 = \int_\Omega \Psi(U_{m,\varepsilon}(t^*))U_{m,\varepsilon}(t^*)dx.$

Both cases are easily reduced to contradiction. □

Using the estimate (15) we can obtain, as in §2, the following:

Theorem 2

Let $u_0 \geqslant 0$ and $u_0^{p+1} \in \mathcal{W}$. Then the problem (1) admits a global solution such that

$$u^{p+1}(t) \in \mathcal{W} \text{ and } \|\nabla u^{p+1}(t)\|_{L^2}^2 + \int_0^T \|\frac{\partial}{\partial t}\left(u^{p/2+1}(t)\right)\|_{L^2}^2 dt < c < \infty$$

for $t \in [0,T]$.

§4 Decay of Solutions for the Case $\alpha > p$.

Here we shall prove the following decay property for the solutions in Th.2.

Theorem 3

The solutions in Th. 2 satisfy the estimate

$$\int_t^{t+1} \left\| \frac{\partial}{\partial s}\left(u^{p/2+1}(s)\right)\right\|_{L^2}^2 ds + \|\nabla u^{p+1}(t)\|_{L^2}^2 \leq c_1(1+t)^{-2(p+1)/p} \tag{17}$$

where c_1 is a constant depending on $d - J(U_0)$ but independent of T.

Proof. It suffices to derive the estimate for $U_{m,\varepsilon}(t)$. For simplicity, however, we treat the equation (1) directly under the assumption that $u(t)$ is a smooth solution on $[0,\infty)$ with $u^{p+1}(t) \in \mathcal{W}$, which does not change the essential feature of the proof. First we observe that

$$\|\nabla u^{p+1}(t)\|_{L^2}^2 \geq \left(\frac{d}{d-\varepsilon_0}\right)^{(\alpha-p)/2(p+1)} \int_\Omega u^{p+\alpha+2} dx \tag{18}$$

for $0 < \forall\, \varepsilon_0 < d - J(u_0{}^{p+1})$. This is derived from the fact $u(t)^{p+1} \in \mathcal{W}$, the inequality (16) and the definition of d. Next, multiplying (1) by $\frac{\partial}{\partial t} u^{p+1}$ we have

$$-\frac{d}{dt} J(u^{p+1}(t) = (p+1)\int_\Omega u^p \left|\frac{\partial}{\partial t} u\right|^2 dx \,. \tag{19}$$

On the other hand, multiplying (2) by u^{p+1}, we have

$$\begin{aligned}\|\nabla u^{p+1}(t)\|_{L^2}^2 &= -\int_\Omega u^{p+1} u_t dx + \int_\Omega u^{p+\alpha+2} dx \\ &\leq \left(\int_\Omega u^p |u_t|^2 dx\right)^{\frac{1}{2}} \left(\int_\Omega u^{p+2} dx\right)^{\frac{1}{2}} \\ &+ \left(\frac{d-\varepsilon_0}{d}\right)^{(\alpha-p)/2(p+1)} \|\nabla u^{p+1}(t)\|_{L^2}^2 \qquad \text{(by (18))}\end{aligned}$$

and hence, with the aid of Sobolev's Lemma and (19),

$$\|\nabla u^{p+1}(t)\|_{L^2}^2 \leq c_1 \left(-\frac{d}{dt} J(u^{p+1}(t))\right)^{\frac{1}{2}} \|\nabla u^{p+1}(t)\|_{L^2}^{(p+2)/2(p+1)}$$

or

$$-\frac{d}{dt} J(u^{p+1}(t) \geq c_1 \|\nabla u^{p+1}(t)\|^{(3p+2)/(p+1)} \geq c_1 J(u^{p+1}(t))^{(3p+2)/2(p+1)} \tag{20}$$

where c_1 denotes constants depending on $d - \varepsilon_0$. The inequalities (20) and (16) yield (17). □

§5 Local Existence and Blow up of Solutions for the Case $\alpha > p$.

In this section we shall show that if $u_0^{p+1} \notin \mathcal{W}$, more precisely if $J(u_0^{p+1}) \leqslant 0$ and $u_0 \neq 0$ we can not expect the global solutions.

Theorem 4

Let $u(t)$ be a solution of (1) such that $J(u^{p+1}(t)) \leqslant 0$ for a.e. $t \in [0,T]$ and

$$\frac{\partial}{\partial t}(u^{p/2+1}) \in L^2([0,T];L^2) \quad \text{and} \quad u^{p+1} \in L^\infty([0,T];H^0_1). \tag{21}$$

Then T must satisfy

$$T < T_0 \equiv \frac{p+2+\alpha}{\alpha(\alpha-p)} \|u_0\|^{-\alpha}_{L^{p+2}}$$

and we have

$$\|u(t)\|_{L^{p+2}} \geqslant c(T_0 - t)^{-1/\alpha} \quad \text{for} \quad \exists c > 0.$$

Remark The condition $J(u^{p+1}(t)) \leqslant 0$ follows formally from $J(u_0^{p+1}) \leqslant 0$ (see (7)).

Proof of Th. 4. Multiplying (1) by $u^{p+1}(t)$ and integrating we get

$$\frac{1}{p+2}\left(\|u(t)\|^{p+2}_{L^{p+2}} - \|u_0\|^{p+2}_{L^{p+2}}\right) + \int_0^t \|\nabla u^{p+1}(s)\|^2_{L^2} ds - \int_0^t\int_\Omega u^{p+\alpha+2} dxds = 0$$

which implies, by virtue of $J(u^{p+1}(t)) \leqslant 0$,

$$\|u(t)\|^{p+2}_{L^{p+2}} \geqslant \left(\|u_0\|^{-\alpha}_{L^{p+2}} - \frac{\alpha(\alpha-p)}{p+\alpha+2} t\right)^{-(p+2)/\alpha}. \qquad \square$$

Finally we give a local existence Theorem of solutions which satisfy $J(u^{p+1}(t) \leqslant J(u_0^{p+1})$.

Theorem 5

Let us assume $0 < \alpha < \{(n+2)p + 4\}/2(n-2)$ if $n \geqslant 3$ and $0 < \alpha < \infty$ if $n = 1,2$. Then, for $\forall\, u_0 \geqslant 0$ with $u_0^{p+1} \in H^0_1$ there exists $T_1 \equiv T_1(u_0) > 0$ such that (2) admits a solution $u(t)$ on the interval $[0,T_1]$ which satisfies (21) and

$$\frac{4(p+1)}{(p+2)^2}\int_0^t \left\|\frac{\partial}{\partial s}(u^{p/2+1}(s))\right\|^2_{L^2} ds + J(u^{p+1}(t) \leqslant J(u_0^{p+1}) \tag{22}$$

for a.e. $t \in [0,T_1]$.

Proof. Let $U_{m,\varepsilon}(t)$ be approximate solutions. Then, from (7) and Sobolev's Lemma (the assumption on α is used here) we have

$$\int_\Omega (\beta_\varepsilon'(U_{m,\varepsilon}) + \varepsilon)\left|\frac{\partial}{\partial t} U_{m,\varepsilon}(t)\right|^2 dx + \frac{d}{dt}\|\nabla U_{m,\varepsilon}(t)\|^2_{L^2}$$

$$\leqslant c\left\{\|\nabla U_{m,\varepsilon}(t)\|^2_{L^2} + \varepsilon^2\right\}^{(p+2\alpha+2)/(p+1)}$$

which shows

$$\|\nabla U_{m,\varepsilon}(t)\|^2_{L^2} \leqslant \left\{\|\nabla U_{m,\varepsilon}(0)\|^2_{L^2} + \varepsilon^2 - \frac{c(2\alpha-p)}{2(p+1)}\, t\right\}^{-2(1+p)/(2\alpha-p)} - \varepsilon^2 .$$

Thus we can conclude that for some $T_1 \equiv T_1\left(\|\nabla u_0^{p+1}\|_{L^2}\right)$

$$\int_0^{T_1}\left\|\sqrt{\beta_\varepsilon'(U_{m,\varepsilon}) + \varepsilon}\,\frac{\partial}{\partial t} U_{m,\varepsilon}\right\|^2_{L^2} dt + \|\nabla U_{m,\varepsilon}(t)\|^2_{L^2} \leqslant c \quad \text{for} \quad t \in [0,T_1], \tag{23}$$

from which we can prove the existence of local solution.

To show (22) we first note that

$$\int_0^t \left\|\sqrt{\beta_\varepsilon'(U_\varepsilon(s)) + \varepsilon}\,\frac{\partial}{\partial s} U_\varepsilon(s)\right\|^2_{L^2} ds + J(U_\varepsilon(t)) \leqslant J(U_0) \quad , \text{a.e.t.} \tag{24}$$

This follows from (7),(23) and the fact that H_1^0 is compactly imbedded in $L^{(p+\alpha+2)/(p+1)}$. Next, by (5) and (23) we can obtain

$$\left\|\frac{d}{dt}\beta_\varepsilon(U_\varepsilon(t))\right\|_{H^{-1}} \leqslant c < \infty \text{ and } \|\beta_\varepsilon(U_\varepsilon(t))\|_{Lip(1/(p+1),2(p+1)} < c < \infty .$$

From this we see $U_\varepsilon(t)^{1/(p+1)} \to U^{1/(p+1)}(t)$ strongly in $L^q([0,T_1];\, L^q)$ with $q = p + \alpha + 2$ and

$$J(U(t)) \leqslant \varliminf_{\varepsilon\to 0} J(U_\varepsilon(t)) \quad \text{for a.e.} \quad t \in [0,T_1]. \qquad \square$$

References

1. N.D. Alikakos, L^p-bounds of solutions of reaction diffusion equations, Comm. P.D.E. 4(8) (1979), 827-868.

2. D.G. Aronson & L.A. Peletier, Large time behavior of solutions of the porous medium equation in bounded domains, J. Diff. Eqs. 39 (1981), 378-412.

3. H. Fujita, On some nonexistence and nonuniqueness theorems for nonlinear parabolic equations, Proc. Symp. in Pure Math. 18, Amer. Math. Soc., Providence, R.I. (1970), 105-113.

4. M. Nakao, On solutions to the initial-boundary value problem fo: $\frac{\partial}{\partial t} u - \Delta\beta(u) = f$, to appear in J. Math. Soc. Japan.

5. D.H. Sattinger, On global solution of nonlinear hyperbolic equations, Arch. Rat. Mech. Anal. 30 (1968), 148-172.

6. M. Tsutsumi, Existence and nonexistence of global solutions for nonlinear parabolic equations, Publ. R.I.M.S. Kyoto Univ. 8 (1972/73), 211-229.

A Survey of Global Properties of Linear Differential Equations of the n-th Order

F. Neuman

I. Introduction

Investigations of linear differential equations started in the last century. In 1834 E. E. Kummer [14] published his results about transformations of second order equations, and then several papers appeared devoted to transformations of higher order equations, their canonical forms and invariants. From many authors let us mention at least E. Laguerre, A. R. Forsyth, F. Brioschi, G. H. Halphen [11], P. Stäckel [33], S. Lie, and E. J. Wilczynski [34]. Among their results there is, e.g., the so called Laguerre-Forsyth canonical form of linear differential equations of the n-th order characterized by vanishing coefficients by the (n-1) st and the (n-2) nd derivatives. However as lately as in 1893 P. Stäckel [33] proved that the transformation considered by Kummer is the most general transformation that converts any linear homogeneous differential equation of the n-th order ($n \geq 2$) into an equation of the same kind.

Anyhow all their investigations were of local character that was pointed out by G. D. Birkhoff [1] already in 1910. He gave an example of the third order linear differential equation that cannot be transformed into the Laguerre-Forsyth form on its whole interval of definition.

Of course, the local nature of results is not suitable for investigations of global problems, like boundedness or periodicity of solutions, their convergence to zero, their asymptotic properties and oscillatory behaviour, or whether they belong to L^2 or L^p, etc.

In the first half of the century some isolated results of global character occurred, as e.g., G. Sansone's example of the third order equation with all oscillatoric solutions, [32].

A systematic study of global properties of second order linear differential equations started about 30 years ago by O. Borůvka. He summarized his deeply developed theory in the monograph [2] and in further works [3, 4, 5].

In the last 10 years there was also found enough general approach to global problems of linear differential equations of arbitrary order, This investigation is based on O. Borůvka's methods, it discovers what is common for the second order and higher order equations, it shows what is specific for the second order, and introduces methods suitable for equations of arbitrary orders.

Here we give a survey of main definitions, methods and some results of the approach to global properties of linear differential equations of the n-th order, $n \geqq 2$.

Algebraic, topological and geometrical tools together with the methods of the theory of dynamical systems and functional equations make it possible to deal with problems concerning global properties of solutions by contrast to the previous local investigations.

For example; the structure of the set of all global transformations of linear differential equations is described by algebraic means (theory of categories-Brandt and Ehresmann groupoids), construction of global canonical forms is given by methods of differential geometry (including E. Cartan's moving-frame-of-reference method).

The theory includes also effective methods for solving several special problems, e.g., from the area of questions concerning distribution of zeros of solutions of linear differential equations.

II. Basic definitions and notations

Let $C^k(I)$ be the class of real scalar or vector functions with continuous derivatives up to order $k \geq 0$ inclusive on an open interval $I \subset \mathbb{R}$.

Consider a linear homogeneous differential equation of the n-th order, $n \geq 2$, with continuous coefficients on I, i.e.

$$y^{(n)}+p_{n-1}(x)y^{(n-1)}+\dots+p_o(x)y = 0,$$

where $p_i \in C^o(I)$, $I \subset \mathbb{R}$, $i = 1,\dots, n-1$. Denote by $P_n(y;I)$ this differential equation.

Similarly

$$Q_n(z;J) \equiv z^{(n)}+q_{n-1}(t)z^{(n-1)}+\dots+q_o(t)z = 0,$$

$q_i \in C^o(J)$, $i = 1,\dots, n-1$. Sometimes we write briefly P_n and Q_n instead of $P_n(y;I)$ and $Q_n(z;J)$, or simply P and Q only.

Let $y_1,\dots,y_n$ be n linearly independent solutions of the equation $P_n(y;I)$. Denote by

$$\underline{y}: = (y_1,\dots,y_n)^T$$

the column vector with coordinates formed by the solutions y_i. Then $\underline{y} \in C^n(I)$ and the Wronskian

$$W[\underline{y}(x)]: = \det (\underline{y}(x), \underline{y}'(x),\dots,\underline{y}^{(n-1)}(x))$$

is different from zero everywhere on I. Analogously for $Q_n(z;J)$ we have $\underline{z} \in C^n(J)$, $W[\underline{z}(t)] \neq 0$ on J.

We say that $P_n(y;I)$ can be globally transformed into $Q_n(z;J)$, if there exist

1° a function $f \in C^n(J)$, $f(t) \neq 0$ on J,

2° a bijection $h \in C^n(J)$ of J onto I,

such that $dh(t)/dt \neq 0$ on J,

3° a constant regular n by n matrix A,

such that

$$\underline{z}(t) = A \cdot f(t) \cdot \underline{y}(h(t)) \qquad \text{for all } t \in J. \tag{1}$$

Due to P. Stäckel [33], the form of the transformation (1) is the most general one that keeps linearity, homogenity and order unchanged. The bijectivity of h guarantees the globality in our definition.

If the coefficient p_{n-1} of $P_n(y;I)$ is identically zero on I, we shall express it by the superscript o as $P_n^o(y;I)$.

Consider

$$P_2^o(y;I) \equiv y''+p_o(x)y = 0 , \tag{2}$$

$p_o \in C^{n-2}(I)$. Let y_1 and y_2 be its two linearly independ solutions. Then $y_1, y_2 \in C^n(I)$, and n functions defined by

$$z_i(x) := y_1^{i-1}(x) \cdot y_2^{n-i}(x), \quad i = 1,\dots,n,$$

belong to the class $C^n(I)$, their Wronskian being nonzero everywhere on I. Hence z_i, $i = 1,\dots,n$, can be considered as n linearly independent solutions of the so-called iterative equation of the n-th order with respect to (2). Let $It_n[p_o]$ denote the differential operator of this equation normalized by unit leading coefficient. It can be shown (see, e.g.,[13]) that

$$It_n[p_o] = z^{(n)}+\binom{n+1}{3}p_o(x)z^{(n-2)}+2\binom{n+1}{4}p_o'(x)z^{(n-3)}+\dots \quad .$$

III. Structure of global transformations

The relation of global transformability is an equivalence relation. Hence the set of all linear differential equations of any order n, $n \geq 2$, is decomposed into classes of globally equivalent equations.

Let D be one of these classes, P, Q and R be equations from the class. Since they are globally equivalent, there exist global transformations, say τ_1 and τ_2, such that τ_1 transforms P globally into Q, and τ_2 transforms Q into R. Denote the fact shortly as

$$\tau_1 P = Q \quad \text{and} \quad \tau_2 Q = R.$$

We may also define the composition of the transformations $\tau_2 * \tau_1$. It is again a global transformation converting P into R. As the identity of P we define the transformation (1) with I = J, A being the unit matrix, f(t) = 1 for all $t \in I$, and $h = id_I$, i.e., the identity of the interval I. Since each transformation τ has its inverse τ^{-1}, we may conclude [26] that

<u>Theorem 1.</u> Linear differential equations of all orders n, $n \geq 2$, as objects and their global transformations as morphisms, form an Ehresmann groupoid. Each class of globally equivalent equations is a Brandt groupoid.

The basic structural notion of every Brandt groupoid D is the so-called stationary group S(P) of its element P formed by all global transformations τ that transforms P into itself, i.e.

$$S(P) := \{\tau;\ \tau P = P\}.$$

It can be shown [12, 26] that stationary groups of two globally equivalent equations are conjugate. Moreover, it holds

<u>Theorem 2.</u> There are 5 possible types of classes of globally equivalent equations with respect to their stationary groups.

Stationary group of a (then any) element of D

(T1) consists of identity only

(T2) is an infinite cyclic group

(T3) is a one-parameter group

(T4) is a two-parameter group

(T5) is a three-parameter group.

The cases (T5) and (T4) occur exactly when D contains an (then only) iterative equation It[p_0]. Then (T5) holds for both-side oscillatoric equation (2), and (T4) when (2) is one-side oscillatoric or of the special kind.

The case (T3) occurs just when D is neither of typ (T5) nor (T4), and a linear differential equation with constant coefficients belongs to D.

D is of the type (T2) if and only if it is not of types (T5), (T4) and (T3), and D contains a linear differential equation with periodic coefficients of the same period.

The proof of the theorem is based on investigations of certain vector functional equations [25] and on the latest results of O. Borůvka [5] about one-parameters groups of transformations of reals. For the definition of the second order equation of the special kind, see [2].

There are several results describing structure of global transformations [26]. E.g.:

<u>Theorem 3.</u> Let τ transform an equation P globally into Q. Then all global transformations of P into Q form the set

$$\tau * S(P) = \{\tau * \delta \ ; \ \delta \in S(P)\} ,$$

where S(P) is the stationary group of P.

IV. Canonical forms

Some linear differential equations of a special form are called canonical. E.g., the Laguerre-Forsyth canonical form is characterized by vanishing coefficients near the (n-1)st and the (n-2)nd derivatives.

A given canonical form is called global, if there is a linear differential equation of the canonical form in each class of globally equivalent equations.

A canonical form is called unique, if there is at most one linear differential equation of the canonical form in each class of globally equivalent equations.

We say that a canonical form is effective, if each linear differential equation can be transformed into the given canonical form effectively from its coefficients. The effectivity must be specified by listing operations which are considered as effective. In this paper quadratures and compositions of known functions are considered effective.

It is known [1, 31] that the Laguerre-Forsyth form is neither global, nor unique, nor effective.

We shall describe one of the possible constructions of a global canonical form [20]. Consider a linear differential equation $P_n(y;I)$ and one of its n-tuple of linearly independent solutions forming a column vector function $\underline{y}$. All such vector functions $\underline{z}$ for all equations globally equivalent with our $P_n(y;I)$ are obtained from formula (1). For getting a special form of differential equations we need to specify f and h in this formula. We shall do it in the following way. We consider the vector function $\underline{y}$ as a curve in the n-dimensional euclidean space E_n with the euclidean norm $|\cdot|$. We take the central projection $\underline{v}$ of $\underline{y}$ onto the unit sphere S_{n-1} in E_n, i.e. $\underline{v} := \underline{y}/|\underline{y}|$. Then we introduce the length parametrization into $\underline{v}$ to get $\underline{u}(t)$; $t \in J$. It can be shown that $\underline{u} \in C^n(J)$, $W[\underline{u}(t)] \neq 0$ on J. Hence coordinates of $\underline{u}$ can be considered as n linearly independent solutions of a certain linear differential equation that may be called canonical. This construction can be done without any restriction on smoothness of coefficients or length of the interval of definition.

For the explicit form of equations in our canonical form, we utilize methods of differential geometry. The Frenet Formulas for our curve $\underline{u}(t)$, $t \in J$, in the group of orthogonal transformations are

$$\begin{aligned}
\underline{u}_1' &= \underline{u}_2 \\
\underline{u}_2' &= -\underline{u}_1 + k_1(t)\underline{u}_3 \\
\underline{u}_3' &= -k_1(t)\underline{u}_2 + k_2(t)\underline{u}_4 \\
&\dots \\
\underline{u}_{n-1}' &= -k_{n-3}(t)\underline{u}_{n-2} + k_{n-2}(t)\underline{u}_n \\
\underline{u}_n' &= -k_{n-2}(t)\underline{u}_{n-1} ,
\end{aligned}$$

where $(\underline{u}_1,\dots,\underline{u}_n)$ is the moving orthogonal frame, $\underline{u}_1 = \underline{u}$, curvatures $k_i(t) \neq 0$ on J, $k_i \in C^{n-i-1}(J)$, $i = 1,\dots,n-2$.

By evaluating $\underline{u}_1 = \underline{u}$ from the system we get our canonical form explicitely. E.g., for $n = 2$ and 3 we have

Theorem 4.

$u'' + u = 0$ on J is a global canonical form for linear differential equations of the second order (it depends on the length of J).

$u''' - k_1'/k_1 . u'' + (1+k_1^2)u' - k_1'/k_1 . u = 0$ on J,

$k_1 = k_1(t) \neq 0$ on J, $k_1 \in C^1(J)$

is a global canonical form for linear differential equations of the third order (it depends on one function, k_1).

There is also another way how to get global canonical forms. It can be proved [31]

Theorem 5. Let $y'' + p(x)y = 0$, $p \in C^{n-2}(I)$, be both-side oscillatoric. Then

$$u^{(n)} + \binom{n+1}{3}p(x)u^{(n-2)} + \sum_{i=3}^{n} r_{n-i}(x)u^{(n-i)} = 0$$

on $J \subset I$, p fixed, $r_{n-i} \in C^0(J)$ for $i = 3,\dots,n$ being n-2 arbitrary functions, is a global canonical form for linear differential equations of the n-th order with coefficients of the class C^{n-1} by the (n-1)st derivative and of the class C^{n-2} by the (n-2)nd derivative.

Especially me have

Corollary.

$$u^{(n)} + u^{(n-2)} + r_{n-3}(x)u^{(n-3)} + \dots + r_0(x)u = 0 \qquad (3)$$

on $I \subset (-\infty,\infty)$, is a global canonical form.

If Laguerre and Forsyth had required the first three coefficients 1, 0, 1 instead of their 1, 0, 0, they would have got our global form (3) instead of their local one.

All introduced here global canonical forms are not, however, unique. Using Cartan's moving-frame-of-reference method [6, 28] for the case of centroaffine (and not only orthogonal) transformations, we get unique canonical forms:

$$u'' + u = 0 \qquad \text{for } n = 2$$

$$u''' + 2\alpha v' + (\alpha' + 1)v = 0 \qquad \text{for } n = 3$$

$$u^{IV} - 10(\alpha^2 + \alpha' + \beta)u'' - 10(\alpha^2 + \alpha' + \beta)'u' - 5u'$$
$$+ (9(\alpha^2 + \alpha' + \beta)^2 - 3(\alpha^2 + \alpha' + \beta)'')u + 3\beta u = 0 \qquad \text{for } n = 4,$$

$\alpha = \alpha(t)$, $\beta = \beta(t)$.

In fact we get special types of differential equations similar to those already considered by G. H. Halphen [11]. It can be shown that the types are local [31].

V. Asymptotic behaviour of solutions

Global canonical forms enable us to "coordinate" each linear differential equation. We can do it in the following way.

Consider an equation $P_n(y;I)$. It belongs to a certain class D of globally equivalent equations. Since our canonical form is global, there exists an equation $K_n(u;J)$ of the canonical form in D. Thus K_n can be globally transformed into P_n, or explicitely

$$\underline{y}(x) = A.f(x).\underline{u}(h(x)), \quad x \in I.$$

Hence each linear differential equation P_n can be determined by a differential equation in a global canonical form, K_n, and by a couple $\langle f, h \rangle$ of a factor f and a change of parameter h that form the transformation converting K_n into P_n.

Using this approach, some global properties of solutions were equivalently expressed in terms of K_n, f, and h. Thus periodic solutions [15, 17, 30], asymptotic properties of solutions in connection with the distribution of their zeros were studied in [16, 21].

A deep study of square integrable solutions of the second order linear differential equations $P_2^0(y;I)$ in connection with the so-called limit circle and limit point classification was done by W. N. Everitt [8, 9].

Let us mention an application of our approach to these area of problems.

We have n = 2. Fix $K_2 \equiv u''+u = 0$ on J; cf. Sect. IV. To each $P_2^0(y;I)$ there correspond a couple $\langle |h'|^{-1/2}, h \rangle$, where $h \in C^3(I)$, $dh(x)/dx \neq 0$ on I, h(I) = J; cf. the notion of phase in [2]. Now, we can prove that a. each solution of P_2^0 is square integrable if and only if $\int_I |h'(x)|^{-1} dx < \infty$,

b. each solution of P_2^0 is bounded on I if and only if $|h'(x)|^{-1}$ is bounded on I.

Hence it holds

Theorem 6. There exists an equation $P_2\ (y;I)$ with all square integrable solutions, i.e., which is in the limit circle case, and still with an unbounded solution.

For more details about these problems and further results, see [18, 19, 23, 27].

VI. Criterion of global equivalence

A criterion for global equivalence of two given linear differential equations is based on the following theorem.

Theorem 7. Let $n \geq 2$ be an arbitrary integer. Consider linear differential equations $P_n^o(y;I)$ and $Q_n^o(z;J)$ with $p_{n-2} \in C^{n-2}(I)$ and $q_{n-2} \in C^{n-2}(J)$.

Then we can write

$$P_n^o(y;I) \equiv It_n[p_{n-2}/\tbinom{n+1}{3}] + r_{n-3}(x)y^{(n-3)} + \dots$$
$$\dots + r_o(x)y = 0,$$

and

$$Q_n^o(z;J) \equiv It_n[q_{n-2}/\tbinom{n+1}{3}] + s_{n-3}(t)z^{(n-3)} + \dots$$
$$\dots + s_o(t)z = 0.$$

The equation P_n^o is globally equivalent to Q_n^o if and only if there exists a function $h \in C^n(J)$, $dh(t)/dt \neq 0$ on J, such that

$$r_{n-3}(h(t)).h'^3(t) = s_{n-3}(t) \qquad \text{on } J,$$

and

$$r_{n-4}(h(t)).h'^4(t) = s_{n-4}(t)$$
$$\text{on } J_{n-4} := \{t \in J;\ s_{n-3}(t) = 0\},$$

and

$$r_{n-5}(h(t)).h'^5(t) = s_{n-5}(t)$$
$$\text{on } J_{n-5} := \{t \in J_{n-4};\ s_{n-4}(t) = 0\}, \text{ etc.,}$$

and moreover, the differential equation of the second order

$$y'' + p_{n-2}(x)/\binom{n+1}{3}.y = 0 \qquad \text{on } I$$

is globally transformable by the transformation $\langle |h'|^{-1/2}, h \rangle$ into

$$z'' + q_{n-2}(t)/\binom{n+1}{3}.z = 0 \qquad \text{on } J.$$

Remarks.

Except the case of iterative equations, i.e., when $s_{n-3} = s_{n-4} = \dots = s_0 = 0$, the function h can be expressed by quadratures, hence except the case the criterion is effective.

Sufficient and necessary condition for global equivalence of two second order linear differential equations was found by O. Borůvka [2].

VII. Zeros of solutions

Perhaps one of the mostly considered problem for linear differential equations is the question of distribution of zeros of solutions (their oscillatoric behaviour, disconjugacy etc.)

The essence of our approach to the problem of distribution of zeros of solutions can be expressed in the following theorem, first given in [20], and then used in recent literature, e.g. [10].

Theorem 8. Let coordinates of a curve $\underline{y}$ in the n-dimensional vector space be linearly independent solutions of a linear differential equation of the n-th order, $P_n(y;I)$. There is a 1-1 correspondence between all solutions of $P_n(y;I)$ and all hyperplanes passing through the origin, in which parameters of intersections of the curve $\underline{y}$ with a particular hyperplane are zeros of the corresponding solution, and vice versa, counting multiplicities that occur as the order of contacts.

If morever, $\underline{y}$ is considered in the n-dimensional euclidean space, the central projection $\underline{v}$ of $\underline{y}$ onto the unit sphere S_{n-1} is taken (without any change of parameter), then the same situation is in the compact space, S_{n-1}, and we have to our disposal strong tools of topology as well.

Using this approach several open problems were solved and the essence of some theorems with complicated proofs became clear, [7, 24, 29]. Constructions of linear differential equations with some preseribed properties of zeros of their solutions can be sometimes seen from drawing a curve on the sphere without lengthy calculations.

As an example let us draw a picture of a curve in 3-dimensional space, ensuring the existence of a linear differential equation without oscillatory solutions, but still having solutions with arbitrary (finite) number of zeros.

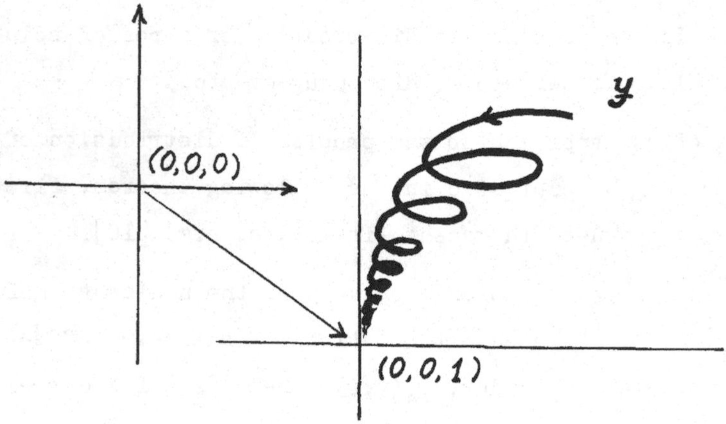

References

1. Birkhoff, G. D.: On the solutions of ordinary linear homogeneous differential equations of the third order, Annals of Math. 12 (1910/11), 103-124.

2. Borůvka, O.: Linear Differentialtransformationen 2. Ordnung, VEB, Berlin 1967; Linear Differential Transformations of the Second Order, The English Univ. Press, London 1971.

3. Borůvka, O.: Teorija global'nych svojstv obyknovennych linejnych differencial'nych uravněnij vtorogo porjadka, Differencial'nyje uravněnija 12 (1976), 1347-1383.

4. Borůvka, O.: Algebraic methods in the theory of global properties of the oscillatory equations Y" = Q(t)Y. Equadiff IV - Proceedings, Prague 1977, Lecture Notes in Mathematics 703, 35-45.

5. Borůvka, O.: Sur une classe des groupes continus à un paramètre formés des fonctions réelles d'une variable, Ann. Polon. Math., to appear.

6. Cartan, É.: La théorie des groupes finis et la géométrie différentielle traitées par la méthode du repère mobile, Gauthier-Villars, 1937.

7. Dolan, J. M.: On the relationship between the oscilatory behavior of a linear third-order differential equation and its adjoint, J. Diff. Equations 7 (1970), 367-388.

8. Everitt, W. N.: A note of the Dirichlet conditions for second-order differential expressions, Canad. J. Math. 28 (1976), 312-320.

9. Everitt, W. N.: On the transformation theory of ordinary second-order linear symmetric differential equations, preprint.

10. Guggenheimer, H.: Distribution of zeros and limit behavior of solutions of differential equations, Proc. AMS 61 (1976), 275-279.

11. Halphen, G. H.: Mémoire sur la réduction des équations différentilles linéaires aux formes intégrables. Mémoires présentés par divers savants à l'académie des sciences de l'institut de France 28 (1884), 1-301.

12. Hasse, M., Michler, L.: Theorie der Kategorien, VEB, Berlin 1966.

13. Hustý, Z.: Die Iteration homogener linearer Differentialgleichungen, Publ. Fac. Sci. Univ. J. E. Purkyně (Brno), 449 (1964), 23-56.

14. Kummer, E. E.: De generali quadam aequatione differentiali tertii ordinis. Progr. Evang. Königl. & Stadtgymnasiums Liegnitz 1834.

15. Neuman, F.: Criterion of periodicity of solutions of a certain differential equation with a periodic coefficient, Ann. Mat. Pura Appl. 75 (1967), 385-396.

16. Neuman, F.: Relation between the distribution of the zeros of the solutions of a 2nd order linear differential equation and the boundedness of these solutions, Acta Math. Acad. Sci. Hungar. 19 (1968), 1-6.

17. Neuman, F.: An explicit form of the differential equations y" = q(t)y with periodic solutions, Ann. Mat. Pura Appl. 85 (1970), 205-300.

18. Neuman, F.: Linear differential equations of the second order and their applications, Rend. Mat. 4 (1971), 559-617.

19. Neuman, F.: A note on Santaló's isoperimetric theorem, Revista Mat. Fis. Teor. Tucuman, 21 (1971), 203-206.

20. Neuman, F.: Geometrical approach to linear differential equations of the n-th order, Rend. Mat. 5 (1972), 579-602.

21. Neuman, F.: Distribution of zeros of solutions of y" = q(t)y in relation to their behaviour in large, Studia Sci. Math. Hungar. 8 (1973), 177-185.

22. Neuman, F.: On n-dimensional closed curves and periodic solutions of linear differential equations of the n-th order, Demonstratio Math. 6 (1973), part I, 329-337.

23. Neuman, F.: On a problem of transformations between limit-circle and limit-point differential equations, Proc. Roy. Soc. Edinburgh, Sect. A. 72 (1973/74), 187-193.

24. Neuman, F.: On two problems about oscillation of linear differential equations of the third order, J. Diff. Equations 15 (1974), 589-596.

25. Neuman, F.: On solutions of the vector functional equation $y(\xi(x)) = f(x).A.y(x)$, Aequationes Math. 16 (1977), 245-257.

26. Neuman, F.: Categorial approach to global transformations of the n-th order linear differential equations, Časopis Pěst. Mat. 102 (1977), 350-355.

27. Neuman, F.: Limit circle classification and boundedness of solutions, Proc. Roy. Soc. Edinburgh, 81 A (1978), 31-34.

28. Neuman, F.: Invarianty linejnych differencial'nych uravnenij 3-go porjadka i metod podvižnogo repera E. Kartana, Differencial'nyje uravnenija XIV (1979), 398-404.

29. Neuman, F.: Global theory of linear differential equations of the n-th order, Proceedings of the Colloquium on Qualitative Theory of Differential Equations, August 79, Szeged-Hungary, Seria Colloquia Mathematica Societatis János Bolyai & North-Holland Publishing Company, 777-794.

30. Neuman, F.: Second order linear differential systems, Ann. Sci, École Norm. Super. (Paris) 13 (1980), 437-449.

31. Neuman, F.: Global canonical forms of linear differential equations, Math. Slovaca, to appear.

32. Sansone, G.: Studi sulle equazioni differenziali lineari omogenee di terzo ordine nel campo reale. Revista Mat. Fis. Teor. Tucuman 6 (1948), 195-253.

33. Stäckel, P.: Über Transformationen von Differentialgleichungen. J. Reine Angew. Math. (Crelle Journal) 111 (1893), 290-302.

34. Wilczynski, E. J.: Projective Differential Geometry of Curves and Ruled Surfaces, Teubner, Leipzig 1906.

Author's address: Mathematical Institute of the
Czechoslovak Academy of Sciences
Branch Brno
Janáčkovo nám. 2a
66295 Brno, Czechoslovakia

STRONGLY NONLINEAR EVOLUTION VARIATIONAL INEQUALITIES

DAN PASCALI *)

Certain free boundary problems related to diffusion processes lead to the evolution variational inequality

$$<\frac{\partial u}{\partial t} + P(t)u\ ,\ v-u> + \int_Q g(t,x,u)(v-u) \geq <f\ ,\ v-u> \quad \forall v \in K$$

in an open cylinder $Q =]0,T[\times\Omega$ corresponding to a bounded domain Ω in $\mathbb{R}^N$ and to a fixed $T>0$. The closed convex set K in $L^\infty(0,T;L^2(\Omega))$ is determined by the heat flux through semi-permeable walls and g is a caloric body function satisfying a sign condition. Here the coercive operator of pseudo-monotone type P(t) is bounded while no growth restriction is imposed on the perturbation g.

The purpose of this report is to give an existence result for a weak solution of the above variational inequality, based on extending the treatment employed by the author [5] to study evolution equations involving this kind of perturbations. Although we use the same compactness criterium as Browder [3], which is essentially a special case of Aubin's lemma for strongly nonlinear parabolic equations [2], the approach here is in a more abstract setting, thus leading to different assumptions.

Let V be a reflexive Banach space compactly and densely imbedded in $H = L^2(\Omega)$ and V^* be the dual space of V. Identify H with its dual and obtain $V \subseteq H \subseteq V^*$. Denote by $|\cdot|_2$ $\|\cdot\|$ and $\|\cdot\|_*$ the norms in H, V and V^* respectively and let $(\cdot,\cdot)$ the pairing between V and V^*, be an extension of the inner product on H.

An operator $A: V \to V^*$ is *monotone* if $(Au - Av, u-v) \geq 0$ for all $u, v \in D(A) \subseteq V$ and *maximal monotone* if it admits no properly monotone extensions in $V \times V^*$. By "$\longrightarrow$" and "$\rightharpoonup$" we denote the strong and

*) *This work was done with the support of an Alexander von Humboldt grant while the author worked at the T H Darmstadt.*

weak convergence respectively. Then A is *demicontinuous* if $u_n \rightharpoonup u$ in V implies $Au_n \rightharpoonup Au$ in V^* while A is *coercive* if

$$\frac{(Au,u)}{\|u\|} \to \infty \quad \text{as} \quad \|u\| \to \infty .$$

Let C be a closed convex set of V. A demicontinuous operator $P : C \to V^*$ is said to be *pseudo-monotone* if for every sequence $\{u_n\}$ in C for which $u_n \rightharpoonup u$ and $\limsup\,(Pu_n, u_n - u) \le 0$ it follows that

$$(Pu, u - v) \le \liminf\,(Pu_n, u_n - v) \quad \text{for all} \quad v \in C.$$

We are given a family $\{P(t) \mid t \in [0,T]\}$ of nonlinear (single-valued) operators from V into V^* such that:

i) For almost every $t \in]0,T[$, the operator $P(t): V \to V^*$ is pseudo-monotone and there are constants $c_1 > 0$ and $p \ge 2$ so that

$$\|P(t)u\|_* \le c_1(1 + \|u\|^{p-1}) \quad \text{for all} \quad u \in V ;$$

ii) There is a positive constant c_2 such that

$$(P(t)u,u) \ge c_2 \|u\|^p \quad \text{for all} \quad u \in V \quad \text{and a.a.} \quad t \in]0,T[.$$

The following hypotheses on the strongly nonlinear perturbing term $g(t,u)$ are imposed:

j) $g : [0,T] \times H \to H$ is continuous at u in H for a.a.$t \in]o,T[$;

jj) g fulfils *the sign condition* in the sense

$$(g(t,u), u) \ge 0 \quad \forall u \in H \quad \text{and} \quad \text{a.a.} \ t \in]0,T[.$$

Note that the perturbation g does not induce a mapping on $V \times V^*$ because no growth restriction is made. In this account, we introduce *the truncated perturbation*

$$g_n(t,u) = \begin{cases} g(t,u) & \text{if } |g(t,u)|_2 \le n , \\ n\dfrac{g(t,u)}{|g(t,u)|} & \text{otherwise} \end{cases}$$

and the semilinear form $a_n(u,v) = (g_n(\cdot,u),v)$. The Hölder inequality yields $|a_n(u,v)| \le n\,|v|_2$ and so $a_n(u,\cdot)$ determines an element $G_n(t)u$ in H. Since V is compactly imbedded in H the sum $(P+G_n)(t): V \to V^*$ is a coercive bounded mapping, for almost all $t \in]o,T[$.

Furthermore, we consider the vector-valued function spaces $C([0,T];H)$ and $\mathcal{H} = L^2(0,T;H) = L^2(Q)$ as well as the dual reflexive spaces $\mathcal{V} = L^p(0,T;V)$ and $\mathcal{V}^* = L^{p'}(0,T;V^*)$, with $\frac{1}{p} + \frac{1}{p'} = 1$, all endowed with the usual norms. We also have $\mathcal{V} \subseteq \mathcal{H} \subseteq \mathcal{V}^*$ and denote by $\langle\cdot,\cdot\rangle$ the pairing between $\mathcal{V}$ and $\mathcal{V}^*$, that is

$$\langle u;v\rangle = \int_0^T (u(t),v(t))\,dt \quad \text{for all} \quad u \in \mathcal{V} \quad \text{and} \quad v \in \mathcal{V}^* ,$$

which is the inner product on $\mathcal{H}$. Suppose, in addition, that:

iii) For every $u \in H$, the function $t \to P(t)u(t)$ is strongly V^*- measurable on $]0,T[$;

jjj) For any $u, v \in V$, the function $t \to (g(t,u(t),v(t))$ is measurable on $]0,T[$ while $g_r(t) = \sup_{\|u\|_\infty \leq r} |g(t,u(t))|_2$ defines a $L^1(0,T)$ - function for $0 \leq r < \infty$, where $\|u\|_\infty = \operatorname{ess\,sup}_{t \in]0,T[} |u(t)|_2$ is the norm in $L^\infty(0,T;H)$.

Set now the nonlinear operators $P: V \to V^*$ and $G_n: H \to H$ defined by

$$(Pu)(t) = P(t)u(t) \quad \text{and} \quad (G_n u)(t) = G_n(t)u(t)$$

for almost all $t \in]0,T[$. The above assumptions guarantes that P is a pseudo-monotone operator defined on V such that

$$\|Pu\|_* \leq c_1'(1 + \|u\|_V^{p-1}) \quad \text{and} \quad \langle Pu,u\rangle \geq c_2 \|u\|_V^p \quad \text{for all} \quad u \in V,$$

with a constant $c_1' > 0$. The same conditions are also satisfied by the sum $P + G_n$.

Further, let $W = \{v \in V \mid \frac{dv}{dt} \in V^*\}$ be a Banach space with respect to the norm $\|v\|_W = \|v\|_V + \|\frac{dv}{dt}\|_{V^*}$, where the derivative $\frac{dv}{dt}$ is taken in the sense of V^*- valued distributions over $]0,T[$. After appropriate modification of $v \in W$ on a subset of $]0,T[$, whose measure is zero, we have $v \in C([0,T];H)$ and the integration by parts formula

$$\int_s^t \{(\frac{dv}{dt}, w) + (v, \frac{dw}{dt})\}dt = (v(t), w(t)) - (v(s), w(s)) \qquad 0 \leq s < t \leq T$$

is valid for all $v, w \in W$. Moreover, the injection of W into $C([0,T];H)$ is continuous and the linear operator $A: V \to V^*$ defined by

$$Aw = \frac{dw}{dt} = w' \quad \text{for} \quad w \in W \quad \text{with} \quad w(0) = 0$$

is maximal monotone in $V \times V^*$, (see e.g. [1,p.65]).

Let C be a closed convex set containing the origin in V. Associate with C *the penalty operator* β, i.e., a demicontinuous bounded monotone operator $\beta: V \to V^*$ with $C = \ker \beta = \{v \in V \mid \beta v = 0\}$. In addition to the set C, consider the following two closed convex sets $K = \{v \in V \mid v(t) \in C \text{ a.e. in }]0,T[,\ v(0) = 0\}$ and $K' = K \cap W$, and denote by B the penalty operator corresponding to K.

As usual, assume that the following *compatibility conjecture* is fulfiled: To every $v \in K$ there corresponds a smoothing sequence $\{w_n\} \in K'$ such that $w_n \longrightarrow v$ in V and $\limsup (Aw_n, w_n - v) \leq 0$.

We are now in the position to state our main result

THEOREM. *Let* f *be a given element of* V^* *and assume that all above hypotheses hold. Then there exists* $u \in K$ *with* $g(\cdot,u) \in L^1(0,T;H)$ *all* $(g(\cdot,u),u) \in L^1(0,T)$ *which satisfies the variational inequality*

$$(1)\quad \langle v',v-u\rangle + \langle Pu,v-u\rangle + \int_0^T (g(t,u(t)),v(t)-u(t))dt \geq \langle f,v-u\rangle \quad \forall v\in K'.$$

Proof. Using an elliptic penalisation method, it follows from a well-known surjectivity result in monotone type operator theory, that the approximante equation

$$Au + (P+G_n)u + \frac{1}{\varepsilon}Bu = f, \quad u(0) = 0$$

admits at least one solution $u_{n\varepsilon} \in W$ for every $n \in \mathbb{N}$ and $\varepsilon > 0$.

Now we multiply by $u_{n\varepsilon} - v$ and integrate over $[0,t]$ to obtain

$$\frac{1}{2}|u_{n\varepsilon}(t) - v(t)|_2^2 + \int_0^t (v'(\tau),u_{n\varepsilon}(\tau) - v(\tau))d\tau +$$
$$+ \int_0^t (P(\tau)u_{n\varepsilon}(\tau),u_{n\varepsilon}(\tau) - v(\tau))d\tau + \int_0^t (g_n(\tau,u_{n\varepsilon}(\tau)),u_{n\varepsilon}(\tau)-v(\tau))d\tau$$
$$+ \frac{1}{\varepsilon}\int_0^t (\beta u_{n\varepsilon}(\tau),u_{n\varepsilon}(\tau)-v(\tau))d\tau = \int_0^t (f(\tau),u_{n\varepsilon}(\tau) - v(\tau))d\tau,$$

for each $0 < t \leq T$ and $v \in K'$. By taking $v \equiv 0$ and since $(\beta u_{n\varepsilon},u_{n\varepsilon}) \geq 0$, we derive

$$\frac{1}{2}|u_{n\varepsilon}(t)|_2^2 + c_2 \|u_{n\varepsilon}\|_V^p + \int_0^t (g_n(\tau,u_{n\varepsilon}(\tau)),u_{n\varepsilon}(\tau))d\tau \leq \|f\|_{V^*}\|u_{n\varepsilon}\|_V$$

whence $\|u_{n\varepsilon}\|_{L^\infty(0,T;H)} + \|u_{n\varepsilon}\|_V \leq$ const. and then $\{(g_n(\cdot,u_{n\varepsilon}),u_{n\varepsilon})\}$ is a bounded sequence in $L^1(0,T)$, for each $\varepsilon > 0$.

Taking first the limit as $\varepsilon \to 0$, we have $u_{n\varepsilon} \rightharpoonup u_n$ in $V \cap L^\infty(0,T;H)$, at least for a subsequence, and, by a standard argument ([4],Chap.3, §6), we can infer that $u_n \in K$ and

$$(2)\quad \langle Av, v-u_n\rangle + \langle (P+G_n)u_n, v-u_n\rangle \geq \langle f, v-u_n\rangle \quad \forall v \in K',$$

for every $n \in \mathbb{N}$. Due to the weakly lower semicontinuity of norms, we have further $\|u_n\|_{L^\infty(0,T;H)} + \|u_n\|_V \leq$ const., while Fatou's lemma assures that $\{(g_n(\cdot,u_n),u_n)\}$ remains a bounded sequence in $L^1(0,T)$.

Let now $u_n \rightharpoonup u$ in $V \cap L^\infty(0,T;H)$ as $n \to \infty$. Since $|u_n|_2 \leq c_3$ and $|G_n u_n|_2 \leq g_{c_3}(t)$ a.e. in $]0,T[$, by *jjj*) we get $G_n u_n \in L^1(0,T;H)$ and so there is a function $\gamma : \mathbb{R}_+ \to \mathbb{R}_+$, with $\gamma(r) \to 0$ as $r \to 0$, such that for any pair (s,t) in $[0,T]$, with $s < t$,

$$\int_s^t \|u_n'(\tau)\|_* d\tau \leq \gamma(t-s) \quad \text{for all } n \in \mathbb{N}.$$

By the above mentioned compactness criterium ([2],theorem 3),the sequence $\{u_n\}$ is contained in a compact set in $L^p(0,T;H)$ and also in $L^2(0,T;H) = L^2(Q)$. Thus,there is a subsequence,again denoted by $\{u_n\}$, such that $u_n \rightarrow u$ a.e. in $Q =]0,T[\times \Omega$ and, consequently

$$g_n(\cdot,u_n(\cdot)) \rightarrow g(\cdot,u(\cdot)) \quad \text{a.e. in }]0,T[.$$

Applying again the Fatou lemma,we deduce that $(g(\cdot,u),u) \in L^1(0,T)$, while the Lebesque dominated convergence theorem yields

$$g_n(\cdot,u_n) \rightarrow g(\cdot,u) \quad \text{in} \quad L^1(0,T;H).$$

Finally,the passage to the limit in (2) concludes that $u \in K$ satisfies the variational inequality (1), which completes the proof of the theorem.

REFERENCES

[1] V.BARBU - TH.PRECUPANU, *Convexity and Optimization in Banach spaces*, *Edit.Academiei Bucuresti - Sijthoff & Nordhooff,1978.*

[2] H.BRÉZIS - F.E.BROWDER, *Strongly nonlinear parabolic initial--boundary valued problems*, *Proc.Natl.Acad.Sci.USA 76 (1979),38-40.*

[3] F.E.BROWDER, *Strongly nonlinear parabolic problems,(abstract) Tagungsbericht 49/1979, Oberwolfach.*

[4] J.L.LIONS, *Quelques Méthodes de Résolution des Problèmes aux Limites Non Linéaires*, *Dunod & Gauthier-Villars,Paris,1979.*

[5] D.PASCALI, *Variational solutions of strongly nonlinear evolution equations*, *Workshop on "Applied Functional Analysis", (Variational methods and ill-posed problems), West-Berlin, September 7-12, 1981.*

A FURTHER RESULT ON THE ESSENTIAL SPECTRUM OF LINEAR 2nth ORDER DIFFERENTIAL EXPRESSIONS WITH COMPLEX COEFFICIENTS

D. Race

1. Introduction

We shall be concerned with the expression

$$\tau y(t) = \sum_{k=0}^{n} (-1)^{n-k} (p_k(t)\, y^{(n-k)}(t))^{(n-k)} \quad , \quad t \in [a,\infty) \tag{1.1}$$

where $a > -\infty$ and the coefficients p_k , $0 \leqslant k \leqslant n$ are complex-valued functions. We assume the coefficients satisfy the minimal conditions, i.e. that $1/p_0, p_1, \ldots, p_n$ are locally Lebesgue integrable on the interval $[a,\infty)$. We define the operator T_0' in $L^2[a,\infty)$ by

$$T_0' f = \tau f \quad , \qquad f \in \mathscr{D}(T_0')$$

where

$$\mathscr{D}(T_0') = \{f \in L^2[a,\infty) : f^{[k]} \text{ is locally absolutely continuous,}$$
$$0 \leqslant k \leqslant 2n-1 \, , \; \tau f \in L^2[a,\infty) \, , \; f \text{ has compact support}$$
$$\text{in } (a,\infty)\}$$

in which $f^{[k]}$ denotes the kth quasi-derivative of f associated with τ . The minimal operator T_0 associated with τ is the closure of T_0'. We also define the essential spectrum of an operator A , $E\sigma(A)$, to be the set of $\lambda \in \mathbb{C}$ for which there exists a non-compact sequence $\{f_n\} \subset \mathscr{D}(A)$ with $\|f_n\| = 1$, satisfying the condition

$$\lim_{n\to\infty} (A - \lambda I) f_n = 0 \; .$$

The general theory of operators associated with τ may be found in [3, 5, 7] . If T_1 is any closed extension of T_0 then $E\sigma(T_1) = E\sigma(T_0)$ so it is clearly of value to locate the set $E\sigma(T_0)$. The significance of this and related results is discussed in [6]. We need the following result here.

Lemma 1.1 (c.f. [2, p.75], [6, §1]) *If* $K(T_o')$ *is the closure of the set of complex numbers*

$$\{(T_o'f, f) : f \in \mathcal{D}(T_o'),\ \|f\| = 1\}$$

then $E\sigma(T_o) \subset K(T_o')$.

In applying Lemma 1.1 it is useful to observe that by integrating by parts repeatedly,

$$(T_o'f, f) = \sum_{k=0}^{n} \int_a^\infty p_k |f^{(n-k)}|^2 . \tag{1.2}$$

We also need the following inequality, which is valid for any $f \in \mathcal{D}(T_o')$ (c.f. [2, p.83])

$$\|f'\| \leqslant \sqrt{C_k}\,(\|f^{(k)}\| + \|f\|) \tag{1.3}$$

where C_k is a constant, $1 \leqslant k \leqslant n$.

We use $\mathrm{Re}(\alpha)$, $\mathrm{Im}(\alpha)$ to denote the real and imaginary parts of α , respectively, and $(f)^-$ to denote the negative part of f , when f is real, i.e.

$$(f)^- = \max(-f,\ 0) .$$

2. Known results

There are few results for $n > 1$, concerning the location of $E\sigma(T_o)$ when the coefficients are real-valued, and even fewer for complex-valued coefficients. Many of the known results for real coefficients were extended to complex coefficients in the case $n = 1$, in [4]. Most of these were then generalized to $n > 1$ in [6], where a brief review of other results for complex coefficients and $n > 1$ may be found. The following is a typical result from [6].

Theorem 2.1 ([6, Theorem 3.1]). *If for some* $N < \infty$, $p_o(t) \geq 1$ *for* $t \in [N,\infty)$, *and for each* $1 \leq k \leq n$,

$$\int_a^\infty (\mathrm{Im}\, p_k)^- < \infty \tag{2.1}$$

and

$$\int_J \mathrm{Re}\, p_k \geq -D_k \quad \text{for some } D_k \geq 0 \text{ and for all intervals } J \text{ of length } \leq 1$$

then there is a constant $\Lambda > -\infty$ *for which*

$$E\sigma(T_o) \subset \{\lambda \in \mathbb{C} : \mathrm{Im}\,\lambda \geq 0 \text{ and } \mathrm{Re}\,\lambda \geq \Lambda\} .$$

A value for Λ is given in [6, (3.17)] , though it is observed there that it is not the best possible value. From Theorem 2.1, the following may be deduced.

Corollary 2.2 ([6, Corollary 3.2]) *If for some* $N < \infty$, $p_o(t) \geq 1$, $t \in [N,\infty)$, *and for each* $1 \leq k \leq n$, $\mathrm{Im}\, p_k$ *satisfies* (2.1), *and there is a real constant* a_k , *such that* $\mathrm{Re}\, p_k$ *satisfies one of the following* :

(i) $\displaystyle\int_{M_{\delta,k}} (\mathrm{Re}(p_k - a_k))^- < \infty$ *for any* $\delta > 0$ *where* $M_{\delta,k} = \{t \in [a,\infty) : (\mathrm{Re}(p_k(t) - a_k))^- \geq \delta\}$;

(ii) $\displaystyle\lim_{t\to\infty} \int_t^{t+\omega} (\mathrm{Re}(p_k - a_k))^- = 0$ *for some (and consequently for any)* $\omega \neq 0$;

(iii) $\displaystyle\int_a^\infty \{(\mathrm{Re}(p_k - a_k))^-\}^{r_k} < \infty$ *for some* $r_k \geq 1$;

then

$$E\sigma(T_o) \subset \left\{\lambda \in \mathbb{C} : \mathrm{Im}\,\lambda \geq 0 \text{ and } \mathrm{Re}\,\lambda \geq \min_{0 \leq x < \infty}\left(\tfrac{1}{4}x^n - \sum_{k=1}^n a_k x^{n-k}\right)\right\} .$$

3. Main Theorem

The purpose of the present work is to give a generalization of [4, Theorem 4.6] and [6, Theorem 3.4]. The former is for $n = 1$ only, and the latter for $n > 1$. In addition to extending the latter, we give a more direct proof than is used there, and hence obtain a more precise bound for $E\sigma(T_o)$.

Theorem 3.1 *If there are real constants* a_k, $1 \leqslant k \leqslant n$, *such that for some* $N < \infty$ *and* $m \in \mathbb{R}$,

$$(\mathrm{Re} + m\ \mathrm{Im})p_o(t) \geqslant a_o > 0\,, \qquad t \in [N, \infty) \tag{3.1}$$

$$\int_a^\infty \{(\mathrm{Re} + m\ \mathrm{Im})(p_k(t) - a_k)\}^- < \infty\,, \qquad 1 \leqslant k \leqslant n \tag{3.2}$$

then $E\sigma(T_o) \subset \{\lambda \in \mathbb{C} : (\mathrm{Re} + m\ \mathrm{Im})\lambda \geqslant \min\limits_{0 \leqslant x < \infty} \sum\limits_{k=0}^{n} a_k x^{n-k}\}$.

Proof In the light of Lemma 1.1 and (1.2) we consider $(T_o'y, y)$ for any $y \in \mathscr{D}(T_o')$ with $\|y\| = 1$ and having compact support in (N, ∞).

$$\begin{aligned}(\mathrm{Re} + m\ \mathrm{Im})(T_o'y, y) &= \sum_{k=0}^{n} \int_N^\infty (\mathrm{Re} + m\ \mathrm{Im})p_k \cdot |y^{(n-k)}|^2 \\ &= \int_N^\infty (\mathrm{Re} + m\ \mathrm{Im})p_o \cdot |y^{(n)}|^2 + \sum_{k=1}^{n} a_k \|y^{(n-k)}\|^2 \\ &\quad + \sum_{k=1}^{n} \int_N^\infty (\mathrm{Re} + m\ \mathrm{Im})(p_k - a_k)|y^{(n-k)}|^2 \\ &\geqslant a_o \|y^{(n)}\|^2 + \sum_{k=1}^{n} a_k \|y^{(n-k)}\|^2 \\ &\quad - \sum_{k=1}^{n} \int_N^\infty \{(\mathrm{Re} + m\ \mathrm{Im})(p_k - a_k)\}^- \cdot |y^{(n-k)}|^2 \end{aligned} \tag{3.3}$$

using (3.1). Now from (3.2), we may choose N sufficiently large so that for each $1 \leqslant k \leqslant n$,

$$\int_N^\infty \{(\mathrm{Re} + m\ \mathrm{Im})(p_k - a_k)\}^- < \alpha$$

where α will be chosen later. Also, for $1 \leqslant k \leqslant n$

$$|y^{(n-k)}(t)|^2 \leqslant 2\left(\int_N^\infty |y^{(n-k)}|^2\right)^{1/2}\left(\int_N^\infty |y^{(n-k+1)}|^2\right)^{1/2}$$

$$= 2\|y^{(n-k)}\| \, . \, \|y^{(n-k+1)}\|, \qquad t \in [N, \infty)$$

by a method used in [1]. Hence in (3.3) we have

$$(\mathrm{Re} + m\,\mathrm{Im})(T_o'y, y) > a_o\|y^{(n)}\|^2 + \sum_{k=1}^{n} a_k\|y^{(n-k)}\|^2$$

$$- \sum_{k=1}^{n} 2\alpha\|y^{(n-k)}\| \, . \, \|y^{(n-k+1)}\|$$

$$\geqslant a_o\|y^{(n)}\|^2 + \sum_{k=1}^{n} a_k\|y^{(n-k)}\|^2$$

$$- \sum_{k=1}^{n} 2\alpha\sqrt{C_k}\,(\|y^{(n)}\| \, . \, \|y^{(n-k)}\| + \|y^{(n-k)}\|^2)$$

using (1.3) with $f = y^{(n-k)}$. Thus for any $0 < \varepsilon < a_o$,

$$(\mathrm{Re} + m\,\mathrm{Im})(T_o'y, y) > (a_o - \varepsilon)\|y^{(n)}\|^2 + \sum_{k=1}^{n}\{\tfrac{\varepsilon}{n}\|y^{(n)}\|^2 - 2\alpha\sqrt{C_k}\|y^{(n)}\| . \|y^{(n-k)}\|$$

$$+ (a_k - 2\alpha\sqrt{C_k})\|y^{(n-k)}\|^2\}$$

$$\geqslant (a_o - \varepsilon)\|y^{(n)}\|^2 + \sum_{k=1}^{n}(a_k - 2\alpha\sqrt{C_k} - n\alpha^2 C_k\varepsilon^{-1})\|y^{(n-k)}\|^2 \tag{3.4}$$

by completing the square. We now choose $\alpha = \min\{\tfrac{1}{4}\varepsilon C_k^{-1/2}, \varepsilon(2nC_k)^{-1/2}\}$ so that

$$2\alpha\sqrt{C_k} + n\alpha^2 C_k \varepsilon^{-1} \leqslant \varepsilon .$$

In (3.4) we now have

$$(\mathrm{Re} + m\,\mathrm{Im})(T_o'y, y) \geqslant \sum_{k=0}^{n}(a_k - \varepsilon)\|y^{(n-k)}\|^2$$

for any $0 < \varepsilon < a_o$ and for $N(\varepsilon)$ sufficiently large. Hence, as in the proof of [2, Theorem 1, p.130] we have

$$(\text{Re} + m\ \text{Im})(T_o{}'y, y) \geqslant \min_{0 \leqslant x < \infty} \sum_{k=0}^{n} (a_k - \varepsilon)x^{n-k} = \mu_\varepsilon , \quad \text{say.}$$

Applying Lemma 1.1 and the decomposition principle (see [2, 5]) this gives

$$E\sigma(T_o) \subset \{\lambda \in \mathbb{C} : (\text{Re} + m\ \text{Im})\lambda \geqslant \mu_\varepsilon\} \tag{3.5}$$

We note that $\mu_\varepsilon > -\infty$ since $\varepsilon < a_o$. Finally, since (3.5) holds for any $\varepsilon > 0$ sufficiently small, we may conclude by continuity that (3.5) holds with $\varepsilon = 0$, as required to complete the proof. □

We observe that the condition $a_o > 0$ guarantees $\mu_o > -\infty$. The case when $a_o = 1$ and $a_k = 0$, $1 \leqslant k \leqslant n-1$ is [6, Theorem 3.4]. The above method of proof may also be used to refine the bound given in Corollary 2.2, by removing the factor of ¼. Geometrically, Theorem 3.1 says that under the conditions stated, the essential spectrum lies to the right of the line $\text{Re}\ \lambda + m\ \text{Im}\ \lambda = \mu_o$ in the complex plane.

4. Concluding remarks

If, in the usual way, we define the regularity field of T_o to be

$$\pi(T_o) = \{\lambda \in \mathbb{C} : \|(T_o - \lambda I)f\| \geqslant k_\lambda \|f\| \text{ for some positive constant } k_\lambda \text{ and for all } f \in \mathcal{D}(T_o)\}$$

then it is known (see [5]) that $\pi(T_o)$ is the complement in the complex plane, of the set $E\sigma(T_o)$. In the theory of J-selfadjoint operators associated with τ, a basic pre-requisite is that $\pi(T_o)$ be non-empty (see [3, 7]). It is therefore useful to have the following result, which is an immediate consequence of Theorem 3.1.

Theorem 4.1 (c.f. [6, Theorem 4.1]) *If the coefficients* p_k *satisfy* (3.1) *and* (3.2) *then* $\pi(T_o)$ *is non-empty.*

References

[1] W.N. Everitt, M. Giertz and J. Weidman. Some remarks on a separation and limit-point criterion of second-order, ordinary differential expressions. *Math. Ann.* 200 (1973), 335-346.

[2] I.M. Glazman. *Direct methods of qualitative spectral analysis of singular differential operators.* (Jerusalem : Israel Program for Scientific Translations, 1965).

[3] I. Knowles. On the boundary conditions characterizing *J*-selfadjoint extensions of *J*-symmetric operators. *J. Differential Equations* 40 (1981), 193-216.

[4] D. Race. On the location of the essential spectra and regularity fields of complex Sturm-Liouville operators. *Proc. Roy. Soc. Edinburgh Sect. A* 85 (1980), 1-14.

[5] D. Race. *The spectral theory of complex Sturm-Liouville operators.* (Ph.D. thesis : University of the Witwatersrand, Johannesburg, 1980)

[6] D. Race. On the essential spectra of linear $2n$th order differential operators with complex coefficients. *Proc. Roy. Soc. Edinburgh Sect. A,* to appear.

[7] N.A. Zhikhar. The theory of *J*-symmetric operators (Russian). *Ukrain. Mat. Z.* 11 (4) (1959), 352-364.

ON ERROR BOUNDS FOR NONSTATIONARY SPECTRAL NAVIER-STOKES APPROXIMATIONS

Reimund Rautmann

Summary: For spectral Galerkin approximations of nonstationary local strong Navier-Stokes solutions on a smoothly bounded 3-dimensional domain, error estimates in suitable Hilbert space are established, which lead to uniform pointwise error bounds in certain Hölder spaces. The estimates follow by application of Fujita's and Kato's results [4].

Since Kiselev's and Ladyženskaya's work [13] the local (in time) existence of unique strong solutions of the Navier-Stokes initial-boundary value problem in three-dimensional regions has been proved in several contextes [9],[18], [12,4] , [23, 22] , [2] . As recently Heywood has shown in [5], the Galerkin procedure based on the eigenfunctions of the linear Stokes problem offers a very natural way to proving local existence, C_∞ -regularity and asymptotic decay of strong solutions.

The question, how good the spectral Galerkin approximations are, has been settled in [2o] and [21] with respect to error estimates in the L^2 - and Dirichlet-norm.Heywood and Rannacher then established related estimates in the framework of finite elements [8] ,[19] , and Heywood [7] with an additional stability assumption inquired estimates uniform in time.

In this note we will strengthen the earlier estimates [2o,21] to bounds in suitable Hilbert spaces, which lead to pointwise uniform error estimates in certain Hölder spaces (locally in time, c.p. Theorem 2.1. below). The essential tool in the following is the application of Fujita's and Kato's results [4] .

1. Local strong Navier-Stokes solutions and spectral Galerkin approximations.

Let Ω be a bounded open set in the (x^1,x^2,x^3)- space $\mathbb{R}^3$, the boundary $\partial\Omega$ being a compact 2-dimensional C_3- submanifold of $\mathbb{R}^3$. The velocity vector $u(t,x)= (u^1,u^2,u^3)$

and the pressure function $p(t,x)$ of a nonstationary incompressible viscous flow in Ω at times $t \geq o$ solve the Navier-Stokes initial-boundary value problem

(1.1) $$u_t - \Delta u + \nabla p = -u\cdot\nabla u \;,\; \nabla\cdot u = o,\; t > o,$$

$$u_{|\partial\Omega} = o; \qquad u = u_o,\; t = o,$$

if we assume the condition of adherence on $\partial\Omega$ and for short neglect external forces, and if distance and time are measured in the appropriate units.
We will consider this problem in the usual framework of the Hilbert spaces H_m containing the vector functions (defined almost everywhere) on Ω, which belong to Lebesgue's class $L^2(\Omega)$ together with their spatial derivatives up to the order m.
The norm on H_m is

$$|f|_{H_m} = \Big(\sum_{n_1+n_2+n_3=o}^{m} \int_\Omega \Big|\frac{\partial^{n_1+n_2+n_3} f(x)}{(\partial x^1)^{n_1}(\partial x^2)^{n_2}(\partial x^3)^{n_3}}\Big|^2 dx\Big)^{\frac{1}{2}},$$

where $|f(x)| = (\sum_{i=1}^{3}(f^i(x))^2)^{\frac{1}{2}}$. For the $L^2(\Omega)$-norm we will write $\|f\| = |f|_{H_o}$. By $\mathcal{H}_m$ we denote the closure in H_m of the linear space of the divergence-free C_∞-vector functions having compact support in Ω. Finally let P be Weyl's projection of H_o on $\mathcal{H}_o$ and A Friedrich's extension of the positive definite, symmetric operator $-P\Delta$ in $\mathcal{H}_o$, D_A (or D_{A^α}) denoting the domain in $\mathcal{H}_o$ of the operator A (or A^α, respectively). We recall that for any $\alpha \in (o,1]$, D_{A^α} is a Hilbert space equipped with the norm
$(\int_\Omega |A^\alpha f|^2 dx)^{1/2}$, because the inverse $A^{-\alpha} : \mathcal{H}_o \to D_{A^\alpha}$ exists as a bounded operator on $\mathcal{H}_o$.

Since P commutes with the strong time derivative ∂_t, with these notations the Navier-Stokes initial-boundary value problem can be transformated in the evolution equation

(1.2) $$(\partial_t + A)u = -P(u\cdot\nabla u) \;,\; t > o,$$
$$u = u_o, \qquad t = o$$

for the $\mathcal{H}_o$-valued function $u(t) = u(t,\cdot)$.

We inquire Galerkin approximations for u in terms of the eigenfunctions of the linear Stokes boundary value problem

$$A\,v = f$$

for the unknown vector function $v \in \mathcal{H}_1 \cap H_2$ and a given $f \in \mathcal{H}_o$.
Since the selfadjoint operator A is positive definite and possesses a compact inverse on $\mathcal{H}_o$, it has the sequence (λ_i) of eigenvalues λ_i, $o<\lambda_1 \leq \lambda_2 \leq .. \leq \lambda_k \to \infty$ with $k \to \infty$ corresponding to the sequence (e_i) of eigenfunctions $e_i \in \mathcal{H}_1 \cap H_3$ in any fixed order. Thus the equation

(1.3) $\quad A e_i = \lambda_i e_i$

holds, and (e_i), $(e_i\lambda_i^{-1/2})$, $(e_i\lambda_i^{-1})$ represent complete orthonormal systems in $\mathcal{H}_o$, $\mathcal{H}_1$, $\mathcal{H}_1 \cap H_2$ equipped with the norms, which are induced by the inner products $\int_\Omega f\cdot g\, d x$, $\int_\Omega (\nabla f)(\nabla g)dx$ or $\int_\Omega (A f)(Ag)dx$, respectively, c.p.[14] p.46 .
Let $\mathcal{H}_{o,m}$ be the m-dimensional subspace of $\mathcal{H}_o$, which is spanned by the first m eigenfunctions $e_1,\dots,e_m$ of A. By P_m we denote the orthogonal projection of H_o on $\mathcal{H}_{o,m}$, i.e. we have

$$(1.4) \qquad P_m f = \sum_{i=1}^{m} e_i \int_\Omega f e_i \, dx$$

for any $f \in H_o$. From (1.3) we get

$$(1.5) \qquad P_k\, A\, f = A\, P_k f \qquad\qquad \text{for } f \in D_A.$$

We require, that the kth Galerkin-approximation

$$u_k(t) = \sum_{i=1}^{k} a_{ki}(t) e_i$$

of u with the unknown coefficients $a_{ki}(t)$ solves the projection of (1.2) in $\mathcal{H}_{o,m}$. Since P_k also commutes with the strong time derivative and because of $PP_k=P_kP=P_k$, we get for u_k the evolution equation

$$(1.6) \qquad \begin{aligned} (\partial_t+A)u_k &= -P_k(u_k\cdot\nabla u_k), \quad t > o, \\ u_k &= P_k u_o, \qquad\qquad t = o. \end{aligned}$$

In terms of the k functions a_{ki}, (1.6) represents the initial value problem of a system of k ordinary differential equations. Due to the energy inequality, the unique solution $u_k(t)$ exists on $[o,\infty)$, u_k belonging to $C_\infty[o,\infty)$ and (with the e_i) to H_3 for any $t\geq o$,[5, 21].
For short we assume $u_o \in \mathcal{H}_1 \cap H_2$. Then on a (possibly small) time interval $[o,T]$, $T > o$, the unique strong solution $u(t) \in \mathcal{H}_1 \cap H_2$ of (1.2) exists and has the following properties:
The norms $\|A\, u(t)\|$ and the Hölder seminorms

$[A^\alpha u]_\mu = \sup\limits_{\substack{t,s\in[o,T] \\ t \neq s}} \{\|A^\alpha u(t)-A^\alpha u(s)\| |t-s|^{-\mu}\}$ are uniformly bounded, i.e. we have

(1.7) (a) $\|A\, u(t)\| \leq c$, (b) $[A^\alpha u]_\mu \leq c$ with a bound $c > o$ depending only on $|u_o|_{H_2}$, Ω, T, α and μ for all $t \in [o,T]$ and any $o < \alpha + \mu < 1$, $\mu > o$. In addition, for any Galerkin approximation u_k from (1.6), the inequalities

(1.8) (a) $\|A\,u_k(t)\| \le c$, (b) $[A^\alpha u_k(t)]_\mu \le c$ [1)]

hold uniformly with respect to $k = 1,2,\ldots$, $t \in [o,T]$, [5,p.666], [4, p. 3o2]. From the moment inequality for fractional powers of the operator A [3,p. 159] and Cattabriga's estimate [1], using (1.7 a),(1.8 a) we find

(1.9) $\|A^\alpha u\,(t)\| \le c(\alpha)$, $\|A^\alpha u_k(t)\| \le c(\alpha)$ for any $t \in [o,T]$,

$k = 1,2,\ldots$, with a bound $c(\alpha) > o$ depending on $\alpha \in [o,1)$.
In the following from the estimates (1.7)-(1.9) we derive a bound for $\|A^\alpha(u-u_k)\|$ firstly in the case $\alpha < 1$. By $c_o, c_1, \ldots$ we will denote positive constants, the value of which may be different in different sections.

2. Error bounds in D_{A^α}, $\alpha < 1$.

Let u be the local strong Navier-Stokes solution with initial value u_o, u_k its kth Galerkin-approximation, $k = 1,2,\ldots$, and $[o,T]$ any existence interval of u, on which (1.7)-(1.9) hold.
We prove

Theorem 2.1: *Assume* $u_o \in \mathcal{H}_1 \cap H_2$.
Then the error estimate

$$\|A^\alpha u(t) - A^\alpha u_k(t)\| \le c^* \lambda_{k+1}^{-\varepsilon}$$

holds with any $\alpha \in (\frac{3}{4},1)$, $\alpha + \varepsilon < 1$, $o < \varepsilon$, *uniformly on* $[o,T]$. *The sequence* (u_k) *converges to* u *also in the Hölder space* $C_\mu(\Omega)$ *with the error bound* $c_* \lambda_{k+1}^{-\varepsilon}$ *for* $\mu \in (o, 2(\alpha - \frac{3}{4}))$ *uniformly on* $[o,T]$.

The positive numbers c^* and c_* depend only on $|u_o|_{H_2}$, Ω, T, α , μ , ε .
$C_\mu(\Omega)$ denotes the Banach space of the continuous vector functions f on Ω , which have a finite Hölder norm

$$|f|_{C_\mu} = \sup_{x\in\Omega} |f(x)| + \sup_{\substack{x,y\in\Omega \\ x\neq y}} \{|f(x)-f(y)|\,|x-y|^{-\mu}\}.$$

For the proof we write the error of u_k from (1.6) in the form

$$u - u_k = Q_k u + w,$$

where Q_k denotes the orthogonal projection $P-P_k$ in H_o and $w = P_k u - u_k$. Because of

---

1) In order to verify (1.8 b), firstly using Cattabriga's estimate [1] we derive from (1.8 a) a uniform bound for $|u_k(t)|_{H_2}$, which leads, by means of Sobolev's inquality $|u|_{L\infty} \le c_o |u|_{H_2}$ to a uniform bound for $\|P(u_k \cdot \nabla u_k)\|$. Then we apply Lemma 2.11. and 2.12. of [4] on (2.3) below.

(1.5) the operator function A^ε commutes with P_k and Q_k for any $\varepsilon \in (o,1]$. Since A^ε is again a positive definite symmetric operator in $\mathcal{H}_o$, A^ε having the eigenvalues λ_i^ε and the eigenfunctions e_i, $i = 1,2,\ldots$, we get error bounds for the Fourier approximations of a function $f \in D_{A^\varepsilon} \subset \mathcal{H}_o$ in terms of the Stokes eigenfunctions e_i from

Lemma 2.1: [1o, p. 38]:

In a Hilbert space H *with inner product* <.,.> *defining the norm* $|\cdot|_H$, *let* A^* *be a symmetric operator which has the complete orthonormal system of eigenfunctions* (e_i^*) *corresponding to the sequence* (λ_i^*) *of eigenvalues* λ_i, $o < \lambda_1^* \leq \lambda_2^* \leq \ldots \leq \lambda_i^* \to \infty$ *with* $i \to \infty$. *Then the error estimate*

$$|f - \sum_{i=1}^{k} < f, e_i^* > e_i^*|_H \leq \lambda_{k+1}^{*\,-1} |A^* f|_H$$

holds for any $f \in D_{A^*}$.

From this using (1.9) we verify

Corollary 2.1:

The estimate

$$(2.1) \quad \|A^\alpha Q_k u\| = \|Q_k A^\alpha u\| \leq c(\alpha+\varepsilon) \cdot \lambda_{k+1}^{-\varepsilon}$$

holds.

On our condition (1.9), the term $P(u\cdot\nabla u)$ in (1.2) is on $[o,T]$ strongly continuous in $\mathcal{H}_o$, as we see from [4], Lemma 1.2. Therefore u from (1.2) and u_k from (1.6) are the unique solutions of the integral equations

$$(2.2) \quad u(t) = e^{-tA} u_o - \int_o^t e^{-(t-s)A} P(u\cdot\nabla u)(s)ds$$

and

$$(2.3) \quad u_k(t) = e^{-tA} P_k u_o - \int_o^t e^{-(t-s)A} P_k(u_k\cdot\nabla u_k)(s)ds,$$

where e^{-tA} denotes the (for Re $t > o$) holomorphic semigroup which is generated by the positive selfadjoint operator A in $\mathcal{H}_o$.

The evolution equation

$$(\partial_t + A) P_k u = -P_k(u\cdot\nabla u), \; t > o,$$

$$P_k u = P_k u_o, \qquad t = o$$

results from (1.2) by the projection P_k commuting with ∂_t and A. The projection P_k being orthogonal in $\mathcal{H}_o$, the estimates in [4], Lemma 1.2. hold with P_k instead of P. Thus $P_k u$ is the unique solution of the integral equation

$$(2.4) \quad P_k u(t) = e^{-tA} P_k u_o - \int_o^t e^{-(t-s)A} P_k(u\cdot\nabla u)(s)ds.$$

Subtracting (2.3) we get

$$(2.5) \quad w(t) = -\int_o^t e^{-(t-s)A} P_k(u\cdot\nabla u - u_k\cdot\nabla u_k)(s)ds.$$

In order to show $w \in D_{A^\alpha}$ and to find a bound for $\|A^\alpha w\|$, we use [4], Lemma 2.1o.

and 2.12. Recalling that $P_k = P_k P$ represents an orthogonal projection in $\mathcal{H}_o$ and using [4] , Lemma 1.2. again and (2.1), by a short calculation we establish the inequality

(2.6) $$\| P_k (u\cdot\nabla u - u_k \cdot \nabla u_k)\| \le c_1 \cdot (\lambda_{k+1}^{-\varepsilon} + \phi)$$

with the function

$$\phi(t) = \| A^{\frac{1}{2}} w(t)\| + \| A^{\gamma} w(t)\| ,$$

ϕ being strongly continuous on $[o,T]$, $\gamma \in (\frac{3}{4},1)$.

Equation (2.5) for w, (2.6) and the lemmata 2.1o. and 2.12. of [4] just mentioned result in the integral inequality

$$\phi(t) \le c_2 \lambda_{k+1}^{-\varepsilon} + c_2 \int_o^t \frac{\phi(s)}{(t-s)^{\gamma}} ds$$

for any $\gamma \in (\frac{3}{4},1)$, $\gamma + \varepsilon < 1$. Because of the monotonicity of the kernel [27,p.21] , the solution formula [29,p. 152] for the corresponding (weakly singular) Volterra integral equation leads to the bound

(2.7) $$\phi(t) \le c_3 \lambda_{k+1}^{-\varepsilon} .$$

This inequality together with (2.1) verifies the first part of Theorem 2.1. The second part follows from [4] , Lemma 1.2. and 2.6.

References

1 Cattabriga, L., Su un problema al contorno relativo al sistema di equazioni di Stokes, Rend.Mat.Sem.Univ. Padova 31 (1961), 3o8-34o.

2 Foias,C., Statistical study of Navier-Stokes equations I, Rend.Mat.Sem.Univ. Padova 48 (1972 219-348.

3 Friedman,A., Partial differential equations, Holt, Rinehart and Winston, New York 1969.

4 Fujita,H., Kato, T., On the Navier-Stokes initial value problem, I. Arch. Rat. Mech. Anal. 16 (1964), 269-315.

5 Heywood, J.G., The Navier-Stokes equations: On the existence, regularity and decay of solutions, Indiana U. Math. J. 29 (198o),639-681.

6 Heywood,J.G., Classical solutions of the Navier-Stokes equations, Proc. IUTAM Symp. Paderborn (W.-Germany) 1979, Springer Lecture Notes in Math. 771, 198o, 235-248.

7 Heywood, J.G., An error estimate uniform in time for spectral Galerkin approximations of a Navier-Stokes problem, Preprint Univ. of British Columbia, Vancouver 198o.

8 Heywood, J.G., Rannacher,R., Finite element approximation of the nonstationary Navier-Stokes problem I, preprint no.385, Sonderforschungsbereich 72, 198o Universität Bonn.

9 Itô, S., The existence and the uniqueness of regular solution of non-stationary Navier-Stokes equation, J. Fac. Sci. Univ. Tokyo, Sect. IA,9(1961) 1o3-14o.

1o Jörgens, K.,Rellich,F., Eigenwerttheorie gewöhnlicher Differentialgleichungen, Springer Berlin 1976.

11 Kato, T., Perturbation theory for linear operators, 2.ed. Springer Berlin 1976.

12 Kato,T., Fujita,H., On the Non-Stationary Navier-Stokes System, Rend.Sem.Math. Univ. Padova 32 (1962) 243-26o.

13 Kiselev, A.A., Ladyženskaya,O.A., On the existence and uniqueness of the solution of the nonstationary problem for a viscous incompressible fluid, Izv. Akad. Nauk SSR Ser. Mat. 21 (1957) 665-68o.

14 Ladyženskaya,O.A., The mathematical theory of viscous incompressible flow, Second Edition, Gordon and Breach. New York 1969.

15 Ladyženskaya,O.A., Solonnikov,V.A., Ural'ceva,N.N., Linear and quasilinear equations of parabolic type, Amer.Math.Soc.Transl. Providence, R.I. 1968.

16 Masuda, K., On the stability of incompressible viscous fluid motions past objects, J.Math.Soc. Japan 27 (1975) 294-327.

17 Oskolkov,A.P., On the Asymptotic Behaviour of Solutions of Certain Systems with a small Parameter Approximating the System of Navier-Stokes Equations, Proc. Steklov Inst. Math. 125 (1973) 137-153.

18 Prodi,G., Teoremi di tipo locale per il sistema di Navier-Stokes e stabilità delle soluzione stazionarie, Rend.Math.Sem. Univ. Padova 32 (1962)374-397.

19 Rannacher, R., On the finite element approximation of the nonstationary Navier-Stokes problem, Proc. IUTAM-Symp. Paderborn (W.-Germany) 1979, Springer Lecture Notes in Math. 771 (198o 4o8-424.

2o Rautmann, R., Eine Fehlerschranke für Galerkinapproximationen lokaler Navier-Stokes-Lösungen. Proc. Conference Oberwolfach 1978, ISNM 48, Birkhäuser Basel (1979) 11o-125.

21 Rautmann,R., On the convergence-rate of nonstationary Navier-Stokes approximations, Proc. IUTAM Symp., Paderborn (W.-Germany) 1979, Springer Lecture Notes in Math. 771 (198o) 425-449.

22 Shinbrot, M., Lectures on fluid mechanics, New York 1973.

23 Shinbrot, M., Kaniel, S., The initial value problem for the Navier-Stokes equations, Arch. Rational Mech. Anal. 21 (1966) 27o-285.

24 Temam, R., Navier-Stokes Equations, North-Holland Publ. Comp., Amsterdam rev. ed. 1979.

25 Temam, R., Behaviour at Time t = o of the Solutions of Semi-Linear Evolution Equations, MRC Technical Summary Report 2162, Univ. of Wisconsin, Madison 198(

26 Velte, W., Eigenwertschranken mit finiten Differenzen beim Stokesschen Eigenwertproblem, ISNM 43, Birkhäuser Basel (1979).

27 Walter, W., Differential and integral inequalities, Springer Berlin 197o.

28 von Wahl, W., Analytische Abbildungen und semilineare Differentialgleichungen in Banachräumen, preprint 229, Sonderforschungsbereich 72, Universität Bonn 1978.

29 Yosida, K., Lectures on differential and integral equations, Interscience Publ. New York 196o.

Sectorial Second Order Differential Operators

Thomas T. Read

We shall be concerned with some properties of differential operators generated by the nonsymmetric second order expression

$$L[y] = -(py' + sy)' + ry' + qy \tag{1}$$

on the half-line $[0,\infty)$ under hypotheses which guarantee that these operators are sectorial. Here p is a positive function, and r, s, and q are complex-valued. For convenience we shall assume that p and s are piecewise C^1 and that r and q are piecewise continuous.

The representation of a given differential expression L in the form (1) is, of course, not unique. Indeed most of our results are concerned either with the construction of a representation with certain properties, or with the consequences of the existence of one.

Our investigation of (1) is based largely upon the properties of the formally symmetric second order expression which is the real part of L. Thus we begin by considering the expression

$$M[y] = -(py' + \bar{u}y)' + uy' + qy \tag{2}$$

where now both p and q are real valued and $\bar{u}$ is the complex conjugate of u. We have the following characterization of when the minimal operator H_0 generated by M is positive definite. This extends Theorem 2.3 of [9].

THEOREM 1. <u>These two properties are equivalent.</u>

(i) $(H_0f,f) > 0$ <u>for all functions</u> f <u>in the class</u> $C_0^\infty(0,\infty)$ <u>of</u> C^∞ <u>functions with compact support in the interior of</u> $(0,\infty)$.

(ii) M <u>has a representation</u> $M[y] = -(py' + \bar{u}_1y)' + u_1y' + q_1y$ <u>on</u> $(0,\infty)$ <u>where</u>

$$q_1 \geq |u_1|^2/p. \tag{3}$$

<u>Proof</u>. It follows easily from an integration by parts that if M has a representation for which (3) holds, then $(H_0f,f) > 0$ for all f in C_0^∞. For the other direction, if $M[y] = 0$ and $z = py' + \bar{u}y$, then y and z satisfy the system

$$y' = -(\bar{u}/p)y + (1/p)z$$
$$z' = (q - |\bar{u}|^2/p)y + (u/p)z.$$

It follows from (i) and Theorem 14, p. 61 of Coppel [3] that the system has a solution for which $y > 0$ on $(0,\infty)$. Then $w = z/y$ satisfies the Riccati equation

$$w' - (2\mathrm{Re}\ u/p)w + w^2/p = q - |u|^2/p$$

on $(0,\infty)$ and if we set $u_1 = u - w$, $q_1 = |u - w|^2/p$ it is easy to see that $q_1 = |u_1|^2/p$.

Remark. It is evident from the proof that the result is still true with the inequality in (3) replaced by equality. The formulation above is far more useful, however, both because it is far easier to construct representations in specific cases satisfying (3) and because, as we shall see below, representations where (3) is replaced by the stronger condition $q_1 \geq (1 + \varepsilon)|u_1|^2/p$ are very easy to work with.

The representation (2) of M may be rewritten as

$$M = N^+N + P$$

where

$$N[y] = \sqrt{p}y' + (\bar{u}/\sqrt{p})y \ ; \ P[y] = (q - |u|^2/p)y,$$

and N^+ is the formal adjoint of N. Note that P involves multiplication by a non-negative function precisely when the representation of M is such that (3) holds.

We shall wish to characterize explicitly the domains of m-sectorial operators generated by L. We do this by applying to the real part of L the following variant of the Dirichlet index theory developed by Kauffman [7] and Bradley, Hinton, and Kauffman [2].

Notation. Suppose that M has a representation satisfying (3). By adding a constant if necessary we may assume $q - |u|^2/p \geq 1$. We then set

$$\mathcal{D} = \{f: f\varepsilon\ AC^{loc} \text{ and } \int_0^\infty(|N[y]|^2 + \bar{y}P[y]) < \infty\}.$$

DEFINITION. If M, N, and P are as immediately above, we say M is NP-minimal if the equation $M[y] = 0$ has exactly one solution which lies in $\mathcal{D}$.

For a given M, the property of being NP-minimal depends in general upon the representation of M being considered, that is upon the choice of N and P. This will be illustrated in Example 1.

When M is NP-minimal, the explicit characterizations of the domains of various operators defined as Friedrichs extensions and of their square roots can be established just as in [2] or [7]. In particular we have

THEOREM 2. <u>The expression</u> M <u>is</u> <u>NP-minimal</u> <u>if and only if the Friedrichs extension</u> H_0 <u>of the minimal operator generated by</u> M <u>has domain</u>

$$\text{domain } H_0 = \{f \in \text{dom } M_{max} : f(0) = 0 \text{ and } f \in \mathcal{D}\}.$$

Moreover the domain of $H_0^{1/2}$ is the set of all elements f in $\mathcal{D}$ such that $f(0) = 0$.

<u>Example 1</u>. Let $M[y] = -(x^4y' - ix^3y)' + ix^3y' - x^2y$, $x \geq 1$. Then for any real constant c we have also

$$M[y] = -(x^4y' + (c-i)x^3y)' + (c+i)x^3y' + (3c-1)x^2y.$$

With $q_c(x) = (3c - 1)x^2$, $u_c(x) = (c+i)x^3$, we have

$$q_c - |u_c|^2/p = (2 - c)(c - 1)x^2.$$

Thus $q_c - |u_c|^2/p \geq 0$ if and only if $1 \leq c \leq 2$ with equality for $c = 1$ and $c = 2$ and $q_c - |u_c|^2/p > 0$ otherwise. The corresponding expressions N_c and P_c are

$$N_c[y] = x^2y' + (c-i)xy; \quad P_c[y] = (2-c)(c-1)x^2y.$$

The solutions of $M[y] = 0$ are x^{-1+i} and x^{-2+i}. It is easy to verify that x^{-2+i} is in the space $\mathcal{D}_c$ for each c, $1 \leq c \leq 2$, while x^{-1+i} is in $\mathcal{D}_c$ only for $c = 1$. Thus M is N_cP_c-minimal for $1 < c \leq 2$. The exceptional case $c = 1$ corresponds to the factorization $M = N_1^+N_1$ where $N_1[x^{-1+i}] = 0$.

We next develop the criterion for NP-minimality that is suitable for our purposes. Niessen [8] has given a characterization of when all solutions of an equation are Dirichlet which when specialized to our situation becomes the following result. (See also Bennewitz [1] for a very simple proof.)

THEOREM 3. (Niessen) *M is NP-minimal if and only if* $p^{-1}\exp(2\mathrm{Re}\int u/p)$ *and* $(q - |u|^2/p)\exp(-2\mathrm{Re}\int u/p)$ *are not both in* $L^1(0,\infty)$.

From this we can obtain the result we want.

THEOREM 4. *If* $q \geq (1 + \varepsilon)|u|^2/p$ *for some* $\varepsilon > 0$, *then* M *is NP-minimal. Moreover, in this case*

$$\mathcal{D} = \{f\varepsilon AC^{loc}: \int_0^\infty (p|f'|^2 + q|f|^2) < \infty.\}$$

Proof. We recall that we have assumed just before the definition of NP-minimality that q is bounded away from 0. In particular, $q \notin L^1(0,\infty)$. Now suppose that both $p^{-1}\exp(2\mathrm{Re}\int u/p)$ and $(q-|u|^2/p)\exp(-2\mathrm{Re}\int u/p)$ lie in $L^1(0,\infty)$. Then $q - |u|^2/p \geq \varepsilon|u|^2/p$ implies that also $(|u|^2/p)\exp(-2\mathrm{Re}\int u/p) \; \varepsilon \; L^1(0,\infty)$. Then by the Schwarz inequality $u/p \; \varepsilon \; L^1(0,\infty)$, since

$$\left(\int_0^\infty |u/p|\right)^2 \leq \int_0^\infty p^{-1} \exp(2\mathrm{Re}\int u/p) \int_0^\infty (|u|^2/p) \exp(-2\mathrm{Re}\int u/p) < \infty.$$

It follows from this that $\exp(-2\mathrm{Re}\int u/p)$ has a finite limit as $x \to \infty$ and thus that $q - |u|^2/p \; \varepsilon \; L^1(0,\infty)$. But $q - |u|^2/p \geq (\varepsilon/(1+\varepsilon))q$ so this is impossible. Thus M must be NP-minimal.

The last assertion follows from the inequality $q - |u|^2/p \geq (\varepsilon/(1+\varepsilon))q \geq \varepsilon|u|^2/p$ since then f in $\mathcal{D}$ implies that $\int_0^\infty q|f|^2$ is finite and also, since

$$\int_0^\infty p|f'|^2 \leq 2\left(\int_0^\infty |N[f]|^2 + \int_0^\infty (|u|^2/p)|f|^2\right),$$

that $\int^\infty p|f'|^2$ is finite.

Remark. In practice, whenever a representation for M can be constructed with $q \geq |u|^2/p$ it is generally true that in fact $q \geq (1+\varepsilon)|u|^2/p$. In Example 1, for instance, this inequality is valid for $1 < c < 2$.

We turn now to the nonsymmetric expression L defined in (1). As stated above, we study this expression by means of its real part. See Kato [6, p. 310] for the general definition of the real part of a form. In our context this becomes the following.

DEFINITION. The *real part*, L_R, of L is the symmetric expression

$$L_R[y] = -(py' + \bar{u}y)' + uy' + (\mathrm{Re}q)y$$

where $u = (r + \bar{s})/2$.

It is easy to verify that for f in $C_0^\infty(0,\infty)$, $(L_R[f],f) = \mathrm{Re}(L[f],f)$.

We recall that if the minimal operator T_0 generated by L is sectorial, then it is possible to adapt the construction of the Friedrichs extension to define an m-sectorial extension T of T_0. The domain of T is a dense subset of the domain of the closure of the form $h_0[f,g] = (H_0 f,g)$ where f and g are in $C_0^\infty(0,\infty)$ and H_0 is the minimal operator generated by the real part of L. See Kato [6, p. 322] for details.

We shall give two main results on the numerical range of m-sectorial operators generated by an expression of the form (1). The second of these (Theorem 6) is in the spirit of the generalization of Molchanov's theorem on discreteness of the essential spectrum given in [9, Theorem 4.1]. The proof consists essentially of constructing a representation of L so that the hypothesis of Theorem 4 holds for the real part of L. In the first result we shall assume that such a representation has already been found.

THEOREM 5. *Let L be as in (1). Suppose that there are positive constants* B, ε, k_1 *and* k_2 *such that*

(a) $\mathrm{Re}\, q + B \geq (1+\varepsilon)|r + \bar{s}|^2/4p$,

(b) $|r - \bar{s}|^2/4p \leq k_1(\mathrm{Re}\, q + B)$,

(c) $|\mathrm{Im}\, q| \leq k_2(\mathrm{Re}\, q + B)$.

Then the minimal operator T_0 *is sectorial and the m-sectorial extension* T *of* T_0 *has the following properties*.

(i) domain $T = \{f \in \text{domain } L_{max} : f(0) = 0$ and

$$\int_0^\infty (p|f'|^2 + (\mathrm{Re}\, q + B)|f|^2) < \infty\}.$$

(ii) $\mathrm{Re}(Tf,f) \geq (\varepsilon \inf \mathrm{Re}\, q - B)/(1+\varepsilon)$.

(iii) *The numerical range of* T *is contained in the sector*

$$|y| \leq ([k_1(1+\varepsilon)/\varepsilon]^{1/2} + k_2(1+\varepsilon)/\varepsilon)(x + B).$$

Here the inf in (ii) is to be taken over all x in $[0,\infty)$.

Remark. This theorem can be converted into a result about the essential spectrum of T simply by replacing inf by lim inf. In particular the essential spectrum is empty if $\mathrm{Re}\, q \to \infty$.

Example 2. Let $L[y] = -y'' + (x^2 + (1+i)k\, x^\alpha \sin x^\beta)y$ where $\alpha < \beta$ or $\alpha = \beta > k$. We can obtain a suitable representation for L by letting $q(x) = x^2$, $r(x) = s(x) = F(x)$ where F is an antiderivative of $(1 + i)k\, x^\alpha \sin x^\beta$ such that $F(x) = O(x^{\alpha-\beta+1})$. Here $q(x) = x^2 \to \infty$ does imply that the essential spectrum is empty.

THEOREM 6. Let $L[y] = -y'' + qy$. Suppose that, for some constants $a > 0$ and $A > 0$, and some bounded region D containing the origin,

(a) for each $x > 0$ there is z, $x < z \le x + a$ such that

$$(1/a)\mathrm{Re} \int_x^z q \ge A,$$

and

(b) $(1/a)\int_x^w q \in D$ whenever $x \le w \le x + a$.

Then the minimal operator T_0 is sectorial, and the m-sectional extension T of T_0 has the following properties.

(i) domain $T = \{f \in \mathrm{dom}\, L_{max}: f(0) = 0$ and $\int_0^\infty (|f'|^2 + q_1|f|^2) < \infty\}$.

(ii) $T \ge A - (Aa + M_1)^2/4 = x_0$.

(iii) The numerical range of T is contained in the set

$$\{x + iy: x \ge x_0,$$

$$|y| \le \begin{cases} M_2(M_1+2A) + 2M_2(x-x_1)^{1/2}, & x - x_1 \le (M_2/M)^2, \\ M_2(M_1+2Aa) + M_2^2/M + M(x-x_1), & x - x_1 > (M_2/M)^2.\} \end{cases}$$

Here q_1 is a mean value of Req so that in particular $q_1 \ge A$, and the constants M, M_1, M_2, x_1 of (ii) and (iii) are defined as follows.

$$\tan M = \sup\{\arg z: z \in D, \mathrm{Re} z = Aa\}$$

$$-M_1 = \sup\{a\mathrm{Re}\, z: z \in D\},$$

$$M_2 = \sup\{a\mathrm{Im}\, z: z \in D\},$$

$$x_1 = Aa + M_1/2.$$

Remark. In general it is not possible to restrict the numerical range of T to a sector with semi-angle smaller than $\theta = \arctan M$, for the hypotheses allow $q(x) = e^{i\theta}f(x)$ where f is any real valued function such that $f \ge A/\cos\theta$.

Remark. Hypothesis (b) may be regarded as a Brinck condition for a complex valued function.

COROLLARY 2. *If* $M_2(x) \to 0$ *as* $x \to \infty$ *where* $M_2(x)$ *is defined as* M_2 *above for the restriction of* L *to* $[x,\infty)$ *of the real line, then the essential spectrum of* T *is contained in the subset* $[x_0,\infty)$ *of the real line. In particular this is true if* $\int_0^\infty q$ *exists as an improper integral with* $-x_0 = \sup_{a,b}[\mathrm{Re}\int_a^b q]^2/4$.

As a corollary of the proof we also obtain

COROLLARY 3. *If* $\int_0^\infty |\mathrm{Re} q| = k_1$, $\int_0^\infty |\mathrm{Im} q| = k_2$, *then the numerical range of* T *is contained in the set*

$$\{x+iy : x \geq -k_1^2/4 = x_0;\ |y| \leq k_1k_2/2 + k_2(x-x_0)^{1/2}\}$$

This result was obtained by Evans [5] with k_1 and k_2 replaced by $2k_1$ and $2k_2$ respectively.

References

1. C. Bennewitz, A generalization of Niessen's limit-circle criterion, Proc. Roy. Soc. Edinburgh 78A(1977), 81-90.
2. J. S. Bradley, D. B. Hinton and R. M. Kauffman, On the minimization of singular quadratic functionals, Proc. Roy. Soc. Edinburgh 87A(1981), 193-208.
3. W. A. Coppel, Disconjugacy (Lecture Notes in Mathematics No. 220, Springer-Verlag, Berlin, 1971).
4. M. S. P. Eastham, Semi-bounded second order differential operators, Proc. Roy. Soc. Edinburgh 72A(1973), 9-16.
5. W. O. Evans, On the spectra of Schrödinger operators with a complex potential, Math. Ann. 255(1981), 57-76.
6. T. Kato, Perturbation Theory for Linear Operators, Second edition (Springer-Verlag, Berlin, 1980).
7. R. M. Kauffman, The number of Dirichlet solutions to a class of linear ordinary differential equations, J. Differential Equations 31(1979), 117-129.
8. H. D. Niessen, A necessary and sufficient limit-circle criterion for left-definite eigenvalue problems, Ordinary and Partial Differential Equations, Lecture Notes in Mathematics 415 (Berlin: Springer-Verlag, 1974).
9. T. T. Read, Factorization and discrete spectra for second-order differntial expressions, J. Differential Equations 35(1980), 388-406.

Spectral Mapping Theorems for Dissipative C_0-Semigroup Generators

Harald Röh

1. Formulation of the problem

Let $G : D(G) \subset H \to H$ be the generator of a C_0-semigroup $T(t) = \exp(Gt)$, $0 \leq t < \infty$, on a complex separable Hilbert space H with inner product $(\cdot,\cdot)$. In order to describe the asymptotic behaviour of solutions $x(t) = T(t)x_0$ of the abstract Cauchy problem

$$\dot{x}(t) = G\,x(t), \qquad x(0) = x_0 \in D(G)$$

we determine conditions for G which guarantee that the spectral mapping theorem

$$\sigma(T(t)) = \overline{\exp(\sigma(G)t)} \tag{1}$$

holds for $t \geq 0$. If G is a normal, self-adjoint or bounded operator or if $T(t)$ is uniformly continuous (e.g. differentiable or analytic), then eq.(1) holds ([5], [3]). However, eq.(1) does not hold in general ([11]). In this paper we consider the case where G is a maximal dissipative operator, i.e. $(G-I)D(G) = H$ and $\operatorname{Re}(Gx, x) \leq 0$ for $x \in D(G)$.

2. Abstract results

Assume that G is a densely defined maximal dissipative operator on the Hilbert space H and satisfies

$$\sigma(G) \subset \{ z \in \mathbb{C} \mid \operatorname{Re} z < 0 \} \tag{2}$$

$$\dim D(G)/D_0(G) = n_0 < \infty \tag{3}$$

where $D_0(G) = \{ x \in D(G) \mid \operatorname{Re}(Gx, x) = 0 \}$. It is straight forward to prove that G^{-1} is a compact operator on H, i.e. G has a compact resolvent. Hence the spectrum $\sigma(G)$ of G consists of at most countable many isolated eigenvalues λ_n with finite algebraic multiplicities ν_n. Assume there exists $\delta > 0$ such that

$$\prod_{\substack{n \\ n\neq k}} \frac{|\lambda_k - \lambda_n|}{|\lambda_k + \overline{\lambda_n}|} \geqslant \delta > 0 \qquad \text{for all } k \tag{4}$$

$$\nu_n \leqslant \nu < \infty \qquad \text{for all } n \tag{5}$$

<u>Theorem 1:</u> Let G be a densely defined maximal dissipative operator on the Hilbert space H satisfying (2) - (5). Then

(a) the generalized eigenspaces of G form an unconditional subspace basis of their closed linear hull.

(b) the generalized eigenspaces of G span the whole Hilbert space H if and only if there exists a sequence $\{z_n\}$ of points in the resolvent set $\rho(G)$ of G such that $\{|\mathrm{Re}\, z_n| \, \|R(z_n,G)\|\}$ is uniformly bounded and $\mathrm{Re}\, z_n \to -\infty$ as $n \to \infty$.

<u>Remark 1:</u> Subspaces $\{H_n\}$ of the Hilbert space H form an (unconditional) subspace basis of H if each vector $x \in H$ has a unique expansion $x = \Sigma\, x_n$ with $x_n \in H_n$, where the series converges (unconditionally) in H.

<u>Remark 2:</u> Elementary calculations show that condition (4) is satisfied if the eigenvalues λ_n lie in a bounded strip parallel to the imaginary axis and are uniformly separated, i.e. there exists some $\varepsilon > 0$ such that $|\lambda_n - \lambda_m| \geqslant \varepsilon$ for all $n \neq m$ ([9]).

<u>Remark 3:</u> Theorem 1 remains valid if instead of (3) we assume that G has a nuclear resolvent. Note that in this case the generalized eigenspaces of G automatically span the whole space H, because G^{-1} is a dissipative nuclear operator ([7]).

<u>Theorem 2:</u> Let $\{H_n\}$ be an unconditional subspace basis of H with $\dim H_n \leqslant n_1 < \infty$ for all n. Let G be a closed operator on H

such that $H_n \subset D(G)$ for all n. Define $G_n := G \mid H_n$ (restriction). If G generates a C_0-semigroup $T(t) = \exp(Gt)$ on H, then

$$\sigma(T(t)) = \overline{\exp(\sigma(G)t)} = \overline{\bigcup_n \exp(\sigma(G_n)t)}$$

An immediate consequence of Theorem 1 and 2 is

Corollary 1: Let G be as in Theorem 1. If the generalized eigenspaces of G span the whole Hilbert space H, then eq.(1) holds.

Remark 4: If the eigenspaces of G do not span the whole space H, then only the point zero must be added to the right side of eq.(1) ([9]).

Remark 5: If the spectrum of G in Corollary 1 is bounded to the left, i.e. $\operatorname{Re} \lambda_n \geqslant \eta > -\infty$ for all n, then (1) implies $0 \in \rho(T(t))$ for all $t \geqslant 0$. Thus in this case G actually generates a C_0-group $T(t)$, $t \in \mathbb{R}$, on H.

Theorem 3: Let G be a densely defined operator on H with nuclear resolvent. Assume that the eigenvalues of G are uniformly separated and that their algebraic multiplicities are uniformly bounded. If G generates a C_0-group $T(t)$, $t \in \mathbb{R}$, on H, then the generalized eigenspaces of G form an unconditional subspace basis of H and eq.(1) holds for all $t \in \mathbb{R}$.

3. Proofs of the Theorems

Theorem 1 is proved in [9], so we only outline the idea of the proof. First we define the Cayley transformation $T := (G+I)(G-I)^{-1}$. The operator T is a contraction on H and has the same generalized eigenspaces as G with respect to its eigenvalues $a_n = (\lambda_n+1)(\lambda_n-1)^{-1}$. In the terminology of [10], condition (2) implies that T is of class C_{00} and (3) implies that the defect indices of T and T^*

both equal $n_0 < \infty$, hence T is even of class C_0. Thus there exists a minimal (inner) function $m_T(\lambda) \in H^\infty$ (bounded analytic functions on the open unit disk) satisfying $m_T(T) = 0$. Because of (2), $m_T(\lambda)$ has the form

$$m_T(\lambda) = a \prod_n \left(\frac{\overline{a_n}}{|a_n|} \frac{a_n - \lambda}{1 - \overline{a_n} \lambda} \right)^{\nu_n} e^{c\frac{\lambda+1}{\lambda-1}}, \quad |a| = 1, \quad c \geqslant 0 \qquad (6)$$

Moreover, the generalized eigenspace of T corresponding to the eigenvalue a_n coincides with $H_n := \{ x \in H \mid m_n(T)x = 0\}$, where $m_n(\lambda)$ is the n-th factor of the Blaschke product appearing in (6). Using condition (4) (which states that $\{a_n\}$ is a H^∞-interpolating sequence), (5) and the corona theorem ([2]), one proves that for any two disjoint subsets σ_1, σ_2 of $\mathbb{N}$ the algebraic sum of H_{σ_1} and H_{σ_2} is direct. Here H_{σ_i} denotes the closed linear hull of all subspaces H_n with $n \in \sigma_i$. It is easily seen that this proves part (a) of Theorem 1. As to part (b) of the theorem, we only remark that the condition stated there is equivalent to the fact $c = 0$ in (6), which characterizes the completeness of the eigenspaces of T.

If G has a nuclear resolvent, then its Cayley transformation T is a weak contraction ([10]), because

$$I - T^*T = -2(G-I)^{-1} - 2(G^*-I)^{-1} - 4(G^*-I)^{-1}(G-I)^{-1}$$

i.e. $I - T^*T$ is a nuclear operator. As by (2) T is of class C_{00}, we again conclude that T is of class C_0. Hence we can as above prove that the eigenspaces of T form an unconditional subspace basis of H. This justifies Remark 3.

In order to prove Theorem 2 note that it is sufficient to prove $\sigma(T(t)) \subset \overline{\cup\, \sigma(\exp(G_n t))}$, because $\sigma(\exp(G_n t)) = \exp(\sigma(G_n)t) \subset \subset \exp(\sigma(G)t)$ and $\exp(\sigma(G)t) \subset \sigma(T(t))$ automatically hold. First recall that if M is any invertible $k \times k$ matrix, then

$$\|M^{-1}\| \leq \gamma_k \|M\|^{k-1} \, |\det M|^{-1} \tag{7}$$

where γ_k is independent of M ([6], chapter I, eq.(4.12)). Next define $T_n(t) := T(t)\,|\,H_n$. Clearly, $T_n(t) = \exp(G_n t)$ and $\|T_n(t)\|$ is uniformly bounded in n. Assume $\lambda \notin \overline{\cup\, \sigma(\exp(G_n t))}$. Regarding $T_n(t)$ as a matrix operator, ineq.(7) implies that $\|(\lambda I - T_n(t))^{-1}\|$ is uniformly bounded in n for fixed t,λ. Hence $\lambda \notin \sigma(T(t))$ and Theorem 2 is proved.

To prove Theorem 3 we assume without loss of generality that G generates a C_0-group $T(t)$ such that $\|T(t)\|$ decays exponentially as $t \to \infty$. Then there exists a new inner product $(\cdot,\cdot)_0$ on H, equivalent to the original one, with respect to which G becomes dissipative ([8]). It is given by

$$(x, y)_0 := \int_0^\infty (T(t)x, T(t)y)\, dt$$

Theorem 3 now easily follows from Corollary 1, observing Remark 2 and 3.

4. Applications

Consider the following parameter dependent damped wave equation

$$\begin{aligned} &u_{tt} + \gamma u_{tx} + u_{xxxx} + \beta u_{xx} = 0, \quad 0 \leq t < \infty, \quad 0 < x < 1 \\ &u(0,t) = u_x(0,t) = u_{xx}(1,t) = u_{xxx}(1,t) = 0 \end{aligned} \tag{8}$$

with real parameters $\beta,\gamma \geq 0$ ([1]). We rewrite (8) in operator form on the Hilbert space $L^2(0,1)$ with inner product $< f, g > = \int f(t)\overline{g(t)}\, dt$ as

$$u_{tt} - \gamma C u_t + (A + \beta B)\, u = 0$$

$$Af = f'''', \quad D(A) = \{ f \in H^4 \mid f(0) = f'(0) = f''(1) = f'''(1) = 0 \}$$

$$Bf = f'' \ , \quad D(B) = \{ f \in H^2 \mid f(0) = f'(0) = 0 \}$$

$$Cf = -f' \ , \quad D(C) = \{ f \in H^1 \mid f(0) = 0 \}$$

(where H^k denotes the Sobolev space $H^k(0,1)$) and in system form on the Hilbert space $H = D(A^{\frac{1}{2}}) \times L^2(0,1)$ as

$$\begin{pmatrix} u \\ u_t \end{pmatrix}_t = G \begin{pmatrix} u \\ u_t \end{pmatrix} , \qquad G = G_0 + \gamma G_1 + \beta G_2$$

$$G_0 = \begin{pmatrix} 0 & I \\ -A & 0 \end{pmatrix}, \quad G_1 = \begin{pmatrix} 0 & 0 \\ 0 & C \end{pmatrix}, \quad G_2 = \begin{pmatrix} 0 & 0 \\ -B & 0 \end{pmatrix}$$

with $D(G) = D(G_0) = D(A) \times D(A^{\frac{1}{2}})$. Note that $A = A^* > 0$ is a positive self-adjoint operator on $L^2(0,1)$ with compact resolvent. The inner product on H is given by

$$\left(\begin{pmatrix} f \\ g \end{pmatrix} , \begin{pmatrix} u \\ v \end{pmatrix} \right) = \langle A^{\frac{1}{2}}f , A^{\frac{1}{2}}u \rangle + \langle g , v \rangle$$

where $\langle A^{\frac{1}{2}}f , A^{\frac{1}{2}}u \rangle = \langle f'' , u'' \rangle$ and $D(A^{\frac{1}{2}}) = D(B)$. Using the compactness of A^{-1}, it is straight forward to prove that $G = G(\beta,\gamma)$ has a compact resolvent for all $\beta,\gamma \geq 0$. An eigenvalue analysis of G ([9]) shows that the eigenvalues of G lie symmetric to the real axis, are simple if their modulus is large enough, and, in the upper complex half plane, admit an asymptotic representation

$$\lambda_n = -\gamma + \frac{1}{16}\gamma^2 i - \frac{1}{2}\beta i + i\pi^2 (n-\tfrac{1}{2})^2 + O(n^{-1}) \tag{9}$$

It is easy to check that the dissipative operator $G(0,\gamma) = G_0 + \gamma G_1$ satisfies conditions (2) - (5) , in particular (3) holds with $n_0 = 1$ because

$$\operatorname{Re} \left(G(0,\gamma)\begin{pmatrix} u \\ v \end{pmatrix} , \begin{pmatrix} u \\ v \end{pmatrix} \right) = -\frac{\gamma}{2} |v(1)|^2 \qquad \text{for } \begin{pmatrix} u \\ v \end{pmatrix} \in D(G).$$

By Theorem 1, the eigenspaces of $G(0,\gamma)$ form an unconditional subspace basis of their closed linear hull. The completeness of the eigenspaces of $G(0,\gamma)$ in H follows by simple perturbation arguments, either by directly verifying the completeness condition in Theorem 1 (b) or by noting that $G(0,\gamma)$ has in fact a nuclear resolvent. Hence the spectral mapping theorem (1) holds for $G = G(0,\gamma)$. Moreover, (9) implies that $G(0,\gamma)$ actually generates a C_0-group on H (Remark 5). As $G = G_0 + \gamma G_1 + \beta G_2$ is a bounded perturbation of $G(0,\gamma)$, we conclude that G also generates a C_0-group $T(t)$, $t \in \mathbb{R}$, on H and that G has a nuclear resolvent. Thus Theorem 3 can be applied to G and proves eq. (1). Finally we note that G is a discrete spectral operator ([4]).

REFERENCES

1. A.K. Bajaj, P.R. Sethna and T.S. Lundgren, Hopf bifurcation phenomena in tubes carrying a fluid, SIAM J. Appl. Math. 39 (1980), 213-230

2. L. Carleson, Interpolation by bounded analytic functions and the corona problem, Ann. of Math. (2) 76 (1962), 547-559

3. E.B. Davies, One-parameter semigroups, Academic Press, London 1980

4. N. Dunford and J.T. Schwartz, Linear operators, Part III: Spectral operators, Pure and applied Mathematics, Vol. VII, Wiley-Interscience, New York, 1971

5. E. Hille and R.S Phillips, Functional analysis and semigroups, AMS Collected publications no. 31, Providence, R.I. 1957

6. T. Kato, Perturbation theory for linear operators, second edition, Springer Verlag, Berlin 1976

7. V.B. Lidskii, Nonselfadjoint operators with a trace, Amer. Math. Soc. Transl. (2) 47 (1965), 43-46

8. A. Pazy, On the applicability of Liapunov's theorem in Hilbert space, SIAM J. Math. Anal. 3 (1972), 291-294

9. H. Röh, Dissipative operators with finite dimensional damping, Proc. Roy. Soc. Edinburgh Sect. A 91 (1982), in print

10. B. Sz.-Nagy and C. Foias, Harmonic analysis of operators on Hilbert space, North-Holland, Amsterdam 1970

11. J. Zabczyk, A note on C_0-semigroups, Bull. Acad. Polon. Sci. Ser. Sci. Tech. 23 (1975), 895-899

Department of Mathematics
Heriot-Watt University
Riccarton, Currie
Edinburgh EH14 4AS / Scotland

Some Preliminary Results on Periodic Solutions of Matrix Riccati Equations

David A. Sánchez

The question to be considered is: given the autonomous matrix Riccati equation

$$\dot{\underline{\bar{X}}} = B\underline{\bar{X}} + \underline{\bar{X}}B^T + \underline{\bar{X}}\,C\,\underline{\bar{X}}, \tag{1}$$

where B and C, are nxn real constant matrices and C is symmetric, together with initial conditions

$$\underline{\bar{X}}(o) = Q, \text{ real, symmetric,} \tag{2}$$

under what conditions does it have periodic solutions? Also, what is the structure of its constant solutions? These questions are of interest in filtering theory as well as the control of periodic systems - for instance knowing the constant solutions one can ask whether periodic solutions will bifurcate from them if (1) is subjected to a periodic forcing term.

Using a formula previously developed by the author in [3], and derived from earlier results by R. Redheffer [1] and W.T. Reid [2], one sees that a pair (ω,Q), $\omega > o$, Q real and symmetric, for which the solution $\underline{\bar{X}}(t)$ of (1) satisfies

$$\underline{\bar{X}}(o) = \underline{\bar{X}}(\omega) = Q,$$

will satisfy the relation

$$e^{\omega B}Q - Qe^{-\omega B^T} + Q(e^{-\omega B^T}\int_0^{\omega} e^{tB^T}Ce^{tB}dt)Q = 0.$$

As it stands the last formula is fairly useless, even expanding the matrix exponential in series, so one makes some assumptions about B and C so as to move the matrix C out of the integral sign. Then the case n = 2 is examined in detail since algebraic equations are formidable but not impossible.

Case 1: B symmetric

If B is symmetric and $\overline{X}(t)$ is a solution (1) defined for $o \leq t \leq \omega$ and satisfying $\overline{X}(o) = \overline{X}(\omega) = Q$, then ω and Q satisfy

$$e^{\omega B}Q - Qe^{-\omega B} + Q(\int_0^{\omega} \cosh tBdt)CQ = 0$$

if BC = CB and

$$e^{\omega B}Q - Qe^{-\omega B} + \omega Qe^{-\omega B}CQ = 0$$

if BC = -CB.

Using these relations one can show for the case n = 2 that there are no periodic solutions.

Case 2: B skew symmetric

If B is symmetric and $\overline{X}(t)$ is a solution of (1) defined for $o \leq t \leq \omega$ and satisfying $\overline{X}(o) = \overline{X}(\omega) = Q$, then ω and Q satisfy

$$e^{\omega B}Q - Qe^{-\omega B} + \omega Qe^{\omega B}CQ = 0$$

if BC = CB, and

$$e^{\omega B}Q - Qe^{\omega B} + Q(\int_0^{\omega} \cosh tBdt)CQ = 0$$

if BC = -CB.

Using these relations for the case n = 2 when B has eigenvalues $\pm ia$, $a \neq o$, these formulas become:

$$\frac{\sin a\omega}{a} [BQ - QB + \omega cQBQ] + \cos a\omega[\omega cQ^2] = 0$$

if BC = CB, and where C = cI, and

$$\frac{\sin a\omega}{a} [BQ - QB + QCQ] = 0$$

if BC = -CB.

For the case BC = CB a straightforward analysis shows that there are no periodic solutions, whereas for the case BC = -CB a possible period is $\omega = \pi/a$. The condition that the solutions exist for $o \leq t \leq \pi/a$ imposes a restriction on the choice of initial values

$$Q = \begin{pmatrix} x & y \\ y & z \end{pmatrix} .$$

The result is:

If B is skew symmetric, n = 2, with eigenvalue $\pm ia \neq o$, and BC = -CB then periodic solutions (possibly constant) of period $\omega = \pi/a$ will exist whenever

$$1 - \frac{1}{2a}(c_1x + 2c_2y - c_1z)\sin 2at$$

$$+ \frac{1}{a}[c_2x - 2c_1y - c_2z + \frac{1}{a}(c_1^2 + c_2^2)(y^2 - xz)]\sin^2 at \neq o,$$

with

$$C = \begin{pmatrix} c_1 & c_2 \\ c_2 & -c_1 \end{pmatrix} .$$

Example:

$$B = \begin{pmatrix} o & 2 \\ -2 & o \end{pmatrix}, \qquad C = \begin{pmatrix} o & -1 \\ -1 & o \end{pmatrix} .$$

The only constant solutions will be the one parameter family

$$Q = \begin{pmatrix} x & o \\ o & \frac{2x}{2-x} \end{pmatrix}, \qquad x \neq 2$$

and periodic solutions will exist whenever x,y,z satisfy

$$1 + \frac{y}{2}\sin 4t + \left\{ -\frac{x}{2} + \frac{z}{2} + \frac{1}{4}y^2 - \frac{1}{4}xz \right\}\sin^2 2t \neq o.$$

All other solutions will be unbounded. This example was only cursorily examined by T. Sasegawa in [4].

Case 3: $B = \begin{pmatrix} p & a \\ -a & p \end{pmatrix}$, $a,p \neq o$

The solution of (1) with initial conditions (2) is given by the expression

$$\underline{X}(t) = A(t)Q[e^{-2pt}I - e^{-2pt}(\int_o^t e^{sB^T}Ce^{sB}ds)Q]^{-1}A(t)^T$$

where

$$A(t) = \begin{pmatrix} \cos at & \sin at \\ -\sin at & \cos at \end{pmatrix} .$$

Now one computes the expression

$$\int_0^t e^{sB^T} C e^{sB} ds = e^{2pt} F(B,C,\sin 2at,\cos 2at) - F(B,C,o,1)$$

and one concludes that the only possible period is $\omega = \pi/a$, and a unique periodic solution exists if and only if

$$F(B,C,\sin 2at,\cos 2at)^{-1}$$

exists for $o \leq t \leq \pi/a$, in which case

$$\overline{X}(o) = Q = -F(B,C,o,1)^{-1} .$$

However, a more detailed analysis shows that such a solution satisfies $\dot{\overline{X}}(t) \equiv$ and hence is a constant. This last assertion is proved by showing that it is a necessary and sufficient consequence of the relation

$$\dot{F}(t) = F(t)A - AF(t), \quad A = \begin{pmatrix} o & a \\ -a & o \end{pmatrix}$$

and then doing some careful bookeeping. We conclude that:

If $B = \begin{pmatrix} p & a \\ -a & p \end{pmatrix}$, $a,p \neq o$, $C = \begin{pmatrix} c_1 & c_2 \\ c_2 & c_3 \end{pmatrix}$, then there are no periodic solutions, $\overline{X}(t) \equiv 0$ is an isolated constant solution and there is an additional isolated constant solution if $F(B,C,o,1)^{-1}$ exists.

The above result holds even in the case where C is not symmetric.

<u>Example:</u>

$$B = \begin{pmatrix} o & 3 \\ -3 & 2a \end{pmatrix}, \quad |a| < 3, \quad C = \begin{pmatrix} o & o \\ o & -1 \end{pmatrix} .$$

A change of variable converts B and C to

$$B = \begin{pmatrix} a & -\theta \\ \theta & a \end{pmatrix}, \quad C = \begin{pmatrix} o & -a/\theta \\ o & -1 \end{pmatrix}, \quad \theta = \sqrt{9-a^2}$$

and for $|a| < 3$ the only nontrivial constant solution is

$$\overline{X}(t) \equiv \begin{pmatrix} 4a & o \\ o & 4a \end{pmatrix}.$$

There are no periodic solutions. This corrects a second example found in the paper by Sasegawa [4/].

The analysis of the structure of constant solutions in the case $n = 2$, as well as the question of the existence of periodic solutions when (1) is subjected to a periodic (possibly nonlinear) small disturbance will be analysed in a forthcoming paper.

Mathematics Department
University of New Mexico

References

1. R. Redheffer, *On solutions of Riccati's equation as a function of initial values*, Jour. of Rational Mech. and Anal. 5 (1956), 835-848.
2. W.T. Reid, *Solutions of a Riccati matrix differential as functions of initial values*, Jour. of Math. and Mech. 8 (1959), 221-230.
3. D.A. Sánchez, *Computing periodic solutions of Riccati differential equations*, Appl. Math. and Comp. 6 (1980), 283-287.
4. ____________, *Periodic and constant solutions of matrix Riccati differential equations: n = 2*, to appear.
5. T. Sasegawa, *A necessary and sufficient condition for the solution of the Riccati equation to be periodic*, I.E.E.E. Trans. Auto. Control, AC-25 (1980), 564-566.

Solitary and Travelling Waves in a Rod

Ralph Saxton

We are concerned with the existence and partial analysis of longitudinal travelling waves in a thin rod, with a later application to a growth condition in incompressible elasticity. We consider only solutions $u(x,t) = \phi(z),\ z = x - ct$, which have the form of steady travelling waves for the equation of motion, which is derived most simply by means of Hamilton's principle ([7]; cp. [4],[5],[6]) applied to the functional

$$\int_0^T (K - I)\,dt. \tag{1}$$

Here, the total kinetic energy is given by

$$K(t) = \frac{1}{2}\int_{-\infty}^{\infty} [u_t^2(x,t) + \beta_t^2(u_x(x,t))]\,dx, \tag{2}$$

where $\beta(.)$ is a measure of the lateral deformation undergone during any elongation or contraction of the rod which is assumed to remain straight at all times. $u(x,t)$ is the longitudinal displacement at time t of a material point at x measured from some chosen point along the rod, which is now taken to be of infinite length. We have taken the density in the current and reference configurations to be unity. In terms of u_x, by adopting the limit of incompressibility (Poisson's ratio $\nu \to \frac{1}{2}$ in the special case of small deformation) β is given here by

$$\beta(u_x) = \frac{1}{(1 + u_x)^{\frac{1}{2}}} - 1, \tag{3}$$

although we need not assume this. To avoid material inversion or flattening in general, it is only necessary that $u_x, \beta > -1$. I, the total strain energy, is of the form

$$I(t) = \int_{-\infty}^{\infty} W(u_x(x,t), \beta(u_x(x,t)))\,dx \tag{4}$$

where $W(.,.)$ is the strain energy.

On carrying out the variations, we obtain from (1) the Euler equation

$$u_{tt} - (\beta'\beta'' u_{xt}^2 + \beta'^2 u_{xtt})_x - (W' + W^{\backprime} \beta')_x = 0 , \tag{5}$$

where subscripts denote differentiation with respect to x and t, and superscripts $'$,$^{\backprime}$ denote diffentiation with respect to dependent variables u_x and β, respectively. Further details of this equation appear in [7].

We now consider the substitution in (5) of

$$u(x,t) = \phi(z) , \quad z = x - ct , \quad c \in \mathbb{R}, \tag{6}$$

giving

$$c^2 \frac{d^2\phi}{dz^2} - \frac{d}{dz}\left(\beta'^2\left(\frac{d\phi}{dz}\right). c^2 \frac{d^3\phi}{dz^3}\right)$$

$$-\frac{d}{dz}\left(\beta'\left(\frac{d\phi}{dz}\right).\beta''\left(\frac{d\phi}{dz}\right). c^2\left(\frac{d^2\phi}{dz^2}\right)^2 + W'\left(\frac{d\phi}{dz}, \beta\left(\frac{d\phi}{dz}\right)\right)\right.$$

$$\left. + W^{\backprime}\left(\frac{d\phi}{dz} , \beta\left(\frac{d\phi}{dz}\right)\right).\beta'\left(\frac{d\phi}{dz}\right)\right) = 0 . \tag{7}$$

On further substituting $\psi = \frac{d\phi}{dz}$, $\chi = c\beta'(\psi)\frac{d\psi}{dz}$ and integrating (7) with respect to z , we obtain

$$-c^2\psi + c\beta'(\psi)\frac{d\chi}{dz} + W' + W^{\backprime}\beta'(\psi) = 0 \tag{8}$$

where we have omitted the constant of integration.

At some point z_0 the initial data for (8) is given by

$$\psi(z_0) = \psi_0 , \quad \frac{d\psi}{dz}(z_0) = \psi_1 , \quad \chi(z_0) = c\beta'(\psi_0)\psi_1 , \tag{9}$$

and on multiplying (8) by $\frac{d\psi}{dz}$ and integrating with respect to z , we have the energy equation $-\frac{c^2}{2}\psi^2 + \frac{1}{2}\chi^2 + W(\psi,\beta(\psi)) = E$, (10)

where E is a constant depending only on the values of ψ_0 , ψ_1 and c.

Our analysis of equation (7) is in the phase plane.

(For a related discussion in another problem, c.p. Ball [1], Calderer [2]). We therefore use (8) and (9) to obtain the equivalent first order system

$$\frac{d}{dz}\underset{\sim}{x}(z) = \underset{\sim}{m}(\Psi, \chi) \tag{11}$$

where
$$\underset{\sim}{x} = \begin{pmatrix}\Psi \\ \chi\end{pmatrix}, \tag{12}$$

$$\underset{\sim}{m} :]-1, \infty[\times \mathbb{R} \to \mathbb{R}^2 , \tag{13}$$

$$\underset{\sim}{m}(\Psi, \chi) = \frac{1}{c\,\beta'(\Psi)}\begin{pmatrix}\chi \\ c^2\Psi - W' - W'\beta'\end{pmatrix}, \tag{14}$$

and

$$\underset{\sim}{x}(z_0) = \begin{pmatrix}\Psi_0 \\ \chi_0\end{pmatrix} = \begin{pmatrix}\Psi_0 \\ c\,\beta'(\Psi_0)\,\Psi_1\end{pmatrix}. \tag{15}$$

We now present a result on local existence of trajectories $\underset{\sim}{x}$:

Proposition 1 Assume $\underset{\sim}{m}(\Psi, \chi)$ is locally Lipschitz continuous on its domain of definition. Then the initial value problem (12) - (15) has a unique maximally defined solution

$\underset{\sim}{x}(z) \in C'(I) \times C'(I)$ where $I =]z_0 - \alpha, z_0 + \alpha[$

for some $\alpha \in \mathbb{R}$. The solution $\underset{\sim}{x}(z)$ satisfies

$\underset{\sim}{x}(z_0) = \begin{pmatrix}\Psi_0 \\ \chi_0\end{pmatrix}$.

Proof For $\underset{\sim}{m}(\Psi, \chi)$ to be locally Lipschitz, it is enough for β (.), W (.,.) and W (.,.) to be locally Lipschitz continuous in each variable, and for β (.) to be sign definite. This implies we require the existence of some $0 < \Gamma < \infty$, $\Gamma = \Gamma(\varepsilon, R)$ such that for any $(\Psi_1, \chi_1), (\Psi_2, \chi_2) \in]-1+\varepsilon, R[\times]-R, R[$, $\varepsilon, R > 0$

$|W'(\Psi_1, \chi_1) - W'(\Psi_2, \chi_2)| \leq \Gamma(\varepsilon, R)(|\Psi_1 - \Psi_2| + |\chi_1 - \chi_2|)$ etc.,

and $\alpha > 0$ such that $|\beta'| > \alpha^{1/2}$.

We may then apply standard results in ordinary differential equations (e.g. Friedrichs [3]) to obtain local existence and uniqueness through $\begin{pmatrix}\Psi_0 \\ \chi_0\end{pmatrix}$.

Referring to equation (10) we define

$$V(\Psi) = -\frac{c^2}{2}\Psi^2 + W(\Psi, \beta(\Psi)) \tag{16}$$

and denote

$$\sigma'(\Psi) = c^2 + \frac{d^2 V}{d\Psi^2}$$

$$\equiv W'' + 2W^{``}\beta' + W^{"}\beta'^2 + W^{`}\beta''. \tag{17}$$

We may characterize the equilibrium points for (11) - (14) by the following proposition.

<u>Proposition 2</u> Let c be given. Then the equilibrium solutions $\chi(z) = 0,\ \Psi(z) = \bar{\Psi},\ z \in \mathbb{R}$ of (11) are the zeros of the function $c^2\Psi - \sigma(\Psi)$, where $\sigma(\Psi) = c^2\Psi + \frac{dV}{d\Psi}$.

Further, for $\bar{\Psi}$ satisfying $c^2 < \sigma'(\bar{\Psi})$, $(\bar{\Psi}, 0)$ is a centre and $V(\bar{\Psi})$ is a relative minimum. When $\bar{\Psi}$ satisfies $c^2 > \sigma'(\bar{\Psi})$, $(\bar{\Psi}, 0)$ is a saddle point and $V(\bar{\Psi})$ is a relative maximum.

<u>Proof</u> Follows from (14) and its linearisation

$$\underset{\sim}{m}(\Psi, \chi) \simeq \frac{1}{c\beta'(\bar{\Psi})}\begin{pmatrix} 0 & 1 \\ c^2 - \sigma'(\bar{\Psi}) & 0 \end{pmatrix}\begin{pmatrix} \Psi - \bar{\Psi} \\ \chi \end{pmatrix}$$

in a neighbourhood of $(\bar{\Psi}, 0)$.

<u>Remark</u> In addition to the above cases we may also have $c^2 = \sigma'(\bar{\Psi})$ which corresponds to a point of inflection of V.

We now assume that for each c, there exists a finite number, n, of zeros

$$\{\bar{\Psi}_i,\ 0 \leq i \leq n\},\ \bar{\Psi}_i < \bar{\Psi}_{i+1} \tag{18}$$

to the function $c^2\Psi - \sigma(\Psi)$.

Further we consider two possible hypotheses for σ :-

(H1) As $\Psi \to +\infty$, $\frac{\sigma(\Psi)}{\Psi} \sim 1$;

(H2) As $\Psi \to +\infty$, $\frac{\sigma(\Psi)}{\Psi} \sim \Psi^{B}$,

where $B > 0$. (H1) and (H2) are compatible with the Lipschitz assumptions made earlier. We use these conditions to investigate the region $\Psi \geq 0$ of the phase plane and obtain the next two results.

<u>Proposition 3</u> Let $\psi \geq 0$ throughout, and let (ψ_0, ψ_1) be chosen so that

$$E(\psi_0, \psi_1) \neq V(\bar{\psi}_i), \quad 0 \leq i \leq n$$

Then a) when $c^2 < 1$ and either (H1) or (H2) hold, the trajectory through (ψ_0, ψ_1) is periodic with period

$$\tau = \sqrt{2} \int_{J_0}^{J_1} \frac{c\beta'(\psi)\, d\psi}{(E - V(\psi))^{1/2}} \tag{19}$$

where $J_0 < J_1$ and $(J_0, 0)$, $(J_1, 0)$ are the intersections of the trajectory with the axis $\chi = 0$.

b) When $c^2 > 1$ and (H2) holds, we again have the result of a).

c) When $c^2 > 1$ and (H1) holds, no periodic orbits exist through any point $(\psi, 0)$ with $\psi > \bar{\psi}_n$.

<u>Proof</u> From (10) and (18).

<u>Corollary</u> When (ψ_0, ψ_1) are chosen so that

$$E(\psi_0, \psi_1) = V(\bar{\psi}_i) \quad \text{for some of } i = 0, 1, \dots, n$$

then either A) $(\psi_0, \psi_1) = (\psi_0, 0)$ is a centre, or when $c^2 < 1$ and (H1) and (H2) hold or $c^2 < 1$ and (H2) holds, then

B) i) for $\psi_1 \geq 0$ we have

$$\lim_{z \to +\infty} \psi(z) = \bar{\psi}_i, \quad \lim_{z \to +\infty} \chi(z) = 0$$

where $\bar{\psi}_i \leq \psi_0$ is the closest saddlepoint to the left of $(\psi_0, 0)$.

ii) for $\psi_1 \leq 0$

$$\lim_{z \to +\infty} \psi(z) = \bar{\psi}_j, \quad \lim_{z \to +\infty} \chi(z) = 0$$

where $\bar{\psi}_j \geq \psi_0$ is the closest saddle point to the right of $(\psi_0, 0)$.

<u>Remark</u> The case B) above is that which provides for the existence of solitary waves. Note that since centres and saddle points alternate (we have excluded the case $c^2 = \sigma'(\bar{\psi})$), given the same ψ_0 in B i) and ii) above implies $\bar{\psi}_j = \bar{\psi}_{i+2}$.

We conclude by looking at a case for $-1 < \psi \leq 0$, with $\beta(\psi)$ given by (3).

Hypothesis (H3) For $p \in]-1,\infty[$ as $q \to +\infty$

$$W(p,q) \sim \delta q^{\gamma}, \quad \gamma \in \mathbb{R}, \ \delta > 0.$$

Proposition 4 Under hypothesis (H3), a sufficient condition preventing flattening of the incompressible rod is $\gamma > 0$.

Proof With β given by (3), (10) becomes

$$-\frac{c^2}{2}\psi^2(1+\psi)^3 + \frac{c^2}{2}\psi'^2 + W(\psi,(1+\psi)^{-3/2})(1+\psi)^3 = E(1+\psi)^2 \tag{20}$$

and so as $\psi \to -1$

$$\frac{c^2}{2}\psi'^2 \sim (1+\psi)^3(E+c^2) - \delta(1+\psi)^{3(1-\gamma/2)} \tag{21}$$

and the result follows since $\delta > 0$ and it is impossible for $u_x = \psi = -1$.

References

[1] J. Ball, Discontinuous Equilibrium Solutions and Cavitation in Nonlinear Elasticity, Preprint.

[2] C. Calderer. The Dynamic Behaviour of Nonlinear Elastic Spherical Shells, J. Elasticity, to appear.

[3] K. O. Friedrichs, " Advanced Ordinary Differential Equations", Gordon and Breach Science Publishers, Inc., New York (1965).

[4] W. A. Green, "Dispersion Relations for Elastic Waves in a Bar", in Progress in Solid Mechanics I, Ed. Sneddon and Hill.

[5] W. Jaunzemis, "Continium Mechanics", New York MacMillan (1967).

[6] A. E. H. Love, Theory of Elasticity (4th ed.) 278, Cambridge Univ. Press (1927).

[7] R. Saxton, "Global Existence for the Nonlinear Pochhammer-Chree Equation", submitted to J.D.E.

INVESTIGATIONS IN THE THEORY OF PARTIAL DIFFERENTIAL EQUATIONS OF INFINITE ORDER

K. Seitz

Technical University of Budapest

I. INTRODUCTION

Consider the following partial differential equation of infinite order

$$(1) \qquad \sum_{n=0}^{\infty} \sum_{k=0}^{n} a_{k,n-k}(z,w) \frac{\partial^n F(z,w)}{\partial z^k \partial w^{n-k}} = 0 ,$$

where z and w are complex variables.

The domain of $a_{k,n-k}(z,w),\ (k=0,\dots,n)$ is $|z| < R$, $|w| < R$ where R is an arbitrary pozitiv constant.

In this short lecture we shall investigate the structure of the

$$(2) \qquad \begin{aligned} z &= \alpha \hat{z} + \beta \hat{w} \\ w &= \gamma \hat{z} + \delta \hat{w} \end{aligned} \qquad (\alpha, \beta, \gamma, \delta \text{ are complex numbers and we suppose that } \alpha\delta - \beta\gamma \neq 0)$$

transformations which map (1), to

$$\sum_{n=0}^{\infty} \sum_{k=0}^{n} a_{k,n-k}(\hat{z}, \hat{w}) \frac{\partial^n \hat{F}(\hat{z}, \hat{w})}{\partial \hat{z}^k \partial \hat{w}^{n-k}} = 0 ,$$

where

$$(4) \qquad \hat{F}(\hat{z}, \hat{w}) = F(\alpha \hat{z} + \beta \hat{w}, \gamma \hat{z} + \delta \hat{w}) .$$

This investigation has some importantce for building up a Lie-like theory of partial differential equations of infinite order.

For the study of the field treated in this paper, the reader is referred to the works [1], [2], [3].

II. THE TRANSFORMATION

It is easy to see that, from [2] and [4] we have

$$\frac{\partial F}{\partial z} = \left(\frac{\delta}{\Delta}\frac{\partial}{\partial \hat{z}} - \frac{\gamma}{\Delta}\frac{\partial}{\partial \hat{w}}\right)\hat{F}, \tag{5}$$

and

$$\frac{\partial F}{\partial w} = \left(-\frac{\beta}{\Delta}\frac{\partial}{\partial \hat{z}} + \frac{\alpha}{\Delta}\frac{\partial}{\partial \hat{w}}\right)\hat{F} \tag{6}$$

where

$$\Delta = \alpha\delta - \beta\gamma \neq 0 . \tag{7}$$

From (5) and (6) it follows that,

$$\frac{\partial^n F(z,w)}{\partial z^k \partial w^{n-k}} = \left(-\frac{\beta}{\Delta}\frac{\partial}{\partial \hat{z}} + \frac{\alpha}{\Delta}\frac{\partial}{\partial \hat{w}}\right)^{(n-k)} \left(\frac{\delta}{\Delta}\frac{\partial}{\partial \hat{z}} - \frac{\gamma}{\Delta}\frac{\partial}{\partial \hat{w}}\right)^{(k)} \hat{F}(\hat{z},\hat{w}) =$$

$$= \sum_{l=0}^{n-k} \binom{n-k}{l} \left(-\frac{\beta}{\Delta}\right)^{l} \left(\frac{\alpha}{\Delta}\right)^{n-k-l} \left[\sum_{j=0}^{k} \binom{k}{j} \left(\frac{\delta}{\Delta}\right)^{j} \left(\frac{-\gamma}{\Delta}\right)^{k-j} \frac{\partial^n \hat{F}(\hat{z},\hat{w})}{\partial \hat{z}^{j+l}\, \partial \hat{w}^{n-j-l}}\right] = \tag{8}$$

$$= \frac{1}{\Delta^n} \sum_{l=0}^{n-k} \sum_{j=0}^{k} \binom{n-k}{l} \binom{k}{j} (-\beta)^{l} \alpha^{n-k-l} \delta^{j} (-\gamma)^{k-j} \frac{\partial^n \hat{F}}{\partial \hat{z}^{j+l}\, \partial \hat{w}^{n-j-l}} =$$

$$= \frac{1}{\Delta^n} \sum_{p=0}^{n} C_{p,n-k,k} \frac{\partial^n \hat{F}(\hat{z},\hat{w})}{\partial \hat{z}^{p}\, \partial \hat{w}^{n-p}} ,$$

where

$$(9)\qquad c_{p,n-k,k} = \sum_{q=0}^{p} \widetilde{\binom{k}{q}} \cdot \widetilde{\binom{n-k}{p-q}} \alpha^{n-k-p+q} (-\beta)^{p-q} (-\gamma)^{k-q} \delta^{q} \qquad (p = 0, 1, \dots, n)$$

and

$$(10)\qquad \widetilde{\binom{k}{q}} = \begin{cases} \binom{k}{q} & \text{if } q \le k, \\ 0 & \text{if } q > k. \end{cases}$$

It is easy to see, that (2) maps (1) to (3) iff,

$$\sum_{k=0}^{n} \Big[\frac{1}{\Delta^n} \sum_{p=0}^{n} c_{p,n-k,k} \frac{\partial^n \hat{F}(\hat{z},\hat{w})}{\partial \hat{z}^p \partial \hat{w}^{n-p}} \Big] a_{k,n-k}(\alpha\hat{z}+\beta\hat{w}, \gamma\hat{z}+\delta\hat{w}) =$$

$$(11)\qquad = \sum_{p=0}^{n} a_{p,n-p}(\hat{z},\hat{w}) \frac{\partial^n \hat{F}(\hat{z},\hat{w})}{\partial \hat{z}^p \partial \hat{w}^{n-p}}$$

Since (11) we have

$$\sum_{p=0}^{n} \Big[\frac{1}{\Delta} \sum_{k=0}^{n} c_{p,n-k,k}\, a_{k,n-k}(\alpha\hat{z}+\beta\hat{w}, \gamma\hat{z}+\delta\hat{w}) \Big] \frac{\partial^n \hat{F}(\hat{z},\hat{w})}{\partial \hat{z}^p \partial \hat{w}^{n-p}} =$$

$$= \sum_{p=0}^{n} a_{p,n-p}(\hat{z},\hat{w}) \frac{\partial^n \hat{F}(\hat{z},\hat{w})}{\partial \hat{z}^p \partial \hat{w}^{n-p}},$$

from which

$$a_{p,n-p}(\hat{z},\hat{w}) = \frac{1}{\Delta^n} \sum_{k=0}^{n} c_{p,n-k,k}\, a_{k,n-k}(\alpha\hat{z}+\beta\hat{w}, \gamma\hat{z}+\delta\hat{w}).$$

$$(p = 0, 1, 2, \dots, n)$$

Thus we have obtained the following theorem:

Theorem If $a_{k,n-k}(z,w)$ are functions of complex variables with domain $|z|<R$, $|w|<R$ where R is an arbitrary pozitiv number, then the transformations (2) which map the following partial differential equations of infinite order

$$\sum_{n=0}^{\infty}\sum_{k=0}^{n} a_{k,n-k}(z,w)\frac{\partial^n F(z,w)}{\partial z^k \partial w^{n-k}} = 0, \qquad \text{to}$$

$$\sum_{n=0}^{\infty}\sum_{k=0}^{n} a_{k,n-k}(\hat{z},\hat{w})\frac{\partial^n \hat{F}(\hat{z},\hat{w})}{\partial \hat{z}^k \partial \hat{w}^{n-k}} = 0,$$

are determined by the following relations

$$a_{p,n-p}(\hat{z},\hat{w}) = \frac{1}{\Delta^n}\sum_{k=0}^{n} C_{p,n-k,k}\cdot a_{k,n-t}(\alpha\hat{z}+\beta\hat{w},\gamma\hat{z}+\delta\hat{w}).$$

It is well known that there is a one to one map between (2) and the groups of 2×2 type matrices $\begin{bmatrix}\alpha & \beta\\ \gamma & \delta\end{bmatrix}$, where $\alpha,\beta,\gamma,\delta$ are complex numbers and $\alpha\delta-\beta\gamma\neq 0$. This group os denoted by $GL(2,K)$ where K is the complex field.

In this way we can see that there is a one to one relation between (1) and a subgroup of $GL(2,K)$, which characterize the (1) partial differential equation of infinite order.

Example. Consider the following Euler-type partial differential equations of infinite order

$$(14)\qquad \sum_{n=0}^{\infty}\sum_{k=0}^{n} S_{k,n-k}\, z^k w^{n-k}\frac{\partial^n F(z,w)}{\partial z^k \partial w^{n-k}} = 0,$$

where $S_{k,n-k}$, $(k=0,1,2,\dots,n)$ are constant numbers.

In this case we have the following transformations

$$(15)\qquad z=\alpha\hat{z} \qquad w=\delta\hat{w}$$

and the set of all matrices $\begin{bmatrix}\alpha & 0\\ 0 & \delta\end{bmatrix}$ which is an abelian subgroups of $GL(2,K)$.

REFERENCES

[1] Einar Hille, "Ordinary Differential Equations in the Complex Domain", Wiley-Interscience, New York, 1976.

[2] Nathan Jacobson, "Lie Algebras", Interscience, New York 1962.

[3] V.S.Varadarajan, "Lie Groups, Lie Algebras, and their representations", Prentice-Hall, Inc., 1974.

WELL-POSED BOUNDARY PROBLEMS FOR HAMILTONIAN SYSTEMS OF LIMIT POINT OR LIMIT CIRCLE TYPE

by

J. K. Shaw[(1)] and D. B. Hinton[(2)]

INTRODUCTION. We consider a $2n \times 2n$ Hamiltonian system of differentia equations

$$(1.1) \qquad J\vec{y}\,' = [\lambda A(t) + B(t)]\vec{y}, \quad -\infty < a < t < b \le \infty,$$

where $A(t)$ and $B(t)$ are $2n \times 2n$ Hermitian matrix functions locally integrable on $[a,b)$, $A(t)$ real and nonnegative definite, λ is a complex parameter, $\vec{y}(t)$ is a $2n \times 1$ vector function and

$$J = \begin{bmatrix} 0 & -I_n \\ I_n & 0 \end{bmatrix} \quad (I_n = n \times n \text{ identity}).$$

We assume Atkinson's "definiteness condition" ([1, p. 253])

$$(1.2) \qquad \int_c^d \vec{y}^{\,*}(t)A(t)\vec{y}(t)dt > 0$$

(1) Research supported in part by NSF Grant No. MCS-8101536.

(2) Research supported in part by NSF Grant No. MCS-8101712.

whenever $a \leq c < d < b$ and $\vec{y}$ is a nontrivial solution of (1.1).

To say that (1.1) is of limit circle type at b means that every solution of (1.1) is of "integrable square" on [a,b), written $\vec{y} \in L^2_A[a,b)$, which is to say $\int_a^b \vec{y}^{\,*} A \vec{y} < \infty$. It is known ([1], [8]) that if (1.1) has only $L^2_A[a,b)$ solutions for any one λ, then this is so for all λ.

The Titchmarsh-Weyl $M(\lambda)$ functions associated with (1.1) are certain $n \times n$ matrix valued analytic functions of λ which generalize the classical $m(\lambda)$ coefficient ([2], [10]). The present authors developed a theory of matrix $M(\lambda)$ functions for "limit point" systems (1.1) in [4], [6]. To continue our study here, we introduce the M functions in the traditional way by setting up a boundary problem on a compact interval [a,d], d < b, and considering the limiting case of the problem as $d \to b$.

Thus consider a compact subinterval [a,d], and let $\delta_1(d)$ and $\delta_2(d)$ be $n \times n$ matrices which may depend on d and which satisfy for all d < b

(1.3) $$\delta_1\delta_1^* + \delta_2\delta_2^* = I \ , \quad \delta_1\delta_2^* = \delta_2\delta_1^* \ .$$

Let $E_\delta = E_\delta(d)$ be the $2n \times 2n$ matrix

(1.4) $$E_\delta = \begin{bmatrix} \delta_1^* & -\delta_2^* \\ \delta_2^* & \delta_1^* \end{bmatrix} .$$

By its defintion E_δ satisfies for all d < b

$$(1.5) \qquad E_\delta^{-1} = E_\delta^*, \quad E_\delta^* J E_\delta = E_\delta J E_\delta^* = I .$$

A matrix in the form (1.4), whose $n \times n$ block components satisfy (1.3), will be termed *admissible.*

At the left endpoint $t = a$ we shall need a fixed admissible matrix

$$E_\alpha = \begin{bmatrix} \alpha_1^* & -\alpha_2^* \\ \alpha_2^* & \alpha_1^* \end{bmatrix}$$

and a $2n \times 2n$ matrix solution $Y_\alpha(t,\lambda)$ of (1.1) with initial value

$$Y_\alpha(a,\lambda) = E_\alpha \quad \text{for all } \lambda.$$

We recall from [1, p. 257] that $Y(t,\lambda)$ is an entire function of λ for each fixed t. We partition Y_α into $n \times n$ blocks by writing

$$(1.6) Y_\alpha(t,\lambda) = \begin{bmatrix} \theta_\alpha(t,\lambda) & \phi_\alpha(t,\lambda) \\ \hat{\theta}_\alpha(t,\lambda) & \hat{\phi}_\alpha(t,\lambda) \end{bmatrix} = \left[\vec{\theta}_\alpha(t,\lambda) \vec{\phi}_\alpha(t,\lambda)\right], \quad \vec{\theta}_\alpha = \begin{pmatrix} \theta_\alpha \\ \hat{\theta}_\alpha \end{pmatrix}, \quad \vec{\phi}_\alpha = \begin{pmatrix} \phi_\alpha \\ \hat{\phi}_\alpha \end{pmatrix}$$

We frequently use the notation of (1.6) for $2n \times k$ matrix quantities, $\vec{f} = \begin{pmatrix} f \\ \hat{f} \end{pmatrix}$, where f and $\hat{f}$ will stand for the first and second $n \times k$ block components, respectively. For complicated expressions we use subscripts instead, $\vec{f} = \begin{pmatrix} f_1 \\ f_2 \end{pmatrix}$.

It is not difficult to prove that the $n \times n$ block components $[E_\delta^* \vec{\phi}_\alpha(d,\lambda)]_1$ and $[E_\delta^* \vec{\phi}_\alpha(d,\lambda)]_2$ are invertible for $a < d < b$ and for $\mathrm{Im}(\lambda) \neq 0$. Indeed, by Atkinson's generalization of Green's formula ([1, p. 253]), and by (1.5) and (1.3),

$$(1.7) \qquad (\lambda - \bar{\lambda}) \int_a^d \vec{\phi}_\alpha^*(s,\lambda)A(s)\vec{\phi}_\alpha(s,\lambda)ds = [\vec{\phi}_\alpha^*(t,\lambda)J\vec{\phi}_\alpha(t,\lambda)]_a^d$$

$$= \vec{\phi}_\alpha^*(d,\lambda)E_\delta J E_\delta^* \vec{\phi}_\alpha(d,\lambda) - [-\alpha_2, \alpha_1]J \begin{bmatrix} -\alpha_2^* \\ \alpha_1^* \end{bmatrix}$$

$$= (E_\delta^* \vec{\phi}_\alpha)_2^* (E_\delta^* \vec{\phi}_\alpha)_1 - (E_\delta^* \vec{\phi}_\alpha)_1^* (E_\delta^* \vec{\phi}_\alpha)_2,$$

where the last expression is evaluated at d,λ. If an $n \times 1$ vector $v \neq 0$ could be found such that $[E_\alpha^* \vec{\phi}_\alpha(d,\lambda)]_1 v = 0$ then (1.7) would imply $(\lambda - \bar{\lambda})v^*(\int_a^d \vec{\phi}_\alpha^* A \vec{\phi}_\alpha)v = 0$, contradicting (1.2) since $\mathrm{Im}(\lambda) \neq 0$. In the same way $[E_\alpha^* \vec{\phi}_\alpha]_2$ is seen to be invertible, and the same can be said for the components of $E_\delta^* \vec{\theta}_\alpha(d,\lambda)$.

Because of invertibility of these components, a unique linear combination

$$\vec{\psi}_\alpha(t,\lambda) = \vec{\theta}_\alpha(t,\lambda) + \vec{\phi}_\alpha(t,\lambda)M_{\alpha,\delta,d}(\lambda)$$

satisfies at $t = d$, and for $\mathrm{Im}(\lambda) \neq 0$, the boundary condition

$$(1.8) \qquad [E_\delta^* \vec{\psi}_\alpha(d,\lambda)]_1 = \delta_1(d)\psi_\alpha(d,\lambda) + \delta_2(d)\hat{\psi}_\alpha(d,\lambda) = 0 ,$$

and in fact

$$(1.9)\qquad M_{\alpha,\delta,d}(\lambda) = -[\delta_1(d)\phi_\alpha(d,\lambda)+\delta_2(d)\hat{\phi}_\alpha(d,\lambda)]^{-1}[\delta_1(d)\theta_\alpha(d,\lambda)+\delta_2(d)\hat{\theta}_\alpha(d,\lambda)]$$

$$= -\,[E_\delta^*\vec{\phi}_\alpha(d,\lambda)]_1^{-1}\;[E_\delta^*\vec{\theta}_\alpha(d,\lambda)]_1\;, \qquad \mathrm{Im}(\lambda) \neq 0\;.$$

The $n \times n$ matrix quantity (1.9) extends the classical Weyl circle formula (see [2, p. 226]).

We prove in [4], [6] that the terms $M_{\alpha,\delta,d}(\lambda)$ are uniformly bounded on compact subsets of the half planes $\mathrm{Im}(\lambda) \neq 0$, using no assumptions whatever on the number of independent $L_A^2[a,b)$ solutions of (1.1). However, in the "limit point" case we show that the limit

$$(1.10)\qquad M_\alpha(\lambda) = \lim_{d\to b} M_{\alpha,\delta,d}(\lambda)$$

exists for all $\mathrm{Im}(\lambda) \neq 0$, and is independent of the values of $\delta(d)$. We call (1.10) the Titchmarsh-Weyl M function, corresponding to initial data E_α, for (1.1).

If the limit circle case prevails at b, the limit in (1.10) fails to exist ([2], [10]). Nevertheless, the subsequential limits as $d \to b$ of the terms $\{M_{\alpha,\delta,d}(\lambda)\}$ play an important role in spectral theory. Thus it is of interest to characterize the set of all such subsequential limits. In [3] C. T. Fulton gave such a characterization for second order scalar equations. One of the aims of this paper is to extend Fulton's results to systems of equations (1.1), and in so doing we will have the necessary structure to generalize the theory of boundary problems at limit circle endpoints due to M. H. Stone ([9, ch. 10]).

2. PARAMETERIZATION OF THE M FUNCTION.

The classical limit circle m coefficients (defined as limits of sequences of the defining quotient of [2, p. 226]) are meromorphic. The analogous result for Hamiltonian systems will now be proved.

THEOREM 2.1. Assume (1.1) is limit circle at b. Let $M(\lambda)$ be the limit, uniform on compact subsets of the half planes $Im(\lambda) \neq 0$, of a sequence of matrices $M_{\alpha,\delta_k,d_k}(\lambda)$ from (1.9). Then $M(\lambda)$ is analytic in the complex λ plane except for poles on the real axis.

PROOF: As a function of λ, the right side of (1.9) is analytic except for the isolated points where the determinant of $[E^*_\delta \vec{\phi}_\alpha(d,\lambda)]_1$ vanishes; i.e., $M_{\alpha,\delta,d}(\lambda)$ is meromorphic. Hence $M(\lambda)$ is analytic at least for $Im(\lambda) \neq 0$.

Let $M_k(\lambda) = M_{\alpha,\delta_k,d_k}(\lambda)$ and $\vec{\psi}_k(t,\lambda) = \vec{\theta}_\alpha(t,\lambda) + \vec{\phi}_\alpha(t,\lambda)M_k(\lambda)$, so that $[E^*_{\delta_k}\vec{\psi}_k(d_k,\lambda)]_1 = 0$ for all k and λ. Fix a complex number λ_1, $Im(\lambda_1) \neq 0$, and use the Green's formula again (cf. (1.7)) to obtain

$$(2.1) \qquad (\lambda_1-\lambda)\int_a^{d_k}\vec{\psi}_k^*(t,\bar{\lambda})A(t)\vec{\psi}_k(t,\lambda_1)dt = [\vec{\psi}_k^*(t,\bar{\lambda})J\vec{\psi}_k(t,\lambda_1)]_a^{d_k}$$

for $\mathrm{Im}(\lambda) \neq 0$. The right side of (2.1) at $t = d_k$ is 0 and at $t = a$ it is $M_k(\lambda_1) - M_k^*(\bar{\lambda})$. Now $M_k^*(\bar{\lambda}) = M_k(\lambda)$ by [4]. Putting $M_k(\lambda_1)-M_k(\lambda)$ on the right of (2.1), substituting $\vec{\psi}_k(t,\bar{\lambda}) = \vec{\theta}_\alpha(t,\bar{\lambda}) + \vec{\phi}_\alpha(t,\bar{\lambda})M_k(\bar{\lambda})$ into the left, and solving for $M_k(\lambda)$ results in

$$(2.2) \qquad M_k(\lambda)\{I - (\lambda-\lambda_1)\int_a^{d_k}\vec{\phi}_\alpha^*(t,\bar{\lambda})A(t)\vec{\psi}_k(t,\lambda_1)dt\} = M_k(\lambda_1)+(\lambda-\lambda_1)\int_a^{d_k}\vec{\theta}_\alpha^*(t,\bar{\lambda})A(t)\vec{\psi}_k(t,\lambda_1)dt,$$

or with obvious notation

$$(2.3) \qquad M_k(\lambda)D_k(\lambda) = N_k(\lambda), \quad \mathrm{Im}(\lambda) \neq 0 .$$

Since the limit circle case prevails, the integrals in (2.2) all converge absolutely as $k \to \infty$. Thus $D_k(\lambda)$ and $N_k(\lambda)$ approach entire functions $D(\lambda)$ and $N(\lambda)$ such that $M(\lambda)D(\lambda) = N(\lambda)$, $\mathrm{Im}(\lambda) \neq 0$. Since $D(\lambda_1) = I$, neither $D(\lambda)$ nor $N(\lambda)$ is identically zero. The determinant of $D(\lambda)$ can vanish at just countably many points, and thus $M(\lambda)$ can be continued analytically onto the real axis, save for those countably many points where a pole will occur. This completes the proof.

We now discuss boundary conditions. We continue to assume the limit circle case. Let $Y_0(t,\lambda)$ be a fundamental matrix solution of

(1.1) satisfying at $t = a$ the condition $Y_0^*(a,\lambda)JY_0(a,\lambda) = J$, and let $Y_0(t) = Y_0(t,\lambda_0)$ where λ_0 is real. A differentiation shows that $Y_0^*(t)JY_0(t) = J$ for all t, since λ_0 is real and the initial condition $Y_0^*(a)JY_0(a) = J$ holds. Let E_β be an admissible matrix (β_1 and β_2 satisfy (1.3)) and define $Y_\beta(t) = Y_0(t)\, E_\beta$, $a \le t < b$.

Let $\vec{f} \in L_A^2[a,b)$ and let $\vec{y}(t)$ satisfy $J\vec{y}' = [\lambda A + B]\vec{y} + A\vec{f}$. By variation of parameters using some fixed τ, $a \le \tau < b$,

$$\vec{y}(t) = Y_\beta(t)\, Y_\beta^{-1}(\tau)\vec{y}(\tau) + Y_\beta(t)J^{-1}\int_\tau^t Y_\beta^*(s)A(s)[(\lambda-\lambda_0)\vec{y}(s)+\vec{f}(s)]ds$$

and consequently

$$(2.4)\qquad Y_\beta^{-1}(t)\vec{y}(t) = Y_\beta^{-1}(\tau)\vec{y}(\tau)+J^{-1}\int_\tau^t Y_\beta^* A[(\lambda-\lambda_0)\vec{y}+\vec{f}]ds.$$

Under the present limit circle hypotheses, the integral above converges absolutely as $t \to b$, and consequently $\lim_{t\to b} Y_\beta^{-1}(t)\vec{y}(t)$ exists. We choose the first $n \times 1$ block component, whose limit also exists, and set as a boundary condition at $t = b$ the requirement

$$(2.5)\qquad (Y_\beta^{-1}\vec{y})_1(b) = \lim_{t\to b} [Y_\beta^{-1}(t)\vec{y}(t)]_1 = 0 .$$

We could use the other component just as well. Since $Y_\beta^{-1} = E_\beta^{-1}Y_0^{-1}$, then

$$(Y_\beta^{-1}\vec{y})_1(b) = \beta_1(Y_0^{-1}\vec{y})_1(b) + \beta_2(Y_0^{-1}\vec{y})_2(b)$$

which extends the analogous formula of Fulton [3, eq. (2.6)].

Reasoning much as in (1.7), we can show that the $n \times n$ matrix $[Y_\beta^{-1}\vec{\phi}_\alpha]_1(b)$ is invertible for $\mathrm{Im}(\lambda) \ne 0$ ($\vec{\phi}_\alpha = \vec{\phi}_\alpha(t,\lambda)$). Thus we can define a matrix function $\tilde{M}_{\alpha,\beta}(\lambda)$ by the equation

$$(2.6)\qquad \tilde{M}_{\alpha,\beta}(\lambda) = -[Y_\beta^{-1}\vec{\phi}_\alpha(\cdot,\lambda)]_1^{-1}(b)[Y_\beta^{-1}\vec{\theta}_\alpha(\cdot,\lambda)]_1(b),\ \mathrm{Im}(\lambda) \neq 0.$$

This is the matrix analog of the representation formula of [3, eq. (1.9)].

Our aim is to show that (2.6) characterizes the set of functions which appear as limits of sequences of terms (1.9). That is, if $M(\lambda)$ satisfies the hypothesis of Theorem 2.1, then $M(\lambda) = \tilde{M}_{\alpha,\beta}(\lambda)$ for some E_β, and vice-versa. In this way, the admissible matrices E_β may be said to "parameterize" the M functions at a limit circle endpoint.

We will not give full details in every instance, but refer the reader instead to the forthcoming article [5]. The first result says that (2.6) introduces no new M functions.

THEOREM 2.2. Let $\tilde{M}_{\alpha,\beta}(\lambda)$ be given by (2.6). Then there exist admissible matrices $E_\delta = E_\delta(d)$, $a < d < b$, such that $\tilde{M}_{\alpha,\beta}(\lambda)$ is the uniform limit on compact subsets of $\mathrm{Im}(\lambda) \neq 0$ of a sequence of terms $M_{\alpha,\beta,d}(\lambda)$ from (1.9).

PROOF: For $a < d < b$ let $\tilde{M}_d(\lambda) = -[Y_\beta^{-1}(d)\vec{\phi}_\alpha(d,\lambda)]_1^{-1}[Y_\beta^{-1}(d)\vec{\theta}_\alpha(d,\lambda)]_1$ so that $\tilde{M}_{\alpha,\beta}(\lambda) = \lim_{d\to b} \tilde{M}_d(\lambda)$. Also

$$(2.7)\qquad [Y_\beta^{-1}\vec{\theta}_\alpha]_1(d) + [Y_\beta^{-1}\vec{\theta}_\alpha]_1(d)\tilde{M}_d = 0 .$$

Now Y_0^{-1} may be calculated from the relation $Y_0^* J Y_0 = J$ and substituted into $Y_\beta^{-1} = E_\beta^{-1} Y_0^{-1}$ to yield

$$Y_\beta^{-1}\vec{\phi}_\alpha = \begin{bmatrix} \beta_1 & \beta_2 \\ -\beta_2 & \beta_1 \end{bmatrix} \begin{bmatrix} \hat{\phi}_0^* & -\phi_0^* \\ -\hat{\theta}_0^* & \theta_0^* \end{bmatrix} \begin{pmatrix} \phi_\alpha \\ \hat{\phi}_\alpha \end{pmatrix}$$

and so $[Y_\beta^{-1}\vec{\phi}_\alpha]_1(d)=[\beta_1(\hat{\phi}_0^*\phi_\alpha-\phi_0^*\hat{\phi}_\alpha) + \beta_2(-\hat{\theta}_0^*\phi_\alpha+\theta_0^*\hat{\phi}_\alpha)](d)$ and similarly for $[Y_\beta^{-1}\vec{\theta}_\alpha]_1$. Substituting these into (2.7), re-arranging and introducing $\sigma_1(d) = (\beta_1\hat{\phi}_0^* - \beta_2\hat{\theta}_0^*)(d)$ and $\sigma_2(d) = (-\beta_1\phi_0^*+\beta_2\theta_0^*)(d)$, (2.7) now becomes

$$(2.8)\qquad (\sigma_1\theta_\alpha+\sigma_2\hat{\theta}_\alpha) + (\sigma_1\phi_\alpha + \sigma_2\hat{\phi}_\alpha)\tilde{M}_d = 0 .$$

By identifying the terms σ_1 and σ_2 in the relation $Y_\beta^{-1}Y_\beta = I$, one can show that $\sigma_1\sigma_2^* = \sigma_2\sigma_1^*$. While it may not be true that $\sigma_1\sigma_1^* + \sigma_2\sigma_2^* = I$, this matrix is certainly positive definite, so we can make a "change of variable" $\delta_1(d) = c(d)\sigma_1(d)$, $\delta_2(d) = c(d)\sigma_2(d)$, where $c(d)$ is a nonsingular matrix, in which

$$E_\delta = E_\delta(d) = \begin{bmatrix} \delta_1^* & -\delta_2^* \\ \delta_2^* & \delta_1^* \end{bmatrix}$$

is admissible. Multiplying (2.8) on the left by $c(d)$, and then taking an inverse, we have

$$(2.9)\qquad \tilde{M}_d(\lambda) = [\delta_1(d)\phi_\alpha(d,\lambda)+\delta_2(d)\hat{\phi}_\alpha(d,\lambda)]^{-1}[\delta_1(d)\theta_\alpha(d,\lambda)+\delta_2(d)\hat{\theta}_\alpha(d,\lambda)]$$

which agrees with (1.9). The right side of (2.9) is uniformly bounded on compact subsets of $\mathrm{Im}\,(\lambda) \neq 0$, so a sequence of its terms converges to an M function. Since the limit must be $\tilde{M}_{\alpha,\beta}$ the proof is complete.

To complete the characterization, we have to take a limit $M(\lambda)$ of a sequence of terms (1.9) and show that it can be represented as in (2.6). First, we require a preliminary result.

LEMMA 2.1. Let $\{d_k\}$ be a monotone sequence, $d_k \to b$, let E_{δ_k} be a sequence of admissible matrices and let $\gamma_{k1}(\lambda) = \delta_{k1}\theta_0(d_k,\lambda) + \delta_{k2}\hat{\theta}_0(d_k,\lambda)$, $\gamma_{k2}(\lambda) = \delta_{k1}\phi_0(d_k,\lambda) + \delta_{k2}\hat{\phi}_0(d_k,\lambda)$, where δ_{k1}, δ_{k2} are the components of E_{δ_k} as in (1.4) and θ_0, ϕ_0, etc. are the components of $Y_0(t,\lambda)$ as in (1.6). Then there exists a real λ_0 and a subsequence $\{k_j\}$ of integers such that either

$$\gamma^{-1}_{k_j2}(\lambda_0)\gamma_{k_j1}(\lambda_0) \quad \text{or} \quad \gamma^{-1}_{k_j1}(\lambda_0)\gamma_{k_j2}(\lambda_0) \tag{2.10}$$

is convergent.

PROOF: Define $M_k^{(0)}(\lambda) = -\gamma^{-1}_{k2}(\lambda)\gamma_{k1}(\lambda)$. Then $M_k^{(0)}$ is of the form of (1.9), so a sequence $M_{k_j}^{(0)}$ of its terms converges uniformly on compact subsets of $\mathrm{Im}(\lambda) \neq 0$ to a limit $M^{(0)}(\lambda)$, which is meromorphic by Theorem 2.1. If λ_0 is a regular point of $M^{(0)}$ then $M_{k_j}^{(0)}$ is also regular there for k_j large enough. This says the left member of (2.10) converges.

Since $[M_{k_j}^{(0)}(\lambda)]^{-1} = -\gamma^{-1}_{k_j1}(\lambda)\gamma_{k_j2}(\lambda)$ and $[M_{k_j}^{(0)}]^{-1} \to [M^{(0)}]^{-1}$, which is also meromorphic, the right member of (2.10) will converge if λ_0 is a regular point of $[M^{(0)}]^{-1}$. This completes the proof.

Thus in the lemma, λ_0 may be taken as a regular point of $M^{(0)}$ or its inverse. In the scalar case this is no restriction on λ_0, for a pole of $m(\lambda)$ is a zero of its inverse and vice-versa. For a Hamiltonian system this need not be the case, and we illustrate this important fact with an example. We can encompass two independent Sturm-Liouville problems $-y'' = [\lambda a_{11}(t)+b_{11}(t)]y$ and $-z'' = [\lambda a_{22}(t) + b_{22}(t)]z$ in a single

4×4 Hamiltonian system by writing

$$\begin{bmatrix} 0 & 0 & -1 & 0 \\ 0 & 0 & 0 & -1 \\ 1 & 0 & 0 & 0 \\ 0 & 1 & 0 & 0 \end{bmatrix} \begin{bmatrix} y \\ z \\ y' \\ z' \end{bmatrix}' = \left\{ \lambda \begin{bmatrix} a_{11} & 0 & 0 & 0 \\ 0 & a_{22} & 0 & 0 \\ 0 & 0 & 0 & 0 \\ 0 & 0 & 0 & 0 \end{bmatrix} + \begin{bmatrix} b_{11} & 0 & 0 & 0 \\ 0 & b_{22} & 0 & 0 \\ 0 & 0 & 1 & 0 \\ 0 & 0 & 0 & 1 \end{bmatrix} \right\} \begin{bmatrix} y \\ z \\ y' \\ z' \end{bmatrix}.$$

Setting $\alpha_1 = \alpha_2 = I_2$, we see that $Y = Y_\alpha(t,\lambda)$ is a 4×4 matrix whose entries $Y_{21} = Y_{41} = Y_{12} = Y_{23} = Y_{32} = Y_{34} = Y_{41} = Y_{43}$ vanish identically because of the initial values they take. Thus $\vec{\theta}_\alpha$ and $\vec{\theta}_\alpha$ have the forms

$$\theta_\alpha = \begin{bmatrix} \theta_y & 0 \\ 0 & \theta_z \end{bmatrix}, \quad \phi_\alpha = \begin{bmatrix} \phi_y & 0 \\ 0 & \phi_z \end{bmatrix}$$

where θ_y, ϕ_y and θ_z, ϕ_z are the classical θ and ϕ solutions of Titchmarsh ([10]) for the separate equations in y and z, respectively. If we take $\delta_{k1} = I_2$, $\delta_{k2} = 0$ then $M^{(0)}$ in the lemma becomes

$$M^{(0)}(\lambda) = \begin{bmatrix} m_y(\lambda) & 0 \\ 0 & m_z(\lambda) \end{bmatrix}$$

where m_y and m_z are subseqeuntial limits of the quotients (θ_y/ϕ_y) and (θ_z/ϕ_z) respectively. That is, m_y and m_z are scalar m-coefficients. It is easy to produce an example where m_y has a zero and m_z a pole at a common point λ_0. Thus $M^{(0)}$ will experience a pole at λ_0, but $[M^{(0)}]^{-1} = \text{diag}[m_y^{-1}, m_z^{-1}]$ has a pole there as well.

THEOREM 2.3. Let $M(\lambda)$ be the limit of a sequence of terms of the form (1.9). Then there exists a real λ_0 and an admissible matrix E_β such that $M(\lambda) = -\lim_{t\to b}[Y_\beta^{-1}(t)\vec{\phi}_\alpha(t,\lambda)]_1^{-1}[Y_\beta^{-1}(t)\vec{\theta}_\alpha(t,\lambda)]_1$.

PROOF: We will sketch the proof; details may be found in [5]. Let $M(\lambda) = \lim_{k\to\infty} M_{\alpha,\delta_k,d_k}(\lambda)$ from (1.9) and define $S_k(\lambda) = E^*_{\delta_k} Y_0(d_k,\lambda)$. Denote the upper $n \times 2n$ portion of $S_k(\lambda)$ by $[S_k(\lambda)]_1 = [\gamma_{k1}(\lambda),\gamma_{k2}(\lambda)]$. Arguing as in Theorem 2.2, we can find a nonsingular matrix $c_k(\lambda)$ such that if $\Gamma_{k1} = c_k\gamma_{k1}$, $\Gamma_{k2} = c_k\gamma_{k2}$ then $\Gamma_{k1}\Gamma^*_{k1} + \Gamma_{k2}\Gamma^*_{k2} = I$. Note that $[S_k(\lambda)]^{-1} = Y_0^{-1}(d_k,\lambda)E_{\delta_k}$. If one writes out the components of $S_k^{-1}S_k = I$, one finds that $\gamma_{k1}\gamma^*_{k2} - \gamma_{k2}\gamma^*_{k1} = 0$, so it follows that $\Gamma_{k1}\Gamma^*_{k2} - \Gamma_{k2}\Gamma^*_{k1} = c_k\gamma_{k1}\gamma^*_{k2}c^*_k - c_k\gamma_{k2}\gamma^*_{k1}c^*_k = 0$; i.e., $E_{\Gamma_k}(\lambda)$ is admissible.

If we introduce the fundamental matrices $Y_k(t,\lambda) = Y_0(t,\lambda)E_{\Gamma_k}(\lambda)$, then a calculation using the above identities leads to

$$(2.11)\quad M_{\alpha,\delta_k,d_k}(\lambda) = -[Y_k^{-1}(d_k,\lambda_0)\vec{\phi}_\alpha(d_k,\lambda)]_1^{-1}[Y_k^{-1}(d_k,\lambda_0)\vec{\phi}_\alpha(d_k,\lambda)]_1$$

for any real λ_0. Using Y_k as the fundamental matrix in (2.4), and putting $\vec{f} = 0$, we have for an arbitrary solution $\vec{y}(t)$

$$(2.12)\quad Y_k^{-1}(t,\lambda_0)\vec{y}(t) = E_{\Gamma_k}^{-1}(\lambda_0)\vec{v}(t,\lambda,\vec{y})\ , \qquad \text{where}$$

$$(2.13)\quad \vec{v}(t,\lambda,\vec{y}) = Y_0^{-1}(\tau,\lambda_0)\vec{y}(\tau)+J^{-1}\int_\tau^t Y_0^*(s,\lambda_0)(\lambda-\lambda_0)A(s)\vec{y}(s)ds.$$

Setting $t = d_k$ in (2.12) gives

$$(2.14)\quad [Y_k^{-1}(d_k,\lambda_0)\vec{y}(d_k)]_1 = \Gamma_{k1}(\lambda_0)v(d_k,\lambda) + \Gamma_{k2}(\lambda_0)\hat{v}(d_k,\lambda).$$

If we put $\vec{u}_k(\lambda) = \vec{v}(d_k,\lambda,\vec{\phi}_\alpha(\cdot,\lambda))$ and $\vec{w}_k = \vec{v}(d_k,\lambda,\vec{\theta}_\alpha(\cdot,\lambda))$ then (2.11) and (2.14) give

$$(2.15)\quad M_{\alpha,\delta_k,d_k}(\lambda) = -[\Gamma_{k1}(\lambda_0)u_k(\lambda)+\Gamma_{k2}(\lambda_0)\hat{u}_k(\lambda)]^{-1}[\Gamma_{k1}(\lambda_0)w_k(\lambda)+\Gamma_{k2}(\lambda_0)\hat{w}_k(\lambda)]$$

Because of the limit circle hypothesis, the integrals in (2.13) converge absolutely as $t \to b$, and this means that $u_k(\lambda)$ and $w_k(\lambda)$ approach finite limits, say $\vec{u}(\lambda)$ and $\vec{w}(\lambda)$, as $k \to \infty$.

Since $\Gamma_{k2}^{-1}\Gamma_{k1} = \gamma_{k2}^{-1}\gamma_{k1}$ and $\Gamma_{k1}^{-1}\Gamma_{k2} = \gamma_{k1}^{-1}\gamma_{k2}$, the lemma tells us that λ_0 can be chosen in (2.15) in such a way that either $\Gamma_{k2}^{-1}(\lambda_0)\Gamma_{k1}(\lambda_0)$ or $\Gamma_{k1}^{-1}(\lambda_0)\Gamma_{k2}^{-1}(\lambda_0)$ has a convergent subsequence. Suppose $\Gamma_{k2}^{-1}(\lambda_0)\Gamma_{k1}(\lambda_0)$ has a convergent subsequence, whose running index we will continue to denote by k to simplify notation. Letting Γ be its limit, we have by (2.15) that

$$(2.16)\qquad M(\lambda) = -[\Gamma u(\lambda) + \hat{u}(\lambda)]^{-1}[\Gamma w(\lambda) + \hat{w}(\lambda)] .$$

By making another "change of variable", it can be shown that an admissible matrix E_β exists for which $\beta_1 = \beta_2\Gamma$. Substituting this into (2.16) there follows

$$(2.17)\qquad M(\lambda) = -[\beta_1 u(\lambda) + \beta_2\hat{u}(\lambda)]^{-1}[\beta_1 w(\lambda) + \beta_2\hat{w}(\lambda)] .$$

Let $Y_\beta(t) = Y_0(t,\lambda_0)E_\beta$. Repeating (2.4), we get $Y_\beta^{-1}(t)\vec{y}(t) = E_\beta^{-1}\vec{v}(t,\lambda,y)$, where $\vec{v}$ is defined above. This expression has first $(n \times 1)$ component $\beta_1 v(t,\lambda,\vec{y}) + \beta_2\hat{v}(t,\lambda,\vec{y})$. Putting first $\vec{y} = \vec{\phi}_\alpha$, then $\vec{y} = \vec{\theta}_\alpha$, and taking an inverse, we obtain on letting $t \to b$

$$\lim_{t\to b}[Y_\beta^{-1}(t)\vec{\phi}_\alpha(t,\lambda)]_1^{-1}[Y_\beta^{-1}\vec{\theta}_\alpha(t,\lambda)]_1 = [\beta_1 u(\lambda)+\beta_2\hat{u}(\lambda)]^{-1}[\beta_1 w(\lambda)+\beta_2\hat{w}(\lambda)].$$

Comparing this with (2.17), the proof is complete in this case. There remains the possibility that $\Gamma_{k1}^{-1}\Gamma_{k2}$, instead of $\Gamma_{k2}^{-1}\Gamma_{k1}$, has a convergent subsequence. The proof for this case is not substantially different, and we omit the details.

3. BOUNDARY VALUE PROBLEMS. The boundary condition (2.5) generalizes the Wronskian codition of M. H. Stone ([9]) used in connection with second order scalar equations. We now consider boundary problems on $[a,b)$ which, in the limit circle case, employ (2.5) as right endpoint conditon.

By "limit point case" at $x = b$ we mean that (1.1) has for all nonreal λ exactly n linearly independent solutions of class $L_A^2[a,b)$. We have recently shown in [7] that this definition of limit point is equivalent to the condition $\lim_{t\to b} \vec{y}^*(t)J\vec{z}(t) = 0$ for all solutions $\vec{y},\vec{z}$ of (1.1) belonging to $L_A^2[a,b)$. The latter was the limit point hypothesis of [4], [6]. Hence the results of [4], [6] may be brought to bear on the present paper.

For the remainder of this paper, the symbol $M_\alpha(\lambda)$ will have the following meaning. If $x = b$ is limit point, $M_\alpha(\lambda)$will be defined by (1.10), the existence of the limit being a consequence of [4]. If $x = b$ is limit circle, $M_\alpha(\lambda)$ will stand for one of the cluster points of the collection $\{M_{\alpha,\delta,d}(\lambda)\}$; i.e., $M_\alpha(\lambda)$ is the limit of a sequence of terms from(1.9). Whichever the case, we set $\vec{\psi}_\alpha(t,\lambda) = \vec{\phi}_\alpha(t,\lambda) + \vec{\phi}_\alpha(t,\lambda)M_\alpha(\lambda)$.

If it is the limit circle case which prevails, select via Theorem 2.3 a real λ_0 and an admissible E_β such that if $Y_\beta(t) = Y_0(t)E_\beta$, $Y_0(t) = Y_0(t,\lambda_0)$, then $M_\alpha(\lambda) = -\lim_{t\to b} [Y_\beta^{-1}(t)\vec{\phi}_\alpha(t,\lambda)]_1^{-1}[Y_\beta^{-1}\vec{\theta}_\alpha(t,\lambda)]_1$. By definition of $\vec{\phi}_\alpha$, we have in this case

$$\lim_{t\to b} [Y_\beta^{-1}(t)\vec{\psi}_\alpha(t,\lambda)]_1 = 0 \tag{3.1}$$

which is to say $\vec{\psi}_\alpha$ satisfies (2.5).

Following [6] we define the Green's function for $\mathrm{Im}(\lambda) \neq 0$ by

$$G(x,t,\lambda) = \begin{cases} \vec{\phi}_\alpha(x,\lambda)\vec{\psi}_\alpha^*(t,\bar{\lambda}), & x < t \\ \vec{\psi}_\alpha(x,\lambda)\vec{\phi}_\alpha^*(t,\bar{\lambda}), & x > t \end{cases}$$

and if $\vec{f}\in L_A^2[a,b]$, let $(G_\lambda\vec{f})(x) = \int_a^b G(x,t,\lambda)A(t)\vec{f}(t)dt$. If the limit point case holds, $\vec{\psi}_\alpha \in L_A^2[a,b)$ ([4]) and so the integral defining G_λ converges.

Now let $\vec{f}\in L_A^2[a,b)$ and consider integrable square solutions of the nonhomogeneous equation

$$J\vec{y}\,' = [\lambda A(t)+B(t)]\vec{y} + A\vec{f}\,, \quad \vec{y}\in L_A^2[a,b), \quad \mathrm{Im}(\lambda) \neq 0. \tag{3.2}$$

<u>THEOREM 3.1</u>. In the limit point case the boundary problem consisting of (3.2) together with $\alpha_1 y(a) + \alpha_2\hat{y}(a) = u$ has a unique solution given by $\vec{y}(t) = \vec{\psi}_\alpha(t,\lambda)u + (G_\lambda\vec{f})(t)$. In the limit circle case, the boundary problem consisting of (3.2) together with $\alpha_1 y(a)+\alpha_2\hat{y}(a) = v$, $(Y_\beta^{-1}\vec{y})_1(b) = \hat{v}$ has a unique solution given by $\vec{y}(t) = Y_\alpha(t,\lambda)\vec{w}+(G_\lambda\vec{f})(t)$

where $\vec{w} = \begin{pmatrix} w \\ \hat{w} \end{pmatrix}$, $w = v$ and $\hat{w} = (Y_\beta^{-1}\vec{\phi}_\alpha)_1^{-1}[\hat{v} - (Y_\beta^{-1}\vec{\theta}_\alpha)_1 v]$.

PROOF: We proved $G_\lambda\vec{f} \in L_A^2[a,b)$ in [4] in the limit point case. Since $G(x,t,\lambda)$ has exactly the same form in the limit circle case, the proof in [4] carries over to that case, too. Certainly the forms for $\vec{y}$ in the statement of the theorem then belong to $L_A^2[a,b)$. Moreover, these forms satisfy the differential equation $J\vec{y}' = [\lambda A+B]\vec{y} + A\vec{f}$ by the argument of [6].

In either case, $\vec{g} = G_\lambda\vec{f}$ has the form

$$\vec{g}(x) = (G_\lambda\vec{f})(x) = \vec{\psi}_\alpha(x,\lambda)\int_a^x \vec{\phi}_\alpha^*(t,\bar{\lambda})A(t)\vec{f}(t)dt + \vec{\phi}_\alpha(x,\lambda)\int_x^b \vec{\psi}_\alpha^*(t,\bar{\lambda})A(t)\vec{f}(t)dt ,$$

and so $(G_\lambda\vec{f})(a) = \begin{pmatrix} -\alpha_2^* \\ \alpha_1^* \end{pmatrix}\int_a^b \vec{\psi}_\alpha^*(t,\bar{\lambda})A(t)\vec{f}(t)dt$. Thus $\alpha_1 g(a)+\alpha_2\hat{g}(a) = (-\alpha_1\alpha_2^* + \alpha_1\alpha_2^*)\int_a^b \vec{\psi}_\alpha^* A\vec{f} = 0$. As for the right endpoint condition in the limit circle case,

$$(Y_\beta^{-1}\vec{g})_1(b) = (Y_\beta^{-1}\vec{\psi}_\alpha)_1(b)\int_a^b \vec{\phi}_\alpha^* A\vec{f} = 0$$

on account of (3.1). Thus $\vec{g}$ satisfies all homogeneous endpoint conditions. Consequently in the limit point case for $\vec{y}(t) = \vec{\psi}_\alpha(t)u + \vec{g}(t)$, $\alpha_1 y(a) + \alpha_2\hat{y}(a) = \alpha_1(\alpha_1^* - \alpha_2^* M_\alpha)u + \alpha_2(\alpha_2^* + \alpha_1^* M_\alpha)u = u$ by conditions (1.3) In the limit circle case, the function $\vec{y}(t) = Y_\alpha(t)\vec{w} + \vec{g}(t)$ satisfies $\alpha_1 y(a) + \alpha_2\hat{y}(a) = \alpha_1(\alpha_1^* w - \alpha_2^*\hat{w}) + \alpha_2(\alpha_2^* w + \alpha_1^*\hat{w}) = w$ by (1.3). At the right endpoint, $(Y_\beta^{-1}\vec{y})_1(b) = (Y_\beta^{-1}\vec{\theta}_\alpha)_1(b)w + (Y_\beta^{-1}\vec{\phi}_\alpha)_1\hat{w} = \hat{v}$ if $\vec{w}$ is as specified in the statement.

Finally, for uniqueness, we must show that if $J\vec{y}' = [\lambda A + B]\vec{y}$ and $\vec{y}$ satisfies the homogeneous end conditions, then $\vec{y} = 0$. In the limit point case $\vec{y}(t) = \vec{\psi}_\alpha(t)c$, where c is $n \times 1$; but $0 = \alpha_1 y(a) + \alpha_2\hat{y}(a) = c$ as above. Similarly, in the other case $\vec{y}(t) = Y_\alpha(t)\vec{k}$, where $\vec{k}$ is $2n \times 1$, and $0 = \alpha_1 y(a) + \alpha_2\hat{y}(a) = k$ as before. Lastly, $0 = (Y_\beta^{-1}\vec{y})_1(b) = (Y_\beta^{-1}\vec{\theta}_\alpha)_1(b)k + (Y_\beta^{-1}\vec{\phi}_\alpha)_1(b)\hat{k}$ implies $\hat{k} = 0$ since $k = 0$ and $(Y_\beta^{-1}\vec{\phi}_\alpha)_1(b)$ is invertible.

REFERENCES

1. F. V. Atkinson, "Discrete and Continuous Boundary Problems," *Academic Press,* New York, 1964.

2. E. A. Coddington and N. Levinson, "Theory of Ordinary Differential Equations," *McGraw-Hill,* New York, 1955.

3. C. T. Fulton, Parameterizations of Titchmarsh's m(λ)-functions in the limit circle case, *Trans. A.M.S.* 229 (1977), 51-63.

4. D. B. Hinton and J. K. Shaw, On Titchmarsh-Weyl m(λ)-functions for linear Hamiltonian systems, *J. Diff. Eqs.* 40 (3) (1981), 316-342.

5. D. B. Hinton and J. K. Shaw, Parameterization of the m(λ)-function for a Hamiltonian system of limit circle type, submitted.

6. D. B. Hinton and J. K. Shaw, Titchmarsh-Weyl theory for Hamiltonian systems, in "Spectral Thoery of Differential Operaotrs," pp. 219-231, I. W. Knowles and R. T. Lewis (editors) *North-Holland*, New York, 1981.

7. D. B. Hinton and J. K. Shaw, On boundary value problems for Hamiltonian systems with two singular points, submitted.

8. V. I. Kogan and F. S. Rofe-Beketov, On square-integrable solutions of symmetric systems of differential equations of arbitrary order, *Proc. Royal Soc. Edin.* 74A (1974), 5-39.

9. M. H. Stone, "Linear Transformations in Hilbert Space and their Applications to Analysis," *Amer. Math. Soc. Colloq. Pub.*, vol. 15, *Amer. Math. Soc.*, Providence, RI, 1932.

10. E. C. Titchmarsh, "Eigenfunction Expansions Associated with Second Order Differential Equations," Part I, 2nd edition, *Clarendon Press,* Oxford, 1962.

ON UNIFORM ASYMPTOTIC EXPANSION OF A CLASS OF INTEGRAL TRANSFORMS

K SONI

ABSTRACT

Asymptotic expansions for the integrals of the type

$$F(x,a) = \int_0^a K(xt)\, f(t)\, dt\ , \quad x \to \infty$$

which hold uniformly in a when either $0 \le a \le \delta$ or when $\delta \le a < \infty$ for some $\delta > 0$, are obtained. It is assumed that f has an algebraic singularity at the origin and $K(t)\, t^{\lambda-1}$, $\lambda > 0$ is locally absolutely integrable in $[0,\infty)$. In general, the asymptotic expansion of $F(x,a)$ when $a \to 0+$ cannot be obtained directly from the corresponding expansion when a is bounded away from zero. In some cases, a similar situation may arise as $a \to \infty$. Analytic continuation of the incomplete Mellin transform of K provides a unified approach to this problem.

1. INTRODUCTION

Let

(1.1) $$F(x,a) = \int_0^a K(xt)\, f(t)\, dt\ , \quad 0 < a < \infty\ , \quad x > 0\ .$$

In recent years a number of techniques have been developed that provide an asymptotic expansion of $F(x,a)$ when $x \to \infty$ and a is a fixed positive number. In general such expansions are valid when $a \ge \delta > 0$ but not when $a \to 0+$. It has also been observed that in some cases, even under appropriate conditions on f, the asymptotic expansion of $F(x,\infty)$ does not follow directly from that of $F(x,a)$ as $a \to \infty$, (see, for example, [5, p.228]). An investigation of these phenomena results in a very simple approach to the problem of uniform asymptotic expansions of a certain class of integral transforms. The basic idea is as follows. Suppose that

(1.2) $$f(t) \sim \sum_{k=0}^{\infty} a_k\, t^{k+\mu-1}\ , \quad t \to 0+\ ,$$

and the incomplete Mellin transform of the kernel K, namely

$$g(K,s\;;\;x,a) = \int_0^a K(xt)\,t^{s-1}\,dt \tag{1.3}$$

has an asymptotic expansion as $x \to \infty$. Then under reasonable assumptions on K and f, the formal expansion

$$F(x,a) \sim \sum_{k=0}^{\infty} a_k\, g(K,\, k+\mu\;;\;x,a) \tag{1.4}$$

invariably provides an asymptotic expansion of $F(x,a)$ as $x \to \infty$. Whether such an expansion holds uniformly in $0 \le a \le \delta$ for some $\delta > 0$ or in $0 < \delta \le a < \infty$ may depend on how we represent the function g.

Our results are summed up in two theorems stated in section 2. These are proved in section 3. In section 4 we give some examples.

2. NOTATION, ASSUMPTIONS AND STATEMENT OF RESULTS

We assume that $f(t)$ belongs to $C^p(0,\infty)$ and write

$$f(t) = f_{m-1}(t) + R_m(t)\,, \quad m = 1,2,\ldots,n\,, \tag{2.1}$$

where

$$f_{m-1}(t) = \sum_{k=0}^{m-1} a_k\, t^{\lambda_k}\,, \quad -1 < \lambda_0 < \lambda_1 < \ldots < \lambda_n\,. \tag{2.2}$$

We also assume that

$$R_n^{(k)}(t) = O(t^{\lambda_n - k})\,, \quad t \to 0+\,, \quad k = 0,1,\ldots,p\,. \tag{2.3}$$

Thus the behaviour of $f(t)$ and its successive derivatives near the origin is effectively provided by $f_{n-1}(t)$.

$F(x,a)$ and $g(K,s\;;\;x,a)$ are defined by (1.1) and (1.3) respectively whenever these integrals converge.

We will need to refer to the following conditions regarding the kernel $K(t)$

C_1 : $K(t)\,t^{s-1}$ is locally absolutely integrable in $[0,\infty)$ for $\mathrm{Re}\, s > 0$.

C_2 : For some α,β, the Mellin transform of K,

$$M[K,s] = \int_0^{\infty} K(t)\,t^{s-1}\,dt\,, \quad \alpha < \mathrm{Re}\, s < \beta\,, \tag{2.4}$$

converges. Furthermore, $M[K,s]$ is an analytic function of s in this strip and can be continued analytically as a meromorphic function into the whole s-plane.

$K^{(-m)}(t)$, $m = 0,1,\ldots$, denotes an mth iterated integral of $K(t)$. We will need to refer to the following conditions for $K^{(-m)}(t)$.

C_3 : For $m \geq 1$, $K^{(-m)}(t)$ is continuous in $[0,\infty)$.

C_4 : There exist numbers $b \geq 0$ and $\gamma < 1$ such that for $m \geq 0$,

$$K^{(-m)}(t) = O(t^{m\gamma+b}) , \quad t \to \infty .$$

Finally, $G(K,s\,;\,x,a)$ is defined by

$$G(K,s\,;\,x,a) = \int_a^\infty K(xt)\, t^{s-1} dt , \quad a > 0 . \tag{2.5}$$

Whenever the condition C_2 is satisfied, $G(K,s\,;\,x,a)$ and $g(K,s\,;\,x,a)$ are complementary functions at least in the strip $\alpha < \text{Re } s < \beta$.

THEOREM 1 Let $\lambda_k = k + \mu - 1$, $0 < \mu \leq 1$. If $K(t)$ and $K^{(-m)}(t)$ satisfy the conditions C_1 and C_3 respectively, then for any $\delta > 0$, $0 \leq a \leq \delta$ and $x > 0$,

$$F(x,a) = \sum_{m=1}^{n} (-1)^{m-1} x^{-m} K^{(-m)}(xa) R_m^{(m-1)}(a) + \sum_{m=0}^{n-1} a_m (-x)^{-m} (\mu)_m\, g\left(K^{(-m)},\mu\,;\,x,a\right) + (-x)^{-n} \int_0^a K^{(-n)}(xt) R_n^{(n)}(t) dt . \tag{2.6}$$

The above expansion is in decreasing powers of x and for $0 < c \leq x \leq d$ where c and d are any positive numbers, $F(x,a) \to 0$ uniformly with respect to x as $a \to 0$. Whether this expansion is an asymptotic expansion which holds uniformly in $0 \leq a \leq \delta$ as $x \to \infty$ depends on the behaviour of $K^{(-m)}(xa)$ and $g\left(K^{(-m)},\mu\,;\,x,a\right)$. The main advantage of this expansion is that we can identify and possibly isolate the expressions which cause nonuniform behaviour. For example, if $K(t) = \exp(it)$, $K^{(-m)}(t) = (-i)^m K(t)$. In this case we obtain an expansion containing the simplest integral with the same critical points as the integral defining $F(x,a)$. Apart from that the expansion holds uniformly in a near the origin as $x \to \infty$ (see Example 1 in section 4).

Before we state Theorem 2, we observe that the condition C_2 implies that $g(K,s\,;\,x,a)$ is analytic in $\text{Re } s > \alpha$ and $G(K,s\,;\,x,a)$ is analytic in $\text{Re } s < \beta$. Now by the relation

(2.7) $$g(K,s\ ;\ x,a) + G(K,s\ ;\ x,a) = x^{-s}\, M[K,s]\ ,\quad \alpha < \mathrm{Re}\ s < \beta\ ,$$

both of them can be continued analytically as meromorphic functions to the whole complex s-plane. Moreover, by applying the condition C_4 to integrate $G(K,s\ ;\ x,a)$ by parts, it follows that each of the functions $G\left(K^{(-m)},s\ ;\ x,a\right)$, $m = 1,2,\ldots$, also can be continued analytically to the whole s-plane.

THEOREM 2 Let $\lambda_n + 1 > p$. If $M[K,s]$ is regular at $s = \lambda_k + 1$, $k = 0,1,\ldots,n-1$, and the conditions $C_1 - C_4$ are satisfied, then for $a \geq \delta > 0$, $x > 0$,

(2.8) $$F(x,a) = \sum_{m=0}^{n-1} a_m x^{-\lambda_m - 1}\, M[K,\lambda_m + 1] + \sum_{m=1}^{p} (-1)^{m-1}\, x^{-m}\, K^{(-m)}(xa)\, f^{(m-1)}(a) + E$$

where

(2.9) $$E = (-x)^{-p} \int_0^{\delta} K^{(-p)}(xt)\, R_n^{(p)}(t)\, dt + (-x)^{-p} \int_{\delta}^{a} K^{(-p)}(xt)\, f^{(p)}(t)\, dt - (-x)^{-p} \sum_{m=0}^{n-1} a_m (\lambda_m + 1 - p)_p\, G\left(K^{(-p)},\lambda_m + 1 - p, x, \delta\right) .$$

The expansion (2.8) is asymptotic as $x \to \infty$ in the sense that if $F(x,a)$ is approximated by the finite sums on the right, the function E represents the error. E is not necessarily of the same order as x^{-p} as $x \to \infty$. It is also possible that some of the terms in the approximating sum may be of lower order. However, the order of E decreases as p increases and the expansion is valid uniformly in $\delta \leq a \leq \Delta$ for any $\Delta > 0$. If $K^{(-m)}(t)$, the iterated integrals of $K(t)$ are bounded, then indeed $E = O(x^{-p})$ as $x \to \infty$. In addition to this, if $f^{(k)}(a) \to 0$ as $a \to \infty$, $k = 0,1,\ldots,p-1$, and the second integral in (2.9) converges uniformly in $x \geq N$ for some N as $a \to \infty$, then the expansion (2.8) is valid uniformly in $a \geq \delta$ as $x \to \infty$. These conditions are sufficient but not always necessary.

Sometimes E is written in a slightly different form,

$$(2.10)\qquad E = (-x)^{-p} \int_0^a K^{(-p)}(xt)\, R_n^{(p)}(t)\, dt$$

$$- (-x)^{-p} \sum_{m=0}^{n-1} a_m (\lambda_m+1-p)_p\, G\left(K^{(-p)}, \lambda_m+1-p\ ;\ x,a\right)\ ;$$

or, when the integral representation for $G\left(K^{(-p)}, \lambda_{n-1}+1-p\ ;\ x,a\right)$ can be used,

$$(2.11)\qquad E = (-x)^{-p} \int_0^a K^{(-p)}(xt)\, R_n^{(p)}(t)dt - (-x)^{-p} \int_a^\infty K^{(-p)}(xt)\, f_{n-1}^{(p)}(t)dt\ .$$

3. PROOF OF THEOREMS

We start with the well known technique in asymptotics. By (2.1) and (2.2) we can write $F(x,a)$ in terms of the functions g and R_n.

$$(3.1)\qquad F(x,a) = \sum_{m=0}^{n-1} a_m\, g(K,\lambda_m+1\ ;\ x,a) + \int_0^a K(xt)\, R_n(t)dt.$$

Next, we apply integration by parts and use (2.3) together with the conditions C_1 and C_3. For $p < \lambda_n + 1$, we obtain

$$(3.2)\qquad \int_0^a K(xt)\, R_n(t)dt = \sum_{k=1}^{p} (-1)^{k-1}\, x^{-k}\, K^{(-k)}(xa)\, R_n^{(k-1)}(a)$$

$$+ (-x)^{-p} \int_0^a K^{(-p)}(xt)\, R_n^{(p)}(t)dt\ .$$

Similarly, by the conditions $C_1 - C_4$, for every positive integer q,

$$(3.3)\qquad G(K,s\ ;\ x,a) = \sum_{k=1}^{q} (-x)^{-k}(s-k+1)_{k-1}\, K^{(-k)}(xa)\, a^{s-k}$$

$$+ (-x)^{-q}(s-q)_q\, G\left(K^{(-q)}, s-q\ ;\ x,a\right)\ .$$

For $\operatorname{Re} s < -b$, (3.3) follows by using integration by parts and for $\operatorname{Re} s \geq -b$, by analytic continuation. This provides an asymptotic expansion for $G(K,s\ ;\ x,a)$ and consequently, by means of the relation (2.7), for $g(k,s\ ;\ x,a)$. However, this is useful only when a is bounded away from the origin. Under the conditions of Theorem 1, an expansion for $g(K,m+\mu\ ;\ x,a)$ which is in decreasing exponents of x and is valid when $a \to 0$ can be given as follows. We use integration by parts.

$$(3.4)\qquad g(K,m+\mu\ ;\ x,a) = \sum_{k=1}^{m} (-1)^{k-1} x^{-k} K^{(-k)}(xa)\, a^{m+\mu-k} (m+\mu-k+1)_{k-1}$$

$$+ (-x)^{-m} (\mu)_m \int_0^a K^{(-m)}(xt)\, t^{\mu-1}\, dt\ .$$

This is a finite term expansion. In general, the behaviour of the last integral near $a = 0$ depends on whether xa approaches zero or moves away from zero.

PROOF OF THEOREM 1 We start with (3.1). By (3.4),

$$(3.5)\qquad \sum_{m=0}^{n-1} a_m\, g(K,m+\mu\ ;\ x,a) = \sum_{m=0}^{n-1} a_m (-x)^{-m} (\mu)_m \int_0^a K^{(-m)}(xt) t^{\mu-1} dt + E_1$$

where

$$(3.6)\qquad E_1 = -\sum_{m=1}^{n-1} a_m \sum_{k=1}^{m} (-x)^{-k} K^{(-k)}(xa)\, a^{m+\mu-k} (m+\mu-k+1)_{k-1}\ .$$

Change the order of summation in E_1. Since

$$(3.7)\qquad f_{n-1}^{(k-1)}(a) - f_{k-1}^{(k-1)}(a) = \sum_{m=k}^{n-1} a_m (m+\mu-k+1)_{k-1}\, a^{m+\mu-k}\ ,\quad 1 \le k \le n-1\ ,$$

$$(3.8)\qquad E_1 = -\sum_{k=1}^{n-1} (-x)^{-k} K^{(-k)}(xa) \{f_{n-1}^{(k-1)}(a) - f_{k-1}^{(k-1)}(a)\}\ .$$

Now let $\lambda_m = m + \mu - 1$, $p = n$ and write (3.1) as follows. For the finite sum, use (3.5) with E_1 as given in (3.8) above and replace the integral by its expansion (3.2). The expansion (2.6) follows by observing that for $1 \le k \le n-1$,

$$(3.9)\qquad R_n^{(k-1)}(a) = R_k^{(k-1)}(a) - f_{n-1}^{(k-1)}(a) + f_{k-1}^{(k-1)}(a)\ .$$

PROOF OF THEOREM 2 We start with (3.1) again. By the relation (2.7) which is valid for all s,

$$(3.10)\qquad \sum_{m=0}^{n-1} a_m\, g(K,\lambda_m+1\ ;\ x,a) = \sum_{m=0}^{n-1} a_m x^{-\lambda_m-1} M[K,\lambda_m+1] - E_2$$

where

$$(3.11)\qquad E_2 = \sum_{m=0}^{n-1} a_m\, G(K,\lambda_m+1\ ;\ x,a)\ .$$

By the condition C_1, $g(K,s\ ;\ x,a)$ has no singularity in $\operatorname{Re} s > 0$. Since $M[K,s]$ is regular at $s = \lambda_m + 1$, $m = 0,1,\ldots,(n-1)$, by (2.7) and (3.3) it follows that $G(K,s\ ;\ x,a)$ and $(s-p)_p\, G\big(K^{(-p)},s-p\ ;\ x,a\big)$ also are regular at these points. By (3.3), E_2 can be written as follows.

$$(3.12)\qquad E_2 = (-x)^{-p} \sum_{m=0}^{n-1} a_m(\lambda_m+1-p)_p\, G\big(K^{(-p)},\lambda_m+1-p\ ;\ x,a\big) + \sum_{k=1}^{p} (-x)^{-k} K^{(-k)}(xa)\, f_{n-1}^{(k-1)}(a)\,.$$

The last sum in (3.12) above is obtained by changing the order of summation and then using (2.2). Now rewrite (3.1) by using (3.2), (3.10) and (3.12). The result can be simplified by observing that

$$R_n^{(k-1)}(a) = f^{(k-1)}(a) - f_{n-1}^{(k-1)}(a)\,.$$

This provides the expansion (2.8) with E as given in (2.10). To obtain (2.9), use

$$\int_\delta^a K^{(-p)}(xt)\, f_{n-1}^{(p)}(t)dt = \sum_{m=0}^{n-1} a_m(\lambda_m+1-p)_p \int_\delta^a K^{(-p)}(xt)t^{\lambda_m-p}dt\,,$$

and

$$\int_\delta^a K^{(-p)}(xt)t^{s-1}dt = G\big(K^{(-p)},s\ ;\ x,\delta\big) - G\big(K^{(-p)},s\ ;\ x,a\big)\,.$$

4. EXAMPLES

EXAMPLE 1. Let $K(t) = e^{it}$ and $K^{(-m)}(t) = e^{-im\pi/2}\, e^{it}$. By Theorem 1,

$$(4.1)\qquad \int_0^a f(t)\, e^{ixt}\, dt = -\sum_{m=1}^{n} x^{-m}\, e^{im\pi/2}\, e^{ixa}\, R_m^{(m-1)}(a) + \Big(\int_0^a e^{ixt}\, t^{\mu-1}\, dt\Big) \sum_{m=0}^{n-1} x^{-m}\, a_m\, (\mu)_m\, e^{im\pi/2} + x^{-n}\, e^{in\pi/2} \int_0^a e^{ixt}\, R_n^{(n)}(t)\, dt.$$

The relation (4.1) actually holds for all complex x, $x \neq 0$. In particular, by setting $x = iz$, we may obtain an asymptotic expansion for the integral

$$(4.2)\qquad \int_a^\infty e^{-zt} f(t)\,dt = \int_0^\infty e^{-zt} f(t)\,dt - \int_0^a e^{-zt} f(t)\,dt$$

which is valid uniformly in a as $a \to 0$. Such an expansion was first given by Erde'lyi [1] in 1974 by a technique based on fractional integrals. It can be shown that the two expansions are identical.

The asymptotic expansions for the Fourier integrals, complete with remainder terms, are well known (see, for example, [3] and [4]). Theorem 2 provides still another way of deriving such expansions.

EXAMPLE 2. Let $K(t) = t^{-\nu/2} J_\nu(\sqrt{t})$, $\mathrm{Re}\,\nu > -1$, and let

$$K^{(-m)}(t) = (-2)^m t^{-(\nu-m)/2} J_{\nu-m}(\sqrt{t}).$$

The Mellin transform of $K(t)$ converges in the strip $0 < \mathrm{Re}\,s < (2\,\mathrm{Re}\,\nu + 3)/4$. By using [2, p.326, (1)],

$$(4.3)\qquad M[K,s] = 2^{2s-\nu}\,\Gamma(s)/\Gamma(\nu-s+1)\ .$$

$M[K,s]$ has no singularity in $\mathrm{Re}\,s > 0$. In particular, it is regular at $\lambda_m + 1$, $m = 0,1,\ldots$. Finally,

$$(4.4)\qquad K^{(-m)}(t) = O\left(t^{(m/2)+b}\right)\ ,\quad t \to \infty\ ,$$

where $b = -(2\nu+1)/4$. Thus $K(t)$ satisfies the conditions $C_1 - C_4$. By Theorem 1, for $0 \le a \le \delta$,

$$(4.5)\qquad \int_0^a f(t)\,(xt)^{-\nu/2} J_\nu(\sqrt{xt})\,dt$$

$$= -\sum_{m=1}^{n} (2/x)^m (xa)^{-(\nu-m)/2} J_{\nu-m}(\sqrt{xa})\, R_m^{(m-1)}(a)$$

$$+ \sum_{m=0}^{n-1} (2/x)^m a_m (\mu)_m \left(\int_0^a (xt)^{-(\nu-m)/2} J_{\nu-m}(\sqrt{xt})\, t^{\mu-1} dt\right)$$

$$+ (2/x)^n \int_0^a (xt)^{-(\nu-n)/2} J_{\nu-n}(\sqrt{xt})\, R_n^{(n)}(t)\,dt\ .$$

Although the above expansion is valid for all a near the origin, computationally it is not very attractive. In contrast to the expansion (4.1), where the approximating sum contains only one integral whose behaviour changes with the magnitude of xa, each term of the second sum in (4.5) contains such an integral. For certain kernels, including the Bessel function kernels, this can be avoided if we modify our technique by applying an appropriate differential operator to integrate by parts.

When $a \geq \delta > 0$, by Theorem 2,

$$(4.6)\quad \int_0^a f(t)\,(xt)^{-\nu/2} J_\nu(\sqrt{xt})dt = \sum_{m=0}^{n-1} a_m\, 2^{-\nu}(4/x)^{\lambda_m+1}\,\Gamma(\lambda_m+1)/\Gamma(\nu-\lambda_m)$$

$$- \sum_{m=1}^{p} 2^m x^{-(\nu+m)/2} a^{(m-\nu)/2} J_{\nu-m}(\sqrt{xa})\, f^{(m-1)}(a) \; + \; E$$

where by (2.9),

$$(4.7)\quad E = (2/x)^p \int_0^\delta (xt)^{-(\nu-p)/2} J_{\nu-p}(\sqrt{xt})\, R_n^{(n)}(t)dt$$

$$+ 2^p x^{-(\nu+p)/2} \int_\delta^a t^{-(\nu-p)/2} J_{\nu-p}(\sqrt{xt})\, f^p(t)dt$$

$$+ (-x)^{-p} \sum_{m=0}^{n-1} a_m(\lambda_m+1-p)_p\, G\left(K^{(-p)},\lambda_m+1-p\,;\,x,\delta\right) .$$

The second term on the right in (4.7) indicates that E is not always of the order of x^{-p} as $x \to \infty$. If $a^{-(2\nu+1)/4 + m/2} f^{(m-1)}(a) \to 0$ for $m = 1,2,\ldots,p$, and the second integral in (4.7) converges uniformly in x for all x sufficiently large as $a \to \infty$, then the above expansion holds uniformly in $a \geq \delta$ as $x \to \infty$. The last sum in (4.7) cannot in general be represented as an integral of the form (2.5) although it can be represented as the sum of a finite number of terms together with such an integral. However, it is of order x^{-1} as $x \to \infty$ and has an asymptotic expansion in decreasing exponents of x (see (3.3)).

REMARK. Recently Wong [5] obtained a uniform asymptotic expansion for the integral

$$(4.8)\qquad I(h) = \int_q^z g(\sqrt{y^2 - q^2}\,) \sin yh\, dy , \quad h \to \infty$$

$$(4.9)\qquad g(x) = x^{-1} f(x) .$$

Presumably such integrals arise in the study of small-angle x-ray scattering from membranes. By a change of variable, we can write

$$(4.10) \qquad I(h) = (\pi h/8)^{1/2} \int_0^a t^{-1/2} f(\sqrt{t})\ (t+b)^{-1/4} J_{1/2}(h\sqrt{t+b})dt \ ,$$

where $a = z^2 - q^2$ and $b = q^2$. It is no longer of convolution type but it is similar to Example 2 and the analysis presented in this paper is still applicable. In particular, under appropriate conditions on f, the expansion which corresponds to (4.6) and (4.7), holds uniformly in $a = z^2 - q^2 \geq \delta > 0$.

REFERENCES

1. A Erde'lyi, Asymptotic evaluation of integrals involving a fractional derivative, SIAM J. Math. Anal., 5 (1974), pp. 159-171.

2. A Erde'lyi, W Magnus, F Oberhettinger and F Tricomi, Tables of Integral Transforms, Volume 1, McGraw-Hill, 1954.

3. F W J Olver, Error bounds for stationary phase approximations, SIAM J. Math. Anal., 5 (1974), pp. 19-29.

4. K Soni, On uniform asymptotic expansions of finite Laplace and Fourier integrals, Proc. Royal Soc. Edin., 85 A (1980), pp. 299-305.

5. R Wong, On a uniform asymptotic expansion of a Fourier-type integral, Quart. Appl. Math., 38 (1980), pp. 225-234.

ON A NEW NUMERICAL METHOD FOR A NEW CLASS OF NONLINEAR PARTIAL DIFFERENTIAL EQUATIONS ARISING IN NONSPHERICAL GEOMETRICAL OPTICS

Robert L. Sternberg

Marvin J. Goldstein and Drew Drinkard

The surfaces S: $z = z(x,y)$ and S': $z' = z'(x',y')$ of a three-dimensional, aperture or volume extremized and ellipticized, bifocal, nonspherical singlet optical, radar, or sonar lens antenna satisfy--by Snell's law and the optical path length condition of Fermat's principle--differential equations of the form

$$S:\quad \frac{\partial z}{\partial x} = F, \quad \frac{\partial z}{\partial y} = G, \qquad S':\quad \frac{\partial(z',y')}{\partial(x,y)} = F'\,\frac{\partial(x',y')}{\partial(x,y)}, \quad \frac{\partial(x',z')}{\partial(x,y)} = G'\,\frac{\partial(x',y')}{\partial(x,y)}, \tag{1}$$

symmetry conditions of the form

$$z(-x,y) = z(x,y), \quad z'(-x',y') = z'(x',y'),$$
$$z(x,-y) = z(x,y), \quad z'(x',-y') = z'(x',y'), \tag{2}$$

and boundary conditions of the form

$$z(x,y) = 0 \quad \text{and} \quad z'(x,y) = 0, \tag{3}$$

on the ellipse

$$\Gamma:\quad (x^2/b_0^2\cos^2\psi_0) + (y^2/b_0^2) = 1\ . \tag{4}$$

Here we are given the lens focal length $|z_0| > 0$, its maximum diameter $2b_0 > 0$, its index of refraction $n_0 > 1$, and its off-axis or bifocal angle

$\psi_0 \neq 0$, while x and y are independent variables and x', y', z, and z' are dependent variables such that $x' = x'(x,y)$, $y' = y'(x,y)$, $z = z(x,y)$, and $z' = z'(x',y') = z'[x'(x,y), y'(x,y)]$. Geometrically, x, y, z and x', y', z' are the intercepts on S: $z = z(x,y)$ and S': $z' = z'(x',y')$ of an arbitrary ray R passing from the primary finite, perfect or stigmatic, focal point F at x_0, y_0, z_0 where

$$x_0 = 0, \qquad y_0 = [b_0^2 + z_0^2 \sec^2 \psi_0]^{1/2} \sin \psi_0, \qquad z_0 = -|z_0|, \tag{5}$$

to the infinite perfect focal point F_∞ located in the direction $-\psi_0$ relative to z-axis and at infinity in the y,z-plane as in the illustration shown in Figure 1. By virtue of the symmetry conditions (2), the lens has, in addition, two symmetrically displaced secondary, perfect or stigmatic, focal

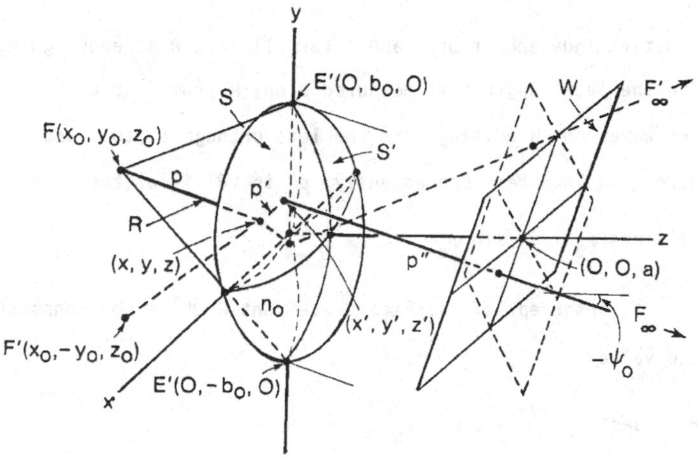

Figure 1

points, one finite, F' at x_0, $-y_0$, z_0, and one infinite, F'_∞, located in the direction $+\psi_0$ relative to the z-axis and also at infinity in the y,z-plane. The secondary focal points F' and F'_∞ are dual to the primary foci F and F_∞.

The coefficient functions F, G, F', and G' in (1) are of the form

$$F = \frac{n_0(x' - x)p - xp'}{(z - z_0)p' - n_0(z' - z)p}, \qquad G = \frac{n_0(y' - y)p - (y - y_0)p'}{(z - z_0)p' - n_0(z' - z)p}, \tag{6}$$

$$F' = \frac{n_0(x' - x)}{p' \cos\psi_0 - n_0(z' - z)}, \qquad G' = \frac{n_0(y' - y) + p' \sin\psi_0}{p' \cos\psi_0 - n_0(z' - z)},$$

where p and p' are the path length elements

$$p^2 = x^2 + (y - y_0)^2 + (z - z_0)^2, \quad p'^2 = (x' - x)^2 + (y' - y)^2 + (z' - z)^2. \tag{7}$$

Here we have suppressed x_0 since, as in (1), we assume that $x_0 = 0$.

Essentially equivalent to either the equations for S or those for S' in (1)--and capable of being substituted for one of those sets of equations--is the optical path length condition that

$$P \equiv p + n_0 p' + p'' = P_0, \tag{8}$$

where P_0 is a constant independent of x and y for all rays R as above going from F to F_∞ when the path length P is measured along R from F to a momentarily fixed wavefront W exiting from the lens orthogonal to R and propagating toward F_∞. Thus here consequently, p" in (8) is of the form

$$p'' = (a - z') \cos\psi_0 + y' \sin\psi_0 , \tag{9}$$

where a is the z-axis intercept of the fixed wavefront W while the constant P_0 in (8) has the value

$$P_0 = [b_0^2 + z_0^2 \sec^2\psi_0]^{1/2} + a \cos\psi_0 . \tag{10}$$

The geometrical arrangement showing a typical ray R and its corresponding wavefront W is perhaps further clarified by the illustration.

For full details of the genesis of the problem and the equations (1) through (10) and information on previously developed and applied methods of their solution, see Sternberg [1, 3 and 4].

In the following--assuming always that the F-number, $\mathcal{F} = |z_0|/2b_0$ is sufficiently large and that the bifocal angle ψ_0 is sufficiently small, in fact that they are substantially larger and smaller, respectively, than the absolute bounds

$$\mathfrak{F} = |z_0|/2b_0 > \tfrac{1}{2} \sin\psi_0 \cos\psi_0 \qquad \text{and} \qquad \psi_0 < \pi/2, \tag{11}$$

noted in the existence conjecture of Sternberg [3] since the index n_0 here is assumed finite--we outline a new numerical method of solving approximately the problem consisting of the equations (1) and the symmetry and boundary conditions (2), (3), and (4) by what may be described as ellipticized and constrained spline approximations for the surfaces S: $z = z(x,y)$ and S': $z' = z'(x',y')$. The new method is based on a new transcendental eigenvalue-like process similar to, but distinct from, the related techniques described in Sternberg [1,4].

To keep matters simple we limit our considerations for the present to solving the problem approximately as described above, first, only locally in two-dimensional neighborhoods of the points E_0 and E'_0 located at 0, $\pm b_0$, 0 at the ends of the principle diameter of the lens and, second, to extending the partial solution thus found to solve the complete problem, still only locally and approximately, but now in a three-dimensional neighborhood of the complete lens boundary curve Γ given by (4) which passes through the points E_0 and E'_0.

To these ends we begin by assuming that the surfaces S: $z = z(x,y)$ and S': $z' = z'(x',y')$ can be approximated in two dimensions near the points E_0 and E'_0 at 0, $\pm b_0$, 0 in the y,z-plane by polynomials of the forms

$$\begin{aligned} S:&\quad z = \gamma_1(y^2 - b_0^2) + \gamma_2(y^2 - b_0^2)^2, \\ S':&\quad z' = \gamma'_1(y'^2 - b_0^2) + \dots + \gamma'_4(y'^2 - b_0^2)^4, \end{aligned} \tag{12}$$

and seek to determine the coefficients γ_k and γ'_k. The first pair are known. In fact

$$\gamma_1 = -\frac{1}{2b_0}\tan\xi_0 \qquad \text{and} \qquad \gamma'_1 = -\frac{1}{2b_0}\tan\xi'_0, \tag{13}$$

where ξ_0 and ξ'_0 are the angles made by S and S' with the y-axis at E_0 and E'_0 and are readily computed as in Sternberg [1, 4]. To evaluate the remaining coefficients in (12) we proceed as follows.

Write $\gamma_2 \equiv \omega$ and treat the latter as an eigenvalue to be determined. With reference to Figure 2, pass a ray R in the y,z-plane from the focal point F at x_0, y_0, z_0 as in (5) to the orthogonal wavefront W via points $x=0$, y, z on S and $x'=0$, y', z' on S' in the y,z-plane, bending the ray at S so as to satisfy Snell's law and computing $y' = y'(y)$, $z' = z'[y'(y)]$ on S' by the path length condition (8).

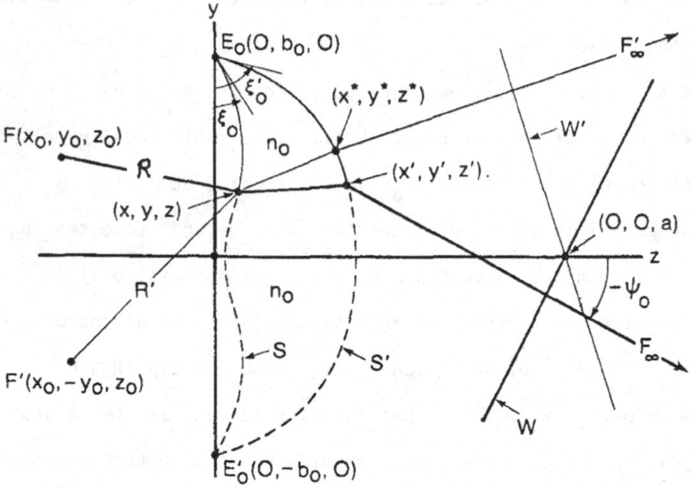

Figure 2

Here y is an independent variable and for fixed values of y, the quantity $z = z(y)$ is to be thought of as a function of ω and is to be computed by the first of equations (12).

By the process of passing the ray R through S and S' we also find that $y' = y'(y)$ and $z' = z'[y'(y)]$ can be computed in terms of ω and can be expressed as elementary transcendental functions of the latter and the several constants of the problem. By applying Snell's law at $y' = y'(y)$,

$z' = z'[y'(y)]$ on S', the derivative dz'/dy' can be similarly expressed. For given fixed y let

$$Z(\omega), \quad Y'(\omega), \quad Z'(\omega), \quad \text{and} \quad Z'_{y'}(\omega), \tag{14}$$

denote these elementary algebraic and transcendental expressions.

Similarly, we trace a ray R' in the y,z-plane from F' at x_0, $-y_0$, z_0 to a corresponding orthogonal wavefront W' via the same point x=0, y, z on S and through a point x*=0, y*, z* on S'. Similarly as before we find that $y^* = y^*(y)$ and $z^* = z^*[y^*(y)]$ can be computed in terms of ω and also that the derivative dz^*/dy^* can be similarly evaluated. For given fixed y let

$$Y^*(\omega), \quad Z^*(\omega), \quad \text{and} \quad Z^*_{y^*}(\omega), \tag{15}$$

denote these elementary algebraic and transcendental expressions.

Keeping in view the fact that each of the quantities $Z(\omega)$, $Y'(\omega)$, $Z'(\omega)$, $Z'_{y'}(\omega)$, $Y^*(\omega)$, $Z^*(\omega)$, and $Z^*_{y^*}(\omega)$ could now be computed explicitly for fixed y as a function of ω if ω were known, we set up the system of equations

$$\begin{aligned}
&(Y'^2 - b_0^2)^2\gamma_2' + \ldots + (Y'^2 - b_0^2)^4\gamma_4' = Z' - \gamma_1'(Y'^2 - b_0^2),\\
&4Y'\,(Y'^2 - b_0^2)\gamma_2' + \ldots + 8Y'(Y'^2 - b_0^2)^3\gamma_4' = Z'_{y'} - 2\gamma'_1Y',\\
&(Y^{*2} - b_0^2)^2\gamma_2' + \ldots + (Y^{*2} - b_0^2)^4\gamma_4' = Z^* - \gamma_1'(Y^{*2} - b_0^2),\\
&4Y^*(Y^{*2} - b_0^2)\gamma_2' + \ldots + 8Y^*(Y^{*2} - b_0^2)^3\gamma_4' = Z^*_{y^*} - 2\gamma_1'Y^*,
\end{aligned} \tag{16}$$

for the unknowns γ_2', γ_3', and γ_4'. These are four equations in three unknowns with coefficients entirely expressible for fixed y in terms of elementary algebraic and transcendental functions of the eigenvalue ω and known constants.

To solve the system (16) for the unknowns γ_2', γ_3', γ_4', we have first to determine the value of ω by solving the eigenvalue-like equation

$$\Delta(\omega) \equiv \begin{vmatrix} (Y'^2 - b_0^2)^2 & \cdots & (Y^{*2} - b_0^2)^4 & \gamma_1'(Y'^2 - b_0^2) - Z' \\ 4Y'(Y'^2 - b_0^2) & \cdots & 8Y'(Y'^2 - b_0^2)^3 & 2\gamma_1' Y' - Z'_{y'} \\ (Y^{*2} - b_0^2)^2 & \cdots & (Y^{*2} - b_0^2)^4 & \gamma_1'(Y^{*2} - b_0^2) - Z^* \\ 4Y^*(Y^{*2} - b_0^2) & \cdots & 8Y^*(Y^{*2} - b_0^2)^3 & 2\gamma_1' Y^* - Z^*_{y^*} \end{vmatrix} = 0. \quad (17)$$

The determinant $\Delta(\omega)$ simplifies to the expression

$$\begin{aligned}\Delta(\omega) \equiv\ & 8Y'Y^*R'R^*(R' - R^*)\,[S'R^{*3}(R^* - 2R') - S^*R'^3(R' - 2R^*)] \\ & + 2R'^2R^{*2}(R^* - R')^2\,[T'R^{*2}Y^* + T^*R'^2Y'],\end{aligned} \quad (18)$$

where

$$\begin{aligned} R' &= Y'^2 - b_0^2, & R^* &= Y^{*2} - b_0^2, \\ S' &= Z' - \gamma_1' R', & S^* &= Z^* - \gamma_1' R^*, \\ T' &= Z_{Y'} - 2\gamma_1' Y', & T^* &= Z^*_{y^*} - 2\gamma_1' Y^*, \end{aligned} \quad (19)$$

in which each capital letter is now a known elementary algebraic or transcendental function of the eigenvalue ω.

The solution of the determinantal equation (17) using (18) and (19) is not difficult computationally and $\omega = \gamma_2$ in (12) is reasonably easily computed following which the remaining coefficients γ_2', ..., γ_4' therein can be found in theory by solving any three of the equations (16), assuming the appropriate lesser minors therein are nonsingular. In practice, in an attempt to reduce round-off errors, we solve each triple of equations (16) and average.

An approximate solution for the surfaces S: $z = z(x,y)$ and S': $z' = z'(x',y')$ valid in a three-dimensional neighborhood of the boundary curve Γ given by (4) which passes through the points E_0 and E_0' at 0, $\pm b_0$, 0 can now be written down in the form

$$\begin{aligned} S:\ & z = \gamma_1[(x^2/\cos^2\psi_0) + y^2 - b_0^2] + \gamma_2[(x^2/\cos^2\psi_0) + y^2 - b_0^2]^2, \\ S':\ & z'\ \gamma_1'[(x'^2/\cos^2\psi_0) + y'^2 - b_0^2] + \ldots + \gamma_4'[(x'^2/\cos^2\psi_0) + y'^2 - b_0^2]^4, \end{aligned} \quad (20)$$

by ellipticizing the two-dimensional solution (12); that is to say by constructing the ellipticized solution (20) from (12). Note that here x' and y' can be thought of as simply new additional independent variables if we wish, their dependence on x and y being of no consequence in using (20) to specify the shape of the surfaces S and S'. To determine the quantities x' and y' as functions $x' = x'(x,y)$ and $y' = y'(x,y)$, if these expressions are actually desired, however, is a simple matter by ray-tracing once the solutions for S and S' have been found. The ray tracing process also provides a simple check on the accuracy of the computed solution in terms of any given wavelength λ_0.

To improve the solutions obtained for S and S' the value of y used in the computation can be taken closer and closer to b_0 or higher degree polynomials can be used in (12) or, alternatively, completion or correction terms, as they have been called, can be added, if desired, to each of the expressions (20) for S: $z = z(x,y)$ and S': $z' = z'(x',y')$ by using a Cauchy-Kovalevsky type process, or even by trial and error, as described in Sternberg [4].

Finally, the entire computational process can be repeated with suitable modifications to step the two-dimensional solution out of neighborhoods of the points E_0 and E'_0 at $0, \pm b_0, 0$ and the three-dimensional solution out of the neighborhood of the boundary curve Γ thereby determining the full form of the lens antenna to high accuracy.

As a numerical example consider the case in which the focal length $|z_0|$, maximum diameter $2b_0$, index of refraction n_0, and off-axis or bifocal angle ψ_0 have the values

$$|z_0| = 9, \quad 2b_0 = 18, \quad n_0 = 1.594 \quad \text{and} \quad \psi_0 = 20°, \tag{21}$$

as in the example discussed in Sternberg [4].

For this case, proceeding as in Sternberg [1, 3, 4] we find for the angles

ξ_0 and ξ'_0 and the coefficients γ_1 and γ'_1 in (13), the values

$$\xi_0 = 4.8738440^\circ, \qquad \xi'_0 = 55.743568^\circ, \tag{22}$$
$$\gamma_1 = -4.7372426 \times 10^{-3}, \qquad \gamma'_1 = -8.1574583 \times 10^{-2},$$

whence, applying our new numerical process with the given value of y set at $y = b_0 - \delta b_0$, we obtain for the eigenvalue $\omega = \gamma_2$ and the remaining coefficients $\gamma'_2, \ldots, \gamma'_4$, the values given in the following table for different assumed values of the step-size-factor δ.

Table of Results

δ	$\omega \times 10^5$	$\gamma'_2 \times 10^4$	$\gamma'_3 \times 10^6$	$\gamma'_4 \times 10^8$
0.1	-1.336	-3.071	-4.094	-9.612
0.01	-1.231	-2.952	-4.705	----
0.001	-1.215	-2.950	-2.655	----

In comparison to these results the true values of $\omega = \gamma_2$ and the coefficients γ'_2 and γ'_3--as calculated by a Cauchy-Kovalevsky like process carried out some time ago in connection with Sternberg [1]--are

$$\gamma'_2 = -2.936 \times 10^{-4},$$
$$\omega = \gamma_2 = -1.220 \times 10^{-5}, \tag{23}$$
$$\gamma'_3 = -2.033 \times 10^{-6}.$$

Although no check on γ'_4 is available, since no accurate calculations of that coefficient have ever been made, it may nevertheless be seen that the new results compare quite well with the old. Hence, in suitably restricted neighborhoods of the boundary curve Γ, the antenna should perform well even at very small values of the wavelength λ_0 and by extending the method step-by-step or by using some of the other refinement methods as described previously, the solutions for the complete surfaces S and S' should be able to be computed to reasonably high accuracy.

Physical and engineering applications of the lens antennas studied here and related devices are discussed in Sternberg [1, 3, 4], Sternberg and Goldberg [2], Corbett, Middleton, and Sternberg [5], and Sternberg and Anderson [6] and in the therein referred to additional references and text books.

Finally, note that our results are computational only, no complete existence, uniqueness or convergence theory having ever been established for the problem.

References

1. R. L. Sternberg, "Successive Approximation and Expansion Methods in the Numerical Design of Microwave Dielectric Lenses," Journal of Mathematics and Physics, Vol. 34, 1955, pp. 209-235.

2. R. L. Sternberg and H. B. Goldberg, "Off-Axis Short Focal Length Dielectric Lens Antennas," Proceedings National Aerospace Electronics Conference, Dayton, Ohio, 1961.

3. R. L. Sternberg, "Surface Series Expansion Coefficients for Numerical Design of Scannable Aspherical Microwave and Acoustic Lenticular Antennas," Proceedings of the IEE (England), Vol. 121, 1974, pp. 1351-1354.

4. R. L. Sternberg, "Optimizing and Extremizing Nonlinerar Boundary Value Problems in Lenticular Antennas in Oceanography, Medicine and Communications: Some Solutions and Some Questions," in Nonlinear Systems and Applications, V. Lakshmikantham, Editor, Academic Press, New York, 1977.

5. K. T. Corbett, F. H. Middleton, and R. L. Sternberg, "Nonspherical Acoustic Lens Study," Journal of the Acoustical Society of America, Vol. 59, May 1976, pp. 1104-1109.

6. R. L. Sternberg and W. A. Anderson, "Frequency Independent Beamforming," Naval Research Reviews, Vol. 33, No. 1, 1981, pp. 34-47.

Local Existence Theorems for Ordinary Differential Equations of Fractional Order

by

Ahmad Zain - Alabedeen Mohammad Tazali

ABSTRACT

In this paper, we prove two local existence theorems, by using both the Picard method and the Schauder fixed-point theorem, for the following initial-value problem:

$$g^{(\alpha)}(x) = f(x,g(x)) \quad \text{(almost all } x \in [a,a+h])$$

with (A)

$$g^{(\alpha-1)}(a) = b\Gamma(\alpha), \quad 0 < \alpha \leq 1 ,$$

where $g^{(\alpha)}$ denotes the derivative of order α of a real-valued function g ; $\Gamma(\alpha)$ is the Gamma function where $\alpha > 0$; b is a real number, and under suitable conditions on the function f.

If $\alpha = 1$ in the initial-value problem (A), then the existence theorems corresponding to this problem are known (sometimes) as the Carathéodory theorem, see Coddington and Levinson (1955) and Hale (1969). Finally, we prove a local existence theorem of the maximum and the minimum solutions for the initial-value problem (A) above; when $\alpha = 1$, this theorem reduces to Theorem 1.2 of Coddington and Levinson (1955).

1. Preliminaries and definitions

Throughout of this work the symbols are:

Γ, β denote the Gamma and Betafunctions respectively; R, R_+ are the sets of all real numbers, non-negative real numbers; L(1) is the class of all real-valued functions which are Lebesgue-integrable on I _ R; AC(J) denotes for the set of all real-valued absolutely continuous functions defined on any compact subinterval J of R; $AC_{loc}(I)$ is the set of all real-valued absolutely continuous functions defined on any compact subinterval of I; C(I) is the set of all real-valued continuous functions defined on interval I _ R; N_o, N_+ are the sets of all non-negative, positive integers respectively; f = f(.) denotes the function f, and f(x) denotes the function value at the point x; $\{f_n(.) : n \in N_+\}$ denotes a sequence of functions f_n; a.e. is to be read as "almost everywhere"; (ppx ε A) is to be read as "almost all x εA)".

2. Definitions and lemmas

For the following definitions 2.1, 2.2 and the proof of lemma 2.2 see Barrett (1954).

Definition 2.1 Let f be a real-valued function which is Lebesgue-measurable and is defined a.e. on [a,b[. For real α J 0, define

$$\underset{a}{\overset{b}{I}}{}^{\alpha} f = \frac{1}{\Gamma(\alpha)} \int_a^b (b - t)^{\alpha-1} f(t)dt,$$

provided that the (Lebesgue) integral exists.

An extension of definition 2.1 is:

Definition 2.2 If $\alpha \leq 0$ and if n is the smallest positive integer such that $\alpha + n > 0$, then define

$$\underset{a}{\overset{b}{I}}{}^{\alpha} f = D_x^n \underset{a}{\overset{x}{I}}{}^{\alpha+n} f \text{ at } x = b, \quad (D_x^n = \frac{d^n}{dx^n})$$

provided that $\underset{a}{\overset{x}{I}}{}^{\alpha+n} f$ and its (n - 1) derivatives exist in a segment $|b - x| < h$ (h > 0) and the nth derivative exists at x = b.

Lemma 2.1 Let p,q ε R such that q > -1. If x > a, then

$$\underset{a}{\overset{x}{I}}{}^{p} \frac{(t - a)^q}{\Gamma(q + 1)} = \begin{cases} \dfrac{(x - a)^{p+q}}{\Gamma(p + q + 1)}, & p + q \neq -n \\ 0, & p + q = -n \end{cases} \qquad (n \in N_+)$$

Proof: The proof is straightforward.

Lemma 2.2 Let α > 0 and f ε L(a,b). Define

$$F_\alpha(x) = \underset{a}{\overset{x}{I}}{}^{\alpha} f \quad (x \in (a,b)).$$

Then the following results hold:

(i) F_α ε L(a,b) and exists a.e on [a,b] when α ε (0,1),

(ii) $\int_a^x F_\alpha(t)dt = \underset{a}{\overset{x}{I}}{}^{1+\alpha} f(.)$ (x ε [a,b]),

(iii) $F_\alpha(x) = D_x \underset{a}{\overset{x}{I}}{}^{1+\alpha} f(.)$ (ppx ε [a,b]).

(iv) $F_\alpha(.)$ ε AC[a,b] when $\alpha \geq 1$ or f ε AC[a,b],

(v) $\underset{a}{\overset{x}{I}}{}^{p} \underset{a}{\overset{t}{I}}{}^{q} f(.) = \underset{a}{\overset{x}{I}}{}^{p+q} f(.)$ (p,q > 0 and ppx ε [a,b]).

<u>Definition 2.3</u> Let $\alpha \in (-\infty,1)$; let I be a compact interval of the real line R containing the point a; then define

$$D^{(\alpha)}(I) = \{y \in L(I) : x \to \overset{x}{\underset{a}{I}}{}^{1-\alpha} |y(.)| (x \in I)\} \in AC(I).$$

<u>Note</u> Clearly $D^{(0)}(I) = L(I)$.

<u>Definition 2.4</u> For $y \in D^{(\alpha)}(I)$, we define for $0 \leq \alpha < 1$,

$$y^{(\alpha)}(x) = \frac{d^\alpha y}{dx^\alpha} = \frac{d}{dx} \overset{x}{\underset{a}{I}}{}^{1-\alpha} y(.) \quad (\text{pp}x \in I)$$

and for $\alpha < 0$,

$$y^{(\alpha)}(x) = \overset{x}{\underset{a}{I}}{}^{-\alpha} y(.) \quad (x \in I),$$

provided that $\overset{x}{\underset{a}{I}}{}^{-\alpha} y(.)$ exists.

<u>Definition 2.5</u> Let $\alpha \in (0,1)$ and $I_\mu = [a,a+\mu]$; we define the space $LT_\alpha(I_\mu)$ as the set of all $m \in L(I_\mu)$ such that:

(i) $\{x \to \overset{x}{\underset{a}{I}}{}^{\alpha} |m(.)| \; (x \in (a,a+\mu]\} \in C(a,a+\mu]$,

(ii) $\lim_{x \to a^+} \overset{x}{I}{}^{\alpha} |m(.)| = 0$.

<u>Lemma 2.3</u> (i) If $\alpha \in (0,1)$, then $L^\infty(a,b) \quad LT_\alpha[a,b] \quad L(a,b)$.

(ii) If $\alpha = 1$, then $L^\infty(a,b) \quad LT_\alpha[a,b] = L(a,b)$.

<u>Proof</u> Clearly $LT_\alpha[a,b] \subseteq L(a,b)$. To prove $LT_\alpha \neq L(a,b)$, we consider the following example:

Let $m : [a,b] \to R$ be defined by $m(x) = (b-x)^{-\alpha}$ for all $x \in [a,b]$ and $0 < \alpha < 1$. Obviously $m \in L(a,b)$, but $m \notin LT_\alpha[a,b]$ since $\lim_{x \to b^-} {}_a\int^x (x-t)^{\alpha-1}(b-t)^{-\alpha}dt > \lim_{x \to b^-} \ln\frac{b-a}{b-x}$ does not exist.

Now we prove that $L^\infty(a,b) \subseteq LT_\alpha[a,b]$ and to do this let $m \in L^\infty(a,b)$, define $g : [a,b] \to R$ by:

$$g(x) = \frac{1}{\Gamma(\alpha)} {}_a\int^x (x-t)^{\alpha-1} |m(t)| dt \quad (x \in (a,b]).$$

Then $m \in LT_\alpha[a,b]$ and the proof is straightforward.

Next, we show that $L^\infty(a,b) \neq LT_\alpha[a,b]$. Let $m : [a,b] \to R$ be such that $m(x) = \ln\frac{1}{b-x}$ $(x \in [a,b))$. Then, by using integration by parts, we obtain that $m \in L(a,b)$; let further $\phi_\alpha : [a,b] \to R$, where $\alpha \in (0,1)$, be defined by

$$\phi_\alpha(x) = {}_a\int^x (x-t)^{\alpha-1} \ln\frac{1}{b-t}\, dt \quad (x \in (a,b]).$$

Thus

$$\phi_\alpha \frac{(x-a)^\alpha}{a} \ln\left(\frac{1}{b-a}\right) + \frac{1}{\alpha} {}_a\int^x \frac{(x-t)^\alpha}{b-t} dt \quad (x \in (a,b]),$$

and hence $\phi_\alpha(x)$ exists ($x \varepsilon (a,b]$) and $\lim_{x\to a^+} \phi_\alpha(x) = 0$. Then, by using sections 10.8, 11.7 of Titchmarsh (1939), one can show that $\phi_\alpha \varepsilon AC[a,b]$. Consequently the definition 2.5 implies that $m \varepsilon LT_\alpha[a,b]$. Furthermore, $m \notin L^\infty(a,b)$ since $m(x) = \ln(\frac{1}{b - x}) \to \infty$ as $x \to b^-$. Clearly (ii) holds and this completes the proof of lemma 2.3

In the following theorem, we prove a result which gives some information about the elements of the space $LT_\alpha[a,b]$.

Theorem 2.1(a) Let $\alpha \varepsilon (0,1)$ and $\phi : [a,b] \to R$ be a Lebesgue measurable function; let $m : [a,b] \to R_+$ be such that

$$|\phi(t)| \leq m(t) \quad (\text{almost all } t \varepsilon [a,b]), \text{ then } \phi \varepsilon LT_\alpha[a,b].$$

(b) For any $x_1, x_2 \varepsilon [a,b]$ such that $x_1 < x_2$, there exists $\delta = \delta_\varepsilon$ depending only on ε such that

$$\int_{x_1}^{x_2}(x_2 - t)^{\alpha-1}m(t)dt < \varepsilon \text{ whenever } x_2 - x_1 < \delta.$$

Proof of (a) Let $y,z : [a,b] \to R$ be defined by

$$y(x) = \frac{1}{\Gamma(\alpha)} \int_a^x (x - t)^{\alpha-1}|\phi(t)|dt \quad (x \varepsilon (a,b])$$

and (2.1)

$$z(x) = \frac{1}{\Gamma(\alpha)} \int_a^x (x - t)^{\alpha-1}m(t)dt \quad (x \varepsilon [a,b]).$$

Now, if $x_1,x_2 \varepsilon (a,b]$ such that $x_2 > x_1$ (the case $x_2 < x_1$ is similar), then from (2.1) we have

$$|y(x_2) - y(x_1)| \leq |z(x_1) - z(x_2)| + \frac{1}{\Gamma(\alpha)} \int_{x_1}^{x_2}(x_2 - t)^{\alpha-1}m(t)dt \quad (2.2)$$

But

$$\frac{1}{\Gamma(\alpha)} \int_{x_1}^{x_2}(x_2 - t)^{\alpha-1}m(t)dt = \frac{1}{\Gamma(\alpha)} \int_a^{x_2}(x_2 - t)^{\alpha-1}m(t)dt$$

$$- \frac{1}{\Gamma(\alpha)} \int_a^{x_1}(x_1 - t)^{\alpha-1}m(t)dt - \frac{1}{\Gamma(\alpha)} \int_a^{x_1}\{(x_2 - t)^{\alpha-1} - (x_1 - t)^{\alpha-1}\}m(t)dt$$

$$\leq |z(x_1) - z(x_1)| - \frac{1}{\Gamma(\alpha)} \int_a^{x_1}\{(x_1 - t)^{\alpha-1} - (x_2 - t)^{\alpha-1}\}m(t)dt \quad (2.3)$$

But $\alpha \varepsilon (0,1)$, $x_2 > x_1$ imply that

$$\{(x_1 - t)^{\alpha-1} - x_2 - t)^{\alpha-1}\} < (x_1 - t)^{\alpha-1}. \quad (2.4)$$

Therefore, from (2.1), (2.3) and (2.4), we obtain

$$|y(x_2) - y(x_1)| \leq 3|z(x_2) - z(x_1)| + \frac{2}{\Gamma(\alpha)} \int_a^{x_1}(x_1 - t)^{\alpha-1}m(t)dt. \quad (2.5)$$

Since $m \varepsilon LT_\alpha[a,b]$, it follows that $z \varepsilon C[a,b]$ and so the integral

${}_{a}\int^{x_1}(x_1 - t)^{\alpha-1}m(t)dt$ exists $(t \in [a,x_1])$, then by the Lebesgue dominated convergence theorem of Titchmarch (1939), we have

$$\lim_{x_2 \to x_1} \int_{x_1}^{x_2} \{(x_1 - t)^{\alpha-1} - (x_2 - t)^{\alpha-1})\}m(t)dt = 0.$$

Thus $\lim_{x_2 \to x_1} \int_{x_1}^{x_2} (x_2 - t)^{\alpha-1}m(t)dt = 0.$ Now, by using (2.2), we see $y \in C(a,b[$. Furthermore $m \in LT_\alpha[a,b]$ implies that $\lim_{x \to a^+} z(x) = 0$ and therefore $\lim_{x \to a^+} y(x) = 0$ since $|y(x)| \leq z(x)$ $(x \in (a,b])$. Consequently, by definition 2.5, it follows that $\phi \in LT_\alpha[a,b]$.

Proof of (b) From part (a) we have

$$\lim_{x_2 \to x_1} \int_{x_1}^{x_2} (x_2 - t)^{\alpha-1}m(t)dt = 0.$$

By using (2.3), (2.4), it is easy to show that the integral $\int_{x_1}^{x_2} (x_2 - t)^{\alpha-1}m(t)dt$ converges uniformly on $a \leq x_1 < x_2 \leq b$. Hence (2.1)(b) holds.

Note It is of interest to note that if in the definition of $LT_\alpha[a,a + \mu]$ we replace condition (i) by

(i) $\{ x \to \frac{1}{\Gamma(\alpha)} {}_{a}\int^{x}(x - t)^{\alpha-1}|m(t)dt \quad (x \in (a,a + \mu]\} \in AC_{loc}(a,a + \mu]$

then it is not known whether theorem 2.1 remains true.

Definition 2.6' Let $\alpha \in (0,1]$ and $I_h' : (a,a + h]$. $h > 0$. Define $E^{(\alpha)}(I_h') = \{g \in C(I_h') : \lim_{x \to a^+} (x - a)^{1-\alpha}g(x)$ exists and is finite$\}$. For any $g \in E^{(\alpha)}(I_h')$, let the norm $||.||_\alpha$ be defined by

$$||g||_\alpha = \sup_{x \in I_h'}\{(x - a)^{\alpha-1}|g(x)|\}.$$

Lemma 2.6

(i) $((E^{(\alpha)}(I_h'), ||.||)$ is a Banach space.

(ii) A sequence $\{g_n(.) : n \in N_+\}$ in $E^{(\alpha)}(I_h')$ converges to $g \in E^{(\alpha)}(I_h')$, if and only if the sequence $\{(x - a)^{1-\alpha}g_n(x) : n \in N_+\}$ is uniformly convergent to $(x - a)^{1-\alpha}g(x)$ on I_h'. The proof is straightforward.

3. Local existence theorem for ordinary differential equation of order α, where $\alpha \in (0,1]$.

Theorem 3.1 (The Picard method)

Let $I,J \_ R$ and $\alpha \in (0,1]$; let $(a,b) \in IXJ$ and choose $\mu,\gamma \in R_+$ so that the set

$$G_{\alpha,\gamma} := \{(x,y) : a < x \leq a + \mu, |(x - a)^{1-\alpha}y - b| \leq \gamma\} \quad IXJ.$$

Suppose $f : IXJ \to R$ satisfies:

(i) $f(x,.) \in C(J) \quad (x \in I)$,

(ii) $f(.,y)$ is Lebesgue-measurable $(y \in J)$,

(iii) there exists $m : I_\mu := [a,a+\mu] \to R_+$ such that $m \in LT_\alpha(I_\mu)$ and $m(x) > 0$ $(x \in I_\mu)$ and $|f(x,y)| \leq m(x) \quad ((m,y) \in G_{\alpha,\gamma})$. Define $0 < h \leq \mu$ such that

$$h^{1-\alpha} \sup_{x\in[a,a+h]} \frac{1}{\Gamma(\alpha)} \int_a^x (x-t)^{\alpha-1} m(t)dt \leq \gamma,$$

then there exists at least one function $g : (a,a+h] \to R$ which satisfies:

(1) $g \in C(a,a+h)$ but $g \notin C[a,a+h]$,

(2) $g \in D^{(\alpha)}[a,a+h]$,

(3) $g^{(\alpha)}(x) = f(x,g(x))$ (almost all $x \in [a,a+h]$),

(4) $g^{(\alpha-1)}(.) \in AC[a,a+h]$,

(5) $g^{(\alpha-1)}(a) = b\Gamma(\alpha)$.

Note Before giving the proof of theorem 3.1, we need to prove the following lemma:

Lemma 3.1 Let the conditions (ii) and (iii) of theorem be given and consider the following equation

$$g(x) = \frac{\lambda(x-a)^{\alpha-1}}{\Gamma(\alpha)} + \overset{x}{I}{}^{\alpha} f(.,g(.)), \quad (x \in (a,a+h]) \tag{3.1}$$

where $\lambda := b\Gamma(\alpha)$. Then (3.1) is equivalent to the following initial-value problem:

$$g^{(\alpha)}(x) = f(x,g(x)) \quad (\text{ppx } \in [a,a+h]) \tag{3.2}$$

with $$g^{(\alpha-1)}(a) = b\Gamma(\alpha) \quad (0 < \alpha \leq 1).$$

Proof of Lemma 3.1 The case $\alpha = 1$ is obvious. Let $0 < \alpha < 1$ and suppose (3.1) is true, then

$$\overset{x}{I}{}^{1-\alpha} g(.) = \underset{a}{\overset{x}{I}}{}^{1-\alpha}\{\frac{\lambda(t-a)^{\alpha-1}}{\Gamma(\alpha)} + \underset{a}{\overset{t}{I}}{}^{\alpha} f(.,g(.))\};$$

applying lemmas 2.1 and 2.2 we obtain that

$$\underset{a}{\overset{x}{I}}{}^{1-\alpha} g(.) = \lambda + \underset{a}{\overset{x}{I}}{}^{1} f(.,g(.)) = \lambda + \int_a^x f(t,g(t))dt \quad (x \in [a,a+h])$$

and this proves that $\overset{x}{I}{}^{1-\alpha} g(.)$ is absolutely continuous on $[a,a+h]$. Since $f \in L(a,a+h)$ (from (ii) and (iii) of theorem 3.1), it follows from definition 2.4 that

$$g^{(\alpha)}(x) = \frac{d}{dx} \underset{a}{\overset{x}{I}}{}^{1-\alpha} g(.) = f(x,g(x)) \quad (\text{ppx } \in [a,a+h]$$

and $$g^{(\alpha)}(a) = \lambda = b\Gamma(\alpha).$$

The proof of the converse is similar.

Now we give the proof of theorem 3.1.

Proof of theorem 3.1 Let $I_h' = (a,a+h]$ and $I_h = [a,a+h]$; define the sequence $\{g_n(.) : n \in N_+\}$ as follows:

$$g_0(x) = \frac{\lambda(x-a)^{\alpha-1}}{\Gamma(\alpha)} \quad (x \in I_h') \tag{3.3}$$

and $g_n(x) = g_o(x) + \frac{1}{\Gamma(\alpha)} {}_a\int^x (x-t)^{\alpha-1} f(t,g_{n-1}(t)dt, \quad (x \in I_h',\ n \in N_+)$.

Clearly, by induction, we have $(x,g_n) \in G_{\alpha,\gamma} \quad (n \in N_+)$.

Next we prove $g_h(.) \in C(I_h') \quad (n \in N_o)$. Obviously $g_o \in C(I_h')$, by using the conditions (ii) and (1) of theorem 3.1 and applying these in Theorem 2.1(a), we obtain that $f(.,g_{n-1}(.)) \in LT_\alpha(I_h) \quad (n \in N_+)$ and this implies that $g_n(.) \in C(I_h')$ $(n \in N_+)$. To prove that $\lim g_n(x) = g_o(x) + \frac{1}{\Gamma(\alpha)} {}_a\int^x (x-t)^{\alpha-1} f(t,g(t))dt$, $(x \in I_h')$, we define the sequence $\{H_n(.) : n \in N_o\}$ as follows:

$H_n(x) = (x - a^{1-\alpha}\ g_n(x) \quad (x \in I_h',\ \ n \in N_o)$, where g_n is as given in (3.3).

Hence $H_0(x) = b \quad (x \in I_h')$ and

$$H_n(x) = b + \frac{(x-a)^{1-\alpha}}{\Gamma(\alpha)} {}_a\int^x (x-t)^{\alpha-1} f(t,g_{n-1}(t))dt, \quad (x \in I_h',\ n \in N_+) \tag{3.4}$$

Since $g_n(.) \in C(I_h') \quad (n \in N_o)$, we have $H_n(.) \in C(I_h') \quad (n \in N_o)$. Also from (3.4), we have $\lim_{x \to a^+} H_n(x) = b \quad (n \in N_+)$, and thus by defining $H_n(a) = \lim_{x \to a^+} H_n(x) = b$ $(n \in N_+)$, we find that $H_n(.) \in C(I_h)$ $(n \in N_+)$. Furthermore, by using (3.4) and theorem 2.1(b), one can show that the sequence $\{H_n(.),\ \ n \in N_+\}$ is uniformly bounded and equicontinuous on I_h. Therefore by the Asceli lemma, (see Coddington and Levinson (1955)), there exists at least one subsequence $\{H_{n_s}(.) : s \in N_+\}$ which converges uniformly on I_h to some limit function H(say) as $s \to \infty$.

i.e. $\lim_{s\to\infty} H_{n_s}(x) = \lim_{s\to\infty}(x-a)^{1-\alpha}\ g_{n_s}(x) \quad (x \in I_h')$.

Now define $g(x) = \lim_{s\to\infty} g_{n_s}(x) = \lim_{s\to\infty}(x-a)^{\alpha-1}\ H_{n_s}(x)$

$$= (x-a)^{\alpha-1} \lim_{s\to\infty} H_{n_s}(x) = (x-a)^{\alpha-1} H(x) \quad (x \in I_h').$$

This means that the sequence $\{g_n(.) : n \in N_+\}$ converges to the function g on I_h' and hence from this fact and the hypothesis (i) of the theorem 3.1, we obtain that the sequence $\{f(t,g_{n_s}(t)) : s \in N_+\}$ converges to the function $f(t,g(t))$ for any fixed $t \in (a,x]$ and for all $x \in I_h'$. Then by using the condition (iii) of theorem 3.1 and the Lebesgue dominated convergence theorem, see Titchmarch (1939), we obtain that

$$g(x) = \lim_{x\to\infty} g(x) = g_o(x) + \frac{1}{\Gamma(\alpha)} {}_a\!\int^x (x-t)^{\alpha-1} f(t,g(t))dt \qquad (x \in I_h'). \qquad (3.5)$$

Some properties of the limit function g

(i) $g \in C(I_h')$ but $g \notin C(I_h)$; this follows from the fact $g_{n_s}(.) \in C(I_h')$ $(s \in N_+)$ and the uniform convergence of the sequence $\{g_{n_s}(.) : s \in N_+\}$, but $g \notin C(I_h)$ because $g_o \notin C(I_h)$.

(ii) $g \in D^{(\alpha)}(I_h)$. In fact $g \in C(I_h')$ implies that $g \in L(I_h)$; furthermore from the proof of lemma 3.1, we have $\overset{x}{\underset{a}{I}}{}^{1-\alpha} g(.)$ is absolutely continuous on I_h, thus by definition (2.3), we find that $g \in D^{(\alpha)}(I_h)$. Finally from the proof of lemma 3.1, we deduce the following properties:

(iii) $g^{(\alpha)}(x) = f(x,g(x))$ $\quad$ (ppx $\in I_h$),

(iv) $g^{(\alpha-1)}(.) \in AC(I_h)$,

(v) $g^{(\alpha-1)}(a) = b\Gamma(\alpha)$.

4. S_1 and S_2 continuities

Definition 4.1 Let $0 < \alpha \leq 1$ and D be an arbitrary region in the strip $G_h := (a,a+h]XR$. A function $f : D \to R$ is said to be S_1-continuous on D, if for any $\varepsilon > 0$, there exists $\sigma = \sigma_\varepsilon > 0$ such that for all $(x,y_1),(x,y_2) \in D$ we have

$$|f(x,y_1) - f(x,y_2)| < \varepsilon \text{ whenever } |y_1 - y_2|(x-a)^{1-\alpha} < \sigma.$$

Definition 4.2 Let α,D be as given in definition 4.1; a function $f : D \to R$ is said to be S_2-continuous on D, if for any $\varepsilon > 0$, there exists $\sigma = \sigma_\varepsilon > 0$ such that for all $(x,y_1),(x,y_2) \in D$ we have $|f(x,y_1) - f(x,y_2)|(x-a)^{1-\alpha} < \varepsilon$ provided that $|y_1 - y_2|(x-a)^{1-\alpha} < \sigma$.

Note When $\alpha = 1$, S_1 and S_2 continuities reduce to ordinary continuity.

Lemma 4.1 Let α,D be as given in definition 4.1; consider the following sets of functions $f : D \to R$ such that

$M_1 = \{f : f$ is uniformly continuous on D$\}$

$M_2 = \{f : f$ is S_1-continuous on D$\}$,

$M_3 = \{f : f$ is S_2-continuous on D$\}$

then the following relations hold:

(1) $M_1 \quad M_2 \quad M_3 \neq \phi$, where ϕ denotes the empty set,

(2) $M_2 \quad M_3$ but $M_3 \quad M_2$,

(3) $M_2 \quad M_1$.

Proof of (1) A constant function $f = C$ (say) belong to $M_1 \quad M_2 \quad M_3$.

Proof of (2) Obviously $M_2 \quad M_3$; to prove $M_3 \quad M_2$ we consider the following example:

Let $f : D \to R$ be defined by $f(x,y) = y$ for all $(x,y) \in D$. Then $f \in M_3$ but $f \notin M_2$ and these can be checked easily.

Proof of (3) Define $f : D \to R$ by:

$$f(x,y) = \frac{1}{x - a} \qquad ((x,y) \in D)$$

let ε and $\delta = \delta_\varepsilon$ be arbitrary positive numbers and choose $x_1 \in (a,a + h]$ such that $x_1 = a + \delta$. If $x_2 \in (a,a + h]$ such that $|x_2 - x_1| < \delta$, $|y_2 - y_1| < \delta$, we have

$$|f(x_1,y_1) - f(x_2,y_2)| = \frac{|x_2 - x_1|}{(x_1 - a)(x_2 - a)} < \frac{\delta}{(x_1 - a)(x_2 - a)} = \frac{1}{x_2 - a} .$$

Now, by choosing x_2 close enough to a, we then find that

$$|f(x_1,y_1) - f(x_2,y_2)| > \varepsilon \text{ although } |x_1 - x_2| < \delta, \ |y_1 - y_2| < \delta.$$

Since this is true for any $\delta > 0$, so we have $f \notin M_1$. Clearly $f \in M_2$, consequently we have $M_2 \quad M_1$.

Definition 4.3 Let α,D be as given in definition 4.1; let

$$A : [a,a + h] \to R_+ \text{ be such that } A \in C[a,a + h].$$

We say f satisfies the Lipschitz condition on D and briefly we write $f \in \underset{\alpha,A}{\mathrm{Lip}}(D)$, if the following condition holds:

$$|f(x,y_1) - f(x,y_2)| \leq A(x)|(x - a)^{1-\alpha}|y_1 - y_2|.$$

Lemma 4.2 If $f \in \underset{\alpha,A}{\mathrm{Lip}}(D)$, then f is S_1-continuous in $y \in R$.

Proof The proof is straightforward.

Theorem 4.1 (Carathéodory theorem, the Schauder method).

Let all the assumption of theorem 3.1 be given except in the condition (i) we assume the given function f is S_2-continuous in $y \in R$ for each fixed $x \in (a,a + \mu]$. Then the conclusions of theorem 3.1 hold.

Proof Let $(E^{(\alpha)}(I_h'), ||.||_\alpha)$ be the space as given in lemma 2.6; let

$$\lambda = b\Gamma(\alpha) \text{ and } g_o(x) = \frac{\lambda(x - a)^{\alpha-1}}{\Gamma(\alpha)}, \qquad (x \in I_h').$$

Clearly $g_o \in E^{(\alpha)}(I_h')$. Define a closed subset $F^{(\alpha)}(I_h')$ of $E^{(\alpha)}(I_h')$ as:

$$F^{(\alpha)}(I_h') = \{g \in E^{(\alpha)}(I_h') : ||g - g_o||_\alpha \leq \gamma\} .$$

Thus $F^{(\alpha)}(I_h'),||\cdot||_\alpha)$ is a Banach space, since every closed subspace of a Banach space is also a Banach space and hence $F^{(\alpha)}(I_h')$ is a convex set. Now, for any $g \in F^{(\alpha)}(I_h')$, define an operator T as follows:

$$(Tg)(x) = Tg(x) = g_o(x) + \frac{1}{\Gamma(\alpha)} {}_a\int^x (x - t)^{\alpha-1} f(t,g(t))dt, \quad (x \in I_h') \tag{4.1}$$

Clearly $T : F^{(\alpha)}(I_h') \to F^{(\alpha)}(I_h')$. Furthermore T is continuous on $F^{(\alpha)}(I_h')$, for if $g_1, g_2 \in F^{(\alpha)}(I_h')$, then from (4.1) we have

$$||Tg_1 - Tg_2||_\alpha \leq \operatorname*{Sup}_{x\varepsilon I_h'} (x - a)^{1-\alpha} {}_a\int^x (x - t)^{\alpha-1} |f(t,g_1(t) - f(t,g_2(t)|(t - a)^{1-\alpha}dt.$$

Hence, by using the S_2-continuity of f in y for each fixed $t \in (a,x]$, it follows from (4.2) that

$$||Tg_1 - Tg_2||_\alpha \leq \varepsilon \operatorname*{Sup}_{x\varepsilon I_h'} \frac{(x - a)^{1-\alpha}}{\Gamma(\alpha)} {}_a\int^x (x - t)^{\alpha-1}(t - a)^{\alpha-1}dt,$$

where ε is a positive number. Then by lemma 2.1, we see that

$$||Tg_1 - Tg_2||_\alpha = \varepsilon \operatorname*{Sup}_{x\varepsilon I_h'} (x - a)^{1-\alpha}\Gamma(\alpha) \, {}_a^x I^\alpha \frac{(t - a)^{\alpha-1}}{\Gamma(\alpha)} \leq \varepsilon \frac{h^\alpha \Gamma(\alpha)}{\Gamma(2\alpha)},$$

and this proves T is continuous on $F^{(\alpha)}(I_h')$.

Next we prove $T(F^{(\alpha)})$ is relatively compact and to do this, we define

$$H_n(x) = (x - a)^{1-\alpha} Tg_n(x) \quad (x \in I_h', \ n \in N_+)$$

where $Tg_n(x)$ as given in (4.1).

It follows that the sequence $\{H_n(.) : n \in N_+\}$ is uniformly bounded and equicontinuous on [1,a + h] and for the proof of this, see the properties of the corresponding sequence defined in (3.4). Consequently, by the Asceli lemma, we find at least one subsequence $\{(x - a^{1-\alpha} Tg_{n_m}(x) : m \in N_+\}$ is convergent in $F^{(\alpha)}(I_h')$, hence $T(F^{(\alpha)})$ is relatively compact. Finally by the Schauder fixedpoint theorem, (see Hale (1969)), there exists $g \in F^{(\alpha)}(I_h')$ such that $Tg = g$. Then from (4.1) we obtain

$$Tg(x) = g(x) = g_o(x) + \frac{1}{\Gamma(\alpha)} {}_a\int^x (x - t)^{\alpha-1} f(t,g(t))dt, \quad (x \in I_h').$$

For the rest of the proof of this theorem, see the proof of theorem 3.1.

5. A local existence theorem for the maximum and minimum solutions of fractional differential equations.

Definition 5.1 Consider the following initial-value problem:

$$g^{(\alpha)}(x) = f(x,g(x)) \text{ (almost all } x \in [a,a + h])$$

with

$$g^{(\alpha-1)}(a) = b\Gamma(\alpha), \quad \alpha \in (0,1] \tag{5.1}$$

where (5.1) is as defined in equation (3.2) if g_M is a solution of (5.1) on $[a,a + h]$,

i.e. $g_M^{(\alpha)}(x) = f(x,g_M(x))$ (almost all $x \in [a,a + h]$ with $g_M^{(\alpha-1)}(a) = b\Gamma(\alpha)$; if g is any other solution of (5.1) such that

$$g(x) \leq g_M(x) \quad (x \in (a,a + h])$$

with $g^{(\alpha-1)}(a) = g_M^{(\alpha-1)}(a) = b\Gamma(\alpha)$, then we say that g_M is a maximum solution of (5.1) on $]a,a + h]$. Similarly, g_m is called a minimum solution of (5.1) on $[a,a + h]$ if any other solution g of (5.1) satisfies the following relation:

$$g(x) \geq g_m(x) \quad (x \in (a,a + h])$$

with

$$g^{(\alpha-1)}(a) = g_m^{(\alpha-1)}(a) = b\Gamma(\alpha) .$$

Theorem 5.1 Let the hypothesis of theorem 3.1 be satisfied. Then there exist maximum and minimum solutions g_m, g_m of (5.1) on $[a,a + h]$ respectively with $g_M^{(\alpha-1)}(a) = g_m^{(\alpha-1)}(a) = b\Gamma(\alpha)$.

Proof From lemma 3.1, we have that any solution g of (5.1) must satisfy the following integral equation:

$$g(x) = g_o(x) + \frac{1}{\Gamma(\alpha)} {}_a\!\int^x (x - t)^{\alpha-1} f(t,g(t))dt, \quad (x \in (a,a + h]) \tag{5.2}$$

where $g_o(x) = \dfrac{\gamma(x - a)^{\alpha-1}}{\Gamma(\alpha)}$ and $\gamma = b\Gamma(\alpha)$. Thus

$$(x - a)^{1-\alpha} g_o(x) = b \quad (x \in (a,a + h])$$

and

$$(x - a)^{1-\alpha} g(x) = b + \frac{1}{\Gamma(\alpha)} {}_a\!\int^x (x - t)^{\alpha-1}(x - a)^{1-\alpha} f(t,g(t))dt, \quad (x \in (a,a+h]) \tag{5.3}$$

$$\left.\begin{aligned} &\text{Let } H(x,t) := (x - a)^{1-\alpha} g(t) \\ &\text{and} \\ &F(t,H(x,t)) = (x - a)^{1-\alpha} f(t,g(t)) \end{aligned}\right\} \quad a < t \leq x \leq a + h \tag{5.4}$$

then using (5.4)m we see that (5.3) is equivalent to the following equation:

$$H_o(x,x) := H_o(x) = b \quad (x \in (a, a + h])$$

and

$$H(x,x) := H(x) = b + \frac{1}{\Gamma(\alpha)} {}_a\!\int^x (x - t)^{\alpha-1} F(t,H(x,t))dt \quad (x \in (a,a+h]) \tag{5.5}$$

It follows from the proof of theorem 3.1 that $H(.) \varepsilon C[a,a+h]$; hence H is uniformly continuous on $[a,a+h]$. Thus for any $\varepsilon > 0$, there exists $\sigma = \sigma(\varepsilon) > 0$, independent of x and H (because of the equicontinuity of set $\{H(.)\}$, and for the proof of this see the proof of the corresponding sequence $\{H_n(.) : n \varepsilon N_+\}$ defined in equation (3.4)) such that

$$|H(x_1) - H(x_2)| < \varepsilon \text{ whenever } |x_1 - x_2| < \sigma. \tag{5.6}$$

Let $$Z(x) := \sup\{H(x) = (x-a)^{1-\alpha} g(x) : x \varepsilon I_h\}, \tag{5.7}$$

taken over all solutions g of (5.2) with $g^{(\alpha-1)}(a) = b\Gamma(\alpha)$. Then Z exists and is continuous on I_h and so it is uniformly continuous on I_h. Hence, given $\varepsilon > 0$, there exists $\sigma = \sigma(\varepsilon) > 0$ such that, no only (5.6) is true for this σ but also for all $x_1, x_2 \varepsilon I_h$,

$$|Z(x_1) - Z(x_2)| < \varepsilon \text{ whenever } |x_1 - x_2| < \sigma \tag{5.8}$$

For a given $\varepsilon > 0$, choose $\sigma = \sigma(\varepsilon) > 0$ so that (5.6) and (5.8) hold. Divide the interval I_h into n intervals by the points $a = x_o < x_1 < x_2 < \ldots < x_n = a + h$ in such a way that $\max(x_{i+1} - x_i) < \sigma$, $(i = 1,2,\ldots,n-1)$.

For $x_i \varepsilon I_h$ $(i = 1,2,\ldots,n-1)$, choose a function H_i from (5.7) such that

$$0 \leq Z(x_i) - H_i(x_i) < \varepsilon \text{ and for } i \geq 1$$

$$H_i(x_i) - H_{i-1}(x_i) \geq 0 \, ;$$

this is possible from the definition of Z. Now, for the given $\varepsilon > 0$, define the function $W_\varepsilon : I_h \to R$ by:

$$W_\varepsilon(x) = H_{n-1}(x) \quad (x \varepsilon [x_{n-1}, x_n]).$$

If $H_{n-1}(x_{n-1}) > H_{n-2}(x_{n-1})$, define W_ε to the left of x_{n-1} as H_{n-1} up to the point a_{n-2} (if it exists) in (x_{n-2}, x_{n-1}) nearest to x_{n-1} such that $W_\varepsilon(a_{n-2}) = H_{n-1}(a_{n-2}) = H_{n-2}(a_{n-2})$. If a_{n-2} does not exist, define W_ε on $[x_{n-2}, x_{n-1})$ as H_{n-1}. If $H_{n-1}(x_{n-1}) = H_{n-2}(x_{n-1})$, define $W_\varepsilon(x) = H_{n-1}(x)$ on $[x_{n-2}, x_{n-1})$. Continuing this way, we can define W_ε on I_h, having the property

$$0 \leq Z(x_i) - W_\varepsilon(x_i) < \varepsilon, \quad (x_i \varepsilon I_h, \ i = 1,2,\ldots,n). \tag{5.9}$$

Since the variation of Z and W_ε in each interval $[x_i, x_{i+1})$ is less than ε, by (5.6), (5.8) the results from (5.9) give us

$$0 \leq Z(x) - W_\varepsilon(x) = |z(x) - W_\varepsilon(x)|$$
$$\leq |Z(x) - Z(x_i)| + |Z(x_i) - W_\varepsilon(x_i)| + |W_\varepsilon(x_i) - W_\varepsilon(x)|$$
$$< 3\varepsilon \quad (x \varepsilon I_h). \tag{5.10}$$

Letting $\varepsilon = \frac{1}{m}$, $(m \varepsilon N_+)$, we obtain a sequence $\{W_{\frac{1}{m}}(.) : m \varepsilon N_+\}$ of functions defined on I_h, which by (5.10) converges uniformly to z on I_h. Thus from this fact and using the same orguments as given in the proof of theorem 3.1 and an application of the Lebesgue dominated convergence theorem to (5.5) with H replaced by $W_{\frac{1}{m}}$. it follows that

$$Z(x) = b + \frac{1}{\Gamma(\sigma)} {}_a\!\int^x (x - t)^{\alpha-1} F(t,Z(x,t))dt \quad (x \varepsilon I_h) \tag{5.11}$$

Then from the definition of Z and H, there exists a function $g_M : (a,a + h] \rightarrow R$ such that

$$Z(x) = (x - a)^{1-\alpha} g_M(x) \quad (x \varepsilon (a,a + h]) \tag{5.12}$$

and $g_M(x) \geq g(x)$, where g satisfies (5.2).

Consequently, from (5.4), (5.11) and (5.12) we obtain

$$(x - a)^{1-\alpha} g_M(x) = b + \frac{1}{\Gamma(\alpha)} {}_a\!\int^x (x - t)^{\alpha-1} (x - a)^{1-\alpha} f(t,g(t))dt \quad (x \varepsilon (a,a + h).$$

Thus

$$g_M(x) = \frac{\gamma(x - a)^{\alpha-1}}{\Gamma(\alpha)} + \frac{1}{\Gamma(\sigma)} {}_a\!\int^x (x - t)^{\alpha-1} f(t,g_M(t))dt, \quad (x \varepsilon (a,a + h]).$$

Finally Lemma 3.1, implies that $g_M^{(\alpha)}(x) = f(x,g_M(x))$ (almost all $x \varepsilon [a,a + h]$) with

$$g_M^{(\alpha-1)}(a) = b\Gamma(\alpha) .$$

Similarly, we can prove the existence of the minimum solution g_M of (5.1) with $g_M^{(\alpha-1)}(a) = b\Gamma(\alpha)$.

Note If $\alpha = 1$ in theorem 5.1, then it reduces to theorem 1.2, pp. 45-46 of Coddington and Levinson (1955).

Remark The uniqueness, global cases and the extension for $\alpha > 1$ of Theorem 3.1 have also been proved; they will be published later.

This paper is a part of a Ph.D. thesis submitted to the Graduate School of the University of Dundee. The author wishes to express his sincere appreciation for the guidance of this work by Professor W.N. Everitt, Baxter Professor in the

Department of Mathematical Sciences, The University of Dundee.

REFERENCES

Barrett, J.H., 1954. Differential equations of non-integer order. Cand. J. Math., 6, 529-541.

Bassam, M.A., 1961. Some properties of the Holmgrem-Riesz transform, Ann. Scuala, Norm. Sup. Pissa, 15, 1-24.

Coddington, E.A. and Levinson, N., 1955. Theory of ordinary differential equations, McGraw-Hill, New York.

Hale, J.K., 1969. Ordinary differential equations, John Wiley & Sons.

Titchmarsh, E.G., 1939. The theory of functions, second edition, Oxford University Press.

APPLICATION OF THE THEOREM OF CONNECTIVITY OF M^7 IN ASTRONOMY

TUNG CHIN-CHU

My talk today is about the application of the theorem of connectivity of M^7. This theorem is a preliminary part of an answer to the G D Birkhoff's seventh problem, which is concerned with the problem of three bodies. The theorem was obtained by myself and by R W Easton independently with minor differences. The idea of the application, however, was suggested to me by my teacher Chin Yuan-Shin.

The problem of three bodies is too long and involved a story to tell, and so I can only restrict myself to a very narrow aspect of it. I have always been of the opinion that the problem of three bodies should be called Newton's problem, because everything is Newtonian here: the equations came from Newton's law of motion with Newton's force of attraction. My interest in the problem was aroused after reading in English Whittaker's treatise on analytical dynamics [15].

It is well-known that Sundman [10] solved the problem of integrating the equations of motion by means of a power series in a fractional power of the time variable t; and in 1907 he obtained further certain qualitative results about the motion from an inequality:

$$\ddot{I} - \frac{1}{4}\dot{I}^2 I^{-1} - \frac{c^2}{3} I^{-1} - 2E \geq 0$$

where $I = m_1 r_1 + m_2 r_2 + m_3 r_3$, (see Siegel [8], p.51).

Siegel [8] observed in his Vorlesungen (1955) that, though Sundman's results are decades old, they are still the newest. And I think this is still true today.

As a concrete problem in dynamics, one wants to know how a system of particles actually behaves. Although Sundman's series gives no indication in general about this, it certainly gives a description of double collisions quantitatively as well as qualitatively. It is intuitively clear that when two bodies collide, their trajectories will suffer a change comparable to a cusp on a curve given by a two-thirds or two-fifths power of a variable at the origin. Sundman's series gives the one-third but not any other power of the time variable. It shows that the singularity of double collisions is removable, while that of triple collisions is irremovable. Here we can see that Siegel's contribution [7] in showing that triple collisions correspond to essential singularities marks a further milestone in the solution of the problem.

The qualitative analysis of motions by means of Sundman's inequality is a method reminiscent of Poincaré's arc-without-contact method or of Lyapunov's function in the qualitative theory of differential equations. The basic idea initiated at the time of Poincaré is to use non-integrals, i.e. inequalities, in obtaining information about the motion, instead of trying to find integrals, which are equalities. And it is, I think, due to the dynamic meaning of the quantity used, i.e. the moment of inertia, that Sundman's inequality still occupies an outstanding position in the general aspect of the problem. Although it is not an integral, it gives an eleventh item of information in addition to the ten classical integrals. One might try to obtain a twelvth item of information by forming an inequality for the evaluation of the rate of change of an angle formed by the three particles. Unfortunately, the expression for such a quantity turns out to be so complicated as to be beyond normal comprehension.

If we consider the motion as a rhythmic variation of a triangle formed by the three particles, the size of the triangle will either increase indefinitely as time increases, or it will decrease in the case of large and positive total energy. Birkhoff therefore made the observation that, qualitatively, we may regard the problem as solved in such a case. In the same sense perhaps we can say that the problem has also been solved for the case of a large negative energy with a non-vanishing angular momentum, since now, by the theorem of connectivity, two of the bodies will always remain closer to each other than to the third.

Let M^7 denote the totality of all motions permitted by the ten classical integrals, let E denote the total energy and c the angular momentum. The theorem can then be stated as follows:

Theorem (Tung-Easton) : For $E \geq 0$, and $E < 0$ with $c = 0$, M^7 is connected. For $E < 0$ and $c \neq 0$, the connectivity of M^7 is as follows. As E decreases from 0 to a large negative value, M^7 will change from one connected piece to three mutually disjoint pieces; and the critical value of E for such changes correspond to constant distance co-linear Lagrangian motions.

The set M^7 mentioned in the theorem is not a manifold unless certain considerations regarding double collisions are taken into account. Easton [3-6] has made the surgery explicit by referring Sundman's result to the sub-manifold of planar three-body motions in order to turn it from a set into a manifold. This surgery is a mathematical reasoning analogous to an operation which turns a set of papers glued along certain lines, such as the binding - rim of a book, into a whole plane sheet of paper.

Before applying the theorem, let us make clear what Birkhoff's seventh problem actually is [2,11].

A natural generalization to higher dimensions of Poincaré's arc without contact joining singular points is Birkhoff's surface of section. That is, to find a surface of dimension $n-1$ in the n-dimensional flow, such that every trajectory intersects the surface many times as the time increases or decreases. The correspondence between consecutive intersections defines a map of points of the surface into itself. For a closed surface of section, the boundary will be a surface of dimension $n-2$, which consists of trajectories.

The ten classical integrals define a manifold M^8 of states of motion in the phase space [13]. The equations of motion can be further reduced one order, so that the manifold can be reduced to a manifold M^7 of dimension 7.

If the surface of section is to be used here, we have to find out such a surface S. Birkhoff made the observation that the sub-manifold M^5 of planar motions might be such a surface in M^7. He remarked in 1927 that not even the connectivity of M^7 had anywhere been discussed. Two years later, he raised the problem of investigating the topological structure of M^7; this is his seventh problem.

The structure of M^5 as given by Easton shows such a complexity that, even if it can be used as a surface of section (which is unlikely, I think), it is still not easy for us to get from it any information concerning the motion of the three bodies.

In case M^7 has three components, notions in one particular component are characterized by the fact that one of the bodies remains relatively remote from the remaining two. If we have an example of motion with a non-vanishing angular momentum and a sufficiently large negative energy as well, and if we are asked to show that some two of the bodies remain as a nearer pair throughout their motion, all that we have to do is to calculate from the data of the example to make sure that the motion corresponds to a trajectory of the flow belonging to a suitable component of M^7. So that the theorem of connectivity can be applied to actual astronomical problems.

Let us now make a survey of two examples carried out for the Sun-Neptune-Triton system [12] respectively, to show that the satellite in each system remains as a satellite in the future as well as in the past.

Let the three masses be denoted by m_1, m_2, m_3.

The real motion as observed in astronomy will be denoted by M_r. This motion has already definite values for the ten classical integrals. Other ficticious possible motions of the same masses will be denoted by M_f. They are to have the same values for all the classical integrals except the energy.

The proof will be along a line which depends on the statements:

(I) If an M_f is a constant-distance collinear Lagrangian motion, its energy can be evaluated. Let the energies be E_1, E_2, E_3 for three such motions respectively. If E_r of M_r is such that $E_r < E_i$, for $i = 1,2,3$, then the M^7 containing M_r will have three components, in accordance with our theorem. Let these be M_{12}, M_{23}, M_{31}.

(II) If, for instance, M_r is in M_{31}, then m_3 and m_1 will remain closer. In order to prove that M_r is in M_{31}, we take a particular M_f, say M_{fo}, with coincidence of m_3 and m_1; and show that M_r and this M_{fo} are in the same component. Whether a component contains both of them depends on whether it contains their initial values, so that we need consider only their initial values.

If M_r has a syzygial position, then we may take it as its initial position and compare it with the initial collineation of other M_f's, and so the consideration is further simplified.

Proof of (I)

The value of c_r and E_r is a matter of the astronomical data or of a calculation from such data.

The value E_f for a constant-distance collinear Lagrangian M_f with the same value of c as c_r is calculated from

$$T - U = \frac{1}{2} I\omega_\tau - \frac{1}{a}\left(\frac{m_2 m_3}{\rho} + \frac{m_3 m_1}{1-\rho} + m_1 m_2\right) = E$$

where $I = \sum_{i=1}^{3} m_i r_i^2$; and a and ρ with $0 < \rho < 1$ are determined from

$$\frac{m_3\rho^{-2} + m_1}{m_3\rho + m_1} = \frac{m_3(1-\rho)^{-2} + m_2}{m_3(1-\rho) + m_2}$$

$$\omega_\tau^2 a^3 = \frac{m_3\rho^{-2} + m_1}{m_3\rho + m_1}(m_1+m_2+m_3) ,$$

where ω_τ is the angular velocity, measured with a certain time scale τ, of a rotating line containing the three bodies. Three values E_1, E_2, E_3 are obtained for three such M_f's. And all of these E_i are greater than E_r.

Proof of (II)

A fundamental inequality for the positional coordinates of the bodies is that

$$c^2 D^2(\sqrt{\mu_1}x_1,\ldots,\sqrt{\mu_2}z_2) \le 2(U+E)$$

for motions determined by the reduced equations of motion, where D is a certain homogeneous rational function of $(x_1,\ldots,z_2)$. It can be transformed into

$$c^2 D^2(\xi_1,\ldots,\eta_2) \le 2(U+E) \qquad (1)$$

by using new variables

$$\xi_i = \sqrt{\mu_i}x_i,\ldots \quad .$$

The projection of points satisfying (1) on the unit sphere

S^5 : $$\xi_1^2 + \xi_2^2 + \eta_1^2 + \eta_2^2 + \zeta_1^2 + \zeta_2^2 = 1$$

is determined by

L^0 : $$U^{0^2} + 2E\, c^2 D^{0^2} \geq 0 \qquad (2)$$

where U^0 and D^0 are values of U and D on S^5 respectively.

The connectivity of M^7 is determined by the connectivity of L^0 on S^5.

Consider the totality Q^0 of points on L^0 corresponding to the initial collineation of the three masses. Suppose M_r has a syzygial initial position $m_2 m_3 m_1$ which has a corresponding point in Q^0.

Since collineation can happen only in the invariable plane, we cant set $\zeta_1 = \zeta_2 = 0$ for such initial values. Thus for Q^0 the sphere S^5 becomes

S^3 : $$\xi_1^2 + \eta_1^2 + \xi_2^2 + \eta_2^2 = 1 .$$

It can be shown that $D^0 = 1$ on S^3, so that (2) simplifies to

$$U^{0^2} \geq -2Ec^2 . \qquad (3)$$

Let r_{c2} be the distance from the centre of gravity to m_2; and let r_{31} be the distance from m_3 to m_1. Take the ratio

$$\alpha = \sqrt{\mu}_2 r_{c2} \;/\; \sqrt{\mu}_1 r_{31}$$

as a parameter. Let the value for M_r, as calculated from the data, be α_0.

Now consider how the value of U^0 is affected by varying the parameter α. Writing down the expression of U^0 and setting all the available data in it, we get a function in α :

$$U^0 = \sqrt{1 + \alpha^2}\; (\sqrt{\mu}_1 m_3 m_1 + \ldots) \equiv F(\alpha) .$$

This is an increasing function of α. As its value does decrease for increasing α, the inequality holds for such α. And the value of α at infinity corresponds to M_{f0}.

For a non-planar motion of the three masses, it may happen that no syzygial position exists. Such a case can be turned into a case with a syzygial position by the following consideration.

Let the angle between the vector m_3m_1 and cm_2 be β when the three masses are in the invariable plane. This instant of time, say t_1, is taken to be the initial time. Consider the set of all motions M_r and M_f with the initial condition that the centre of gravity c of m_3m_1 is in the invariable plane. Let θ be the angle between the same vectors as above for all such motions. Non-existence of a syzygial position means that $\theta = \beta \neq 0$ for M_r. Substituting the numerical data into U^0, we get a function of θ : $U^0 = G(\theta)$. This function $G(\theta)$ does not decrease for monotonic variation of θ in a suitable direction from $\theta = \beta$ to 0 or to π . So that (3) holds in such a process. But $\theta = 0$ or $\theta = \pi$ means a collinear case.

In conclusion let me make a remark as follows.

The calculation has nothing to do with the differential equations except that sometimes we argue with the trivial fact that the phase space of a flow contains a path if it contains its initial point, or conversely. For motions of a system in reality, the energy and angular momentum are determined values. But now they are calculated from the astronomical data. A set of data from simultaneous observation at an instant is quite enough, since we need only a set of initial values. An observation without error is most desirable. But such an ideal set of data is not available in reality. We have to make use of data from various sources. Two values of the same quantity, say the distance between two celestial bodies, observed at different time, might differ greatly. On the other hand, some quantities may have a value too small, such as the mass ratio m_ψ/m_Θ . We shall be helpless if we always stick to astronomical errors in a mathematical treatment of a problem. If we consider the mass ratio as negligible, i.e. m_ψ/m_Θ as negligible, then we shall come up with a new mathematical proble, the restricted problem of three bodies. Therefore we have used the data as if they were exact values.

Still another point that I'd like to call attention to is concerned with the difference of a mathematical model and the physical reality. The moon will not escape from the earth if it is a member of an ideal three-body system; just as we may say that the earth's orbit is an ellipse if the earth is a member of an ideal two-body problem.

In conclusion, I wish to express my sincere thanks to the Organizing Committee for inviting me to attend this notable Conference, and to the Royal Society of Great Britain and to the Academy of Science in the People's Republic of China for their financial support which enabled me to do so.

References

1. Birkhoff, G.D. : Dynamical systems, Amer. Math. Soc. Coll. Publ. IX (1927).
2. Birkhoff, G.D. : Einige Problem der Dynamik, Jahresbericht der Deutschen Mathematikers Vereinigung 38, Heft 1/4 (1929), 1-11. Collected works 2778-2793
3. Easton, R.W. : Some topology of the 3-body problem, J. of Diff. Equations 10, 2 (1971).
4. Easton, R.W. : Some topology of n-body problems, J. of Diff. Equations 19, 2 (1975).
5. Easton, R.W. : The topology of the regularized integral surfaces of the 3-body problem, J. of Diff. Equations 12, 2 (1972).
6. Easton, R.W. : Some qualitative aspects of the 3-body flow, Dynamical Systems, No 2, An international symposium (1976).
7. Siegel, C.L. : Der Dreierstoss, Ann. of Maths. 42 (1941), 127-168.
8. Siegel, C.L. : Lectures on celestial mechanics (in Germ.) 1955.
9. Siegel, C.L. and Moser, : Lectures on celestial mechanics, Springer-Verlag, 1971.
10. Sundman, K.F. : Mémoire sur le problème des trois corps, Acta. Math. 36 (1912), 105-179.
11. Tung, Chin-Chu : A brief account of G D Birkhoff's problem in the problem of 3 bodies, Appl. Math. & Mech. (Engl. ed. 1, No 2, Dec 1980).
12. Tung, Chin-Chu : The motion of Triton as a satellite of Neptune, Publ. of the Beijing Astron. Observatory, No 15 (1978).
13. Tung, Chin-Chu : Some properties of the classical integrals of the general 3-body problem, Scientia Sinica, 17, No 3 (1974).

14. Tung, Chin-Chu : A property of the motion of the moon, Publ. of the Beijing Astron. Observatory, No 1 (1979).

15. Whittaker, E.T. : A treatise on the analytical dynamics of particles and rigid bodies, Cambridge University Press, 4th ed., 1937.

Perturbations of Periodic Boundary Conditions

Lawrence Turyn

Section 1: Liapunov-Schmidt; Linear Perturbations

We consider boundary value problems which are perturbations of a linear boundary value problem with periodic boundary conditions. Specifically, we are interested in

$$(1.1)\qquad \begin{cases} \tau x \overset{\text{defn}}{=} -x'' + q(\cdot)x = \lambda r(\cdot)x + f_0(\cdot,\varepsilon,\lambda,x,x') \\ Mx \overset{\text{defn}}{=} \begin{pmatrix} x(0) - x(1) \\ x'(0) - x'(1) \end{pmatrix} = \varepsilon L(\varepsilon,x) + \begin{pmatrix} f_1(\varepsilon,\lambda,x,x') \\ f_2(\varepsilon,\lambda,x,x') \end{pmatrix} \end{cases}$$

where $L(\varepsilon,\cdot)$ is linear and f_i is non-linear in x, x', $i = 0,1,2$. We will apply the method of Liapunov-Schmidt. In Turyn [8] we also were able to use Hill's discriminant and classical perturbation theory of Rellich to achieve results on linear perturbations, i.e. $f_i \equiv 0$, $i = 0,1,2$, but Liapunov-Schmidt has the advantage of discussing both linear and non-linear perturbations.

We assume that λ_0 is a double eigenvalue of the linearisation

$$(1.2)\qquad \begin{cases} \tau x = \lambda r(\cdot)x \\ Mx = 0 \,, \end{cases}$$

i.e. there are two linearly independent solutions x_1, x_2 of the periodic boundary value problem (1.2). Without loss of generality x_1, x_2 are normalised to satisfy

$$x_1(0) = 1, \quad x_2(0) = 0$$
$$x_1'(0) = 0, \quad x_2'(0) = 1.$$

For the linear perturbation we take

$$(1)\qquad L(\varepsilon,x) = H\begin{pmatrix} x(1) \\ x'(1) \end{pmatrix} + \int_0^1 W(t)\begin{pmatrix} x(t) \\ x'(t) \end{pmatrix} + O(|\varepsilon|),$$

where H is a constant 2×2 real matrix, i.e. $H \in \mathbb{R}^{2\times 2}$, and $W(\cdot)$ is a continuous $\mathbb{R}^{2\times 2}$ valued function. In Turyn [8] $W \equiv 0$ although we did allow distributed terms in the non-linearities.

Define Banach spaces $Y = C^2[0,1]$, $Z = C[0,1] \times \mathbb{R}^2$ with norms $\|x\| = |x|_\infty + |x'|_\infty + |x''|_\infty$, $\|(v;c,d)\| = |v|_\infty + |c| + |d|$, respectively, where $|x|_\infty = \sup_{0\le t\le 1}|x(t)|$. Then (1.1) is equivalent to the problem

$$(1.4) \qquad (B-\nu A-\varepsilon C)x = G(\varepsilon,\lambda_0+\nu,x) + O((|\nu| + |\varepsilon|)^2\|x\|)$$

where $\nu = \lambda-\lambda_0$, $x \in Y$, $Bx = (\tau x-\lambda_0 r(\cdot)x;Mx)$, $Ax = (r(\cdot)x;\underset{\sim}{0})$

$$Cx = \left(0;\ H\begin{pmatrix} x(1) \\ x'(1)\end{pmatrix} + \int_0^1 W(t)\begin{pmatrix} x(t) \\ x'(t)\end{pmatrix}\right),\quad \underset{\sim}{0} = \begin{pmatrix} 0 \\ 0\end{pmatrix},\quad \underset{\sim}{f} = \begin{pmatrix} f_1 \\ f_2\end{pmatrix},\ \text{and}$$

$G(\varepsilon,\lambda,x) = (f_0(\cdot,\varepsilon,\lambda,x,x');\ \underset{\sim}{f}(\varepsilon,\lambda,x,x'))$. We assume that $G(\cdot,\cdot,\cdot)$ is twice continuously Fréchet differentiable and furthermore satisfies $G(\varepsilon,\lambda_0+\nu,x) = O((|\varepsilon| + |\nu|)^2\|x\| + \|x\|^2)$.

Denote $Y_0 = N(B)$, $Z_1 = R(B)$. By assumption $Y_0 = \text{span}\{x_1,x_2\}$. One can show that

$$R(B) = \{(v;c,d):\ \ell_j(v;c,d) = 0 \text{ for } j = 1,2\}$$

where $\ell_1(v;c,d) = -d + \int_0^1 vx_1$, $\ell_2(v;c,d) = c + \int_0^1 vx_2$ are linear functionals on Z Let $c_1 = -\int_0^1 x_1x_2 = -d_2$, $\alpha_j = \int_0^1 x_j^2$, and $z_1 = \alpha_1^{-1}(x_1;c_1,0)$, $z_2 = \alpha_2^{-1}(x_2;0,d_2)$. Then

$$Q:\ Z \to R(B):\ z \mapsto z - \ell_1(z)z_1 - \ell_2(z)z_2$$

is a projection. Define another projection

$$P:\ Y \to Y_0:\ x \mapsto \alpha_1^{-1}\left(\int_0^1 xx_1\right)x_1 + \alpha_2^{-1}\left(\int_0^1 xx_2\right)x_2$$

and corresponding subspace $Y_1 = (I-P)Y$.

The method of Liapunov-Schmidt consists of replacing (1.4) by the system

$$(1.5)\qquad Q(B-\nu A-\varepsilon C)(Px+(I-P)x) = QG(\varepsilon,\lambda_0+\nu,x) + O((|\varepsilon|+|\nu|)^2\|x\|)$$

$$(1.6)\qquad (I-Q)(B-\nu A-\varepsilon C)(Px+(I-P)x) = (I-Q)G(\varepsilon,\lambda_0+\nu,x) + O((|\varepsilon|+|\nu|)^2\|x\|).$$

Rewrite $Px = u_1x_1 + u_2x_2$ for real u_1, u_2. Since $Bx_i = 0$ for $i = 1,2$ and $QB\colon Y_1 \to R(B)$ has a bounded inverse, the "auxiliary equation" (1.5) can be solved by $(I-P)x = w^*(\varepsilon,\nu,\underset{\sim}{u})$ for all sufficiently small $|\varepsilon|$, $|\nu|$, $|\underset{\sim}{u}|$, where $\underset{\sim}{u} = (u_1,u_2)$, $|\underset{\sim}{u}| = |u_1| + |u_2|$. Furthermore, $w^* = O((|\varepsilon|+|\nu|)|\underset{\sim}{u}| + |\underset{\sim}{u}|^2)$ as $|\varepsilon|$, $|\nu|$, $|\underset{\sim}{u}| \to 0$. Substitute into (1.6) to arrive at the "bifurcation equation"

$$(1.7)\qquad \begin{aligned}&(I-Q)(B-\nu A-\varepsilon C)(u_1x_1+u_2x_2+w^*) - (I-Q)G(\varepsilon,\lambda_0+\nu,u_1x_1+u_2x_2+w^*)\\ &\quad = O((|\varepsilon|+|\nu|)^2|\underset{\sim}{u}| + (|\varepsilon|+|\nu|)|\underset{\sim}{u}|^3) \overset{\text{defn}}{=} R,\end{aligned}$$

R standing for "remainder". Note $(I-Q)B \equiv 0$. Since z_1, z_2 are linearly independent, (1.7) separates into the system

$$(1.8)\qquad [\nu\Sigma - \varepsilon(JH + \int_0^1 JW(t)X(t))]\underset{\sim}{u} + \underset{\sim}{g}(\varepsilon,\nu,\underset{\sim}{u}) = R\begin{pmatrix}1\\1\end{pmatrix}$$

where $X(t) = \begin{pmatrix}x_1 & x_2\\ x_1' & x_2'\end{pmatrix}$ is the principal fundamental matrix for the linear o.d.e. system $\begin{pmatrix}x\\ x'\end{pmatrix}' = \begin{pmatrix}0 & 1\\ q-\lambda_0 r & 0\end{pmatrix}\begin{pmatrix}x\\ x'\end{pmatrix}$, $\sigma = (\sigma_{ij})_{i,j=1,2}$, $\sigma_{ij} = \int_0^1 rx_ix_j$, $J = \begin{pmatrix}0 & 1\\ -1 & 0\end{pmatrix}$, $\underset{\sim}{g}(\varepsilon,\nu,\underset{\sim}{u}) = O((|\varepsilon|+|\nu|)^2|\underset{\sim}{u}|+|\underset{\sim}{u}|^2)$, and $H, W(\cdot)$ are as in (1.3).

We remark that $W(\cdot)$ need not be continuous for all of this to make sense. One could include finite jumps at interior points, leading to perturbations of the boundary conditions such as

$$\begin{pmatrix}x(0) - x(1)\\ x'(0) - x'(1)\end{pmatrix} = \varepsilon\left[H\begin{pmatrix}x(1)\\ x'(1)\end{pmatrix} + \sum_{i=1}^{\infty} H_i\begin{pmatrix}x(t_i)\\ x'(t_i)\end{pmatrix} + \int_0^1 W(t)\begin{pmatrix}x(t)\\ x'(t)\end{pmatrix}\right] + O(|\varepsilon|^2)$$

as long as $\sum_{i=0}^{\infty} |H_i| < \infty$.

For linear perturbations of (1.1), i.e. $f_i \equiv 0$ for $i = 0,1,2$, (1.8) can be rewritten as

(1.9) $$0 = \sigma\nu^2 + \gamma\varepsilon\nu + \delta\varepsilon^2 + \ldots$$

where ... represents terms of degree three or higher. Always $\sigma = \det \Sigma$. Denote $\Omega = H + \int_0^1 W(t)X(t) = (\omega_{ij})_{i,j=1,2}$. Then

$$\delta = \det \Omega,\ \gamma = -\omega_{11}\sigma_{12} + \omega_{22}\sigma_{12} + \omega_{12}\sigma_{11} - \omega_{21}\sigma_{22}\ .$$

Σ is always symmetric; upon assuming $r(\cdot) \geq r_0 > 0$ it is also positive definite.

<u>Theorem 1.1</u>: For linear perturbations, i.e. $f_i \equiv 0$ for $i = 0,1,2$, a sufficient condition for local splitting of the double eigenvalue λ_0 into two C^1 curves parametrised by ε is $\gamma^2 - 4\delta\sigma > 0$.

<u>Remark 1.2</u>: Assume $f_i \equiv 0$ for $i = 0,1,2$ and $r(\cdot) \geq r_0 > 0$. If $tr\ \Omega = 0$ then $\gamma^2 - 4\delta\sigma \geq 0$.

<u>Proof</u> (of the remark): Σ is positive definite and symmetric. Since $tr\ \Omega = 0$ implies $J\Omega$ symmetric, the solutions β^0 of $0 = \det(\Sigma - \beta J\Omega)$ are all real. □

For $W = 0$ the boundary conditions are self-adjoint iff $1 = \det(I + \varepsilon H + O(|\varepsilon|^2))$, from Coddington & Levinson [3, p. 297]. Thus $tr\ H = 0$ is necessary for self-adjointness. For the more general boundary conditions with $W(\cdot) \not\equiv 0$, any concept of self-adjointness should imply that eigenvalues λ are real; if $tr\ \Omega \neq 0$ then for some positive $r(\cdot)$ the solutions β^0 of $\det(\Sigma - \beta J\Omega)$ will be neither real nor zero, leading to λ_0 splitting into curves which are not real. Thus, for the more general boundary conditions with $W(\cdot) \not\equiv 0$ a necessary condition for self-adjointness will be $tr\ \Omega = 0$.

<u>Example 1.3</u>: When $r \equiv 1$, $q \equiv 0$, $tr\ \Omega = 0$, $\Omega \not\equiv 0$ and $\lambda_0 = 4\pi^2 n^2$ for some integer n, we calculate $\sigma_{12} = 0$ and $\gamma^2 - 4\delta\sigma = \frac{1}{4}((\omega_{12} + \lambda_0^{-1}\omega_{21})^2 + \lambda_0^{-1}\omega_{11}^2) > 0$, so Theorem 1.1 is applicable.

As a specific sub-example, take $r \equiv 1$, $q \equiv 0$, $W \equiv 0$, $H = \begin{pmatrix} 1 & 0 \\ 0 & -1 \end{pmatrix}$. Then use of the Hill's discriminant (see Turyn [8]) gives explicit curves $\lambda = (\lambda_0^{\frac{1}{2}} \pm 2 \arcsin(\varepsilon/2))^2$. These boundary conditions do not remain self-adjoint for $\varepsilon \neq 0$, but the curves exist anyway for $|\varepsilon| \leq 2$.

Application 1.4: Consider a ring of metal obtained by joining the endpoints $\xi = 0$, $\xi = 1$. If the joining is not perfect then there will be some "contact resistance". Appropriate boundary conditions for the temperature $u(\xi)$ are then (see e.g. Özişik [6, p. 283])

$$\begin{cases} u(0) - u(1) = \varepsilon u'(1) \\ u'(0) - u'(1) = 0 \end{cases}$$

where $\varepsilon = k/h = (\text{Biot number})^{-1}$ can be taken to be small and positive if the heat transfer coefficient h is large. These boundary conditions have as a consequence a temperature drop across the join, this phenomenon being well-known in practise. See Holman [4, pp. 45-48] for more details on the causes of contact resistance. These boundary conditions are self-adjoint for all ε, with $H = \begin{pmatrix} 0 & 1 \\ 0 & 0 \end{pmatrix}$. Since $\gamma = \sigma_{11} > 0$, Theorem 1.1 guarantees the existence of two curves $\lambda = \lambda_0 + \nu_{\pm}^*(\varepsilon)$. In fact, for the example $r \equiv 1$, $q \equiv 0$ one can see that $\nu_+^*(\varepsilon) = 0$ for all ε, since the second row of H is trivial, and further one can calculate that $D_\varepsilon \nu_-^*(0) = -2\lambda_0$

Section 2: Nonlinear Perturbations

Liapunov-Schmidt allowed us to reduce the nonlinear boundary value problem (1.1) to the nonlinear system (1.8). Re-write the latter as

$$(2.1) \qquad (\alpha_1 L_1 + \alpha_2 L_2)\underset{\sim}{u} + \underset{\sim}{Q}(\underset{\sim}{u}) + \underset{\sim}{R}(\underset{\sim}{\alpha},\underset{\sim}{u}) = \underset{\sim}{0} \in \mathbb{R}^2,$$

where $\underset{\sim}{\alpha} = (\alpha_1, \alpha_2)$, $|\underset{\sim}{\alpha}| = |\alpha_1| + |\alpha_2|$, $L_1 = \Sigma$, $L_2 = \Omega$, $\alpha_1 = \nu$, $\alpha_2 = \varepsilon$, and $\underset{\sim}{R}(\underset{\sim}{\alpha},\underset{\sim}{u}) = O(|\underset{\sim}{\alpha}|^2|\underset{\sim}{u}| + |\underset{\sim}{u}|^3)$. $\underset{\sim}{Q}(\underset{\sim}{u})$ is assumed to be a homogeneous quadratic, specifically

$$(2.2) \qquad \underset{\sim}{Q}(\underset{\sim}{u}) = (q_{20}u_1^2 + 2q_{11}u_1u_2 + q_{02}u_2^2, r_{02}u_1^2 + 2r_{11}u_1u_2 + r_{20}u_2^2).$$

$\alpha_1 = \alpha_2\beta_1$. Then (2.1) is equivalent to

$$h(w,\beta_1,\alpha_2) \overset{\text{defn}}{=} (\beta_1 L_1 + L_2)w + Q(w) - O(|\alpha_2|) = 0. \tag{2.3}$$

Examination of (2.3) will produce bifurcation curves in the α-plane for the unscaled system (2.1).

We remark that using (H2) in practise relies on distinguishing between L_1 and L_2: Since $L_1 v^0 + Q(v^0) = 0$ always has at least the solution $v^0 = 0$, (H2) requires that the 2×2 matrix L_1 be invertible.

The bifurcation curves will be approximately straight lines $\alpha_1 = \beta_1^0\alpha_2$ where β_1^0 is such that there is a $w^0 \neq 0$ satisfying

$$\begin{cases} h^0(w,\beta_1^0) \overset{\text{defn}}{=} (\beta_1^0 L_1 + L_2)w + Q(w) = 0 \\ \Delta(w,\beta_1^0) \overset{\text{defn}}{=} \det\left[\dfrac{\partial h^0}{\partial w}(w,\beta_1^0)\right] = 0 . \end{cases} \tag{2.4}$$

To ensure that such β_1^0 do give rise to bifurcation curves one must assume some further generic hypotheses.

Mallet-Paret [5] and Chow, Hale, and Mallet-Paret [2] discuss a problem similar to (2.1), and Chow and Hale [1, Chapter 7, Section 4] discuss (2.1) in generality. Our view will now be narrowed to the matrices L_1, L_2 from Application 1.4: Take $L_1 = \begin{pmatrix} 1 & 0 \\ 0 & \lambda_0^{-1} \end{pmatrix}$, $L_2 = \begin{pmatrix} 0 & 0 \\ 0 & 2 \end{pmatrix}$. For this specific choice of L_1, L_2 we will denote the corresponding nonlinear system by (2.1'), so as to distinguish it from the general system (2.1). A prime after an equation number will always indicate this specific choice of L_1, L_2. Corresponding to system (2.4) is

$$\begin{cases} 0 = h_1^0(w,\beta_1) = \beta_1 w_1 + q_{20}w_1^2 + 2q_{11}w_1w_2 + q_{02}w_2^2 \\ 0 = h_2^0(w,\beta_1) = \chi w_2 + r_{02}w_1^2 + 2r_{11}w_1w_2 + r_{20}w_2^2 \\ 0 = \Delta(w,\beta_1) = \beta_1\chi + 2(\chi q_{20} + \beta_1 r_{11})w_1 + \\ \qquad + 2(\chi q_{11} + \beta_1 r_{20})w_2 + 4d_{21}w_1^2 + 4d_{22}w_1w_2 + 4d_{12}w_2^2 \end{cases} \tag{2.4'}$$

where

$$\begin{cases} \chi = 2 + \lambda_0^{-1}\beta_1 \\ d_{21} = q_{20}r_{11} - q_{11}r_{02},\ d_{22} = q_{20}r_{20} - q_{02}r_{02},\ d_{12} = q_{11}r_{20} - q_{02}r_{11}. \end{cases} \tag{2.5}$$

Note that χ depends on β_1; given β_1^0, we denote $\chi_1^0 = 2 + \lambda_0^{-1}\beta_1^0$.

The immediate task is to find the solutions $(\underset{\sim}{w}^0,\beta_1^0)$ of (2.4').

Remark 2.2: (a) If $q_{02} \neq 0 \neq r_{02}$ then the only solutions $(\underset{\sim}{w}^0,\beta_1^0)$ of (2.4') satisfying $w_1^0 = 0$ or $w_2^0 = 0$ are

(2.6) $$(\underset{\sim}{0},0),\quad (\underset{\sim}{0},-2\lambda_0).$$

(b) The only solutions $(\underset{\sim}{w}^0,\beta_1^0)$ satisfying (2.4') and $w_1^0 \neq 0 \neq w_2^0$ are those for which $\underset{\sim}{w}^0$ satisfies

(2.7) $$\begin{cases} 0 = 2w_1w_2+r_{02}w_1^3+(2r_{11}-\lambda_0^{-1}q_{20})w_1^2w_2+(r_{20}-2\lambda_0^{-1}q_{11})w_1w_2^2 \\ \qquad -\lambda_0^{-1}q_{02}w_2^3 \\ 0 = q_{20}r_{02}w_1^4+4q_{11}r_{02}w_1^3w_2+(-q_{20}r_{20}+4q_{11}r_{11}+3q_{02}r_{02})w_1^2w_2^2 \\ \qquad +4q_{02}r_{11}w_1w_2^3+q_{02}r_{20}w_2^4 \end{cases}$$

with correspondingly

(2.8) $$\beta_1^0 = -(w_1^0)^{-1}(q_{20}(w_1^0)^2+2q_{11}w_1^0w_2^0+q_{02}(w_2^0)^2)\ .$$

The proofs of these results are relatively simple, but sometimes tedious, calculations, as found in Turyn [8]. Lemma 2.1, the discussion which followed it, Remarks 2.2 - 2.4 and other calculations below will all serve to prove the existence of bifurcation curves. The results are summarised in Theorem 2.5.

For any solution $(\underset{\sim}{w}^0,\beta_1^0)$ of (2.4') it is important to know the Jacobian matrix

$$\Phi(\underset{\sim}{w}^0,\beta_1^0) \overset{\text{defn}}{=} \frac{\partial(\underset{\sim}{h}^0,\Delta)}{\partial(\underset{\sim}{w},\beta_1)}(\underset{\sim}{w}^0,\beta_1^0)$$

because we will use solutions $(\underset{\sim}{w}^0,\beta_1^0)$ of (2.4') to obtain the bifurcation curves $\alpha_1 = \beta_1^0\alpha_1 + o(|\alpha_2|)$ for the original system (2.1').

We calculate $\Phi(\underset{\sim}{w},\beta_1) =$

We call $\underset{\sim}{u}$ the "solution" and $\underset{\sim}{\alpha}$ the "parameters". Our goal is to find out for which values of $\underset{\sim}{\alpha}$ there are non-trivial solutions $\underset{\sim}{u} \neq \underset{\sim}{0}$ and how the number of solutions changes as $\underset{\sim}{\alpha}$ is varied. Scaling and "generic bifurcation" methods are the techniques we will use to address these questions.

Scaling will effectively reduce the number of parameters from two to one. To do so one makes generic hypotheses which insure that no qualitative features are overlooked in the process.

Lemma 2.1 (Scaling): Assume

(H1) $$\underset{\sim}{Q}(\underset{\sim}{u}) = \underset{\sim}{0} \text{ implies } \underset{\sim}{u} = \underset{\sim}{0}.$$

Then there is a neighbourhood of $(\underset{\sim}{\alpha},\underset{\sim}{u}) = (\underset{\sim}{0},\underset{\sim}{0}) \in \mathbb{R}^4$ and a constant c in which all solutions of (4.1) satisfy

$$|\underset{\sim}{u}| \leq c|\underset{\sim}{\alpha}|$$

Proof: See Chow, Hale and Mallet-Paret [2, p. 222] for the proof of a similar result. □

If $\underset{\sim}{u} = \alpha_1\underset{\sim}{v}$, $\alpha_2 = \alpha_1\beta_2$ then solving (2.1) in a neighbourhood of $(\underset{\sim}{0},\underset{\sim}{0})$ is equivalent to solving

$$(L_1 + \beta_2 L_2)\underset{\sim}{v} + \underset{\sim}{Q}(\underset{\sim}{v}) = O(|\alpha_1|)$$

for small $\underset{\sim}{\alpha}$. We assume

(H2) $$L_1\underset{\sim}{v}^0 + \underset{\sim}{Q}(\underset{\sim}{v}^0) = 0 \text{ implies } \det\left[L_1 + \frac{\partial \underset{\sim}{Q}}{\partial \underset{\sim}{v}}(\underset{\sim}{v}^0)\right] \neq 0.$$

Since (H2) holds, every solution of $L_1\underset{\sim}{v}^0 + \underset{\sim}{Q}(\underset{\sim}{v}^0) = \underset{\sim}{0}$ corresponds to a solution of (2.1) which remains simple for $|\beta_2|$, $|\alpha_1|$ sufficiently small.

We will say that $\underset{\sim}{\alpha}^0$ is a "bifurcation point" if the number of solutions $\underset{\sim}{u}$ of (2.1) is not locally constant as a function of $\underset{\sim}{\alpha}$ near $\underset{\sim}{\alpha}^0$. A curve of bifurcation points will be called a "bifurcation curve". Hypothesis (H2) assures that there are no bifurcation curves in the intersection of a double cone $\{|\alpha_2| \leq \delta|\alpha_1|\}$ with a neighbourhood of the origin in $\mathbb{R}^2$. As for what remains in a neighbourhood of $\underset{\sim}{0}$ in the $\underset{\sim}{\alpha}$-plane we introduce a different scaling for (2.1): Let $\underset{\sim}{u} = \alpha_2\underset{\sim}{w}$,

$$\left[\begin{array}{c|c|c}
\beta_1+2q_{20}w_1+2q_{11}w_2 & 2q_{11}w_1+2q_{02}w_2 & w_1 \\
\hline
2r_{02}w_1+2r_{11}w_2 & \chi+2r_{11}w_1+2r_{20}w_2 & \lambda_0^{-1}w_2 \\
\hline
\begin{array}{c} 2(\chi q_{20}+\beta_1 r_{11}) \\ +8d_{21}w_1+4d_{22}w_2 \end{array} & \begin{array}{c} 2(\chi q_{11}+\beta_1 r_{20})+ \\ +4d_{22}w_1+8d_{12}w_2 \end{array} & \begin{array}{l} 2+2\lambda_0^{-1}\beta_1+ \\ +2(\lambda_0^{-1}q_{20}+r_{11})w_1 \\ +2(\lambda_0^{-1}q_{11}+r_{20})w_2 \end{array}
\end{array}\right].$$

When $(\underset{\sim}{w}^0,\beta_1^0)$ is a solution of (2.4') for which $\det \Phi(\underset{\sim}{w}^0,\beta_1^0) \neq 0$ the Implicit Function Theorem implies the existence of a local curve $\beta_1 = \beta_1^0 + o(1)$, $\underset{\sim}{w} = \underset{\sim}{w}^0 + o(1)$ of solutions of (2.3') for $|\alpha_2|$ small, with corresponding curve of solutions $\alpha_1 = \beta_1^0\alpha_2 + o(|\alpha_2|)$, $\underset{\sim}{u} = \alpha_2\underset{\sim}{w}^0 + o(|\alpha_2|)$ of (2.1') for $|\alpha_2|$ small. Furthermore, $\alpha_1 \sim \beta_1^0\alpha_2$ is in fact a bifurcation curve since in a neighbourhood of $\underset{\sim}{\alpha} = \underset{\sim}{0}$ it separates regions between which the number of solutions $\underset{\sim}{u}$ changes by at least two. The proof of this can be found in Chow and Hale [1, Chapter 7, Section 4].

It is interesting when $(\underset{\sim}{w}^0,\beta_1^0)$ solves (2.4') but $\det \Phi(\underset{\sim}{w}^0,\beta_1^0) = 0$. In fact, the two solutions (2.6) both have $\det \Phi(\underset{\sim}{w}^0,\beta_1^0) = 0$. But, by analysing higher-order terms one can establish the existence of bifurcation curves $\alpha_1 = \beta_1^0\alpha_2 + O(|\alpha_2|^2)$ for (2.1') when $\beta_1^0 = 0$ and $\beta_1^0\ -2\lambda_0$, upon assuming some generic hypotheses. Those values of β_1 are exactly the "simple eigenvalues" for $(L_2;L_1)$, as exposed by Mallet-Paret [5, p. 390] and Chow and Hale [1].

<u>Remark 2.3</u>: There are solutions of (2.4') given by

$$\underset{\sim}{w} = -m_0(\beta_1-\beta_1^0)\underset{\sim}{z}^0 + O(|\alpha_2| + (|\alpha_2|+|\beta_1-\beta_1^0|)^2)$$

for all $|\alpha_2|$, $|\beta_1-\beta_1^0|$ sufficiently small, where

(a) $\qquad m_0 = r_{20}^{-1}\lambda_0^{-1}, \quad \underset{\sim}{z}^0 = \begin{pmatrix}0\\1\end{pmatrix}, \quad \beta_1^0 = -2\lambda_0$, assuming $r_{20} \neq 0$

(b) $\qquad m_0 = q_{20}^{-1}, \quad \underset{\sim}{z}^0 = \begin{pmatrix}1\\0\end{pmatrix}, \quad \beta_1^0 = 0$, assuming $q_{20} \neq 0$.

Moreover, those solutions are unique in some neighbourhood of $\underset{\sim}{w} = \underset{\sim}{0}$, $\beta_1 = \beta_1^0$ in $\mathbb{R}^3$ for each small, fixed $\alpha_2 \neq 0$. Hence, upon assuming $r_{20} \neq 0 \neq q_{20}$ there are bifurcation curves $\alpha_1 = \beta_1^0\alpha_2 + O(|\alpha_2|^2)$ for (2.1'), since on such a curve there is one less solution $\underset{\sim}{u}$ for (2.1') than for $\underset{\sim}{\alpha}$ nearby, but not on, the curve.

Define

$$(2.9) \quad \begin{cases} c_4 = q_{20}r_{02},\ c_3 = 4q_{11}r_{02},\ c_2 = -q_{20}r_{20}+4q_{11}r_{11}+3q_{02}r_{02}, \\ c_1 = 4q_{02}r_{11},\ c_0 = q_{02}r_{20}, \\ b_3 = r_{02},\ b_2 = 2r_{11}-\lambda_0^{-1}q_{20},\ b_1 = r_{20}-2\lambda_0^{-1}q_{11},\ b_0 = -\lambda_0^{-1}q_{02}. \end{cases}$$

<u>Remark 2.4</u>: Assume that $q_{20} \neq 0 \neq r_{20}$ and $q_{02} \neq 0 \neq r_{02}$ both hold. Then there can be at most four solutions $\underset{\sim}{w}^0$ of (2.4') satisfying $w_1^0 \neq 0 \neq w_2^0$: $w_1^0 = mw_2^0$ where $m \in \{m: c_4m^4+c_3m^3+c_2m^2+c_1m+c_0 = 0,\ m \neq 0 \neq b_3m^3+b_2m^2+b_1m+b_0\}$ and $w_2^0 = -2m(b_3m^3+b_2m^2+b_1m+b_0)$.

We can now state the main result of this section.

<u>Theorem 2.5</u>: For the system (2.1') let us assume (H1), (H2), and two other hypotheses:

(H3) If $(\underset{\sim}{w}^0,\beta_1^0)$ satisfies (2.7 - 2.8) then $\det \Phi(\underset{\sim}{w}^0,\beta_1^0) \neq 0$

(H4) $q_{20} \neq 0 \neq r_{20}$ and $q_{02} \neq 0 \neq r_{02}$.

Then there are only a finite number of local bifurcation curves in the $\underset{\sim}{\alpha}$-plane and they are of the form $\alpha_1 = \beta_1^0\alpha_2 + o(|\alpha_2|)$ where β_1^0 is given in (2.6) and (2.8). The curves corresponding to β_1^0 given in (2.8) separate a neighbourhood of the origin in the $\underset{\sim}{\alpha}$-plane into regions between which the number of solutions $\underset{\sim}{u}$ of (2.1') changes by at least two; if the β_1^0 are distinct then "at least two" can be replaced by "exactly two". The curves corresponding to β_1^0 in (2.6), i.e. $\beta_1^0 = 0$ and $-2\lambda_0$, separate regions with the same number of solutions, but since the number of solutions is one less on the curve itself the number of solutions is not locally constant.

The simplicity of the eigenvalues of $(-J\Omega;\Sigma)$ is required for the application of the results of Chow and Hale [1] for the general problem (2.1). The general concept of simple eigenvalue was motivated by considerations of transversality; in our particular problem we recall that $\gamma^2 - 4\delta\sigma > 0$ implies that the linear perturbation problem has locally two distinct curves of eigenvalues passing through $\lambda_0 = 0$, $\varepsilon = 0$. From Remark 1.2 we know that $\gamma^2 - 4\delta\sigma \geq 0$ when tr $H = 0$, so that the local existence of two such distinct curves will usually be the case when the boundary conditions are self-adjoint for $\varepsilon \neq 0$.
Shearer [7] discusses (2.1) when the linearisation has locally two distinct curves of eigenvalues and the nonlinear problem has symmetry or symmetries.

In Turyn [8] one finds more of the proofs, some further remarks on simple eigenvalues, and some numerical examples.

Acknowledgements: I would like to thank Professors Chow and Hale for making available to me portions of their book, which will be published soon. I would like to thank Professors Binding, Browne, and Hale for helpful discussions and Professor Rostamian for asking about distributed terms in linear perturbation.

Bibliography:

[1] S.-N. Chow and J.K. Hale, Methods of Bifurcation Theory, Springer-Verlag, Berlin and New York (to appear).

[2] S.-N. Chow, J.K. Hale, and J. Mallet-Paret, Applications of generic bifurcation. II, Arch. Rational Mech. Anal., 62 (1976), pp. 209-235.

[3] E.A. Coddington and N. Levinson, Theory of Ordinary Differential Equations, McGraw-Hill, New York, 1955.

[4] J.P. Holman, Heat Transfer, McGraw-Hill, New York, 1976 (fourth edition).

[5] J. Mallet-Paret, Buckling of cylindrical shells with small curvature, Quart. Appl. Math., 35 (1977), pp. 383-400.

[6] M.N. Özişik, Boundary Value Problems of Heat Conduction, International Textbook Company, Scranton, Pennsylvania (1968).

[7] M. Shearer, Secondary bifurcation near a double eigenvalue, S.I.A.M. Math. Anal., 11 (1980), pp. 365-389.

[8] L. Turyn, Perturbation of periodic boundary conditions, S.I.A.M. Math. Anal.. submitted December, 1981.

BIFURCATION PROBLEMS WITH TRIANGULAR SYMMETRY

A. VANDERBAUWHEDE

1. INTRODUCTION

In this contribution we want to discuss generic bifurcation for problems having the symmetry of an equilateral triangle. A prototype of such a problem is the following nonlinear eigenvalue problem :

$$\begin{aligned} \Delta v + \sigma f(z,v,\lambda) &= 0 \quad , \quad z \in \Omega , \\ v(z) &= 0 \quad , \quad z \in \partial\Omega , \end{aligned} \tag{1}$$

where we make the following assumptions :

(i) $\Omega \subset \mathbb{R}^2$ is open, smooth, bounded, and has the symmetry of an equilateral triangle, i.e. $\Omega = \gamma(\Omega)$, for each $\gamma \in \Delta_3$, where Δ_3 is the 6-element subgroup of $O(2)$ generated by the operators $\delta \in L(\mathbb{R}^2)$ and $\tau \in L(\mathbb{R}^2)$ given by :

$$\delta z = \exp(\tfrac{2\pi}{3} i) z \quad , \quad \tau z = \bar{z} \quad , \quad \forall z \in \mathbb{R}^2 \; ; \tag{2}$$

(for notational convenience, we identify $\mathbb{R}^2$ with the complex plane $\mathbb{C}$);

(ii) $\sigma \in \mathbb{R}$ and $\lambda \in \Lambda$, an arbitrary Banach space;

(iii) $f : \bar{\Omega}\times\mathbb{R}\times\Lambda \to \mathbb{R}$ is sufficiently smooth, $f(0,\lambda) = 0$ for all λ, and

$$f(\gamma z,v,\lambda) = f(z,v,\lambda) \quad , \quad \forall \gamma \in \Delta_3 . \tag{3}$$

Now fix some $\lambda_0 \in \Lambda$; we will assume that $D_v f(z,0,\lambda_0)$ is not identically zero. For each λ near λ_0 we want to study the bifurcation of nontrivial solutions of (1) from the line $(v,\sigma) = (0,\sigma)$ of trivial solutions.

Bifurcation can only take place at those σ-values for which the linearized problem :

$$\begin{aligned} &\Delta u + \sigma D_v f(z,0,0)u = 0 \quad , \quad z \in \Omega \\ &u(z) = 0 \quad , \quad z \in \partial\Omega \end{aligned} \tag{4}$$

has nontrivial solutions. We know from standard theory that (4) has a countable set of eigenvalues $\{\sigma_j \mid j \in \mathbb{N}\}$, with $|\sigma_j| \to \infty$ as $j \to \infty$; the corresponding eigenspaces are finite-dimensional. Because of the symmetry some of these eigenvalues may generically be multiple eigenvalues. In section 2 we will use group representation theory to classify the eigenvalues according to the symmetry properties of the corresponding eigenfunctions. In section 3 we apply a generalization of the Crandall-Rabinowitz theorem for bifurcation from a simple eigenvalue [1] to obtain the bifurcation equations. We discuss the solution set in section 4.

2. THE LINEAR PROBLEM

Let $\alpha \in]0,1[$, $W = C^{0,\alpha}(\bar{\Omega})$ and $V = C_0^{2,\alpha}(\bar{\Omega}) = \{v \in C^{2,\alpha}(\bar{\Omega}) \mid v(z) = 0, \forall z \in \partial\Omega\}$. Define $M : V \times \Lambda \times \mathbb{R} \to W$ by $M(v,\lambda,\sigma)(z) = \Delta v(z) + \sigma f(z,v(z),\lambda)$, $\forall z \in \bar{\Omega}$, and let $L(\sigma) = D_v M(0,0,\sigma)$, $\forall \sigma \in \mathbb{R}$. Then problem (1) takes the form

$$M(v,\lambda,\sigma) = 0 \tag{5}$$

while the linear problem (4) corresponds to the determination of $N(L(\sigma))$.

We can define a representation $\Gamma : \Delta_3 \to L(W)$ of the group Δ_3 over the space W, as follows :

$$(\Gamma(\gamma)w)(z) = w(\gamma^{-1}z) \quad , \quad \forall \gamma \in \Delta_3 . \tag{6}$$

Then the mappings M and $L(\sigma)$ are equivariant with respect to this representation, i.e. we have :

$$M(\Gamma(\gamma)v,\lambda,\sigma) = \Gamma(\gamma)M(v,\lambda,\sigma) \quad , \quad \forall\gamma \in \Delta_3 \tag{7}$$

and

$$L(\sigma)\Gamma(\gamma) = \Gamma(\gamma)L(\sigma) \quad , \quad \forall\gamma \in \Delta_3 . \tag{8}$$

It is clear from (8) that $N(L(\sigma))$ is invariant under the symmetry operators $\Gamma(\gamma)$, $\gamma \in \Delta_3$. If σ_j is an eigenvalue, then generically $N(L(\sigma_j))$ will be irreducible under this group action, i.e. $N(L(\sigma_j))$ will have no proper invariant subspace. If $N(L(\sigma_j))$ should happen to be reducible, then an arbitrary small perturbation of the domain Ω or the function f such that the triangular symmetry is maintained will split σ_j in several eigenvalues for which the corresponding eigenspaces are irreducible.

Lemma 1. Let U be a finite-dimensional real vectorspace, and $\Gamma : \Delta_3 \to L(U)$ an irreducible representation of Δ_3 over U. Then one of the following three possibilities occurs :

(1) $\dim U = 1$ and $\Gamma(\gamma)u = u$, $\forall\gamma \in \Delta_3$, $\forall u \in U$;

(2) $\dim U = 1$, and :

$$\Gamma(\delta)u = u \quad , \quad \Gamma(\tau)u = -u \quad , \quad \forall u \in U ; \tag{9}$$

(3) $\dim U = 2$, and U has a basis $\{u_1,u_2\}$ such that if we define $\psi : \mathbb{C} \to U$ by $\psi(x+iy) = xu_1 + yu_2$, then :

$$\Gamma(\gamma)\psi(z) = \psi(\gamma z) \quad , \quad \forall\gamma \in \Delta_3 , \forall z \in \mathbb{C} . \tag{10}$$

P r o o f. Since $\Gamma(\tau)^2 = \Gamma(\tau^2)$ is the identity on U, $\Gamma(\tau)$ will have eigenvalues ±1; let $u_0 \neq 0$ be such that $\Gamma(\tau)u_0 = \varepsilon u_0$, $\varepsilon = \pm 1$. Then $\Gamma(\tau)\Gamma(\delta)u_0 = \Gamma(\delta^{-1})\Gamma(\tau)u_0 = \varepsilon\Gamma(\delta^{-1})u_0$, since $\tau\circ\delta = \delta^{-1}\circ\tau = \delta^2\circ\tau$. Let $\bar{u}_0 = u_0 + \Gamma(\delta)u_0 + \Gamma(\delta^2)u_0$; then it is immediate that the subspace $\mathrm{span}\{\bar{u}_0\}$ of U is invariant under $\Gamma(\delta)$ and $\Gamma(\tau)$. Since Γ is irreducible on U, it follows that either $U = \mathrm{span}\{\bar{u}_0\}$ or $\bar{u}_0 = 0$. The first possibility gives us the cases (1) and (2), depending on whether $\varepsilon = +1$ or $\varepsilon = -1$.

Now suppose $\bar{u}_0 = 0$. Let $v_0 = (\sqrt{3})^{-1}(\Gamma(\delta)u_0 - \Gamma(\delta^2)u_0)$. Then $\Gamma(\tau)u_0$

$= \varepsilon u_0$, $\Gamma(\tau)v_0 = -\varepsilon v_0$, and

$$\Gamma(\delta)u_0 = -\frac{1}{2}u_0 + \frac{\sqrt{3}}{2}v_0 \quad , \quad \Gamma(\delta)v_0 = -\frac{\sqrt{3}}{2}u_0 - \frac{1}{2}v_0 \ . \tag{11}$$

This shows that $v_0 \neq 0$ and that $\text{span}\{u_0,v_0\}$ is a two-dimensional subspace of U, invariant under Γ. From the irreducibility hypothesis it follows that $U = \text{span}\{u_0,v_0\}$. To obtain (10) we take $u_1 = u_0$, $u_2 = v_0$ if $\varepsilon = +1$, and $u_1 = -v_0$, $u_2 = u_0$ in case $\varepsilon = -1$. □

Under the hypothesis that each eigenvalue σ_j of (4) corresponds to an irreducible eigenspace, we can use lemma 1 tc divide the set of eigenvalues into three disjoint classes $\{\sigma_j^{(i)} \mid j \in \mathbb{N}\}$, where for each $i = 1,2,3$ the eigenspace $U_j^{(i)}$ corresponding to $\sigma_j^{(i)}$ transforms under the symmetry group Δ_3 according to case (i) of lemma 1.

3. THE BIFURCATION EQUATION

Since $L(\sigma_j^{(i)})$ is a Fredholm operator, we can apply the Liapunov-Schmidt method on equation (5). Because of the equivariance of M this can be done in a symmetry-compatible way (see [4]). More precisely we can apply the following theorem ([3],[4]) which is a generalization of the classical theorem of Crandall and Rabinowitz [1] on bifurcation from simple eigenvalues.

Theorem 2. Let V, W and Λ be real Banach spaces, and G a compact group with representations $\Gamma : G \to L(V)$ and $\tilde{\Gamma} : G \to L(W)$ over V, respectively W. Let $M : V\times\Lambda\times\mathbb{R} \to W$ be a mapping of class C^{r+2} $(r \geq 0)$ such that $M(0,\lambda,\sigma) = 0$ for all (λ,σ), and which is equivariant with respect to $(G,\Gamma,\tilde{\Gamma})$:

$$M(\Gamma(\gamma)v,\lambda,\sigma) = \tilde{\Gamma}(\gamma)M(v,\lambda,\sigma) \quad , \quad \forall\gamma \in G \ . \tag{12}$$

Let $(\lambda_0,\sigma_0) \in \Lambda\times\mathbb{R}$ be such that :

(i) $L = D_vM(0,\lambda_0,\sigma_0)$ is a Fredholm operator with index zero ;

(ii) $N(L)$ is irreducible under the representation Γ ;

(iii) $D_\sigma D_vM(0,\lambda_0,\sigma_0).u \notin R(L)$ for some $u \in N(L)$.

Then all solutions of (5) in a neighbourhood of $(0,\lambda_0,\sigma_0)$ belong to a C^{r+2}-submanifold M of the form :

$$M = \{(\Phi(u,\lambda,\sigma),\lambda,\sigma) \mid (u,\lambda,\sigma) \in \Omega\} ,$$

where Ω is a symmetry-invariant neighbourhood of $(0,\lambda_0,\sigma_0)$ in $N(L)\times\Lambda\times\mathbb{R}$, $\Phi : \Omega \to V$ is of class C^{r+2} and equivariant (i.e. $\Phi(\Gamma(\gamma)u,\lambda,\sigma) = \Gamma(\gamma)\Phi(u,\lambda,\sigma)$), $\Phi(0,\lambda,\sigma) = 0$ for all (λ,σ), $D_u\Phi(0,\lambda_0,\sigma_0) = I$ on $N(L)$, and $P\Phi(u,\lambda,\sigma) = u$, where P is an equivariant projection in V onto $N(L)$.

Moreover, on the submanifold M the equation (5) is equivalent to an equation of the form :

$$F(u,\lambda,\sigma) = 0 , \tag{13}$$

where $F : \Omega \to N(L)$ is of class C^{r+2}, and has the following properties :

(a) $F(0,\lambda,\sigma) = 0 \quad , \quad \forall(\lambda,\sigma)$;

(b) $D_uF(0,\lambda_0,\sigma_0) = 0$, $D_\sigma D_uF(0,\lambda_0,\sigma_0) = I$ on $N(L)$;

(c) F is equivariant with respect to (G,Γ) :

$$F(\Gamma(\gamma)u,\lambda,\sigma) = \Gamma(\gamma)F(u,\lambda,\sigma) \quad , \quad \forall\gamma \in G . \quad \square \tag{14}$$

It is also possible to give an explicit definition of the mapping F, but for the qualitative discussion which follows the properties (a)-(c) above are sufficient. For our particular problem (1) we replace σ_0 by any eigenvalue $\sigma_j^{(i)}$ of (4); then $L = L(\sigma_j^{(i)})$ and the hypotheses (i) and (ii) of theorem 2 are immediately satisfied; (we take $G = \Delta_3$ and Γ defined by (6)). As for the hypothesis (iii), we have

$$R(L(\sigma_i^{(i)})) = \{w \in W \mid \int_\Omega w(z)u(z)dz = 0, \ \forall u \in N(L(\sigma_j^{(i)}))\} . \tag{15}$$

Also, if $u \in N(L(\sigma_j^{(i)}))$, then :

$$(D_\sigma D_V M(0,\lambda_0,\sigma_j^{(i)})u)(z) = D_V f(z,0,\lambda_0)u(z) = -(\sigma_j^{(i)})^{-1}\Delta u(z) ,$$

(remark that $\sigma_j^{(i)} \neq 0$), and :

$$\int_\Omega \Delta u(z).u(z)dz = -\int_\Omega (\nabla u(z))^2 dz \neq 0 .$$

Consequently also (iii) is satisfied, and our problem reduces to that of studying the bifurcation equation (13) for (u,λ,σ) near $(0,\lambda_0,\sigma_j^{(i)})$ in $N(L(\sigma_i^{(i)}))\times\Lambda\times\mathbb{R}$; in the next section we analyze how the solution set of (13) depends on the value of $i \in \{1,2,3\}$.

4. THE SOLUTION SET

When $i = 1$, then $N(L(\sigma_j^{(1)})) = \text{span}\{u_j^{(1)}\}$ for some $u_j^{(1)} = 0$. It follows from the properties (a) and (b) of F that

$$F(\rho u_j^{(1)},\lambda,\sigma) = \rho F_1(\rho,\lambda,\sigma) \ , \tag{16}$$

where $F_1(0,\lambda_0,\sigma_j^{(1)}) = 0$ and $D_\sigma F_1(0,\lambda_0,\sigma_j^{(1)}) \neq 0$. Consequently, the equation $F_1(\rho,\lambda,\sigma) = 0$ can be solved for σ, and the bifurcation equation has for each λ near λ_0 a unique branch of nontrivial solutions, given by

$$\{(\rho u_j^{(1)},\sigma^*(\rho,\lambda)) \mid |\rho| < \rho_0\} \ , \tag{17}$$

where σ^* is of class C^{r+1} and $\sigma^*(0,\lambda_0) = \sigma_j^{(1)}$. Moreover, since $\Gamma(\gamma)u_j^{(1)} = u_j^{(1)}$, $\forall\gamma \in \Delta_3$, all solutions on the corresponding solution branch of (1) will have the full Δ_3-symmetry of the original problem. We conclude that near each eigenvalue $\sigma_j^{(1)}$ of (4) there bifurcates, for each λ near λ_0, a branch of symmetry invariant solutions of (1).

Also in the case $i = 2$ the bifurcation equation will have a branch of nontrivial solutions of the form (17) (just replace $u_j^{(1)}$ by $u_j^{(2)}$). However, since $\Gamma(\tau)u_j^{(2)} = -u_j^{(2)}$, we will have $\sigma^*(-\rho,\lambda) = \sigma^*(\rho,\lambda)$ (i.e. the bifurcating branch is symmetric in ρ), and $\Gamma(\tau)\Phi(\rho u_j^{(2)},\lambda,\sigma^*(\rho,\lambda)) = \Phi(-\rho u_j^{(2)},\lambda,\sigma^*(\rho,\lambda))$. We conclude that near each eigenvalue $\sigma_j^{(2)}$ of (4) there bifurcates, for each λ near λ_0, a symmetric branch of nontrivial solutions of (1), which are invariant under $\Gamma(\delta)$, and where the two symmetric parts of the branch are obtained one from the other by application of $\Gamma(\tau)$.

Finally, we come to the case $i = 3$. We have $\dim N(L(\sigma_j^{(3)})) = 2$, and we can use the isomorphism ψ given by lemma 1 to identify $xu_1 + yu_2 \in N(L(\sigma_j^{(3)}))$ with the complex number $z = x+iy \in \mathbb{C}$. Then we can write the bifurcation equation (13) in the form

$$F(z,\lambda,\sigma) = 0 \ , \tag{18}$$

where $F : \mathbb{C}\times\Lambda\times\mathbb{R} \to \mathbb{C}$ is of class C^{r+2} (we consider $\mathbb{C}$ as a *real* vector-space), $F(0,\lambda,\sigma) = 0$ and

$$F(\gamma z,\lambda,\sigma) = \gamma F(z,\lambda,\sigma) \quad , \quad \forall\gamma \in \Delta_3 . \tag{19}$$

Lemma 3. Let $F : \mathbb{C} \to \mathbb{C}$ be of class $C^{\ell+4}$, and equivariant with respect to Δ_3. Then there exist unique functions $g : \mathbb{C} \to \mathbb{R}$ and $h : \mathbb{C} \to \mathbb{R}$ such that :

$$\text{(i)}\ F(z) = g(z)z + h(z)\bar{z}^2 \quad , \quad \forall z \in \mathbb{C} ; \tag{20}$$

$$\text{(ii)}\ g(\gamma z) = g(z)\ ,\ h(\gamma z) = h(z)\ ,\ \forall z \in \mathbb{C}\ ,\ \forall\gamma \in \Delta_3 . \tag{21}$$

The functions g and h are of class C^{ℓ}; if F depends smoothly on parameters, then also g and h depend smoothly on these parameters.

P r o o f. Let $G(z) = \text{Im}\ \bar{z}F(z)$; then $G(\delta z) = G(z)$ and $G(\bar{z}) = -G(z)$. Therefore $G(z) = 0$ if $z = \bar{z}$, and consequently :

$$G(z) = (2i)^{-1}(z-\bar{z})G_1(z) \quad , \quad \forall z \in \mathbb{C} ,$$

where $G_1 : \mathbb{C} \to \mathbb{C}$ is of class $C^{\ell+3}$. From $G(\delta z) = G(z)$ it then follows that $G(z) = (2i)^{-1}(\delta z-\bar{\delta}\bar{z})G_1(\delta z)$, and consequently $G_1(\delta z) = 0$ if $z = \bar{z}$. This implies that $G_1(\delta z) = (2i)^{-1}(z-\bar{z})G_2(z)$, where $G_2(z)$ is of class $C^{\ell+2}$. Another application of the same argument gives us finally :

$$G(z) = h(z)\ \text{Im}\ \bar{z}^3 , \tag{22}$$

where $h : \mathbb{C} \to \mathbb{R}$ is of class $C^{\ell+1}$, and clearly satisfies $h(\gamma z) = h(z)$, $\forall\gamma \in \Delta_3$.

Let now $\tilde{F}(z) = F(z) - h(z)\bar{z}^2$; then $\text{Im}\ \tilde{F}(z)\bar{z} = 0$ for all $z \in \mathbb{C}$, which immediately implies that $\tilde{F}(z) = g(z)z$, with $g : \mathbb{C} \to \mathbb{R}$ of class C^{ℓ}. The invariance property of g follows then from the equivariance of $\tilde{F}$. This proves the lemma. □

Using this result, we can rewrite the bifurcation equation (18) in the form :

$$g(z,\lambda,\sigma)z + h(z,\lambda,\sigma)\bar{z}^2 = 0 , \tag{23}$$

where the functions g and h are real valued and symmetry invariant; it also follows from the property (b) of the bifurcation function tha

$$g(0,\lambda_0,\sigma_0) = 0 \quad , \quad D_\sigma g(0,\lambda_0,\sigma_0) = 1 \; . \tag{24}$$

For $z \neq 0$ we can multiply (23) by $\bar{z}$, write $z = \rho e^{i\theta}$, split into real and imaginary parts, and divide by ρ^2, respectively ρ^3; we obtain the equivalent system :

$$\begin{aligned} & g(z,\lambda,\sigma) + \rho h(z,\lambda,\sigma) \cos 3\theta = 0 \; , \\ & h(z,\lambda,\sigma) \sin 3\theta = 0 \; . \end{aligned} \tag{25}$$

Assuming enough differentiability, it follows from (24) that the first equation of (25) can be solved for $\sigma = \tilde{\sigma}(\rho,\lambda,\sigma)$, with $\tilde{\sigma}(0,\theta,\lambda_0) = 0$ for all θ, and $\tilde{\sigma}(\rho,\pm\theta+\frac{2\pi}{3},\lambda) = \tilde{\sigma}(\rho,\theta,\lambda) = \tilde{\sigma}(-\rho,\theta+\pi,\lambda)$. The second equation is satisfied for $\theta = \frac{\pi}{3}k$, $k \in \mathbb{N}$; returning to the original bifurcation equation (13), this gives us for each λ near λ_0 three branches of solutions of (13), of the form

$$\{(\rho\Gamma(\delta^j)u_1,\tilde{\sigma}(\rho,0,\lambda) \mid |\rho| < \rho_0\} \quad , \quad j = 0,1,2. \tag{26}$$

The three branches are obtained from each other by application of the symmetry operator $\Gamma(\delta)$, while along each of these branches the solutions are invariant under an appropriate reflection in Δ_3; for example, along the branch $j = 0$ the solutions satisfy $\Gamma(\tau)u = u$. A similar result then holds for the corresponding solution branches of (1).

All other solutions of (25) have to satisfy the equations :

$$g(z,\lambda,\sigma) = 0 \quad , \quad h(z,\lambda,\sigma) = 0 \; . \tag{27}$$

Again we can solve the first of these equations for $\sigma = \sigma^*(z,\lambda)$, with $\sigma^*(0,\lambda_0) = \sigma_0$ and $\sigma^*(\gamma z,\lambda) = \sigma^*(z,\lambda)$, $\forall\gamma \in \Delta_3$. Of course, for each solution of (27) we will have $\sigma^*(z,\lambda) = \tilde{\sigma}(\rho,\theta,\lambda)$, with $z = \rho e^{i\theta}$. Bringing the solution $\sigma^*(z,\lambda)$ of $g = 0$ in the second equation $h = 0$, we obtain the final bifurcation equation

$$p(z,\lambda) = 0 \;, \tag{28}$$

with $p(\gamma z,\lambda) = p(z,\lambda)$, $\forall\gamma \in \Delta_3$. This symmetry invariance implies that

$$p(z,\lambda) = \alpha(\lambda) + \beta(\lambda)\rho^2 + O(\rho^3) \quad , \quad z = \rho e^{i\theta} \;. \tag{29}$$

If $\alpha(\lambda_0) \neq 0$, then (27) has no small solutions. If $\alpha(\lambda_0) = 0$ and $\beta(\lambda_0) \neq 0$, then a simple scaling argument shows that (36) has no non-trivial solutions if $\alpha(\lambda)\beta(0) \geqslant 0$, while if $\alpha(\lambda)\beta(0) < 0$ there is a small "circle" of solutions of the form

$$\rho = \rho_\lambda(\theta) = |\alpha(\lambda)|^{1/2}\eta(\theta,\lambda) \;,$$

with $\eta(\theta,0) = |\beta(0)|^{-1/2}$ and $\rho_\lambda(\pm\theta + \frac{2\pi}{3}k) = \rho_\lambda(\theta)$. This gives us a closed curve of solutions of (1), which connects the three branches which we found already; so, if $\alpha(\lambda_0) = 0$ and $\beta(\lambda_0) = 0$, there will be 6 points of secondary bifurcation for those λ near λ_0 for which $\alpha(\lambda)\beta(0) < 0$.

It is interesting to compare these last results with some of the results obtained by Golubitsky and Schaeffer [2] using singularity theory.

REFERENCES

[1] Crandall M.G. and P.H. Rabinowitz, Bifurcation from simple eigenvalue. J. Funct. Anal. 8 (1971) 321-340.

[2] Golubitsky M. and Schaeffer D., Bifurcation with O(3)-symmetry including applications to the Bénard problem. Preprint 1980.

[3] Vanderbauwhede A., Symmetry and bifurcation from multiple eigenvalues. Lecture Notes in Math. Vol. 846, 1981, 356-365.

[4] Vanderbauwhede A., Local bifurcation and symmetry. Research Notes in Math., Pitman, London. To appear.

Instituut voor Theorctische Mechanica
Rijksuniversiteit Gent
Krijgslaan 281
B-9000 Gent, Belgium.

On the Foundations of Thermodynamics

Johann Walter

Wer kann das ansehen, ohne leise zu pfeifen?

1. Technical Remark: Since the present paper is rather closely connected with our previous paper [30] (and also because of lack of space) we take accross the denotation and terminology of [30] without explaining it again. The only modification is that in the discussions to follow a certain postulate occurring in the definition of the set $\mathfrak{M}$ of all thermodynamical systems (cf. [30; p.90,(iii)]) will be replaced successively by one of the "axioms" (1), (2), (3), (4) corresponding to the "lower corners" of a certain tetrahedron to be described below. We also use the additional abbreviation $\vartheta_A := (\tilde{A} + \hat{A}_\alpha, \hat{A}_t)$ and ask the reader to replace the "A" after "denotated by" in the seventh line of [30; p.88] by "$\tilde{A}$".

2. Aim of the Paper: It is a commonplace that even today thermodynamics has not reached yet a generally accepted mathematical status as, e.g., electrodynamics. In spite of numerous widespread complaints concerning this point there does not seem to exist much plausible knowledge as to the deeper reasons for the obstinacity of this state of affairs. "The recent tendency to apply more and more rigorous mathematical methods to physical investigations" [22] rather tends to aggravate than to improve the situation - as will become clear in the following. It is the aim of the present paper to pick out a small sector of this bunch of problems, viz., the problem of derivation of the Fundamental Theorem from some appropriately chosen mathematical representative of the Second Law of classical equilibrium thermodynamics and to try to give a "geometric" overview of a certain set $\mathfrak{D}$ of such derivations to be found in the literature. Even in this restricted area of research we have not succeeded in finding a simple Venn-diagram like that used in [29]. We have developed instead a three-dimensional "graph-theoretic" representation. Besides serving its main purpose, viz., providing "a critical analysis of various axiomatic systems" [18; p.129] as well as a sound basis for the choice of an "optimal" element of $\mathfrak{D}$, this geometric model works in a way like the "Periodic System of Elements" in bringing to light new aspects overlooked so far. It is in this way that we have identified a new variant of such a derivation which from the point of view of elementary vector analysis might be of independent interest (cf. Remark 2 below).

3. The "Foundational" Tetrahedron: The main assertions or axioms used in the derivations belonging to $\mathfrak{D}$ will be enshrined into the corners (one exception) of a three-dimensional tetrahedron in such a way that every element of $\mathfrak{D}$ may be represented as a certain sequence of such "corners" whose "edges" correspond

to the theorems making up that special derivation:

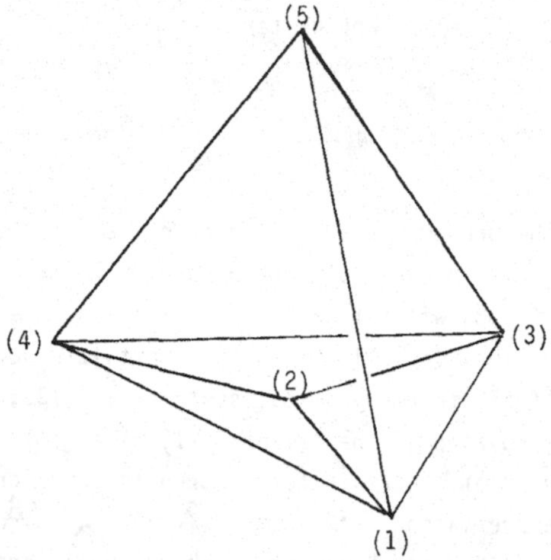

4. The Corners:

(1) $$\mathfrak{N} = \emptyset$$

where $\mathfrak{N} := \bigcup_{n(A) \geq 2} \mathfrak{A}$ and for any $A \in \mathfrak{M}$ with $\hat{A}_t > 0$ and $n(A) \geq 2$, $\mathfrak{A}$ denotes the set of all solution curves of $Q_A = 0$ whose projection onto the a-space is closed but which are not closed themselves.[1)] Bringing into play the "composition" of a special "Hilfskörper" E (cf. postulates (i), (ii) in [30; p.88,89]) with any $A \in \mathfrak{M}$ we get our own version of axiom (1) (which is bound to replace (1) in our derivation[2)] being especially well-suited for applying elementary vector analytic methods), viz.,

(1') $$\mathfrak{N}_1 = \emptyset \ , \ \mathfrak{N}_2 = \emptyset$$

where $\mathfrak{N}_1 := \bigcup_{n(A) \geq 1} \mathfrak{A}_1 \quad [\, \mathfrak{N}_2 := \bigcup_{n(A) \geq 2} \mathfrak{A}_2 \,]$

and for each $A \in \mathfrak{M}$ with $n(A) \geq 1 \ [n(A) \geq 2]$ $\mathfrak{A}_1 \ [\mathfrak{A}_2]$ denotes the set of solution curves of $Q_{E \dot{+} A} = 0$

----

1) " "closed" but not closed " in the terminology of [18].

2) It is interesting to note that we have, in fact, (1)⇔(1'). But we will not make use of this equivalence and, therefore, omit its proof. It is further interesting that the very formulation of (1) or (1') renders pointless an objection raised in [2; Aufgabe 34 (p.84) and Lösung (p.232)].

whose projection onto the (a,t)-space is closed but which are not closed themselves [the set of closed isothermal curves $\mathcal{L}$ with $\int_{\mathcal{L}} Q_A \neq 0$].

The plausibility of (1) or (1') as a mathematical representative of the Second Law derives from its high degree of falsibility: each one of the relations $\mathcal{N} \neq \emptyset$, $\mathcal{N}_1 \neq \emptyset$ or $\mathcal{N}_2 \neq \emptyset$ amounts to the existence of a perpetuum mobile of the second kind!

Remark 1: In this connection by indulging in some rather strange terminology (already hinted at in footnote 1) the author of [18; p.122] (cf. also [17; p.329] and [30; p.90, Remark 2]) puts forward a Postulate whose meaning is that of the First or the Second Law of thermostatics according as to which of two possible interpretations of certain quotation marks occuring in it is preferred (compare the use of the graphems " "cyclic" " and " "closed" " in [31; p.237] or [18; p.122], respectively). -

(2) For every $A \in \mathcal{M}$ with $\hat{A}_t > 0$ and every x in the domain of definition of A no neighborhood of x exists all points of which can be joined with x by a solution curve of $Q_A = 0$.

There is a lot of controversy connected with this axiom of Carathéodory [9]. On the one hand it has been hailed as "more elegant and at the same time more rational" [25; p.39], "logically more economic" [15; p.37], "Das Axiom von Carathéodory beschränkt die axiomatischen Aussagen auf das logische Minimum" [28; p.96], "Tatsächlich reicht die Erfahrung, daß es überhaupt gewisse unausführbare Prozesse gibt" [5; p.172], whereas on the other hand its physical plausibility has been depreciated as "not based directly on experience" [32; p.915], "There has always been some difficulty in motivating Carathéodory's principle convincingly" [20; p.485] "no obvious physical interpretation" [18; p.125]. The persistence up to the present day of this controversy seems to be due to the fact that there are no specific reasons available to kick the beam in favor of one of the two groups of these rather general judgements. It is, in the opinion of the present writer, the second group that can be made precise; the reduced degree of falsibility of (2) (as compared with that of (1) or (1')) will become evident in the following way: Since (2) cannot be made to assert the emptiness of a set each element of which embodies some immediately desirable physical process, at the first sight it might seem for the task of falsifying (2) to be necessary "Versuche anzustellen in der Absicht, alle Nachbarzustände irgendeines bestimmten Zustandes auf adiabatischem Wege zu erreichen" [23; p.619 and will become clear after using an additional mathematical theorem only (cf. [20; p.485],[21; p.42])[3] that again the existence of a single perpetuum mobile of the second kind will do the job.

----

3) This theorem corresponds to the edge (1) (2) of our tetrahedron, cf. next section.

It seems to be a rather interesting question which are the reasons for the choice of an axiom of this kind. -

(3) $$\bigwedge_{n(A)\geq 2} \left\{ Q_A \wedge d Q_A = 0 \right\}^{4)}.$$

This axiom looks reasonably compact and simple at the first sight. Also rather sympathetic is its tendency to exclude some widespread misuse of " dQ " as a synonym for " Q " ([4], however, still writes $\delta Q \wedge d\delta Q$). But on closer inspection it reveals itself as highly redundant and we have been able to dig out an unexpectedly small basis of (3), viz.,

(3') $$\bigwedge_{n(A)=1} \left\{ Q_{E\dot{+}A} \wedge d Q_{E\dot{+}A} = 0 \right\}, \bigwedge_{n(A)\geq 2} \left\{ d_a W_A = 0 \right\}.$$

Since no use will be made of the equivalence (3)⇔(3') we omit its proof (cf. footnote 2) which, however, is quite simple algebra. Owing to its lack of redundancy (3') fits more readily into the logical texture as precursor as well or as successor than does (3). Most important in this respect: For $n(A)=1$ the "integrability condition" $Q_{E\dot{+}A} \wedge d Q_{E\dot{+}A} = 0$ already guarantees the existence of an integrating denominator (in fact, the "absolute temperature") depending on t only and being given as a solution of a simple ordinary differential equation (cf. [30; p.91,(2)]). Replacing (3) by (3') thus amounts to the complete avoidance of a theorem of Frobenius[5] which is difficult to prove even in the most simple nontrivial case (for such proofs cf.[14; p.318],[19], [24; p.225],[26; p.346]). Another advantage of (3') which I suppose will be welcomed not by the least part of my readers is that it can be written in the more familiar form

(3") $$\bigwedge_{n(A)=1} \left\{ v_{E\dot{+}A} \operatorname{rot} v_{E\dot{+}A} = 0 \right\}, \bigwedge_{n(A)\geq 2,\, t\in\mathbb{R}} \left\{ \tilde{A}\, da \text{ exact} \right\}^{*)}$$

thus throwing out exterior calculus altogether. -

(5) $$\bigvee_{\tau(t)} \bigwedge_{n(A)\geq 1} \left\{ \frac{Q_A}{\tau} \text{ exact} \right\}.$$

The function τ is uniquely determined by this property up to a constant factor and is "after normalization in the usual manner" [17; p.329] denoted by T and

---

4) "$\wedge$" or "d" denotes the exterior product or derivative, respectively.

5) We do not share the point of view expressed by [1; p.220] in a similar context: "To try to avoid using some theorems of mathematics is to impose an absurd handicap on one's technique". The theorem of Frobenius, by the way, is represented as the edge (3) (4) of our tetrahedron, cf. next section.

*) See additional note at the end of the paper.

called "absolute temperature". Theorem (5) is the natural endpoint of every sequence of corners corresponding to an element of $\mathfrak{D}$ and is, therefore, called the Fundamental Theorem of thermostatics. -

(4) $$\bigwedge_{n(A)\geq 2} \bigvee_{\tau(a,t)} \left\{ \frac{Q_A}{\tau} \text{ exact} \right\}.$$

This assertion (not being chosen very often as the initial point) seems to be affected by a rather strong tendency to appear as the endpoint of an element of $\mathfrak{D}$: The integrating denominator τ although by no means uniquely defined by (4) (compare, e.g., the different positions of the two quantifications occurring in (5) or (4)) is frequently denoted by T too. Bound variables, it is true, can be denoted ad libitum - but then T cannot be called "absolute" temperature any longer as is done in [4; p.401],[25; p.41],[31; p.191]. This equivocation does'nt give a fair chance of understanding to the uninitiated because certain relations which look rather familiar at the first sight (cf. [4; p.404,(6)],[15; p.26, (3.13)],[31; p.191, (4.44)]) (but are nevertheless valid for every integrating denominator!) do not enjoy their fundamental physical meaning except for the appropriate choice of T . This fact has already been criticized more than a hundred years ago by R. Clausius [10; p.367] in a similar context (cf. also [8]).

5. The Edges:

(1) (2): This edge has already been mentioned in footnote 3.

(2) (4): This is Carathéodory's famous "accessibility theorem". It is interesting to note that Carathéodory's proof of this theorem essentially falls into two parts containing the first or second edge of the sequence (2) (1) (4), respectively. On the other hand (2) (4) is equivalent (modulo the trivial edge (4) (3)) to (2) (3) which on its part is equivalent to (i.e. contraposition of) a rather difficult theorem of Chow [11](cf. [13; p.947/948],[16; p.249],[19]; a sketch of (2) (3) is also given in [15; p.38/39] where, however, (2) (4) is asserted). Under these circumstances it does not seem to be astonishing that together with (2) itself the edge (2) (4) originating in (2) has provoked disagreement too: "encumbered by the emphasis which is laid on an abstract mathematical theorem which is otherwise of little use in physics" [12; p.246], "This "simple" proof, however, is not simple for the ordinary physicist" [27; p.781], "The usual proofs of which are often so unpalatable to the physicist" [6; p.41] (proofs for the case n=3 in [25;p.35 and 37] and for the general case in [3], [7]).

(1') (5): This edge is given in [30] where, however, only $(1')_1$ is used.

6. The new sequence (1') (3") (5):

(3") (5): Existence of T as in [30; Section 4, First Step]. Exactness of $\frac{Q_A}{T}$ by inspection.

(1') (3"): Let A be a system with $n(A) = 1$ and ξ a point such that $(3")_1$ is violated at ξ. Setting $\varepsilon := (1,0,0)$ we have $v_{E+A}(\xi)\cdot\varepsilon \neq 0$. According to our Lemma (cf. Appendix) on a suitable cylinder parallel to the e-axis there exists a solution curve of $Q_{E+A} = 0$ contradicting $(1')_1$ (note that $(1')_1$ is used for the case $n(A) = 1$ only). Let now A be a system with $n(A) \geq 2$ and ξ a point such that $(3")_2$ is violated at ξ. Then two indices i, k exist such that $(\tilde{A}_{ia_k} - \tilde{A}_{ka_i})(\xi) \neq 0$. Applying Gauss'theorem in the a_i, a_k-plane there exists a closed curve $\mathcal{L}$ near ξ such that $\int_{\mathcal{L}} \tilde{A}\, da \neq 0$ which according to the First Law contradicts $(1')_2$.

7. Appendix: Let ξ be a point, ε a vector and $v(x)$ a vectorfield in $\mathbb{R}^3$ such that $v(\xi)\cdot\varepsilon \neq 0$. Let $\mathcal{L}$ be a circle of length ℓ around ξ orthogonal to $v(\xi)$, let $c(s)$, $s \in [0,\ell]$, s = arclength be a parametrization of $\mathcal{L}$ and $\mathfrak{Z}$ the cylinder through $\mathcal{L}$ parallel to ε. Then $p(s,h) := c(s) + h\varepsilon$, $h \in \mathbb{R}$ is a parametrization of $\mathfrak{Z}$ and

(6)
$$\frac{dh}{ds} = f(s,h) := -\frac{v(p)\cdot\dot{c}}{v(p)\cdot\varepsilon}$$

a parametrization of the directional field induced in $\mathfrak{Z}$ by $v(x)\cdot dx = 0$. The solution of (6) with $h(0) = 0$ then corresponds (for small ℓ) to a "flat screw" on $\mathfrak{Z}$ with "pitch" $h(\ell)$.

Lemma $h(\ell) = -\left.\frac{\ell^2 v\cdot \operatorname{rot} v}{4\pi\, v\cdot\varepsilon}\right|_{\xi}$ to first order as $\ell \to 0$.

The proof follows by simple estimations from the identity

(7)
$$h(\ell) = \left[\int_0^\ell f(s,h(s))\,ds - \int_0^\ell f(s,0)\,ds\right] + \left[\int_{\mathcal{L}} \frac{v\cdot dx}{v(\xi)\cdot\varepsilon} - \int_{\mathcal{L}} \frac{v\cdot dx}{v\cdot\varepsilon}\right]$$

$$+\frac{1}{v(\xi)\cdot\varepsilon}\left[\left.\frac{\ell^2 v\cdot\mathrm{rot}\,v}{4\pi}\right|_\xi-\int_{\mathcal{L}} v\,dx\right]-\left.\frac{\ell^2 v\cdot\mathrm{rot}\,v}{4\pi v\cdot\varepsilon}\right|_\xi .$$

Remark 2: According to (7) the "circulation" of $v(x)$ along $\mathcal{L}$ can be interpreted to first order as the "pitch" of the above-mentioned "screw". A rather special result of this type is contained in [26; p.352, Anmerkung 1].

References

1 Arens, R.: J. Math. Anal. Appl. 6 (1963), 207.

2 Basarov, J.P.: Thermodynamik, VEB Deutscher Verlag der Wissenschaften, Berlin 1974.

3 Bernstein, B.: J. Math. Phys. 1 (1960), 222.

4 Bleuler, K.: Reports on Math. Phys. 12 (1977), 395.

5 Born, M.: Ausgewählte Abhandlungen 1, Vandenhoeck & Ruprecht, Göttingen 1963.

6 Buchdahl, H.A.: Am. J. Phys. 17 (1949), 41.

7 Buchdahl, H.A.: Am. J. Phys. 17 (1949), 212.

8 Budde, E.: Wied. Ann. 45 (1892), 751.

9 Carathéodory, C.: Gesammelte mathematische Schriften 2, C.H. Beck, München 1955.

10 Clausius, R.: Die mechanische Wärmetheorie, Vieweg, Braunschweig 1876.

11 Chow, W.L.: Math. Ann. 117 (1940), 98.

12 Eisenschitz, R.: Sci. Progr. 43 (1955), 246.

13 Flanders, H.: Bull. Amer. Math. Soc. 1 (1979), 944.

14 Fleming, W.: Functions of several variables, Springer, New York-Heidelberg-Berlin 1977.

15 Finkelstein, R.J.: Thermodynamics and statistical physics, W.H. Freeman, San Francisco 1969.

16 Hermann, R.: Differential geometry and the calculus of variations, Academic Press, New York 1968.

17 Jauch, J.M.: Found. Phys. 2 (1972), 327.

18 Jauch, J.M.: Found. Phys. 5 (1975), 111.

19 Kirchhart, K.J. und Walter, J.: Eine elementare Studie über die Sätze von Carathéodory, Chow und Frobenius in der Thermodynamik, Institut für Mathematik der RWTH Aachen 1982.

20 Landsberg, P.T.: Nature 201 (1964), 485.

21 Leontowitsch, M.A.: Einführung in die Thermodynamik, Deutscher Verlag der Wissenschaften, Berlin 1953.

22 Mrugała, R.: Reports on Math. Phys. 14 (1978), 419.

23 Planck, M.: Physikalische Abhandlungen und Vorträge 2, Vieweg, Braunschweig 1958.

24 Petrovski, J.G.: Ordinary differential equations, Prentice-Hall, Inc., Englewood Cliffs (N.J.) 1966.

25 Sneddon, I.N.: Elements of partial differential equations, McGraw-Hill, New York-Toronto-London 1957.

26 Stepanov, W.W.: Lehrbuch der Differentialgleichungen, VEB Deutscher Verlag der Wissenschaften, Berlin 1956.

27 Turner, L.A.: Am. J. Phys. 28 (1960), 781.

28 Waals, J.D. van der: Lehrbuch der Thermostatik, J.A. Barth, Leipzig 1927.

29 Walter, J.: J. Math. Anal. Appl. 59 (1977), 587.

30 Walter, J.: Proc. Royal Soc. Edingburgh 82 A (1978), 87 (cf. also Autorreferat in Zentralblatt f. Mathematik 395 (1979) 35074).

31 Westenholz, C. von: Differential forms in mathematical physics, North-Holland, Amsterdam 1978.

32 Zemanski, M.W.: Am. J. Phys. 34 (1966), 914.

Additional Note: After definition of T relation $(3")_1$ can be looked at as an ordinary linear first order differential equation in t for each component of $\widetilde{A}$. From this we deduce that $(T(0)\cdot\widetilde{A}(a,t) - T(t)\cdot\widetilde{A}(a,0))\,da$ is exact for each t. Thus there is still some redundancy left in (3").

On some conjectures on the deficiency index for symmetric differential operators

A.D. Wood and R.B. Paris

Introduction

We consider the symmetric differential expression

$$M[y] = \sum_{r=0}^{m} (-1)^r (p(x)y^{(r)})^{(r)} \qquad x \in [X,\infty),\ X \geq 0 \qquad (1)$$

where p_r are sufficiently continuously differentiable real-valued functions and $p_m(x) \neq 0$ on $[X,\infty)$. The deficiency indices of (the minimal symmetric operator T_0 defined in $L^2(X,\infty)$ from) M are the dimensions of the linear subspaces of $L^2(X,\infty)$ functions y satisfying $M[y] = \lambda y$ for im$\lambda > 0$ and im $\lambda > 0$ respectively. It is well-known [6] that the deficiency indices of a symmetric differential operator of even order 2m are (N,N) where $m \leq N \leq 2m$. We shall be concerned with two recent conjectures on the possible values of N, given certain conditions on the coefficients p_s of M.

The first of these is the positive coefficient conjecture. It assumes that $p_m(x) > 0$ and $p_r(x) \geq 0$, $0 \leq r < m$, for all $x \in [x,\infty]$. It was originally thought that this would force the deficiency indices to lie in the minimal position N = m, but this was disproved by R. Kauffman in 1976 [4]. The method depended on showing that a certain necessary condition for the minimal position was broken and it was not possible to say how large N might become. The following conjecture was then made by J.B. McLeod [3].

Conjecture 1 For positive coefficients all values of N except the maximal case N = 2m are possible. We shall construct below a sixth-order operator for which N = 2m − 1, the largest value suggested by McLeod.

The second is the zero coefficient conjecture made by Eastham and Grudniewicz at the end of a long paper [1] in 1981. They suggested that if some of the coefficients p_s were identically zero, this restricted the scope for moving away from the minimal case $N = m$.

Conjecture 2 If exactly K of the coefficients p_s in M are identically zero in $[X,\infty]$, then $m \leq N \leq 2m - K$. We shall exhibit below a sixth-order operator of form (1) with $p_1(x) = p_2(x) \equiv 0$ for which $N = 5$, thus showing conjecture 2 to be incorrect.

2. Transformation to a generalised hypergeometric equation

We consider the interval $[1,\infty)$ with $m = 3$ and $p_r(x) = \alpha_r x^{2r+\beta}$ in (1), where $\beta > 0$, $\alpha_3 = 1$, $\alpha_2 = \alpha_1 = 0$, $\alpha_0 = \alpha \geq 0$.

$$\text{Then } M[y] = -(x^{6+\beta}y^{(3)})^{(3)} + \alpha x^{\beta} y \qquad (2).$$

If we write $\Theta \equiv x\frac{d}{dx}$, by standard manipulations given in Watson [8,109] we may express $(x^{6+\beta}y^{(3)})^{(3)}$ as

$$x^{\beta}\,(\Theta+\beta+1)\,\Theta\,(\Theta+\beta+2)\,(\Theta+1)\,(\Theta+\beta+3)\,(\Theta+2)y$$

Let $-\beta_r$ $(r = 1,2,..,6)$ denote the zeros of the polynomial in μ

$$(\mu+\beta+1)\,\mu\,(\mu+\beta+2)\,(\mu+1)\,(\mu+\beta+3)\,(\mu+2) - \alpha. \qquad (3)$$

Then the equation $M[y] = \lambda y$ may be written as

$$x^{\beta}\prod_{r=1}^{6}(\Theta+\beta_r)y + \lambda y = 0 \qquad (4)$$

If we set $z = \lambda\beta^{-6}x^{-\beta}$ and $\delta \equiv z\frac{d}{dz}$, then $\Theta = -\beta\delta$ and (4) becomes

$$\prod_{r=1}^{6}(\delta - \frac{\beta_r}{\beta})\,y + zy = 0 \qquad (5)$$

Multiplying this equation by $z^{-\beta_1/\beta}$ and noting $z^{\nu}\delta y = (\delta-\nu)(z^{\nu}y)$ for arbitrary ν we obtain

$$\delta(\delta - \frac{\beta_2-\beta_1}{\beta})(\delta - \frac{\beta_3-\beta_1}{\beta})\dots(\delta - \frac{\beta_6-\beta_1}{\beta})Y + zY = 0 \qquad (6)$$

where $Y = z^{-\beta_1/\beta}y$. This is the generalised hypergeometric equation giver by Luke [5, p181] which may be solved in terms of the hyper-Bessel functions ${}_0F_{p-1}(z)$ whose asymptotic properties are given in [2, p.182]. Reverting to the original variable x, we see that one solution of $M[y] = \delta y$ is

$$\phi_1(x) = x^{-\beta_1} {}_0F_{p-1}\left(1 - \frac{\beta_2-\beta_1}{\beta}, \ldots, 1 \frac{\beta_6-\beta_1}{\beta} ; - \frac{\lambda}{\beta^6 x^\beta}\right)$$

$$\sim x^{-\beta_1} \quad \text{as } x \to \infty.$$

Assuming that α can be chosen so that the roots $-\beta_r$ of (3) are all distinct and no two differ by an integer multiple of β, we may repeat the above process, multiplying (5) successively by $z^{-\beta_k/\beta}$ for $k = 2,3,\ldots,6$ to obtain a fundamental set for $M[y] = \lambda y$ as

$$\phi_k(x) = x^{-\beta_k} {}_0F_{p-1}\left(1 - \frac{\beta_1-\beta_k}{\beta}, \ldots * \ldots, 1 - \frac{\beta_6-\beta_k}{\beta} ; - \frac{\lambda}{\beta^6 x^\beta}\right)$$

$$\sim x^{-\beta_k} \quad \text{as } x \to \infty, \; k = 1,2,\ldots,6$$

where * denotes the omission of the factor $1 - \frac{\beta_r-\beta_k}{\beta}$ with $r = k$.

The $L^2(1,\infty)$ nature of these solutions is governed by the disposition of the zeros $-\beta_r$ of the polynomial (3) with regard to the line re $\mu = -\frac{1}{2}$ (in the complex μ plane). Zeros lying to the left of this line will give rise to $L^2(1,\infty)$ solutions.

3. The location of the zeros $-\beta_r$

If we define a new variable $s = \mu + \frac{1}{2}\beta + \frac{1}{2}$ and write $s = \sigma + it$ the condition for $L^2(1,\infty)$ solutions becomes $\sigma < \frac{1}{2}\beta$. The polynomial (3) which characterises the asymptotic behaviour of the solutions becomes

$$G(s) = (s^2 - a^2)(s^2 - b^2)(s^2 - c^2) - \alpha$$

where $a = \frac{1}{2}(\beta+1)$, $b = \frac{1}{2}(\beta+3)$, $c = \frac{1}{2}(\beta+5)$. (7)

When $\alpha = 0$, the characteristic polynomial $G(s)$ has the real zeros $s = \pm a, \pm b, \pm c$. It is clear that all the positive zeros are greater than $\frac{1}{2}\beta$ and that there are only 3 L^2 solutions of $M[y] = \lambda y$.

We now consider the locus of these positive zeros in the s-plane as α is allowed to increase from 0. We shall show that for suitable values of β two of these zeros move across the line $\sigma = \frac{1}{2}\beta$ to give rise to two additional L^2 solutions.

We note that the zeros of $G(s)$ are symmetric in the σ and t axes and observe that $G(s)$ has zeros when

$$H(s) \equiv (s^2 - a^2)(s^2 - b^2)(s^2 - c^2) = \alpha e^{2\pi ik} \quad (k=0, \pm 1, \pm 2, \ldots).$$

The locus of these zeros consists of those points s which satisfy

$$\arg H(s) \equiv \arg(s^2 - a^2) + \arg(s^2 - b^2) + \arg(s^2 - c^2) = 2\pi k \tag{8}$$

This can only be satisfied for real s if $a < s < b$ or $s > c$. By considering $\alpha = |H(s)|$ we see that, as α increases from 0, the zero initially at c moves outwards along the σ-axis while those initially at a and b move towards each other, eventually meeting and branching off the axis as a complex conjugate pair. Does this branch enter the strip $0 < \sigma < \frac{1}{2}\beta$ corresponding to L^2 solutions of $M[y] = \lambda y$?

For $s = it$ we find $\arg H(s) = 3\pi$, hence the locus cannot cross the imaginary axis. Also, for non-real s, we see that as $\sigma \to \infty$, $\arg H(s) \to 0$ monotonically. Thus, for given t, there is a unique value $\sigma = \sigma(t)$ at which $\arg H(s) = 2\pi$. The locus of such points is obtained by setting $k = 1$ in (8), taking tangents and separating real and imaginary parts to give

$$3(\sigma^2 - t^2)^2 - 4\sigma^2 t^2 - 2(a^2+b^2+c^2)(\sigma^2-t^2) + a^2b^2+b^2c^2+c^2a^2 = 0. \tag{9}$$

This quartic crosses into the strip $0 < \sigma < \frac{1}{2}\beta$ if the quartic in t resulting when we substitute $\sigma = \frac{1}{2}\beta$ has real roots. This condition will be satisfied if (9), regarded as a quadratic in t^2, has distinct positive roots. Substituting for a,b and c in terms of β from (7), this will occur when

$$\beta^4 - 18\beta^3 - 23\beta^2 + 72\beta + 112 > 0$$

which holds if $\beta > 18.99495$ with a corresponding α value of 1.68112×10^6 Since the locus has asymptotics args $= \pm\,\pi/3$, it must cross over the line $\sigma = \frac{1}{2}\beta$ for some higher value of α. Recalling the symmetry in the σ-axis, we obtain a range of values of α for which there are two additional L^2 solutions of $M[y] = \lambda y$, a total of five.

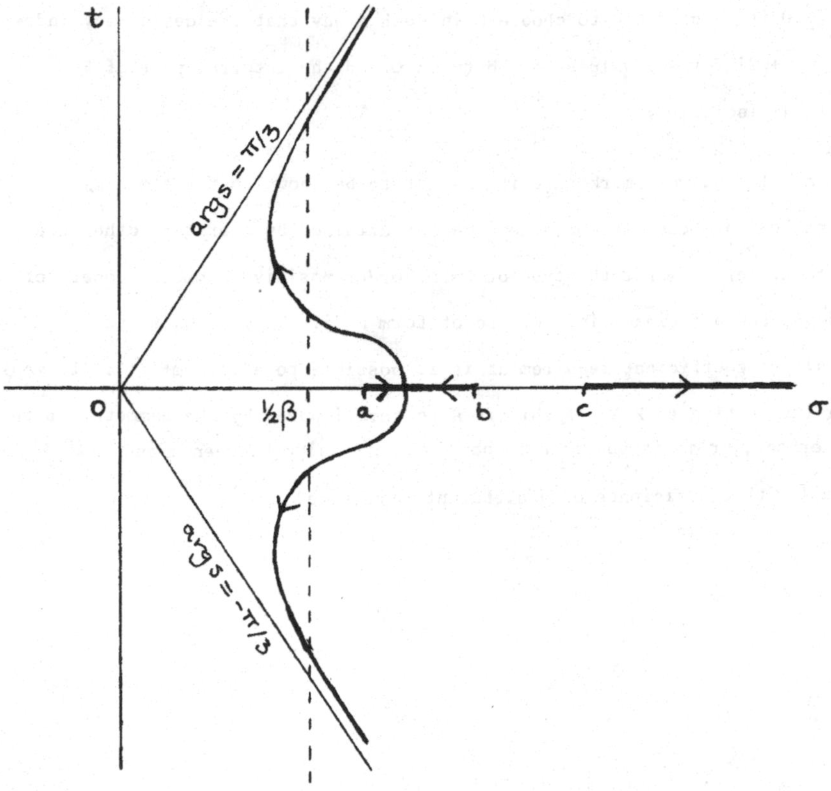

The root-locus diagram for (3). a, b and c are the roots when $\alpha = 0$. The arrows denote the direction of α increasing.

4. Concluding Remarks

This single example shows both that the case 2m-1 of Conjecture 1 is attainable and that Conjecture 2 is incorrect. Since $k = 2$, Conjecture 2 would require $N = 3$ or 4.

These results have been extended in [7] to equations of general even order 2m of form

$$(-1)^m y^{(2m)} + \sum_{r=0}^{q} (-1)^r (\alpha_r x^{2r+\beta} y^{(r)})^{(r)} = \lambda y \quad \text{on} \quad (0,\infty)$$

where $m \geq q > 0$, α_r real and $\beta > 0$. It is shown that when $\alpha_q = 1$ and $\alpha_r \geq 0$ it is possible to choose β in such a way that the deficiency index $N = m + 2j$ for all values of j between 0 and the integral part of $\frac{1}{4}(q+1)$ inclusive.

In conclusion we remark that it has yet to be shown that the maximal case 2m-1 of McLeod's conjecture can be attained by operators other than of sixth-order. For the type of 2mth. order positive symmetric operator above, the deficiency indices are of form m+2j. Only if we relax the positive coefficient requirement is it possible to attain m+2j+1. It would be interesting to know if this is a feature imposed by the symmetry of the operator, or is it possible to obtain every value between m and 2m-1 with coefficients of a different structure?

References

1. M.S.P. Eastham and C.G.M. Grudniewicz, Asymptotic theory and deficiency indices for higher-order self-adjoint differential equations, J. London Math.Soc. (2), 24 (1981), 255-271.

2. A. Erdelyi (Ed.), Higher Transcendental Functions, Vol. I, (New York: McGraw Hill, 1955).

3. W.N. Everitt, On the deficiency index problem for ordinary differential operators 1910-1970, Proceedings of Uppsala 1977 International Conference on Differential Equations, 62-81.

4. R.M. Kauffman, On the limit-n classification of ordinary differential operators with positive coefficients, Springer Verlag Lecture Notes in Mathematics, 564 (1976), 259-266.

5. Y.L. Luke, Mathematical functions and their approximations, (New York: Academic Press, 1975).

6. M.A. Naimark, Linear Differential Operators, Part II (London: Harrap, 1968).

7. R.B. Paris and A.D. Wood, On the L^2 nature of solutions of nth. order symmetric differential equations and McLeod's conjecture, Proc. Roy. Soc. Edin., 90A (1981), 209-236.

8. G.N. Watson, A Treatise on the Theory of Bessel Functions (Cambridge University Press, 1944).

On a second-order differential expression and its Dirichlet integral

S. D. Wray

Introduction

We consider the symmetric second-order differential expression M defined by, for suitable functions f,

(1) $$M[f] = w^{-1}(-(pf')' + qf) \quad \text{on } [a,\infty) \quad ('\equiv d/dx) ,$$

where p, q and w are real-valued coefficients on the interval $[a,\infty)$ of the real line R. We also consider the quadratic form

$$\int_a^\infty \{p|f'|^2 + q|f|^2\} ,$$

this being the Dirichlet integral associated with M. For recent work on the Dirichlet integral, see [4], [8] and the other references listed therein.

Conditions will be imposed on the coefficient functions, p, q and w, of M that will put M in the so-called limit-circle condition at ∞ (see the next section). A self-adjoint operator derived from M is considered, as is a "spectral" identity involving the Dirichlet integral, from which we obtain information about its lower bound. Use is made of a unitary change of independent variable.

Notation

AC denotes absolute continuity and 'loc' a property to be satisfied on all compact sub-intervals of the interval in question. A symbol such as '$(f\in\mathcal{D}(T))$' is to be read as 'for all f in the set $\mathcal{D}(T)$'.

Statement of results

We work on the half-line $[a,\infty)$, with $a\in R$. Throughout, the coefficients p, q and w are real-valued, Lebesgue measurable on $[a,\infty)$ and satisfy the basic conditions:

(i) $p(x)>0$ (almost all $x\in[a,\infty)$) and $p^{-1}\in L_{loc}[a,\infty)$;

(2) (ii) $q\in L_{loc}[a,\infty)$;

(iii) $w(x)>0$ (almost all $x\in[a,\infty)$) and $w\in L_{loc}[a,\infty)$.

The space $L_w^2[a,\infty)$, where w satisfies (2)(iii), is the Hilbert space of Lebesgue measurable functions $f:[a,\infty)\to C$ (the complex field) that satisfy

$$\int_a^\infty w|f|^2<\infty.$$

We consider the linear differential equation

(3) $M[f] = \lambda f$ on $[a,\infty)$,

where M is defined by (1) and λ is a complex parameter. M is in a singular case because $[a,\infty)$ is an infinite interval. Let ϕ, θ be the fundamental solutions of (3) determined by the boundary conditions

$$\phi(a,\lambda) = \sin\alpha \quad , \quad p(a)\phi'(a,\lambda) = -\cos\alpha \quad ,$$

$$\theta(a,\lambda) = \cos\alpha \quad , \quad p(a)\theta'(a,\lambda) = \sin\alpha \quad ,$$

for all $\lambda\in C$, where α is a real number in the interval $[0,\pi)$. The general solution of (3) is of the form $\theta + \ell\phi$; if c is any point in (a,∞), we consider those solutions which satisfy the boundary condition

$$\{\theta(c,\lambda) + \ell(c,\lambda)\phi(c,\lambda)\}\cos\beta +$$
$$+ \{p(c)\theta'(c,\lambda) + \ell(c,\lambda)p(c)\phi'(c,\lambda)\}\sin\beta = 0 \ ,$$

for all $\lambda\in C$, where β is a real number in $[0,\pi)$ again. For given c and β, this condition defines $\ell(c,\cdot)$ as a meromorphic function, whose poles are all on the real axis. For the remainder of this paper, we suppose that λ has positive imaginary part.

Now $\ell(c,\lambda)$ depends upon β: for fixed c and λ, $\ell(c,\lambda)$ describes a circle in the complex plane as β varies. H. Weyl [7] showed that as $c\to\infty$ these circles converge either to a limit-point or to a limit-circle. Let $m(\lambda)$ be the limit-point or a point on the limit-circle. Weyl also proved that the solution

$$\psi(x,\lambda) = \theta(x,\lambda) + m(\lambda)\phi(x,\lambda)$$

of (3) is in $L^2_w[a,\infty)$ as a function of x, for fixed λ (Weyl has $w\equiv 1$, but his method extends to the present situation without difficulty). For this theory, see [6,Ch.2].

Now, it is known, see [5], that there is a real non-decreasing function σ defined on R which induces a unitary transformation from $L^2_w[a,\infty)$ onto $\mathcal{L}^2$, which is the space of all Lebesgue measurable functions $F:R\to C$ such that

$$\int_{-\infty}^{\infty} |F|^2 d\sigma < \infty \ .$$

The unitary transform, F, of f is given by

$$(4) \qquad F(t) = \underset{s\to\infty}{\ell im} \int_a^s f(x)\phi(x,t)w(x)dx \ ,$$

where the limit is in $\mathcal{L}^2$ norm. We shall call σ a spectral function of M.

Finally, we introduce a self-adjoint operator and a quadratic form associated with M. Suppose that M is in the limit-circle condition at ∞. Let $\mathcal{D}(T)\subset L^2_w[a,\infty)$ be given by $f\in\mathcal{D}(T) \Leftrightarrow f\in L^2_w[a,\infty)$, f, $pf'\in AC_{loc}[a,\infty)$, $M[f]\in L^2_w[a,\infty)$, and f satisfies the boundary conditions $f(a)\cos\alpha+(pf')(a)\sin\alpha=0$ and $\underset{c\to\infty}{\ell im}\{f(c)p(c)\psi'(c,\lambda) - p(c)f'(c)\psi(c,\lambda)\}=0$. Here, $c\to\infty$ through a special sequence: see [5] or [2,Ch.9,Theorem 4.1]. It is assumed henceforth that c always tends to ∞ through this sequence. We then define the operator T by

$$Tf = M[f] \quad (f \in \mathcal{D}(T)).$$

It is known that T is self-adjoint [2,p.246], and it will follow from our results below that its domain $\mathcal{D}(T)$ is independent of λ (although this is also a known result, see [3]).

Let $\mathcal{D}(\tau)$ be given by $f \in \mathcal{D}(\tau) \Leftrightarrow f \in L^2_w[a,\infty)$, $f \in AC_{loc}[a,\infty)$, $p^{\frac{1}{2}}f'$, $|q|^{\frac{1}{2}}f \in L^2[a,\infty)$, $f(a)=0$ if $\alpha=0$, $f(\infty) = \lim_{c\to\infty} f(c)$ exists, and $f(\infty)=0$ if $\beta=0$. In the one case $\beta=\pi/2$, the requirement that $f(\infty)$ exist is dropped. The quadratic form τ is then defined by

$$\tau[f] = \int_a^\infty \{p|f'|^2 + q|f|^2\} - |f(a)|^2\cot\alpha +$$

$$+ |f(\infty)|^2\cot\beta \qquad (f \in \mathcal{D}(\tau)) ,$$

where we omit the $\cot\alpha$ ($\cot\beta$) term if $\alpha=0$ ($\beta=0$). The (absolutely convergent) integral is the Dirichlet integral associated with M.

Now we state the results of this paper as

Theorem Let the differential expression M on $[a,\infty)$ be defined by (1) and let its coefficients p, q and w satisfy the basic conditions (2). Let the operator T and quadratic form τ be as above.

Suppose additionally that p, q and w are so chosen that

(5) $\quad p^{-1}, q, w \in L[a,\infty)$.

Then

(i) M is in the limit-circle condition at ∞ (Amos[1]);

(ii) the operator T has discrete spectrum $\lambda_0 < \lambda_1 < \lambda_2 < \ldots$, tending to ∞ and having no finite cluster point; its domain $\mathcal{D}(T)$ is a sub-set of $\mathcal{D}(\tau)$;

(iii) $f \in \mathcal{D}(\tau)$ if and only if $|t|^{\frac{1}{2}}F(t) \in L^2$, where $F \in L^2$ is the unitary transform of f (see (4)), and if $f \in \mathcal{D}(\tau)$ then we have

$$\tau[f] = \int_{-\infty}^\infty t|F(t)|^2 d\sigma(t) ;$$

(iv) *if $f \in \mathcal{D}(\tau)$ then*

$$f(x) = \int_{-\infty}^{\infty} F(t)\phi(x,t)d\sigma(t) \qquad (x \in [a,\infty)),$$

the convergence being uniform on $[a,\infty)$, and

$$\lim_{s\to\infty} \int_a^{\infty} p|f' - f_s'|^2 = 0 ,$$

where

$$f_s(x) = \int_{-\infty}^{s} F(t)\phi(x,t)d\sigma(t) \qquad (x \in [a,\infty),\ s \in R).$$

Proof See the next section.

Corollary *Under the conditions of the Theorem, we have*

$$\tau[f] \geq \lambda_0 \int_a^{\infty} w|f|^2 \qquad (f \in \mathcal{D}(\tau)),$$

with equality if and only if f is an eigenfunction of T corresponding to the least eigenvalue, λ_0.

Proof This follows the lines of the proof of the Corollary in [4,§6].

Proof of the Theorem

We outline the method of proof here. The principal tool is the following change of independent variable. We define $\hat{x}(x)$ by

$$\hat{x}(x) = \int_a^x p^{-1} \qquad (x \in [a,\infty)) ,$$

so that $\hat{x} \in [0,\hat{b})$, where

$$\hat{b} = \int_a^{\infty} p^{-1} < \infty .$$

Since $p(x)>0$ (almost all $x \in [a,\infty)$), the function $\hat{x}:[a,\infty) \to [0,\hat{b})$ is continuous and strictly increasing, and hence has a continuous inverse $x:[0,\hat{b}) \to [a,\infty)$. Then, any function f on $[a,\infty)$ is transformed to a function $\hat{f}$ on $[0,\hat{b})$ via the relation

$$\hat{f}(\hat{x}) = f(x(\hat{x})) \qquad (\hat{x}\in[0,\hat{b})).$$

The transformation $f \mapsto \hat{f}$ is one-to-one; it is also unitary in the following sense. If $f\in L^2_w[a,\infty)$ and $\hat{f}$ is its transform, then $\hat{f}\in L^2_{\hat{w}}[0,\hat{b})$ and

$$\int_0^{\hat{b}} \hat{w}(\hat{x})\,|\hat{f}(\hat{x})|^2\, d\hat{x} = \int_a^{\infty} w(x)\,|f(x)|^2\, dx .$$

Under this transformation, the coefficient p is removed from the differential expression. If primes on letters without caps denote differentiation with respect to x, and on letters with caps with respect to $\hat{x}$, then the original differential equation (3) on $[a,\infty)$ is transformed to

$$\hat{w}^{-1}(\hat{x})\{-\hat{f}''(\hat{x})+\hat{q}(\hat{x})\hat{f}(\hat{x})\} = \lambda\hat{f}(\hat{x}) \quad \text{on } [0,\hat{b}) , \tag{7}$$

where $\hat{q}, \hat{w}\in L[0,\hat{b})$ and $\hat{w}(\hat{x})>0$ (almost all $\hat{x}\in[0,\hat{b})$). The same spectral parameter, λ, is used in both equations (3) and (7); it follows that both equations are associated with the same spectral function σ and the same associated unitary transformation (as in (4)). The equation (7) is actually in the <u>regular</u> case because of the integrability of its coefficients $\hat{q}$ and $\hat{w}$, and σ is a step function, derived from the Sturm-Liouville problem on the closure $[0,\hat{b}]$ of $[0,\hat{b})$ defined by (7) and the boundary conditions

$$\hat{f}(0)\cos\alpha + \hat{f}'(0)\sin\alpha = 0 ,$$

$$\hat{f}(\hat{b})\cos\beta + \hat{f}'(\hat{b})\sin\beta = 0 , \tag{8}$$

where we use the same α and β as formerly (see [8,Section 2] where σ is denoted by σ_c). The usual corresponding self-adjoint operator $\hat{T}$ is transformed to T via the transformation $\hat{x} \mapsto x$; it is straightforward to check most of this; we show here only that the boundary condition at ∞ (for T) is equivalent to the condition (8) (for $\hat{T}$).

Let $c \in [a,\infty)$ and let $\psi_c(x,\lambda) = \theta(x,\lambda) + \ell(c,\lambda)\phi(x,\lambda)$. If f is in $\mathcal{D}(T)$, then

$$0 = \lim_{k\to\infty}\{f(k)p(k)\psi'(k,\lambda) - (pf')(k)\psi(k,\lambda)\}$$

(9) $$= \lim_{k\to\infty}\lim_{c\to\infty}\{f(k)p(k)\psi_c'(k,\lambda) - (pf')(k)\psi_c(k,\lambda)\}$$

(10) $$= \lim_{c\to\infty}\{f(c)p(c)\psi_c'(c,\lambda) - (pf')(c)\psi_c(c,\lambda)\}$$

$$= \lim_{c\to\infty} - \{f(c)\cos\beta + (pf')(c)\sin\beta\} ;$$

if $\hat{f}$ is the transform of f then it now follows that (8) holds if we define $\hat{f}(\hat{b})$ and $\hat{f}'(\hat{b})$ to be the limits of $\hat{f}(\hat{c})$ and $\hat{f}'(\hat{c})$ as $\hat{c} \to \hat{b}-$. (Here, we rely on the Lemma of [1].) The step from (9) to (10) above is justified by a uniform convergence argument.

It is clear now that the transform of $\mathcal{D}(T)$ is a sub-set of the domain of $\hat{T}$. But both T and $\hat{T}$ are self-adjoint, and so their domains must be in one-to-one correspondence. For each $f \in \mathcal{D}(T)$, the transform of Tf is $\hat{T}\hat{f}$.

The results (ii) - (iv) of the Theorem now follow from their analogues in the regular case on the interval $[0,\hat{b}]$; see [8, Section 3]. Note that

$$\int_a^\infty \{p|f'|^2 + q|f|^2\} = \int_0^{\hat{b}} \{|\hat{f}'|^2 + \hat{q}|\hat{f}|^2\} \qquad (f \in \mathcal{D}(\tau)) ;$$

and that in part (iii)

$$\int_{-\infty}^\infty t|F(t)|^2 \, d\sigma(t) = \sum_{n=0}^\infty \lambda_n |f_n|^2 ,$$

where

$$f_n = \int_a^\infty wf\psi_n \qquad (n \geq 0) ,$$

ψ_n being the normalised eigenfunction corresponding to the eigenvalue λ_n.

REFERENCES

1 Amos, R. J.: On a Dirichlet and limit-circle criterion for second-order ordinary differential expressions. Quaestiones Mathematicae 3 (1978), 53-65.

2 Coddington, E. A. and Levinson, N.: Theory of Ordinary Differential Equations. McGraw-Hill, New York, 1955.

3 Everitt, W. N.: A note on the self-adjoint domains. Quart. J. Math. Oxford (2) 14 (1963), 41-45.

4 Everitt, W. N. and Wray, S. D.: A singular spectral identity and inequality involving the Dirichlet integral of an ordinary differential expression. To appear.

5 Sears, D. B.: Integral transforms and eigenfunction theory (I). Quart. J. Math. Oxford (2) 5 (1954), 47-58.

6 Titchmarsh, E. C.: Eigenfunction Expansions associated with Second-order Differential Equations, Part I, 2nd Edn. Oxford University Press, Oxford, 1962.

7 Weyl, H.: Über gewöhnliche Differentialgleichungen mit Singularitäten und die zugehörigen Entwicklungen willkürlicher Funktionen. Math. Ann. 68 (1910), 220-269.

8 Wray, S. D.: An inequality involving a conditionally convergent Dirichlet integral of an ordinary differential expression. Utilitas Mathematica, in press.

AN INVERSE EIGENVALUE PROBLEM FOR THE LAPLACE OPERATOR

E M E Zayed

§1 Introduction

Let $D \subseteq \mathbb{R}^2$ be a smooth bounded domain with finite connectivity $1 \leq p < \infty$. We wish to consider the oscillations, $\psi(x,y;t)$, of a membrane stretched across D and clamped along the boundary ∂D of D.

It is well known that $\psi(x,y;t)$ satisfies the wave equation

$$\frac{\partial^2\psi}{\partial x^2} + \frac{\partial^2\psi}{\partial y^2} = \frac{\partial^2\psi}{\partial t^2} \quad \text{in} \quad D \times \mathbb{R}^+ \tag{1.1}$$

together with the Dirichlet boundary condition

$$\psi(x,y;t) = 0 \,, \quad \text{on} \quad \partial D \,. \tag{1.2}$$

If we make the substitution $\psi(x,y;t) = u(x,y)e^{ikt}$, we arrive at a consideration of the problem

$$\Delta u + k^2 u = 0 \quad \text{in} \quad D \,, \tag{1.3}$$

$$u = 0 \quad \text{on} \quad \partial D \,, \tag{1.4}$$

where Δ is the Laplace operator in xy-plane. This is a well known eigenvalue problem for which non-trivial solutions ϕ_n (eigenfunctions) exist provided $k^2 = \lambda_I$ (eigenvalues).

These eigenvalues form an increasing sequence

$$0 < \lambda_1 \leq \lambda_2 \leq \ldots \leq \lambda_n \leq \ldots \to \infty \quad \text{as} \quad n \to \infty \,.$$

At the beginning of this century the principle problem was that of investigating the asymptotic behaviour of the eigenvalues $\{\lambda_n\}$. If $N(\lambda)$ is the number of these eigenvalues less than or equal to a bound λ, then

$$N(\lambda) \sim \frac{|D|}{4\pi}\lambda \quad \text{as} \quad \lambda \to \infty \quad \text{(H Weyl 1912)} \,, \tag{1.5}$$

and

$$N(\lambda) \sim \frac{|D|}{4\pi}\lambda + O(\lambda^{\frac{1}{2}} \log \lambda) \quad \text{as} \quad \lambda \to \infty \quad \text{(R Courant 1920)}, \tag{1.6}$$

where $|D|$ is the area of the domain D.

For either (1.5) or (1.6) the presence of the quantity $|D|$ leads one to ask whether, for example, the order term in (1.6) contains further information about the geometry of D and its connectivity. Such a question was discussed and put rather nicely by M Kac [1] who simply asked "Can one hear the shape of a drum?". To explore this question, one studies certain functions of the spectrum. The most useful to date comes from the heat equation or the wave equation.

Our study in this paper is concerned only with the heat equation method. If $e^{-t\Delta}$ denotes the heat operator, then we can construct the trace function

$$\Theta(t) = \mathrm{tr}(e^{-t\Delta}) = \sum_{n=1}^{\infty} e^{-\lambda_n t} , \tag{1.7}$$

which converges for all positive t.

Now the following question is pertinent: suppose that $\{\lambda_n\}$ is the spectrum of eigenvalues for the Laplace operator, can the shape of the domain D and its attendant boundary conditions be uniquely determined from the asymptotic expansion of the trace function $\Theta(t)$ for small positive t ? It has been shown by Kac [1] that some progress is possible on establishing the first three terms of the asymptotic expansion of $\Theta(t)$ for small positive t, viz

$$\Theta(t) \sim \frac{|D|}{4\pi t} - \frac{|\partial D|}{8(\pi t)^{\frac{1}{2}}} + a_0 + O(t^{\frac{1}{2}}) \quad \text{as} \quad t \to \infty , \tag{1.8}$$

where $|D|$ is the area of D and $|\partial D|$ is the total length of its boundary ∂D. Furthermore, the constant term a_0 has geometric significance, e.g., if D is smooth and convex, then $a_0 = \frac{1}{6}$, if D is permitted to have a finite number "r" of smooth convex holes, then $a_0 = (1-r)\frac{1}{6}$ and finally if D is a piecewise smooth domain (without cusps), then

$$a_0 = \frac{1}{12\pi} \int_{\partial D} K(\sigma)d\sigma + \sum_i c(\alpha_i) , \tag{1.9}$$

where the summation is taken over all the corners of D. $K(\sigma)$ denotes the curvature of the boundary ∂D and $c(\alpha_i)$ may be called the corner number corresponding to the angle α_i at the ith corner of D.

A simple method for calculating $c(\alpha)$, $0 < \alpha < 2\pi$ is due to D B Ray and is described by McKean and Singer [2] and the explicit result is

$$c(\alpha) = \frac{\pi^2 - \alpha^2}{24\pi\alpha} \quad . \tag{1.10}$$

The result (1.8) is also deduced from an analysis by Pleijel [3], who essentially determined the first four terms of the asymptotic form of $\Theta(t)$ for small positive t by the iterative solution of an appropriate Fredholm integral equation.

We merely note that aspects of the question of Kac, namely, "Can one hear the shape of a drum?" have been discussed recently in [6] for the equation (1.3) together with the impedance boundary condition $\frac{\partial u}{\partial n} + \gamma u = 0$ on ∂D, where γ is a positive constant.

Our objective of this paper is to determine the length of a uniform vibrating string and its Sturm-Liouville end constraints from a complete knowledge of its frequencies of vibration. The problems to be discussed here can be formulated as follows:

PI

Suppose the eigenvalues $\lambda_1 \le \lambda_2 \le \ldots \le \lambda_n \le \ldots \to \infty$ as $n \to \infty$ are known exactly for the Sturm-Liouville problem

$$\frac{d^2y}{dx^2} + \lambda y = 0 \qquad \text{for} \quad x \in (0,a) \ , \tag{1.11}$$

$$\left.\begin{aligned} y(0)\cos\alpha - y'(0)\sin\alpha = 0 \ , \quad & \alpha \in [0,\pi) \ , \\ y(a)\cos\beta - y'(a)\sin\beta = 0 \ , \quad & \beta \in (0,\pi] \ . \end{aligned}\right\} \tag{1.12}$$

Determine the unknown length "a" of the vibrating string and the unknown angles α and β .

PII

Suppose the eigenvalues $\lambda_1 \le \lambda_2 \le \ldots \le \lambda_n \le \ldots \to \infty$ as $n \to \infty$ are known exactly for the τ-periodic problem comprising (1.11) and the boundary conditions

$$\left.\begin{aligned} y(a) &= y(0)e^{i\pi\tau} \ , \\ y'(a) &= y'(0)e^{i\pi\tau} \ , \end{aligned}\right\} \tag{1.13}$$

where τ is a real parameter such that $\tau \in (-1,1]$ and $i = \sqrt{-1}$. Determine the period "a" and the parameter τ .

These results extend to eigenvalue problems for rectangles and cuboids. In fact we obtain the Roe result [4] easily and certain generalizations of it. In fact the results extend easily to n-dimensions. These problems are attacked by a careful analysis of the asymptotic behaviour of the trace function $\Theta(t) = \sum_{n=1}^{\infty} e^{-\lambda_n t}$ for small positive t.

§2 Construction of the trace function for problem I

Kac [1] has shown that the trace function $\Theta(t)$ can be expressed in terms of the simple formula

$$\Theta(t) = \int_0^a G(x,x;t)dx , \tag{2.1}$$

where $G(x,\xi;t)$ is the Green's function for the heat equation

$$\left(\frac{\partial^2}{\partial x^2} - \frac{\partial}{\partial t} \right) G(x,\xi;t) = 0 , \tag{2.2}$$

subject to the boundary conditions (1.12) and the initial condition

$$\lim_{t \to 0} G(x,\xi;t) = \delta(x-\xi) , \tag{2.3}$$

where $\delta(x-\xi)$ is the Dirac delta function located at the source point $x = \xi$.

Let us write

$$G(x,\xi;t) = G_0(x,\xi;t) + \chi(x,\xi;t) , \tag{2.4}$$

where

$$G_0(x,\xi;t) = (4\pi t)^{-\frac{1}{2}} \exp\{ - \frac{|x-\xi|^2}{4t} \} \tag{2.5}$$

is "the fundamental solution" of the heat equation (2.2), while $\chi(x,\xi;t)$ is a "regular" solution chosen in such a way that $G(x,\xi;t)$ satisfies the boundary conditions (1.12).

From (2.1), (2.4) and (2.5) we deduce that

$$\Theta(t) = \frac{a}{(4\pi t)^{\frac{1}{2}}} + K(t) , \tag{2.6}$$

where

$$K(t) = \int_0^a \chi(x,x;t)dx . \tag{2.7}$$

The problem now centres on providing an asymptotic estimate, for small positive t, of the integral term (2.7). In what follows we shall use Laplace transforms with respect to t, with s^2 as the Laplace transform parameter. That is define

$$\bar{G}(x,\xi;s^2) = \int_0^\infty e^{-s^2 t} G(x,\xi;t)dt . \tag{2.8}$$

An application of the Laplace transform to the heat equation (2.2) shows that $\bar{G}(x,\xi;s^2)$ satisfies the string equation

$$\left(\frac{\partial^2}{\partial x^2} - s^2 \right) \bar{G}(x,\xi;s^2) = -\delta(x-\xi) . \tag{2.9}$$

The asymptotic expansion of $K(t)$, for small positive t, may then be deduced directly from the asymptotic expansion of $\bar{K}(s^2)$, for large positive s, where

$$\bar{K}(s^2) = \int_0^a \bar{\chi}(x,x;s^2)dx . \tag{2.10}$$

Lengthy but otherwise straight forward calculations show that the following cases can be considered

(i) $\alpha = 0,\ \beta = \pi$ (Dirichlet problem)

$$\Theta(t) \sim \frac{a}{(4\pi t)^{\frac{1}{2}}} - \tfrac{1}{2} + O(t^{\frac{1}{2}}) \quad \text{as} \quad t \to 0 . \tag{2.11}$$

(ii) $\alpha = 0,\ \beta \neq \pi$

$$\Theta(t) \sim \frac{a}{(4\pi t)^{\frac{1}{2}}} + \frac{\cot\beta}{\sqrt{\pi}}\, t^{\frac{1}{2}} + O(t) \quad \text{as} \quad t \to 0. \tag{2.12}$$

(iii) $\alpha \neq 0,\ \beta = \pi$

$$\Theta(t) \sim \frac{a}{(4\pi t)^{\frac{1}{2}}} - \frac{\cot\alpha}{\sqrt{\pi}}\, t^{\frac{1}{2}} + O(t) \quad \text{as} \quad t \to 0 . \tag{2.13}$$

(iv) $\alpha \neq 0,\ \beta \neq \pi$ (Mixed problem)

$$\Theta(t) \sim \frac{a}{(4\pi t)^{\frac{1}{2}}} + \tfrac{1}{2} + (\cot\beta - \cot\alpha)\left(\frac{t}{\pi}\right)^{\frac{1}{2}} + \tfrac{1}{2}(\cot^2\beta + \cot^2\alpha)t + O(t^{\frac{3}{2}}) \quad \text{as} \quad t \to 0 . \tag{2.14}$$

(v) $\alpha = \beta = \frac{\pi}{2}$ (Neumann problem)

$$\Theta(t) \sim \frac{a}{(4\pi t)^{\frac{1}{2}}} + \tfrac{1}{2} + O(t^{\frac{1}{2}}) \quad \text{as} \quad t \to 0 . \tag{2.15}$$

An examination of the results (2.11) and (2.15) shows that the coefficient of $t^{-\frac{1}{2}}$ determines the length "a" of the vibrating string, while the sign of $\frac{1}{2}$ determines whether we have a Dirichlet or a Neumann problem. In the cases (ii) and (iii) we observe that the constant term is zero.

If $\alpha \neq 0$, $\beta \neq \pi$ then the coefficient of $t^{\frac{1}{2}}$ and that of t can be solved to determine the angles α and β.

§3 Construction of the trace function for problem II

As we have done in §2 we deduce that the trace function for the problem II can be written in the form

$$\Theta(t) = \frac{a}{(4\pi t)^{\frac{1}{2}}} \sum_{m=-\infty}^{\infty} \cos(m\pi\tau)\, e^{-\frac{m^2a^2}{4t}} , \quad t > 0 . \tag{3.1}$$

There are two cases to be considered.

(i) $\tau = 0$ (periodic problem)

$$\Theta(t) = \frac{a}{(4\pi t)^{\frac{1}{2}}} \sum_{m=-\infty}^{\infty} e^{-\frac{m^2a^2}{4t}} \quad \text{for} \quad t > 0 , \tag{3.2}$$

which is an example of the classic summation formula, viz, the celebrated Poisson-Jacobi inversion formula (see [5]).

(ii) $\tau = 1$ (semi-periodic problem)

$$\Theta(t) = \frac{a}{(4\pi t)^{\frac{1}{2}}} \sum_{m=-\infty}^{\infty} (-1)^m\, e^{-\frac{m^2a^2}{4t}} \quad \text{for} \quad t > 0 .$$

§4 Applications to eigenvalue problems for rectilinear domains

A convenient point for an anlysis of rectilinear domains is to examine the structure of $\Theta(t)$ for

(i) a rectangle D of sides a, b where the boundary conditions (1.12) are holding on x-axis and y-axis respectively,

(ii) a cuboid D of sides a, b, c where the boundary conditions (1.12) are holding on x-axis, y-axis and z-axis respectively.

§§4.1 <u>If $\alpha = 0$, $\beta = \pi$ (Dirichlet problem)</u>

For a rectangle D of sides a, b ,

$$\Theta(t) \sim \frac{ab}{4\pi t} - \frac{2(a+b)}{8(\pi t)^{\frac{1}{2}}} + \tfrac{1}{4} \quad \text{as} \quad t \to 0 , \tag{4.1.1}$$

while for a cuboid D of sides a, b, c ,

$$\Theta(t) \sim \frac{abc}{(4\pi t)^{\frac{3}{2}}} - \frac{2(ab+bc+ca)}{16\pi t} + \frac{a+b+c}{8(\pi t)^{\frac{1}{2}}} - \tfrac{1}{8} \quad \text{as} \quad t \to 0 . \tag{4.1.2}$$

§§4.2 <u>If $\alpha = 0$, $\beta \neq \pi$</u>

For a rectangle D of sides a, b ,

$$\Theta(t) \sim \frac{ab}{4\pi t} + \frac{2(a+b)}{4\pi} \cot\beta + \frac{\cot^2\beta}{\pi} t \quad \text{as} \quad t \to 0 , \tag{4.2.1}$$

while for a cuboid D of sides a, b, c ,

$$\Theta(t) \sim \frac{abc}{(4\pi t)^{\frac{3}{2}}} + \frac{2(ab+bc+ca)}{8\pi^{\frac{3}{2}} t^{\frac{1}{2}}} \cot\beta + \frac{(a+b+c)\cot^2\beta}{2\pi^{\frac{3}{2}}} t^{\frac{1}{2}}$$

$$+ \left(\frac{t}{\pi}\right)^{\frac{3}{2}} \cot^3\beta \quad \text{as} \quad t \to 0 . \tag{4.2.2}$$

§§4.3 <u>If $\alpha \neq 0$, $\beta = \pi$</u>

For a rectangle D of sides a, b ,

$$\Theta(t) \sim \frac{ab}{4\pi t} - \frac{2(a+b)}{4\pi} \cot\alpha + \frac{\cot^2\alpha}{\pi} t \quad \text{as} \quad t \to 0 , \tag{4.3.1}$$

while for a cuboid D of sides a, b, c ,

$$\Theta(t) \sim \frac{abc}{(4\pi t)^{\frac{3}{2}}} - \frac{2(ab+bc+ca)}{8\pi^{\frac{3}{2}} t^{\frac{1}{2}}} \cot\alpha + \frac{(a+b+c)\cot^2\alpha}{2\pi^{\frac{3}{2}}} t^{\frac{1}{2}} - \left(\frac{t}{\pi}\right)^{\frac{3}{2}} \cot^3\alpha$$

$$\text{as} \quad t \to 0 . \tag{4.3.2}$$

§§4.4 If $\alpha \neq 0$, $\beta \neq \pi$ (Mixed problem)

For a rectangle D of sides a, b,

$$\Theta(t) \sim \frac{ab}{4\pi t} + \frac{2(a+b)}{8(\pi t)^{\frac{1}{2}}} + \frac{1}{4} + \frac{2(a+b)}{4\pi}(\cot\beta - \cot\alpha)$$

$$+ \left\{\frac{2(a+b)(\cot^2\beta+\cot^2\alpha)}{8\pi^{\frac{3}{2}}} + \frac{(\cot\beta-\cot\alpha)}{\pi^{\frac{1}{2}}}\right\} t^{\frac{1}{2}}$$

$$+ \left\{\frac{(\cot\beta-\cot\alpha)^2}{\pi} + \frac{\cot^2\beta+\cot^2\alpha}{2}\right\} t$$

$$+ \frac{(\cot\beta-\cot\alpha)(\cot^2\beta+\cot^2\alpha)}{\pi^{\frac{1}{2}}} t^{\frac{3}{2}}$$

$$+ \tfrac{1}{4}(\cot^2\beta+\cot^2\alpha)^2 t^2 \quad \text{as} \quad t \to 0, \qquad (4.4.1)$$

while for a cuboid D of sides a, b, c,

$$\Theta(t) \sim \frac{abc}{(4\pi t)^{\frac{3}{2}}} + \frac{2(ab+bc+ca)}{16\pi t} + \left\{\frac{2(ab+bc+ca)(\cot\beta-\cot\alpha)}{8\pi^{\frac{3}{2}}} + \frac{a+b+c}{8\pi^{\frac{1}{2}}}\right\}\frac{1}{t^{\frac{1}{2}}}$$

$$+ \tfrac{1}{8} + \frac{2(ab+bc+ca)(\cot^2\beta+\cot^2\alpha)}{16\pi} + \frac{(a+b+c)(\cot\beta-\cot\alpha)}{2\pi}$$

$$+ \left\{\frac{(a+b+c)(\cot\beta-\cot\alpha)^2}{2\pi^{\frac{3}{2}}} + \frac{(a+b+c)(\cot^2\beta+\cot^2\alpha)}{4\pi^{\frac{1}{2}}} + \frac{3(\cot\beta-\cot\alpha)}{4\pi^{\frac{1}{2}}}\right\} t^{\frac{1}{2}}$$

$$+ \left\{\frac{(a+b+c)(\cot\beta-\cot\alpha)(\cot^2\beta+\cot^2\alpha)}{2\pi} + \frac{3(\cot\beta-\cot\alpha)^2}{2\pi} + \frac{3(\cot^2\beta+\cot^2\alpha)}{8}\right\} t$$

$$+ \left\{\frac{(a+b+c)(\cot^2\beta+\cot^2\alpha)}{8\pi^{\frac{1}{2}}} + \frac{3(\cot\beta-\cot\alpha)(\cot^2\beta+\cot^2\alpha)}{2\pi^{\frac{1}{2}}} + \frac{(\cot\beta-\cot\alpha)^3}{\pi^{\frac{3}{2}}}\right\} t^{\frac{3}{2}}$$

$$+ \left\{\frac{3(\cot^2\beta+\cot^2\alpha)^2}{8} + \frac{3(\cot\beta-\cot\alpha)^2(\cot^2\beta+\cot^2\alpha)}{2\pi}\right\} t^2 \quad \text{as} \quad t \to 0.$$

(4.4.2)

§§4.5 If $\alpha = \beta = \frac{\pi}{2}$ (Neumann problem)

For a rectangle D of sides a, b,

$$\Theta(t) \sim \frac{ab}{4\pi t} + \frac{2(a+b)}{8(\pi t)^{\frac{1}{2}}} + \frac{1}{4} \quad \text{as} \quad t \to 0 , \qquad (4.5.1)$$

while for a cuboid D of sides a, b, c,

$$\Theta(t) \sim \frac{abc}{(4\pi t)^{\frac{3}{2}}} + \frac{2(ab+bc+ca)}{16\pi t} + \frac{a+b+c}{8(\pi t)^{\frac{1}{2}}} + \frac{1}{8} \quad \text{as} \quad t \to 0 \qquad (4.5.2)$$

An examination of the results (4.1.1) and (4.5.1) shows that the coefficient of t^{-1} determines the area of the rectangle D and the coefficient of $t^{-\frac{1}{2}}$ determines the total length of its boundary, while the sign of the coefficient of $t^{-\frac{1}{2}}$ determines whether we have a Dirichlet or a Neumann problem.

We note also that the first three terms in (4.1.2) and (4.5.2) are in agreement with the result of Roe [4] and show that the coefficient of $t^{-\frac{3}{2}}$ determines the volume of the cuboid D, the coefficient of t^{-1} determines its surface area, and the coefficient of $t^{-\frac{1}{2}}$ determines the sum of the total length of its sides, while the sign of the coefficient of t^{-1} determines whether we have a Dirichlet or a Neumann problem.

References

1. M Kac, Can one hear the shape of a drum?, Amer. Math. Monthly, 73 No 4 part II (1966), 1-23.

2. H P McKean and I M Singer, Curvature and the eigenvalues of the Laplacian. J. Diff. Geometry 1 (1967), 43-69.

3. A Pleijel, A study of certain Green's function with applications in the theory of vibrating membranes. Arki Matematik 2 (1952), 553-569.

4. G M Roe, Frequency distribution of normal modes. J. Acoust. Soc. Amer. 13, No 1 (1941), 1-7.

5. T Sunada, Trace formula for Hill's operator, Duke. Math. J. 47 No 3 (1980), 529-546.

6. B D Sleeman and E M E Zayed, An inverse eigenvalue problem for a general convex domain. J. Math. Anal. Applics (to appear).